The Origins of Genome Architecture

THE ORIGINS OF GENOME ARCHITECTURE

Michael Lynch
Indiana University

Sinauer Associates, Inc. Publishers
Sunderland, Massachusetts

Cover art courtesy of Lewis Rifkowitz, Guam.

The Origins of Genome Architecture

For information, address
Sinauer Associates, Inc., 23 Plumtree Road, Sunderland, MA 01375 USA
Fax: 413-549-1118
E-mail: publish@sinauer.com
Internet: www.sinauer.com

Library of Congress Cataloging-in-Publication Data

Lynch, Michael, 1951-
The origins of genome architecture / Michael Lynch.
p. ; cm.
Includes bibliographical references and index.
ISBN 978-0-87893-484-3 (alk. paper)
1. Evolutionary genetics. 2. Genomes. 3. Molecular evolution. 4. Population genetics. I. Title.
[DNLM: 1. Genome—genetics. 2. Eukaryotic Cells. 3. Evolution, Molecular. QU 470 L987o 2007]

QH390.L96 2007
572.8'38—dc22

2007000012

Printed in China

5 4 3 2 1

I am not prone to moralize
  In scientific doubt
On certain facts that Nature tries
  To puzzle us about,
For I am no philosopher
  Of wise elucidation,
But speak of things as they occur,
  From simple observation.

From *Natural Perversities* by James Whitcomb Riley, in *The Complete Poetical Works of James Whitcomb Riley*, 1993, Indiana University Press, Bloomington, IN.

Contents

Preface

We live in one of the most amazing periods of intellectual and technological development that has thus far been experienced by the biological sciences. The emergence of methods for the rapid characterization of genomes has led to a rate of data proliferation that was literally unimaginable just 25 years ago, presenting enormous opportunities and challenges for the field of evolutionary biology. For several decades, we have had a fairly mature theory of evolution in terms of expected gene/genotype frequency changes in response to various evolutionary forces, but owing to the absence of detailed information on the raw materials of evolution, much of the theory was couched in terms too abstract for the tastes of the average biologist. Now, with a picture of the full spectrum of DNA-level resources in hand, there is real potential for making quantum-level progress in our understanding of the mechanisms of evolution at both the molecular genetic and population genetic levels.

This is a book of advocacy on several fronts. Contrary to common belief, evolution is not driven by natural selection alone. Many aspects of evolutionary change are indeed facilitated by natural selection, but all populations are influenced by the nonadaptive forces of mutation, recombination, and random genetic drift. These additional forces are not simple embellishments around a primary axis of selection, but are quite the opposite—they dictate what natural selection can and cannot do. Although this basic principle has been known for a long time, it is quite remarkable that most biologists continue to interpret nearly every aspect of biodiversity as an outcome of adaptive processes. This blind acceptance of natural selection as the only force relevant to evolution has led to a lot of sloppy thinking, and is probably the primary reason why evolution is viewed as a soft science by much of society.

A central point to be explored in this book is that most aspects of evolution at the genomic level cannot be fully explained in adaptive terms, and moreover, that many genomic features could not have emerged without a near-complete disengagement of the power of natural selection. This contention is supported by a wide array of comparative data, as well as by well-established principles of population genetics. However, even if such support did not exist, there is an important reason for pursuing nonadaptive (neutral) models of evolution. If one wants to confidently invoke a specific adaptive scenario to explain an observed pattern of comparative data, then an ability to reject a hypothesis based entirely on the nonadaptive forces of evolution is critical.

Because evolution is a population-genetic process, the litmus test for any evolutionary argument is the demonstration of its compatibility with basic population-genetic principles. Nevertheless, it is still common to hear molecular, cellular, and developmental biologists arguing that we evolutionary biologists have missed the boat entirely. Such a view derives in part from the mistrust that most biologists have for all things mathematical and the consequent failure to appreciate the accomplishments of evolutionary theory, an inherently quantitative enterprise. Nothing of a productive nature has yet come from such posturing. If it ever does, biology will have been confronted with a crisis of major proportions.

Such tensions are not confined to a single-lane on the bridge between molecular and evolutionary genetics. In the field of evolutionary genetics, vague references to molecular/developmental constraints are often invoked to explain the limited distributions of observed phenotypes. Such uncertainty is largely a consequence of our rudimentary understanding of the genetic architecture of complex traits. At the level of the genome, however, we can no longer hide from the facts. Over the past twenty-five years, molecular geneticists have revealed a huge amount of information about the structural and functional aspects of gene architecture and expression, to the point that we really do know the raw material upon which evolutionary forces operate. As in population genetics, there are still some gaps in the field of molecular genetics, but the knowledge base is so well established that if we are to concoct arguments on the matter of genome evolution, they must also be compatible with basic principles of molecular genetics.

The goal of this book is to help bring about a synthesis of our understanding of genomic evolution from the standpoint of both population genetics and molecular biology, and to outline the major challenges to the transformation of the descriptive field of comparative genomics into a more explanatory enterprise. The central issues to be explored revolve around the question of how the striking architectural diversity within and among prokaryotic and eukaryotic genomes came to be. This is not to say that the biggest questions in evolutionary biology are being addressed here. Those questions would include how phenotypic diversity came to be. But given that the complexities of cellular and developmental biology are constrained by the basic architectural features of genes and genomes, genomic repatterning underlies all aspects of evolution at higher levels of organization. Natural history is important to evolution, and in a very real sense, the cell and its contents define the natural historical setting within which the architectural features of genomes evolve.

There have been two major challenges to composing this book. First, although most biologists find mathematics unpalatable, evolutionary biology is one of the most mathematically sophisticated areas in the life sciences. Throughout, I have tried to clarify why the quantitative details matter, while also resisting the temptation to reveal all the gory details. In contrast, many evolutionary biologists find the reductionistic details of molecular biology to be arcane and irrelevant. But as noted above, because phenotypic evolution derives from change at the molecular level, many of the details matter here as well—the key is to figure out just which ones are indeed relevant to evolutionary processes. Second, the rapidity at which the field of genomic biology continues to emerge is daunting. I wrote this book over a period of about six years, but each time a chapter draft was set down for a few months, dozens of key findings had surfaced. Although I believe that a synthetic theory of genomic evolution is now possible, we can also be certain that many significant embellishments of our understanding will emerge in the very near future.

Hardcore population geneticists know that I am not a particularly good mathematician, and hardcore molecular biologists will easily see my weak understanding of subcellular processes. There is an intrinsic element of beauty in both of these fields, making it easy for the specialist to become sidetracked by the details, so there may be a small advantage of partial ignorance. I am hopeful that I know just enough about both fields to begin to see which details are essential to building a field of evolutionary genomics. As I embarked on this project, my intention was simply to develop a series of neutral models to evaluate whether selective explanations for genomic evolution were broadly justified. It quickly became clear that they often are not. We will soon know if I am right.

Acknowledgments

I come from a lineage of janitors, plumbers, cooks, factory workers, and clerks, so not a day goes by that I do not appreciate the unusual and unexpected opportunities that life has provided to me—the time to think about what got us here, and to do so through interactions with an enormous number of very smart people from a wide number of fields. The work reviewed on the following pages is derived from the labors of thousands of investigators in molecular, cellular, and developmental biology, genomic analysis and bioinformatics, molecular evolution, and population genetics. Although the names of many of these individuals are displayed in the Literature Cited section, countless lab technicians, undergraduate students, etc., have labored in the background of the more visible side of science. I am grateful to all of these people. Throughout the entire period of writing, I have been generously supported by the National Institutes of Health, the National Science

Foundation, and the Lilly Foundation funding to Indiana University. Such financial support has enabled me to maintain a laboratory containing a remarkable series of undergraduate and graduate students and postdoctoral fellows, all of whom have played a central role in my education.

Numerous specific acknowledgments are in order. I entered the field of genomic evolution largely as a consequence of early conversations with Allan Force on gene duplication. He regularly pushed me to the limits, and it is highly likely that this book would have never originated had such a Force not been with me at this point in my career. Matt Hahn, Harmit Malik, and Sally Otto read the first draft of the entire manuscript and provided enormously helpful suggestions regarding content, interpretation, and presentation. Their critical insights are woven throughout the final product. Extremely helpful input on one or more chapters was provided by Charlie Baer, Nicole Crown, Ben Evans, Mario dos Reis, Laurent Duret, Eric Haag, Laura Higgins, Alex Kondrashov, John Logsdon, Tomoko Ohta, Dmitri Petrov, Anthony Poole, Aaron Richardson, Sarah Schaack, Doug Scofield, Arlin Stoltzfus, and Greg Wray. Earlier comments by Vincent Daubin and Nancy Moran prompted me to think more deeply about prokaryotic evolution. Many of these people do not agree with everything that I have written, but I greatly appreciate their dissent (and that of others, who will remain nameless), as numerous issues regarding genomic evolution certainly remain to be resolved.

Once again, it has been a great pleasure to work with the extraordinarily talented staff at Sinauer Associates, Inc.: Andy Sinauer (editor), Sydney Carroll and Chelsea Holabird (production editors), Chris Small (production manager), Janice Holabird (composition and design), Norma Roche (copy editing), Marie Scavatto (advertising), and The Format Group (art production). Sitting on top of a hill in Guam, Lewis Rifkowitz (a potter) devised the cover.

Finally, I am especially grateful to Jeff Palmer for providing me with a job in the Department of Biology at Indiana University. Hermann Muller spent the last half of his productive career in this department, formulating among other things his seminal ideas on the role of deleterious mutations in evolution, and it has been a very special treat to work in the building in which he once walked. The intellectual setting for pursuing cross-disciplinary questions in the life sciences at Indiana University is truly exceptional. Many of the ideas that will be encountered on the following pages crystallized after I arrived here, and I doubt they would have ever come to light if I had not made such a move. Every aspect of my day-to-day academic life has been facilitated by the tolerance and support of the people at IU, from the administrators at the top to the office staff working behind the scenes to my numerous faculty, graduate student, and postdoctoral colleagues. Most of all, I wish to thank Emília Martins and our son Gabe for their enduring patience, provisioning of distractions, and the privilege of being in their company.

Michael Lynch
Bloomington, IN.

1 The Origin of Eukaryotes

Approximately 4.6 billion years ago (BYA), a cloud of cosmic gas and dust condensed into the solar system that contains our planet (Nisbet and Sleep 2001). For the next 0.8 billion years, Earth was heavily bombarded by interstellar debris, with some of the more massive impacts generating enough heat to sterilize the entire planet (Sleep et al. 1989). Thus, although early steps in the origin of life may have been taken prior to 3.8 BYA, the roots of modern biology are probably younger than this. The earliest stages of life presumably involved simple polymers, perhaps capable of replicating only on time scales of days or months and probably doing so quite inaccurately. Unfortunately, for obvious physical and chemical reasons, no fossils of noncellular life from this period are likely to be found.

Successful life forms must be capable of acquiring energy while also harboring a heritable genotype containing the information for perpetuating such abilities. All of today's cellular life employs DNA for information storage, proteins for enzymatic activity and cell structure, and RNA for various aspects of DNA processing and protein production. It is implausible that all three types of biomolecules appeared simultaneously, but which came first? With its complementary double-stranded structure, DNA provides a superb substrate for information storage and replication, but is for the most part catalytically inert. In contrast, although proteins carry out a bewildering diversity of tasks, self-replication is not one of them. This leaves RNA as the only reasonable candidate for a starting point in evolution, and with its dual capacity for information storage and processing, a consensus has gradually emerged that the initial informational component of the biosphere consisted entirely of RNA (Woese 1983; Gilbert 1986; Gesteland et al. 1999).

With no fossil record to restrict our imagination, this earliest phase of molecular evolution has inspired much creative thought, often generating testable predictions. For example, numerous laboratory evolution experiments involving diverse populations of RNA molecules have successfully selected for a wide variety of catalytic activities, including some key steps toward self-replication (Wilson and Szostak 1999; Joyce 2004). Such demonstrations of the enzymatic potential of RNA support the idea that a primitive metabolism may have existed prior to the evolution of protein synthesis. The central role of RNA in transcript processing and translation in modern organisms is presumably a reflection of this early era, and the diverse assemblage of viruses with RNA-based genomes provides direct evidence that RNA can serve as a reliable information storage molecule.

If a biosphere consisting entirely of RNA existed, it was relatively quickly displaced by life forms that rely on cooperative activities of RNA, DNA, and a broad assemblage of proteins. According to one view, a diverse assemblage of microbes, including cyanobacteria-like cells, had colonized the seas by ~3.5 BYA (Schopf 1993; Schopf et al. 2002). Some aspects of this interpretation of the fossil record have been questioned (Brasier et al. 2002), but other signs of biological activity have been found in rocks from 3.4–3.8 BYA (Rosing 1999; Furnes et al. 2004; Tice and Lowe 2004), and unambiguous fossils of filamentous organisms deposited around hydrothermal vents have been dated to 3.2 BYA (Rasmussen 2000). Thus, cellular life (as we know it) appears to have emerged from inorganic materials within a window of just a few hundred million years.

The lack of distinctive morphological features renders the detailed taxonomic composition of the early fossil record quite uncertain, but the first 0.5–1.0 billion years of life appear to have been dominated by prokaryotes (simple cells lacking membrane-enclosed organelles, often referred to as bacteria), if not entirely restricted to them. Given the presence of methanogenesis as early as 3.0 BYA (Nisbet 2000) and photosynthesis as early as 2.8 BYA (Des Marais 2000), we can conclude that the microbial world was quite sophisticated biochemically by this time, perhaps harboring the full repertoire of metabolic/molecular processes from which all subsequent cellular lineages were built. The first evidence of eukaryotes (cells with membrane-enclosed organelles, including a nucleus) appears in the form of putative diagnostic biomarkers of membrane components deposited in shale from ~2.7 BYA (Brocks et al. 1999). The first presumptive algal fossils date to ~2.1 BYA (Han and Runnegar 1992), and many other fossils of unicellular eukaryotes with well-developed cytoskeletons date to 1.7–1.5 BYA (Knoll 1992; Shixing and Huineng 1995; Javaux et al. 2001). But despite this gradual addition of eukaryotic diversity, the biosphere continued to be dominated by unicellular and oligocellular (a few cell types) species for at least another billion years.

A dramatic shift occurred ~0.55 BYA, when all of the major groups of multicellular animals appear essentially simultaneously in the fossil record, in what is popularly known as the Cambrian explosion. Shortly thereafter, jawed vertebrates arose (~0.44 BYA), as did land plants (~0.40 BYA). The origin of the major angiosperm (flowering plant) groups came somewhat later (~0.14 BYA), followed by the radiation of the major orders of mammals (~0.10

BYA). The only species capable of thinking about things like this, *Homo sapiens*, is a very recent arrival (~0.002 BYA), but now dominates the planet ecologically to such an extent that much of the global legacy of biodiversity is on the verge of extinction. An enormous amount of literature chronicles these kinds of events on a finer scale in the context of long-term changes in climate, atmospheric composition, and landmass locations (a highly readable overview is provided by Knoll 2003).

The main goal of this chapter is to establish a general phylogenetic setting for the diverse assemblage of organisms that will be encountered in succeeding chapters. First, we will consider how a DNA-based genome might have emerged out of an RNA world. This issue raises numerous other critical questions, including whether true cells arose prior to the reliance on DNA for information storage, and if so, whether DNA-based cellular genomes evolved more than once. Because these questions address singular events, they are not subject to traditional hypothesis testing, and some may never be fully settled. However, what is not in question is the remarkable degree to which the descendants of the primordial DNA-based cells have colonized Earth (Box 1.1). Second, we will examine the degree of phylogenetic affinity between the major functional groups of cellular life: the prokaryotes and the eukaryotes. Despite the centrality of this issue to all of biology, many questions remain unanswered, including the nature of the

Box 1.1 How much DNA?

A crude estimate of the amount of DNA within currently living organisms can be made by noting that the length spanned by one base of DNA is ~0.3×10^{-12} km (Cook 2001).

The number of viral particles in the open oceans is ~10^{30} (Suttle 2005). Assuming that there are twice as many viruses on land and in fresh water does not change the global estimate very much at the order-of-magnitude level. Thus, assuming an average viral genome size of 10^4 bp, the total length of viral DNA if all chromosomes were linearized and placed end to end is ~10^{22} km.

The estimated global number of prokaryotic cells is ~10^{30} (Whitman et al. 1998), and assuming an average prokaryotic genome size of 3×10^6 bp (see Chapter 2) yields an estimated total DNA length of 10^{24} km.

With a total population size of 6×10^9 individuals, 10^{13} cells per individual (Baserga 1985), and a diploid genome size of 6×10^9 bp, the amount of DNA occupied by the human population is ~10^{20} km. Assuming there are ~10^7 species of eukaryotes on Earth (~6 times the number that have actually been identified), that the average eukaryotic genome size is ~1% of that of humans (see Chapter 2), and that all species occupy approximately the same amount of total biomass (see Chapter 4), total eukaryotic DNA is ~10^5 times that for humans, or ~10^{25} km.

Given the very approximate nature of these calculations, any one of these estimates could be off by one to two orders of magnitude, but it is difficult to escape the conclusion that the total amount of DNA in living organisms is on the order of 10^{25} km, which is equivalent to a distance of 10^{12} light-years, or 10 times the diameter of the known universe.

prokaryote(s) that gave rise to eukaryotes. Third, drawing from a wide array of comparative studies, we will find that eukaryotes are monophyletic (share a most recent common ancestor, the stem eukaryote, to the exclusion of any prokaryotic group). The shared presence of a wide array of genomic features across all eukaryotes suggests that the stem eukaryote was much more complex genomically than any known prokaryote, although many of the individual peculiarities of the eukaryotic cell can be found in isolated prokaryotic species. Finally, we will find that the major phylogenetic lineages of eukaryotes diverged within a fairly narrow window of time, and some speculation will be offered on the mechanisms that may have precipitated such a rapid radiation.

Entry into the DNA World

Despite its broad appeal, the RNA world hypothesis leaves open the question of how and why an RNA-based genome was eventually displaced by DNA. There are compelling reasons to think that proteins evolved prior to this transition. For example, an early RNA–protein world would imply the existence of the genetic code prior to the arrival of DNA, which is consistent with the ubiquitous use of transfer RNAs (tRNAs) and messenger RNAs (mRNAs) in translation. In addition, the nucleotide building blocks from which DNA is built in today's organisms are derived from RNA precursors by use of enzymatic proteins: ribonucleotide reductases are used in the production of dAMPs, dCMPs, and dGMPs, while thymidylate synthase catalyzes the production of dTMPs by methylation of dUMPs (Figure 1.1). This additional step in dTMP synthesis suggests that the initial transition to a DNA-based genome might have involved an intermediate form of DNA in which U was used instead of T (as is the case in RNA). Remarkably, the genomes of some bacteriophage (viruses that infect prokaryotes) have such a structure (Takahashi and Marmur 1963). Equally notable, the fact that two unrelated thymidylate synthases with substantially different mechanisms of operation are distributed among the primary prokaryotic lineages (Myllykallio et al. 2002) suggests the intriguing possibility that DNA originated more than once. Although such a claim may seem rather fantastic, it is consistent with the use of two apparently nonhomologous sets of DNA replication proteins in different prokaryotic lineages (Edgell and Doolittle 1997; Olsen and Woese 1997; Leipe et al. 1999).

Given the early success of the RNA world, why would the transition to a DNA world be so complete as to eradicate all RNA-based genomes (except those of RNA viruses)? The answer is probably related to the reduction in mutational vulnerability afforded by a DNA-based genome. As an information storage molecule, DNA has two significant chemical advantages. First, the additional –OH group on ribose renders RNA much less structurally stable than DNA. Second, one of the most common sources of mutation is the production of uracil via the deamination of cytosine (see Chapter 6). In thymine-bearing DNA, uracil can be recognized as aberrant and

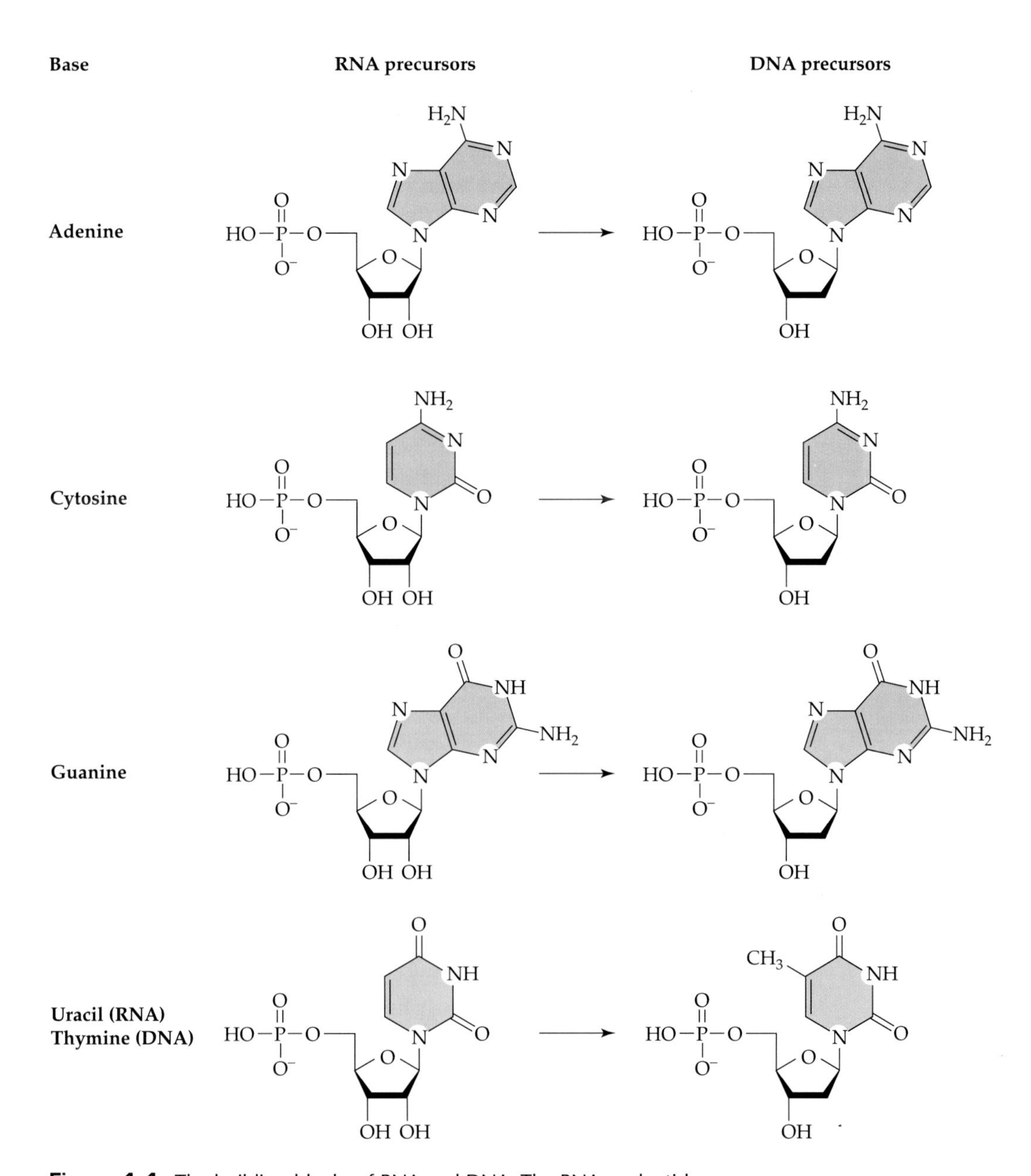

Figure 1.1 The building blocks of RNA and DNA. The RNA nucleotide precursors are enzymatically converted into those for DNA (arrows).

corrected prior to replication, but such a distinction is impossible in RNA. Thus, in an RNA world, an organism that discovered a way to store its genome as DNA would have had a substantial advantage in terms of the reliable production of progeny genotypes.

The arrival of DNA, with its reduced mutational vulnerability, was a revolutionary event for evolution, as it provided a more permissive environment for genomic expansion and hence for the emergence of more complex biological functions. Because the vast majority of mutations are deleterious, an excessively high mutation rate can overwhelm the power of purifying natural selection, resulting in a deterministic decline in fitness and eventual population extinction. This general principle was first promoted by Eigen and Schuster (1977) in their error catastrophe theory, which was later expanded in more formal population genetic terms to describe the conditions leading to a mutational meltdown (Lynch and Gabriel 1990; Lynch et al. 1995a,b). Depending on the reproductive mode and the distribution of mutational effects, a population is unable to sustain itself if the deleterious mutation rate exceeds one to five per individual per generation, and once a genome is close to this meltdown threshold, further expansion of genome size is impossible without a mechanism for reducing the per-nucleotide mutation rate. As will be seen in many contexts in the following pages, a variant of this principle also applies to the structural features of individual genes: in populations of sufficiently small size with sufficiently low per-nucleotide mutation rates, changes in gene architectural features with mild mutational costs (such as increases in the length of coding or regulatory regions) can accumulate with little opposition from natural selection and without imperiling the survival of the species; and although such features need not be of immediate adaptive significance, they may eventually contribute to adaptive evolution (Lynch 2002b, 2006a). Thus, once the technology for reliable maintenance and utilization of DNA evolved, entirely novel pathways for adaptive evolution would have opened up, leading to the rapid displacement of the RNA world by DNA-based genomes.

A viral origin of DNA?

With their peculiar genomic architectures and their lack of structural features universal to cellular species, the phylogenetic positions of viruses in the tree of life have remained enigmatic, and many biologists regard them as nothing more than products of ill-defined mechanisms of reductive evolution from cellular species. Forterre (2005, 2006a,b) has challenged this view, arguing not only for the early existence of a diverse virosphere, but for a central role of viruses in the invention of DNA. The plausibility of this hypothesis derives from the fact that viruses exploit hosts from all domains of cellular life and can have either RNA- or DNA-based single- or double-stranded genomes. Under Forterre's hypothesis, DNA first evolved in viruses as a means for avoiding recognition in a world of RNA host cells, and once established in such cells, began to accumulate host cell gene copies until the entire host genome had been transferred. The physical possibility of such transfers is not in doubt, as today's viruses commonly encode reverse transcriptase, an enzyme that produces complementary strands of DNA from RNA templates. However, Forterre takes things a step further, arguing that DNA takeovers by entirely different viruses occurred in independent RNA host lineages, eventually driving them all into the DNA world. An extraordinary diversity of DNA processing machinery exists among viral lineages, including many mechanisms with no known counterparts in cellular species, so one attractive feature of this argument is its potential to explain the use of radically different DNA processing machinery in the major lineages of cellular life (noted above).

Any hypothesis for the transition to a DNA world requires that such a change be either neutral or immediately beneficial, but is it necessary to invoke an initial role for DNA as a viral defensive mechanism? The hypothesis that a DNA-based viral genome provided a means for avoiding detection presupposes the existence in the RNA world of host mechanisms for discriminating self from foreign RNA, raising the question of why DNA would not be recognized as foreign as well. In addition, the absorption of an entire host genome by a parasite imposes some rather stringent requirements, not the least of which is the need for the rapid transformation of an initially negative host–parasite interaction into a more benign relationship so as to keep the liaison stabilized during the potentially millions of years of transfer of genetic information. In principle, this could be accomplished by the transfer of a single essential host gene to the viral genome, but the ability of a cell lineage containing such a rearrangement to spread throughout the host population would require an advantage to the host exceeding the cost of the parasite load.

Because a reduction in the deleterious mutation rate provides an immediate advantage to any DNA-based genome, there is no clear need to invoke the initial establishment of DNA as a viral defensive mechanism, although this does not negate the possibility of a viral origin for DNA, with secondary transfer to host genomes occurring by other mechanisms. At the very least, Forterre has forced us to reevaluate the role of viruses in early genomic evolution, and the recent discovery of numerous DNA viral genomes containing hundreds of genes affirms this view, blurring the distinction between viruses and prokaryotes (Shackelton and Holmes 2004; Claverie et al. 2006).

Membranes early or late?

A final significant issue regarding early evolution concerns the stage at which individuality evolved (Figure 1.2). Individualization (a strict one-to-one linkage between genotype and phenotype) is essential for the successful operation of adaptive evolution, as it ensures the transmission of genotypes by selectively favored members of the population. If a genome led to a useful product that diffused broadly, then most members of the population (including those not encoding the product) would receive equal benefits, greatly reducing the efficiency of selection. One view is that the isolation of genomes behind true cell membranes evolved prior to the emergence

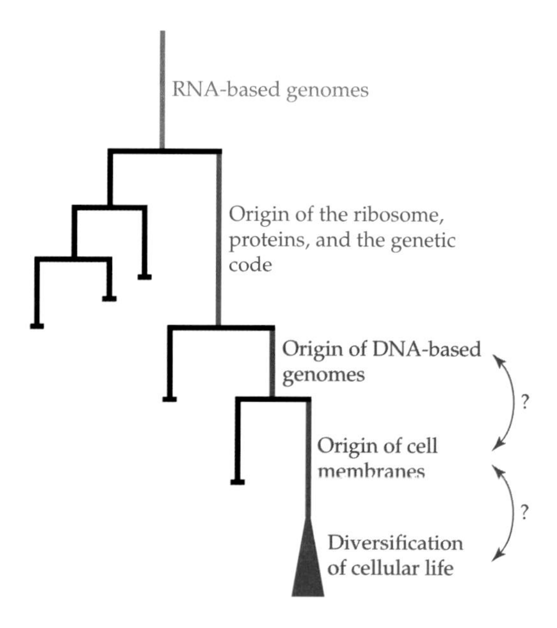

Figure 1.2 A schematic for the early transition to DNA-based genomes and cellularity. Although the hypothesis that the biosphere initially was composed entirely of RNA, with proteins evolving secondarily, has gained wide acceptance, the order of events in the transition to cellular life is less clear (as denoted by the question marks). According to one view, the major cellular domains of life arose prior to the end of the RNA world (Forterre 2005, 2006a,b), but another view postulates that DNA-based genomes were well established prior to the evolution of cell membranes (Koonin and Martin 2005). Black lines ending in crossbars denote hypothetical extinct lineages.

of the DNA world (Forterre 2005, 2006a,b), but a major challenge to this membranes-early hypothesis is the presence of radically different pathways for cell membrane lipid biogenesis in the two major groups of prokaryotes. The archaea use isoprene ethers, whereas the eubacteria use fatty acid esters (Boucher et al. 2004). It is not immediately obvious how these states could evolve from a common ancestor or how one state could make a transition to another. However, both difficulties are eliminated under Koonin and Martin's (2005) hypothesis that cell membranes evolved in independent lineages subsequent to the establishment of the DNA world. Under this membranes-late view, the earliest DNA-based genomes were sequestered within abiogenic, hydrophobic membranes (which are known to be generated in environments such as hydrothermal vents). Although the shared presence of several key membrane-associated enzymes across all domains of cellular life has been viewed as inconsistent with the membranes-late hypothesis (Forterre 2005), Koonin and Martin (2005) make the case that such enzymes could have been effective in abiogenic membranes.

The Three Domains of Cellular Life

Although we can only speculate on the earliest steps in the evolution of precellular life, information on the genealogical relationships of surviving lineages is contained within the genomes of extant taxa. Unlike hard fossils, which provide an effectively permanent snapshot of past history, DNA acquires nucleotide substitutions and rearrangements over time, which lead to gradual divergence of DNA sequences among sister taxa. Quantification of such change provides an entrée into the relationships of different lineages as, on average, species with higher levels of DNA sequence similarity are more closely related phylogenetically. Based on this logic, numerous methods have been developed for estimating phylogenetic trees and dating evolutionary events (e.g., Feng et al. 1997; Li 1997; Nei and Kumar 2000; Hedges et al. 2001; Felsenstein 2003).

One of the first applications of such methods led to one of the greatest breakthroughs in systematics of the last century: the demonstration of an entirely unexpected level of phylogenetic diversity in the microbial world. Noting that all cellular organisms use ribosomes to translate messenger RNAs (mRNAs), Woese and Fox (1977) reasoned that the entire tree of life might be resolved via comparative analysis of ribosomal RNA (rRNA) sequences. (Although rRNAs are encoded at the DNA level, at the time it was easier to isolate and sequence RNA than DNA). Up to this point, all prokaryotes had been viewed as one large, ill-defined, monophyletic group, deeply separated from the eukaryotes. However, Woese and Fox found a deep phylogenetic furrow within the prokaryotes, implying the existence of two species clusters as distinct from each other as they are from eukaryotes (Figure 1.3). Following Woese et al. (1990), these two prokaryotic groups have come to be known as the archaea and the eubacteria. The frequent occupation of extreme environments (e.g., hot springs and hypersaline brine pools) by archaea has inspired many successful searches for life in environments previously thought to be biological deserts (Howland 2000). Because the term "prokaryote" is no longer phylogenetically informative, it has been argued that its usage should be abandoned entirely (Pace 2006). However, as a matter of convenience, the term will be retained in the following text whenever the collective groupings of eubacteria and archaea are being referred to.

The division of prokaryotes into two distantly related domains raises several questions about the base of the tree of life. Are the eubacteria, archaea, and eukaryotes fully monophyletic, or is one or more clades embedded within another? Is the eukaryotic lineage more closely related to the archaea or the eubacteria, or does it have affinities with both? Can the possibility that eukaryotes are ancestral to prokaryotes be ruled out? The key to answering these questions is a correctly rooted phylogenetic tree denoting the most recent common ancestor from which all species in the tree descend. But, herein lies a significant problem: placing a root on a phylogenetic tree requires the inclusion of an outgroup (a lineage that is clearly outside the

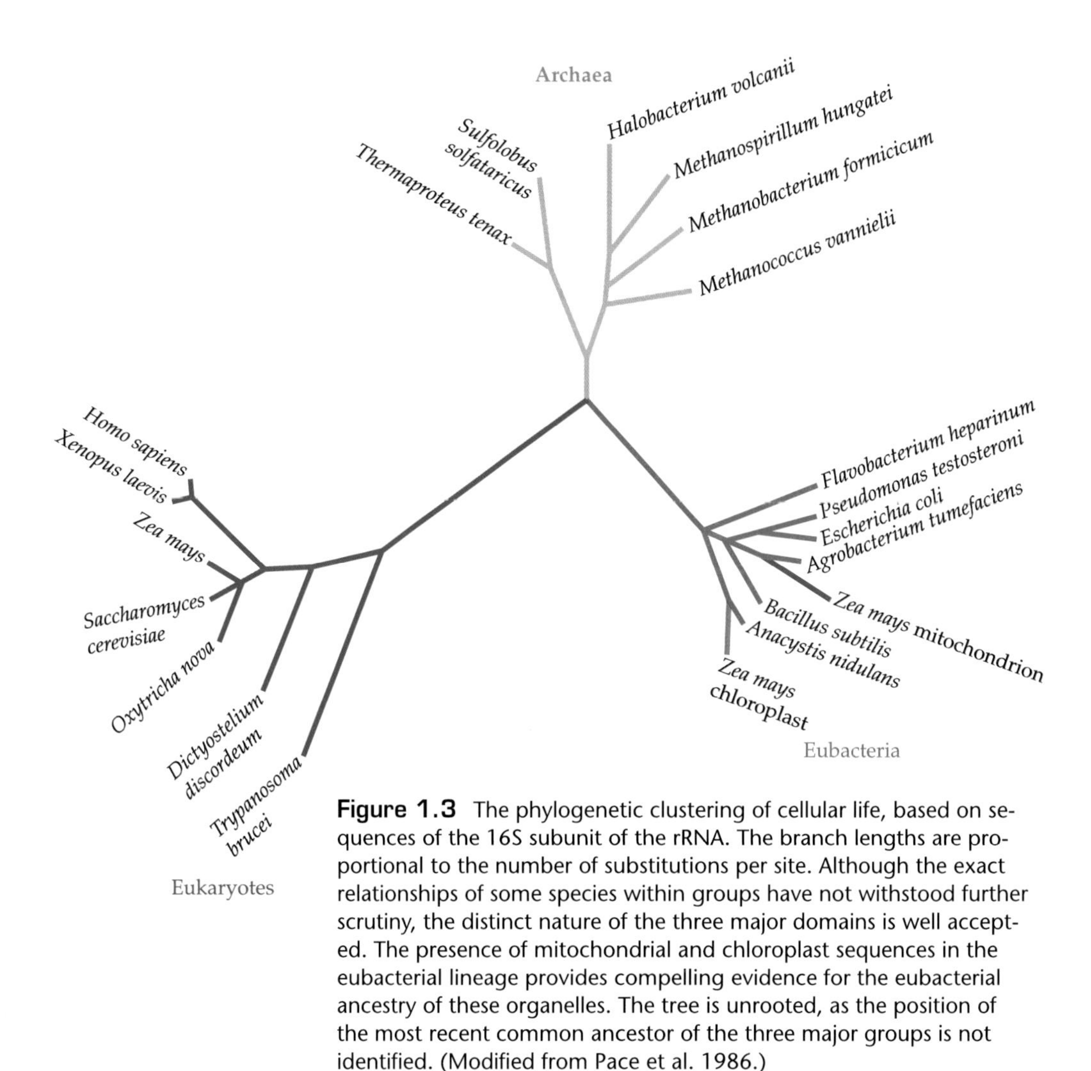

Figure 1.3 The phylogenetic clustering of cellular life, based on sequences of the 16S subunit of the rRNA. The branch lengths are proportional to the number of substitutions per site. Although the exact relationships of some species within groups have not withstood further scrutiny, the distinct nature of the three major domains is well accepted. The presence of mitochondrial and chloroplast sequences in the eubacterial lineage provides compelling evidence for the eubacterial ancestry of these organelles. The tree is unrooted, as the position of the most recent common ancestor of the three major groups is not identified. (Modified from Pace et al. 1986.)

group under consideration, such as a bird for a mammalian phylogeny) in the analysis, but that is not an option when the entire tree of life is under consideration.

The only known way to solve this problem is to focus on a pair of ancient duplicate genes. If the duplicates are present in each of the three major domains of life, then they almost certainly were present in the last common ancestor of all life. Each gene can then serve to root the phylogeny of its partner, and the topologies for the two copies are expected to be identical. Following this approach, Gogarten et al. (1989) and Iwabe et al. (1989) used anciently duplicated subunits of membrane ATPase to show that archaea and eukaryotes consistently group together (as in Figure 1.4). The same result was obtained with several other pairs of ancient duplicate genes

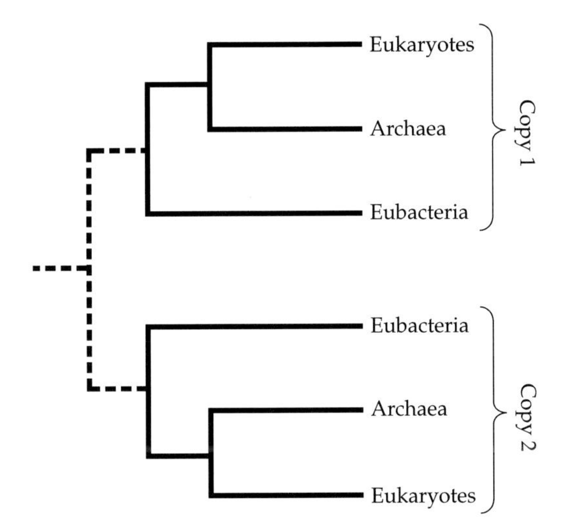

Figure 1.4 The reciprocal rooting of a phylogenetic tree using a pair of ancient duplicate genes. Provided the duplication event occurred prior to the divergence of the species under consideration (here depicted as single members of the eukaryotes, archaea, and eubacteria) and all lineages retain it, two identical topologies are expected for each gene, with the connecting branches (dashed lines) reflecting the divergence of the two copies within the ancestral species.

(Iwabe et al. 1989; Brown and Doolittle 1995; Baldauf et al. 1996; Lawson et al. 1996; Gribaldo and Cammarano 1998), leading to the conclusion that of the two branches emanating from the base of the tree of life, one contains only eubacteria, while the other contains the common ancestor that later diverged into the archaea and eukaryotes.

Unfortunately, subsequent analyses have raised significant concerns about this view. Most notably, the two phylogenies associated with some duplicate genes are discordant (Philippe and Forterre 1999). Occasional excursions of an archaeal gene into the eubacterial clade, or vice versa, may result from horizontal gene transfer (gene acquisition from distantly related lineages), which is known to be important in microbes (Ochman et al. 2000; Gogarten et al. 2002; Jain et al. 2002), so we should not be overly concerned with a few exceptions to a general pattern that associates eukaryotes and archaea. However, horizontal transfer is not the only source of ambiguity in these kinds of analyses. Reciprocal rooting of phylogenetic trees with duplicate genes rests critically on the assumption that no additional gene family members existed in the most recent common ancestor of today's organisms. If that is not the case, and different gene copies have been lost in different lineages, all sorts of illusions can appear (Figure 1.5), and the situation is exacerbated if lineage-specific duplications have also occurred. The common occurrence of gene duplication in all genomes (see Chapter 8) makes this a nontrivial issue.

Many additional problems can arise in phylogeny reconstruction as a simple consequence of sampling error (inadequate numbers of informative sites) or of the extraordinary phylogenetic distances among the three major domains (long enough to allow for multiple substitutions per site). Unequal rates of evolution among phylogenetic groups are particularly sinister in this regard, as they encourage branches with exceptionally high rates of evo-

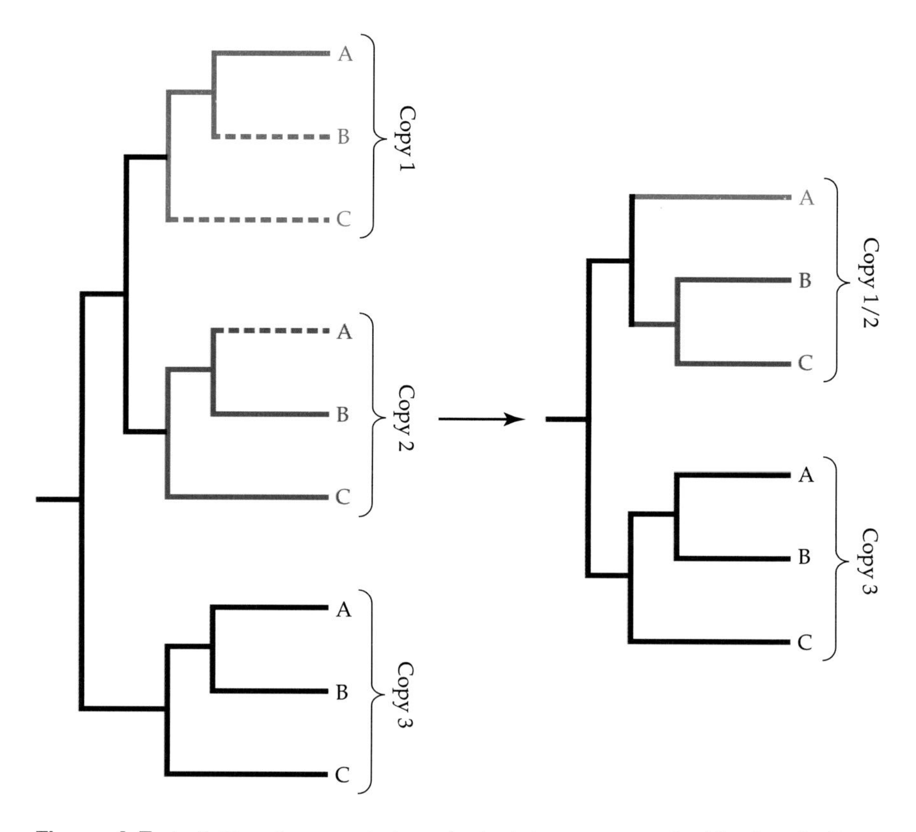

Figure 1.5 *Left*: Here the ancestral species had three gene copies (the first duplication led to copy 3 and the ancestor to 1 and 2; and the second duplication created copies 1 and 2). Copy 1 was lost in species B and C, and copy 2 was lost in species A (dashed lines). *Right*: Using just the surviving gene copies, a conflicting signal is obtained on the phylogenetic relationships of species A, B, and C, even though the overall topology of extant gene relationships is correct. The top cluster incorrectly implies a phylogeny in which species B and C are grouped together (as a consequence of an incorrect mixture of copy 1 and 2 genes), whereas the bottom cluster correctly groups A and B.

lution to group together due to chance parallel character changes (Felsenstein 1988). If, for example, eubacterial genes evolve at higher average rates than those of archaea and eukaryotes, the reciprocal rooting technique will be biased toward the production of duplicate-gene trees with eubacterial outgroups (as in Figure 1.4, where the bases of the two eubacterial lineages connect directly to the root of the tree). Ordinarily, one can test whether two lineages evolve at different rates by comparing their divergence from a common outgroup (Figure 1.6), but we cannot rely on this logic here, as it is the outgroup that we are trying to discover.

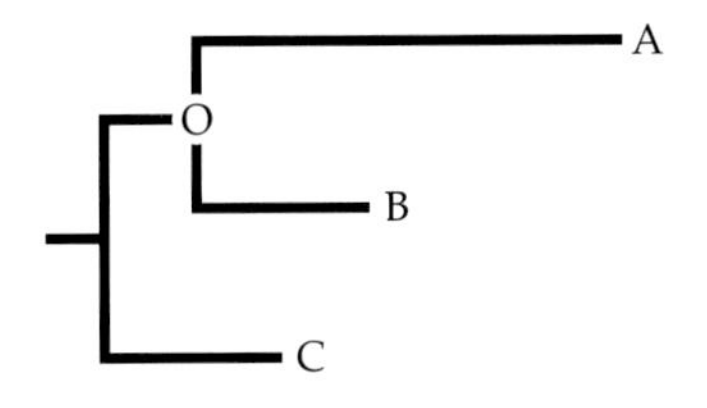

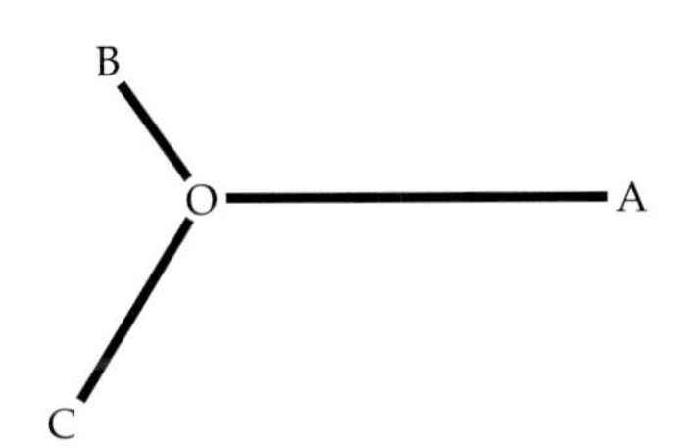

Figure 1.6 A three-species phylogeny with unequal rates of evolution in different lineages (branch lengths are proportional to the amount of sequence divergence). *Top*: When a known outgroup species (C) is included, the hypothesis of equal rates of evolution in the lineages leading to A and B can be tested by comparing the total divergence between A and C with that between B and C, as both A and B share the path from C to common ancestor O. *Bottom*: The topology here is identical to that above, except that the location of the root has been removed. With this view, the most similar species pair in terms of sequence divergence is B and C, even though the true genealogical relationship between A and B is closer.

Tree, ring, or web of life?

Conflicting phylogenetic signals from different genes would also be expected if any of the three major domains was a hybrid between members of the remaining two lineages. Numerous analyses of single-copy genes suggest that this is the case for eukaryotes (Brown and Doolittle 1997; Feng et al. 1997; Koonin et al. 1997; Golding and Gupta 1995; Gupta 1998a,b; Rivera et al. 1998; Doolittle et al. 2003). Unlike the duplicate-gene studies noted above, these surveys rely on the relative similarities of eukaryotic genes to those of archaea versus eubacteria as measures of phylogenetic affinity. This approach carries the risk that with an unrooted phylogeny, slowly evolving lineages can appear to be more similar than they actually are in a genealogical sense (see Figure 1.6). Nevertheless, an emergent pattern from these analyses is that eukaryotic genes involved in information processing (e.g., transcription and translation) tend to be more similar to those in archaea, while those involved in housekeeping functions (e.g., metabolism) tend to be more similar to those in eubacteria (Brown and Doolittle 1997; Rivera et al. 1998; Leipe et al. 1999; Brown et al. 2001; Horiike et al. 2001; Esser et al. 2004). In addition, some genes appear to be present in only two of the three major domains, and in this context, the pattern of presence and absence for information processing versus housekeeping genes is consistent with the sequence-based dichotomy (Lake et al. 1999).

Such observations strongly suggest that the nuclear genome of the stem eukaryote arose, at least in part, as an archaeal-eubacterial chimera, as first postulated by Sogin (1991) and Zillig (1991). Under this interpretation, the base of the tree of life is not a stem but a ring, and further analysis implies that the ring was closed by the mixture of eubacterial and archaeal lineages to form eukaryotes, rather than by the mixture of eukaryotes and one of the

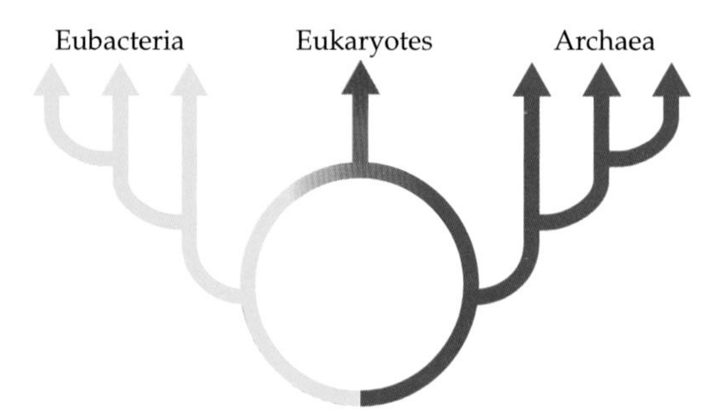

Figure 1.7 The "ring-of-life" hypothesis for the origin of eukaryotes. Yellow and blue lineages denote branches in the phylogenetic trees for eubacteria and archaea, respectively. Members of two such lineages fused to form the eukaryotic domain (green). (Modified from Rivera and Lake 2004.)

prokaryotic lineages (Rivera and Lake 2004) (Figure 1.7). Less clear is whether the emergence of the eukaryotic lineage was a simple consequence of a fusion between single eubacterial and archaeal species or a more gradual result of incremental horizontal transfers from multiple contributors (Lester et al. 2006; Doolittle et al. 1998). Doolittle et al. (2003) offer the extreme view that the earliest phases of cellular evolution involved such rampant horizontal gene transfer, both within and among major domains, that the base of the tree of life is more appropriately viewed as a web.

Eukaryotic gene acquisition from endosymbionts

There is broad agreement on one source of eubacterial genes in the eukaryotic domain: almost all eukaryotes harbor one or more minor genomes within organelles (mitochondria and chloroplasts) originally derived from endosymbiotic eubacteria (Delwiche 1999; Lang et al. 1999), as revealed by phylogenetic analysis (see Figure 1.3). Although a few amitochondriate eukaryotes exist, they all exhibit vestiges of the prior presence of a mitochondrion (in the form of endosymbiont genes transferred to the nucleus; Roger et al. 1998; Mai et al. 1999; Dyall and Johnson 2000; McArthur et al. 2001; Tachezy et al. 2001; Silberman et al. 2002; Dolezal et al. 2005). Thus, we can be confident that the stem eukaryote contained a genome-bearing mitochondrion. Phylogenetic analysis strongly suggests an α-proteobacterial origin for the mitochondrion, which is consistent with the double membrane surrounding this organelle. This means that from the very beginning, eukaryotes had a built-in mechanism for importing eubacterial genes into the nuclear genome. Radical genomic incompatibilities would be expected to result from full-scale contact between two highly divergent genomes, but this would have been avoided by the incremental seeding made possible by the sequestration of incipient organellar genomes behind membrane barriers.

Several other significant issues regarding the chimeric origin of the nuclear genome, including the identity of the host genome(s) and the nature of the hybridizing event(s), remain unclear (Gupta 1998a,b; Lake and Rivera 1994; Margulis et al. 2000; Cavalier-Smith 2002b,c; Hartman and Fedorov 2002). The simplest explanation for the existing data is that the initial contents of the nuclear genome were derived from an archaeal host cell (Martin and Russell 2003). Otherwise, one must invoke massive horizontal transfer of archaeal information processing genes in the face of evidence that the transfer of such genes among microbes occurs at a much lower rate than that of housekeeping genes (Jain et al. 1999).

How might the ancestral eukaryote have acquired the primordial mitochondrion? Under the hypothesis that acquisition involved ingestion, it has long been thought that the predatory host cell must have had a cytoskeleton (to support phagocytosis), a vesicle transport system (for processing prey), and a nuclear membrane (to protect the genome from the shearing effects of intracellular movement) (Stanier 1970). The fact that at least one eubacterium harbors a eubacterial endosymbiont (von Dohlen et al. 2001) raises questions about the necessity of such a morphology, although phylogenetic analysis supports it (Richards and Cavalier-Smith 2005). Moreover, although the nuclear membrane is generally viewed as a eukaryotic innovation, its presence at the time of mitochondrial colonization cannot be ruled out. Some members of the eubacterial Planctomycetes house their DNA within a double membrane–enclosed nucleus (Lindsay et al. 2001; Fuerst 2005), and although this structure may not be homologous to the eukaryotic nuclear membrane (Mans et al. 2004), its mere presence clearly indicates that the ancestral eukaryote could have had significant cell structure prior to the origin of organelles.

The unity of the eukaryotes

Although the preceding results strongly support the hypothesis that the stem eukaryote was some kind of chimera, even this view has a few detractors. For example, Poole et al. (1998, 1999) have argued that the idea that the nuclear genome is derived from a prokaryote is entirely backward, postulating instead that the species at the base of the tree of life was a complex eukaryote. The logic underlying this argument derives from the idea that all functional RNAs must have emerged in the RNA world. Under this assumption, classes of RNA molecules found only in eukaryotes (e.g., those involved in the processing of introns, ribosomes, and the telomeres of linear chromosomes) must be relics of the RNA world that were lost from the secondarily derived prokaryotic lineages. The power of this argument would be eliminated entirely if the three major domains of life emerged prior to the transition to the DNA world, which, as noted above, cannot be ruled out. Moreover, the idea that new RNA functions cannot evolve in a DNA world is inconsistent with the exploitation of numerous small RNAs in animal and land plant development (see Chapter 3).

In summary, although the phylogenetic reality of the three-domain view of life continues to be a contentious issue (Margulis 1996; Gupta 1998a,b; Mayr 1998; Woese 1998; Cavalier-Smith 2002b,c), a good deal of the controversy is a consequence of our uncertainty as to whether the unique morphology of the eukaryotic cell arose before or after the unique features of the eukaryotic genome. Resolution of the remaining issues will ultimately have to come from genealogical information recorded at the DNA level, but at this point, no evidence of this kind supports the eukaryotes-first hypothesis. In contrast, the idea that all eukaryotes are contained within a single monophyletic lineage is fully consistent with the data (e.g., Ciccarelli et al. 2006).

The Stem Eukaryote

From the shared characters of all extant species, we can be confident as to the genomic and cellular characteristics that accumulated in the line of descent leading up to the most recent common ancestor of all eukaryotes. The assumption here is that highly complex cellular features are unlikely to have arisen independently in dozens of deeply branching eukaryotic lineages, and therefore must have been present in the ancestor of all eukaryotes. Such an exercise tells us that the stem eukaryote was quite sophisticated. The most celebrated eukaryotic attributes are physical ones: a membrane-enclosed nucleus, cellular organelles, and a cytoskeleton. But eukaryotes also distinguish themselves in numerous ways at the level of gene structure, genome organization, replication, and transcript processing.

First, as noted above, some genome size expansion must have occurred in the stem eukaryote by horizontal transfer from the primordial mitochondrion. In principle, such an event could have roughly doubled the gene content of the nuclear genome, although many redundant genes were probably lost at an early stage in the transfer process (see Chapter 8). The stable maintenance of a genome-bearing organelle requires a reliable mechanism for coordinating the transmission of organellar and nuclear genomes. The nuclear genome came to be inherited by mitotic and meiotic mechanisms, perhaps even before the arrival of the proto-mitochondrion, and as a by-product, meiosis also provided a novel means for reassembling genotypes by segregation and recombination (Dacks and Roger 1999; Ramesh et al. 2005). The mitochondrial genome continued to be inherited in the manner of its immediate prokaryotic ancestor, with a single circular genome and a single origin of replication (see Chapter 11).

Second, subdivision of a circular ancestral genome with a single origin of replication into a nuclear genome containing a series of linear fragments raised structural challenges for eukaryotic genome inheritance (see Chapter 5). Chromosome linearity did not necessarily originate in eukaryotes, as several eubacteria have single linear chromosomes, including *Streptomyces coelicolor* (Bentley et al. 2002), the plant pathogen *Agrobacterium tumefaciens* (Goodner et al. 2001; Wood et al. 2001), and the agent of Lyme disease, *Bor-*

relia burgdorferi (Fraser et al. 1997). Nevertheless, chromosome linearity imposes requirements for the maintenance of chromosome ends, while chromosome multiplicity imposes the need for a mechanism for transmitting a balanced set of chromosomes to each daughter cell. The capping of chromosomes with telomeres and the evolution of the mitotic machinery played central roles in solving these problems.

Third, the processing of transcripts underwent considerable modification in the stem eukaryote. Most, if not all, prokaryotes have operons—cassettes of often functionally related genes that are cotranscribed—but such polycistronic transcription constitutes a significant challenge for a membrane-enclosed genome: a multigene transcript must either be exported from the nucleus in its entirety or processed into single-gene fragments that can be individually recognized by the ribosome. Although most of today's eukaryotes harbor no operons at all, there are a few exceptions in distantly related lineages (e.g., trypanosomes and nematodes). In all cases, the same *trans*-splicing mechanism is used to process the transcript prior to export from the nucleus: the polycistron is snipped into pieces, and a small leader sequence is spliced to the front end of each individual transcript (see Chapter 10). *Trans*-splicing is also applied to some non-operon-inhabiting genes in a few eukaryotic groups (including cnidarians, flatworms, rotifers, and hemichordates). The complexity of this process and its wide phylogenetic distribution implies that the stem eukaryote likely used *trans*-splicing to process at least some genes. By the same reasoning, the stem eukaryote probably also initiated transcription well upstream of translation initiation sites, as eukaryotic mRNAs typically have 5′ untranslated regions (UTRs, usually 100 bp or more in length; see Chapter 10). In contrast, prokaryotic transcription generally starts just a few base pairs upstream of the translation initiation site. In addition, the stem eukaryote processed at least some genes by the direct addition of a 5′ cap (a modified base) and a 3′ poly(A) tail (a string of adenines) to mRNAs. Both features are broadly used by eukaryotes, but generally not by prokaryotes (see Chapter 10).

Fourth, the stem eukaryote contained intragenic spacers in some protein-coding genes. Eukaryotic genes are often fragmented into coding exons separated by noncoding introns. Because introns are transcribed along with their surrounding exons, this genes-in-pieces architecture imposes another significant challenge for information processing: introns must be neatly excised and exons spliced back together (i.e., *cis*-spliced) prior to the export of mature mRNAs to the cytoplasm. This splicing is carried out by a complex molecular machine unique to eukaryotes, the spliceosome, consisting of five small RNA subunits and 150 or more proteins. Not only do all well-studied eukaryotes harbor the spliceosomal machinery, but the fact that numerous distantly related lineages utilize two distinct spliceosomes (see Chapter 9) suggests that the stem eukaryote also had two such units. In striking contrast, nearly all prokaryotic genes consist of a single uninterrupted coding region, and in the very few instances in which this is not the case, the introns are self-splicing (Belfort et al. 1995; Dai and Zimmerly 2002). Eukaryotes generally deposit an exon junction complex just upstream of each exon–exon junction in mRNAs, and these proteins assist in the flagging of premature termination codons in aberrant transcripts via the nonsense-mediated decay pathway, another unique eukaryotic innovation (see Chapter 9).

Fifth, although the origin of the nuclear membrane may have greatly reduced the incidence of horizontal gene transfer among species, the stem eukaryote very likely harbored genetic elements capable of self-mobilization and colonization of new genomic locations. The two major classes of mobile genetic elements, the cut-and-paste transposons and the copy-and-paste retrotransposons, are found in nearly all major eukaryotic lineages (see Chapters 3 and 7). If not resisted by some physical mechanism or natural selection, runaway colonization by such parasitic DNAs would eventually impose a substantial genetic load on the host population. Meiotic recombination and independent chromosomal segregation helped reduce this problem by enabling parents to produce progeny with a reduced genetic load, although they also provide a vehicle for the transfer of mobile elements among host population members.

The net implication of all of these observations is that an enormous remodeling of genome structure and organization occurred in the lineage leading to the most recent common ancestor of all eukaryotes. As at least vaguely similar traits can be found in isolated prokaryotic lineages (Table 1.1), some aspects of early eukaryotic design may have arisen via early horizontal transfers, perhaps even from viruses (Bell 2001; Villareal 2005). Moreover, the acquired embellishments of the eukaryotic genome may not have been independent events. For example, introns may have been introduced into the nuclear genome via the primordial mitochondrion (see Chapter 8), and Martin and Koonin (2006) have argued that this event promoted the evolution of the nuclear membrane as a mechanism to ensure the complete splicing of precursor mRNAs prior to their exposure to the translation apparatus. A central theme of the following chapters is that many of the peculiar features of the eukaryotic genome are unlikely to have arisen by natural selection, but once established, they would have provided the substrate for an evolutionary revolution: the explosion in cell architectural diversity that is the hallmark of eukaryotes.

The Eukaryotic Radiation

So enormous are the differences in the gross features of the major eukaryotic groups that traditional approaches to revealing phylogenetic relationships based on morphology inspire little confidence. Thus, as in investigations of the prokaryote–eukaryote divide, progress on these issues has largely depended on comparative gene sequence analysis. Although molecular data have begun to reveal some rigorous phylogenetic groupings, a variety of issues (including idiosyncratic changes in rates of evolution, divergent nucleotide compositions across lineages, possibilities of early horizontal gene transfer, gene duplications, and inadequate taxon sampling) still con-

TABLE 1.1 Some of the features that set eukaryotic genomes apart from those of prokaryotes, and their exceptions

EUKARYOTES	PROKARYOTES
Presence of a nuclear membrane	Also present in the Planktomycetes
Organelles derived from endosymbionts	Also present in the β-proteobacteria
Cytoskeleton and vesicle transport machinery	Tubulin-related proteins, but not microtubules
Trans-splicing	Absent
Introns in protein-coding genes, and a complex spliceosomal apparatus for excising them	Rare self-splicing introns, but almost never in coding DNA
Expansion of the untranslated regions of transcripts	Untranslated regions are generally very short
Addition of poly(A) tails to all mRNAs	Rare and nonessential polyadenylation of transcripts
Translation initiation by scanning for start codon	Ribosome binds directly to a Shine-Dalgarno sequence
Messenger RNA surveillance	The nonsense-mediated decay pathway is absent
Multiple linear chromosomes capped with telomeres	Single linear chromosomes in a few eubacteria
Mitosis and meiosis	Absent
Expansion in gene number	The largest prokaryotic genomes contain more genes than the smallest eukaryotic genomes
Expansion of cell size and number	A few have very large cell sizes (e.g., *Thiomargarita*), and several produce multiple cell types

spire to cloud our view of the basal structure of the eukaryotic phylogeny (Graur and Martin 2004; Hedges and Kumar 2004). Two things can be agreed on: the primary eukaryotic lineages are deeply branching in time, and the major groups on which most biological research is performed (animals, fungi, and plants) constitute only a small fraction of eukaryotic phylogenetic diversity. In addition, although these three well-studied sets of taxa are sometimes viewed as members of a "crown group" of eukaryotes (Knoll 1992), there is now fairly compelling evidence that they do not constitute a monophyletic lineage.

An attempt to summarize what is known about eukaryotic phylogeny is presented in Figure 1.8, with two caveats. First, this description is by no means complete, as it contains only the groups that will be encountered in the following chapters, and even if all of the major known groups of eukaryotes were included, the story would be an abstract at best. In a recent search for novel eukaryotes, Dawson and Pace (2002) used degenerate PCR

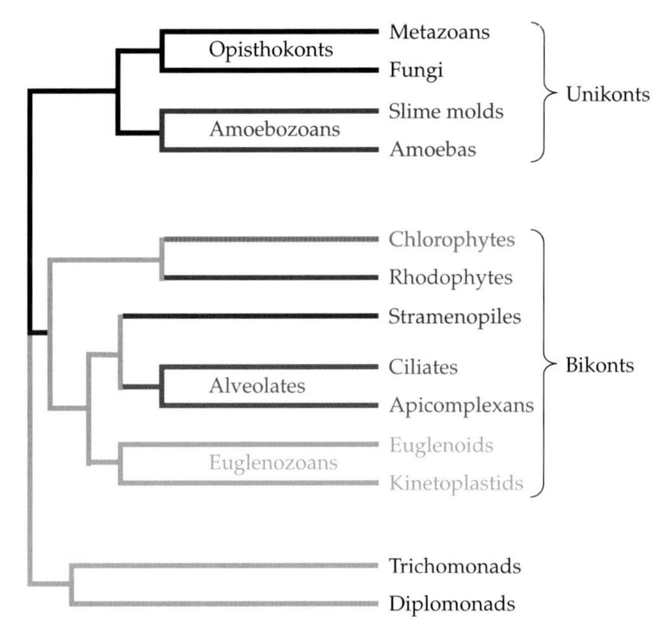

Figure 1.8 An approximate phylogenetic tree for some of the major eukaryotic groups, based on the arguments in the text. The branch lengths are not proportional to time, although all external branches are expected to be in excess of 700 million years in length. Gray lines denote areas of uncertainty.

primers to isolate rRNA sequences from several anaerobic environments. (PCR, the polymerase chain reaction, is a method used to amplify DNA from small samples.) As in previous studies with prokaryotes, this survey revealed a number of novel sequences far removed from any others ever recorded within the eukaryotic tree. Such results indicate that many novel lineages of microbial eukaryotes, never before visualized, reside in our midst. Contained within these lineages may be the secrets of the origin of eukaryotes.

Second, although the phylogeny presented in Figure 1.8 is largely supported by analyses based on multiple protein-coding genes (e.g., Burger et al. 1999; Baldauf et al. 2000; Forget et al. 2002; Bullerwell et al. 2003; Bhattacharya et al. 2004; Richards and Cavalier-Smith 2005), it is not entirely congruent with earlier phylogenies of rRNA sequences. A broader sample of genes should reduce the effects of biases that might be associated with individual loci, and a number of recent studies have raised concerns about the utility of rRNA sequences in reconstructing phylogenies. As an example of such difficulties, consider the phylogenetic affinity of nematodes and arthropods. Based largely on the absence of a true coelom, nematodes were traditionally thought to reside in a lineage basal to arthropods, chordates, and most other metazoan phyla. However, phylogenetic analyses based on rRNA sequences place nematodes firmly in a clade with arthropods

(Aguinaldo et al. 1997; Van de Peer et al. 2000), leading Aguinaldo et al. (1997) to proclaim that all molting animals constitute a single monophyletic group (the Ecdysozoa or "molting animals"). This new grouping continues to be presented as fact in many major textbooks, even though phylogenies based on large numbers of protein-coding genes generally either place nematodes in their traditional position or are equivocal on the matter (Mushegian et al. 1998; Wang et al. 1999; Baldauf et al. 2000; Blair et al. 2002; Dacks et al. 2002; Y. I. Wolf et al. 2004; Rokas et al. 2005; Ciccarelli et al. 2006). Given such uncertainties regarding the relationships of invertebrate lineages separated by just 0.7 billion years, difficulties in deciphering phylogenetic relationships dating to the dawn of eukaryotes should come as no surprise.

The unikonts and bikonts

The vast majority of eukaryotes reside in two major clades. The first of these groups, the unikonts, is united at the morphological level by the presence of cells with single flagella at some stage of the life cycle (Cavalier-Smith 1998; Steenkamp et al. 2006) and contains the opisthokonts, the now universally recognized assemblage of animals and fungi (at the top of Figure 1.8). The microsporidia, a group of animal parasites with reduced genomes, once thought to be a lineage deep in the eukaryotic phylogeny, appear to be members of the fungal lineage (and hence opisthokonts) (Katinka et al. 2001). Filling out this first major clade is the amoebozoan group, which contains most of the familiar lobose amoebas as well as the slime molds (Bapteste et al. 2002).

The second major eukaryotic clade is the bikonts (at the middle of Figure 1.8), all members of which are believed to be ancestrally biflagellate (Cavalier-Smith 1998). This large assemblage contains several subgroups, one encompassing the green plants (chlorophytes, including the green algae) and the red algae (rhodophytes). Another bikont sublineage that will be encountered frequently in subsequent chapters is the diverse alveolate subclade (united by the presence of alveoli, a system of sacs underlying the cell surface), which contains the ciliates (e.g., *Paramecium* and *Tetrahymena*), the dinoflagellates (a diverse group of aquatic flagellates), and the obligately parasitic apicomplexans (e.g., the malarial parasite *Plasmodium*) (Fast et al. 2002). A third large subclade, the stramenopiles, includes the diatoms, brown algae, and oomycetes, and still another, the euglenozoans, unites the euglenoids (e.g., *Euglena*) and the parasitic kinetoplastids (e.g., the trypanosomes *Trypanosoma* and *Leishmania*). At least two amoeboid lineages (which include the heliozoans, radiolarians, and foraminiferans) also appear within the bikonts (Nikolaev et al. 2004). With the dual nuclei of the ciliates, the condensed chromosomes of the dinoflagellates, and the massive RNA editing in trypanosome mitochondria, these latter groups harbor some of the most bizarre forms of genomic architecture in all of life, highlighting the pronounced diversification that occurred prior to the origin of multicellularity.

Monophyly of the entire bikont group is supported by a unique fusion between the genes encoding two key enzymes (dihydrofolate reductase and thymidylate synthase), which are encoded separately in all unikonts and prokaryotes (Stechmann and Cavalier-Smith 2002). However, two groups of protists, the amitochondriate diplomonad (including *Giardia*) and trichomonad lineages, appear not to contain either gene and so cannot be assigned to either of the two major eukaryotic clades on this basis. Most molecular phylogenies place these two lineages at the very base of the eukaryotic tree, but the long branches associated with both groups may cause statistical artifacts, and significant uncertainty remains over their exact position (Keeling and Palmer 2000; Philippe et al. 2000; Bapteste and Philippe 2002; Dacks et al. 2002; Arisue et al. 2005).

A eukaryotic big bang?

The very deep and frequently unresolved nature of the relationships among the major eukaryotic groups has inspired a "big bang" hypothesis, under which most of the major lineages became established in a period of 10–100 million years (Philippe et al. 2000; Cavalier-Smith 2002c). On the basis of fossil and paleoclimatological evidence, Cavalier-Smith (2002c) argues for a major radiation around 800–850 (MYA). However, molecular estimates for the divergence times of plants, metazoans, and fungi are twofold deeper, falling between 1.6 and 1.7 BYA (Wang et al. 1999; Nei et al. 2001; Yoon et al. 2004). Given this enormous discrepancy, which type of data is most reliable?

The molecular dates could be inflated if molecular clocks calibrated with recent species greatly underestimate the rates of molecular evolution that prevailed in the deeper past (Cavalier-Smith 2002c). However, the early rates necessary to yield dates in accordance with the fossil record approach levels that defy explanation (Bromham and Hendy 2000). Moreover, molecular clocks calibrated with recent data are generally more, not less, rapid than those based on ancient divergences (Ho and Larson 2006), which should downwardly bias divergence time estimates. Thus, there is no compelling reason to think that the molecular dates are greatly inflated, although they are certainly subject to large inaccuracies. On the other hand, given that the initial eukaryotes were probably unicellular, the silence of the early fossil record is certainly not unexpected, although recent discoveries of microfossils of algae and bilaterian embryos raise hope that signs of even the earliest phases of eukaryotic evolution will eventually be found in rocks (Xiao et al. 1998; Chen et al. 2002).

Discordant fossil and molecular dates have also been found in analyses of the emergence of the major animal lineages. As noted above, nearly all of the animal phyla appear in the Cambrian fossil record within a window of about 10 million years, focused around 545 MYA (Knoll and Carroll 1999; Valentine et al. 1999; Conway Morris 2000). However, although there is substantial variation among sequence-based dates of the animal radiation, all such estimates greatly predate the animal fossil record. This point was first made by Wray et al. (1996), who used multiple protein-coding sequences to date the origin of triploblasts (animals with mesoderm) to 1.2 BYA. This study generated considerable debate, but although more recent molecular estimates date the emergence of metazoans to 0.6–1.0 BYA, this still substantially predates the Cambrian (Ayala et al. 1998; Gu 1998; Lynch 1999; Wang et al. 1999; Nei et al. 2001; Aris-Brosou and Yang 2003; Peterson et al. 2004). Interestingly, molecular dates of the emergence of the major orders of placental mammals, a group with an unusually good fossil record, precede fossil-based dates by about 30% (Kumar and Hedges 1998), thereby approximating the situation observed with the major animal phyla.

In principle, these mixed messages from molecules and fossils might be reconciled if the earliest members of radiating lineages tend to be low in abundance and/or geographically isolated and hence unlikely to be found in the fossil record. Two key events in Earth's history could have induced such conditions. Both the emergence of the major eukaryotic groups and the subsequent radiation of animals roughly coincide with two of the most extreme climatological shifts experienced by Earth's biota. Major episodes of global glaciation in which mean surface temperatures approached –50°C and the oceans were frozen over with up to a kilometer of ice (so-called "snowball Earth" effects) occurred ~2.4 BYA and ~0.7 BYA. For the ~35-million-year duration of each episode, few habitats with significant sunlight would have been coincident with liquid water and oxygen, other than possible hot springs (Hoffman et al. 1998; Kirschvink et al. 2000; Hoffman and Schrag 2002). Such prolonged, intense conditions must have had a major effect on Earth's evolutionary history, although details on the matter remain controversial (Runnegar 2000; Corsetti et al. 2003).

Genome Repatterning and the Eukaryotic Radiation

Eukaryotic diversity is reflected not just in the large number of deep lineages, but also in the astounding morphological divergence within lineages. In total, there are fewer than 10,000 described species of eubacteria and archaea (Oren 2004), in contrast to the roughly 1,500,000 named species of eukaryotes, about half of which are insects. One might argue that the small size and general absence of obvious morphological disparities in prokaryotes results in their estimated number being biased dramatically downward. However, similar arguments can be made for many unicellular eukaryotes, and yet there are already about 30,000 described species of protists. Thus, although huge uncertainties surround the numbers, one gets the impression that the eukaryotic domain may be substantially more subdivided than either the eubacteria or the archaea.

Assuming that the major eukaryotic groups arose in a temporally explosive manner, as suggested above, what might have precipitated such an active phase of lineage isolation? Most attempts at explaining evolutionary radiations resort to ecological arguments, invoking either a dramatic change in the environment or the chance appearance of an evolutionary novelty that allows the exploitation of new ecological niches. A prolonged snowball Earth certainly qualifies as a dramatic (and global) environmental shift, and the new "body plan" of the eukaryotic cell in itself may have served as a potential launching pad for subsequent diversification. However, an adaptive radiation requires more than just ecological opportunity. In particular, a mechanism is needed to keep new lineages distinct in the face of potential gene flow during and after their period of phenotypic diversification.

Opportunities for speciation arise when populations are isolated for long periods, as this allows the accumulation of mutational changes in independent lineages. Should secondary contact occur and cross-recognition still be possible, the viability and fertility of hybrid individuals will then depend on the compatibilities of the two parental genomes. Although postreproductive isolating barriers may arise by many different mechanisms, microchromosomal rearrangements resulting secondarily from gene duplication are especially worth exploring in the context of the eukaryotic radiation, as the basal eukaryotic lineages may have experienced an unusually high level of such activity.

Consider, for example, the primordial mitochondrion. Because most prokaryotic genomes contain a few thousand genes, while most mitochondrial genomes contain no more than a few dozen, it is clear that hundreds, if not thousands, of organelle-to-nuclear gene transfers occurred early in the establishment of mitochondria, although many genes were probably also simply lost. The different mitochondrial genome contents across various eukaryotes (see Chapter 11) indicate that such events continued to unfold stochastically as the major lineages of eukaryotes were developing. Indeed, a low level of such gene movement still continues in some organisms (Ricchetti et al. 1999; K. L. Adams et al. 1999, 2000, 2001, 2002; Blanchard and Lynch 2000; Millen et al. 2001; Rujan and Martin 2001).

Intergenomic transfer may have passively promoted the origin of genomic incompatibilities at unusually high rates, in part because organellar genomes are generally inherited uniparentally (here we assume through the mother). Consider two closely related but geographically separated lineages, both initially with an organelle copy of a key gene. One lineage then experiences a duplication of this gene to the nuclear genome, followed by its subsequent silencing in the mitochondrion (Figure 1.9). Assuming a diploid nucleus and letting small letters denote absentee alleles, the respective parental genotypes are ***aa* / *M*** and ***AA* / *m***, and a hybrid cross involving a female of the second species would yield ***Aa* / *m*** progeny, half of whose gametes would lack a functional gene. Although a single genomic transfer of this sort is not sufficient to produce complete reproductive isolation, just a few independent transfers would have a powerful effect. Imagine an incipient pair of species experiencing n independent organelle-to-nuclear gene

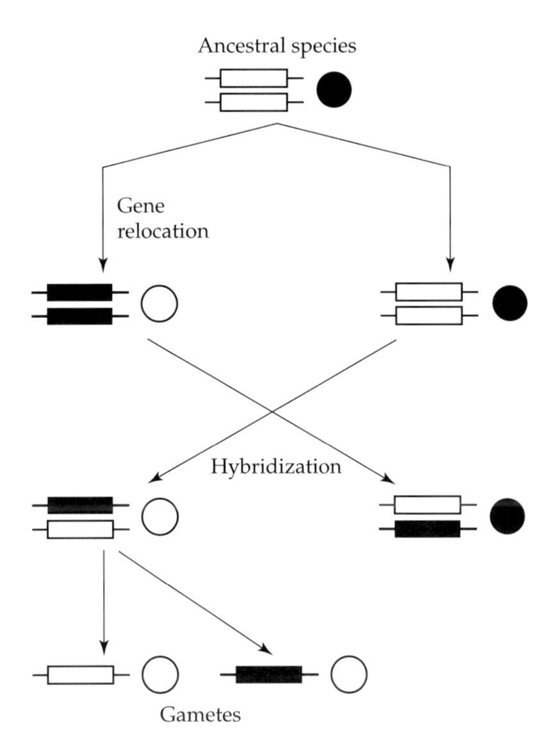

Figure 1.9 The development of a reproductive incompatibility following the relocation of an organelle gene. Rectangles and circles denote autosomal and organellar gene copies, respectively, with open symbols indicating gene absence. Following a geographic isolating event, the incipient species on the left experiences an organellar gene transfer to the nucleus. Subsequent hybridization yields presence/absence heterozygotes at the autosomal locus, with the status at the organellar locus depending on maternal identity. As a consequence of Mendelian segregation of the diploid autosomal locus, half of the gametes of the individual in the lower left will lack the gene entirely.

transfers in each lineage. Assuming independent assortment of the nuclear genes during meiosis, the fraction of F_1 gametes that are entirely lacking in a functional gene at one or more loci would be $1 - 0.5^n$, which is 0.969 for $n = 5$.

Thus, when one considers the hundreds of organelle-to-nuclear gene transfers that may have occurred soon after the colonization of the primordial mitochondrion (see Chapter 11), it is plausible that such gene traffic played a significant role in the passive development of isolating barriers among the earliest eukaryotes. Moreover, because most transfer events involving organellar DNAs generally encompass only a gene or two at a time (Blanchard and Lynch 2000), following the first isolation events of this kind, substantial substrate would still exist for further nested events. Finally, as will be discussed in Chapter 8, these kinds of gene relocations are by no means restricted to organelle-to-nuclear gene transfers, but are expected under any type of gene duplication event, including those involving only nuclear genes (Lynch and Force 2002a). A key point here is that these processes yield reproductive isolating barriers only in species with multiple chromosomes and sexual reproduction, as both are necessary for the independent segregation of unlinked loci. Thus, although the influence of an altered adaptive landscape on the dramatic radiation of the primary eukaryotic lineages is not precluded, the simple structural consequences of two of the defining cytological attributes of the stem eukaryote, a genome-bearing mitochondrion and meiotic recombination, may have played a central role in the passive emergence of reproductive isolation.

A Synopsis of the First 2 Billion Years of Biology

Given the magnitude of ground that has been covered in the previous pages, a brief summary is in order. Although several uncertainties remain, the most parsimonious explanation for the earliest steps in evolution involves four major stages:

1. At the dawn of life's history, information storage and all forms of catalysis most likely relied entirely on RNA. In this RNA world, replication would have been slow and highly error prone.
2. Soon thereafter, the basic translation machinery for protein synthesis emerged. In this RNA–protein world, RNA continued to be the substrate for information storage, but proteins began to supplement the toolbox of catalytic activities, expanding the range of metabolic diversity. With an increased ability to acquire energy and the refinement of catalytic capacity, both the rate and accuracy of genome replication would have been enhanced.
3. Following the origin of the proteins necessary for producing and assembling DNA precursors, RNA was probably rapidly replaced by DNA as a more reliable mechanism for information storage, enabling still further genomic expansion.
4. Biogenic pathways for the construction of cell membranes then appeared, giving rise to the era of cellular life, and further expanding the potential for morphological and metabolic diversification. The origin of cell membranes would have marked a profound turning point in evolution, as the sequestration of genomically encoded products would have greatly increased the efficiency of natural selection.

The emergence of eukaryotes involved at least one substantial admixture event between the two major prokaryotic lineages: the addition of the eubacterially derived mitochondrion to an archaeal host, which may have provided the seeds for horizontal gene transfer to the primordial nuclear genome over a period of hundreds of millions of years. Although many eukaryotic lineages may have coexisted during this early period, only one eventually gave rise to today's eukaryotes—the stem eukaryote. This keystone species was extraordinarily unique in terms of its cellular architecture and DNA transaction mechanisms, having acquired a genome littered with introns and mobile elements and an array of new methods for processing transcripts, all of which are difficult to explain with adaptive arguments. Although the nuclear membrane, mitosis, and meiosis may have been relatively late innovations in the stem eukaryote, they were probably well established prior to the divergence of the major eukaryotic lineages. It remains unclear whether any of the physical modifications associated with these latter features owe their origins to adaptive processes. However, once established, these new mechanisms of chromosome management and assortment may have facilitated the rapid diversification of the eukaryotes.

2 Genome Size and Organismal Complexity

When the key insights of Crick, Franklin, Watson, and Wilkins led to the elucidation of the structure of DNA, half a century ago, little was known about the molecular aspects of genome structure. We now know that chromosomes, the vehicles of DNA, are enormously versatile in terms of content and sequence malleability. Recurrent mutation generally ensures that most homologous chromosomes within a species are unique in multiple ways, and this variation provides the fuel for evolutionary divergence among species, as revealed in striking detail by whole-genome comparisons. For example, the 250 or so fully sequenced prokaryotic genomes contain between 350 and 8000 genes packed into 0.5–9.0 megabases (Mb), while all well-characterized genomes of animals and land plants contain more than 13,000 genes and are at least 100 Mb in size. As will be detailed in Chapter 3, most of the increase in genome size in multicellular eukaryotes is a consequence of the expansion of noncoding forms of DNA, including introns and mobile elements. The phylogenetic positions of animals and land plants suggest the independent emergence of this complex genomic syndrome in both lineages (Meyerowitz 2002), but what is cause and what is effect?

Across the entire domain of life, there is a moderate positive scaling between organism size and number of cell types (Bell and Mooers 1997; Bonner 2004). However, although specific genes play a central role in cellular differentiation, there is little evidence that a substantial increase in genome size or gene number is essential for the evolution of multicellularity. For example, numerous cyanobacteria (Meeks et al. 2002), myxobacteria (Goldman et al. 2006), streptomycetes (Bentley et al. 2002), methanogens (Galagan et al. 2002), and other prokaryotic lineages are capable of producing multiple cell types, despite having moderate numbers of genes (4000–8000) and relatively little noncoding DNA. A number of eukaryotes with complex cell structures and multiple cell types harbor 10,000 or fewer genes, whereas the genomes of some unicellular eukaryotes (e.g., *Paramecium*) harbor more genes than those of vertebrates.

This weak relationship between gene number and organismal complexity suggests that the increased structural innovation and developmental flexibility of the eukaryotic cell must largely be a consequence of the unique ways in which genes are deployed. But were new ways of expressing genes (such as complex spatial and temporal patterns of transcription regulation and alternative splicing) promoted as a direct response to selection for new cell types in large organisms? Or did the evolution of large size and/or multicellularity induce side effects that provoked nonadaptive changes in genomic architecture, which then secondarily paved the way for the adaptive origin of new cell functions? In the following chapters, the case will be made that the roots of many aspects of eukaryotic genomic complexity are likely to reside in nonadaptive processes (in particular, mutation pressure and random genetic drift) that are particularly potent in eukaryotes, especially in multicellular lineages. This chapter reviews some of the historical background leading up to this argument.

First, a broad phylogenetic survey will demonstrate the continuity of scaling of genome content with genome size across the transitions from prokaryotes to unicellular eukaryotes to multicellular species. This observation leads to the conclusion that aspects of cell structure and metabolism are not the central determinants of genomic architecture. Second, previous hypotheses for the evolution of genome size will be evaluated in this context and their limitations outlined. Although it is commonly argued that microbial genomes are kept streamlined by efficient selection against the negative metabolic costs of replicating excess DNA, there appear to be no data in support of this contention. Nor is there compelling evidence that an intrinsic bias toward deletion mutations is sufficient to prevent runaway genome growth. Finally, a brief verbal description of the mutational hazards of excess DNA will be given, setting the theme for many of the topics to be considered in subsequent chapters.

Genome Size and Complexity

Many aspects of genomic architecture exhibit continuous transitions within and across all cellular domains of life, extending even to DNA viruses (Figure 2.1). In viruses and prokaryotes, the amount of coding DNA scales nearly linearly with total genome size, occupying 80%–95% of the latter, and a similar allocation is found in the smallest eukaryotic genomes. However, the expansion of coding DNA progressively slows in genomes with total sizes in excess of 10 Mb, eventually leveling off at about 100 Mb in vertebrates and land plants, in which 90%–98% of the genome is allocated to noncoding DNA. As a consequence, over the 10,000-fold range in total genome sizes for well-studied cellular species, there is only a 100-fold range in the amount of DNA devoted to protein coding.

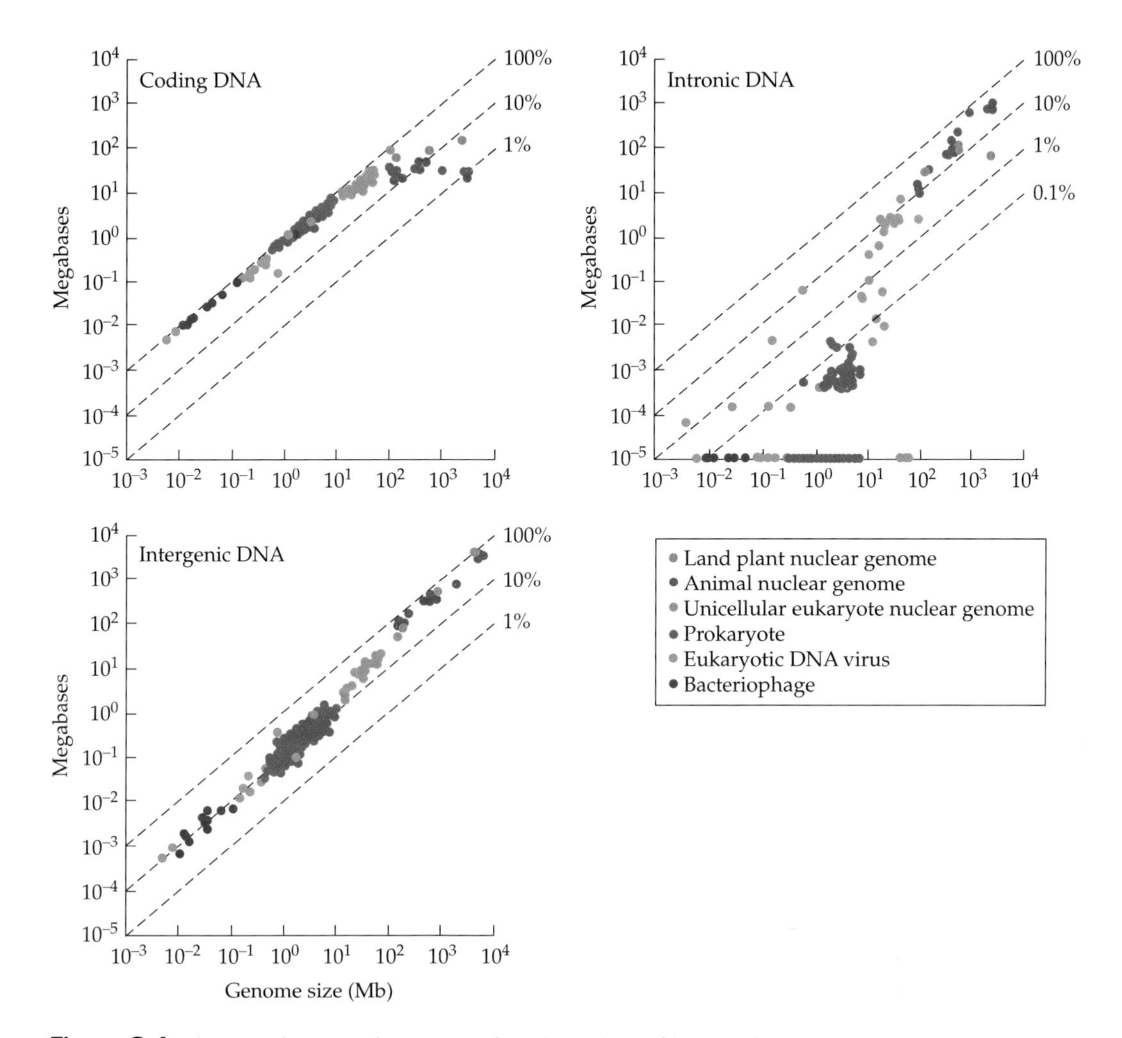

Figure 2.1 The contributions of protein-coding, intronic, and intergenic DNA to total genome size in various organismal groups. The diagonal lines define points of equal proportional contribution to total genome size. The intronic DNA depicted here does not include introns in noncoding exons (UTRs). Intergenic DNA comprises all of the genome exclusive of spans between translation initiation and termination points for protein-coding genes. Points residing on the *x* axis denote situations in which the component contribution is zero. (Modified from Lynch 2006b.)

The two most easily identified classes of noncoding DNA, introns and mobile elements, scale similarly with genome size. Although spliceosomal introns are confined to eukaryotes, prokaryotes sometimes harbor small numbers of self-splicing introns (see Chapter 9), but never at levels exceeding 0.2% of the entire genome (see Figure 2.1). The smallest eukaryotic genomes also contain very little intronic DNA, but above total genome sizes of 10–20 Mb, there is an abrupt and progressive increase in investment in introns. Genomes that are about 100 Mb in size (all of which are eukaryotes) have nearly equal amounts of DNA allocated to introns and exons, whereas about 95% of the total length of protein-coding genes is intronic in the large (>2500 Mb) mammalian genomes. A qualitatively similar transition is seen with the fraction of the genome occupied by intergenic DNA: typically less than 20% for genomes smaller than 1 Mb and progressively increasing to more than 80% for genomes beyond 10 Mb. Residing within intergenic regions are sequences involved in transcription, chromatin packaging, and replication initiation (see Chapters 3, 5, and 10), but in species with large genomes, the majority of intergenic DNA consists of active mobile elements (transposons and retrotransposons) and other debris associated with their past activities.

Two general conclusions emerge from the enormous phylogenetic breadth of the patterns in Figure 2.1. First, the common assertion that there is essentially no correlation between genome size and organismal complexity (Thomas 1971; Cavalier-Smith 1978; Gregory 2005a), appears to derive from a focus on extreme outliers rather than on measures of central tendency. Although there is considerable variation in genomic features among species with similar levels of cellular/organismal complexity, there is a clear ranking from viruses to prokaryotes to unicellular eukaryotes to multicellular eukaryotes in terms of genome size, gene number, mobile element number, intron number and size, size of intergenic spacer DNA, and complexity of regulatory regions (Lynch and Conery 2003b; Lynch 2006a). Second, despite this gradient, there are no abrupt discontinuities in the scaling of genome content with genome size across radically different groups of organisms. This smooth transition in patterns of genome content scaling across all forms of life provide compelling evidence that the primary forces influencing the evolution of genomic architecture are unlikely to be direct consequences of organismal differences in cell structures or physiologies.

The Selfish-DNA and Bulk-DNA Hypotheses

The early idea that genome sizes vary wildly among organisms with similar levels of cellular and developmental complexity became known as the C-value paradox (where the C value denotes the total amount of DNA in a haploid genome). Depending on one's point of view, the puzzle was either solved or deepened as it became clear that a substantial fraction of many eukaryotic genomes consists of noncoding and putatively nonfunctional DNA. Two general classes of hypotheses emerged to explain this odd set of observations.

On the one hand, Doolittle and Sapienza (1980) and Orgel and Crick (1980) promoted the idea that a good deal of noncoding DNA consists of "selfish" elements capable of proliferating until the cost to host fitness becomes so prohibitive that natural selection prevents their further spread.

This selfish-DNA hypothesis, under which genome size expansion is a simple pathological response to internal genomic upheaval, draws support from the ubiquity of mobile elements across the eukaryotic domain (see Chapter 7). However, other major contributors to genome size, such as spliceosomal introns, small repetitive DNAs, and random insertions, are not self-replicable and hence not subject to selection for proliferative ability within host genomes. Thus, a central challenge for the selfish-DNA hypothesis is the need to explain why *all* types of excess DNA mutually expand (or contract) in some genomes and not in others.

In striking contrast to the view that much of noncoding DNA is expendable junk, Commoner (1964), Bennett (1972), and Cavalier-Smith (1978) had argued earlier that the total content of the noncoding DNA within a genome (independent of its information content) is a direct product of natural selection. This bulk-DNA hypothesis postulates that genome size has a direct effect on nuclear volume, cell size, and cell division rate, all of which in turn influence life history features such as developmental rate and size at maturity. The supporters of this hypothesis have pointed out an impressive number of correlations between genome size and cell properties in a diversity of phylogenetic groups (Figure 2.2 contains two examples), although the evolutionary mechanisms responsible for such statistical relationships are unclear. Cavalier-Smith (1978) suggested that the evolution of large cell size imposes secondary selection on nuclear genome size as a physical mechanism for modulating the area of the nuclear envelope and hence regulat-

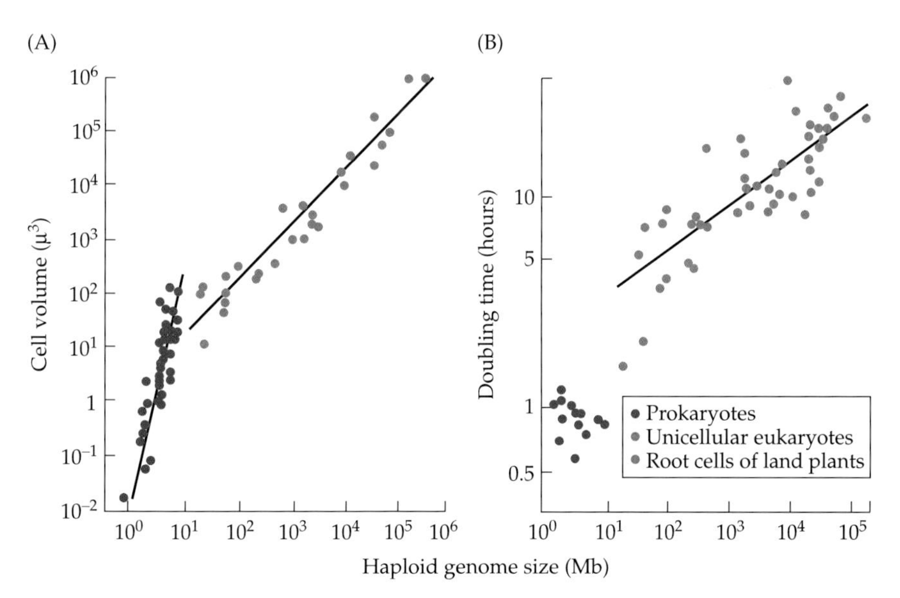

Figure 2.2 Correlations of two cell biological features, (A) cell volume and (B) doubling time, with genome size. Doubling times are measured for cells at 23°C. (Modified from Shuter et al. 1983.)

ing the flow of transcripts to the cytoplasm. One concern with this view is that many additional mechanisms for achieving elevated transcript numbers exist (e.g., increases in nuclear membrane porosity, gene copy number, ribosome number, and transcript longevity), all of which appear to be less risky than the expansion of bulk DNA. Indeed, the most striking challenge for any adaptive hypothesis for the expansion of excess DNA is the nature of the filler material itself, predominantly mobile elements, which are known to impose a heavy mutational burden in eukaryotes (see Chapter 7). Remarkably, there is also a strong positive correlation between cell size and genome size in prokaryotes (see Figure 2.2), which cannot be a consequence of either cytoskeletal effects (given the absence of nuclear membranes) or of noncoding DNA expansion (given its near constant proportion; see Figure 2.1).

Because evolution is a population-level process, any evolutionary explanation for a pattern of variation must be consistent with basic population genetic mechanisms (e.g., mutation, random genetic drift, recombination, and natural selection), but the failure of prior studies to directly confront these issues in a quantitative manner has been a major impediment to sorting out cause and effect in genome size evolution. For example, the logic underlying the bulk-DNA hypothesis will remain unconvincing until it is demonstrated that: (1) heritable within-population variation in genome size significantly covaries with cellular features that are mechanistically associated with individual fitness, and (2) mobile element proliferation is an easy means of achieving such variation with minimal negative side effects. The absence of population-level thinking from much of the ongoing debate about genome size evolution has fostered the impression that unknown evolutionary mechanisms remain to be discovered, leading some to invoke undefined "macroevolutionary" phenomena (Gregory 2005c). However, the logical problems with arguments that abandon established microevolutionary principles are well known (e.g., Charlesworth et al. 1980), and a central goal of this book is to demonstrate that there are very few, if any, aspects of genomic evolution that cannot be explained with well-accepted population genetic mechanisms.

The Metabolic Cost of DNA

Microbial species pose a special challenge for both the bulk-DNA and selfish-DNA hypotheses. With its adherence to adaptive arguments, the bulk-DNA hypothesis invokes a premium on energetic efficiency as an explanation for the diminutive genomes of prokaryotes (Cavalier-Smith 2005), whereas adherents to the selfish-DNA hypothesis have argued that small genomes are products of strong selection for high replication rates (Doolittle and Sapienza 1980; Orgel and Crick 1980; Pagel and Johnstone 1992). With both competing hypotheses conceptually aligned on at least this one matter, the metabolic expense of DNA is widely cited as the explanation for the streamlining of microbial genomes (e.g., Rogozin et al. 2002; Giovannoni et al. 2005; Ranea et al. 2005).

Is the cost of maintaining and replicating an additional DNA segment of a few base pairs (the typical size of an intergenic insertion/deletion; see below) significant enough to be perceived by natural selection? Because the large population sizes of unicellular species magnify the efficiency of natural selection (see Chapter 4), this possibility cannot be ruled out entirely. However, there is no direct evidence that cell replication is ever limited by DNA metabolism, and there are several reasons to think otherwise. First, within and among prokaryotic species, there is no correlation between cell division rate and genome size (Bergthorsson and Ochman 1998; Mira et al. 2001). Second, during rapid growth phases, prokaryotic chromosomes are often present in a nested series of replication stages (Casjens 1998), with some species harboring tens to hundreds of chromosomal copies at various stages of the life cycle (e.g., Maldonado et al. 1994; Komaki and Ishikawa 2000). Third, in *E. coli* and other eubacteria, DNA replication forks progress 10–20 times faster than mRNA elongation rates (Bremer and Dennis 1996; Cox 2004; French 1992). Fourth, DNA constitutes 2%–5% of the total dry weight of a typical prokaryotic cell (Cox 2003, 2004), and the estimated cost of genomic replication relative to a cell's entire energy budget is even smaller (Ingraham et al. 1983). Similar conclusions emerge for eukaryotic cells (Rolfe and Brown 1997).

Directional Mutation Pressures on Genome Size

Genome size evolution ultimately depends on two factors: the relative rates of mutational production of insertions and deletions, and the ability of natural selection to promote or eliminate such changes. Thus, if the energetic consequences of noncoding DNA are not great enough to be perceived by natural selection, species with small genomes must be subject to unusual deletional mutation pressures, and/or excess DNA must be disadvantageous in some other way.

Several types of mutational activity encourage genome size expansion. For example, mobile elements are capable of self-replicating and inserting copies elsewhere in the genome at high rates (in excess of 10^{-5} per element per generation; see Chapter 7), and their activities also result in the insertion of pseudogenes (dead-on-arrival copies of otherwise normally functioning genes) (see Chapter 3). In addition, segmental duplications involving stretches of hundreds to thousands of kilobases (kb) are universal among eukaryotes (see Chapter 8), and strand slippage during replication can also lead to small-scale insertions (Chen et al. 2005).

Double-strand breaks of chromosomes are another common source of insertions and deletions. Such breaks occur spontaneously in nonreplicating cells and are also produced when a replication fork encounters a single-strand nick, severing the entire chromosome. In mammals, 5%–10% of somatic cells acquire at least one double-strand break per cell division

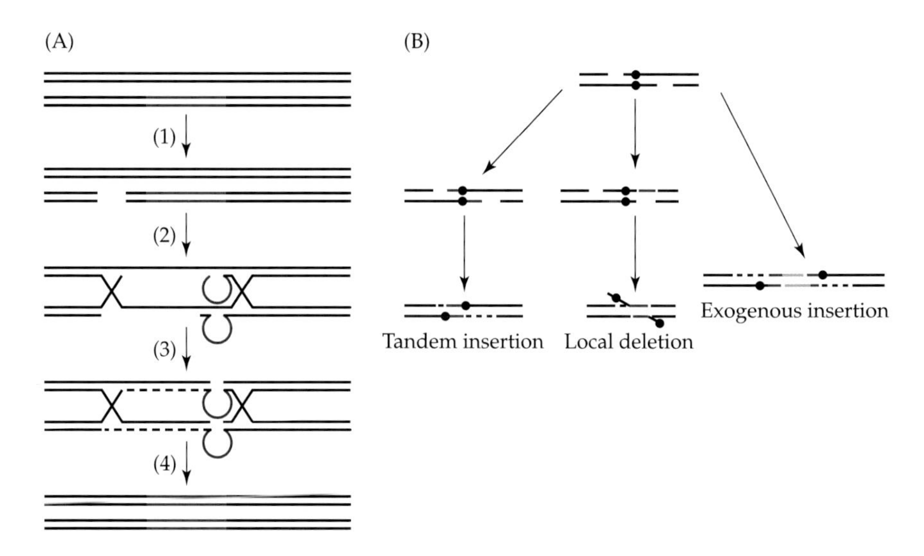

Figure 2.3 Some ways in which insertions and deletions can be created at double-strand breaks. (A) Repair by homologous recombination can result in a local conversion of one chromosome type to the other. (1) A break appears in one chromatid adjacent to an insertion (red) that is absent from the homologous chromosome. (2) The free ends of the lower chromosome invade the upper chromosome to initiate formation of a recombination junction. Complementary DNA from the two chromatids aligns, leaving loops in each recombination intermediate. (3) This mismatch is resolved by cutting the non-looped strands. (4) Synthesis of the missing complementary DNA results in the conversion of the top chromatid to the insertion-bearing form. (If the loop-containing strands were cut instead, the invading strands could be converted to the insertion-free form). (B) Repair by nonhomologous end joining proceeds in the absence of a homologous chromosome. Such repair requires small regions of microhomology (illustrated in red), which are relatively free to align at multiple sites. Black dots serve as points of reference, and the starting point is a staggered cut. On the left and in the center, the selected regions of homology on the two strands (red) determine whether there is a local duplication or deletion. Occasionally, double-strand breaks capture foreign DNA (green lines), as shown on the far right.

(Lieber et al. 2003), and germ line cells that have experienced multiple divisions in the progression toward gamete production can be expected to incur even more. To maintain cell lineage viability, such breaks must be stitched back together by one of two mechanisms.

First, if a homologue (in a diploid species) or a sister chromosome (in the early stages of mitosis or meiosis) is available as a template, homologous recombination can restore the original state of the broken chromosome, provided the regions involved have near complete sequence identity. However, if in the region of the break, one of the chromosomes happens to have an insertion that is absent from the partner chromosome, recombination can alter the state of one of the chromosomes to that of the other by the process of gene conversion (Figure 2.3A). Some evidence suggests that insertions are retained more often than lost (Lamb 1985), although the degree to which this is the case depends on the size and structure of the insertion (Bill et al. 2001). Such biased gene conversion, which is purely a physical process, can encourage the spread of insertions throughout a population in a manner that is indistinguishable from positive selection (Nagylaki 1983; Walsh 1983).

Second, in nondividing cells, where templates are less likely to be available, the error-prone process of nonhomologous end joining, which directly ligates the two edges of a break back together, must be relied on (Moore and Haber 1996a; Heidenreich et al. 2003; Daley et al. 2005; Puchta 2005). Nonhomologous end joining is initiated by complementary base pairing in regions of microhomology (at least 2–3 bp), and the way in which this is done can lead to small insertions or deletions (Figure 2.3B). Double-strand break repairs can also be accompanied by the capture of exogenous DNA derived from the mitochondrial genome, retrotransposons, or microsatellites (small repetitive DNAs, such as dinucleotide repeats) (Moore and Haber 1996b; Teng et al. 1996; Ricchetti et al. 1999; Yu and Gabriel 1999; Lin and Waldman 2001b; Decottignies 2005).

With large-scale insertion events operating on a recurrent basis, the prevention of runaway genome expansion requires direct selection to prevent the fixation of insertions at the population level and/or mutational mechanisms for their subsequent deletion. To evaluate whether deletion mutations alone are capable of putting a cap on genome size without any assistance from selection, Petrov and colleagues (Petrov 2001, 2002a,b; Petrov et al. 1996, 2000) have performed comparative surveys of the numbers and sizes of insertions and deletions in various types of pseudogenes in insects. Their studies and others (Table 2.1) suggest that the rate of small-scale nucleotide losses exceeds that of gains, yielding a net erosion in the length of large inserts of nonfunctional DNA over time. Taken at face value, these data imply a half-life of nonfunctional DNA in the nematode *Caenorhabditis* and the fruit fly *Drosophila* on the order of the time required for neutral DNA to acquire 0.15–0.25 substitutions per site (10 million years or so), whereas that for orthopterans (grasshoppers and crickets), mammals, and birds (all of whose genomes are much larger) is 15–50 times longer (see Table 2.1). Mira et al. (2001) also document substantially higher rates of small-scale nucleotide losses than gains from pseudogenes in a variety of prokaryotes. However, a dramatically different picture emerges in rice (*Oryza*), where the rate of nucleotide gain by pseudogenes exceeds that of loss by a factor of 16 (Noutsos et al. 2005).

These kinds of observations have encouraged the view that interspecific variation in the mutational tendency to delete excess DNA is a primary determinant of genome size, with species with the highest rates of deletion having the smallest genome sizes (Petrov et al. 2000; Mira et al. 2001; Ochman and Davalos 2006). However, a number of uncertainties remain. Why, for example, should insertion/deletion rates differ so dramatically among animals, given the high degree of conservation of their DNA repair

TABLE 2.1 Rates and average sizes of deletions and insertions derived from observations of nonfunctional DNA in various animals

	RATE		SIZE (BP)		NET	HALF-
	DELETION	INSERTION	DELETION	INSERTION	CHANGE	LIFE
Caenorhabditis	0.034	0.019	166	151	–2.8	0.25
Drosophila	0.115	0.028	42	12	–4.5	0.15
Laupala	0.070	0.020	7	7	0.0	—
Podisma	0.060	0.030	2	1	–0.1	6.93
Birds	0.043	0.007	12	4	–0.3	2.31
Mammals	0.033	0.017	5	6	–0.1	6.93

Sources: *C. elegans*, Witherspoon and Robertson 2003; *Drosophila*, average from Blumenstiel et al. 2002 and Petrov 2002b; *Laupala* (Hawaiian cricket) and *Podisma* (grasshopper), Petrov 2002b; birds (pigeons and doves), Johnson 2004; mammals (mouse, rat, and human), average from Ophir and Graur 1997 and Zhang and Gerstein 2003.

Note: All rates are given relative to the time required for the accumulation of one nucleotide substitution per silent site. Net change is defined to be the difference between (insertion rate × size) and (deletion rate × size), so, for example, –2.8 for *C. elegans* implies that by the time an average surviving nucleotide site has acquired a single substitution, an average net loss of 2.8 nucleotides per site is expected to occur. Half-life is the number of substitutions per silent site that are expected to accrue by the time a nonfunctional stretch of DNA experiences a 50% erosion in length, assuming exponential decay.

machinery (Eisen and Hanawalt 1999)? And if species with small genomes have evolved increased deletion rates, as Lawrence et al. (2001) have suggested as an adaptive mechanism for the streamlining of prokaryotic genomes, how is the increased burden on coding DNA avoided?

Central to these issues is the matter of whether the long-term behavior of pseudogenes provides an unbiased view of the de novo mutation spectrum or whether deletions and insertions in pseudogenes are subject to selection (Charlesworth 1996a). This is a concern because insertion-associated disadvantages and/or deletion-associated benefits will tilt the observed spectrum of effects toward deletions relative to the mutational distribution, in which case the negative association between observed net deletion rates and genome size could simply reflect interspecific variation in the efficiency of selection rather than intrinsic differences in mutational properties. As noted above, it is unclear whether the energetic advantages of small deletions are ever substantial enough to cause perceptible fitness differences, but as will be discussed in the following section, excess DNA can impose additional disadvantages.

Evidence that deletions may not outnumber insertions at the mutational level derives from observed excesses of insertions over deletions in several laboratory experiments. In *Drosophila melanogaster*, spontaneous insertions greater than 4 kb in length are fourfold more abundant than deletions (Yang et al. 2001), and reporter construct experiments in the yeast *Saccharomyces cerevisiae* suggest a similar insertion/deletion disparity (Kunz et al. 1998; Ohnishi et al. 2004; Hawk et al. 2005). However, although these direct assays imply an innate mutational tendency for genome size *expansion*, above and beyond that caused by mobile element activity and segmental duplications, such a bias may not be universal. For example, estimates of the human mutational spectrum derived from de novo mutations for genetic disorders suggest that microdeletions are 2.5 times more common than microinsertions, with both exhibiting very similar size distributions (Kondrashov 2003; Ball et al. 2005). Studies involving reporter constructs in *E. coli* also reveal a deletion bias (Schaaper and Dunn 1991; Sargentini and Smith 1994).

In principle, these indirect assays could be biased if deletions and insertions are not equally likely to produce a detectable phenotype, and the only truly unambiguous way to ascertain the insertion/deletion spectrum is to randomly sequence genomic regions after a long period of complete relaxation of selection. Such a study has been performed with the nematode *Caenorhabditis elegans* by using long-term mutation accumulation lines taken through single-individual bottlenecks each generation to eliminate the effectiveness of natural selection against all mutations except those causing complete lethality or sterility (Denver et al. 2004). This study revealed a 15:4 insertion/deletion ratio (both types of mutations were of similar size, and none were associated with mobile element activity), a dramatically different pattern from the 1:1 ratio derived from phylogenetic analysis (see Table 2.1).

Thus, despite the clear need for more data of the type procured for *C. elegans*, these observations, along with the enormous half-life estimates in Table 2.1, raise significant questions as to whether mutational deletion processes are *universally* sufficient to prevent the runaway growth of genome size. If they are not, then some form of natural selection is necessary for genome size stabilization, and lineages with a reduced ability to selectively promote deletions and/or purge insertions can be expected to experience nonadaptive expansions in genome size. Bennetzen and Kellogg (1997) refer to species in this kind of evolutionary situation as having acquired a "one-way ticket to genome obesity." However, arguments presented in the following chapters suggest that genomic expansion and contraction is really a two-way street, with the prevailing direction of traffic depending on the current population genetic conditions. A key question that remains to be resolved is whether the large genomes of multicellular eukaryotes are still in active phases of expansion.

Population Size and the Mutational Hazard of Excess DNA

Although DNA without a function is often assumed to be neutral, this view ignores a fundamental genetic observation: that the operation of every gene depends on its local physical environment. Thus, even if inert spacer DNA is immune to selection against loss-of-function mutations—i.e., is totally expendable—it need not be immune to harmful gain-of-function mutations. Many lines of evidence support this view. First, noncoding regions are known to be depauperate in short motifs with the potential for generating inappropriate transcription factor binding (Hahn et al. 2003), posttranscriptional silencing (Farh et al. 2005), and translation initiation (Rogozin et al. 2001; Lynch et al. 2005). Selection against mutations causing inappropriate gene expression is the likely cause of the maintenance of such sequences below levels expected by chance. A dramatic example of this point is a human blood disorder in which a single nucleotide substitution creates a novel regulatory element in an otherwise inert segment of intergenic DNA (De Gobbi et al. 2006). Second, insertions of mobile elements into coding exons will virtually always inactivate a gene, whereas those in noncoding regions can influence the regulation of adjacent genes (Sorek et al. 2002; Lev-Maor et al. 2003; Kreahling and Graveley 2004; Shankar et al. 2004). Third, introns are a mutational burden for their host genes, as the splicing of each intron requires a specific set of local sequences for proper spliceosome recognition (Lynch 2002b). Fourth, the fact that the majority of eukaryotic genomic DNA may be transcribed (Cawley et al. 2004; Kampa et al. 2004), at least at low levels, raises the question as to whether any segment of nonfunctional DNA is truly neutral.

All of these issues will be explored in further detail in the following chapters. The central point to be understood here is that a primary cost of excess DNA is its mutational liability (Lynch 2002b, 2006a; Lynch and Conery 2003b). Each embellishment of the structure of a gene or of its surrounding area increases the risk that the gene will be rendered defective by subsequent mutational processes.

This matter becomes important in the context of comparative genomics because the mutational burden associated with most excess DNA is quite small, but not so small as to be effectively neutral in all phylogenetic contexts. A key theme that will appear repeatedly in the following pages (particularly in Chapter 4) is that population size is a central determinant of the efficiency of natural selection: by magnifying the power of random genetic drift, fluctuations in allele frequencies caused by small population size can overwhelm the ability of natural selection to influence the dynamics of mutations of small effect. A second key point is that although random genetic drift is often viewed as simple noise that causes variation in evolutionary outcomes around expectations under selection alone, this is a false caricature. The size of a population specifically defines the kinds of genomic evolution that can and cannot proceed, with small population size facilitating the accumulation of deleterious mutations and inhibiting the promotion of beneficial changes. Finally, the tendency for mutationally hazardous DNA to accumulate depends on both the population size and the mutation rate: the latter defines the burden of excess DNA, while the former defines the ability of natural selection to eradicate it. These simple ideas provide a potentially unifying explanation for a wide range of observations on phylogenetic variation in gene structure and genomic composition.

The basic argument can be summarized with the following example. Suppose that natural selection favors an increase in body size in a particular lineage. A general observation from population ecology is that an increase in body size results in a reduction in the number of individuals within a species (see Chapter 4). By reducing the efficiency of natural selection, diminished population size magnifies the tendency for mildly deleterious insertions to accumulate in a genome, while also reducing the ability of selection to promote advantageous deletions. Thus, genome size is expected to expand in organisms of increasing body size, not necessarily because of an intrinsic tolerance of excess DNA or because of an increased need for bulk DNA, but because of a reduced ability to eradicate it. In contrast, genome size contraction can be expected in organisms selected for small size, again not because of direct selection for rapid genome replication, but because purifying selection more efficiently eliminates deleterious genomic elements from large populations. Thus, contrary to the assertion that "the limited variation in prokaryotic genome sizes ... presents an important puzzle in its own right ... just as important a question as the enormous genome size variation in eukaryotes" (Gregory 2005c), from a population genetic perspective, the uniformly simple genomes of prokaryotes are not surprising at all—they are the expectation.

Before we proceed, three fundamental points need to be emphasized. First, an increase in body size is not the only mechanism that results in population size reductions. Thus, the theory to be presented below does not rule out the possibility of a large genome in a unicellular species, provided that the latter has experienced an unusually high level of random genetic drift for reasons associated with the breeding system, degree of linkage in the genome, or historical long-term population bottlenecks (see Chapter 4). Random genetic drift is a fundamental attribute of all species, which in effect operates like a rheostat for modulating genomic growth versus contraction. Once this is understood, it is relatively straightforward to move beyond the rather vague description of excess DNA as being selfish or inert junk to a more mechanistic theory of genomic evolution capable of explaining why specific phylogenetic lineages are prone to the establishment of certain kinds of genetic elements and gene structures.

Second, because evolution is a stochastic process, subject to numerous probabilistic events, we should not expect to find tight deterministic relationships between all genomic attributes in all lineages. Indeed, when random genetic drift is a prominent evolutionary force, substantial deviations around overall patterns are *expected*. For a particular genome size, there can often be an order-of-magnitude range of variation among species with respect to genomic composition, and near the threshold for intron/mobile element expansion, the dispersion can be even greater (see Figure 2.1). Thus, although there is a tendency for those embroiled in the debate over genome size evolution to treat every deviation from an overall pattern as a definitive argument for or against a particular point of view (e.g., Cavalier-Smith 2005; Gregory 2005b), this strategy can be quite misleading. Returning to Figure 2.1, it is fairly easy to pick out two or three points that appear contrary to the overall trend, but much more difficult to do so with five to ten points. Consistent patterns are observed both within and among major phylogenetic groups, and it is this level of variation that merits explanation. This is not to say that outliers to an overall trend are uninteresting. Indeed, as will be seen in the following chapters, once the key biological features of such oddities are understood, they often provide deeper insights into evolution than might otherwise be possible.

One final point on the fundamental mechanisms driving genomic evolution merits further attention. As noted above, striking correlations exist between genome size and various aspects of cell size, metabolism, and division rate. In particular, within many phylogenetic groups, organisms with larger cells and lower metabolic rates generally have larger genomes (e.g., Gregory 2001, 2002a,b; Vinogradov and Anatskaya 2006). Is there any way to reconcile such patterns with the hypothesis that drift- and mutation-associated phenomena are the primary drivers of genomic evolution? The tentative answer is yes, in that a plausible case can be made that the evolved correlations between genome size and cytological features may be by-products of the shared involvement of intervening factors, rather than outcomes of direct causal connections.

As noted above (and argued more formally in Chapter 4), small population sizes and low mutation rates independently encourage the expansion of genome complexity. It is noteworthy that both population size and mutation rates tend to decline with increasing cell size. There is, of course, a one-to-one relationship between cell size and organism size in unicellular species. However, increases in cell size often accompany selectively driven increases in body size in laboratory experiments with multicellular species (Falconer et al. 1978; Riska and Atchley 1985; Stevenson et al. 1995; Calboli et al. 2003), and such associations also exist within a variety of phylogenetic groups, e.g., nematodes (Wang et al. 2002; Watanabe et al. 2005; Lozano et al. 2006), and sea anemones (Francis 2004). Thus, given the nearly universal inverse relationship between population size and body size (see Chapter 4), a negative correlation between cell size and population size is expected within many major taxonomic groups, with the causal intermediary being body size. In addition, metabolic rates tend to increase with decreasing cell size (Kozlowski et al. 2003), and because mutation rates scale positively with metabolic rates (Martin and Palumbi 1993; Gillooly et al. 2005), we can expect a negative association between cell size and mutation rate within broad taxonomic groups, driven by the intermediate factor of metabolic rate. Thus, although population size and mutation rate need not be inversely related to cell size in all phylogenetic groups, such scaling is likely for at least one of these parameters in most groups. This implies that although correlations between genome size and cytological features appear to support the causal connections postulated by the bulk-DNA hypothesis, they are also compatible with the mutational-hazard hypothesis.

3 The Human Genome

As first highlighted in the landmark papers of Lander et al. (2001) and Venter et al. (2001), enormous international effort has led to a refined understanding of the basic architectural features of the human genome. At about 3000 megabases (Mb), the human genome is larger than that of any other well-characterized species (although plenty of more massive genomes await exploration). Despite the unique aspects of human biology, our genome is not particularly noteworthy with respect to gene structure or organization, at least from the perspective of multicellular species, and it harbors most of the features that are the hallmarks of eukaryotic genomic evolution. Thus, a broad overview of our own genetic blueprint and its deviations from those of other eukaryotes will provide a useful entrée to many of the concepts to be explored in the pages to follow. This chapter will also provide some insights into the challenges of exploiting genomic information in attempts to understand organismal evolution. As will be discussed at the close of the chapter, we are still a long way from identifying the molecular-level changes responsible for the unique human phenotype.

Gene Number

One of the greatest surprises of the Human Genome Project was the discordance between the count of protein-coding genes (~24,000) and expectations based on perceived phenotypic and behavioral complexity. Only about 1% of the human genome consists of protein-coding DNA (Table 3.1), and although the number of human protein-coding genes exceeds that in most well-characterized unicellular species, it is only moderately greater than that in nematodes and insects and substantially less than that in pufferfish and rice (Table 3.2). Because the gene counts reported in Table 3.2 are

TABLE 3.1 Approximate fractional composition of the human genome

TYPE OF DNA	FRACTION
Coding exons	0.008
Internal introns	0.308
5′ Untranslated regions	
Exons	0.045
Introns	0.002
3′ Untranslated regions	
Exons	0.006
Introns	0.001
Intergenic DNA	0.683
Conserved noncoding DNA	0.016
Pseudogenes	0.007
Mobile genetic elements	0.446

Note: Derived from various references given in the text. Intergenic DNA is all DNA except coding exons and internal introns. The fractions do not sum to one because mobile elements, pseudogenes, and transcription factor binding sites reside in introns, UTRs, and/or intergenic DNA.

substantially dependent on gene prediction programs, they should be viewed as approximations, and may in some cases be off by as much as 20%, but it is clear that the evolution of multiple cell types is only weakly associated with an expansion in the number of protein-coding genes. There are several hundred noncoding RNA genes with well-understood features (specifying transfer RNAs, ribosomal RNA subunits, and small nucleolar and nuclear RNAs), but the numbers for humans are not much different from those in other eukaryotes. Nor does there appear to be any relationship between organismal complexity and average protein length. Although the total coding DNA associated with individual eukaryotic genes is 10%–15% greater than that for orthologous prokaryotic genes (Wang et al. 2005), there is no obvious pattern among eukaryotic lineages (see Table 3.2).

One conceptual matter regarding gene number merits consideration up front. Because nearly all eukaryotic genomes contain substantial numbers of duplicate genes, many of which may be functionally redundant, gene number may be a poor predictor of functional gene diversity. In the human genome, about 4000 pairs of duplicate genes (not including members of large multigene families) can be readily identified (Lynch and Conery 2003a,b), and about 5% of the genome consists of relatively recent segmental duplications ranging up to 200 kb in size (Bailey et al. 2002; Samonte and Eichler 2002). Using the number of substitutions between silent sites as an estimate of the age of a pair of duplicates (the rationale for which is discussed

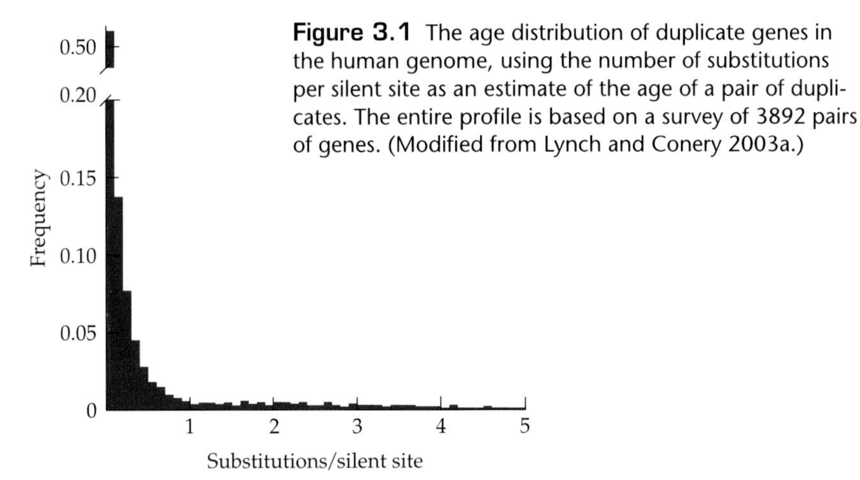

Figure 3.1 The age distribution of duplicate genes in the human genome, using the number of substitutions per silent site as an estimate of the age of a pair of duplicates. The entire profile is based on a survey of 3892 pairs of genes. (Modified from Lynch and Conery 2003a.)

below), the age distribution is found to be highly L-shaped in humans (Figure 3.1) and in several other species (Lynch and Conery 2000, 2003a,b). As will be described in more detail in Chapter 8, such age distributions can be used to estimate the rates of birth and loss of gene duplicates. For humans, the average rate of duplication is ~0.01/gene/million years (MY), while the average half-life of a duplicate is ~7.5 MY (Lynch and Conery 2003a,b). Thus, on a time scale of ~100 MY, most genes within the lineage leading to the human genome will have had an opportunity to duplicate, but almost all duplicates experience mutational silencing within ~10 MY.

Such observations imply that a significant fraction of genes is nonessential and simply a reflection of stochastic expansion and contraction processes. And this implication in turn raises significant questions about the common assumption that the size of a gene family provides information on its adaptive significance (Rubin et al. 2000; Lespinet et al. 2002). For example, expansions in gene families associated with reproduction and immunity have been taken to imply diversifying selection for new gene functions in the mouse (Waterston et al. 2002). Yet the total numbers of genes in the human and mouse genomes are nearly equal (Waterston et al. 2002), suggesting that expansions and contractions of gene family sizes may be balanced in both lineages, perhaps as a simple consequence of a stochastic equilibrium birth/death process (e.g., Nei 2005). Similarly, the reduced number of active olfactory receptor genes in humans relative to mice has been taken as evidence for reduced selection on olfaction in humans (Dehal et al. 2001; Gilad et al. 2003), but again, the simple possibility that such changes have arisen in a neutral fashion has not been ruled out.

With a fully sequenced genome, it is possible to investigate these issues formally by organizing the entire set of protein-coding genes into families using a sequence similarity criterion. Family sizes can range from singletons with no obvious similarity to any other component of the genome to

TABLE 3.2 Haploid genome size, number of protein-coding genes, and average number of nucleotides per gene for some well-characterized eukaryotic genomes

	GENOME SIZE (MB)	GENE NUMBER	KILOBASES/GENE		
			TOTAL	CODING	NON-CODING
Unicellular species					
Encephalitozoon cuniculi	2.90	1997	1.45	1.01	0.44
Saccharomyces cerevisiae	12.05	6213	1.94	1.44	0.50
Schizosaccharomyces pombe	13.80	4824	2.86	1.43	1.43
Cyanidioschyzon merolae	16.52	5331	3.10	1.55	1.55
Cryptococcus neoformans	19.05	6572	2.89	1.62	1.27
Plasmodium falciparum	22.85	5268	4.34	2.29	2.05
Entamoeba histolytica	23.75	9938	2.39	1.14	1.25
Leishmania major	33.60	8600	3.91	2.15	1.76
Thalassiosira pseudonana	34.50	11242	3.07	0.99	2.08
Trypanosoma spp.	39.20	10000	3.92	1.96	1.96
Oligocellular species					
Ustilago maydis	19.68	6572	2.99	1.84	1.15
Aspergillus nidulans	30.07	9541	3.15	1.57	1.58
Dictyostelium discoideum	34.00	9000	3.78	2.45	1.33
Neurospora crassa	38.64	10082	3.83	1.44	2.39
Land plants					
Arabidopsis thaliana	125.00	25498	4.90	1.80	3.10
Oryza sativa	466.00	60256	7.73	1.18	6.55
Lotus japonicus	472.00	26000	18.15	1.35	16.80
Animals					
Caenorhabditis elegans	100.26	21200	4.73	1.25	3.48
Drosophila melanogaster	137.00	16000	8.56	1.66	6.90
Ciona intestinalis	156.00	16000	9.75	0.95	8.80
Anopheles gambiae	278.00	13683	20.32	1.64	18.68
Fugu rubripes	365.00	38000	9.61	0.93	8.68
Bombyx mori	428.70	18510	23.16	1.66	21.50
Gallus gallus	1050.00	21500	48.84	1.44	47.40
Mus musculus	2500.00	24000	83.33	1.30	82.03
Homo sapiens	2900.00	24000	96.67	1.33	95.36

Source: Lynch 2006a.

Note: The total number of kilobases per gene is equivalent to the reciprocal of the gene density, i.e., total genome size divided by gene number, and does not necessarily reflect functional significance. The number of noncoding nucleotides per gene contains all noncoding material (including introns and all intergenic DNA). The average length of a protein in amino acids is equal to one-third the number of coding nucleotides per gene. All numbers are subject to minor revision as the various genome annotation projects mature.

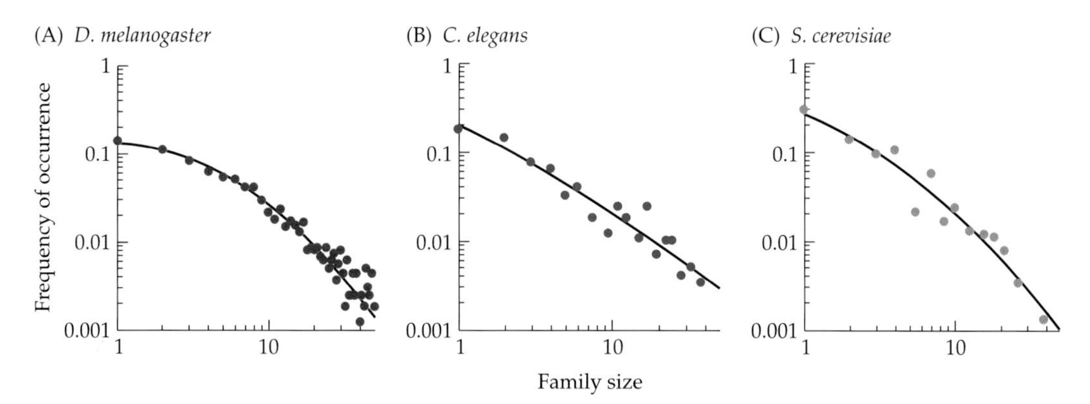

Figure 3.2 The frequency distributions of family sizes for genes in *D. melanogaster* (A) and different protein folds (structural subunit types) in *C. elegans* (B) and *S. cerevisiae* (C). (Data from Qian et al. 2001 and Luscombe et al. 2002.)

large multigene lineages. Although the matching criterion deployed in such analyses is quite arbitrary, regardless of the stringency of family definition, the number of families of a given size virtually always declines with the size of the family, with the most common family size being the singleton class (Figure 3.2). This sort of scaling holds for entire genes, functional domains of such genes (based on protein features), families of pseudogenes, and even various oligomers (specific small-nucleotide sequences) in both eukaryotes and prokaryotes (Huynen and van Nimwegen 1998; Yanai et al. 2000; Qian et al. 2001; Harrison et al. 2002; Koonin et al. 2002; Luscombe et al. 2002).

Several authors have argued that these kinds of plots obey a power law, such that the number of families of size x (i.e., containing x members) is proportional to x^{-b}, with b for protein family sizes typically being ~1.5–2.0 (Luscombe et al. 2002). Such behavior can be seen in Figure 3.2, where a tenfold increase in family size is accompanied by a decline in frequency of one to two orders of magnitude. Similar patterns exist in linguistic analysis, in which the number of words used in a text is usually inversely related to the number of times the words are used (most words are used just once or twice in a literary tract, whereas a few words such as "the" and "a" are used many times), a relationship generally referred to as Zipf's (1949) law. Although such universal patterning has invited speculation that long strings of DNA sequence are structured much like a language, there are substantial differences between the syntax of language and the structure of genomic sequences, not the least of which is the constant upheaval in the latter caused by duplication, deletion, and substitution events. Moreover, the patterns shown in Figure 3.2 do not strictly obey a power law, as there is a clear tendency for group size frequency distributions to bow downward (a power law would generate straight lines in logarithmic plots).

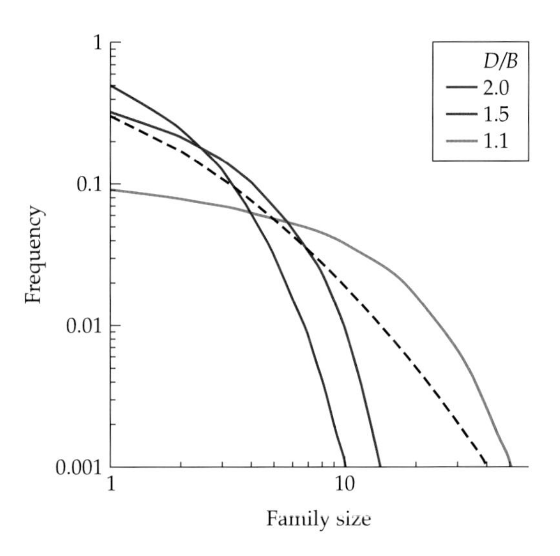

Figure 3.3 Equilibrium probability distributions of gene family sizes generated by computer simulations of the stochastic birth/death process described in the text. The form of the distribution depends only on the ratio of *D* and *B*, not on their absolute values. The patterns describe either the temporal distribution of family sizes experienced by a single gene over evolutionary time or the steady-state distribution of family sizes for a large number of genes with equal *D*/*B*. The dashed curve is the average of the three solid curves.

How might we account for this type of pattern? Consider a simple birth/death model, with each gene having a per-generation probability B of giving rise to a new copy, such that the average birth rate of a family of x members is Bx. If it is assumed that the presence of at least one member of the gene family is essential (i.e., complete loss of the gene family is not possible), but that all excess copies have a probability D of being eliminated, the expected number of losses per unit time is zero if $x = 1$, but Dx otherwise. Because these are just the *expected* numbers of gains and losses per time interval, and true changes come in discrete units, the number of copies (x) will wander over evolutionary time, with an effective upper limit being imposed provided that $D > B$. Under this model, the equilibrium probability of a gene being a member of a family of a certain size at any particular time is a simple function of D/B (Figure 3.3). Consistent with the observed data, this model predicts a bowed negative relationship between family number and family size, with the maximum number of families being in the singleton class. Although the curvature of the distribution for each D/B ratio may appear stronger than that seen in Figure 3.2, it should also be noted that the curves for individual D/B ratios intersect. This means that if data were averaged over a heterogeneous pool of genes subject to different rates of birth and death, the average frequency distribution would be more linear than that for an individual D/B ratio (dashed curve in Figure 3.3).

Although more complicated models for gene family size evolution include gene accrual by horizontal transfer or assume fixed initial and/or final genome sizes (Huynen and van Nimwegen 1998; Yanai et al. 2000; Qian et al. 2001; Karev et al. 2002, 2005), the need for such embellishments remains unclear. The key point here is that with D/B ratios consistent with actual estimates for eukaryotic genes (see Chapter 8), a simple random birth/death model, with no direct selection on copy number (other than the maintenance of a single copy), readily generates distributional forms like those seen in Figure 3.2. Thus, although a number of gene families have almost certainly expanded in specific phylogenetic lineages for adaptive reasons (e.g., immune system genes in vertebrates), many multigene families may simply be transient outcomes of the universal stochastic processes of gene duplication and elimination, with the total gene content being of little or no adaptive relevance. As a guard against overinterpreting large gene family size as a definitive indicator of unusual biological significance, statistical methods for formally evaluating the null hypothesis of stochastic expansion/contraction have been proposed (Hahn et al. 2005).

Finally, a remarkable one-third of the predicted genes in the human genome have no identifiable homologue in any other well-characterized non-primate species, even other mammals (Dehal et al. 2001). This poorly understood genomic feature is not unique to humans (or primates), as it has been repeatedly found in the sequenced genomes of other metazoans. The source and functionality of these orphan genes remains unresolved. Many might simply be derivatives of ancestral coding sequences subject to unusually rapid rates of substitution and/or rearrangement (Long 2001; W. Wang et al. 2004), and relatively short proteins may fortuitously arise de novo from chance open reading frames in noncoding DNA that happen to be located near spurious transcription factor binding sites (Begun et al. 2006). Still other novel genes originate when partial duplications of preexisting genes create chimeric open reading frames at sites of insertion (Katju and Lynch 2003, 2005). Although it was initially suggested that 100–200 human genes have been acquired by lateral transfer from prokaryotes (Lander et al. 2001), the number is closer to 40 after various sampling and computational biases are accounted for, and even these are uncertain (Salzberg et al. 2001).

Introns and Exons

Although most eukaryotic lineages produce proteins with similar average lengths, dramatic variation exists among species in investments in the noncoding intragenic portions of genes. The average coding region of a human gene is subdivided by 7.7 introns, yielding an average exon length of about 0.15 kb. Thus, with an average intron length of 4.66 kb, most exons in human genes are tiny islands in a vast sea of intronic DNA. In contrast, almost all *Saccharomyces* genes are intron-free, and the average *Caenorhabditis* gene contains 5.2 introns of length 0.12 kb. With fewer introns, the average invertebrate exon is larger than that in humans, but more striking is the strong reduction in the variance of exon size in humans, with almost none exceeding 300 bp in length (Figure 3.4). The precise localization of unusually long introns by mammalian spliceosomes appears to rely on a scanning mechanism for exon (rather than intron) boundaries, which becomes increasingly inaccurate with long exons (see Chapter 9).

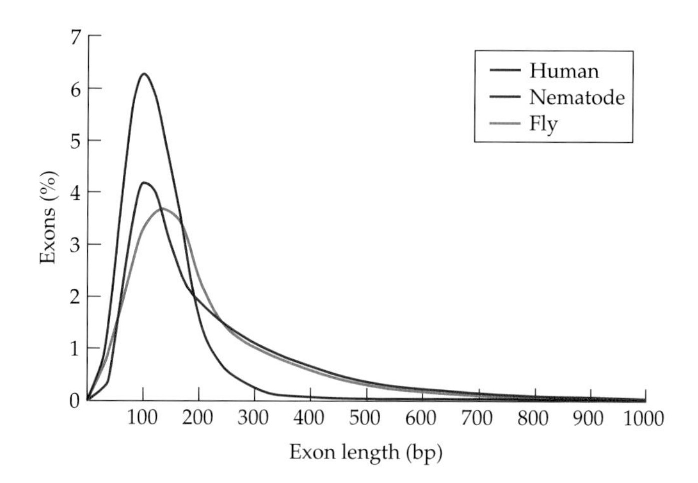

Figure 3.4 The distribution of exon sizes in the genes of humans, *Caenorhabditis* (nematode), and *Drosophila* (fly). (Modified from Lander et al. 2001.)

Introns provide potential opportunities for diversifying gene functions without increasing gene number. For example, if alternative intron–exon junctions can be recognized, it becomes possible to modify the products of a gene by including/excluding or extending/truncating specific exons. More than half of human genes are subject to alternative splicing, with an average of 2.6 transcript variants per gene (Lander et al. 2001; Modrek and Lee 2003), and some human genes have dozens of alternatively spliced forms (Maniatis and Tasic 2002; Roberts and Smith 2002). Approximately 20% of alternatively spliced variants appear to be expressed in a tissue-specific pattern in humans (Xu et al. 2002), and the variants produced involve the insertion/deletion of complete functional domains of proteins more often than expected by chance (Kriventseva et al. 2003). Thus, there is little question that the repertoire of functional human proteins exceeds the number of genes.

Could an enhanced amount of alternative splicing contribute to the greater level of cellular/developmental complexity in mammals relative to invertebrates? Only about 20% of the genes in *C. elegans* and *D. melanogaster* are alternatively spliced, with an average of 1.3 transcripts per gene (Misra et al. 2002). However, although this suggests a reduction in alternative splicing in these invertebrate species, the depth of cDNA coverage (libraries of mRNAs transformed into more readily sequenced DNA by reverse transcription) for humans and mice tends to be much greater than that for invertebrates, and hence more likely to reveal rare transcript types (Brett et al. 2002). In addition, it is unclear how many alternatively spliced transcripts are actually products of adaptive evolution as opposed to being simple accidents of an imperfect splicing process. Modrek and Lee (2003) found that about 70% of minor variants of alternatively spliced genes in humans and mice contain exons that have arisen since the divergence of these species. Such splicing variants may be simple consequences of poorly refined sets of splice signals, unless they represent lineage-specific adaptations. These issues will be covered in greater depth in Chapter 9.

Summing up to this point, the human genome contains between 1.5 and 2.0 times more protein-coding genes than the genomes of nematodes and flies, and after accounting for the increased number of splicing variants per gene, this potentially translates into more than 50,000 additional proteins. In terms of protein diversity, the proportional scaling may be even greater: in proceeding from yeasts to invertebrates (fly and nematode) to mammals, the number of functional domains used as building blocks of proteins increases from 1000 to 1200 to 1900, respectively (Li et al. 2001); and human proteins have about twice the number of distinct domain arrangements as those of invertebrates and nearly six times as many as those of yeasts (Lander et al. 2001). Thus, when one adds to these structural differences the severalfold increase in DNA allocated to the regulation of human genes (see below), the potential diversity of ways in which gene products can be deployed in mammals appears to greatly exceed that in well-studied invertebrates, although it is far from clear how much of that potential has been promoted or exploited by adaptive mechanisms.

Regulatory DNA

Given the general lack of correlation of gene number and protein size with phenotypic complexity, the roots of morphological diversification may reside largely in changes in noncoding DNA that influence the regulation of gene expression, alternative splicing being one manifestation of such evolution. As noted above, the average amount of noncoding DNA per gene varies dramatically among species (see Table 3.2), from 0.5 to 2.0 kb in most unicellular eukaryotes to more than 3.0 kb for most multicellular species and up to nearly 100 kb for humans, so the potential certainly exists for more complex organisms to harbor more regulatory DNA. However, because a substantial fraction of the noncoding DNA in most species is derived from mobile genetic elements, the extent to which multicellular species have accumulated regulatory DNA embellishments is not immediately clear.

The direct identification of regulatory DNA is difficult because in contrast to coding regions, which can be identified by continuous open reading frames (adjacent sets of nucleotide-triplet codons for amino acids), there is no general language for regulatory elements. However, functionally significant noncoding DNA can be inferred indirectly by comparing orthologous sequences from different species, capitalizing on the idea that regions under purifying selection will appear as islands of conserved sequence on a background of more random interspecific divergence. This strategy will yield minimal estimates of the amount of functionally relevant noncoding DNA for at least two reasons: first, the regions used as measures of background variation are assumed to be neutral; and second, noncoding DNA that is under directional selection and hence rapidly evolving will be missed entirely. Application of this approach to mouse and human sequences yields

TABLE 3.3 Average amount of DNA per gene (in kilobases) associated with coding exons, internal introns, and intergenic spacers (outside points of translation initiation and termination)

			INTERGENIC	
	EXON	INTRON	REGULATORY	OTHER
Saccharomyces	1.44	0.02	0.11	0.37
Aspergillus	1.57	0.27	0.03	1.55
Plasmodium	2.29	0.25	0.04	1.76
Caenorhabditis	1.25	0.64	0.43	2.41
Drosophila	1.66	2.93	1.37	2.60
Homo/Mus	1.32	32.27	1.95	61.14

Note: Exonic and intronic DNA includes only that associated with the coding region, i.e., excludes UTR regions, which are included in the intergenic categories. Estimates for the intergenic regulatory DNA category are based on islands of observed intergenic sequence conservation among closely related species: *Saccharomyces* (Kellis et al. 2003); *Aspergillus* (Galagan et al. 2005); *Plasmodium* (van Noort and Huynen 2006); *Caenorhabditis* (Webb et al. 2002); *Drosophila* (Bergman and Kreitman 2001; Andolfatto 2005); *Homo/Mus* (Shabalina et al. 2001). Intergenic other refers to all DNA between the stop codon of an upstream gene and the start codon of the following gene that is not discernable as intergenic regulatory. Qualitatively similar results have been obtained with other methods (e.g., Siepel et al. 2005).

a conservative estimate of more than 2.0 kb of intergenic DNA reserved for the regulation of the average mammalian gene (Shabalina et al. 2001). This suggests that the amount of DNA devoted to gene regulation in mammals exceeds that associated with coding DNA by about 50%. Similar analyses for two fungi (*Aspergillus* and *Saccharomyces*), *Plasmodium*, and two invertebrates (*Caenorhabditis* and *Drosophila*) suggest smaller investments in conserved intergenic DNA (Table 3.3). Although data are available for only five taxa, the overall pattern is consistent with a gradient of increasing amounts of regulatory DNA with increasing organismal complexity.

As will be discussed extensively in Chapter 10, the nature of several forms of regulatory DNA is fairly well understood. All genes depend on transcription factor binding sites to recruit the machinery for initiating transcription, internal motifs for locating intron–exon boundaries, and downstream signals for terminating transcription. However, because most sequences of these types consist of only a dozen or so bases, they cannot fully account for the more than 2.0 kb of regulatory DNA associated with an average mammalian gene. Many regulatory DNA sequences may be so diffuse and nonspecific as to be entirely recalcitrant to computational identification, but some additional aspects of mammalian gene regulation appear to be associated with large sequence tracts located deep within intergenic regions. For example, comparative analysis of the mouse and human genomes reveals about 66,000 highly conserved blocks of intergenic DNA, averaging 150 bp in length (Dermitzakis et al. 2002, 2003a, 2004). Several hundred such segments in the human genome have homologous sequences that are 100% identical in rodents, 95% identical in chickens, and 90% identical in fish (Bejerano et al. 2004; Woolfe et al. 2005). Despite the general absence of assignable functions for these elements, they are clearly under much more stringent selective constraints than most coding DNA.

Because conserved islands of noncoding DNA do not lead to protein production, it is likely that they play some sort of role in gene expression, either attracting proteins associated with transcription and serving as long-range tissue-specific enhancers or repressors (Woolfe et al. 2005; Kamal et al. 2006) or being transcribed into functional RNAs. It is well known that not all RNA transcripts are destined to become protein. For example, ribosomal RNAs and transfer RNAs are essential for translation in the cytoplasm, five small nuclear RNAs constitute the heart of the spliceosome used for removing introns from precursor mRNAs, several small nucleolar RNAs play key roles in directing base modifications during ribosome synthesis, and the enzyme that maintains chromosome ends contains a key RNA component. (See Table 3.4 for a set of definitions for various families of RNAs.) Although none of the islands of conserved DNA mentioned above have these particular functions, genome-wide transcription profiles suggest that many of them yield products with important cellular roles.

TABLE 3.4 A glossary of terminology for RNAs

mRNAs	Messenger RNAs; mature gene transcripts, after introns have been processed out of the mRNA precursor
miRNAs	Micro RNAs; generally 20–30 bp in length, and processed from transcribed "hairpin" precursor RNAs; used in the regulation of gene expression by complementary binding to nearly identical motifs in the 3′ UTRs of transcripts
ncRNAs	Noncoding RNAs; loosely defined as any transcript that does not encode protein
rRNAs	Ribosomal RNAs; the RNA subunits of the ribosome
sRNAs	Small RNAs; a generic term that encompasses miRNAs and siRNAs
siRNAs	Small interfering RNAs; generally 20–30 bp in length, and processed from longer double-stranded RNAs by the RNA interference pathway; deployed in posttranscriptional gene silencing
snRNAs	Small nuclear RNAs; a heterogeneous group of small RNAs whose functions are confined to the nucleus, including those involved in splicing introns out of precursor mRNAs and in telomere maintenance
snoRNAs	Small nucleolar RNAs; involved in the chemical modifications made in the construction of ribosomes; often encoded within the introns of ribosomal protein genes
tRNAs	Transfer RNAs; serve as vehicles for delivering amino acids during the translation of an mRNA

Further insights into these issues have been acquired from microarray technology, which allows the investigator to deposit thousands to millions of small bits of genomic DNA onto a microscope slide–sized platform, each spot pertaining to a specific genomic region. The pool of RNA transcripts can then be extracted from an organism, turned into cDNAs by reverse transcription, and hybridized to the array. The use of fluorescently labeled dyes provides a means for quantifying transcript levels associated with each DNA sequence on the array. Traditionally, microarrays have harbored only the DNA from predicted open reading frames, but arrays that include noncoding DNA (as well as other methods) have revealed the shortcomings of this approach (Carninci 2006). For example, Kampa et al. (2004) found that about 50% of the transcripts associated with the well-annotated human chromosomes 21 and 22 are derived from DNA entirely outside of known protein-coding genes and that about 20% are from antisense strands. Many of these noncoding transcripts exhibit differential tissue-specific expression (Cawley et al. 2004). Although it is fair to ask whether putative noncoding RNAs are simple artifacts of spurious transcription factor binding sites, the fact that homologs of a substantial fraction of human noncoding RNAs are also observed in the mouse (Kapranov et al. 2002; Okazaki et al. 2002) suggests that this is not universally the case, and it is now generally accepted that an enormous number of functional RNAs are transcribed from the noncoding regions of mammalian genomes (Eddy 2001; Mattick and Gagen 2001; Mattick 2003; Bertone et al. 2004).

Although the precise functions of the vast majority of noncoding RNAs are unknown, numerous noncoding transcripts appear to form stem–loop structures (with the stems held together by complementary base pairing) and probably have functions associated with the regulation of protein-coding genes (Washietl et al. 2005). A diversity of roles in cellular maintenance and development appears likely (Lim et al. 2005). One class of small RNAs in particular is centrally involved in gene regulation. Micro RNAs (miRNAs) are single-stranded, 20–22 bp products cut from transcribed foldback RNA precursors (double-stranded RNA stems with a small end loop) by the enzyme Dicer (Figure 3.5A). In animals, miRNAs are typically expressed in a stage- or tissue-specific manner retain, often at very high levels, and their sequences have strong complementarity to regions in the 3′ UTRs (untranslated regions) of the transcripts of many key developmental genes (Eddy 2001; He and Hannon 2004; Zamore and Haley 2005). By pairing with complementary motifs of about 8 bp in 3′ UTRs, miRNAs cause translational silencing of the mRNA and in some cases tag the mRNA for destruction. By one account, about 20% of human genes are regulated in this manner (Xie et al. 2005). The number of miRNAs encoded in the human genome may be as high as 800 (Bentwich et al. 2005), and at least 100 such genes are highly conserved among nematodes, arthropods, and vertebrates (Lim et al. 2003a,b).

Why and how a system for gene silencing that depends on gene expression would evolve remains to be determined, but the general mechanism may be ancient, perhaps dating to the stem eukaryote. Like animals, all land

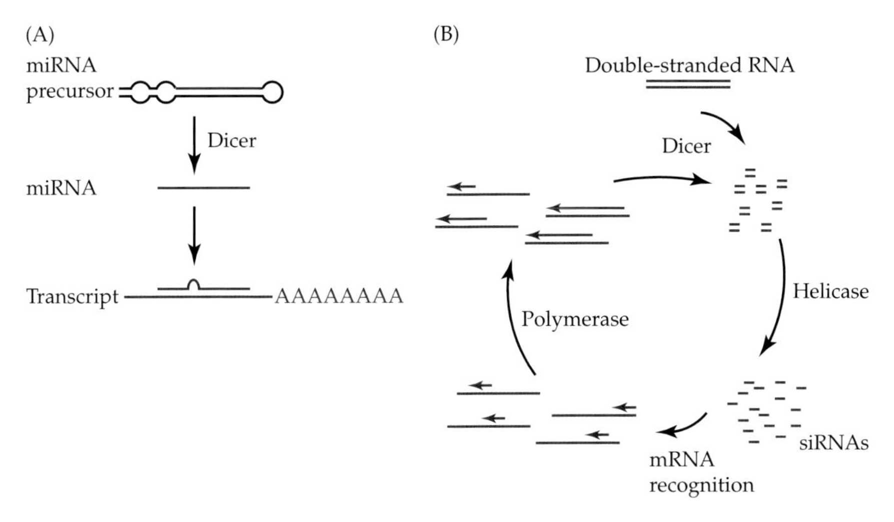

Figure 3.5 The effects of small RNAs on gene expression. (A) Gene silencing by micro RNA. After developing from a transcript, a micro RNA precursor is folded into a hairpin structure (loops denote noncomplementary base positions), which the enzyme Dicer recognizes as containing double-stranded RNA. The released single strand is largely complementary to a messenger RNA, and by binding to it, blocks translation. (B) The RNA interference pathway. An initial target double-stranded RNA is spliced into small interfering RNAs by Dicer, which are then separated into single strands. The siRNAs have perfect complementarity to larger single-stranded mRNAs and serve as primers for converting them into new double-stranded RNAs, which again elicit the action of Dicer, leading to a self-sustaining loop.

plants apparently deploy miRNAs in cell differentiation (Bartel and Bartel 2003; Floyd and Bowman 2004). Although no miRNAs have yet been discovered in unicellular species, the more general use of small antisense RNAs (processed by other means than miRNAs) in gene silencing appears to be ubiquitous. Antisense RNAs play important functions in eubacteria (Selinger et al. 2000; Wagner and Flärdh 2002; Gottesman 2005) and archaea (Tang et al. 2002). Small complementary RNAs of the same approximate size as miRNAs are also deployed in gene silencing in the ciliates *Paramecium* (Garnier et al. 2004) and *Tetrahymena* (Lee and Collins 2006), and, although their functions are unknown, antisense RNAs derived from both genic and intergenic regions are common in *Plasmodium* (Gunasekera et al. 2004) and yeasts (Havilio et al. 2005; Samanta et al. 2006).

Whereas miRNAs simply repress translation of mRNAs that have already been transcribed, many of these other small antisense RNAs play a role in silencing at the transcriptional level, either by blocking transcription entirely or by tagging mRNAs for destruction via the RNA interference (RNAi) pathway (Figure 3.5B). The same mechanism used in the processing of miRNAs is central to RNAi: upon encountering double-stranded RNA, Dicer splices the target molecule into pieces of essentially the same size as typical miRNAs (~21 bp). In their single-stranded form, the resultant small interfering RNAs (siRNAs) can then anneal to complementary regions of homologous transcripts, marking them for cleavage (again by Dicer), generating still additional siRNAs used in further rounds of targeting. Thus, once an RNAi cascade is initiated, an intracellular environment for the efficient destruction of newly arising complementary single-stranded transcripts is maintained, thereby silencing the target gene (Klahre et al. 2002; Timmons et al. 2003; Agrawal et al. 2003; Novina and Sharp 2004). The distinction between the functioning of miRNAs in transcript silencing and siRNAs in transcript elimination is related to the degree of complementarity of the small RNA. In general, perfect base complementarity leads to target cleavage by RNAi and hence mRNA removal, whereas the slight mismatches usually found with miRNAs are insufficient to elicit RNAi and instead lead simply to translational repression.

Although short on technical details, the preceding points make it clear that we are still a long way from having an accurate quantitative view of the fraction of the genome that is involved in gene regulation. After conservatively accounting for putative regulatory DNA under purifying selection, over 95% of the human genome still appears to be associated with mobile elements, pseudogenes, introns, and other ill-defined categories of potentially nonfunctional DNA. However, with substantial transcription occurring across the entire genome, on both sense and antisense strands, and with the central role played by the RNAi pathway in the silencing of mobile elements (see Chapter 7) and the maintenance of centromeric regions (see Chapter 5), the definition of "functional DNA" is by no means straightforward. Noncoding RNAs also impose substantial challenges to the enterprise of quantifying gene numbers. Arriving at a universally accepted definition of a gene has never been easy, but from a molecular perspective, a gene can be viewed as a segment of DNA that is transcribed to provide a functional unit. If one adheres to this definition, then the protein-centric gene counts outlined in Table 3.2 may be substantial underestimates of the true totals.

Mobile Genetic Elements

One of the most striking aspects of the human genome is the extraordinary abundance of mobile genetic elements—numbering about 100 for every protein-coding gene and together constituting nearly half of the genome. If we add in the large pool of processed pseudogenes that have arisen as byproducts of mobile element activity (see below) and additional intergenic DNA that is presumably derived from ancient mobilization events but is no longer recognizable as such, then 75% or more of the human genome may be a product of past mobile element activities. Because of their mutagenic side effects (as agents of insertions and substrates for ectopic recombination), most mobile elements must have negative consequences for their

TABLE 3.5 Abundances of various general classes of mobile genetic elements in the human genome

CLASS	NUMBER OF COPIES (1000s)	FRACTION OF GENOME (%)
LINEs	868	20.4
SINEs	1558	13.1
LTR elements	443	8.3
Transposons	294	2.8

Source: Lander et al. 2001.

hosts. However, the maintenance of a population of such elements within a host genome is guaranteed so long as the rate of element self-proliferation offsets the rate of selective removal (see Chapter 7).

Mobile elements are generally subdivided into two major categories: retrotransposons that travel via an RNA intermediate, and transposons that move via genomic excision. The human genome contains plenty of both, although the former vastly outnumber the latter (Table 3.5). The retrotransposons in the human genome can be subdivided further into three classes: LINEs (*l*ong *in*terspersed *e*lements), SINEs (*s*hort *in*terspersed *e*lements), and LTR (*l*ong *t*erminal *r*epeat) elements (Deininger and Batzer 2002). We will consider these three classes in turn, with special attention being given to LINEs, which are capable of both self-replication and inadvertent mobilization of nonautonomous SINEs. The discussion will be rather generic in nature, as all of the major groups can be subdivided into numerous subfamilies (see Chapter 7).

LINEs and SINEs

The dominant retrotransposons in the human genome are the interrelated families of LINEs, together constituting nearly 870,000 total recognizable copies and about 20% of total genomic DNA (Smit 1999; Ostertag and Kazazian 2001). Despite these enormous numbers, only about 100 LINEs are thought to be retrotranspositionally competent. Approximately 6 kb in length, these autonomous elements contain an internal 5′ promoter (which directs transcription initiation to the upstream start of the element), two open reading frames (one encoding an RNA-binding protein and the other a dual-function protein that acts as an endonuclease and a reverse transcriptase), and a downstream poly(A) tail. Although some LINEs colonize via integration into preexisting breaks in the DNA—that is, in an endonuclease-independent manner (Morrish et al. 2002), most replicate by a process known as target-primed reverse transcription (Luan et al. 1993; Weiner 2002). Following the export of the transcript of an active element to the cytoplasm, the two translated proteins bind to their associated mRNA molecule. This ribonucleoprotein complex is then imported back into the nucleus, where the endonuclease has an opportunity to nick the genomic DNA, generally at a TT | AAAA target site (the bar denotes the insertion point). The exposed nick then serves as a primer site for reverse transcription of the mRNA, after which the cDNA of the element is integrated into the genome.

Retrotransposition of LINEs is known to be a sloppy process, with new insertions deviating from their parental elements in numerous ways. First, because the polyadenylation signal built into autonomous LINEs is weak, occasional transcriptional read-throughs proceed into the 3′ flanking DNA. This excess terminal sequence is then subject to reinsertion at a new location along with its linked LINE, providing a potential mechanism for gene duplication (Moran et al. 1999). Second, because reverse transcription is initiated at the 3′ end of the LINE mRNA and frequently fails to reach the 5′ terminus prior to integration, truncated insertions lead to dead-on-arrival elements. Third, for reasons that are not yet understood, new LINE insertions are often accompanied by local rearrangements, including large deletions (up to several kilobases) of adjacent genomic DNA, inversions, and even the insertion of chimeric elements derived from two different mRNAs (Gilbert et al. 2002; Symer et al. 2002). Fourth, because LINE-encoded proteins do not bind exclusively to the mRNA from which they arise, other defective LINEs can be mobilized as long as they are transcribed (Wei et al. 2001). By this means, nonautonomous elements can continue to spread throughout the genome.

Because LINEs are incapable of cleaving themselves from the host DNA, once fixed at the population level, they are permanently retained until they happen to be encompassed by a deletion event. Lander et al. (2001) argued that most of the retrotransposons within the human genome are ancient relics, and that a dramatic reduction in retrotranspositional activity occurred some 50 million years ago. However, this conclusion was reached by deriving the age distribution of elements from the divergence of each individual element from an overall consensus sequence thought to approximate the ancestral sequence. Because all members of a gene genealogy are equally old with reference to the base of the lineage, this technique could be misleading. A more informative strategy is to consider the pairwise divergences of members of the extant pool of elements, as a pair of highly similar elements must denote a recent colonization regardless of the divergence from the putative ancestral sequence. Focusing just on the LINE-1 subfamily, numerous pairs of elements are less than 2% divergent (Boissinot et al. 2000; Myers et al. 2002; Ovchinnikov et al. 2002), and many of the younger elements reside at sites exhibiting presence/absence polymorphisms (Bennett et al. 2004; Boissinot et al. 2004; Salem et al. 2003), which means that they have not yet gone to fixation within the human population. The absence of the younger elements from orthologous sites in the genomes of African apes further implies an age of less than 5 million years (Hedges et al. 2004). Thus, although there may have been an ancient period of explosive LINE colonization in the ancestor leading to apes, this family of elements is still quite active.

Even more numerically abundant than LINEs are SINEs, which total about 1,500,000 copies, 70% of which belong to the primate-specific *Alu* family. With a typical length of only 300 bp, SINEs are much smaller than LINEs. Their small size is due to the fact that they do not encode proteins, and there is no evidence that they ever have. Thus, SINEs are incapable of self-mobilization, and their abundance is entirely a consequence of their ability to hijack the endonuclease/reverse transcriptase activity of LINEs. *Alu* elements specifically parasitize the retrotranspositional apparatus of LINE-1 family members and exhibit the same TT | AAAA insertion-site preference. Like LINEs, a substantial fraction of SINEs are defective in ways that prevent transcription, and these elements have no future retrotranspositional possibilities. Also like some LINEs, many *Alu* elements are recent enough arrivals in the human genome to exist as presence/absence polymorphisms (Carroll et al. 2001; Roy-Engel et al. 2001). Such polymorphic insertions are present at a density of about 0.7 per million base pairs (Heilig et al. 2003), and the total rate of insertion of new *Alu* elements is estimated to be 0.05/host genome/generation (Cordaux et al. 2006).

LTR elements

The main class of human LTR elements is the *HERVs* (*h*uman *e*ndogenous *r*etro*v*iruses), which, like LINEs, carry genes encoding proteins for reverse transcription and integration. The long terminal repeats of these elements contain key regulatory sequences for transcription initiation and termination, as will be described more fully in Chapter 7. Unlike the situation with LINEs, however, HERV mRNA contains its own primer site for reverse transcription, which is carried out in the cytoplasm to produce a double-stranded DNA molecule that is then transported back to the nucleus. This unique mechanism results in the production of newborn elements with identical LTR sequences (see Chapter 7), so these retrotransposons carry a built-in record of their age, in the form of sequence divergence between the 5′ and 3′ LTRs. The rate of nucleotide substitution per site, about 1.25×10^{-9} per year (estimated from a comparative analysis of homologous human and chimpanzee sequences), is virtually the same as that seen elsewhere in the human genome for putative neutral sites (see below) (Sverdlov 2000; Hughes and Coffin 2001). Using this rate as a calibration, it appears that a major expansion of the HERV family began in the Old World primate lineage, with a substantial acceleration of insertions occurring in the great apes, especially humans.

A remarkable feature of HERVs is the presence of an *env* gene, which is used only for movement among cells. Analysis of sequence data indicates that this gene is under strong purifying selection, which would not be expected for a mobile element experiencing only vertical transmission (Belshaw et al. 2004). Although a fully active HERV element has not yet been discovered, these elements may be capable of moving horizontally, which would require transmission from somatic to germ line cells. Such movement is known in *Drosophila* (see Chapter 7).

Transposons

Much less abundant than the "copy-and-paste" retrotransposons in the human genome are the "cut-and-paste" DNA-based transposons, which comprise a dozen or so families (Smit and Riggs 1996). Such elements generally have 10–500 bp terminal inverted repeats (TIRs) and are flanked by small (< 10 bp) target-site duplications. Autonomous transposons encode a transposase that, after translation in the cytoplasm, returns to the nucleus, where it binds specifically to TIRs, catalyzing a process that moves elements to new genomic locations. Of course, if each gain of an insert were accompanied by an excision, it would not be possible for a DNA transposon to spread numerically. However, excision of an element results in a double-strand break, which must be repaired, and this is often accomplished by using the homologous chromosome as a template. Transposons are often activated during cellular replication, and if mobilization occurs at a late enough stage of replication that a sister chromatid is available, the proliferating element can simply be back-copied into the gap in the parental chromosome (Engels et al. 1990; Gloor et al. 1991). Alternatively, if in an early stage of germ cell division a copy transposes from a replicated to a nonreplicated portion of a chromosome, even if the donor site of one chromatid were to lose its element, this would be offset by the gain of a new element in both descendant chromosomes (in the not yet replicated site) (Chen et al. 1992).

In contrast to LINEs, where the proteins encoded by an element are intimately associated with its mRNA from the outset, the transposase of a DNA element is imported back into the nucleus with no memory of its genomic source. Thus, whether they contain a functional transposase gene or not, all genomic copies of the element are equally susceptible to mobilization so long as they have recognizable TIRs. In this sense, the genomic consequences of transposons are similar to those of retrotransposons: transposition is usually, though not always, duplicative, and nonautonomous elements are able to expand at the expense of their autonomous sister elements. As defective elements accumulate in the genome by this mechanism, hijacking more and more of the replicative proteins of active elements, transposition of the autonomous members must become less efficient, which may provide a natural barrier to runaway element proliferation.

Pseudogenes

Embedded within the noncoding DNA of the human genome are the decomposing remnants of a huge number of failed gene duplication events. These pseudogenes can be roughly subdivided into two major classes, defined by the mechanism of origin.

Processed pseudogenes arise when reverse transcription of an mRNA intermediate is followed by reintegration of the cDNA into the genome. Although such insertions often cover the entire coding region of the ancestral gene, they lack introns (which will have been removed from the mRNA prior to reverse transcription) and most (if not all) key regulatory elements, and are generally thought to be dead on arrival. Processed pseudogenes are also born with a poly(A) tract at the 3′ end (added at the time of prior transcription) and flanking direct repeats (gained at the time of insertion). Because of its reliance on reverse transcriptase in germ line cells, processed pseudogene production is intimately connected with retrotransposons, which, as noted above, can inadvertently supply the enzymatic machinery for mobilizing other mRNAs (Esnault et al. 2000; Wei et al. 2001).

Nonprocessed pseudogenes arise at the DNA level, often via local events that result in tandem duplications. Because such duplications arise from stretches of DNA with arbitrary truncation points that may lie within or between gene boundaries, they too will often be dead on arrival. Nonprocessed pseudogenes may also simply arise from the inactivation of a vertically inherited gene, in some cases promoted by positive selection (Wang et al. 2006).

Working with the relatively well-curated human chromosomes 21 and 22, Harrison et al. (2002) estimated that there are 0.4 pseudogenes for every annotated protein-coding gene, whereas parallel estimates for chromosomes 7 (Hillier et al. 2003) and 14 (Heilig et al. 2003) are 0.8 and 0.4, respectively. Altogether, there are about 15,000 pseudogenes in the entire human genome, at least half of which are of the processed type and nearly all of which are younger than the rodent–primate divergence (Zhang and Gerstein 2004). Because there are about 0.5 pseudogenes per active gene in the mouse (Waterston et al. 2002), about the same number as in humans, this high density of pseudogenes appears to extend to other mammals. However, ratios of pseudogenes to active genes tend to be lower in nonmammalian species: just 0.01 for *D. melanogaster* (Harrison et al. 2003); between 0.10 and 0.20 for *C. elegans* (Echols et al. 2002; Mounsey et al. 2002); 0.03 and 0.01, respectively, for the yeasts *S. cerevisiae* (Echols et al. 2002) and *S. pombe* (Wood et al. 2002); and 0.03 for *Arabidopsis* (Echols et al. 2002).

The likelihood of pseudogenization can vary substantially among gene classes. For example, a remarkable 13% of all human pseudogenes are derived from ribosomal protein-coding genes, with an average of 26 per gene (Zhang, Harrison et al. 2002). The age distribution of the 2000 or so ribosomal protein pseudogenes, based on their nucleotide divergence from still active ancestral genes, peaks at about 9% sequence divergence and thereafter exhibits a nearly exponential decline (Figure 3.6). A similar pattern has been found in the mouse genome, except that the intermediate peak is much less pronounced (Z. Zhang et al. 2004). Stable age distributions resulting from a steady-state birth/death process necessarily have the form of an exponential decay curve (as in Figure 3.1), provided that the probability of eradication is independent of age. Thus, one interpretation of these data is that all pseudogenes older than 9% divergence experienced an environment in which pseudogene birth and death rates were roughly constant, and that a decline in the birth rate occurred more recently.

When might this transition in behavior have occurred? From comparative data on pseudogenes of humans and chimpanzees, Nachman and

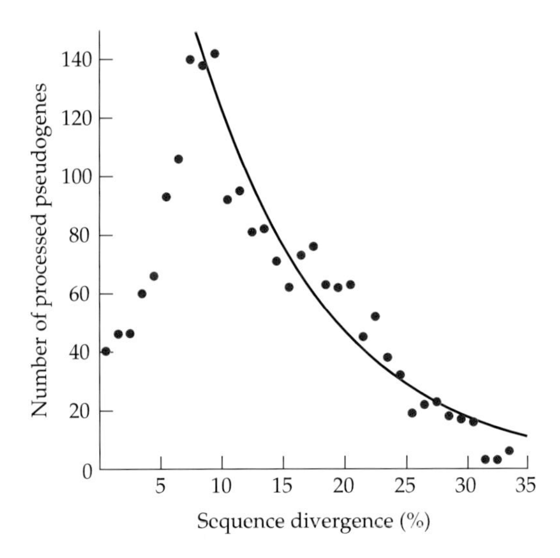

Figure 3.6 The age distribution of pseudogenes for the full set of 79 ribosomal protein genes in the human genome. The dashed line denotes the fitted exponential decay curve $y = 322\ e^{-0.096x}$. (Data from Zhang, Harrison et al. 2002.)

Crowell (2000) estimate the rate of substitution of silent sites to be 1.25×10^{-9} per year. Noting that active ribosomal proteins are nearly invariant across all animals at the amino acid level, we can roughly assume that only the 25% silent fraction of their nucleotide sites is free to accumulate nucleotide substitutions, whereas all sites within pseudogenes are free to evolve. The approximate rate of divergence of a ribosomal protein pseudogene from its active relative is then $(1.00 + 0.25) \times (1.25 \times 10^{-9}) = 1.56 \times 10^{-9}$ per year, so 9% divergence translates to about 58 MYA. Because pseudogene production depends critically on the intracellular availability of reverse transcriptase, it is notable that this date is close to the 50 MYA point of curtailed retrotransposon activity suggested by Lander et al. (2001). Remarkably, Bensasson et al. (2003) found that there was also a substantial pulse of insertion of mitochondrial DNA fragments into the human nuclear genome during this period. Thus, three entirely independent sets of analyses point to a dramatic phase of genomic upheaval in the lineage leading to humans and other primates 50–60 MYA, not long after the mass extinction event that marks the boundary between the Cretaceous and the Tertiary (65 MYA).

Because of the unstable age distribution of human pseudogenes, only a rough approximation can be made of the possible rate of pseudogene origin. However, given that there are 40 ribosomal protein–derived pseudogenes with divergence of less than 1% (0.5% being equivalent to 3 MY) and 79 active ribosomal protein-coding genes in the human genome, the current rate of origin of pseudogenes for this particular class of genes appears to be about 17%/gene/MY. Extrapolating the exponential curve in Figure 3.6 to time zero, a crude estimate of the birth rate prior to 58 MYA is 68%/gene/MY. These are enormous rates of pseudogene production, 10–100 times greater than other estimated rates of gene duplication (see Chapter 8). However, because ribosomal proteins are very highly expressed, constituting about 5%–10% of the protein in a typical cell (and probably much more in an egg, where ribosomes need to be prepackaged for use before transcriptional competence sets in), they provide an unusually large amount of mRNA substrate for reverse transcription.

How long do pseudogenes remain in the human genome before they are rendered unrecognizable by mutational substitution or deletion events? From the decline in the right side of the age distribution in Figure 3.6, for the period subsequent to 58 MYA, the half-life of pseudogenes is estimated to be equivalent to 7.2% divergence at silent sites, or 37 MY. This is severalfold longer than the estimated half-life of putatively functional duplicate genes in humans, 7.5 MY (Lynch and Conery 2003a,b), but more than 100 times shorter than estimates derived from dead-on-arrival retrotransposons (see Chapter 2). All of these estimates are quite rough and are based on numerous assumptions, but the latter contrast again points to the uncertainty of the generality of estimates of DNA eradication rates based on mobile elements (see Chapter 2).

Finally, it is worth considering whether processed mRNA insertions are always simply nonfunctional husks of previously active genes. Few formal investigations of this question have been made at the molecular level, but a clear case has been uncovered in the mouse in which a transcribed (but untranslated) retroprocessed gene has an important function in regulating the expression of the homologous parental gene (Hirotsune et al. 2003). In addition, several human genes with testis-specific functions appear to have been produced by the retroprocessing of more ubiquitously expressed genes (Marques et al. 2005). Such genes have apparently retained function by fortuitously acquiring transcription factor binding sites at the time of origin. These examples and others (Balakirev and Ayala 2003) provide another reminder that what superficially appears to be nonfunctional DNA may be quite the opposite.

The Unique Trajectory of Human Evolution

Although we humans are fairly ordinary mammals from a metabolic perspective, we are also strikingly unique in several key ways, such as our braininess and use of language and tools, hairlessness, and bipedalism. Yet we differ from our closest living relative, the chimpanzee (Ruvolo 1997), by just 1% at the nucleotide level (Mikkelsen et al. 2005), less than the level of divergence among reproductively compatible individuals within many eukaryotic species (see Chapter 4). Equally astounding is the substantial amount of human variation in morphology, behavior, physiology, and disease resistance when, at the molecular level, we are one of the most uniform species on Earth, with two random chromosomes differing by only 0.1% at the nucleotide sequence level (Cargill et al. 1999; Halushka et al. 1999; Yu et al. 2004).

Two of the primary goals of human genomics are to identify the molecular basis of functionally relevant human genetic variation and the key genomic changes that led to the unique human condition. Given the exceptionally low amounts of human variation involved, achieving these goals should be simpler than with more diverse species, but the challenges are still enormous. With a total haploid genome size of ~3 Mb, two individual humans are expected to differ at ~6,000,000 nucleotide sites, whereas ~30,000,000 fixed nucleotide differences exist between humans and chimpanzees. A large fraction of these differences are likely to involve inconsequential neutral variants, and as discussed below, many are undoubtedly maladaptive, so the number of candidate sites of functional significance could be in the hundreds or low thousands. It is mind-boggling to think that such a tiny number of nucleotides may make all the difference between a species that utterly dominates the global ecosystem and a small hairy ape inhabiting a few remnant forests in Africa.

Although a detailed evaluation of what makes us human is beyond the scope of this book, a few general points are worth making. One key question regarding human evolution concerns the rate of morphological change and the extent to which it has been driven by natural selection. It is frequently argued that mammals evolve at unusually rapid rates, with primates being exceptionally rapid for mammals and humans being extraordinary for primates (e.g., Cherry et al. 1978, 1982; Van Valen 1985). To account for this putative gradient in morphological evolution, Wilson and colleagues (Wilson et al. 1975; Wyles et al. 1983; Larson et al. 1984) suggested two possibilities: (1) the shift of a behaviorally sophisticated organism to a new habitat could induce secondary selection on morphology to yield a more favorable phenotype in the altered environment; or (2) modifications of gene regulation mechanisms, perhaps engendered by chromosomal rearrangements, could open up new pathways for adaptive evolution. Although a number of the problems with these arguments have been pointed out (Lynch 1990), here we simply ask just how impressive the rate of anatomical evolution in mammals is.

An objective means of evaluating this question is to contrast the observed rate of evolution with the expectation in the absence of natural selection. Under this neutral model, the only evolutionary forces are the recurrent introduction of variation by mutation and the gradual fixation of mutant alleles by random genetic drift. The application of such a model requires information on the per-generation rate at which mutation introduces new genetic variation for morphological traits, σ_m^2. Although this rate has not been measured directly in humans, it has been determined for mice, and very similar rates have also been found for a wide array of morphological traits in a variety of other animals (Lynch and Walsh 1998). Letting t be the number of generations that have elapsed since two species diverged from a common ancestor (estimated from the fossil record and generation times of extant species), $2\sigma_m^2 t$ is the expected amount of divergence between the two species in the absence of selection (where divergence is measured as the average squared deviation (variance) between the mean phenotypes of two species; Lynch and Hill 1986).

When this test is applied to the skeletal measures of a wide variety of mammals, the rate of observed evolution is almost always found to be 10–100 times slower than the neutral expectation. The only exception is the human lineage, which is evolving at a rate consistent with neutrality (Lynch 1990; Ackermann and Cheverud 2004). This tells us, first, that the long-term pattern of morphological evolution in nearly all mammals is relative stability, and second, that although skeletal traits do evolve more rapidly in humans than in other mammals (including chimpanzees and gorillas), the rate is far lower than would be possible with long-term directional selection.

These types of observations may seem far removed from the field of genome biology, but they provide a useful perspective by demonstrating that the types of genomic changes leading to our unique human morphology are unlikely to be out of the ordinary and may not even have been entirely promoted by natural selection. To account for patterns of primate morphological evolution, there is no need to invoke either a behavioral-drive process (which presumably would elevate the rate of evolution beyond the neutral rate) or extreme mutational changes in gene regulation mechanisms. In addition, the quantitative genetic methodology developed for morphological traits is directly relevant to comparisons of gene expression between chimpanzees and humans, which provide more direct insight into matters of regulatory evolution. Because of the high degree of sequence similarity between primate genomes, it is possible to use the same microarray platform to evaluate relative mRNA levels from different species (at least for gene spans that are 100% identical), treating transcript levels as an ordinary quantitative trait. Applying this approach to the transcriptome of blood leukocytes, livers, and brains, Khaitovich et al. (2004, 2005) argued that the data for humans, chimpanzees, orangutans, and macaques are consistent with a neutral model of evolution.

Although this conclusion appears to be contrary to that derived from morphology, the analytical model deployed by Khaitovich et al. is based on ideas that are only loosely connected with neutral expectations: an expected positive correlation between levels of within- and between-species variation for the expression of different genes, and a linear increase in between-species expression divergence with time. Because the *genetic* components of within-species and between-species variance are both driven by mutation, they are indeed expected to be correlated under the neutral model (Lynch and Hill 1986). However, measures of gene expression are functions of both the genetic and the environmental backgrounds of individuals, and unless the latter is factored out, they provide little information on the more relevant levels of genetic divergence, because genes whose expression is strongly influenced by the environment will naturally exhibit higher levels of variation both within and among samples. In addition, linear patterns of evolutionary diversification are entirely compatible with models in which selection randomly drives divergence. The only objective way to test for neutrality in gene expression divergence is to follow the same procedure outlined above for morphology; that is, to compare quantitative levels of divergence (measured in a controlled environment) with those expected given the mutation rate and time of divergence. This has been done for both *C. elegans* (Denver et al. 2005) and *D. melanogaster* (Rifkin et al. 2005), and in both cases the neutral hypothesis was strongly rejected in favor of a model in which purifying selection eliminates most newly arising mutations influencing gene expression. Thus, it is premature to conclude that primate gene expression is evolving in a neutral fashion, although this could be specifically true in the human lineage.

Because the vast majority of mutations with phenotypic effects are deleterious, a genome-wide elevation in the rate of evolution can simply result from an enhanced rate of fixation of mildly deleterious mutations, and there are two reasons to think that the human species is unusually vulnerable to such events. First, the exceptionally low level of nucleotide variation in humans reflects a very small long-term effective population size (see Chapter 4), and hence a high vulnerability to random genetic drift. Second, contrary to the situation envisioned under the behavioral-drive hypothesis, behavioral sophistication may result in a relaxation of the power of natural selection. Tool use, dietary flexibility, medical procedures, agriculture, and use of domesticated animals are just a few of the ways in which human behavior may reduce the intensity of selection against maladaptive morphological and/or physiological changes. Although the possibility that the relatively rapid rate of many aspects of morphological evolution in the human lineage is simply a consequence of the reduced efficiency of stabilizing selection may be hard to swallow for those with a high affinity for adaptive arguments, several lines of evidence at the molecular level support this contention.

First, from a genome-wide analysis of the amounts of divergence at silent (synonymous) and replacement (nonsynonymous) sites in protein-coding genes, Mikkelsen et al. (2005) found that 77% of amino acid–altering mutations are prevented from going to fixation in the human–chimpanzee lineage, as compared with 87% in the mouse–rat lineage (see Box 3.1 for the rationale underlying such analyses). Second, although a substantial fraction of human molecular evolution is associated with changes in gene regulation, contrary to the suggestions of Wilson and colleagues, such alterations, appear to have accumulated largely by nonadaptive processes. As in the case of replacement sites, regions immediately upstream and downstream of coding regions exhibit a substantial reduction in the efficiency of selection in the human–chimpanzee lineage, suggesting a degradation of regulatory elements (Keightley et al. 2005), although the magnitude of this effect is uncertain (Bush and Lahn 2005). In addition, human transcription factor genes have acquired roughly 50% more amino acid changes than have their counterparts in the chimpanzee (Mikkelsen et al. 2005), and the expression of such genes has evolved particularly rapidly in the human lineage (Gilad et al. 2006). Third, several lines of evidence suggest an expanded accumulation of excess DNA in the human lineage. For example, in a comparative study of about 11,000 lineage-specific mobile element insertions in humans and chimpanzees, 73% were found to be unique to humans (Mills et al. 2006), although it is uncertain whether this finding represents differential

Box 3.1 Inferring mutation rates and patterns of selection from DNA sequences of protein-coding genes

Because of the redundancy of the genetic code, nucleotide sites within the protein-coding regions of genes can be subdivided into two categories: silent (or synonymous) sites, where substitutions have no influence on the encoded amino acid, and replacement (or nonsynonymous) sites, where nucleotide substitutions lead to amino acid changes. For example, alanine is encoded by four triplets—GCA, GCC, GCG, and GCT—so the first two positions are replacement sites and the third is silent. Complications arise at sites where only a fraction of substitutions are silent, but these matters can be accommodated by an appropriate weighting of such sites—for example, as one-third silent and two-thirds replacement. As a consequence of the structure of the genetic code, approximately 25% of the nucleotide sites in coding DNA are silent.

Silent sites are often assumed to be neutral, given their lack of effect on the encoded protein (but see Chapters 4 and 6 for other potential selective effects), in which case they provide an entrée into the mutation rate when sequences are compared between two species. Imagine a diploid population of N individuals, with u being the rate of mutation per nucleotide site per generation. There are then an expected $2Nu$ new mutations entering the population per generation at any particular site. Because the initial frequency of each mutation is $1/(2N)$, and there is no directional force operating on the mutation subsequent to its origin, the probability that a new neutral mutation will eventually drift to fixation (i.e., to frequency 1.0) is also equal to $1/(2N)$. The expected rate of nucleotide substitution at a site within a given lineage is simply the product of the mutation rate and the probability of fixation, which in the case of neutrality simplifies to $2Nu \times 1/(2N) = u$. Thus, if two species have been separated from a common ancestor for t generations, each lineage will have acquired an expected ut mutations per neutral site, and the total number of substitutions separating the two lineages will be $2ut$ at such sites.

As sequences diverge, the observed frequency of differences among nucleotide sites will eventually approach saturation, due to the accumulation of multiple changes per site, but numerous methods exist for transforming observed levels of divergence into the total expected number of substitutions per site (Graur and Li 2000). Because they are linear in time (t), such measures provide surrogate estimates of relative divergence times (proportional to u), as used in Figures 3.1 and 3.6. In addition, letting S be the number of substitutions per silent site that have accumulated in the divergence of two species, then if t is known, $S/(2t)$ provides an estimate of u.

For replacement sites, the rate of accumulation of nucleotide substitutions is modified to $2Nu \times p_f = 2Np_f u$, where p_f is the average probability of fixation. If such sites are neutral, $p_f = 1/(2N)$, but p_f will be above or below the neutral expectation if average mutations are beneficial or deleterious, respectively. Because the expected cumulative amount of divergence between replacement sites (R) is $4Np_f ut$, dividing by the expected divergence of silent sites at the same locus ($S = 2ut$) yields the ratio R/S with expected value $2Np_f$, which is equivalent to the ratio of the average fixation probabilities at the two types of sites, $p_f /[1/(2N)]$.

If most mutations are deleterious or neutral, which is usually the case, the quantity $1 - (R/S)$ provides an estimate of the fraction of mutations that are opposed by natural selection. (The quantity R/S is often referred to as the width of the selective sieve). In contrast, in extreme cases in which selection favors amino acid substitutions, R/S can be positive. Aside from the assumption of neutrality for silent sites, such analyses have other limitations; for example, $R/S \approx 1.0$ is consistent with neutrality, but can also come about if the pool of changes at replacement sites involves a mixture of deleterious and positively selected mutations.

expansion in the human genome, deletion in the chimpanzee genome, or both. Relative to the chimpanzee genome, the human genome also appears to have accumulated twice as many gene duplications (Fortna et al. 2004; Cheng et al. 2005), although the adaptive (or maladaptive) significance of this finding remains to be determined. While by no means ruling out the operation of adaptive evolution on the human lineage, all of these observations suggest that the fraction of positively selected mutations separating humans from chimpanzees may be very small.

As noted in Box 3.1, one approach to identifying genes under unusual modes of selection is to examine the ratio of rates of change at replacement (R) versus silent (S) sites in protein-coding genes. An R/S ratio significantly greater than one implies strong selection for change at the protein level (because changes are accumulating at the amino acid level at a rate that exceeds the rate of mutation), and genes with ratios significantly greater than the average are also candidates for such selection in particular protein domains. When applied to the human–chimpanzee clade, such analyses suggest that the pool of genes under strong directional selection is enriched in functions associated with transcription regulation, spermatogenesis, olfaction, nucleotide metabolism, nuclear hormone receptors, and immune responses (Bustamante et al. 2005; Mikkelsen et al. 2005; Nielsen et al. 2005). A case has also been made that several key proteins associated with the nervous system have evolved at unusually high rates in the great ape lineage (Dorus et al. 2004; Evans et al. 2005). There are, however, several caveats with respect to the interpretation of these results. First, because the types of genes that evolve rapidly in the great ape lineage often exhibit rapid changes in other animals as well, it is uncertain whether these observations tell us anything about unique human attributes. Second, evaluations of the prevailing form of selection based on ratios of R and S are necessarily confined to the coding regions of genes, and do not necessarily reveal anything about evolution at the level of gene regulation. Third, an elevation of R/S may simply reflect a reduction in the efficiency of selection against mildly deleterious mutations, as noted above.

A novel set of approaches that may prove more useful in resolving some of these problems involves the use of information on intraspecific variation (Bamshad and Wooding 2003). Loci that have recently been under positive selection should tend to be associated with fairly long spans of DNA with exceptionally low diversity (as a consequence of an adaptive variant having expanded into the population by a selective sweep; see Chapter 4). Using this strategy, Voight et al. (2006) have found the pool of candidate loci in the human lineage to be enriched for functions associated with spermatogenesis, skin pigmentation, olfaction, and alcohol metabolism. Analyses of specific genes known to be associated with language development (Enard et al. 2002; Zhang, Webb et al. 2002) and perception and memory (Rockman et al. 2005) have also revealed the signature of strong directional selection. These types of studies, combined with follow-up functional analyses, represent the earliest stages of what will soon become a substantial investigative enterprise.

4 Why Population Size Matters

Basic principles of Mendelian genetics, supported by decades of empirical observation, tell us that long-term evolutionary change is a consequence of the origin of new variation by mutation and subsequent modifications of allele frequencies. This fundamental fact has provided a solid foundation for the development of a mechanistic framework for understanding evolutionary processes with a level of mathematical rigor that has few rivals in the life sciences. Indeed, the general principles of population genetics are now so well established that the credibility of any proposed scenario for genomic evolution must remain in doubt until it has survived this theoretical gauntlet. Given our primary concern with genomic differences among distantly related organisms, the focus here will be on the factors that influence species divergence. However, because the roots of evolutionary change reside in within-species variation, the two issues are by no means independent: no matter how beneficial a mutational change might be in the long run, it will never contribute to evolutionary divergence unless it is capable of first surviving as a transient polymorphism.

Evolutionary change involves the complex interplay of mutation, recombination, selection, and the chance sampling of gametes from generation to generation, raising some of the most difficult theoretical challenges that biologists have had to confront. The goal of this chapter is to show why these issues must be addressed if comparative genomics is to mature beyond its descriptive "natural history" phase into an explanatory framework based on known evolutionary mechanisms. An attempt has been made to do this in a way that is mathematically rigorous but also transparent from a biological perspective. In many cases this has meant the presentation of just the final results of what may be an intricate tapestry of analytical solutions, first-order approximations, and computer simulations. The theoretician reading this book will realize that an enormous number of details have been glossed over. He/she who abhors all things mathematical will be glad that this is the case, but should not lose track of the fact that the technical details do matter.

The key point in the following pages is that the types of evolution that can occur within a species depend strongly on population size. Natural selection is ubiquitous, but for many aspects of genomic evolution, the forces of selection are quite weak. If the population size is sufficiently small, the noise associated with random gamete sampling can completely overwhelm weak selective forces, yielding a situation in which the population evolves as though selection were entirely absent. This idea has long been appreciated in the field of molecular evolution (Kimura 1983; Ohta 1997). However, a general theme that will emerge in the following chapters is that although small population size promotes the accumulation of mutations that are mildly deleterious in the short term, the resultant alterations to gene and genomic architecture can provide a potential setting for secondary adaptive changes that are unattainable in large populations. Central to this argument is a general reduction in the efficiency of selection across the transitions from prokaryotes to unicellular/oligocellular eukaryotes to multicellular species, which results from the confluence of three factors: a decrease in population size, a decrease in the amount of recombination between functionally relevant genomic sites, and an increase in the deleterious mutation rate.

Once one understands these basic issues, the menu of plausible mechanisms for genomic evolution takes on a new complexion. For example, although it is almost universally assumed that an expansion of genomic complexity was a necessary prerequisite to the origin of organisms with multiple cell types and mechanisms of cell signaling (Raff 1996; Gerhart and Kirschner 1997; Davidson 2001; Carroll et al. 2001), the extent to which eukaryotic genome expansion was essential for phenotypic diversification or even adaptive in nature is far from clear. Although mobile elements may occasionally spawn an adaptive mutation, it is unlikely that the millions of such elements and their remnants in the human genome have been retained for their beneficial effects. Likewise, the massive increase in average intron size in some multicellular eukaryotes has no obvious advantage, and we will see in the following chapters that even expansions in gene number and regulatory region complexity can arise in the absence of any direct selection for these features.

Random Genetic Drift at a Neutral Locus

Evolution is an inherently stochastic process, starting from the chance events that produce single mutations in single individuals and proceeding through a series of fortuitous steps that gradually lead to the spread of those mutations to every member of the descendent population. The likelihood that a new mutation will successfully complete such a journey is almost never very high, although the chances are certainly improved for advantageous mutations. However, before invoking the guiding hand of natural selection in evolution—as most biologists are prone to do—it helps to understand what to expect in the absence of selection. At the very least, such neutral models serve as formal null hypotheses for interpreting evolutionary observations. But as we will soon see, even non-neutral mutations will behave in an effectively neutral fashion provided the population size is sufficiently small that the magnitude of selection intensity is overwhelmed by the stochastic force of random genetic drift. In the following pages, this situation will be seen to be common in eukaryotic genomic evolution.

Our point of departure is the classic Wright–Fisher model of population genetics, named after two of the founding fathers of the field. We assume a population consisting of N diploid individuals, with each member of the population contributing equally and synchronously to an effectively infinite gamete pool, from which pairs of gametes are drawn randomly to produce the next generation. The mating system is idealized in that we assume hermaphroditic individuals and ignore complexities associated with overlapping generations, spatial structure, environmentally induced variation in family sizes, and so on. However, most of these complications can be dealt with by redefining N to be a measure of the genetic effective size of the population; that is, the size of the ideal Wright–Fisher population that would most closely reflect the evolutionary behavior of the nonideal natural population. In the following pages, we will denote the genetic effective population size by N_e.

An enormous amount of technical theory has been devoted to the concept of effective population size. Although the specific mathematical details need not concern us here, one fundamental issue merits emphasis. Most of the features of natural populations that violate the assumptions of the Wright–Fisher model conspire to ensure that N_e will be less than the total number of potentially reproductive individuals (N), often considerably less. For example, if adults differ in the number of gametes produced, the effective population size will be reduced simply because some individuals contribute little or nothing to the following generation. Variable gamete production can result from selection or purely from chance ecological events (e.g., from some individuals inhabiting microhabitats with more resources than others). If a population is spatially structured into semi-isolated patches, N_e will be further decreased because the tendency to mate with locally related individuals will cause geographic concentrations of particular alleles, which can then become lost by chance extinction of local demes. A sex ratio that deviates from 1:1 will also reduce N_e, because the rarer sex (which necessarily contributes half of the next generation's genes) acts as the limiting factor. Similarly, population size fluctuations reduce N_e via the long-lasting effects of bottlenecks on genetic variation. Readers wishing to make side excursions into the more technical side of N_e can find useful reviews by Caballero (1994), Whitlock and Barton (1997), and Rousset (2003).

The main point here is that the power of random genetic drift in a species is almost always greater than that expected on the basis of actual numbers of individuals. How large might this deviation be? A number of different procedures have been developed to estimate N_e by relating the magnitude of generation-to-generation fluctuations in allele frequencies to the expectations based on the sampling of $2N_e$ effective gametes. In a survey of studies of mostly low-fecundity vertebrate populations, Frankham (1995) found that average N_e is ~10% of the total number of breeding adults (N), although high-fecundity species in spatially variable environments may have an N_e/N < 0.001 (Hedrick 2005). Few reported studies have incorporated the effects of long-term fluctuations in population size, and almost all have involved obligate outcrossers. As many unicellular species have conspicuous phases of asexual reproduction, which can encourage the rapid proliferation of a small number of clones, N_e/N ratios even lower than 0.001 are expected to be quite common in such species. Indeed, the following discussion will demonstrate that even the linkage of genes in sexually reproducing populations under selection can have effects of comparable magnitude.

To return to the point at hand, now imagine a newly arisen neutral mutation (Figure 4.1). Because a diploid population contains $2N$ gene copies at each locus (the 2 entering because there are two gene copies per individual), the initial frequency of this allele is simply $p_0 = 1/(2N)$. Any mutation must eventually suffer one of two fates, either being lost entirely from the population (declining to frequency 0.0) or gradually drifting to fixation (achieving a frequency of 1.0). Usually, a new mutation will be quickly lost. For example, the probability that the new mutant allele will not be among the $2N$ genes randomly incorporated into the very next generation is $(1 - p_0)^{2N}$, as there are $2N$ independent draws each with probability $(1 - p_0)$ of being the nonmutant. Provided N is greater than ten or so, this quantity is closely approximated by $e^{-1} \approx 0.368$, where $e = 2.718$ is the base of natural logarithms. Thus, the probability of immediate loss of a newly arisen mutation is essentially independent of population size, and in the event that the mutant allele does happen to be transmitted to the next generation, it will most likely still be present as a single copy, recreating the same scenario. In rare cases, however, a mutant allele will drift to higher and higher frequencies, entirely by a series of chance events, eventually becoming fixed in the population and totally displacing the ancestral allele.

The probability of ultimate fixation of a neutral allele is quite simple. Provided there are no external forces favoring one allele over another, the probability that a newly arisen neutral mutation will eventually drift to frequency 1.0 is always equal to its initial frequency, $1/(2N)$. The probability that the mutant allele will be lost from the population is $1 - 1/(2N)$. Thus, only a very tiny fraction of new neutral mutations becomes permanently incorporated into a population, and this fraction is inversely proportional to N. While at first glance this result might suggest that neutral changes will accumulate more slowly in larger populations, this is not the case. Because the number of mutational targets in a diploid population is proportional to $2N$, if μ is the rate of origin of neutral mutations per gene, then the average number of new mutant alleles arising per generation is $2N\mu$, each of which has the same prob-

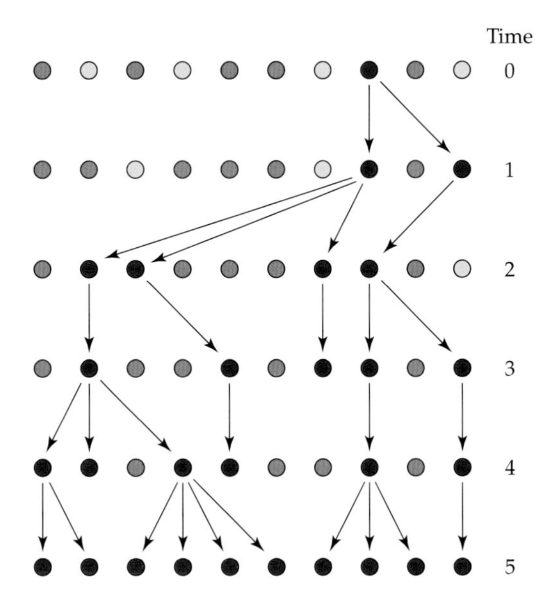

Figure 4.1 Random genetic drift of a single mutation to fixation (denoted in red). The population initially contains two additional alleles (denoted in yellow and green). After three generations, the yellow allele has been lost by chance, and the green allele is lost soon thereafter. The arrows trace the full gene genealogy of the population at generation 5 back to the single mutation at time 0. The different numbers of descendants of each gene, in this case between 0 and 4 per generation, are purely a matter of chance.

ability of ultimate fixation, $1/(2N)$. The total rate of fixation of neutral alleles is equal to the product of the number arising per generation and the fixation probability of individual mutations, and because these two quantities depend on $2N$ in opposing ways, the long-term rate of neutral evolution is simply equal to the genic mutation rate μ. Thus, for neutral changes, neither the absolute nor the effective population size plays a role in the long-term rate of evolution. This simple result, which holds for virtually any breeding system regardless of the degree of linkage or the mode of gene action, is the hallmark of the neutral model of molecular evolution (Kimura 1983).

One additional issue with respect to the neutral model merits consideration. Although the long-term rate of evolution under neutrality is independent of population size, the time that it takes each individual mutation to become fixed is not. Because N_e determines the size of sampling fluctuations in allele frequencies across generations, the magnitude of random genetic drift per generation is diminished and the time to fixation is extended in populations of larger effective size. In one of their many mathematical achievements with the neutral model, Kimura and Ohta (1969) showed that a neutral mutation that successfully makes a sojourn from gene frequency $1/(2N)$ to 1.0 does so in $4N_e$ generations, on average. As will be shown below, the increased residence times of neutral alleles in larger populations result in elevated levels of within-population variation.

The Effects of Selection

Intuition tells us that the probability of fixation of a beneficial allele must be greater than $1/(2N)$ and of a detrimental allele must be less than $1/(2N)$, but how much so? To see that a mutant allele, no matter how favorable, is never guaranteed to go to fixation, we will follow the logic developed above for a neutral allele. Suppose that the mutant allele improves the fitness of its initial carrier by a fraction s (the selection coefficient, with $s = 0$ denoting neutrality), such that the allele has expected frequency $p_0 = (1 + s)/(2N)$ in the gamete pool leading to the next reproductive episode. At this point, the favorable allele can do nothing to further guarantee its transmission to the next generation, and the probability that it is not acquired by at least one offspring is again $(1 - p_0)^{2N}$, which is roughly equal to $e^{-(1+s)}$, or $(1 - s)\,e^{-1}$ for small s. Thus, relative to the situation for a neutral allele, selection only reduces the probability of an initial exit for a beneficial allele by a fraction s. Even for a high selection coefficient of 0.1, the probability of loss in the first generation is 0.333 (compared with 0.368 for a neutral mutation). Until a favorable mutant allele has successfully avoided chance elimination in the first few generations and increased its frequency in doing so, there is little assurance that it will successfully go to fixation.

For the case in which an allele has additive effects on fitness (adding s to the fitness of heterozygotes and $2s$ to that of homozygotes), diffusion theory can be used to make a precise statement about the probability of fixation, p_f (Kimura 1962). Following the modification of Bürger and Ewens (1995),

$$p_f \approx \frac{1 - e^{-(2N_e s/N)/(1-s)}}{1 - e^{-4N_e s/(1-s)}} \tag{4.1}$$

To simplify the following discussion, we will assume that $s < 0.1$ (which in practice is almost always true), in which case $(1 - s) \approx 1$. Because N_e/N is usually much less than 1, the preceding formula then simplifies to

$$p_f \approx \frac{(2sN_e / N)}{1 - e^{-4N_e s}} \tag{4.2}$$

As $4N_e s \rightarrow \infty$, the denominator $\rightarrow 1$, showing that even in very large populations, the high degree of stochasticity in the early phase of establishment of a beneficial mutation still causes the probability of fixation to reach an upper limit of about $2sN_e /N$. Thus, the probability of fixation of a beneficial allele with additive effects is virtually always less than twice its selective advantage in the heterozygous state.

It is useful to consider the ratio of p_f and the neutral probability of fixation, $1/(2N)$, which is a simple function of $4N_e s$, $\theta_f \approx 4N_e s/(1 - e^{-4N_e s})$, graphed in Figure 4.2, where negative values of s denote deleterious mutations. Note that if the strength of positive selection is sufficiently great rel-

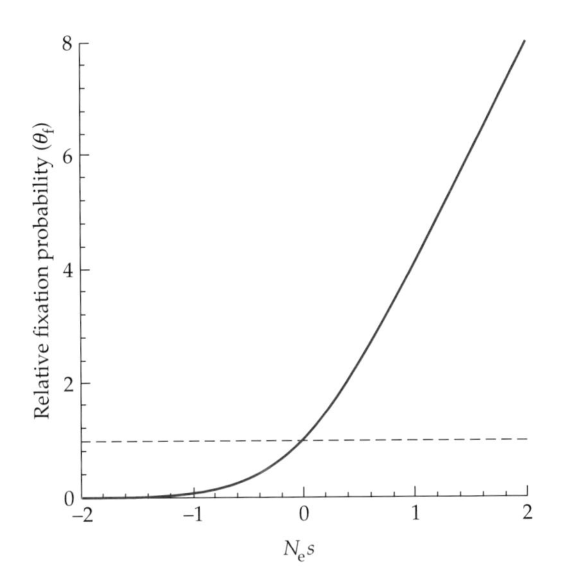

Figure 4.2 Probability of fixation of a new mutant allele relative to the neutral expectation of $1/(2N)$, given as a function of the product of the effective population size (N_e) and the selection coefficient (s). The dashed line denotes the neutral expectation.

ative to the power of random genetic drift (i.e., $4N_e s >> 1$), then the fixation probability is inflated by a factor of $4N_e s$ relative to neutrality. However, if the absolute value of $N_e s$ is less than 0.05, the probability of fixation is within 10% of the neutral expectation, and the deviation reduces to less than 1% if $|N_e s| < 0.005$, where the vertical lines denote absolute value. Thus, for any selection coefficient, there is a critical population size, $N_e \approx 0.005/|s|$, below which the probability of fixation is essentially the same (within a factor of 1%) as that for a neutral mutation. Stated in another way, for any population size, there exists a range of selection coefficients for which the behavior of mutant alleles is nearly identical to that of neutral alleles. Such a condition is generally referred to as effective neutrality.

In general, expressions for the time to fixation of a mutation under selection are quite complicated mathematically. However, a remarkable result was obtained by Maruyama and Kimura (1974): for a given population size, the time to fixation of a beneficial mutation, conditional on the mutation actually going to fixation, is identical to the time to fixation of a deleterious allele with the same selection coefficient (but of opposite sign). This rather counterintuitive result comes about because although a deleterious allele has a reduced *probability* of fixation, if such an allele is going to fix, it must do so as a consequence of some rather extreme and clumped sampling errors in gene frequency. For $|4N_e s| >> 1$ in an idealized population with mutations with additive effects, the mean time to fixation is about $(4/|s|)\ln(2N_e)$ generations (Crow and Kimura 1970). Thus, mutations under efficient selection have much shorter periods of polymorphism than neutral mutations, which, as noted above, require an average of $4N_e$ generations to become fixed.

To summarize, in contrast to neutral mutations, whose fixation probability, $1/(2N)$, declines with population size, the probability of fixation for beneficial mutations increases with N_e, eventually reaching an asymptote at $2sN_e/N$. Fortuitously, this same expression applies to a population that is changing in size, provided appropriate modifications are made in the definition of N_e (Otto and Whitlock 1997). For a subdivided population in which offspring production is constant for all demes (subpopulations), the limit needs to be multiplied by $(1 - F)$, where the index of subdivision F ranges from zero to one, because restricted migration makes it more difficult for a mutant allele to diffuse through the entire metapopulation (Whitlock 2003). If, on the other hand, deme productivity is a function of local mean fitness, the limit needs to be multiplied by $(1 + F)$, because a highly fit deme can come to dominate the entire metapopulation (Whitlock 2003).

Despite the need for some of these modifications, letting μ_b be the rate of beneficial mutation per gene, such that $2N\mu_b$ is the beneficial mutation rate at the population level, the above results imply that the upper limit to the rate of incorporation of beneficial mutations scales as $(2sN_e/N) \times 2N\mu_b = 4N_e\mu_b s$, which, unlike the situation for neutral mutations, increases with the effective population size. In addition, because deleterious mutations with $-0.3 > N_e s > 0.0$ go to fixation at rates at least half as great as that for neutral alleles, if the rate of origin of mutations in this range of effects is sufficiently high, we can expect the accumulation of a considerable load of fixed mildly deleterious mutations in populations of sufficiently small size (Ohta 1973, 1974). Taken together, these results suggest that the ability of a population to incorporate beneficial mutations and to purge deleterious mutations should scale positively with absolute population size, assuming that N_e scales positively with N. However, we will now see that something beyond the demographic features of the population—the physical structure of the genome itself—generally limits the growth of N_e with N in the largest of populations, perhaps even imposing an upper limit to N_e and hence to the evolutionary potential of a population.

The Importance of Linkage

Because tightly linked nucleotide sites are transmitted across generations as a unit, to a degree that depends on the rate of recombination, the fate of any new mutation depends on the selective forces operating on all loci physically close to its chromosomal location. The resultant selective interference will almost always cause the fixation rates of beneficial mutations to be lower and of detrimental mutations to be higher than suggested by the single-locus approximations in the previous section. Consider, for example, a beneficial mutation that rapidly sweeps through a population to fixation. Such a mutation will necessarily drag along any deleterious alleles at tightly linked loci with which it happened to be associated at the time of origin (Figure 4.3). Linked beneficial mutations can also interfere with each other's fixation. To see how, consider a beneficial allele $\boldsymbol{A}$ segregating at one locus,

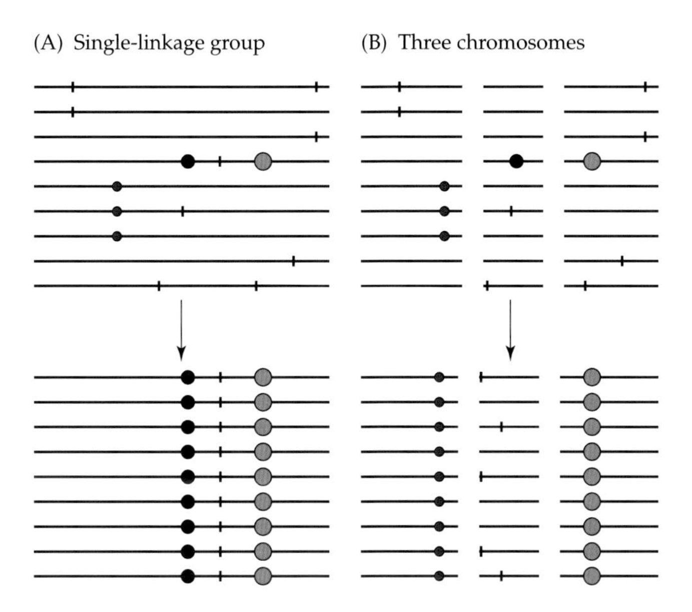

Figure 4.3 The effects of linkage on the fates of alternative alleles. The small red and larger green balls denote beneficial mutations at two loci, whereas the black balls denote deleterious mutations at a third locus. Tick marks denote segregating neutral markers. (A) All three selected loci are completely linked, and it is assumed that the net advantage of the chromosome with the black and green alleles exceeds that of the chromosome with the red allele. In the process of the former chromosome sweeping to fixation, the variation at other linked loci is cleansed from the population, one beneficial mutation (red) is lost, and one deleterious mutation (black) is fixed. (B) Each of the three mutations is located on a separate chromosome. Independent segregation allows the two beneficial mutations to be promoted and the deleterious mutation to be purged, while some variation is retained on the middle chromosome.

with a second beneficial mutation B arising at a tightly linked locus on an a chromosome. If the advantages of each mutation were the same, then the Ab and aB linkage groups would compete with each other in the fixation process, and in the absence of any recombination between them, one would eventually exclude the other. Similar logic leads to the conclusion that linked deleterious mutations will interfere with each other's removal, thereby increasing their probability of fixation. Linkage need not be absolute for these effects to be important, but the stronger the degree of linkage, the greater the degree of selective interference.

Before we explore these issues in more detail, it should be emphasized that selection on linked loci has no influence on the long-term rate of fixation of evolutionary changes at neutral sites (although the fixation times of neutral mutations will be altered) (Birky and Walsh 1988). Consider a neutral allele A with current frequency p, and suppose that a beneficial mutation destined to fixation arises at a completely linked locus. With probability p, the beneficial mutation will arise in association with linked allele $\boldsymbol{A}$, driving it to fixation, and with probability $(1 - p)$, it will arise in association with the alternative allele, preventing the fixation of $\boldsymbol{A}$. The net effect is that the probability of fixation of the $\boldsymbol{A}$ allele remains at p, the neutral expectation.

Now consider a detrimental mutation destined to be purged from the population. Such a mutation again arises in association with neutral allele $\boldsymbol{A}$ with probability p, in which case one $\boldsymbol{A}$ chromosome is ultimately removed from the population and replaced by a random deleterious mutation–free chromosome, reducing the expected frequency of $\boldsymbol{A}$ to $(2pN - 1)/(2N - 1)$. Alternatively, the detrimental mutation will arise in association with the $\boldsymbol{a}$ allele with probability $(1 - p)$, in which case the frequency of $\boldsymbol{A}$ increases, on average, to $2pN/(2N - 1)$. The probability of fixation of $\boldsymbol{A}$ again remains at the neutral rate, as

$$\left[p \cdot \frac{2pN - 1}{2N - 1}\right] + \left[(1 - p) \cdot \frac{2pN}{2N - 1}\right] = p \tag{4.3}$$

Thus, for fixations that are effectively neutral, the complications added by selection on linked sites are no cause for concern. This makes sense because, as will be shown below, selective interference from linked loci is essentially equivalent to a reduction in the effective population size in the region of selected loci, and as noted above, the rate of neutral evolution is completely independent of the effective population size.

The theory of genetic draft

The reduced efficiency of selection on linked loci is often referred to as the Hill–Robertson effect, after the authors who first pointed out the issue (Hill and Robertson 1966), and the selective sweep phenomenon is alternatively referred to as a hitchhiking effect (Maynard Smith and Haigh 1974; Kaplan et al. 1989) or genetic draft (Gillespie 2000). Gillespie (2000) provided an elegant explanation for the influence of linkage on the local effective population size of a reference locus, which although relying on a rather idealized model, captures the essence of the problem in an intuitive way. We start by noting that the effective size of a population defines the variance of neutral allele frequency change from generation to generation, $p(1 - p)/(2N_e)$, where p is the current allele frequency and $2N_e$ is the effective number of genes sampled per locus per generation in a diploid population. Now imagine a neutrally evolving site completely linked to another site experiencing selective sweeps at rate δ. As noted above, such a sweep will result in the fixation of the neutral allele with probability p and its loss with probability $(1 - p)$. If we assume that sweeps cleanse a population of linked variation essentially instantaneously, then conditional on a sweep occurring, the variance in allele frequency change is $p(1 - p)$. Thus, for a neutral locus in an ideal randomly mating population subject to periodic selective sweeps, the variance in allele frequency change is approximately $p(1-p) \cdot \{[(1-\delta)/(2N_e)] + \delta\}$.

Equating the term in brackets to $1/(2N_l)$, we obtain

$$N_l \approx \frac{N_e}{1+2N_e\delta} \tag{4.4}$$

where N_l denotes the long-term effective population size, taking into consideration the constraints imposed by linkage. Maruyama and Birky (1991) obtained essentially the same result via a different method.

The preceding formula shows how the degree of linkage can reduce the long-term effective population size (N_l), relevant to genomic evolution, below the short-term measure (N_e) defined by demographic features alone. It also makes a rather nonintuitive point: that different genes within the same genome will have different effective population sizes if they are subject to varying levels of influence from selection on linked loci.

A key question concerns the way in which δ scales with population size. If δ were to be completely independent of population size, then N_l would increase with N at a decreasing rate, eventually reaching an upper limit of $1/(2\delta)$ (top curve in Figure 4.4). However, because larger populations contain more mutational targets, the rate of sweeps is likely to increase with increasing population size. For the extreme situation in which fixations occur at the selected locus at rate $4N_e\mu_b s$, as predicted by the single-locus theory, then

$$N_l \approx \frac{N_e}{1+8N_e^2\mu_b s} \tag{4.5}$$

which asymptotically approaches $1/(8N_e\mu_b s)$ as $N_e \rightarrow \infty$, leading to a peculiar situation in which the long-term effective population size eventually

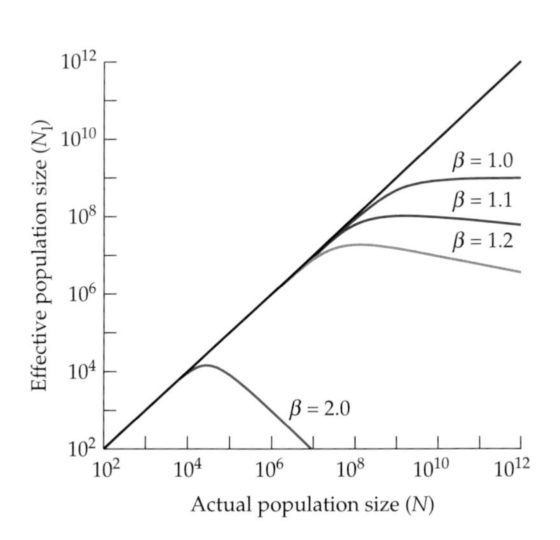

Figure 4.4 Hypothetical relationship between long-term effective population size and actual population sizes for ideal randomly mating populations with short-term $N_e = N$. The diagonal line denotes the situation in which $N_l = N$. The remaining four functions make various assumptions about the relationship between population size and the incidence of selective sweeps, and are defined by the general formula $N_l = N/(1 + \alpha N^{\beta})$, with $\alpha = 10^{-9}$. When $\beta = 1$, the rate of selective sweeps (δ in the text) is independent of population size. For $\beta > 1$, the rate of sweeps increases with population size, as defined by $\delta = (\alpha/2)N^{(\beta-1)}$. When $\beta = 2$, the qualitative behavior of the model is equivalent to the situation in which $\delta = 4N\mu_b s$, as described in the text; that is, the rate of selective sweeps increases linearly with N.

declines to zero at large N_e (bottom curve in Figure 4.4). There are at least three reasons why reality must lie between these two extremes.

First, the model just described is idealized in that it considers only two loci, with a single selected locus dominating the allele frequency dynamics at the neutral hitchhiking locus. However, because the selected locus itself may be subject to selective interference from still other linked loci, the rate of sweeps is necessarily less than the single-locus prediction, $4N_e\mu_b s$. Second, the single-locus prediction for the incidence of sweeps continuously grows with increasing N_e, eventually exceeding one per generation. Because each selective sweep cumulatively requires $10N$ to $100N$ selective deaths in a population of approximately constant size regardless of the selection intensity (Haldane 1957), and because the number of linked sites subject to adaptive sweeps is limited, such a condition leads to a biologically impossible situation. Third, in a sexual population, there is always some probability of recombination between any two linked sites, and there are more opportunities for such events in larger populations. Recombination reduces the likelihood that a selective sweep will completely purge the variation at a linked locus, as it breaks down the statistical associations between the alleles at linked sites. Gillespie (2000) showed that recombination modifies the expression for the effective population size to

$$N_1 \approx \frac{N_e}{1+2N_e C\delta} \tag{4.6}$$

where C is the average squared frequency of the neutral hitchhiking allele after the completion of a sweep at the selected locus. We previously assumed that $C = 1$, but C can be substantially smaller, to a degree that depends on the recombination rate and the magnitude of selection at the linked locus, which mutually determine the breadth and rapidity of the sweep.

These qualitative arguments do not rule out the possibility that after initially scaling linearly with increasing N_e and then leveling off, N_1 eventually begins to decline, but we cannot expect N_1 to decline indefinitely with increasing N_e. Rather, N_1 must eventually reach an asymptote at some level in the vicinity of $1/(2C\delta_{max})$, where δ_{max} is the limiting value of δ reached at large N_e. In the absence of any direct observations on this matter, a more general approximation is to treat $C\delta$ as a power function of N_e, which yields a relationship of the form

$$N_1 \approx \frac{N_e}{1+\alpha N_e^{\beta}} \tag{4.7}$$

where α and β are arbitrary scaling coefficients (see Figure 4.4). Here, $\beta = 1$ describes the case in which $C\delta$ is independent of N_e, as we initially assumed, whereas $\beta = 2$ describes the situation in which the rate of selective sweeps is proportional to N_e, as in the single-locus result described above. Although β is likely to fall in the range of 1 to 2, it must be emphasized that the functions outlined in Figure 4.4 are simply qualitative approximations of expected behavior. Nevertheless, despite the uncertainties of the patterns in Figure 4.4, one conclusion cannot be avoided: stochastic evolutionary change is by no means restricted to small populations, but in very large populations, such noise is more a consequence of genetic draft, a function of the chromosomal nature of the genome, than of genetic drift, a function of population size.

Empirical considerations

Some insight into the frequency of adaptive substitutions (and a rough approximation of δ) may be acquired by considering the distribution of replacement-site and silent-site mutations in protein-coding genes. As noted in Chapter 3, the rate of substitution per replacement site (R) is expected to equal that per silent site (S) in the absence of selection. Thus, the quantity $1 - (R/S)$ for interspecific divergence is often taken to be a measure of the fraction of mutations at replacement sites with substantial enough deleterious effects to prevent fixation (see Chapter 3). The R/S ratio can also be computed for variation within populations (polymorphisms), and the contrast between the two measures provides some insight into the degree to which replacement-site mutations that do become fixed are promoted by positive selection (as opposed to being effectively neutral and simply drifting to fixation).

For example, average R/S for the divergence between humans and Old World monkeys is 0.35 (Fay et al. 2001), whereas that for moderately common (frequencies of at least 0.15) single-nucleotide polymorphisms (SNPs) *within* the human population is only 0.20. Because the latter ratio is influenced by the residence time of segregating mutations, which is depressed by directional forms of selection, the within-population results suggest that 80% of replacement mutations are either deleterious enough to be rapidly eliminated from the population or beneficial enough to be rapidly fixed. The remaining 20% of segregating replacement-site mutations must then be either neutral or under very mild selection. Some may be so mildly deleterious as to be capable of going to fixation, but because selection only promotes the fixation of beneficial mutations, the excess of replacement substitutions between species (0.35 – 0.20 = 0.15 per fixed silent substitution) most likely reflects the action of positive selection. Assuming a 20-year generation time and extrapolating to a protein of average size, Fay et al. (2001) suggested an average fixation rate of 2×10^{-6} adaptive substitutions per protein-coding gene per generation in this pair of primate lineages. Application of a similar approach to two *Drosophila* species yielded an estimated rate of adaptive substitution per protein-coding gene of 2×10^{-7} per generation (Smith and Eyre-Walker 2002).

Although these analyses are unable to partition fixed mutations between the two lineages being compared, some further insight into this matter is possible for primates. In contrast to the analysis of humans and Old World monkeys, a human–chimpanzee comparison reveals virtually identical average R/S ratios at the within- and between-species levels (Mikkelsen et al.

2005). Thus, unless a substantial fraction of segregating replacement mutations is beneficial, which seems unlikely, this observation implies a relatively small amount of adaptive evolution within the great apes, which is consistent with the idea that the human lineage has experienced a considerable relaxation in the efficiency of selection (see Chapter 3).

These observations crudely, but conservatively, suggest that δ for a gene-sized fragment within an animal genome is about 10^{-7}–10^{-6} per generation. However, extrapolating these results to yield an estimate of the upper bound to the long-term effective population size—using, for example, $1/(2C\delta_{max})$—is fraught with difficulties. Some recombination must occur between selected sites and flanking neutral sites, and as will be discussed below, unicellular species can have much higher rates of recombination per physical distance than do multicellular species. Although this difference could cause C (the extent of the sweep) to be lower in unicellular species, with their larger N_e and greater sensitivity to selection, such species are also expected to experience higher rates of selective sweeps (larger δ). Despite these considerable uncertainties, empirical results to be discussed below suggest that the upper limit to the long-term effective population size is at least 10^9, the value assumed for the upper curve in Figure 4.4, although it may not be much greater than that.

These largely theoretical arguments may seem rather unconvincing to some readers, but more direct insight into the power of genetic draft can be acquired from direct observations on nucleotide polymorphisms. If hitchhiking effects have a significant influence on genomic evolution, they should leave a strong signature in chromosomal regions that are maximally sensitive to genetic draft. In particular, genes in regions of low recombination should exhibit exceptionally low levels of segregating polymorphisms within individual species. This hypothesis can be tested by taking advantage of the fact that the rate of recombination per physical distance varies by at least two orders of magnitude among chromosomal regions in typical eukaryotic species. Population surveys of nucleotide diversity in *Drosophila* (Begun and Aquadro 1992), humans (Nachman 2001; Lercher and Hurst 2002), nematodes (Cutter and Payseur 2003), tomatoes (Stephan and Langley 1998; Roselius et al. 2005), and goatgrass (Dvořák et al. 1998) uniformly point to reduced levels of variation in regions of low recombination.

Although these general observations would seem to provide strong support for the hitchhiking hypothesis, an alternative explanation for such patterns is that the production of new variation is at least in part a mutagenic side effect of recombination (Yeadon and Catcheside 1999; Lercher and Hurst 2002). Because the rate of silent-site divergence is equal to the mutation rate under the assumption of neutrality, it was initially argued that the lack of a correlation between recombination rate and silent-site divergence between species is inconsistent with the mutagenesis hypothesis (Begun and Aquadro 1992; Nachman 2001). However, if silent sites are not truly neutral, but rather subject to weak selection (see Chapter 6), and if regions of high recombination generate mutations while also serving as relatively effective environments for selection (Hey and Kliman 2002), the two effects would tend to cancel each other out. Moreover, studies in humans (Filatov and Gerrard 2003; Hellmann et al. 2003), mice (Huang et al. 2005), and *Caenorhabditis elegans* (Cutter and Payseur 2003) reveal positive correlations between interspecies divergence and the recombination rate, and correlations between intra- and interspecies silent-site diversity across loci in maize (Tenaillon et al. 2004) and tomatoes (Roselius et al. 2005) also suggest a mutual dependence on heterogeneous mutation rates. Some direct observations point to the mutagenicity of recombination as well. In yeasts, mutation rates during meiotic cell divisions are about 13 times those during mitosis (in which recombination is suppressed) (Magni and von Borstel 1962), apparently because of the error-prone nature of the DNA polymerase used in the repair of double-strand breaks (Strathern et al. 1995; Rattray et al. 2002).

Although recombination may serve to both reduce sensitivity to individual selective sweeps and increase the frequency of such events, numerous additional observations from *Drosophila* support the idea that the net effect of linkage is to constrain the potential for adaptive evolution. For example, *Drosophila* genes with higher rates of putative adaptive amino acid substitutions tend to reside in regions of high recombination (Betancourt and Presgraves 2002), and the frequency spectrum for segregating polymorphisms is consistent with expectations under the hitchhiking model (Andolfatto and Przeworski 2001). In *D. miranda*, one of the autosomes has become fused to the Y chromosome, and because male *Drosophila* lack recombination, this neo-Y chromosome is completely non-recombining. Relative to the neo-X (which undergoes recombination in females), the neo-Y exhibits numerous signs of deleterious mutation accumulation, including the fixation of mobile element insertions and the silencing of many previously active genes (Bachtrog 2003). Further attention will be given to the evolutionary consequences of recombinational quiescence in organelle genomes and sex chromosomes in Chapters 11 and 12.

The Three Genomic Perils of Evolving Large Body Size

In terms of biomass, a typical single-celled prokaryote is ~10^5–10^7 times smaller than the average single-celled eukaryote, and a similar disparity exists between unicellular and multicellular eukaryotes (Bonner 1988). Such massive differences in size impose numerous ecological and physiological constraints and opportunities, but their implications for the population genetic environment are equally as great. As noted above, theory tells us that small population size, tight linkage, and high background mutational activity reduce the efficiency of natural selection. We will now see that the confluence of all three of these conditions in organisms of increasing body size induces substantial reductions in average effective population sizes across the boundaries from prokaryotes to unicellular eukaryotes to multicellular eukaryotes. This general pattern will be central to many of the issues discussed in subsequent chapters.

Smaller population size

As noted above, all other things being equal, the genetic effective size of a population should generally increase with the actual number of breeding adults (N), possibly leveling off at high N as a consequence of constraints from linkage. One of the few established laws in ecology is that a primary determinant of N is the average body size of members of the population. In general, a doubling of organism size leads to a 50% reduction in population density, although ecological factors unique to individual species can cause local deviations around this pattern (Damuth 1981; Schmid et al. 2000; Enquist and Niklas 2001; Carbone and Gittleman 2002; Finlay 2002). The extremes of average densities (individuals per square meter) for eukaryotic species are 10^{-7} for the largest vertebrates and 10^{11} for the smallest unicellular species, and with the addition of prokaryotic species, the range of variation in population density exceeds 20 orders of magnitude.

An inverse scaling between population density (per unit area) and organism size need not reflect the pattern for total population size, as it does not account for total species ranges. However, the total area occupied by vertebrate species is negligibly to weakly positively correlated with average body size (Gaston and Blackburn 1996; Diniz-Filho and Torres 2002; Housworth et al. 2003), and the geographic ranges of unicellular species appear to be substantially greater than those of multicellular taxa (Finlay et al. 2001, 2002; Green et al. 2004; Horner-Devine et al. 2004). Thus, in a broad phylogenetic sense, there is little doubt that the total number of individuals within a species declines with increasing body size, with the exact scaling probably exceeding that of a strictly inverse relationship.

The numbers for microbes are particularly staggering. The total number of prokaryotic cells inhabiting Earth is about 10^{30} (Whitman et al. 1998). Fewer than 10^4 such species have actually been identified (Oren 2004), so if we conservatively estimate that this represents just 0.1% of the true number of species (Hammond 1995), the total global population size (N) for an average prokaryotic species is 10^{23} cells. One unnamed marine species has an estimated N of 10^{28} (Morris et al. 2002).

Higher mutation rates

Given the central role that mutation plays in evolution and in the generation of genetic disorders, one would think that the rate of origin of replication errors would be well understood. However, because of the enormous technical challenges in identifying very rare events at individual nucleotide sites, the data are extremely sparse and largely confined to indirect observations. Most estimates of the per-nucleotide mutation rate are derived either from surveys of visible mutations at reporter loci (which enhance the detectability of mutations) or from nucleotide sequence comparisons of silent sites in distantly related species (which magnify the accumulation of mutations). Neither approach is without problems, the first requiring assumptions about the fraction of total mutations with observable phenotypic effects, and the second requiring assumptions about absolute times of interspecific divergence, species-specific generation times, and neutrality of the monitored nucleotide sites. The most reliable data obtained from laboratory studies are summarized in Figure 4.5.

Across a phylogenetically diverse set of species, there is a strong correlation between the base-substitution mutation rate per generation and genome size, with average rates for prokaryotes and vertebrates of 5.0×10^{-10} and 5.4×10^{-8}/bp/generation, respectively. Despite the uncertainties in each estimate contributing to this pattern, the validity of the overall relationship is supported by two observations. First, the estimate for the nematode *C. elegans* obtained by direct sequence analysis (highest blue point in Figure 4.5; Denver et al. 2004) is reasonably consistent with the remaining data obtained via reporter constructs. Second, estimates of the human mutation rate obtained from observations of dominant genetic disorders (Kondrashov 2003) are very similar to those obtained from comparisons of pseudogene sequences in humans and chimpanzees (Nachman and Crowell 2000): 2.6×10^{-8} and 2.2×10^{-8}/bp/generation, respectively. An earlier analysis of a more limited database led to the conclusion that the mutation rate per nucleotide site per generation is inversely related to genome size in microbial species (including viruses) (Drake 1991), but the results in Figure 4.5 are consistent with such scaling in cellular species.

The causal determinants of the increase in the per-generation mutation rate in larger organisms are not entirely clear, although genome size itself is

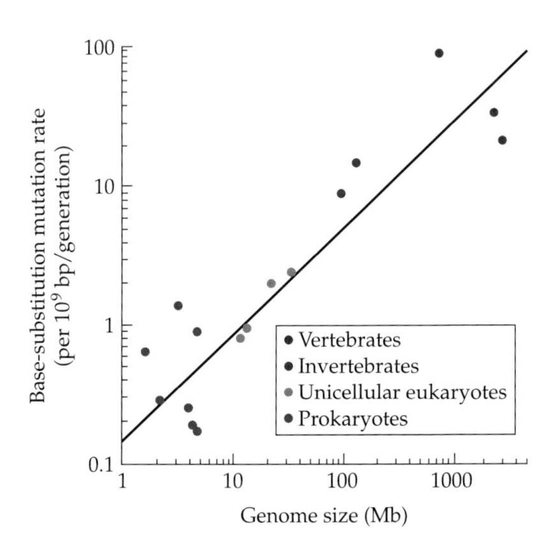

Figure 4.5 The relationship between the rate of base-substitution mutations per nucleotide site per generation (u) and genome size (G) scales as $u = 0.14G^{0.78}$ ($\times 10^{-9}$), with the exponent having a standard error of 0.10 and with genome size accounting for 79% of the variance in the mutation rate. (From Lynch 2006a.)

unlikely to be directly involved. Numerous indirect estimates of the mutation rate for quantitative characters from mutation accumulation experiments support the idea that the per-generation mutation rate scales positively with generation time (Lynch et al. 1999), and organisms with larger genomes tend to have larger numbers of germ line cell divisions per generation: just one for unicellular species, about nine in *C. elegans*, about twenty-five in *D. melanogaster*, and a few hundred in mammals (Drake et al. 1998; Crow 2000). The scaling of the per-generation mutation rate with generation time may be a simple consequence of the greater number of opportunities for germ line replication errors in species with long generation times, although such scaling would also be expected if most mutations derive from processes unassociated with replication, such as spontaneous double-strand breaks arising at a relatively constant rate per absolute time.

Reduced recombination in large genomes

Surveys of high-density molecular markers provide a basis for establishing refined genetic maps from observations of meiotic crossover events, which in turn allow the estimation of average rates of recombination per physical distance. This measure of crossover rate per nucleotide per generation scales negatively with genome size, ranging from 3×10^{-10} in the gymnosperm *Pinus* to 3×10^{-6} in the yeast *S. cerevisiae* (Figure 4.6). Such scaling is due in large part to the simple fact that meiosis in virtually all species employs just one to two crossover events per chromosome. Because chromosome number is largely independent of genome size, it naturally follows that there is a greater intensity of recombination per nucleotide position in species with smaller genomes (whose chromosomes are also smaller). For the same reason, this inverse scaling extends to chromosomes of different sizes within species (Jensen-Seaman et al. 2004).

Because there is only a tenfold range in gene number across all eukaryotes, it follows that species with small genomes also experience increased levels of recombination on a per-gene basis. For example, the rate of recombination over the entire physical distance associated with an average gene (including intergenic DNA) is about 0.0072 in *S. cerevisiae* and about 0.0012 in *H. sapiens*. As noted above, however, superimposed on these species-specific differences is up to a hundredfold range of variation in regional recombination rates within chromosomes. The scale of such spatial variation ranges from thousands to millions of bases, and highly localized recombinational hot spots exist in some species (Petes 2001; Nachman 2002; de Massy 2003; Jeffreys et al. 2004; McVean et al. 2004; Myers et al. 2005). Less clear is why the decline in average recombination frequency with increasing genome size is twice as rapid in unicellular as in multicellular species (see Figure 4.6). This altered scaling is not a consequence of higher numbers of chromosomes in unicellular species, as the average (12.2 haploid chromosomes) is no greater than that for invertebrates (12.3), vertebrates (25.2), or land plants (14.0). One possibility is that mitotic crossing-over

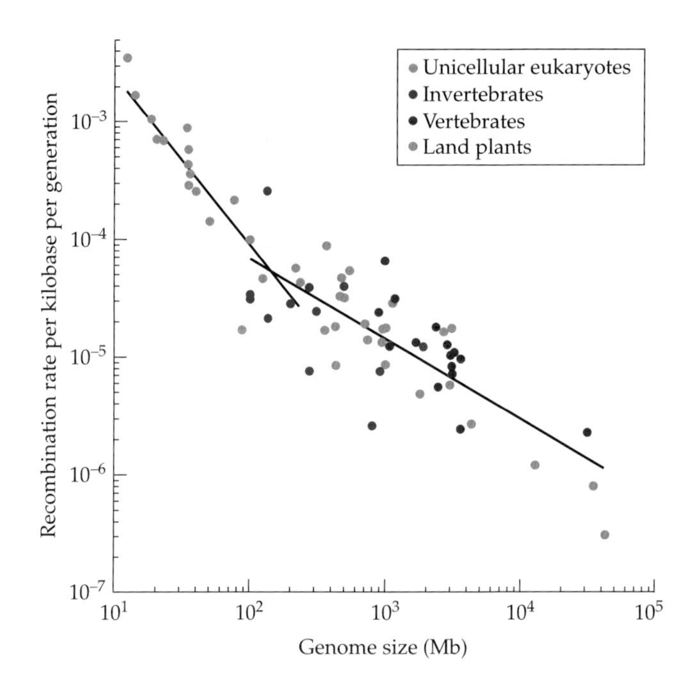

Figure 4.6 The inverse relationship between recombination rate per physical distance and total genome size in eukaryotes. For animals and land plants, this relationship scales as $y = 0.0019x^{-0.71}$, with the exponent having a standard error of 0.06; for fungi and other unicellular species, it scales as $y = 0.045x^{-1.32}$, with the exponent having a standard error of 0.12. Genome size accounts for 70% and 80% of the variance in recombination rate in these two cases, respectively. (From Lynch 2006a.)

prior to gametogenesis contributes to per-generation recombination rates in multicellular species. Although mitotic recombination rates are much lower than those during meiosis, as noted above, there can be dozens to hundreds of mitotic cell divisions during germ line development in multicellular species.

The results in Figure 4.6 imply that, depending on the species, chromosomal segments of 10^4–10^6 nucleotides may remain intact through hundreds to thousands of meioses. The best evidence for such conserved segments derives from surveys of SNPs (single *n*ucleotide *p*olymorphisms) in the human population, which reveal many large regions, tens to hundreds of kilobases long, that appear to have experienced little internal recombination (Daly et al. 2001; Reich et al. 2001; Dawson et al. 2002; Gabriel et al. 2002). The largest haplotype blocks tend to be associated with chromosomal regions with low rates of recombination (Greenwood et al. 2004), and in principle might owe their origins to partial selective sweeps like those hypothesized above. However, with an average rate of recombination of 1.3×10^{-8}/bp/generation, a 10 kb stretch of human DNA has just a 13% chance of experiencing a recombination event over a 1000-generation period, and this probability could easily be one or two orders of magnitude lower in regions of low recombination (McVean et al. 2004; Myers et al. 2005). Thus, it is unclear whether selection needs to be invoked at all in the case of humans. Computer simulations demonstrate that the level of current haplotype structure in the human population can be generated entirely by random genetic drift, assuming a bottleneck of $N_e \approx 50$ or so individuals 1000–2000 generations ago (Reich et al. 2001; Wall and Pritchard 2003), or even assuming a constant $N_e \approx 10^4$ over prolonged periods (K. Zhang et al. 2003).

Although prokaryotes are incapable of participating in conventional meiotic recombination and are often viewed as reproducing in a purely clonal fashion, they do have a number of mechanisms for uptake and exchange of exogenous DNA. A critical issue is whether such activities are rare enough to render typical prokaryotic genomes exceptionally sensitive to hitchhiking events. A combination of two types of observations on polymorphic sites in natural populations contribute to the idea that they are not. First, as will be described more fully in the following section, under the assumption of neutrality, the level of nucleotide heterozygosity at silent sites in protein-coding genes is a function of the composite parameter N_1u, with the rate of input of variation being governed by the base-substitution mutation rate per generation (u) and the rate of loss by the power of drift ($1/N_1$). Second, the statistical association between nucleotide polymorphisms across linked sites is a function of N_1c, where c is the rate of recombination per nucleotide site per generation. Such associations, known as linkage disequilibria, develop stochastically as a consequence of random genetic drift and subsequently break down via recombination (Ohta and Kimura 1971; Hill 1975). Under the assumption of an equilibrium between the forces of mutation, drift, and recombination, N_1c can be estimated from random samples of gene sequences within a species by evaluating the rate of decline of the average level of disequilibrium with the physical distance between linked nucleotide sites. Numerous technical details are being glossed over here, but the important point is that when joint estimates of N_1u and N_1c are available, their ratio eliminates N_1, providing an estimate of relative rates of recombination and mutation per nucleotide site (c/u). A lower c/u ratio implies greater vulnerability to hitchhiking effects resulting from the inability of recombination to separate independently arising mutations, and c/u ratios between selected sites greater than 4 or so can largely alleviate problems with Hill–Robertson interference (Marais and Piganeau 2002).

The few available estimates for c/u derived from polymorphism data for animals and land plants yield averages of 1.9 (SE = 0.8) and 1.2 (0.4), respectively, somewhat lower than the average for prokaryotes, 3.3 (1.3) (Table 4.1). Although such estimates are subject to numerous sources of error, those for eukaryotes are quite consistent with inferences that can be derived from the more direct estimates of c and u outlined in Figures 4.5 and 4.6. Extrapolation of the fitted functions implies that c/u in eukaryotes should generally decrease with increasing genome size (G in Mb), with $c/u \approx 321{,}000G^{-2.1}$ for unicellular species and $13{,}600G^{-1.5}$ for multicellular species. These expressions yield expected values of $c/u \approx 2550$, 17, and 0.4 for genomes of 10, 100, and 1000 Mb, respectively, the latter two being in the range of genome sizes for most animals and land plants.

TABLE 4.1 Estimates of c/u, the ratio of the recombination rate to the mutation rate per base pair, for various species, derived from polymorphism surveys of natural populations

Animals		
Homo sapiens	0.6	Ptak et al. 2004
Caenorhabditis elegans	0.5	Cutter et al. 2006
Chorthippus parallelus	2.5	Ibrahim et al. 2002
Drosophila spp.	3.8	Hey and Wakeley 1997; Machado et al. 2002
Land plants		
Arabidopsis spp.	0.7	Wright et al. 2003b
Brassica nigra	0.3	Lagercrantz et al. 2002
Cryptomeria japonica	3.0	Kado et al. 2003
Hordeum vulgare	1.5	Morrell et al. 2006
Pinus taeda	0.3	Brown et al. 2004
Zea mays	1.6	Tenaillon et al. 2004
Prokaryotes		
Bradyrhizobium spp.	0.9	Vinuesa et al. 2005
Neisseria gonorrheae	1.0	Posada et al. 2000
Neisseria meningitidis	4.8	Feil et al. 2001
Pseudomonas syringae	0.3	Sarkar and Guttman 2004
Rhizobium spp.	0.5	Silva et al. 2005
Staphylococcus aureus	6.5	Feil et al. 2001
Streptococcus pneumoniae	8.9	Feil et al. 2001

Thus, relative to the background rate of mutation, recombination at the nucleotide level does not appear to be exceptionally low in prokaryotes when compared with that in multicellular species. Although c/u in prokaryotes is certainly not as high as the theoretical estimate of 2550 for an average unicellular eukaryote, it should be noted that the latter calculation assumes an obligately sexual life cycle. As most unicellular eukaryotes have extended periods of clonal propagation, the true value of c/u for a natural population needs to be scaled by the number of asexual generations between meiotic events; for example, becoming 25.5 with 100 mitotic cell divisions per life cycle (assuming negligible mitotic recombination). Much additional evidence of a less quantitative nature supports the idea that prokaryotes experience substantial intraspecific genetic exchange (Jiggins 2002; Papke et al. 2004; Hughes and Friedman 2005; Baldo et al. 2006; Nesbø et al. 2006).

The Global Effective Population Sizes of Species

Despite the compelling reasons to expect the long-term effective population size (N_l) to scale positively with absolute population size (N) and recombination rate and negatively with mutation rate, there are many uncertainties about the ways in which these three factors jointly define N_l. For example, fluctuations in population size can result in a substantial depression of N_l below average N, but it is unclear whether the magnitude of such fluctuations typically increases or decreases with organism size. All other things being equal, hitchhiking effects should depress the N_l/N ratio more in large populations, where selective sweeps are more common, and as noted above, linkage might even cause a reduction in N_l at very large N. But because the rate of recombination (per meiosis) is substantially higher and the mutation rate substantially lower in smaller species with large N, the scaling of N_l with N could be quite different from the theoretical expectations in Figure 4.4. Given the many additional factors that can influence N_l, quantitative insight into this key population genetic parameter requires direct empirical observation.

One approach to estimating N_l is to consider the amount of nucleotide sequence variation at silent sites in protein-coding genes within natural populations. Under the assumption that mutations at such sites escape the eyes of natural selection, the amount of silent-site variation has a simple interpretation. The rate of divergence per nucleotide site in two randomly compared alleles is $2u$ (twice the mutation rate per nucleotide), while the expected rate of loss of variation by genetic drift is $1/(2N_l)$ in a diploid population. At equilibrium, the average number of nucleotide substitutions per neutral site in two randomly sampled alleles is simply the ratio of these two rates, $4N_lu$. For a haploid species, the rate of loss of variation is $1/N_l$, and the equilibrium divergence among neutral nucleotide sites becomes $2N_lu$. Both results have the same meaning: at mutation–drift equilibrium, the amount of within-species nucleotide variation at silent sites is equal to twice the effective number of gene copies at the locus times the per-nucleotide mutation rate to base substitutions. To unify the following discussion, we will rely on the concept of the effective number of genes at a locus (N_g), which is equivalent to $2N_l$ in sexual diploids and N_l in haploids (see Table 4.2 for a summary of definitions).

An estimate of the composite parameter $2N_gu$ can be acquired by directly observing levels of silent-site variation among random samples of genes within a species (hereafter π_s). The complex nature of the definition of N_l introduces some interpretive issues with π_s (Laporte and Charlesworth 2002), but a fully general definition can be stated in terms of the ancestry of alleles in a sample: under the assumption of neutrality, π_s is equal to the average age of random pairs of sequences times twice the base-substitution mutation rate per nucleotide site. The average number of generations to the common ancestor of a random pair of genes is equal to N_g (Hein et al. 2004), and multiplying by $2u$ (to account for mutations down both lineages) recov-

TABLE 4.2 A glossary of population size definitions

N	Absolute number of individuals in a species
N_e	Short-term effective population size; takes into account features of demographic and population structure
N_l	Long-term effective population size; takes into account all of the factors that define N_e, as well as the influence of the linked nature of the genome on the efficiency of selection
N_g	Long-term effective number of genes at a locus; equal to N_l for a haploid species, and about $2N_l$ for a diploid species, depending on the mating system

ers $2N_gu$. Although an estimate of $2N_gu$ may seem undesirable in the sense that the separate contributions from N_g and u are not immediately transparent, we will see below that the product $2N_gu$ is actually a more critical determinant of the alternative pathways of genomic evolution than are its individual components.

Estimates of silent-site diversity exist for a wide enough phylogenetic range of species that some general statements can be made. Across a diverse assemblage of over 100 eukaryotic and prokaryotic species, there is an inverse relationship between organism size and π_s (Figure 4.7). For prokaryotes, π_s lies in the broad range of 0.007–0.388, with an average value of 0.104. This

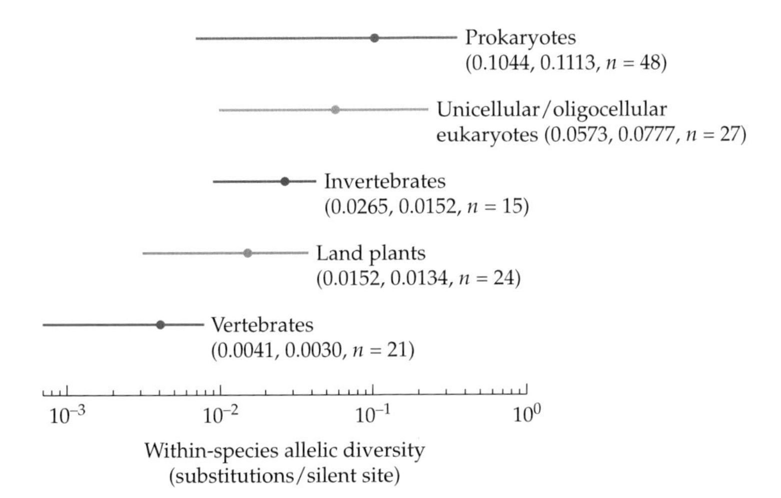

Figure 4.7 Average levels of within-species nucleotide divergence for silent sites among sampled alleles for protein-coding genes from a phylogenetically diverse assemblage of eukaryotic and prokaryotic species. Means, standard deviations, and number of genera (n) involved in the analyses are given in parentheses; horizontal bars provide estimates of the 5%–95% limits to the ranges of values for each group. (From Lynch 2006a.)

is nearly twice the average value for unicellular eukaryotes (0.057), although the range of values among the latter taxa is again very broad (0.010–0.252). For the still larger invertebrates, there is a further reduction in average π_s to 0.026, with a range of 0.009–0.047. The average value of π_s for land plants (0.015) is still lower, and that for vertebrates (0.004) is even lower.

As a consequence of sampling error at the gene, individual, and population levels, the mean values for these five functional groups are certainly more reliable than the extreme values, and there are other limits to these data as well. For the unicellular groups, the bulk of existing surveys involve pathogenic species, and because the N_g of pathogens may be largely dictated by that of their multicellular host species (Hartl et al. 2002), this probably leads to a downward bias of average π_s relative to that for free-living species. Two of the three prokaryotes with π_s less than 0.01 are *Serratia macrescens* (a human pathogen) and *Buchnera aphidicola* (an obligate endosymbiont of aphids), and the six lowest estimates of π_s for unicellular eukaryotes are derived from pathogens (*Candida*, *Coccidioides*, *Encephalitozoon*, *Fusarium*, *Phytophthora*, and *Plasmodium*).

Although π_s is a function of the product of N_g and u, the two factors can be disentangled by applying the mutation rate estimates described in Figure 4.4. Assuming the average observed value of $u \approx 0.5 \times 10^{-9}$ for prokaryotes, the average N_g for prokaryotes is 10^8. When parasitic species are removed, application of the average mutation rate for unicellular eukaryotes (1.6×10^{-9}) yields an average N_g of 10^7 for this group. Similar analyses for invertebrates and vertebrates yield average estimates of $N_g \approx 10^6$ and 10^4. Finally, a phylogenetically based mutation rate estimate of 7.3×10^{-9}/bp/year for plants (Lynch 1997) yields an average estimate of $N_g \approx 10^6$ for annual plants and $N_g \approx 10^4$ for trees (assuming a generation time of 20 years). Given their approximate nature, the preceding calculations are intentionally given to just the nearest order of magnitude.

Sources of bias

Although we now have a well-established theoretical framework for estimating the power of two fundamental population genetic processes, drift and mutation, from measures of within- and between-species variation at neutral sites, the degree to which silent sites actually fulfill the assumption of neutrality is less clear. One notable concern, to be discussed in detail in Chapter 6, is that the nucleotide composition of silent sites can be influenced by selection operating on translation efficiency (codon usage) and/or other features associated with mRNA processing and stability, as well as by biased gene conversion. By favoring specific nucleotides, any one of these processes can reduce standing levels of variation below the neutral expectation, thereby causing π_s to be a downwardly biased estimate of $2N_gu$. Such bias might ordinarily be expected to be greater in species for which N_g is high, which will increase the efficiency of selection and/or biased gene conversion, but this need not always be the case. In fact, if selection favors a particular nucleotide at a site where mutation is biased in the opposite direction—a situation that cannot be entirely ruled out (Kondrashov et al. 2006)—the standing level of variation resulting from the balance between opposing pressures could approximate or even exceed that expected under the neutral model (McVean and Charlesworth 2000).

A second way in which selection might reduce standing levels of silent-site variation involves the influence of nucleotide heterozygosity on recombinational capacity. Because of its role in repairing double-strand breaks and ensuring meiotic chromosome disjunction, recombination is critical to the production of viable gametes. Recombination events are initiated by the invasion of an end of a double-strand break into the homologous region of an intact chromosome (see Chapters 2 and 6), and nucleotide sequence variation is significant in this context because mismatches in the heteroduplex DNA can lead to the abortion of recombinant intermediates (Chambers et al. 1996; Datta et al. 1997; Vulić et al. 1999; Majewski et al. 2000; Emmanuel et al. 2006). The most compelling evidence for such effects derives from experiments in which silencing of the mismatch repair pathway leads to elevated levels of exchange in crosses that are otherwise recombinationally inactive or suppressed: *Escherichia* and *Salmonella* (Rayssiguier et al. 1989); divergent strains of yeast (Hunter et al. 1996); and congeneric species of mice (Shao et al. 2001). Such observations led Stephan and Langley (1992) to suggest that high nucleotide heterozygosity might inhibit the repair of double-strand breaks enough to have significant effects on individual fitness. If this were the case, then selection against highly divergent alleles (independent of their functional attributes) could impose an upper bound on nucleotide heterozygosity.

Are such effects likely to be important at the within-species level? In the yeast *S. cerevisiae*, successful recombination events appear to require at least 60–250 bp stretches of 100% sequence identity between donor and recipient molecules, and recombinational efficiency increases substantially with further increases in the length of identical sequence tracts (Ahn et al. 1988; Mézard et al. 1992; Sugawara and Haber 1992; Jinks-Robertson et al. 1993). The minimum tract lengths for efficient processing in *Leishmania* (Papadopoulou and Dumas 1997) and *Trypanosoma* (Bell and McCulloch 2003) are 140–180 bp, and animals and land plants have fairly similar requirements (Rubnitz and Subramani 1984; Waldman and Liskay 1988; Gloor 2002; Opperman et al. 2004), again with recombination potential increasing with further increases in the length of homology.

These tract length requirements can be compared with the likely situation in natural populations by extrapolating from observations on π_s in coding regions. As a consequence of the redundancy of the genetic code, about 25% of coding sites are silent, and levels of variation at replacement sites are almost always less than a third of that at silent sites (Graur and Li 2000). This observation implies a total level of average within-population sequence divergence of less than $[0.25 + (0.75/3)]\pi_s = 0.5\pi_s$. Thus, with an average π_s of 0.06 for unicellular eukaryotes (see Figure 4.7), the average length of sequence identity is at least $1/(0.5\pi_s) = 33$ bp, whereas average identity tract lengths for multicellular species are about 140 bp. Because there will be substantial variation around these average values, these observations imply that mismatch inhibition of double-strand break repair is likely to occur in eukaryotes, particularly in unicellular species. Without explicit information on the fitness effects of double-strand breaks, the degree to which such effects might inhibit the accumulation of otherwise neutral variation cannot be quantified. However, the consequences of unrepaired breaks must be greater in species in which a single cell constitutes the entire organism, and because of the elevated levels of π_s in unicellular species, the selective effects associated with mismatch repair inhibition must be highest in such species.

These kinds of repair deficiency effects should have less day-to-day relevance to haploid prokaryotes, in which repair of a divergent homologous chromosome would be an unusual event. However, high levels of sequence divergence might still compromise the recombinational cohesiveness among the most extreme genotypes within prokaryotic populations, eventually leading to complete reproductive isolation (Cohan 1995; Vulić et al. 1997). The minimum identity tract length for efficient recombination in eubacteria is smaller than that in eukaryotes—only 20–40 bp (Watt et al. 1985; Majewski and Cohan 1999; Majewski et al. 2000; Springer et al. 2004)—and yet by the above reasoning, for an average prokaryotic $\pi_s = 0.10$, the average length of identical coding sequence tracts between random isolates is just 20 bp.

Although these calculations suggest that mismatch repair inhibition may contribute to a downward bias in estimates of $2N_g u$ in prokaryotes, the extraordinary potential for horizontal transfer in such species by mechanisms other than homologous recombination should also be noted (Ochman et al. 2000). Horizontal transfer across species boundaries expands the genetic resources accessible to prokaryotes in ways that are not accessible to eukaryotes, further magnifying the efficiency of selection on alternative alleles. This source of diversity is not included in the preceding analyses, and the degree to which its inclusion would increase the long-term effective population sizes of prokaryotes remains unclear.

The preceding arguments, as well as those to be presented in Chapter 6, leave little doubt that all species are subject to some forms of variation-reducing selection on silent sites. However, the impact of such effects on the magnitude of deviations of π_s from $2N_g u$ in various lineages is less clear. In mammals, average rates of interspecific divergence for intron sequences are about the same as those for silent sites in coding exons, encouraging the view that both are evolving in a neutral fashion, but the fact that the molecular spectrum of substitutional changes differs significantly among the two regions suggests otherwise (Chamary and Hurst 2004; Mikkelsen et al. 2005). As it is unlikely that both types of sequences would evolve at the same rate if only one were under selection, this observation strongly suggests that different forces of selection and/or biased gene conversion fortuitously lead to similar divergence rates.

Unfortunately, without an unambiguously defined neutral region of DNA, the impact of such effects on silent-site diversity cannot be easily quantified. A comparison of human and chimpanzee sequences indicates a 15% slowdown in the rate of divergence for silent sites relative to that for intergenic DNA (Hellmann et al. 2003), but this discrepancy must be an underestimate of the level of constraint on silent sites because not all intergenic DNA is neutral (see Chapter 3). More telling, perhaps, is the 30% reduction in the rate of divergence for silent sites in human, mouse, and rat genes relative to that for pseudogenes (Bustamante et al. 2002). However, because even pseudogene sequences may be subject to gene conversion and some forms of selection, and because levels of polymorphism are influenced less by selection than are levels of interspecific divergence, it is difficult to say whether the reported estimates of $2N_g u$ for mammals are downwardly biased by a factor greater or less than 3. Things are not much clearer in *Drosophila*, in which average levels of both intraspecific and interspecific silent-site variation are 2–3 times those for intronic and intergenic DNA, and it has again been suggested that both regions are constrained by selection (Halligan et al. 2004; Andolfatto 2005). Finally, for prokaryotes, some evidence suggests that the divergence of silent sites is at least 10 times lower than the expectation based on mutation rates (Eyre-Walker and Bulmer 1995; Ochman 2003), and on this basis alone, the $2N_g u$ estimates given above for this group could be as much as 10 times too low.

One final caveat with respect to estimates of N_g derived from polymorphism data is that they apply only to the past $2N_g$ generations for each taxon—the average time to fixation of a neutral mutation. Because this time scale necessarily increases in species with larger N_g, whereas many of the gross features of genomes may require tens to hundreds of millions of years to emerge, one may question the degree to which polymorphism-based estimates of N_g inform the field of comparative genomics, at least for species with N_g on the order of 10^5 or smaller. However, because they reduce sampling idiosyncrasies, it is reasonable to expect that averages of species-specific estimates of N_g within broad taxonomic groups are representative of long-term evolutionary conditions experienced by group members.

Resolution of the "paradox of variation"

In the earliest days of molecular population genetics, when variation was almost always measured at the level of allozymes (variants of proteins revealed by mobility differences on electrophoretic gels), Lewontin (1974) noted a peculiar pattern: within-species levels of allozyme variation appeared to be nearly independent of population size, contributing to what he broadly viewed as the "paradox of variation." Nei (1983) later pointed out that allozyme heterozygosity levels are actually positively correlated with N, but only weakly so, with the levels of variation in the largest populations falling far below the neutral expectation. Because we can expect a significant fraction of protein variants to be under selection, these types of observations are difficult to interpret. For example, populations with large N would be expected to harbor more neutral variation (higher $2N_g u$) but less variation involving selectively disadvantageous alleles, and these contrasting patterns might roughly cancel each other out, yielding the weak association between allozyme variation and N. The use of DNA sequence data allows this type of analysis to be restricted to silent sites in protein-coding genes, alleviating but not entirely removing these types of interpretative problems.

With the above caveats in mind, three general conclusions, whose validity is critical to many of the ideas outlined in the following chapters, appear to be warranted:

- Average long-term effective population sizes vary by at least five orders of magnitude between the largest multicellular and the smallest unicellular species. Assuming that silent sites in low-N_g species are nearly neutral, N_g should be about 10^4 in many large vertebrates and perennial plants. In contrast, the largest estimate of N_g for a prokaryote derivable from the results in Figure 4.7 is 3×10^8 for *Helicobacter pyogenes*, a highly recombining member of the eubacteria. Thus, because of the potentially strong selective biases operating on silent sites in high-N_g species, it is virtually certain that N_g is greater than 10^9 in some prokaryotes. If the downward bias in estimates of prokaryotic N_g induced by selection is as much as tenfold, then the upper limit to N_g is about 10^{10}.
- Unless the effects of selection on silent sites are beyond our wildest dreams, the scaling of $2N_g u$ with the absolute population size (N) is much less than linear. This pattern is partly a consequence of an increased per-generation mutation rate (u) in larger organisms, which compensates somewhat for the reduction in N. However, even taking this factor into account, the disparity in N_g across all domains of cellular life is nearly 20 orders of magnitude less than the disparity in N. Although once viewed as a paradox, such behavior appears to be quite consistent with the significant variance-reducing effects of stochastic hitchhiking events in large populations.
- Given these results, all deleterious mutations with selection coefficients of 10^{-6} or less are expected to accumulate in the genomes of large multicellular eukaryotes in an effectively neutral fashion (unless accompanied by substantially stronger biased gene conversion). In contrast, the critical selection coefficient for effective neutrality in a prokaryote may commonly be in the vicinity of 10^{-10}–10^{-9}. These numerical estimates are relevant because, as will be shown in the following chapters, many mutational modifications of gene and genomic architecture have intrinsically deleterious effects with selection coefficients in the range of 10^{-8}–10^{-6}. This implies that various forms of cellular life experience radically different population genetic environments—many forms of mutant alleles that are able to drift to fixation in multicellular eukaryotes are efficiently eliminated by selection in prokaryotic species. Without the theoretical and empirical results of population genetics, we would be unable to make such statements.

Finally, it should be recalled that many unicellular species have life cycles with prolonged (and in some cases, continuous) phases of clonal reproduction. Because their entire genome acts effectively as a single supergene in the absence of recombination, such species can be expected to be extremely vulnerable to selective interference. Given that a leading hypothesis for the evolution of sex is that recombination (and segregation) allows populations to avoid the accumulation of deleterious mutations while enhancing the promotion of beneficial alleles (Muller 1932; Lynch et al. 1993; Peck 1994; Barton and Charlesworth 1998; Orr 2000b; Otto and Barton 2001; Johnson and Barton 2002; Otto 2003; Martin et al. 2006), why should clonal propagation be most pronounced in microbial species? If the hypothesis that linkage and hitchhiking events impose an upper limit on the long-term effective population size of 10^{10} is correct, then asexual populations with absolute sizes well beyond the level at which draft overwhelms drift might experience little or no change in evolutionary potential after a transition to sexual reproduction. Other considerations, including the enhanced likelihood that parallel mutation alone can produce advantageous multilocus genotypes in sufficiently large populations, lead to a similar conclusion (Kim and Orr 2005). Thus, the general association of sexual reproduction with multicellular species may largely be a consequence of their reduced effective population sizes. That is, like many aspects of genomic architecture that we will encounter below, the evolution of sex may be a "small population size" phenomenon.

Mutation as a Weak Selective Force

Population geneticists have historically treated selection and mutation as independent forces in the dynamics of evolutionary change, with mutation producing the variation on which natural selection acts, but having no further influence on the fates of alleles. However, in the context of the evolution of the architectural features of genes and genomes, there are numerous ways in which mutation can act indirectly as a selective agent. As discussed at the close of Chapter 2, alleles with more elaborate structural embellishments (e.g., more or larger introns, larger UTRs, and/or larger intergenic regions) are larger targets for degenerative mutations. The resultant burden of an excess mutation rate operates in a fashion analogous to natural selection, as it results in a smaller fraction of successful progeny alleles. The selective disadvantage associated with any single aspect of gene complexity need not be very large, as it is roughly equivalent to the product of the mutation rate per nucleotide per generation (u) and the excess number of nucleotide sites in the more complex allele that are critical to gene function (n) (Lynch 2002b). As we have already seen, u is of the magnitude 10^{-9}–10^{-7}, and arguments presented in later chapters suggest that n often falls in the range of 1–50. As a consequence, expansions of gene architectural complexity are unusually vulnerable to fixation by random genetic drift in populations with small genetic effective sizes. In contrast, in populations with sufficiently large N_g, a differential rate of mutation to defective alleles can act as a selective deterrent to the expansion of gene architectural complexity.

The degree to which a differential mutation rate might influence the likelihood of fixation of a newly arisen allele can be obtained by simply expanding the theory for selection discussed above. Denoting the excess degenerative mutation rate of a newly arisen allele as u_d, and assuming no intrinsic adaptive (or maladaptive) effect of such an allele, u_d can simply be substituted for the deleterious selection coefficient (negative s) to determine the probability of fixation. From the left side of Figure 4.2, recalling that $N_g = 2N_e$ for a diploid, it is clear that if $N_g u_d > 2$, there is essentially no chance of fixation, but if $0 < N_g u_d < 0.5$, such an allele will be capable of drifting to fixation in a nearly neutral fashion. Letting $u_d = nu$ as noted above, the criterion for near-neutrality becomes $2N_g u < 1/n$.

This relationship is a highly useful one for two reasons. First, the product $N_g u$ has a simple intuitive interpretation, as it is equivalent to the ratio of the power of mutation (u) to the power of random genetic drift ($1/N_g$). If this ratio is too low, the negative force of degenerative mutation will be so overwhelmed by the noise in the evolutionary process that selection against a mutationally harmful genomic addition will be ineffective. Second, because both $2N_g u$ and n are estimable parameters, the first as π_s and the second from the key molecular features associated with a modified gene structure, explicit predictions can be made regarding the population genetic conditions that are conducive to the accumulation of slightly deleterious modifications in gene and genomic features. This also means that hypotheses can be tested with empirical data.

For example, the number of nucleotides critical for proper spliceosome recognition of an intron is about 25 (see Chapter 9), which implies a threshold $2N_g u$ for intron colonization of about 0.04. Recalling the average estimates of π_s and their likely downward biases (see Figure 4.7), it is clear that the population genetic environment of prokaryotes may only rarely be conducive to intron colonization, whereas the extremely low levels of π_s for multicellular eukaryotes imply conditions that are incapable of preventing intron proliferation. With their wide range of π_s, the various lineages of unicellular eukaryotes are expected to straddle these two extremes. As previewed in Chapters 2 and 3, these predictions are quite consistent with the phylogenetic distribution of introns.

This fundamental approach can also be extended to other aspects of genomic complexity. For example, tandemly linked groups of duplicate genes, with each member specializing in restricted sets of functions of the ancestral gene (see Chapter 8), are discouraged from proceeding to fixation in sufficiently large populations, because such duplications greatly expand the amount of coding DNA that must avoid mutations. The modularization of regulatory elements can also be inhibited in populations with sufficiently large $2N_g u$ (see Chapter 10), as can the expansion of mobile element families (see Chapter 7).

That the qualitative predictions of population genetic theory combined with quantitative estimates of $2N_g u$ jointly point to the effectiveness of large population size as a barrier to the evolution of complex gene and genomic architectural features raises significant challenges to purely adaptive explanations for the evolution of the eukaryotic genome. Indeed, as will be repeatedly pointed out in the following chapters, it is difficult to reject the hypothesis that the emergence of many aspects of eukaryotic gene structure and genomic complexity was largely a consequence of nonadaptive mechanisms operating in a direction contrary to that expected under natural selection. This is not to say that the repatterning of the eukaryotic genome had no adaptive consequences. Quite the contrary; as will also be emphasized in subsequent chapters, there are good reasons to believe that once the peculiar features of eukaryotic genes were broadly established, entirely novel pathways were opened up for adaptive evolution by descent with modification.

As an empirical preview to these matters, the negative association between genome size and π_s across a broad array of phylogenetic groups is noted in Figure 4.8. Caution should be exercised in overinterpreting the quantitative scaling of the relationship in this figure because errors in the

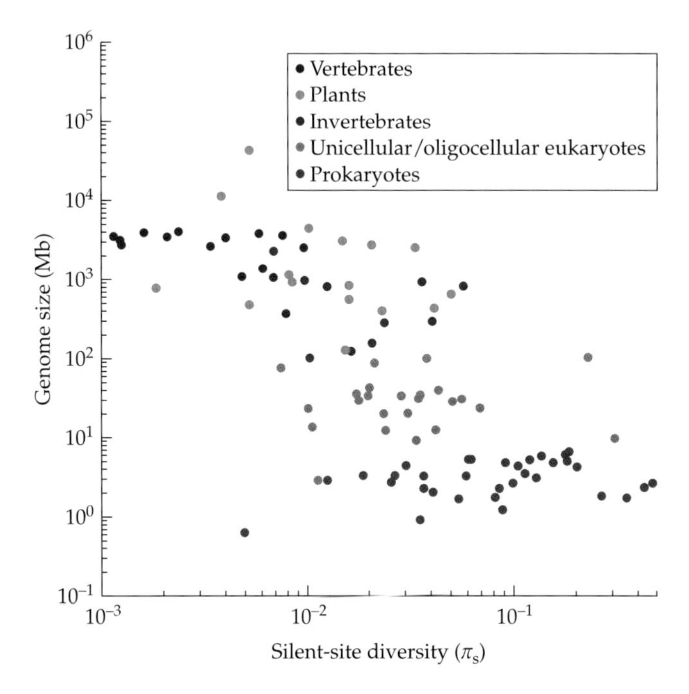

Figure 4.8 The relationship between genome size and π_s for members of various taxonomic groups. (From Lynch 2006.)

estimates of π_s will cause the slope to be less pronounced (by spreading the points out along the x axis), whereas the expected downward bias in π_s resulting from selection is likely to have the opposite effect (by pushing the points with highest π_s more to the left). Nonetheless, no amount of reasonable transformation of the x axis in Figure 4.8 would alter the conclusion that there is a broad negative association between genome size and $2N_g u$. Because most aspects of genomic evolution develop on time scales of millions of years, clear associations of this nature are unlikely to be observed at the level of species within genera, but a remarkably similar pattern to that in Figure 4.8 has been revealed within the ray-finned fishes (Yi and Streelman 2005). A major challenge for evolutionary genomics—and the goal of the remaining chapters—is to explain why such patterns exist.

5 Three Keys to Chromosomal Integrity

Having achieved a broad overview of the potential population genetic processes driving genomic evolution, our remaining goal is to more fully explore the origins and maintenance of specific aspects of genomic architecture from a mechanistic perspective. Because the adaptive morphologies, physiologies, and behaviors on which all organisms depend require the proper functioning of thousands of gene products, it is easy to lose sight of the fact that, regardless of a gene's phenotypic consequences, transmission to the next generation critically depends on the structural features of the chromosome on which it resides. Successful chromosomes must be capable of maintaining structural stability, replicating in entirety, and properly assorting into daughter cells. Moreover, these processes must be coordinated among all members of a chromosome set to ensure the inheritance of complete genomic complements by daughter cells. Three noncoding features of chromosomes—origins of replication, centromeres, and telomeres—are essential for these functions.

This chapter is focused on the gross anatomical aspects of these vital chromosomal components and the special mechanisms by which they evolve. The coverage here is not simply a side excursion from the main course of the book. Because the processes of chromosome maintenance and replication also determine the types of mutations that arise, they are intimately related to patterns of nucleotide composition (see Chapter 6). In addition, chromosomal segments that are physically involved in delivery to gametes are subject to selection for features that maximize their likelihood of inheritance at the expense of homologous chromosomes. The resultant biased transmission can create a drive-like process, reducing the effective population size of a chromosome and compromising the efficiency of selection operating on linked genes, in the same manner as an adaptive selective sweep (see Chapter 4).

Before proceeding, it should be noted that eukaryotic chromosomes are not simply limp and naked molecules of DNA, but exist as chromatin complexes with a number of chromosome-specific proteins, most notably histones (Van Holde 1989). During the nonreplicating phase of the cell cycle, chromatin consists of core subunits of spools of DNA wound around octamers of histone proteins (nucleosomes), which are in turn aggregated into a nested hierarchy of architectural compactions to form visible chromosomes. Given our focus on genomic architecture, the following discussion considers primarily only DNA-level aspects of chromosomal architecture, although the influential role of DNA-binding proteins will become apparent in a number of contexts.

Origins of Replication

The faithful inheritance of genes across generations requires accurate replication of the complete chromosomes on which they reside. Jacob et al. (1963) thought that this process must be achieved by the recognition of a specific DNA element (the replicator) by a replication protein (the initiator), and the limited details that have subsequently emerged on chromosome replication provide general support for this model (DePamphilis 1996). First, replication is usually initiated at well-defined origin of replication sites (ORIs), which the cell detects by use of an origin recognition protein complex. Second, replication is bidirectional and semiconservative, proceeding outward in both directions from each ORI, with each parental DNA strand specifying its complement. Third, except in organelles, replication is semi-discontinuous, with continuous synthesis proceeding along "leading" strands, and discontinuous production of Okazaki fragments and their subsequent annealing occurring on "lagging" strands (Figure 5.1). This latter feature is an unavoidable consequence of the bidirectional synthesis of both strands in a 5′ to 3′ direction, which causes newly synthesized lagging strands to grow in the opposite direction of parental strand exposure. Remarkably, these three ancestral features of replication are shared by all prokaryotes and eukaryotes, despite the fact that the replication proteins (DNA polymerases, primases, and helicases) in archaea and eukaryotes appear to be nonhomologous with those in eubacteria (see Chapter 1).

Most prokaryotes have single ORIs located close to the locus of the initiator protein gene (Matsunaga et al. 2001; Kelman and Kelman 2004), with replication proceeding bidirectionally around the circular chromosome until a terminator sequence is reached at the opposite pole. A region near the ORI then attracts the proteins necessary for guiding the daughter chromosomes to opposite ends of the dividing cell (Sherratt 2003), functioning much as the centromeres of eukaryotic chromosomes do (see below). Under normal growth conditions, unidirectional rates of replication fork progression are about 21, 35, and 50 kb/min, respectively, in the eubacteria *Caulobacter*

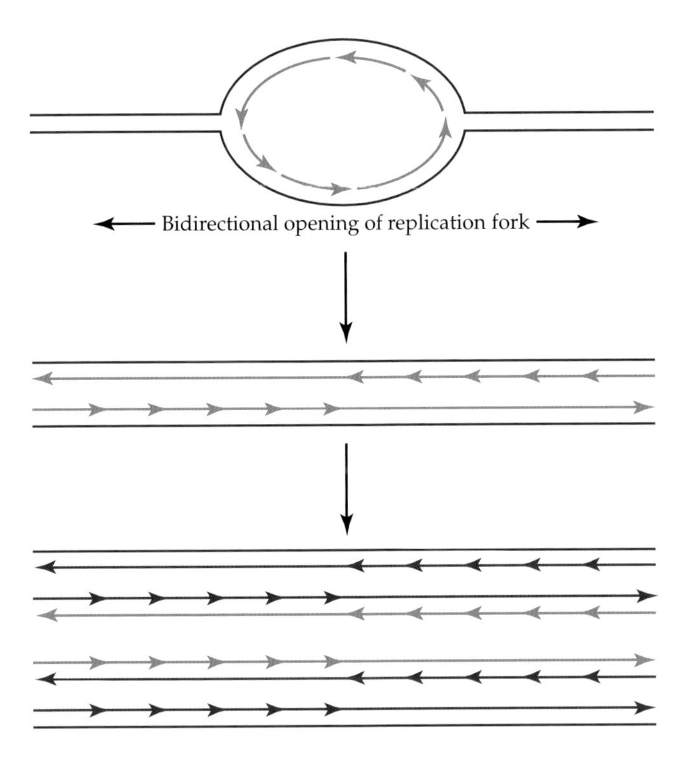

Figure 5.1 *Top*: Opening of an origin of replication with simultaneous, bidirectional synthesis of the continuous leading (long arrows) and discontinuous lagging (short arrows) strands. *Middle*: The discontinuous Okazaki fragments are eventually ligated together in the two daughter chromosomes. *Bottom*: The four chromosomes resulting from the next round of replication, with the newly synthesized strands represented as blue lines. The lagging strands in each cell division become substrates for leading strand synthesis in the following division, and vice versa.

crescentus, *Mycoplasma smegmatis*, and *Escherichia coli* (Hiriyanna and Ramakrishnan 1986; Stillman 1996), whereas the rate for the slow-growing *Mycoplasma tuberculosis* is just 3 kb/min (Hiriyanna and Ramakrishnan 1986). The rates for archaeal species in the genera *Sulfolobus* (Lundgren et al. 2004) and *Pyrococcus* (Myllykallio et al. 2000) are about 5 and 20 kb/min, respectively, although *Sulfolobus* species have three ORIs, bringing the total rate up to about 15 kb/min. Thus, with most prokaryotes having genomes in the range of 1.0–8.0 Mb, the time to replicate an entire genome under good growth conditions is on the order of 10 minutes to 4 hours.

Chromosome sizes in multicellular eukaryotes tend to be larger than those in prokaryotes, averaging 10–250 Mb for invertebrates, 40–200 Mb for vertebrates, and 30–2000 Mb for land plants (see Chapter 2). In contrast, eukaryotic fork progression rates are about an order of magnitude slower than in prokaryotes, usually in the range of 0.2–3.0 kb/min (Table 5.1). With

TABLE 5.1 Properties of replicons in various eukaryotes

	RATE (KB/MIN)	LENGTH (KB)	REFERENCES
Saccharomyces cerevisiae	2.9	30	Raghuraman et al. 2001
Schizosaccharomyces pombe	—	36	Segurado et al. 2003
Physarum polycephalum	1.2	35	Pierron and Bénard 1996
Drosophila melanogaster	0.6	25	Ananiev et al. 1977; MacAlpine et al. 2004
Human	1.7	144	Jackson and Pombo 1998
Mouse	1.8	46	Brown et al. 1987; Takebayashi et al. 2001
Angiosperms	0.2	56	Van't Hof and Bjerknes 1981; Francis et al. 1985

just a single ORI, replication of a 10 Mb chromosome at a fork rate of 1.5 kb/min would require more than 2 days, and a 100 Mb chromosome would require more than 3 weeks. Eukaryotes compensate for this problem by populating their chromosomes with multiple ORIs, but this poses an additional challenge. For efficient genomic replication, a full set of ORIs must be recognizable on each chromosome, but with hundreds of ORIs subject to independent mutations, how can the origin recognition complex cope with what is likely to be substantial sequence variation?

ORI specification

The simplest known eukaryotic ORIs are those of the budding yeast *S. cerevisiae*, each of which is only about 200 bp in length and contains at least one copy of an 11 bp binding site sequence for the origin recognition complex (Newlon and Theis 2002; Nieduszynski et al. 2006). The ORIs of *S. cerevisiae* are autonomously replicating sequences (ARSs), in the sense that they will support replication when placed in artificial constructs. Altogether, there are about 300 such elements per haploid yeast genome, with an average interval between them (replicon length) of about 30 kb. Although the average time to replicate a replicon in this species is about 5 minutes, the time to completely replicate the genome is considerably longer, about 55 minutes, because not all ORIs are activated simultaneously. ORIs near centromeres replicate earlier than others, and only a subset of ORIs appears to be employed during any cell division (Raghuraman et al. 2001).

No other well-studied species has ORIs approaching the simplicity of those in the genus *Saccharomyces*. The next simplest known situation is in the fission yeast *S. pombe*, in which ORIs are about 0.5–1.0 kb in length, sometimes containing an 11 bp ARS-like sequence along with at least two 30–55 bp A/T-rich regions and other ill-defined stimulatory sequences (Dubey et al. 1996). Remarkably, about 50% of random intergenic sequences from *S. pombe* with lengths and A/T contents similar to verified ORIs exhibit potential origin activity (Dai et al. 2005).

The ORIs of animals are even more loosely defined than those in *S. pombe*, sharing only the attribute of A/T richness. Again, many random sequences of animal DNA can act as ARSs, although it is unclear how many of these are actually used in the cell (DePamphilis 1996). *Drosophila* and mammalian ORIs have lengths of 1–10 kb, individually contain multiple discrete initiation sites, and generally reside within intergenic regions (Ina et al. 2001; Altman and Fanning 2004). The numbers of ORIs per animal genome are in the thousands, and as in yeasts, only subsets of the potential pool of such regions appear to be deployed in individual cell divisions (Anglana et al. 2003), although many are used very reliably (Jackson and Pombo 1998). In some animal tissues, replication and cell division are uncoupled, leading to the production of endopolyploid cells, and cases exist in which restricted genomic intervals are over-replicated by repeatedly licensing the same ORIs within the cell. Such activities provide a mechanism for increasing the transcriptional substrate for a number of highly active genes, including ribosomal RNA genes in amphibian oocytes and chorion (eggshell) protein genes in the follicle cells of *Drosophila* ovaries (Claycomb et al. 2004).

The lack of well-defined core sequences for animal ORIs and the approximately random distribution of replicons have led to the suggestion that epigenetic markers (heritable genomic modifications involving factors other than the specific DNA sequence, such as DNA methylation and histone acetylation) facilitate ORI identification (Hyrien et al. 2003; Aggarwal and Calvi 2004). Such effects remain elusive, although there is a positive association between replication and transcription (MacAlpine et al. 2004; Kohzaki and Murakami 2005). Some transcriptional control regions have the potential to operate as ORIs when moved to ectopic locations (Danis et al. 2004), and some transcription factors may directly recruit the origin recognition complex.

Evolutionary consequences

The process of replication influences genomic evolution in several ways. First, the distribution of semi-discrete ORIs imposes constraints on the sequences within intergenic DNA (where ORIs almost universally reside). In the absence of comparative data on specific ORIs across species, it is not possible to estimate the magnitude of such constraints, but they need not be trivial. In vertebrates, for example, the total amount of DNA associated with ORIs probably exceeds the amount allocated to coding exons, although the requirements at any particular nucleotide site might be quite diffuse. In addition, the development of Okazaki fragments appears to depend on specific priming sites. In *E. coli*, Okazaki fragments average 1.5 kb in length, and the priming sites generally contain a CAG triplet, possibly accompanied by contextual requirements in surrounding sequence (Zechner et al.

1992; Bhattacharyya and Griep 2000). Okazaki fragments in the archaebacterium *Pyrococcus abyssi* are shorter, averaging 100 bp (Matsunaga et al. 2003). Unfortunately, such details are lacking for eukaryotic species.

The second way in which replication can influence genomic architecture concerns the interference between jointly replicating and transcribing regions of the chromosome (French 1992). Because transcription always proceeds down an antisense strand in the 3′ to 5′ direction, DNA polymerases (replication) and RNA polymerases (transcription) will incur codirectional versus head-on collisions depending on whether they are proceeding along leading or lagging strands. Nevertheless, because rates of transcription are substantially slower than rates of replication [about 20 times slower in *E. coli*, according to Bremer and Dennis (1996), and about 100 times slower in eukaryotes, based on the data in Table 5.1 and in Reines et al. (1996) and Jackson et al. (2000)], collision rates are nearly independent of the strand being transcribed. All such events can interrupt gene expression by dislodging RNA polymerases, but head-on collisions have the additional effect of reducing the rate of replication fork progression (Mirkin and Mirkin 2005). Although the negative consequences of collisions between the transcription and replication machinery can be reduced by localizing genes on leading strands, the quantitative advantage of such gene positioning may be quite small, as experimental inversions of genes in prokaryotes are not generally known to cause measurable changes in fitness (Price, Alm et al. 2005). Nevertheless, there is a tendency for essential genes to be located on leading strands in prokaryotic genomes (Rocha and Danchin 2003; Price, Alm et al. 2005). The matter is entirely unresolved in eukaryotes, for which information on the locations of ORIs is sparse.

Finally, the semi-discontinuous nature of replication can impose different mutational spectra on genes residing on leading versus lagging strands. The mechanistic details will be explored in Chapter 6, but for now, the salient point is that strand-specific mutational biases generally lead to excesses of G over C and of T over A on leading strands. As a consequence, genes experience different types of mutation pressures depending on the strands on which they reside; for example, Gs and Ts are more stable in genes on leading strands. The consequences of such mutational asymmetries become especially relevant if a gene is moved from one strand to another, as this usually transiently increases the mutation rate of the gene. For example, a gene residing on a leading strand for a sufficiently long time is expected to have a G+T-rich coding sequence, but relocation of the gene to the lagging strand would yield an excess of more highly mutable Cs and As on the leading strand, leading to an elevation in the genic mutation rate until the local equilibrium state of nucleotide usage has been reestablished.

Such a process could have significant implications for the ability of genes to successfully move to opposite strands in various lineages. As a simple example, consider a situation in which the numbers of Gs and Cs in a leading-strand gene are n_G and n_C, with leading-strand Cs being the predominant sites of mutations (see Chapter 6), which arise at a rate of u per C. If

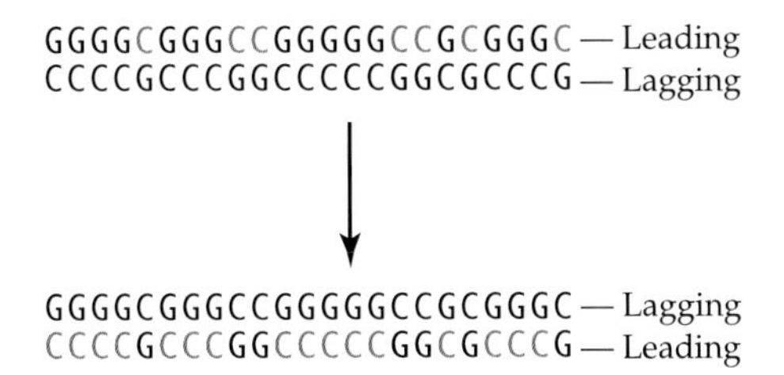

Figure 5.2 The mutational consequences of switching the DNA strands on which a gene resides. For simplicity, only complementary G:C base pairs are shown, and it is assumed that the vast majority of mutations are C→G changes initiated at leading-strand Cs (denoted in orange). If the gene switches strands (e.g., by an inversion), the large number of Cs previously on the lagging strand are now vulnerable to mutation, leading to a transient increase in the gene-wide mutation rate (until a new equilibrium nucleotide composition evolves, with high G content again on the leading strand).

such a gene were to relocate to a lagging strand, the genic mutation rate would initially increase by an amount $(n_G - n_C)u$, because the gene's high number of Gs would place the complementary Cs in a more mutable (leading-strand) environment (Figure 5.2). If this directional mutation pressure exceeded the power of random genetic drift (see Chapter 4) and the fitness consequences of such mutations were sufficiently negative, fixation of the strand-shifted gene copy would be unlikely. This leads to the prediction that strand switching by genes should be less common in prokaryotes (with large effective population sizes) than in eukaryotes (assuming that the physical initiation of such events is not greatly reduced in the latter). An analysis of this matter remains to be done.

Centromeres

One of the most poorly understood aspects of eukaryotic genomic evolution involves the centromere, a semi-discrete region on each chromosome that is essential to chromosome transmission to daughter cells. After chromosome duplication, the replicated sister chromatids remain attached at the centromere until the initiation of cell division. At this point, each sister centromere serves as an assembly site for a proteinaceous kinetochore, which captures the spindle microtubules along which the sister chromatids will be dragged to opposite poles of the dividing cell. In most animals and land plants, centromeres are observable as constrictions in each metaphase chromosome. Although such structures are not visible in the tiny chromosomes of most unicellular species, the replication-associated functions of centromeres and the core set of proteins that carry them out are highly conserved. Thus, the centromere appears to be an evolutionary legacy of the stem eukaryote (Tyler-Smith and Floridia 2000). (In a few exceptional species, including *C. elegans*, discrete centromeres appear to have been lost, and the chromosomes are holocentric, with microtubules attaching at many sites.)

How the cell division machinery recognizes the centromere is largely a mystery. The discovery that *S. cerevisiae* centromeres consist of a specific 125 bp signature (Ng and Carbon 1987) initially encouraged the view that the functional cores of centromeres might be quite conserved, but the simplicity of the *S. cerevisiae* centromere turns out to be exceptional, even for a fungus. For example, the centromeres of the fission yeast *S. pombe* are 35–110 kb in length, containing a central core that is only about 50% identical across nonhomologous chromosomes (Chikashige et al. 1989). The centromeres of *Candida albicans* are intermediate in size to those of budding and fission yeasts, occupying 4–16 kb regions, and centromeres from different chromosomes appear to harbor unique sequences (Sanyal et al. 2004). Still larger are the 200–400 kb centromeres of *Neurospora crassa*, which appear to be graveyards of several mobile element families despite the general absence of active mobile elements in this species (Cambareri et al. 1998; Borkovich et al. 2004). The few centromeres that have been tentatively characterized in other unicellular species follow this general theme of sequence complexity. For example, a putative 2 kb centromere in *Plasmodium falciparum* has 97% A/T content and contains a number of repetitive DNAs (Bowman et al. 1999); and the one centromere characterized in *Trypanosoma cruzi* is about 16 kb in length, highly enriched with G+C, and contains several degenerate retrotransposons (Obado et al. 2005).

Even more poorly defined are the centromeres of animals and land plants. Functional studies indicate that *D. melanogaster* centromeres are about 420 kb in length, containing numerous incapacitated mobile elements and about 8% AATAT and TTCTC tandem repeats, but harboring no other obvious centromere-specific sequences (Murphy and Karpen 1995; Sun et al. 2003). Human centromeric regions, which range in length from 0.5 to 5.0 Mb, are also densely occupied by remnants of mobile elements and repetitive DNAs, including thousands of copies of a 171 bp repeat (Yang et al. 2000; Sullivan 2001; Rudd et al. 2006), as well as abundant gene-containing duplicative transpositions, many younger than a few million years (Eichler et al. 1999; She et al. 2004). The short satellite repeats tend to have a nested age structure, with a substantial expansion of the youngest elements in the interior of the centromere and descendants of older elements near the edges (Schueler et al. 2001, 2005). Similar observations have been made with centromeres from the frog *Xenopus* (Edwards and Murray 2005). Land plant centromeres generally fall into the range of 0.5–3.0 Mb in length, and again contain numerous arrays of repetitive DNAs (with 150–180 bp element lengths) and mobile element insertions, many of which appear to be centromere-specific (Kaszas and Birchler 1998; Copenhaver et al. 1999; Gindullis et al. 2001; Jin et al. 2004; Nagaki et al. 2004; Ma and Bennetzen 2006).

Although the data are still rather limited phylogenetically, these observations strongly suggest that centromere lengths increased with the transition from unicellularity to multicellularity, largely as a consequence of the expansion of repetitive DNA families and the accumulation of mobile elements. The large size and complexity of centromeres make it very difficult to localize functionally relevant regions, and despite considerable effort, no specific sequence requirement for centromere activity has been identified in any species other than *S. cerevisiae*. In addition, the boundary between centromeric and noncentromeric regions is by no means abrupt, and in multicellular species, generally extends over a long, ill-defined gradient referred to as the pericentromeric region. One commonality does seem to exist: although the sequences of centromeric satellite repeats in different multicellular species are widely divergent, they are almost universally 150–180 bp long, putting them in the range of segment lengths that wrap around histones to form the units of chromosomal compaction (nucleosomes) (Henikoff et al. 2001). Remarkably, however, the occasional emergence of new centromeres in regions free of satellite DNAs (below) indicates that these sequences are not essential to centromere definition.

The centromeric-drive hypothesis

The preceding observations illustrate one of the central paradoxes in chromosome biology—the tendency of centromeric sequences to diverge rapidly among species despite their fundamental role in the survival of every chromosome. DNA sequences that are under strong selective constraints usually exhibit very slow rates of molecular evolution, but divergence in putative centromeric core elements well in excess of the neutral expectation has been documented in *Oryza* (rice), *Arabidopsis*, and *Drosophila* (Malik and Henikoff 2001; Kawabe and Nasuda 2005; Lee et al. 2005; Ma and Bennetzen 2006). A potential explanation for such rapid evolution was provided by Henikoff et al. (2001), who argued that selection should favor centromeric features that enhance their own chances of their being loaded into successful gametes (Figure 5.3). The key underpinning of this hypothesis is the peculiar nature of female meiosis in animals and land plants: only one of the four meiotic products becomes an oocyte nucleus, setting up an opportunity for competition between alternative centromeres striving for delivery into the oocyte. Whether a particular centromere will win such a contest depends on its position in the female tetrad (the array of four meiotic products), its strength of binding to the spindle fibers, and so forth (Pardo-Manuel de Villena and Sapienza 2001).

Such a competitive arena might cause centromeric sequences to eventually evolve to a state that maximally exploits the meiotic machinery, bringing further evolutionary change to a standstill. However, Henikoff et al. (2001) suggested that driving centromeres impose significant selection pressure for the restoration of centromeric parity, setting up a sort of coevolutionary loop that makes such stasis unlikely. The assumption here is that although disparities in centromeric strength should have negligible effects on female fitness (as each egg will still acquire a full haploid set of chromosomes), male gamete production (which normally uses all four meiotic

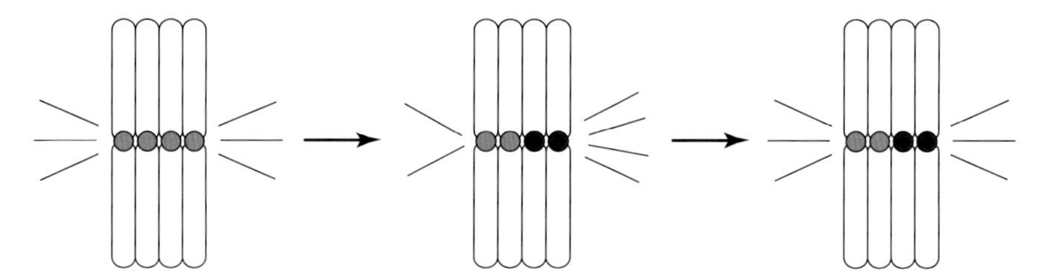

Figure 5.3 The centromeric-drive hypothesis. *Left*: Two pairs of sister chromatids, all with functionally identical centromeres (colored balls), are lined up at metaphase. Each chromatid has an equal probability of being delivered to the oocyte. *Center*: One centromere (red) has acquired an alteration that attracts a larger number of spindle fibers, enhancing the chance of delivery to the oocyte. *Right*: A compensatory mutation in a centromeric binding protein restores a balanced number of spindle fibers. (From Henikoff et al. 2001.)

products) may be compromised by centromeric competition. The centromeric-drive hypothesis postulates that such problems impose selection for changes in centromeric (or heterochromatin-binding) proteins that restore parity in microtubule attachment, resulting in cycles of fixations of driving centromeres and restoring modifiers. Likening this series of events to an intragenomic arms race, Henikoff et al. (2001) further suggested that the process might increase rates of speciation by creating incompatibilities between the centromeric proteins of one species and the centromeric sequences of another. In the case of *Drosophila*, the group in which the genetics of speciation has been studied most closely, reproductive incompatibility does not appear to be associated with the major centromeric protein (Sainz et al. 2003). However, a factor that restores viability in interspecies hybrids has been mapped to the centromeric region of the X chromosome in *D. melanogaster* (Sawamura and Yamamoto 1993).

Attempts to demonstrate the operation of directional selection on the centromeric binding proteins, which would be expected under a drive hypothesis, have yielded mixed results. One target of interest is the centromeric histone protein (usually referred to as CenH3). Unlike the DNA-binding regions of histone H3, which interacts with the full lengths of all chromosomes and is one of the most highly conserved eukaryotic proteins, those of CenH3 evolve at rates in excess of the neutral expectation in both *Arabidopsis* and *Drosophila* (Malik and Henikoff 2001; Cooper and Henikoff 2004). In contrast, in mammals and grasses, CenH3 is under stabilizing selection, although another centromeric protein (CENP-C) does appear to be under positive selection (Talbert et al. 2004). Centromeric drive is not expected to occur in species in which all four tetrad products of meiosis yield gametes (e.g., fungi), and at least in yeasts, the centromeric proteins do appear to be under purifying selection (Talbert et al. 2004). The obvious systems to search for the process are animals with haplodiploid mating systems, as no counter-selection is expected in haploid males, which produce sperm by mitosis.

Despite the seductive simplicity of the centromeric-drive hypothesis, the population genetic conditions required for its operation may be quite restrictive. The key requirement for such a coevolutionary drive process is the maintenance of functionally significant polymorphisms in centromeric regions for a sufficiently long time to enable a compensatory response on the part of the centromeric proteins. In the case of an aggressive centromere that rapidly achieves high frequency, the likelihood of the selective promotion of modifier mutations is small because homozygotes for driving centromeres are not subject to differential spindle-binding strength. On the other hand, if the deleterious effects of a driving centromere on male fitness are sufficiently greater than the power of the drive process, the driving centromere will be eliminated from the population too rapidly to allow the arrival of compensatory mutations. Small population size will facilitate stochastic increases in the frequencies of mildly deleterious centromeres, but will also inhibit the population-level rate of mutational origin of modifiers and the ability of natural selection to promote them. Although a quasi-equilibrium frequency of the driving centromere might arise under some conditions, with the selective advantage through females being balanced by the disadvantage through males (Feldman and Otto 1991), there would then be no selective advantage of modifiers for the transmission of one versus the other centromeric type (although, in principle, a modifier that simply ensures "fairness" might emerge). Such equilibria appear to exist with naturally occurring "knobbed" chromosomes in maize, one of the first known examples of preferential chromosome segregation (Buckler et al. 1999; Birchler et al. 2003). Finally, a fundamental constraint on the centromeric drive process is imposed by the fact that centromeric proteins must recognize the full set of centromeric sequences across all chromosomes. A successful modifier protein would have to restore parity at the problematic chromosome without generating new difficulties elsewhere. Until these issues have been addressed formally in a population genetic framework, the mechanisms responsible for centromere evolution will remain uncertain. A substantial body of theory on the fixation of compensatory mutations could be adapted to this end (Kimura 1985; Hartl and Taubes 1996; Michalakis and Slatkin 1996; Phillips 1996; Stephan 1996; Higgs 1998; Carter and Wagner 2002).

The presence of small, apparently expendable B chromosomes in a wide variety of multicellular species may have additional bearing on the issue of centromeric drive (Camacho et al. 2000, 2005), as may the parasitic plasmids of yeast (Hajra et al. 2006). These diminutive chromosomes appear to be derived from the centromeric regions of ordinary chromosomes and largely owe their success to masquerading as such. Like centromeres, B chromosomes are virtually always saturated with mobile elements and other forms of repetitive DNA and are largely devoid of coding sequences (Camacho 2005; Page et al. 2001). They need not segregate as bivalents, and the rate of transmission by presence/absence heterozygotes often exceeds 50%, as expected for a drive process. B chromosome abundance is positively correlated with genome size (Palestis et al. 2004; Levin et al. 2005), with only a few examples known outside of animals and land plants (all of which are fungi). This pattern may reflect the fact that the mildly deleterious effects that are frequently associated with B chromosomes (Burt and Trivers 2006) tend to be overwhelmed by the forces of random genetic drift in the typically small populations of multicellular species—the same mechanism that probably drives the expansion of noncoding DNA on the primary chromosomes. And this, in turn, reemphasizes the point that consideration of population size issues may help solve the many remaining mysteries of centromeric evolution.

Centromere positioning

Pardo-Manuel de Villena and Sapienza (2001) have argued that some sort of drive process also underlies patterns of centromere positioning in various vertebrates. Within most such species, there is a strong tendency for the full set of chromosomes to have centromeres in the same relative locations, being either mostly metacentric (centromeres centrally located) or mostly acrocentric (centromeres near chromosome tips). Experiments in the mouse demonstrate that a single metacentric chromosome tends to lose to two otherwise homologous acrocentrics during meiosis, whereas the opposite occurs in humans and chickens. That is, the oocyte appears to attract centromeres in the mouse, but to repel them in humans and chickens. The predominant forms of chromosome structure in these different lineages, primarily acrocentric in the mouse and primarily metacentric in humans and chickens, are consistent with such behavior having played a central role in the evolution of centromere locations in these lineages.

Centromere repositioning (the appearance of a novel centromere at an ectopic location concurrent with the loss of the original centromere) can be quite rapid, with phylogenetic analyses implying such changes on time scales of a few million years or less within primates (Ventura et al. 2001, 2004; Eder et al. 2003), rodents (Armengol et al. 2003), and horses (Yang et al. 2004; Carbone et al. 2006). However, the mechanisms by which a neocentromere-bearing chromosome spreads through a population remain unclear. Until such a chromosome has risen to high frequency, it will face a significant problem with recombination, as an odd number of intercentromeric crossovers will yield one daughter chromosome with two centromeres and another with none (Figure 5.4), almost certainly leading to nonfunctional gametes and reduced fertility. Under this scenario, a minority chromosome will always be at a selective disadvantage because it will essentially always be present in heterozygous carriers. Thus, successful centromere repositioning requires a mechanism whereby the novel chromosome can rise to a high enough frequency to become the most advantageous state (because the original chromosome will now mostly be found in chromosomal heterozygotes).

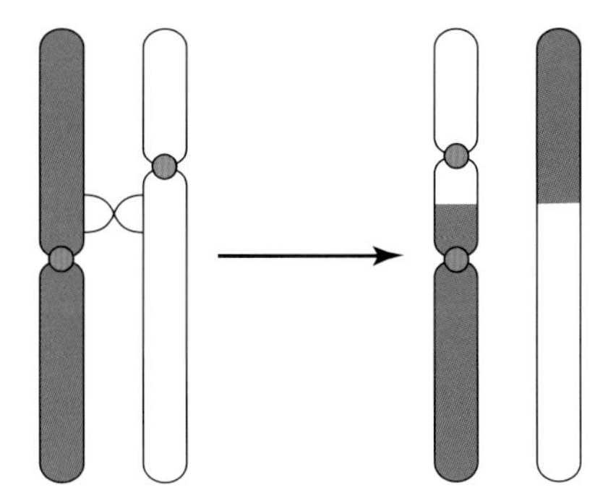

Figure 5.4 The consequences of a crossover (thin connecting lines) in an individual heterozygous for centromere positioning. One daughter chromosome is dicentric, whereas the other is acentric.

In principle, the recombinational load associated with centromere repositioning might be avoided entirely if the centromere moved gradually enough along the chromosome to minimize the chances of crossovers within the intercentromeric region. The early colonization burden could also be overcome during a population bottleneck of significant enough magnitude and intensity that the power of random genetic drift approached or exceeded the selective disadvantage of chromosomal heterozygotes (Lande 1979; Walsh 1982), and such transitions would be facilitated further if neocentromeres were endowed with a driving capacity. Nonetheless, why certain genomes are biased toward one type of karyotypic profile over the other is a mystery.

Centromere definition

Perhaps the central unsolved question regarding centromeres is how they are defined at the molecular level. The emergence of neocentromeres without any obvious rearrangements at the DNA level (e.g., Amor et al. 2004; Heun et al. 2006) implies a de novo origin and the involvement of regional epigenetic modifications of the chromatin, rather than a physical relocation of ancestral centromeric DNA. But how might a heritable and informative set of epigenetic markings be acquired?

As noted above, most centromeric regions contain a high density of lineage-specific tandem repeats and mobile element insertions. Numerous processes can lead to rapid interspecies divergence of tandem arrays of noncoding DNA while maintaining within-array homogeneity (Figure 5.5), and many mobile elements appear to be specialized for residence in centromeric regions, where low gene densities ensure minimal fitness effects of new insertions (Wong and Choo 2004). Thus, a key question is whether the typical contents of centromeric regions are simple reflections of chromosomal environments with exceptional vulnerability to excess DNA accumulation or whether, once established, repetitive DNAs and mobile elements become reliable enough defining features to be exploited for centromere identification.

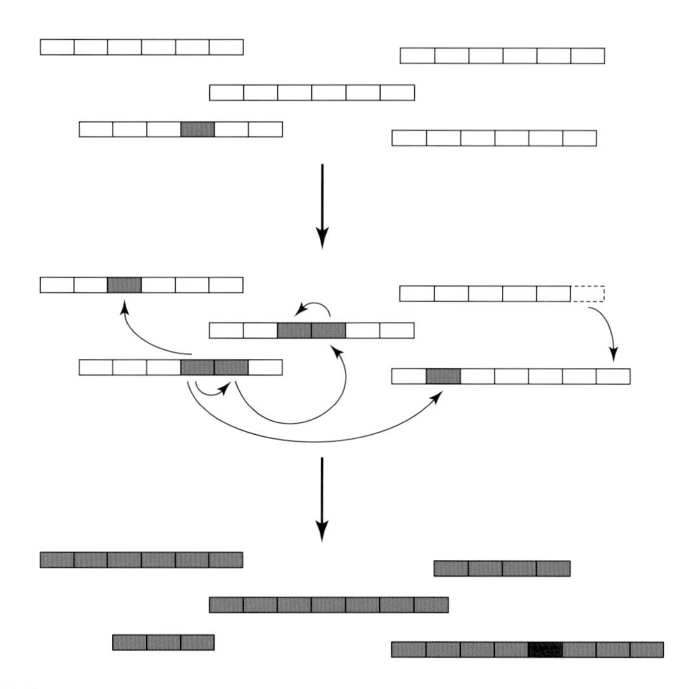

Figure 5.5 Homogenization of a tandem array of repetitive DNAs. A random sample of five chromosomes is illustrated for each of three points in time. *Top*: Initially, all arrays are of equal length, but one has acquired a variant repeat type (green). *Middle*: During replication, various DNA-level mechanisms (gene conversion, unequal crossing over, and replication slippage) can lead to the stochastic gain/loss of various repeat types, as well as expansion/contraction of array lengths. In some cases (e.g., biased gene conversion), one of the repeat types may expand at the expense of another. *Bottom*: Eventually, one of the repeat types will be lost by random processes, but the cycle will continue as new (red) variants arise. This process is often called concerted evolution.

Two sets of observations implicate the involvement of mobile elements in centromere definition. First, the extensive sequence similarity between CENP-B, one of the major centromeric binding proteins, and the transposase of *Tigger* elements (members of the mariner transposon family; see Chapter 7) suggests the possible evolutionary recruitment of centromeric proteins from mobile elements. Because mobile element–encoded transposases bind to their own terminal sequences (see Chapters 3 and 7), the plausibility of this argument is bolstered by the strong homology between the terminal inverted repeat sequences of *Tigger* and the consensus DNA recognition sequence of the CENP-B protein (Kipling and Warburton 1997; Volpe et al. 2002). Second, there is a remarkable similarity between the mechanisms by which centromeres acquire their epigenetic markings and by which mobile elements become silenced in eukaryotic genomes (see Chapter 7). Although the general rarity of protein-coding genes fosters the view that centromeric regions are largely transcriptionally silent, a low level of bidirectional transcription occurs from centromeric sequences in *S. pombe*, including those associated with mobile elements (Volpe et al. 2002; Hall et al. 2003). The double-stranded RNAs resulting from the annealing of the resultant complementary transcripts are subject to cleavage into small interfering RNAs (siRNAs) by the RNA interference (RNAi) pathway (Reinhart and Bartel 2002) (see Chapter 3). These siRNAs then appear to recruit enzymes that methylate centromeric histones specifically in the regions of original transcription, thereby providing an epigenetic signature for centromere location (Sullivan and Karpen 2004; Pidoux and Allshire 2005). Analogous processes appear to occur in plants (e.g., May et al. 2005; Martienssen 2003).

Despite these enticing leads, many questions regarding the epigenetic specification of the centromere remain unanswered, not the least of which concerns the mechanism by which the epigenetic markings of centromeres are stably inherited across cell divisions. One idea is that the tension between sister centromeres imposes a topological change that facilitates the loading of centromeric proteins and their markings (Mellone and Allshire 2003; Henikoff and Dalal 2005). This hypothesis also provides a potentially simple explanation for chromosome repositioning, as an accidental cellular misplacement of the kinetochore assembly would immediately result in a new stable centromeric location. If such accidents were quite local, they might allow for gradual centromeric migrations without substantial negative recombinational effects.

Although many details have been glossed over here, the key point is that the specific contents of a centromeric region may often be of secondary importance with respect to centromeric definition. Instead, provided that an adequate supply of transcription promoters and repetitive sequences is present, as would be the case in a region enriched in mobile elements, the maintenance of centromere location may be largely a self-reinforcing process. If this is the case, then the specific sequences associated with the repetitive tandem arrays commonly found in centromeres may be largely inconsequential, although, as noted above, the near constancy in size of the repeat units (at about 180 bp) implies a significant physical constraint.

The possibility that mobile elements play a key role in centromere definition also provides a potential solution to some of the conceptual challenges confronting the centromeric-drive hypothesis. First, if mobile elements are the governing force, then centromeric sequences do not necessarily evolve so as to take advantage of spindle attachment mechanisms. Rather, by the stochastic expansion and contraction of mobile element families and adjacent tandem repeats, the centromere could simply passively acquire sequence changes that then secondarily drive the evolution of the centromeric proteins. If this hypothesis is correct, then species with limited mobile element activity (generally, those with very large effective population sizes; see Chapter 7) should harbor centromeres with unusually stable contents. Second, mobile element colonization provides a powerful mechanism for rapidly homogenizing the centromeric regions of the full set of a species' chromosomes, thereby minimizing cytogenetic problems that might arise with chromosome-specific centromeres.

Telomeres

Whereas most prokaryotic genomes consist of single circular molecules, all known eukaryotic nuclear chromosomes are linear. This type of chromosomal architecture imposes two substantial challenges. First, chromosome ends are intrinsically vulnerable to gradual loss during replication. Once a replicon reaches the end of a chromosome, there is no more substrate for initiating an Okazaki fragment (see Figure 5.1), so the final nucleotides on lagging strands cannot be replaced without a special mechanism for end sequence replication. Second, as noted in Chapters 2 and 4, eukaryotic chromosomes commonly experience double-strand breaks, which are often repaired by the process of nonhomologous end joining (Lieber et al. 2003; Daley et al. 2005). The accidental fusion of true chromosome ends by this repair pathway would lead to immediate genomic instability, as some chromosomes would have multiple centromeres and others none. Thus, a mechanism is required to distinguish telomeres from chromosome breaks. Muller (1938) demonstrated the existence of such protective properties by showing that chromosomes without telomeres are never recovered after double-strand breaks are imposed in *Drosophila*.

Eukaryotic telomeres generally consist of repetitive DNA sequences, with one strand of the duplex extending tens to hundreds of base pairs beyond the other. In many species (including vertebrates, trypanosomes, and ciliates, but apparently not yeasts), this single strand invades the upstream double-strand repeats to form a loop (Murti and Prescott 1999; de Lange 2004). Such structures, along with the key proteins that bind them, play a central role in end sequence identity and the maintenance of telomere integrity, and their absence generally leads to rapid telomere loss and/or chromosome fusion (Baumann and Cech 2001; Loayza and de Lange 2003; Pardo and Marcand 2005; Shakirov et al. 2005).

The telomeric arrays of unicellular species tend to be fairly short (less than a few hundred bp), whereas those of animals and plants are generally 5–150 kb long (Louis and Vershinin 2005). Such arrays often consist of a wide variety of repeat types, although some commonalities have emerged. Telomeric repeat units are invariably very G+T rich and often small in size. For example, GGGTTA is generally used in vertebrates and some filamentous fungi (Pryde et al. 1997); GGGGTT or GGGGTTTT in ciliates (Murray 1990); GGGTTCA in *Plasmodium* (Carlton et al. 2002); GGTTA(A) in some arthropods (Pryde et al. 1997; Sahara et al. 1999); GGCTTA in *C. elegans* (Wicky et al. 1996); and GGGTTTT(TT)A in *Chlamydomonas* (Hails et al. 1995). This widespread reliance on G+T-rich telomeric repeats invites comparisons to the situation noted above for leading versus lagging strands of DNA. Because exposure of DNA in the single-stranded state imposes strong mutation pressure in the G/T direction, the commonalities in the typical telomeric repeats noted above may be as much a product of shared mutational forces as shared inheritance. A+C-rich telomeres are expected to be highly unstable.

Despite the fairly high degree of conservation of telomeric repeat sequences across a diversity of distantly related lineages, small repeat units are by no means universal. For example, although most plants utilize either GGGTTTA or GGGTTA telomeric repeats, a number of monocot lineages rely on larger, ill-defined sequences (S. P. Adams et al. 2001; Fajkus et al. 2005). The telomeres of the fungus *Ashbya gossypii* consist of 24 bp G+T-rich repeats (Dietrich et al. 2004), and numerous variants are found in other fungi (Teixeira and Gilson 2005). The telomeres of the fly midge *Chironomus* consist of 150–350 bp repeats, the exact sequence of which differs significantly among species (Cohn and Edstrom 1992), and those of the slime mold *Dictyostelium* are even more dramatically divergent, consisting of rDNA-like repeats (Eichinger et al. 2005).

In most species, end sequence maintenance is accomplished by the telomerase enzyme, which is part protein and part RNA. The RNA component contains a sequence complementary to the telomeric repeat, which enables it to bind to single-stranded ends and repetitively prime for the addition of the telomeric repeat sequences by a slippage mechanism (Figure 5.6). Observations on human somatic cells, most of which do not express this enzyme, powerfully demonstrate the critical contribution of telomerase to end sequence maintenance. Such cell lineages experience progressive telomere loss, up to 200 bp per cell division, eventually succumbing to cellular senescence. Mammalian telomere elongation ordinarily occurs only during early embryogenesis (Schaetzlein et al. 2004).

Given that chromosome ends in vertebrate cancer cells are maintained by the activation of an alternative telomere maintenance pathway (Teng and Zakian 1999; Lundblad 2000) and that linear bacterial and viral chromosomes also maintain their integrity without telomerase (Kobryn and Chaconas 2001), it is clear that the earliest eukaryotic chromosomes need not have been telomerase-dependent. On the other hand, the wide phylogenetic distribution of this enzyme suggests an early takeover, quite likely in the stem eukaryote itself. The telomerase protein appears to be related to the reverse transcriptase of non-LTR retrotransposons (Eickbush 1997), suggesting the possibility that the ancestral telomerase was a domesticated mobile element. But the enzymatic features of telomerase activity raise another intriguing issue. If reverse transcriptase is essential for chromosome maintenance, then residing within every telomerase-containing eukaryote is a source of the key protein for retrotransposon proliferation. Can this kind of takeover operate in reverse; that is, can the duplication and divergence of telomerase genes lead to the origin of new retrotransposon families?

A remarkable case for a retrotransposon–telomerase connection comes from the genus *Drosophila*, which appears to be devoid of telomerase, relying instead on arrays of two telomere-homing non-LTR retrotransposon families, HeTA and TART (Sheen and Levis 1994; Pardue and DeBaryshe 2003;

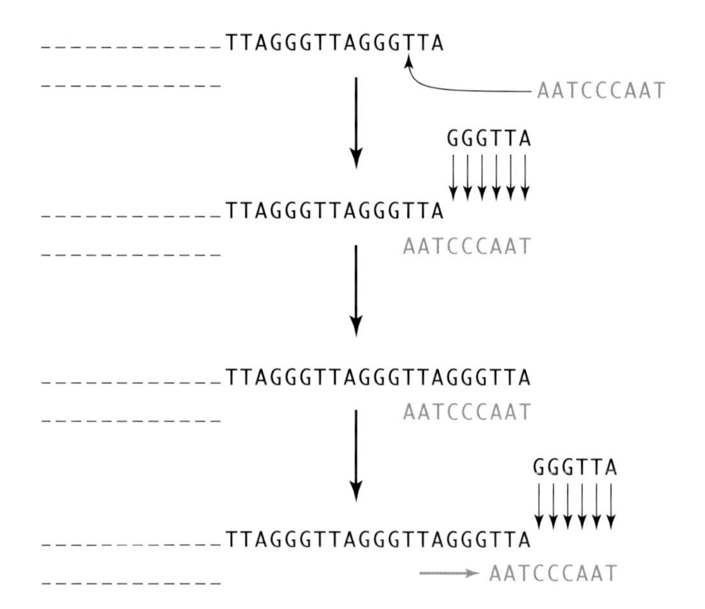

Figure 5.6 Maintenance of a telomere, consisting of GGGTTA repeats, by repeated primer extensions. *Top*: After conventional replication, the chromosome (black) has a single-strand overhang, to which a telomerase containing a complementary RNA template (orange) is recruited. *Middle*: The top strand is extended by the addition of bases complementary to the RNA template. *Bottom*: The template dislodges and slides to the right, and the process is repeated. The number of extension cycles is regulated by cellular processes. (From Chan and Blackburn 2004.)

Abad et al. 2004; Pardue et al. 2005). Both types of elements cooperate in telomere maintenance in a way that is quite similar to the telomerase pathway—the transcripts of each element are diverted to chromosome tips, where they serve as primers for new insertions. Only TART is autonomous, but although HeTA does not encode reverse transcriptase, it does appear to be primarily responsible for the homing process. Phylogenetic evidence suggests an origin of this liaison prior to the divergence of members of the *Drosophila* genus (Pardue and DeBaryshe 2003), and Pardue et al. (2005) make a plausible case that the entire system is derived from telomerase (which is present in many other arthropod genera).

Any attempt to understand telomere evolution must confront the same kinds of issues discussed above with respect to centromeres. Because end maintenance is critical to chromosome survival, chromosomes with telomeres better able to sequester telomerase should experience greater longevities. This idea raises the specter of a potential telomeric drive situation, with a tight coevolutionary loop between telomeric repeat structure and telomerase sequence. However, this issue is further complicated by the presence of the RNA component in telomerase. With each telomere being grown from the same RNA template, telomere homogeneity would seem to be ensured, thereby inhibiting a drive process unless the RNA template itself changed (which, in principle, could cause a rapid transformation of all telomeres, but probably would cause major chromosome maintenance problems until established).

Consistent with this view, most of the rare radical alterations in telomeric constitutions that have occurred across phylogenetic lineages appear to have been followed by long periods of stability. Quite contrary to the situation noted above for centromeres, all species within a genus, or even a family, tend to have very similar telomeric repeat units. Only a few examples of within-species variation among telomeric sequences have been found; for example, on the linear mitochondrial chromosomes of *Tetrahymena* (Morin and Cech 1988) and on the autosomes of *Chironomus* (C. C. Lopez et al. 1996). One mechanism that might initiate a genome-wide shift in telomeric structure with few intermediate negative repercussions is duplication of the telomerase molecule itself, as this might allow one paralog (and/or its RNA moiety) to gradually acquire features leading to the displacement of the ancestral telomeric sequence. As in the case of centromeric drive, this is clearly an area of evolutionary genomics that warrants further population genetic analysis and detailed comparative study.

6 The Nucleotide Composition Landscape

If the nucleotide composition of DNA were totally driven by mutation and the rates of all twelve base-substitution mutations (e.g., A→C, A→G, ..., T→G) were identical, the genome-wide frequencies of all four nucleotides would equal 0.25 on both strands. This is virtually never the case (Sueoka 1962). Because of the Watson–Crick nature of double-stranded DNA (G binding with C, and A with T), when both strands are counted, the genome-wide numbers of Gs and Cs must be equal, and the same is true for As and Ts. However, this is the only universal form of parity in nucleotide usage. For example, the fractional A+T content of prokaryotic genomes ranges from about 0.25 to about 0.85 (Figure 6.1); species with A+T content greater than 0.70 tend to be intracellular pathogens or endosymbionts, whereas those with the lowest values are much more functionally diverse. The A+T contents of the genomes of unicellular eukaryotes are also wide-ranging (0.40–0.80), but with no obvious relationship to lifestyle. The genomes of animals and land plants are almost invariably A+T rich, although not greatly so, falling in the narrow range of about 0.50–0.70. In addition to these dramatic levels of variation in overall nucleotide composition among species, within-strand asymmetry with respect to complementary bases is common (i.e., A ≠ T and/or G ≠ C), and some chromosomes exhibit substantial regional heterogeneity in nucleotide composition.

In principle, such disparities in nucleotide usage could be consequences of biases driven by natural selection. However, before invoking adaptive explanations for the patterning of nucleotide composition, it is appropriate to consider the physical processes associated with mutation and recombination alone. We will first examine the directional effects of mutation, which commonly (if not universally) impose pressure toward A+T content. Opposing this baseline mutational pattern, gene conversion activity asso-

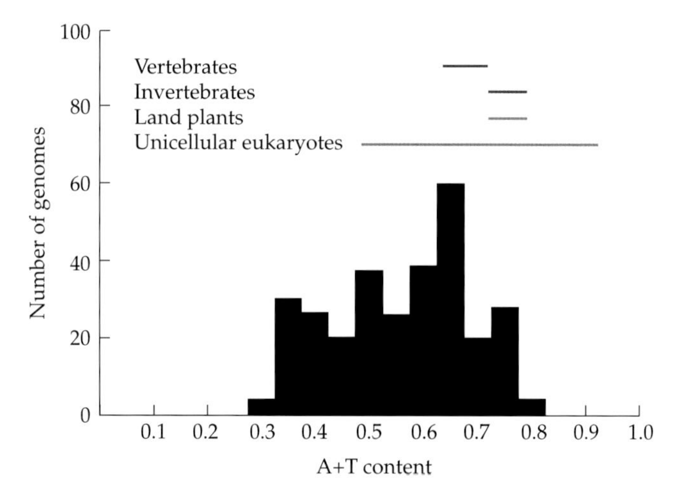

Figure 6.1 Distribution of genome-wide A+T content for prokaryotes with completely sequenced genomes (histogram) and ranges for eukaryotes (colored lines). The prokaryotic mean and SD are 0.54 and 0.13 (sample size = 293). Note that because of the Watson–Crick base complementarity of As and Ts, the A+T content is identical for both strands.

ciated with recombination is generally biased in the direction of G+C content. However, unlike mutation, which operates independently of the genotypic background, gene conversion occurs only in heterozygous individuals and so is most effective in large populations harboring high levels of nucleotide variation (see Chapter 4). After considering the relative strengths of these two physical processes, which can be substantial, we will examine the potential role of selection, which becomes especially relevant in the context of protein-coding genes. Because the net effects of physical forces operating on nucleotide composition may be either parallel to or opposite to the directional forces of selection, an understanding of the determinants of nucleotide composition is fundamental to the full spectrum of issues in genomic evolution.

The Molecular Spectrum of Mutational Effects

Natural cellular processes can induce a wide spectrum of base modifications (premutations) that, if unrepaired at the time of replication, ultimately result in inherited mutations. Although twelve types of base-substitution mutations are possible, observations on *E. coli*, yeast, and human cell cultures suggest that a small subset of the possible changes dominate the distribution of mutational types.

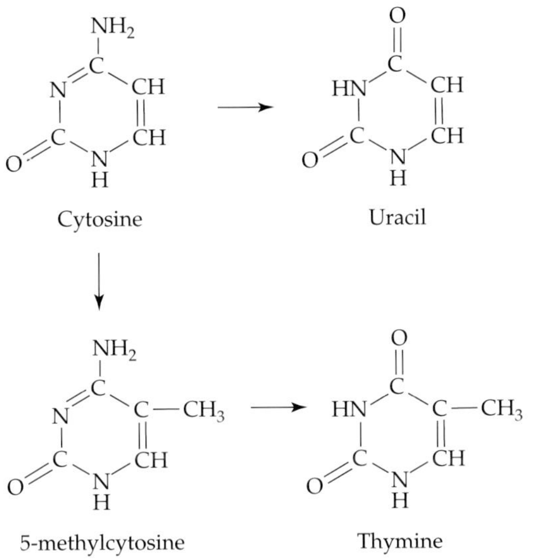

Cytosine

Uracil

5-methylcytosine

Thymine

Figure 6.2 *Top*: Deamination of cytosine produces uracil, which becomes thymine after postreplicative repair. *Bottom*: Deamination of 5-methylcytosine leads directly to thymine.

One of the most common sources of mutation is spontaneous deamination of cytosine or 5-methylcytosine, which promotes C:G→T:A transitions (Figure 6.2), where the colon denotes a bond between DNA strands. In a number of species (mammals and land plants in particular), the cytosines in CG dinucleotides (a C followed by a G on the same strand) are frequently methylated and dominate the total mutational landscape, having mutation rates 2–10 times greater than other sites (Ehrlich and Wang 1981; Krawczak et al. 1998; Siepel and Haussler 2004; Hwang and Green 2004; Huang et al. 2005; Arndt et al. 2005; Mikkelsen et al. 2005; Morton et al. 2006). A CG on one strand will be paired with a GC on the other, so depending on the strand on which C is methylated, a CG doublet will be converted to TG or CA. Although such changes are commonly initiated on single-stranded DNA exposed during replication or transcription, leading to strand asymmetry in G and C content (see below), they can also arise symmetrically from both strands as DNA "breathes" during normal cellular processes. Moreover, the CG mutation rate is higher in A+T-rich regions (Fryxell and Zuckerkandl 2000; Fryxell and Moon 2005; Morton et al. 2006). This effect presumably arises because G and C are linked by three hydrogen bonds, whereas A and T are linked by two, rendering A+T-rich regions less stable and more likely to transiently enter a single-stranded state.

A second common source of premutations, the oxidative conversion of guanine to 8-oxo-guanine (GO), encourages mutations in two ways (Grollman and Moriya 1993; Miller 1996; Wang et al. 1998). First, when a resident GO base is present at the time of replication, an A can be incorporated into the replicating complementary strand (Figure 6.3). Although an A:GO mismatch can still be repaired by first replacing the A with a C and then restoring a C:G state by base excision repair, GO can also be replaced by T, lead-

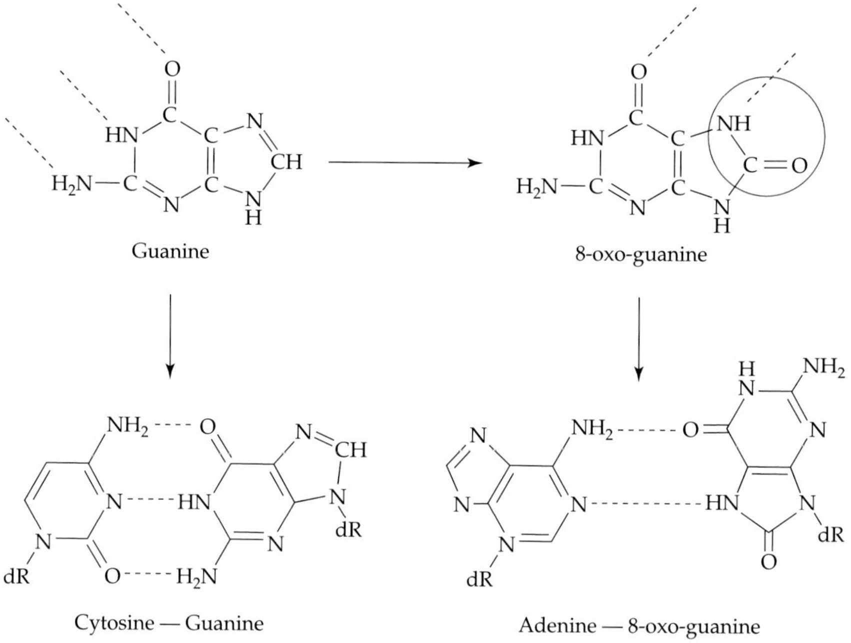

Guanine

8-oxo-guanine

Cytosine — Guanine

Adenine — 8-oxo-guanine

Figure 6.3 Oxidation of guanine to 8-oxo-guanine. Like guanine, 8-oxo-guanine can bond with cytosine, but it bonds much more frequently with adenine (as shown at the bottom right). The red circle demarcates the chemical modifications of guanine, and the dotted lines denote hydrogen bonds. dR denotes a deoxyribose in the sugar–phosphate backbone of the DNA molecule.

ing to an inherited C:G→A:T transversion. Second, GOs can be directly incorporated into replicating strands of DNA when a GO base from the precursor nucleotide pool is inserted across from an A or a C on the template strand, encouraging A:T→C:G transversions. The relative rates of these two types of GO-induced transversions are unknown, but because the first mechanism occurs in both replicating and nonreplicating DNA, it may predominate. Although the rate of origin of GO bases in the germ line is unknown, the ratio of GOs to Gs in the DNA of mammalian somatic tissues is about 10^{-7} to 10^{-5}, depending on the age of the individual (Helbock et al. 1998; Hamilton et al. 2001), so the opportunity for GO-induced mutations may be quite high.

Although the preceding observations suggest two potent biochemical mechanisms that could bias the prevailing mutation pressure in the direction of A+T richness, the full spectrum of mutations potentially depends on numerous additional types of DNA error propagation. An unbiased perspective on the mutational spectrum of effects requires direct observations of heritable changes at the DNA level prior to the operation of natural selection, but owing to the extreme rarity of mutations, such analyses are nearly nonexistent. However, several attempts have been made to indirectly derive the mutational spectrum from laboratory studies of reporter constructs or from phylogenetic analyses of sequence changes in pseudogenes (assumed to be neutral). Despite their mixed nature and potential for bias, these studies have yielded a fairly consistent set of results. For example, across prokaryotes, yeast, and several animals, G/C→A/T mutations (where the slash denotes "or") outnumber those in the opposite direction by factors of 1.6–2.8, with G:C→A:T transitions almost always being the most common substitution type (Table 6.1). Concordant observations have been obtained from phylogenetic comparisons in numerous species (McVean and

TABLE 6.1 Distribution of base-substitution mutations estimated from direct observations of reporter constructs (rc) or phylogenetic analyses of pseudogenes (pa)

SPECIES:	MP	EC	SC	CG	HS	HS	DM
	(pa)	(rc)	(rc)	(rc)	(rc)	(pa)	(pa)
Transitions:							
A:T→G:C	0.14	0.12	0.13	0.13	0.13	0.17	0.18
G:C→A:T	0.42	0.44	0.24	0.31	0.38	0.37	0.30
Transversions:							
A:T→C:G	0.07	0.14	0.05	0.19	0.12	0.09	0.10
A:T→T:A	0.05	0.13	0.10	0.05	0.14	0.14	0.09
G:C→C:G	0.16	0.06	0.22	0.12	0.14	0.10	0.13
G:C→T:A	0.16	0.11	0.26	0.20	0.09	0.13	0.20
A+T bias:							
Numbers	2.76	2.11	2.77	1.59	1.88	1.92	1.78
Rates	3.81	2.12	4.54	2.21	2.70	2.77	2.57
A+T genome content:							
Observed	0.58	0.50	0.62	0.58	0.59	0.59	0.59
Expected	0.79	0.68	0.82	0.69	0.73	0.73	0.72

Sources: *Mycobacterium leprae* (Mp): Mitchell and Graur (2005); *Escherichia coli* (Ec): Schaaper and Dunn (1991), Sargentini and Smith (1994), Yamamura et al. (2000, 2002), Wolff et al. (2004); *Saccharomyces cerevisiae* (Sc): Kunz et al. (1998), Ohnishi et al. (2004); *Cricetus griseus* (hamster, Cg): Nalbantoglu et al. (1987), de Jong et al. (1988), Zhang et al. (1992), Xu et al. (1995), Kimura et al. (1998); *Homo sapiens* (Hs, rc): Podlutsky et al. (1998); *Homo sapiens* (Hs, pa): Petrov and Hartl (1999); *Drosophila melanogaster* (Dm): Petrov and Hartl (1999).

Note: A:T and G:C denote Watson–Crick base pairs across strands. Observed frequencies of each type of mutation, which sum to one for each species, are functions of both the number of sites with a particular parental type and the mutation rate conditional on that type. The A+T number bias is the ratio of observed G/C→A/T to A/T→G/C mutations; e.g., for Mp, (0.42 + 0.16)/(0.14 + 0.07) = 2.76. The A+T mutation rate bias is the ratio of observed G/C→A/T to A/T→G/C mutations, each weighted by the abundance of source nucleotides (given in the second from the last row), which gives an estimate of the relative mutation rates in both directions (m in the text); e.g., for Mp, [(0.42+0.16)/0.42]/[(0.14+0.07)/0.58] = 3.81. The expected A+T content is computed using the expression in Box 6.1.

Vieira 2001; Green et al. 2003; Hudson et al. 2003; Arndt et al. 2005; Singh et al. 2005; Morton et al. 2006). Single-site analyses do not fully capture the mutational substitution spectrum, as the changes at individual sites can depend on the immediately surrounding nucleotides (Siepel and Haussler 2004; Hwang and Green 2004; Arndt et al. 2005), but with the available data, it is hard to escape the conclusion that the prevailing mutation pressure is generally in the direction of A+T. Indeed, no species has yet been identified in which mutation pressure favors G+C richness.

While suggesting that the spectrum of substitution mutations may be roughly similar across diverse life forms, the results in Table 6.1 are quite peculiar in another respect. For a genome that has reached an equilibrium nucleotide composition driven by mutation alone, the numbers of G/C→A/T and A/T→G/C mutations should be exactly equal. Thus, the excess mutations in the direction of A/T imply that each of the genomes in Table 6.1 has a substantially lower A+T composition than expected under a process entirely driven by mutations. The equilibrium A+T composition is a function of the relative rates of the two classes of mutations (Box 6.1). For example, in yeast, the ratio of mutation rates involving G/C→A/T and A/T→G/C changes is 4.54 (see Table 6.1), which, when applied to Equation 6.1 in Box 6.1 yields an expected equilibrium A+T content of 4.54/(4.54 + 1.00) = 0.82, whereas the actual genomic content is 0.62. Given the diversity

Box 6.1 Equilibrium Nucleotide Composition under a Mutation-Driven Process

Here we consider the expected A+T composition of a genome under the assumption that the only relevant processes are the relative mutation rates from A or T to G or C, and vice versa. Because we are considering the entire nucleotide content of the genome on both strands, and because each A is balanced by a T on the opposite strand (and similarly for the other bases), it is sufficient to lump As with Ts and Gs with Cs. For this same reason, A↔T and G↔C mutations are irrelevant to overall A+T composition. We further assume that the mutation rate per site is independent of the status of neighboring sites.

Suppose the mutation rate from A/T to G/C is u, whereas that from G/C to A/T is v. Letting p be the current frequency of A+T, the rate of change in p caused by mutation is equal to $(1 - p)v - pu$. At equilibrium, the number of G/C→A/T mutations, $(1 - p)v$, must equal the number of A/T→G/C mutations, pu, so the equilibrium A+T composition is obtained by solving $(1 - p)v - pu = 0$,

$$\tilde{p} = \frac{v}{(u+v)} = \frac{m}{(1+m)} \tag{6.1}$$

where $m = v/u$ is the mutational bias toward A+T. Because we have assumed neutral sites, $\tilde{p}$ can be interpreted here as the probability that a particular site is of type A or T or as the expected A+T content of the entire genome. In the absence of mutational bias ($m = 1$), the equilibrium A+T content is 0.5. As shown in Table 6.1, m generally appears to be in the range of 2.1–4.5.

of systems analyzed, this consistent pattern suggests the existence of some general force operating in the opposite direction; that is, favoring newly arisen G/Cs more than A/Ts. A leading candidate for such a force is the subject of the next section.

Biased Gene Conversion

Aside from their involvement in codons, it is difficult to imagine a universal selective advantage for G/C nucleotides, unless heteroduplex stability is involved. However, there is a potentially powerful nonadaptive mechanism that might promote such changes: biased gene conversion. Although it is common to think of recombination as a process of reciprocal exchange (crossing-over), the point of recombination almost always experiences a patch of nonreciprocal exchange (Figure 6.4). Recombination events are initiated when a broken strand of DNA is repaired by using the homologous region of an intact chromosome as a template for strand resynthesis. If the sequences of the homologous chromosomes differ in this region, gene conversion will result as mismatched (noncomplementary) nucleotides between invading and recipient strands are restored to Watson–Crick pairs by the

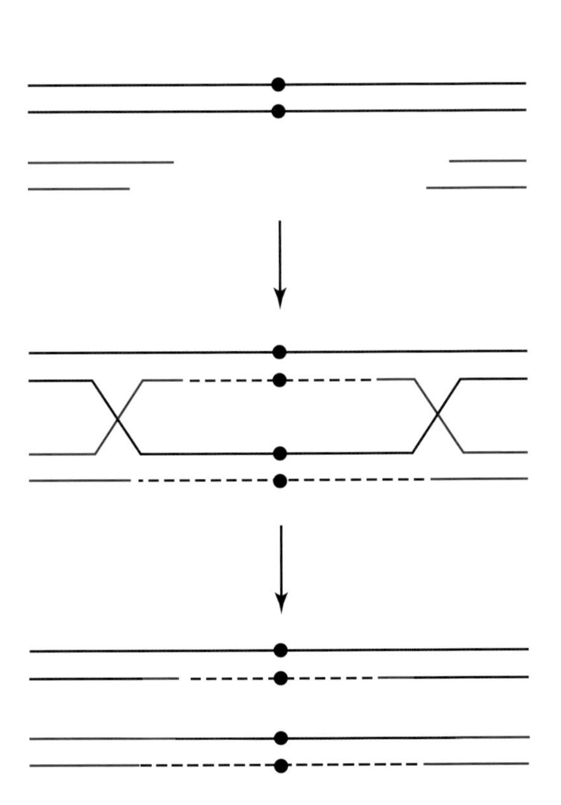

Figure 6.4 The double-strand break repair model of recombination. Parallel sets of lines of the same color denote the two complementary strands of DNA. *Top*: The broken red chromosome has been resected by an exonuclease. The black dots denote a nucleotide position on the black chromosome that had a different state on the red chromosome. *Middle*: Using sequence homology as guide, the red chromosome invades the black chromosome, forming a heteroduplex and utilizing the latter as a template for strand resynthesis (dashed lines). *Bottom*: The intertwined chromatids are cut and rejoined, resulting in the restoration of two complete chromosomes. Note that the red chromosome has acquired the black sequence from the homologous chromosome (the gene conversion tract). In addition, the restored chromosomes may contain small regions of heteroduplex DNA (complementary strands of black and red), which may contain base mismatches resulting from additional nucleotide heterozygosity between the parental chromosomes; in that case, mismatch repair may lead to further gene conversion. Depending on how the strand junctions are cut, the nonreciprocal conversion may or may not be accompanied by reciprocal exchange (crossing-over), although only a noncrossover event is shown here. (Modified from Szostak et al. 1983.)

mismatch repair pathway. This double-strand break repair model describes just one of several physical mechanisms of recombination (Haber 2000), but all result in gene conversion regardless of whether recombination is accompanied by crossing-over.

If the points of chromosome breakage were entirely independent of local nucleotide composition and the repair of mismatches in heteroduplex DNA were entirely random, gene conversion would be unbiased. That is, although individual conversion events would cause nonreciprocal exchange among homologous chromosomes, there would be no tendency for a particular sequence to drive itself through the population. However, a substantial amount of evidence from fungi and animals indicates a general conversion bias toward G/C. The results come from both direct observation (Brown and Jiricny 1988; Bill et al. 1998; Birdsell 2002) and assessments of allele frequency distributions showing that presumptively neutral A/T→G/C mutations are capable of rising to higher frequencies than G/C→A/T mutations (Webster et al. 2003; Webster and Smith 2004; Galtier et al. 2006). A reflection of this behavior is seen in yeasts, in which there is a strong positive correlation between the recombination rate within a chromosomal region and the G+C content at silent sites (Birdsell 2002). Such correlations have also been seen in nematodes and *Drosophila* (Marais et al. 2001; Marais and Piganeau 2002; Singh et al. 2005), humans (Fullerton et al. 2001; Kong et al. 2002), and mice (Huang et al. 2005).

A number of observations support the idea that recombinational activity is driving G+C composition, and not the other way around. First, in humans, the recombination rate of a region is much more highly correlated with the types of mutations being produced than with the current composition of the region (Meunier and Duret 2004). Second, large, recombinationally quiescent genomic regions, such as Y chromosomes and the fourth chromosome of *D. melanogaster*, tend to be very A+T rich (Perry and Ashworth 1999; Marais 2003). Third, large repetitive DNAs that engage in frequent gene conversions, such as those encoding rRNAs and tRNAs, tend to have unusually high G+C contents (Galtier et al. 2001), as do small chromosomes (Webster et al. 2006), which tend to experience more recombination per nucleotide site because of the near constancy of the number of crossover events per chromosome (see Chapter 4).

Can biased gene conversion be a powerful enough force to alter the equilibrium nucleotide composition of a genome relative to that expected with biased mutation alone? To answer this question, two fundamental issues must be considered: the power of the physical pressure of gene conversion and the sufficiency of the population genetic context for the operation of the biasing process. There is little question that conversion tracts arise at a high rate relative to that of mutation. The recombination rate per base pair associated with crossing-over is often on the order of the mutation rate (see Chapter 4), and there are several reasons to expect the rate of conversion to be substantially higher. First, although all recombination events are accompanied by gene conversion (only those involving nonidentical sequences having consequences), just a fraction of recombination events result in crossing-over: about 25% in *S. cerevisiae* (Malkova et al. 2004), 9% in *Arabidopsis* (Copenhaver et al. 2002; Haubold et al. 2002), 15% in *Drosophila* (Hilliker et al. 1991), and 12% in humans (Frisse et al. 2001). Second, meiotic gene conversion tract lengths appear to follow a geometric distribution of the form $p(n) = \phi^n$, where n is the length of the tract in nucleotides and ϕ is the per-base probability of extending a conversion tract (Ahn et al. 1988; Hilliker et al. 1994). Under this distribution, the probability of a particular tract length declines with n, but because ϕ is generally very close to 1.0, the average tract length, $\phi/(1-\phi)$, is generally quite high: about 400 bp in eubacteria (Santoyo et al. 2005), 500 bp in *S. cerevisiae* (Ahn and Livingston 1986), and 350 bp in *Drosophila* (Hilliker et al. 1994). Third, gene conversion can occur during mitotic as well as meiotic divisions (Ahn et al. 1988; Mézard et al. 1992; Sugawara and Haber 1992; Gloor 2002; Omilian et al. 2006), and the average tract lengths of mitotic conversions can be 10 times larger than those arising during meiosis (Judd and Petes 1988). The combination of these three factors suggests that conversion rates per nucleotide site must commonly be at least a thousand times greater than mutation rates, which is consistent with direct estimates of conversion rates of about 10^{-5}/site/generation in multicellular species and as high as 10^{-2} in yeasts (Marais 2003).

The key to understanding the ability of biased gene conversion to drive genomic nucleotide composition away from the expectation under mutation alone is the realization that although gene conversion is an internal physical process like mutation, it is also a selection-like process, whose operation is entirely dependent on the presence of heterozygotes. The power of biased gene conversion can be summarized in the following way. Letting κ be the conversion rate and β be the frequency of G/C transmission from a heterozygote (containing a G or C at a position on one chromosome and an A or T at the same position on the homologous chromosome), the change in G/C frequency $(1-p)$ through a single heterozygote is $\kappa(\beta - 0.5)$, where $(\beta - 0.5)$ is the deviation of the segregation ratio from the random expectation of 0.5. The change at the population level, through the entire pool of heterozygotes (assumed to be in Hardy–Weinberg equilibrium), is $\kappa(\beta - 0.5)[2p(1-p)]$. Thus, if newly arisen mutations with conversion advantages are to have an influence on the nucleotide composition of a population, they must remain segregating in the population for a sufficiently long time to experience conversion events within heterozygous individuals. If the population size is too small, so that most newly arisen mutations are lost or fixed faster than the time scale of conversion ($\sim 1/\kappa$ generations), the process will be ineffective.

These verbal arguments can be made more explicit by taking advantage of the observation that biased gene conversion has an influence on the dynamics of new mutations that is indistinguishable from that of natural selection (Nagylaki 1983; Walsh 1983), which allows the formal use of the theory developed in Chapter 4 for mutations with effects on fitness. In this case, the selection coefficient is $s = \kappa(2\beta - 1)$, which is simply twice the expression derived in the preceding paragraph. Recalling the theory introduced in Chapter 4, mutant alleles with conversion advantages or disadvantages will behave in an essentially neutral fashion if $|2N_g s| << 1$, where N_g is the effective number of genes at the locus in the population. Thus, regardless of the absolute intensity of biased gene conversion, nucleotide composition will approach the expectations under mutational pressure alone if $|\kappa(2\beta - 1)| << 1/(2N_g)$.

Unfortunately, direct empirical estimates of $\kappa(2\beta - 1)$ are scarce except in fungi, which have unusually high recombinational activity (see Chapter 4). In *S. cerevisiae*, for which the data are most reliable, $\kappa(2\beta - 1)$ has an average value of 0.004 during meiotic cell divisions (Lamb 1998). If κ scales with genome size in the same manner as the crossing-over rate (see Figure 4.6), then assuming that the average value of $(2\beta - 1)$ is roughly comparable

Box 6.2 Equilibrium Nucleotide Composition under the Joint Pressures of Mutation and Gene Conversion

As in Box 6.1, we consider the expected A+T composition of a genome in the absence of natural selection, but here we allow for both differential mutation rates and biased gene conversion. Again, we denote the mutation rates from A/T to G/C as u and from G/C to A/T as v, and their ratio as $m = v/u$ (the mutational bias toward A+T). As discussed in the text, biased gene conversion operates like selection by causing heterozygotes to segregate alleles at ratios deviating from 1:1, so now both the rates of origin of different mutations and their subsequent probabilities of fixation may differ. We assume here that As and Ts convert to Gs and Cs by recombination in an equivalent manner, and vice versa.

From Chapter 4, the probabilities of fixation of newly arisen mutations with conversion advantages and disadvantages are, respectively,

$$p_f^+ = \frac{S/(2N)}{1 - e^{-S}}$$

$$p_f^- = -\frac{S/(2N)}{1 - e^{S}}$$

where $S = 2N_g s$, with N_g being the effective number of genes at a locus, N being the absolute population size of reproductive adults, and s being the selection coefficient [equivalent to $\kappa(2\beta - 1)$, where the conversion rate is κ and the conversion bias is $(2\beta - 1)$, as described in the text]. Here we assume that G/C bases have a conversion advantage over A/T bases, which is consistent with most observations, although the following derivations are generalizable by simply shifting the sign on S.

To obtain the equilibrium A+T composition, $\tilde{p}$, we must find the frequency at which A/T→G/C fixations are exactly balanced by G/C→A/T fixations. At this point, the number of A/T→G/C mutations arising per generation is $2N\tilde{p}u$, as there are $2N\tilde{p}$ A/T alleles in the diploid population, each mutating to G/C at rate u, and each of these has a probability of fixation of p_f^-. Similarly, $2N(1 - \tilde{p})v$ G/C→A/T mutations arise per generation, each with a across species, the average $\kappa(2\beta - 1)$ for multicellular eukaryotes is expected to be in the range of 10^{-7}–10^{-5}. This rough calculation is consistent with κ being about 10^{-5} in multicellular species (as noted above) and with the conversion bias toward C-G or G-C bonds for some mismatches (C-A, T-G, and G-T) being as high as $\beta = 0.7$ in some cases (Bill et al. 1998).

The expected equilibrium A+T content, taking into consideration the joint pressures of mutation, random genetic drift, and gene conversion, is derived in Box 6.2. Even with $2N_g s$ as small as 1.0, biased gene conversion can substantially reduce the equilibrium A+T composition relative to the situation expected with mutation alone (Figure 6.5). However, with N_g on the order of 10^4–10^6 for multicellular eukaryotes (see Chapter 4) and s probably in the range of 10^{-7}–10^{-5}, this high a level of $2N_g s$ may only rarely be exceeded in such species. This may explain why the A+T content of nearly all animal

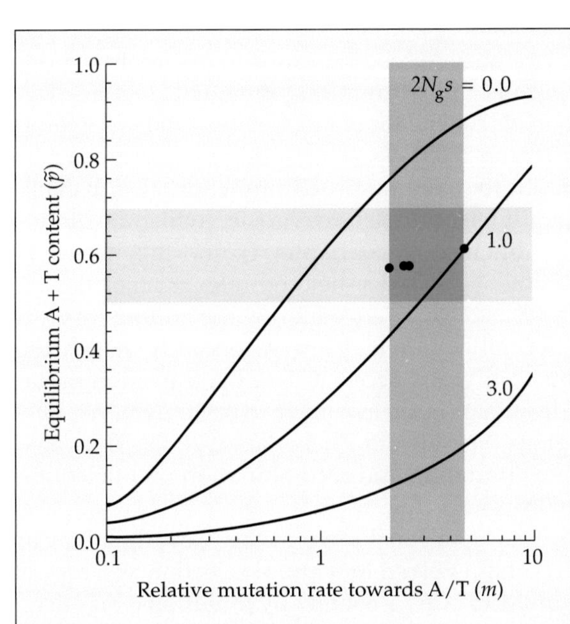

Figure 6.5 The equilibrium A+T compositions of genomes under the joint forces of differential mutation pressure (m), biased gene conversion (s), and random genetic drift ($1/N_g$), given for three values of $2N_g s$ applied to Equation 6.2. The upper curve denotes the situation in which gene conversion is unbiased, such that the equilibrium genome content is determined solely by mutation pressure, as described in Box 6.1. The vertical green bar denotes the range of observed estimates of m from Table 6.1, and the horizontal blue bar denotes the range of A+T composition in observed animal and land plant species from Figure 6.1. The four points are taken from the data on m and p for eukaryotes in Table 6.1, all animals except for the farthest point to the right for the yeast *S. cerevisiae*.

fixation probability of p_f^+. The equilibrium A+T composition is then obtained by solving $2N\tilde{p}u \cdot p_f^- = 2N(1 - \tilde{p})v \cdot p_f^+$. Using the useful relationship

$$\frac{p_f^+}{p_f^-} = e^s$$

the equilibrium A+T composition is found to be

$$\tilde{p} = \frac{me^s}{me^s + 1} \tag{6.2}$$

which reduces to $\tilde{p} = m/(m+1)$ when conversion is unbiased or ineffective ($S << 1$).

and land plant genomes is within the range of expectations for $0.0 < 2N_g s < 1.0$, assuming the mutational biases outlined in Table 6.1 (Figure 6.5). The expectations for unicellular eukaryotes are less clear, because although such species may have substantially higher values of $s = \kappa(2\beta - 1)$ than multicellular species during meiotic divisions, unicellular species also often have prominent haploid and/or clonal phases of reproduction during which s is likely to be very low. Even the lowest observed A+T composition in unicellular eukaryotes, about 0.40, is compatible with $2N_g s$ only slightly greater than 1.0 (see Figure 6.5).

These theoretical results, combined with the existing empirical observations on mutation and gene conversion biases, suggest that subtle variation in these features and in the power of random genetic drift can account for a substantial fraction of the variation in genome-wide nucleotide compositions among species, as well as the general tendency for A+T content to be lower than that expected on the basis of mutation pressure alone. Birdsell (2002) further argues that the joint operation of mutation and biased conversion might explain the tendency for coding exons to be higher in G+C content than intervening introns and surrounding intergenic regions (Kliman and Hey 1994; Green et al. 2003; Zhang and Gerstein 2003; Comeron 2004; Arndt et al. 2005; Mitchison 2005). Because coding regions are generally under selective constraint to maintain protein function, they usually have relatively low levels of nucleotide heterozygosity, which facilitates homology-dependent recombination (see Chapter 4), and this in turn is expected to drive the G+C content upward as heteroduplex mismatches are converted in favor of G/Cs. Moreover, by encouraging the buildup of G+C content in coding regions, recombination may also secondarily modify the mutational spectrum because, as noted above, C:G→T:A transition mutations induced by cytosine deamination are less likely to occur in more physically stable, high-G+C regions (Fryxell and Zuckerkandl 2000). Thus, two of the predominant effects of mutation and recombination on nucleotide composition may be mutually reinforcing.

Finally, although it is tempting to conclude that the prevailing bias of gene conversion toward G/C has evolved by natural selection as a means for avoiding an inevitable buildup of A/T content by mutational bias (e.g., Fryxell and Zuckerkandl 2000), Nagylaki and Petes (1982) and Walsh (1983) articulated a simple reason why opposing patterns of mutation and conversion bias should emerge naturally without any direct involvement of selection. When conversion bias is effective, selectively neutral sites in a particular genomic region are expected to become enriched with nucleotides with a conversion advantage. Mutations arising from such nucleotides will then tend to be sensitive to conversion back to the original state; that is, to be at a conversion disadvantage. This simple logic also has important implications for our understanding of molecular evolution. As discussed in Chapter 4, it is often argued that a universal prediction of the neutral theory is that the long-term rate of nucleotide substitution at a neutral site is equal to the mutation rate. Biased gene conversion is the one nonadaptive mechanism that can lead to a violation of this principle, and the widespread occurrence of this phenomenon may provide a simple explanation for why long-term rates of evolution at silent sites tend to be lower than laboratory estimates of the mutation rate (see Chapter 4).

Evolutionary Consequences of Replication

Because mutations are frequently instigated when DNA is in a single-stranded state, there is substantial opportunity for the development of nucleotide composition heterogeneity in association with the semi-discontinuous nature of replication. As seen in the previous chapter (see Figure 5.1), to the left of an origin of replication (ORI), top-strand DNA is replicated discontinuously, whereas bottom-strand DNA is replicated continuously. The opposite occurs to the right of an ORI. As a consequence, parental strands on which Okazaki fragments are built spend significant time in the mutagenic single-stranded state, whereas the opposite strands are nearly constantly in a more protected double-stranded state. The different mechanisms of processivity of the two strands may also promote variation in repair processes. For example, Okazaki fragments may serve as entry points for mismatch repair enzymes, which require DNA nicks to proceed (Radman 1998).

If the types of mutations incurred during lagging strand and leading strand synthesis differ, an asymmetry in base composition is expected (Lobry 1995; Sueoka 1995), with the top strand to the left of an ORI tending to evolve toward the same equilibrium nucleotide frequencies as the bottom strand to the right, and vice versa. The signature of such mutational variation is readily seen in prokaryotic genomes with a single ORI. When viewed at the level of the entire genome, most prokaryotes exhibit intrastrand parity, with the abundance of A equaling that of T, and of C equaling that of G (Lobry 1996). However, this symmetry breaks down when chromosomes are viewed on a more local scale. Most prokaryotes have enrichments of G over C and T over A on leading strands (Lobry and Sueoka 2002), the main exception being species utilizing an alternative DNA polymerase that reverses the polarity of A/T bias (Worning et al. 2006). With just a single ORI on a circular chromosome, this kind of asymmetry results in the typical subdivision of prokaryotic genomes into two regions, with each strand exhibiting G+T richness on one side of the chromosome and C+A richness on the other. This pattern is generally so pronounced that it provides a reliable indicator of ORI location (Grigoriev 1998; Tillier and Collins 2000) (Figure 6.6).

Two of the leading candidates for the mutational events contributing to such asymmetry have been noted above (see Figures 6.2 and 6.3). Deamination of cytosine to uracil (or 5-methylcytosine to thymine) occurs more than a hundred times faster in single- than double-stranded DNA (Beletskii and Bhagwat 1996; Frank and Lobry 1999; Lutsenko and Bhagwat 1999), and mismatches involving 8-oxo-guanine are also repaired less efficiently on single-stranded DNA (Pavlov et al. 2003). Also encouraged on single-

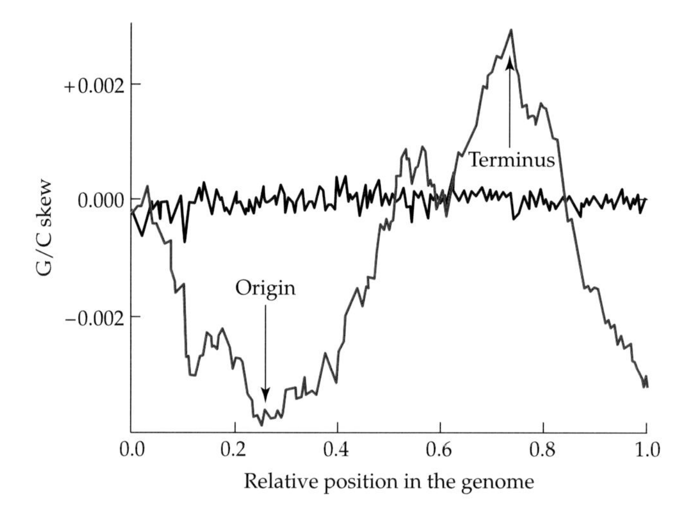

Figure 6.6 G/C skew for the genome of the prokaryote *Mycoplasma pneumoniae*, measured as (G – C)/(G + C), where (G – C) and (G + C) denote the difference and sum of G and C nucleotides in windows of about 10 kb. The horizontal black jagged line plots the noncumulative measures, which when summed from left to right yield the much more obvious cumulative pattern given in red. The initiation of an increase in the cumulative skew to the right of the ORI (Origin) demarcates an abrupt change in (G – C) content. Linear trends in the cumulative skew reflect nearly constant G/C levels to the right and left of the ORI, with local discontinuities and sampling error causing the jagged edges. More complex analyses (Frank and Lobry 2000) that simultaneously utilize the frequencies of all four nucleotides yield smoother profiles. (Modified from Grigoriev 1998.)

stranded DNA is the deamination of adenine to produce hypoxanthine, which base-pairs with C rather than T, resulting in an A→G transition on the affected strand (Lindahl 1993; Ames et al. 1995). The joint action of all three mechanisms causes (G – C) and (T – A) differences to increase on exposed strands.

Mutational asymmetries associated with single-strand vulnerability are also apparent in the mitochondrial genomes of eukaryotes. Although mitochondria generally have circular chromosomes, the mechanism of genome replication differs from that of prokaryotes in that two ORIs are utilized (Figure 6.7). Replication is initiated at ORI_H, building a new G+T-rich heavy strand using the light strand as a template and displacing the parental heavy strand in the process. The latter then remains single-stranded until ORI_L is reached about two-thirds of the way around the molecule, at which point light-strand replication is initiated in the opposite direction. In contrast to the situation in prokaryotes (and nuclear genomes), no Okazaki fragments are made during this process, so the parental heavy strand remains single-stranded for up to 2 hours, with an essentially linear decline in exposure

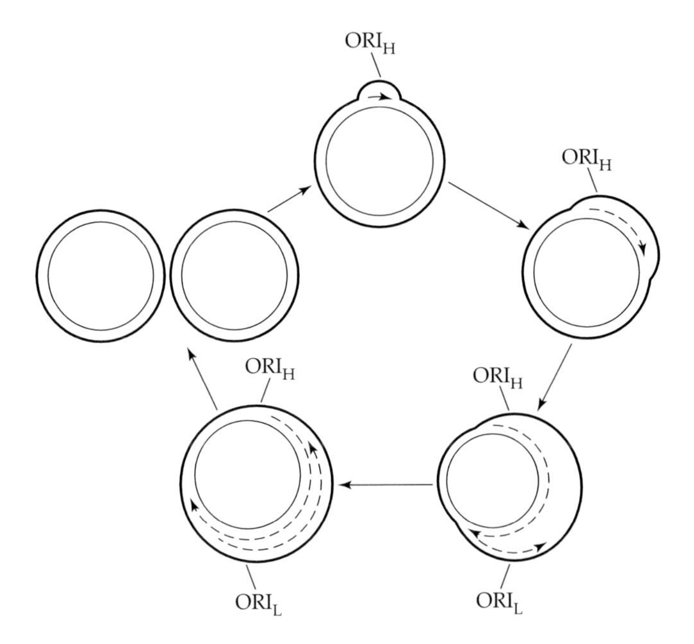

Figure 6.7 Replication of a mitochondrial genome. The inner and outer solid circles denote the light and heavy strands on the parental genome; the dashed lines denote newly synthesized strands. Replication is initiated at ORI$_H$ (*upper right*) and proceeds to build a new heavy strand. Once ORI$_L$ is exposed, synthesis begins in the opposite direction, using the original heavy strand as a template (*lower right*). Note that the parental heavy-strand DNA closest to ORI$_H$ remains single-stranded for the longest time. (Modified from Brown et al. 2005.)

time between the positions closest to ORI$_H$ and those closest to ORI$_L$. This decline is reflected in a substantial increase in the abundance of G and T at silent sites on the heavy strand from ORI$_L$ to ORI$_H$, driven by elevated rates of C→T and A→G transitions on the heavy strand (Reyes et al. 1998; Faith and Pollock 2003).

Do replication-associated mutational biases extend to the nuclear genomes of eukaryotes? On the one hand, the shorter Okazaki fragments and replicon lengths in eukaryotes relative to prokaryotes will tend to reduce the time of exposure of single-stranded DNA, and the possibility that not all eukaryotic ORIs are uniformly used in all cell divisions (see Chapter 5) will blur any expected pattern. On the other hand, these conditions may be mitigated by the greatly reduced rates of strand extension in eukaryotes. Analysis of the problem is further complicated by the fact that the locations of ORIs are generally unknown in most eukaryotes. Nevertheless, regions surrounding known ORIs in the human genome reveal the same kinds of patterns seen in prokaryotes, with very sharp bends in cumulative skew profiles occurring at ORI locations and a significant bias toward G and T appearing on leading strands (Touchon et al. 2005). Parallel analyses in mice and dogs yield almost identical profiles, suggesting that the locations of mammalian ORIs are stable on time scales of 50–100 million years. Application of the cumulative skew approach has also had some success with yeasts (Gierlik et al. 2000).

Coding regions are also subject to single-strandedness during transcription, and Francino and Ochman (1997) have argued that this may introduce mutational asymmetries in two ways. First, during transcription, the transcribed strand is kept protected by the RNA polymerase and the growing mRNA chain, while the coding strand is exposed. Second, transcription-coupled repair corrects bulky lesions in the transcribed strand, but not in the coding strand (Hanawalt 1994). There is little question that transcription causes an overall elevation in the mutation rate (Datta and Jinks-Robertson 1995; Green et al. 2003; Yoshiyama and Maki 2003), but are transcription-associated asymmetries in the spectrum of mutational effects significant enough to overwhelm those associated with replication? The tendency for genes that have been transposed from one strand to the other to adopt the local nucleotide usage pattern (Lafay et al. 1999; Tillier and Collins 2000; Szczepanik et al. 2001) strongly suggests that transcription-associated biases are secondary to those associated with replication. In addition, in the animal mitochondrion, almost all transcription is off the heavy strand (the coding sequences reside on the light strand), and yet (as noted above) the heavy strand is enriched with G and T, contrary to the pattern expected under the transcription-bias hypothesis.

Isochores

Whereas spatial heterogeneity in nucleotide usage along chromosomes can be found in every species, it has long been thought that the magnitude of such variation is unusually pronounced in tetrapods. Extensive regions of approximate homogeneity, known as isochores, were first discovered by density centrifugation of fractionated bovine DNA (Filipski et al. 1973; reviewed in Bernardi 2000), which revealed alternating G+C- and A+T-rich tracts. Although isochores do not exist in the strict sense of having sharp boundaries, there is no question that the level of heterogeneity in nucleotide composition far exceeds that expected by chance. For example, the G+C content of 100 kb regions in the human genome ranges from 35% to 65% (Lander et al. 2001), whereas the expected range associated with sampling error is well under 1%. Such regional variation affects both coding and noncoding DNA, although coding regions tend to be more G+C rich than introns and intergenic DNA (Bernardi 2000). Comparative analyses suggest a relatively A+T-rich vertebrate ancestor, with G/C isochores arising about 350 MYA in the ancestor leading to mammals, birds, and other reptiles (Bernardi et al. 1997; Duret et al. 2002).

The mechanisms that might contribute to such patterns include selection, mutation, and biased gene conversion, but why such processes would vary on spatial scales of 100 kb in these particular lineages remains uncertain (Eyre-Walker and Hurst 2001). Given the presence of isochores in both birds and mammals, Bernardi et al. (1985; Bernardi 1993, 2000) have argued that G+C-rich isochores increase the thermostability of DNA in homeotherms. However, this hypothesis fails to explain why G+C richness should be confined to some regions as opposed to being genome-wide, and it is also challenged by circumstantial evidence suggesting the presence of isochores in snakes and crocodilians, both of which are ectotherms (Hughes et al. 1999; Hamada et al. 2003). Vinogradov (2005) suggested that G+C content influences the potential for nucleosome formation and hence the accessibility of chromosomal regions to gene expression. But again, this hypothesis does not explain why such a phenomenon should be restricted to a subset of vertebrates. If selection is responsible for the origin and maintenance of isochores, it must be capable of promoting single-base substitutions on the basis of their effects over megabase-sized regions. Moreover, any such advantages would have to exceed the potential influence on gene function, because even the amino acid composition of proteins is influenced by the isochore context of the genes that encode them (D'Onofrio et al. 1991). In light of these difficulties and the fact that isochores are present in precisely the lineages in which selection is expected to be least efficient (vertebrates with low N_g), nonadaptive mechanisms for the origin and maintenance of isochores must be taken seriously.

If isochores are not sculpted by selection, then some kind of regional bias in mutational and/or recombinational activity must be involved. At least three candidate mechanisms can be envisioned, although none have achieved indisputable support. One idea relies on the observation that base misincorporation errors are influenced by the composition of the nucleotide pool, which can vary during the cell cycle (Wolfe et al. 1989). Because variation also exists in the timing of replication of different chromosomal regions, such conditions could lead to base compositional heterogeneity along chromosomes. If this is to be a viable hypothesis, two key issues must be resolved: first, whether significant temporal variation in nucleotide pools occurs in germ line cells, and second, why such variation should be peculiar to (or most effective in) mammals and birds.

A second nonadaptive hypothesis for the origin of isochores is that stochastic variation in nucleotide composition along a chromosome nucleates localized changes in susceptibility to mutations, thereby leading to self-sustaining patterns of heterogeneity. For example, as noted above, regions with elevated levels of A+T spend more time in single-stranded form, encouraging the mutational production of Ts associated with cytosine deamination, which reinforces the local A+T content.

Third, isochores could be a consequence of regional variation in biased gene conversion (Holmquist 1992; Galtier et al. 2001). However, if isochores are to be confidently attributed to gene conversion, it will have to be resolved why such activity is more intense in some regions than in others and how it can be spatially restricted for long enough periods to create an isochore.

In principle, this might happen if recombinational activity is encouraged by G+C content, as this would impose a self-reinforcing process like that envisioned above for the elevated rate of mutations toward As and Ts in A+T-rich regions. In yeasts, for example, recombination hot spots tend to be G+C rich (Borts and Kirkpatrick 2006), and circumstantial evidence suggests a similar situation in mammals (Petes 2001), with the motif CCTCCCCT (AGGGGAGG on the opposite strand) being a likely hot spot (Myers et al. 2005). Runs of Gs frequently lead to structural modifications of DNA involving quartets of Gs within or between strands (G4 DNA), which might serve as nuclei for initiating recombination events (Maizels 2006).

Attempts to determine the factors responsible for isochores generally assume that the same forces that gave rise to them are operable today; that is, that isochore compositions more or less reflect an equilibrium situation. Superficially, this assumption appears reasonable, because different lineages of mammals have roughly comparable isochore structures (Bernardi 2000). Moreover, phylogenetic analyses suggest that both the types of mutations that arise and their probabilities of fixation differ between G+C- and A+T-rich regions (Duret et al. 2002; Arndt et al. 2003, 2005; Belle et al. 2004; Comeron 2006; Lipatov et al. 2006). Remarkably, however, the latter studies demonstrate that regional differences in nucleotide substitution patterns are insufficient for the maintenance of isochores. Instead, G+C-rich isochores appear to be slowly disappearing from mammalian lineages due to an excess of G/C→A/T substitutions over the reverse. Even though newly arisen G/Cs have higher probabilities of fixation than A/Ts, consistent with the expectations of biased gene conversion, this fixation bias is being overwhelmed by mutation pressure in the direction of A/T. Thus, the comparable isochore features among different mammals appear to be a simple consequence of parallel evolution and common ancestry. As noted above, the directional force of biased gene conversion can be overwhelmed by random genetic drift in populations of sufficiently small size, so these observations are consistent with mammalian lineages having a small N_g. However, the question as to how isochores originated remains unresolved, and equally puzzling is the observation that substitutional biases in the chicken lineage appear to be enforcing isochore maintenance (Webster et al. 2006).

Finally, it is worth considering whether the attention focused on vertebrate isochores is entirely justified. All genomes exhibit some level of heterogeneity in nucleotide composition, and when scaled appropriately, the pattern does not necessarily deviate greatly from that seen in mammals. In yeasts, for example, the G+C content of 5 kb windows ranges from 35% to 45% (Figure 6.8A). Although the human chromosome in Figure 6.8B is about 470 times the length of that in yeasts, the recombination rate per physical distance in yeasts is about 270 times greater than that in humans (see Chapter 4). Thus, if regional variation in nucleotide composition is driven by recombination processes, the finer scaling and reduced magnitude in yeasts could be a simple consequence of more frequent recombination and higher efficiency of selection.

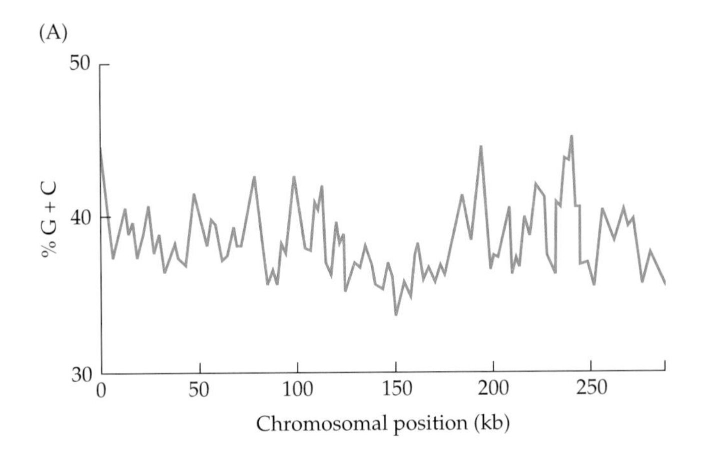

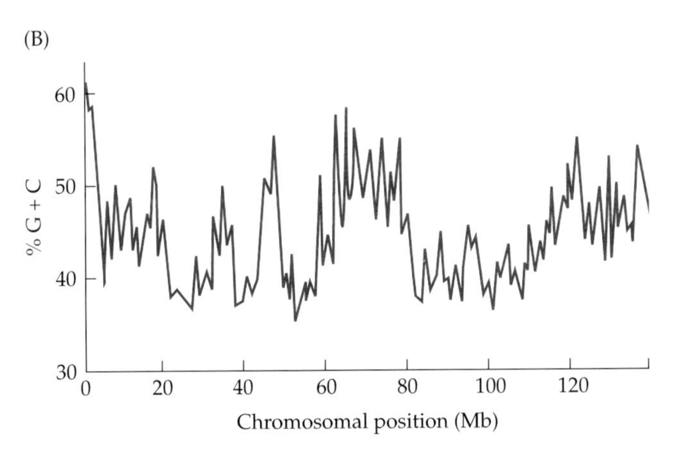

Figure 6.8 Regional patterns of nucleotide composition. (A) Yeast chromosome 6; 5 kb sliding averages, shifted by 1 kb; horizontal scale in kb. (Modified from Petes 2001.) (B) Human chromosome 11; a redrawn version of a 100 kb window plot; horizontal scale in Mb. (Modified from Costantini et al. 2006.)

The Nonadaptive Basis of Genome-Wide Nucleotide Composition Variation

Although selection plays a role in constraining small-scale intergenic sequences in the vicinity of genes (e.g., transcription factor binding sites; see Chapters 3 and 10), the preceding observations provide little, if any, support for the hypothesis that variation in nucleotide composition in putatively nonfunctional genomic regions is a product of natural selection. Rather, the data suggest that most species are subject to an intrinsic mutational bias toward A+T richness, which is counteracted by the tendency of gene conversion to favor G+C composition. Because biased gene conversion is promoted by recombination, its effects are expected to be most pronounced in small eukaryotic genomes (where the rate of crossing-over per physical distance is high; see Chapter 4), especially in those that regularly engage in meiosis. Moreover, because biased gene conversion operates in a selection-like manner, all other things being equal, G+C richness is expected to be greatest in species with high N_g. In principle, a formal test of this hypothesis would be possible if accurate estimates of both long-term N_g and the per-generation recombination rate were available for a wide range of species, but information on the latter is especially sparse for natural populations.

Given their haploid nature, prokaryotes may be effectively immune to gene conversion activity, but given their large effective population sizes, they should be maximally sensitive to the modification of nucleotide composition in noncoding regions by direct forces of selection. However, an examination of the two selective mechanisms thought to influence nucleotide usage does not inspire confidence in adaptive explanations for prokaryotic G+C composition. First, as noted above, the stability of DNA increases with G+C content because of the three versus two hydrogen bonds involved in G:C versus A:T bonds. Thus, if DNA stability is a significant target of selection, species inhabiting warm environments, where DNA is more susceptible to melting, would be expected to have relatively G+C-rich genomes. However, there is no correlation between optimal growth temperatures and G+C contents of prokaryotic genomes (Galtier and Lobry 1997; Hurst and Merchant 2001; Marashi and Ghalanbor 2004), although the G+C richness of the stems of structural RNAs does increase with optimal growth temperature (Galtier and Lobry 1997).

Second, the metabolic costs of production of the four nucleotides are not equal: among pyrimidines, C is more costly than T, and among purines, G is more costly than A. Thus, under the metabolic hypothesis for genomic streamlining (see Chapter 2), prokaryotic genomes should be biased in the direction of A+T richness (Rocha and Danchin 2002). However, as noted in Figure 6.1, prokaryotic genomes exhibit a broad range of nucleotide compositions, with many having higher G+C contents than any eukaryote. The fact that endosymbiotic eubacteria, which are entirely dependent on their hosts for survival, have A+T-rich genomes has been cited as support for the metabolic hypothesis (Rocha and Danchin 2002), but this pattern may also be a simple consequence of an altered mutational spectrum, as such species have lost numerous DNA repair pathway genes (Moran 2002). Indeed, phylogenetic analysis of one such genus, *Buchnera*, suggests very strong mutational bias in the direction of A+T and provides no evidence of a selection-driven process (Wernegreen and Funk 2004). Even among free-living prokaryotes, there is a significant tendency for G+C-rich genomes to encode proteins with elevated usage of amino acids with G+C-rich codons (Lobry 1997; Gu et al. 1998; D'Onofrio et al. 1999; Wilquet et al. 1999), an observation that suggests a mutationally driven process.

This disparate set of observations suggests that even the large effective population sizes of prokaryotes are inadequate to enable selection to significantly influence genome-wide nucleotide composition. If the strength of selection per noncoding nucleotide site is overwhelmed by the power of directional mutation, even the complete absence of random genetic drift would not change this situation. In any case, the data suggest that phylogenetic diversity in G+C composition is largely a consequence of variation in the physical forces of mutation and/or recombination. Although the specific mechanisms responsible for such variation are not yet understood, the underlying changes in such forces may be largely random from a phylogenetic perspective. Indeed, Haywood-Farmer and Otto (2003) find that a simple Brownian motion model with a constant diffusion rate readily explains essentially the entire dispersion of nucleotide compositions across all eubacteria, the only exceptions being the obligate intracellular species, *Rickettsia* and *Buchnera*, which, as noted above, probably have peculiar mutational properties.

Summing up to this point, we have reached the rather surprising conclusion that the relative frequencies of the basic building blocks from which most species construct their genomes may be largely driven by internal physical processes of mutation and gene conversion, with external forces of natural selection playing, at most, a minor role at the genome level. This claim appears to be quantitatively consistent with what we know about the nucleotide composition biases associated with both mutation and gene conversion. Nevertheless, as there are many ways to construct a genome with the same pool of nucleotides, it is important to keep in mind that the lack of a significant role of selection in defining *genome-wide* nucleotide contents need not imply a complete absence of selective influence on the precise sequences residing in particular genomic locations. The next section explores this matter by examining one of the most intensely studied but poorly understood aspects of genome architecture: the usage of alternative nucleotides at the silent sites of protein-coding genes.

Codon Usage Bias

As noted in Chapter 4, population geneticists frequently rely on the assumption of neutrality of silent sites to derive inferences about population structure, mutation, and recombination, as well as to infer levels of selection operating on replacement sites. The logic underlying this assumption draws from the fact that silent-site substitutions have no effect at the protein sequence level. This is an oversimplification, however, as there are numerous other ways by which selection can influence nucleotide usage at silent sites:

1. Alternative codons for a particular amino acid might be more efficient at recruiting appropriately charged transfer RNAs (tRNAs), thereby increasing the rate and/or accuracy of translation.
2. Codons consisting of mononucleotide runs (e.g., AAA) may be mutationally or selectively eliminated as a consequence of their vulnerability to replication and/or transcription slippage (and the resultant production of frameshifts) (Bennetzen and Hall 1982; Ackermann and Chao 2006).
3. Eukaryotic genes often harbor exon-splicing enhancer and silencer sequences for guiding the spliceosome to proper locations for intron removal (see Chapter 9).

4. Specific exonic sequences may yield secondary structures that contribute to mRNA stability.
5. As noted above, biased gene conversion can cause nucleotide frequencies to deviate from expectations based on mutation and drift alone.

Codon usage bias is universal. All but two amino acids (methionine and tryptophan) are encoded by two to six codons in nearly all nuclear eukaryotic and prokaryotic genomes, but in almost no case are the alternative codons for a particular amino acid used in identical frequencies. The central question is whether such biases simply reflect inequities in the physical pressures of mutation and/or gene conversion or whether some additional form of selection needs to be invoked. To address this issue, numerous investigators either have assumed that the most commonly used codon is the one preferred by natural selection or have defined optimal codons to be those that best match the anticodon sequences of the most abundant cognate tRNAs. Indices of overall evenness of codon usage are often relied upon, with deviations from parity being interpreted as the signature of selection. The diversity of approaches, along with the circular reasoning underlying some of them, has led to much confusion in the literature on codon usage bias.

To see the potential perils of measures of codon bias that do not explicitly take into account the underlying population genetic processes, consider the expected usage of alternative codons in the face of the joint pressures of selection, mutation, and random genetic drift. (We will add in the additional complexity of gene conversion shortly.) To simplify the discussion, imagine an amino acid encoded by two alternative third-position nucleotides, and denote the alternative allelic states as **B** and **b**, with **B** denoting the optimal codon state, **B** mutating to **b** at rate u, and **b** mutating to **B** at rate v. The same mathematical machinery developed in Box 6.2 can then be used to obtain the equilibrium usage of optimal codon **B** over the entire genome ($\tilde{p}$):

$$\tilde{p} = \frac{e^{2N_g s}}{e^{2N_g s} + (u/v)} \tag{6.3}$$

where $e^{2N_g s}$ is the selection bias toward **B** and (u/v) is the mutational bias toward **b** (Li 1987; Shields 1990; Bulmer 1991).

Although this derivation is restricted to the two-codon situation and ignores mutations that impose amino acid replacements (under the assumption that selection at the protein level rapidly eliminates such mutations from the population), it nicely encapsulates the problems in interpreting the mechanisms underlying biased codon usage. First, the theory demonstrates that if selection is sufficiently weak and biased mutation sufficiently strong, the most common codon need not be the optimal codon (lower right region of Figure 6.9). Second, unless selection on silent sites is quite strong ($2N_g s$ > 1) and mutation is strongly biased in the direction of the optimal codon

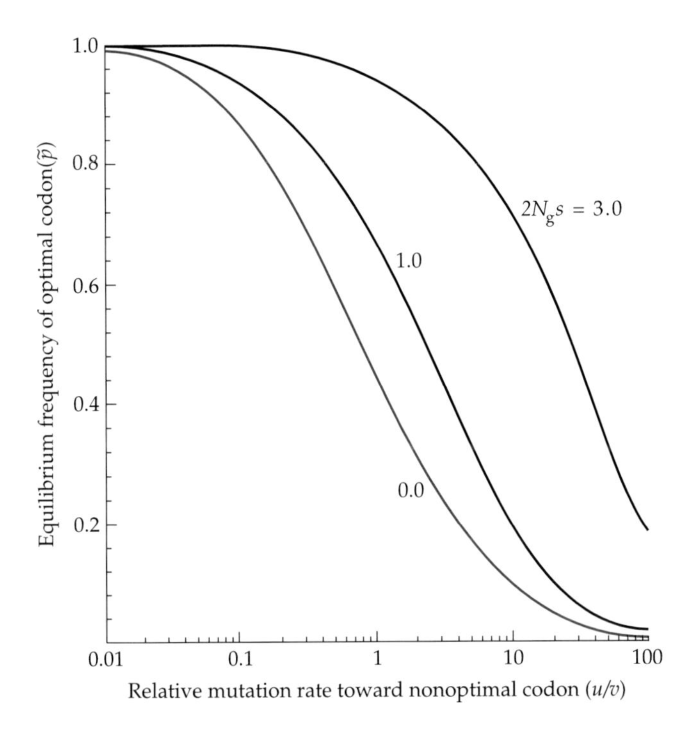

Figure 6.9 The expected genomic frequency of the optimal codon in the case of a twofold redundant silent site under the assumption of mutation-selection-drift equilibrium, derived using Equation 6.3. The bottom (red) curve gives the result under neutrality ($s = 0$).

(upper left region of Figure 6.9), suboptimal codons are expected to be present at appreciable frequencies. Third, if there is a conflict between the directions of mutation and selection bias (right half of Figure 6.9), genes under different intensities of selection for codon usage are expected to exhibit biases toward different codons ($\tilde{p} > 0.5$ for large s, and $\tilde{p} < 0.5$ for small s). This point highlights the precarious nature of the frequent argument that the elevated frequency of "nonoptimal" codons in certain genes is a direct response to natural selection on the regulation of gene expression (e.g., Grosjean and Fiers 1982; Lavner and Kotlar 2005). Finally, to be fully inclusive, s should be viewed as the net directional pressure on a particular nucleotide resulting from the joint forces of natural selection and biased gene conversion. In other words, if s_c and s_n denote the selection coefficients associated with gene conversion and natural selection, respectively, then $s = s_c + s_n$. In principle, these forces could be in opposite directions (e.g., conversion bias is in the direction of the nonoptimal codon, but s_c is negative), rendering s closer to zero than expected based on natural selection alone.

Prokaryotes

In principle, the preceding theory can be used to infer the magnitude of selection operating on codon usage. Rearranging Equation 6.3, $2N_g s$ can be estimated as

$$\ln \frac{\tilde{p}(u/v)}{(1-\tilde{p})}$$

by substituting the genome-wide estimate of the frequency of usage of the optimal codon for p. Such treatment assumes fairly homogeneous evolutionary forces operating on all sites, but a more central issue is the need for an estimate of u/v. If the nucleotide usage from truly neutral regions ($s = 0$) of DNA (p_n) were known, u/v could be estimated from Equation 6.1 as $(1 - p_n)/p_n$, but the problem, of course, is that we are attempting to determine whether such sites actually exist. Sharp et al. (2005) attempted to circumvent this difficulty by assuming that genome-wide codon usage in prokaryotes essentially reflects mutational bias (providing an estimate of p_n), with just the small subset of highly expressed genes being subject to selection on silent sites. They then used the estimate of u/v derived by the genome-wide estimate of p_n to infer $2N_g s$ by applying the preceding expression. For four two-codon amino acids, the optimal codon could be identified by physical arguments alone; for example, of the two codons for phenylalanine (TTT and TTC), the latter has the greatest affinity for the single phenylalanine tRNA with anticodon GAA. Application of this methodology to forty highly expressed genes in eighty prokaryotes yielded estimates of $2N_g s$ that were uniformly ≤ 2.5, similar to earlier results obtained by other methods (e.g., Hartl et al. 1994). Because of the haploid nature of prokaryotic genomes, these results are presumably free of the effects of biased gene conversion. The idea that translation-associated selection drives codon usage in prokaryotes is further supported by the strong correlation between species-specific estimates of $2N_g s$ obtained by Sharp et al. and numbers of tRNA and ribosomal RNA gene copies per genome, two presumptive indicators of the rate of translation (Figure 6.10). A similar conclusion was reached by dos Reis et al. (2004) using substantially different methods.

The estimates of $2N_g s$ in Figure 6.10 will be downwardly biased if codon usage by genes with moderate to low expression is actually molded by natural selection, rather than providing a clean estimate of u/v. These estimates are also derived under the assumption that the mutational spectrum is essentially the same for genes with high and low expression, an assumption that might be questioned given the aforementioned evidence for the influence of transcription on the mutation process. If transcription-associated mutations are biased toward nonoptimal codons, $2N_g s$ will be underestimated, perhaps even enough to yield negative estimates (as seen in Figure 6.10); but if they are biased toward optimal codons, $2N_g s$ will be overestimated (because the elevated usage of optimal codons is due to mutation rather than selection). Given the uncertainties in the magnitude of these potential biases, and recalling from Chapter 4 that the average estimate of the prokary-

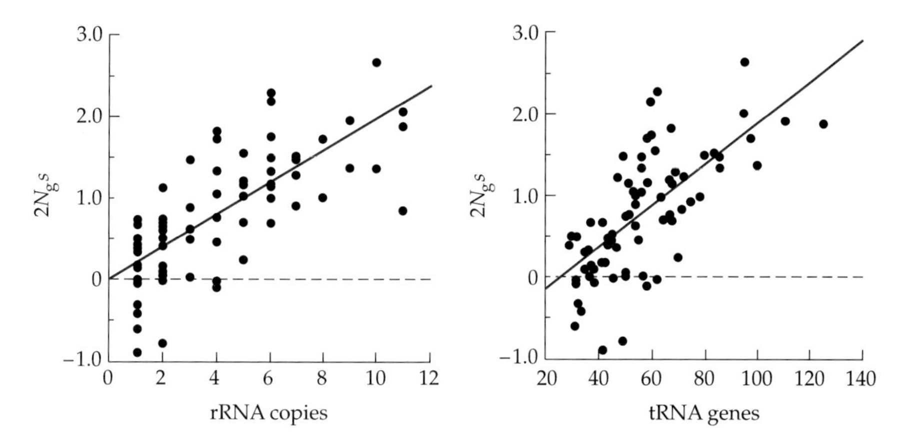

Figure 6.10 Estimates of $2N_g s$ from genomic data for 40 highly expressed genes in 80 species of prokaryotes, and their relationship to the number of genomic copies of rRNA and tRNA genes. The analyses are restricted to four amino acids that utilize two codons, but just one anticodon at the tRNA level (whose affinity to the alternative codons is understood at the molecular level). Negative estimates are obtained when the frequency of usage of the codon assumed to be optimal is less than that predicted from mutational bias alone. (Modified from Sharp et al. 2005.)

otic $2N_g u \approx 0.10$ is likely to be downwardly biased, these results tentatively suggest that the strength of selection associated with codon bias in prokaryotes is of the same order of magnitude as the per-nucleotide mutation rate (i.e., $s \approx u$, because $2N_g s \approx 2N_g u$). Recalling the estimates of u for prokaryotes (see Chapter 4), this further implies s on the order of 10^{-10}–10^{-9} in such species.

With $2N_g s = 0.1$ and 1.0 (and even as large as 3.0), interspecific differences in codon usage are still expected to closely reflect mutational bias differences among species (see Figure 6.9); that is, the effect of selection is statistically significant, but not so large as to eliminate silent-site variation. Thus, if it is correct that most loci in prokaryotes are negligibly influenced by translation-associated selection, it is not surprising that a strong correlation exists between nucleotide usage at silent sites in protein-coding genes and in noncoding regions of such species (Chen et al. 2004).

Two additional sets of observations support the idea that translation-associated selection is effective enough to influence patterns of nucleotide usage at least at some loci in prokaryotes. First, consistent with the implicit assumption in the Sharp et al. analysis, many studies have noted that the degree of bias toward usage of presumptively optimal codons increases in highly expressed genes (e.g., Ikemura 1981, 1985; Gouy and Gautier 1982; Grosjean and Fiers 1982; Karlin et al. 1998). A reduction in the rate of interspecific sequence divergence at silent sites with increasing levels of gene expression in prokaryotes is also consistent with the operation of selection for the maintenance of particular codons (Sharp and Li 1987). Second, correlations often exist between the abundances of alternative tRNAs and the frequency of usage of their most favored codons (the anticodons of the tRNAs for several amino acids vary in nucleotide composition in the third position, and such variants have unequal affinities for alternative codons) (Ikemura 1981, 1985; Kanaya et al. 1999). In addition, prokaryotes with shorter generation times have more but less diverse tRNAs, with the most common tRNA type closely matching the codon most frequently used for its amino acid (Rocha 2004). A fundamental unresolved issue with respect to tRNA usage relationships concerns the question of causality. Does the tRNA repertoire of a species serve as a relatively constant selection pressure on codon usage, or do mutational biases drive nucleotide composition to the extent that selection primarily operates at the level of tRNAs, modifying their relative abundances by gene duplication and/or anticodon-sequence mutations? Bulmer (1987) makes a compelling case that the simultaneous operation of both forces leads to coevolution between codon usage and tRNA abundance.

Finally, it should be noted that some evidence suggests that the nucleotide composition at silent sites of some prokaryotic genes is molded by selective forces unassociated with tRNA recruitment. For example, the bias toward translationally optimal codons is weaker toward the ends of prokaryotic genes than in the centers (Bulmer 1988; Eyre-Walker 1996; Qin et al. 2004). Although such a pattern could result from an overall relaxation of selection on silent sites in such regions, this does not appear to be the case, as interspecific divergence rates (which should be inversely related to the intensity of purifying selection) are actually reduced at such sites, at least in the contrast between *E. coli* and *Salmonella* (Eyre-Walker and Bulmer 1993). In principle, constraints on silent sites at the ends of genes might result from the selective promotion of codons via their effects on the recruitment of the transcriptional machinery and/or the stability of the resultant mRNAs. In addition, the sequences of some closely packed genes may be constrained by the placement of ribosome-binding sites (Shine-Dalgarno sequences; see Chapter 10) within the end sequences of their immediately upstream gene neighbors.

Eukaryotes

Given that the effective population sizes of eukaryotes are typically one to five orders of magnitude smaller than those of prokaryotes (see Chapter 4), the preceding results suggest that codon usage–associated values of $2N_g s$ in eukaryotes are likely to be substantially less than one and perhaps so small as to be entirely inconsequential. Thus, it is surprising that many of the patterns of codon usage bias seen in prokaryotes have also been observed in eukaryotes. Increased levels of codon bias in highly expressed genes have been observed in *Giardia* (Lafay and Sharp 1999), *Saccharomyces* (Bennetzen and Hall 1982; Akashi 2003), *Arabidopsis* (Duret and Mouchiroud 1999; Wright et al. 2004), *Drosophila* (Duret and Mouchiroud 1999; Hey and Kliman 2002; Marais and Piganeau 2002), and *Caenorhabditis* (Castillo-Davis and Hartl 2002; Marais and Piganeau 2002), and a weak pattern even exists in mammals (Urrutia and Hurst 2003; Comeron 2004; Lavner and Kotlar 2005). In most of these species, the most frequently used codon for a given amino acid corresponds with the most abundant tRNA in the genome (Bennetzen and Hall 1982; Moriyama and Powell 1997; Duret 2000; Kanaya et al. 2001; Wright et al. 2004; Comeron 2004), although the correlation is not always strong. In addition, the rate of silent-site divergence is lower in genes with greater codon bias in *Caenorhabditis* (Castillo-Davis and Hartl 2002), *Chlamydomonas* (Popescu et al. 2006), and *Drosophila* (Powell and Moriyama 1997), although the effect has not been seen in *Arabidopsis* (Wright et al. 2004) or mammals (Duret and Mouchiroud 2000; Urrutia and Hurst 2001; Mikkelsen et al. 2005).

Taken at face value, this collective set of results appears to support the idea that translation-associated selection is effective in at least some eukaryotic genes. But this raises a significant paradox, in that such selection would require *s* operating on silent sites in eukaryotic genes to be orders of magnitude greater than that in prokaryotes, whereas there is no direct evidence that this is the case. One potential resolution to this puzzle is that codon usage bias in eukaryotes is driven by nonadaptive mechanisms that are largely inoperable in prokaryotes. For example, as noted above, the power of biased gene conversion is great enough to drive nucleotide composition in the direction of G+C content in recombining genomes of some outcrossing eukaryotes. If this force is effective, then genes in regions of high recombination are expected to exhibit elevated G+C content at silent sites, and such a pattern is indeed seen in *Saccharomyces* (Birdsell 2002), *Caenorhabditis* (Marais et al. 2001), *Drosophila* (Akashi 1995; Akashi et al. 1998; Comeron et al. 1999; Kliman and Hey 2003; Singh et al. 2005), and humans (Galtier et al. 2001; Comeron 2004). In all of these species, for nearly all amino acids, the optimal codon (corresponding to the most abundant tRNA type) ends in G or C, encouraging the view that increased usage of such codons in regions of high recombination is a consequence of an increased efficiency of selection (Hey and Kliman 2002). However, the G+C contents of introns and intergenic DNA are also correlated with recombination rates (Marais et al. 2001; Marais and Piganeau 2002), and in nematodes, optimal codons ending in A or T decline in frequency with increasing recombination rates (Marais et al. 2001).

This broad collection of observations suggests that biased gene conversion, rather than translation-associated selection, may be the primary determinant of the G+C content at silent sites in eukaryotes. If this is the case, the predominance of tRNAs with affinities for G/C-ending codons might be promoted by secondary selection imposed by genome-wide G+C content at silent sites driven by physical forces. However, such tRNAs might also simply accumulate spontaneously, as tRNA genes will be just as susceptible to the conversion process as protein-coding loci (if not more so, as a consequence of their elevated copy numbers). Biased gene conversion may also be responsible for the reduced rate of silent-site substitutions at G+C-rich loci, which will predominantly produce mutations in the A/T direction whose fixation is opposed by the conversion process.

Many other eukaryotic patterns of codon usage bias that superficially appear to be adaptive are plausible outcomes of biased gene conversion. For example, rapidly evolving *Drosophila* proteins tend to be produced by loci with low frequencies of optimal codons (Betancourt and Presgraves 2002; Marais et al. 2004), leading these authors to suggest that selective interference resulting from strong directional selection on replacement sites overwhelms the much weaker translation-associated selection on linked silent sites. A second explanation for this pattern is that genes that are under strong purifying selection at the amino acid level are also under strong selection for translational accuracy, yielding an elevated abundance of "optimal" G/C-ending codons (Akashi 1994; Bierne and Eyre-Walker 2006). However, an alternative exists to both of these arguments: if rapidly evolving proteins experience periodic selective sweeps, the opportunities for biased gene conversion in heterozygous individuals will be reduced, resulting in an elevated abundance of "nonoptimal" A/T-ending codons. A parallel explanation can potentially explain the reduced rates of evolution at both replacement and silent sites in highly expressed genes of eukaryotes (Subramanian and Kumar 2004; Drummond et al. 2005; Lemos et al. 2005): if such genes only rarely experience selective sweeps, as a consequence of a predominant pattern of purifying selection, the susceptibility of newly arisen mutations to biased gene conversion will be enhanced, thereby encouraging elevated levels of relatively stable G/C-ending codons.

Using methods related to those outlined above for prokaryotes, but interpreting nucleotide composition in introns as a reflection of the mutational pattern expected under neutrality, Kondrashov et al. (2006) estimated $2N_g s$ to be about 1.0 in the great ape lineage, and estimates for *Drosophila* species generally fall in the range of 3–18 (Akashi 1995; Akashi and Schaeffer 1997; Maside et al. 2004). These estimates are of the same order of magnitude if not higher than those obtained for prokaryotes (see Figure 6.10), which, given the elevated levels of N_g in prokaryotes, implies that s associated with codon bias in eukaryotes is orders of magnitude higher than in prokaryotes. This pattern can be explained by recalling that s is a composite index of bias imposed by both gene conversion (s_c) and translation-associated selection (s_n); that is, $s = s_c + s_n$. Assuming that the magnitude of translation-associated selection per nucleotide site is roughly comparable between prokaryotes and eukaryotes, then to compensate for the disparity in N_g and yield comparable $2N_g s$ in both groups, s_c associated with eukaryotic biased gene conversion must be several orders of magnitude greater than s_n associated with translation-associated selection. This is a plausible situation, given that s_n is about 10^{-9} in prokaryotes and that s_c is about 10^{-3} in yeasts and in the range of 10^{-7}–10^{-5} in multicellular species (as discussed above).

One other explanation has been offered for the comparable levels of $2N_gs$ associated with silent sites observed in prokaryotes and eukaryotes: if codons exhibit synergistic epistasis, such that consecutive suboptimal codons accumulating in a genome impose increasingly deleterious effects on fitness, the load from such mutations would freely accumulate in a genome until s associated with additional suboptimal codons satisfied the condition for efficient selection ($2N_gs_n > 1$) (Li 1987). The major limitation of this argument is its implications for the accumulation of a much greater absolute fitness load in multicellular eukaryotes than in prokaryotes, which raise significant questions about the maintenance of species viability (Kondrashov 1995). This type of cost is not imposed by biased gene conversion, which need not have any direct effect on fitness.

Unsolved problems

In summary, although diverse sources of circumstantial evidence support the view that silent sites are subject to selection for translation efficiency and accuracy, the magnitude of such selection is so weak that even prokaryotic populations with extremely large effective population sizes exhibit only moderate deviations of nucleotide usage from the expectations based on mutation pressure alone. Eukaryotes, with smaller effective population sizes than prokaryotes, appear to provide an inadequate population genetic environment for the operation of translation-associated selection, but they still present physical opportunities for the biasing effects of gene conversion that may be largely absent from prokaryotes.

In closing, one additional caveat must be given with respect to the interpretation of codon bias. Although a deviation of genome-wide codon usage from the neutral expectation provides compelling evidence for the operation of selection on silent sites, the reverse is not necessarily true; that is, the absence of codon bias need not imply the absence of selection. For example, selection operating on silent sites for reasons other than translation efficiency (e.g., mRNA architecture or exon identification signals) could promote certain codons in some gene locations and their alternatives in others, yielding an overall pattern with little deviation from the mutation/gene-conversion expectation on a genome-wide level. Support for this concern derives from computational and empirical work showing that randomization of the synonymous codons within mammalian and drosophilid genes, while keeping both the protein sequence and the overall codon usage constant, reduces the structural stability of mRNAs (Seffens and Digby 1999; Carlini et al. 2001; Duan and Antezana 2003; Carlini 2005; Chamary and Hurst 2005).

An additional trend bearing on this matter is a gradient of decreasing usage of putatively "optimal" G/C-ending codons from the edges to the interiors of exons in a diversity of species (Comeron et al. 1999; Comeron and Kreitman 2002; Comeron and Guthrie 2005). These authors suggest that this pattern is a consequence of selective interference (with simultaneous selection operating on linked sites reducing the overall efficiency of selection; see Chapter 4). For translation-associated selection, such interference is expected to be highest in exon interiors, where all neighboring sites are coding sequence. However, given the absence of compelling evidence for high $2N_g s_n$ in eukaryotic genes, alternative explanations need to be considered. Willie and Majewski (2004) and Qin et al. (2004) point out another puzzling aspect of the gradient of codon usage bearing on this issue: it is not just the frequency of optimal codons, but the overall diversity of codons, that declines in exon interiors. Such a pattern appears to be inconsistent with a reduction in the efficiency of translation-associated selection in exon interiors, where "suboptimal" codons should be able to drift to higher frequencies, but it can be explained by supposing that codon usage in such regions is driven largely by biased mutation/conversion processes. More even codon usage toward the edges of exons ($\tilde{p}$ closer to 0.5) would be expected if the selectively advantageous codons were discouraged by mutation/conversion processes (see Figure 6.9).

Such conflicts may arise from the constraints imposed by the splicing process. To provide guidance for the spliceosome during precursor mRNA processing (see Chapter 9), the edges of exons generally harbor short sequence motifs that operate as exon-splicing enhancers, whereas intron-splicing enhancers are avoided in these same regions. For example, the exon-splicing enhancer motif GAA is more abundant, and the intron-splicing enhancer GGG less abundant, near intron–exon boundaries (Willie and Majewski 2004). The signature of purifying selection, in the form of reduced levels of interspecific divergence and intraspecific polymorphism, is readily seen in such motifs (Carlini and Genut 2006; Parmley et al. 2006). In a similar vein, the dinucleotides GT and AG, which form the end sequences of most introns, are avoided at the edges of *Drosophila* and *Caenorhabditis* exons (Eskesen et al. 2004). Although these observations do not fully eliminate translation-associated selection as a contributor to spatial patterns of codon usage bias within exons, they do raise significant questions about both the necessity and sufficiency of such an explanation. Perhaps the best evidence for the effects of selective interference on exonic gradients of codon usage bias derives from a study focused entirely on intron-free genes in *Drosophila*, which showed an increase in silent-site divergence rates in gene interiors (Comeron and Guthrie 2005), but even here, caution is required in interpreting this pattern as a consequence of translation-associated selection.

7 Mobile Genetic Elements

Virtually all organisms are at least periodically confronted with intracellular parasites. When the infective agent is a bacterium or a virus, the debilitating effects of such interactions can often be minimized by behavioral avoidance and/or immunological responses, but this is not an option in the case of a mobile genetic element (retrotransposon or transposon) residing in the host genome. Such a niche not only guarantees an element's transmission from host parent to offspring, but also provides access to new colonization sites in other chromosomal locations, and in sexual species, in other individuals.

Mobile elements have been spectacularly successful at exploiting cellular life. Indeed, they are so ubiquitous, so diverse, and have such a profound effect on eukaryotic chromosomal architecture that one can reasonably argue that an overview of genomic evolution ought to start with them, before moving on to the host genes themselves. That is the approach taken here. We will first consider the key molecular attributes of the main classes of mobile elements, focusing on matters relevant to their mobilization within host genomes, and then move on to explore population biological issues, concentrating on the key determinants of mobile element proliferation and persistence within host genomes. Most mobile elements have predominantly negative effects on their hosts, but at the close of the chapter, we will see that the table has been turned many times, with the host procuring a major benefit from a prior genomic parasite. This need not imply that mobile elements are retained for their advantageous effects. Almost certainly they are not, but as a major source of genomic mutation, mobile elements are bound to occasionally spawn beneficial changes, as is the case with virtually any environmental mutagen.

As briefly outlined in Chapter 3, two broad classes of mobile elements are found in eukaryotes, both of which rely on mobility factors encoded within the elements themselves. Retrotransposons (sometimes referred to as class I elements) proliferate by a copy-and-paste mechanism, using an RNA intermediate for dispersal. After the mobility factors for such elements are transcribed and then translated in the cytoplasm, they generally remain associated with their parental (or related) mRNA molecules, using them as templates for reverse transcription to a DNA product that can be inserted elsewhere in the genome. In contrast, transposons (sometimes referred to as class II elements) proliferate by a cut-and-paste mechanism that directly mobilizes DNA-based elements. After the mobility factors encoded by a transposon have been transcribed and translated, they return to the nucleus (without the mRNA), where they have an opportunity to interact with a random element and insert a copy of it elsewhere. Because the life cycles of all mobile elements involve the export of an element-associated mRNA to the cytoplasm followed by the migration of the translated mobility factors back to the nucleus, the barrier presented by the nuclear envelope may have been a challenge to the early success of eukaryotic mobile elements. However, given the enormous numbers of mobile elements in the genomes of today's multicellular eukaryotes, it is clear that the nuclear membrane is no longer a major impediment.

Most mobile elements insert themselves randomly into host genomes, without regard to the relevance of the landing site to the host's well-being, so their potential to generate deleterious mutations is high. Their negative effects can range from frameshifts and truncations caused by insertions into coding DNA to changes in gene expression resulting from insertions into regulatory regions to large-scale chromosomal rearrangements. Not surprisingly, such activities have earned mobile elements the label of "selfish DNAs" (Doolittle and Sapienza 1980; Orgel and Crick 1980). Although the earliest speculations about the selfish nature of mobile elements were based on compelling verbal arguments, the development of formal population genetic models played an essential role in validating this conclusion (Charlesworth 1985). Selection at the level of a mobile element family will then tend to favor lineages with greater proliferative abilities. In contrast, selection against hosts with high mobile element loads should reduce the extent of mobile element proliferation while also encouraging the evolution of host-encoded resistance factors. The degree to which an element family can expand within a species will then depend on the relative strengths of these opposing forces, which might be modified through time by a coevolutionary arms race between both players.

The range of apparent susceptibilities of different host lineages to mobile element expansion is profound. For example, well over half of the human and maize nuclear genomes consist of mobile elements and their remnants, whereas several unicellular eukaryotes appear to lack mobile elements entirely. Given that the ancestors of the major classes of mobile elements were probably present near the very base of the eukaryotic tree, and that at least some transposons are capable of horizontal transfer, a central goal of evolutionary genomics is to clarify the mechanisms responsible for this uneven phylogenetic distribution.

Our overview of the natural history of mobile elements will start with the retrotransposons, which, as noted in Chapter 3, are generally subdivided into two groups with substantially different mobility mechanisms. Like most retroviruses, LTR retrotransposons are flanked by long terminal repeats (LTRs), which play a key role in element proliferation, whereas non-LTR retrotransposons lack this attribute. It may seem a bit odd to name a group for something that it lacks, and an attempt has been made to rectify this situation (Eickbush and Malik 2002), but this nomenclature will serve our purposes. The vast majority of retrotransposons fall neatly into these two major categories, but a few other eukaryotic classes have been suggested (Cappello et al. 1985; Goodwin and Poulter 2001; Duncan et al. 2002), and additional groups reside within prokaryotes (Curcio and Derbyshire 2003; Lampson et al. 2005). Although we will focus on the LTR and non-LTR groups, these exceptions serve as a reminder that other retrotransposon classes probably await discovery. Following a generic description of the two major retrotransposon groups, similar coverage will be given to transposons. A detailed account of the biology of a single element in *Drosophila melanogaster* will provide more in-depth understanding of each major type.

Non-LTR Retrotransposons

A key contributor to the proliferative success of retrotransposons is the replicative nature of their insertion mechanism. Each birth of a new element leaves the parental copy intact, and once in place, a retrotransposon essentially never leaves its insertion site unless it happens to be contained within a deletion event at the level of the host genome. Although some non-LTR insertions arise by integration into preexisting breaks in the host DNA (Morrish et al. 2002), most are initiated by the elements' own endonuclease activity (Luan et al. 1993; Weiner 2002; Eickbush and Malik 2002). The process, known as target-primed reverse transcription (Figure 7.1), begins with transcription of an autonomous element. By directing transcription initiation to an upstream region, the internal promoters in such elements (red box in figure) ensure their own inclusion in the resultant mRNAs. After translation in the cytoplasm, the element-encoded endonuclease and reverse transcriptase become intimately associated with their parent mRNA, and with luck, the aggregate then reenters the nucleus. If at this point the endonuclease is able to produce a single-strand nick in the nuclear DNA, the close association of the reverse transcriptase with the 3′ tail of the element's transcript can initiate DNA synthesis at the nick, using the mRNA as a template. Several molecular aspects of the final integration steps are poorly understood, but most non-LTR element insertions are flanked by variable-length target-site duplications, implying that staggered cuts are made in the recipient chromosome. Synthesis of the second DNA strand of the inserting element

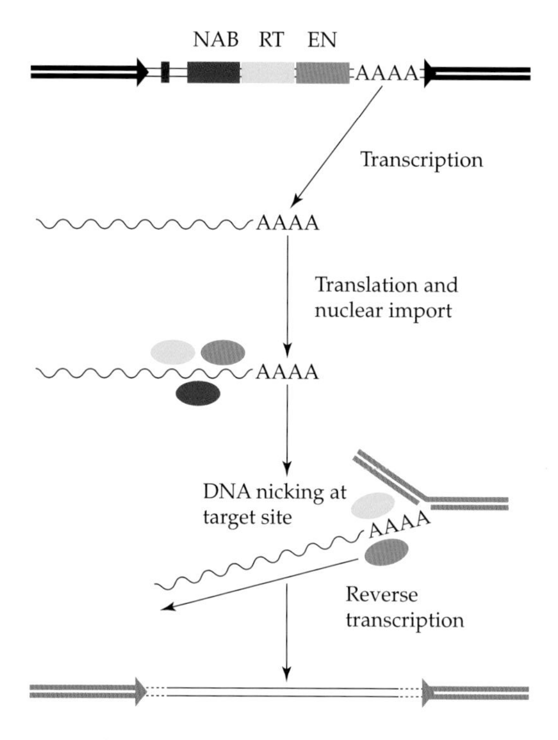

Figure 7.1 Insertion of a non-LTR element into a new target site. Thick lines denote DNA flanking the element, black for the original element and light blue for the new insertion. The original autonomous element encodes three protein domains, which are not necessarily in the order shown or independently transcribed: NAB (nucleic acid–binding protein), RT (reverse transcriptase), and EN (endonuclease). Also included are an internal transcription promoter (red) and a 3′ poly(A) tail. The transcript is shown as a wavy line, and the proteins derived from the transcript as color-coded ovals. Following the creation of a single-strand nick in the recipient DNA, reverse transcription produces single-stranded DNA starting at the poly(A) end of the transcript. It is thought that host-encoded DNA repair enzymes are involved in second-strand synthesis. Potential errors at the 5′ and 3′ ends of the new inserts are shown as dotted lines, and target-site duplications as triangles.

requires removal of the mRNA primer template by an RNase H, usually derived from the host cell, except in the *I* family of elements (Malik 2005).

The sloppy nature of non-LTR element replication causes new insertions to deviate from their parental elements in numerous ways. First, because the polyadenylation signal built into some retrotransposons is weak, transcription occasionally proceeds into the 3′ flanking DNA. Such cotranscription can result in the hitchhiking of parental flanking DNA to new insertion sites, potentially leading to the insertion of regulatory elements, individual exons, or even fully functional genes. Second, because the mRNA template is not directly annealed to the priming site, slippage errors are sometimes made at the initiation of insertion, resulting in the further accumulation of As in the poly(A) region of such elements. Third, 5′-truncated insertions can arise when integration occurs prior to processing of the entire transcript. Fourth, new insertions can be accompanied by other rearrangements, including deletions or inversions of genomic DNA at the insertion site. Finally, due to the absence of a proofreading mechanism, reverse transcription is a relatively noisy process, with a rate of base misincorporation on the order of 10^{-5} per nucleotide site per replication (Mansky and Temin 1995; Boutabout et al. 2001).

The net effect of these aberrations is that many retrotranspositional events result in the insertion of nonautonomous (dead-on-arrival) elements incapable of producing functional proteins. However, not all such incapacitated elements are immobile. Although retrotransposon-encoded proteins bind preferentially to the mRNA from which they arise, they do not do so exclusively. This means that defective elements can still be mobilized as long as they are transcribed and other autonomous (protein-producing) elements remain in the genome (Wei et al. 2001). Such nonautonomous elements effectively operate as secondary genomic parasites by hijacking the mobility apparatus of their autonomous relatives. One of the most dramatic examples of defective element proliferation is the case of the *L1*-dependent *Alu* element in the human genome (see Chapter 3).

As a consequence of their reliance on an endonuclease, all non-LTR elements appear to have at least some insertion-site specificity, although most insertion-site sequence requirements are so minimal that potential sites are widely distributed throughout the host genome. For example, primate LINE elements are preferentially inserted at TT | AAAA sites, which should appear roughly once every $4^6 = 4096$ genomic positions, assuming a random distribution of nucleotides. In dipterans, *Waldo* elements are preferentially inserted into ACAY repeats (where Y denotes an arbitrary pyrimidine), and *Mino* elements are preferentially inserted into AC repeats (Kojima and Fujiwara 2003), both of which are quite common. Many other examples could be cited. On the other hand, a few non-LTR lineages appear to have very specific target-site requirements. Most such lineages rely on highly repetitive and conserved sequences, which presumably offer a reliable and somewhat expendable resource. For example, members of the *R2* clade are inserted specifically into ribosomal RNA gene spacers in arthropods (Eickbush 2002), planarians (Burke et al. 2003), and vertebrates (Kojima and Fujiwara 2004). The *Genie* element is inserted near the telomeres of *Giardia* (Arkhipova and Morrison 2001; Burke et al. 2002). The *NeSL-1* element is inserted into spliced leader sequences in nematodes, and the *CZAR* and *SLAC* elements behave similarly in trypanosomes (Aksoy et al. 1990; Villanueva et al. 1991) (repetitive spliced leader genes encode the leading exon that is *trans*-spliced to the transcripts in these species; see Chapter 9).

The non-LTR elements comprise an enormous phylogenetic group. Through a comparative analysis of reverse transcriptase sequences shared by all elements, Malik et al. (1999) initially suggested the existence of eleven distinct clades. Although subsequent discoveries of new lineages (Malik and Eickbush 2000; Arkhipova and Morrison 2001; Burke et al. 2002) have raised questions about the general validity of this early classification scheme, a case can still be made for the presence of five major groupings based on reverse transcriptase sequence divergence and gene content and arrangement (Figure 7.2). Even so, each of these five groups contains such deeply diverging lineages that only a small fraction of the sequence variation among non-LTR elements separates the major lineages.

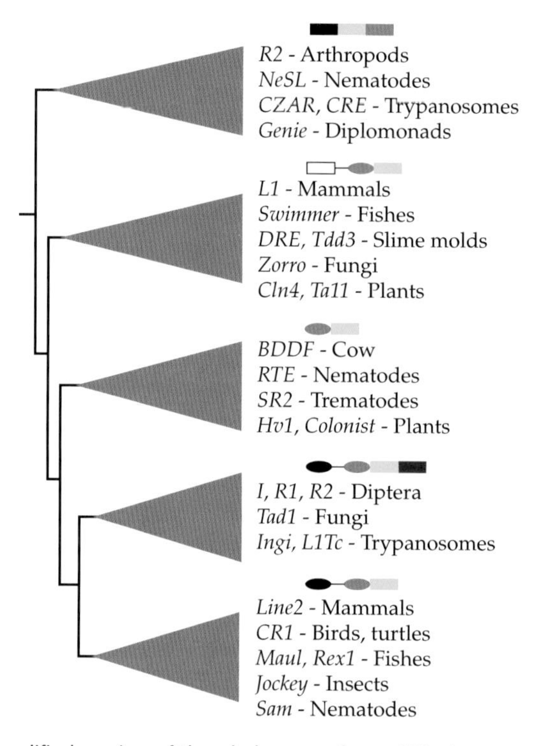

Figure 7.2 Simplified version of the phylogeny of non-LTR elements. The five major clades can be further subdivided into many deep lineages, some representatives of which are given on the right. All autonomous elements carry a coding domain for reverse transcriptase (yellow rectangle). Distinct open reading frames are separated by narrow lines (second, fourth, and fifth clades). The *R2* (top) clade employs an endonuclease that appears to be entirely unrelated to that used by all other clades (green rectangle vs. ovals), although a clade containing both types of endonuclease has been found in the green alga *Chlamydomonas* (Kojima and Fujiwara 2005). Distinct nucleic acid–binding protein genes are illustrated in blue and with an open box. Members of the *RTE*-containing clade uniquely lack a nucleic acid–binding protein (absence of blue), whereas members of the *I*-containing clade uniquely carry a gene for ribonuclease H (red), presumably used in the degradation of mRNA subsequent to reverse transcription. The mechanisms by which the polycistronic open reading frames give rise to multiple proteins are poorly understood. (Modified from Eickbush and Malik 2002.)

One of the greatest enigmas concerning mobile elements concerns their ultimate phylogenetic roots. Although numerous prokaryotes contain small to moderate numbers of mobile elements, no non-LTR-like prokaryotic elements have been found. A plausible case can be made for a relationship between non-LTR retrotransposons and prokaryotic group II introns (Xiong and Eickbush 1990; Eickbush and Malik 2002; Belfort et al. 2002; see Chapter 9), as both employ an endonuclease and a reverse transcriptase to insert new copies via target-primed reverse transcription. However, unlike group II introns, non-LTR retrotransposons are incapable of splicing themselves out of precursor mRNAs, raising the question as to whether non-LTR elements are descendants of group II introns that have lost the ability to self-splice, or vice versa, or whether both are derived from an unknown common ancestor with rather different abilities. Non-LTR inserts within noncoding DNA are under no selection pressure to retain a capacity for RNA splicing, and this, combined with their apparent absence from prokaryotes and that of group II introns from nuclear genomes, suggests that non-LTR elements may be simplified group II introns, but the matter remains unresolved.

Might non-LTR elements be relatively recent arrivals that somehow infectiously spread across various eukaryotic lineages? In contrast to the situation in many transposon families (see below), there is little evidence of phylogenetically distant horizontal transfer in any non-LTR lineage, although a few such events have been suggested (Župunski et al. 2001), and some introductions by inadvertent interspecific matings are likely. The rarity of such movement may be a consequence of cytoplasmic copies of non-LTR retrotransposons being in the form of RNA, which is much less stable than DNA. To complete a successful journey across a species barrier, a non-LTR element's unprotected mRNA molecule would need to be transported intact, most likely along with functional copies of the proteins necessary for integration into the alternative host's genome. In any event, although a number of issues remain unresolved, the deep phylogenies indicated in Figure 7.2 are consistent with the hypothesis that the major non-LTR lineages diversified early, perhaps even prior to the establishment of the major eukaryotic groups.

The *IR* system of hybrid dysgenesis

The first mobile element to be identified in *D. melanogaster*, the *I* factor, was discovered when Picard and L'Héritier (1971) found that crosses between some flies produced a high incidence of genetic abnormalities (a phenomenon known as hybrid dysgenesis, although in this case the participants were members of the same species). All isolated strains of *D. melanogaster* can be characterized as *I* (inducer) or *R* (reactive) (Bucheton et al. 2002; Chambeyron and Bucheton 2005). Most *D. melanogaster* individuals contain 20 to 30 defective *I* elements per haploid genome, but only *I*-strain members carry active elements. These active elements are largely silent in crosses between *I* strains, but when *I* males mate with *R* females, the elements begin to mobilize at very high rates (a 100-fold or so increase) in the germ lines of the hybrid daughters (mobilization has never been seen in males), causing a high incidence of infertility from chromosomal defects. The daughters of the reciprocal cross exhibit a much lower degree of *I* element activation and have nearly normal fertility. *I* does not appear to be active in somatic tissues, and activation in the female germ line appears to be due to the presence of an ovary-specific enhancer within the 5′ UTR of the element. As in the case of LINEs and *Alu*s, nonautonomous elements can spread by exploiting the mobilization apparatus of active elements (Robin et al. 2002).

Remarkably, all individuals in current natural populations of *D. melanogaster* are *I*-strain members, while isolates sequestered in the laboratory prior to 1930 are *R*-strain members. The presence of remnants of inactivated elements in even the oldest isolates of *D. melanogaster* implies significant prior *I* activity followed by element family extinction. Thus, shortly after 1930, either a previously defective element was somehow reactivated within the species or an active element was introduced by horizontal transfer, rapidly spreading throughout the entire geographic range of *D. melanogaster*. Under this hypothesis, the old *R* lab strains have retained their status by being shielded from gene flow from the rest of the species. Although there is little direct evidence for distant horizontal transfers of non-LTR elements in *D. melanogaster*, rare hybridization events with the closely related *D. simulans*, which carries active elements, are possible (Sezutsu et al. 1995; Bucheton et al. 2002).

The rapid proliferation of a mobile element throughout an entire species poses the obvious question as to what, if anything, prevents runaway expansion of copy number within the host genome. In the case of *I*, retrotransposition appears to be tightly regulated, because shortly after the initiation of a dysgenic cross, the number of active elements stabilizes at 10 to 15 per haploid genome. Although the mechanisms by which this is accomplished remain unclear, they go beyond simple natural selection against hosts with high copy numbers. For example, insertion of fragments of either the 5′ UTR or the coding region of an active *I* element into a fly genome greatly reduces the retrotransposition rate in a copy-number-dependent manner. The underlying mechanism appears to involve some sort of homology-dependent suppression mechanism associated with mRNA intermediates (Chaboissier et al. 1998; Jensen et al. 1999a,b; Malinsky et al. 2000). Females that have lost the DNAs associated with introduced transgenes (by segregation), but not their mRNAs, continue to exhibit the suppression effect for at least a generation, pointing to the maternal inheritance of this RNAi-like phenomenon.

The most remarkable feature of *I* element cosuppression is its ability to operate via nonhomologous sequences (Jensen et al. 2002). If a reporter construct with a small portion of *I* is introduced into the fly genome, and then another construct with a nonoverlapping fragment of *I* is added, active transcription of the latter results in the transcriptional silencing of the former. The leading hypothesis to explain such behavior invokes the presence of "relay" sequences containing regions of homology to both introduced constructs, and the obvious candidates are the defective elements residing in the genome. Despite the many remaining mysteries regarding the mechanics of cosuppression, these results strongly suggest that "self-regulation" of *I* element number is an indirect effect of general host cell processes. The enhanced *I* activity in *I*/*R* hybrids is then presumably explained by the 50% reduction in the number of interfering copies in the cytoplasm of hybrid offspring.

LTR Retrotransposons

Like those of non-LTR elements, transcripts of autonomous LTR elements serve as both the information-bearing message necessary for translation into the mobility apparatus and the template for reverse transcription. However, the mobility mechanisms of LTR elements are much more similar to those of eukaryotic retroviruses than to those of non-LTR elements. As in retroviruses, the information critical for the transcription of an LTR element is contained within the terminal direct repeats (the LTRs), and of the three genes encoded by retroviruses (*gag*, *pol*, and *env*), two are also present in LTR elements. The *gag* gene encodes the capsid protein, which helps sequester the key element components during reverse transcription, whereas *pol* encodes the three functional domains necessary for mobilization: protease (PR), reverse transcriptase (RT), and integrase (IN). The RT protein is further subdivided into two domains: a DNA polymerase that can use DNA or RNA as a template, and an RNase H used in the elimination of RNA during the replication reaction. Finally, as in retroviruses, most LTR elements use a host transfer RNA (tRNA) to initially prime reverse transcription during replication, with different element families relying on different tRNAs. The primary distinction between infectious retroviruses and LTR elements is the presence of an *env* gene in the former, the product of which facilitates entry into host cells and hence horizontal transfer.

Following transcription, some LTR element transcripts are translated in the cytoplasm, where the protease activity of the primary translation product subdivides *pol* into its individual protein components. However, most transcripts are destined to be packaged into capsids (generally two per capsid), where RT and IN assemble into viruslike particles along with the host tRNA necessary for priming reverse transcription. In contrast to the situation in non-LTR elements, reverse transcription is carried out within these cytoplasmic vesicles, after which daughter cDNA molecules enter the nucleus in the form of preintegration complexes containing the *pol* products. Insertion into the host chromosome is carried out when IN makes staggered cuts in the host DNA, resulting in short direct repeats flanking the integrated element.

The production of an LTR element's cDNA is a remarkably intricate process. Given that transcription is initiated within the 5′ LTR and terminated within the 3′ LTR, full-length copies of LTR elements are never present at the mRNA level (Figure 7.3). How, then, can a complete daughter ele-

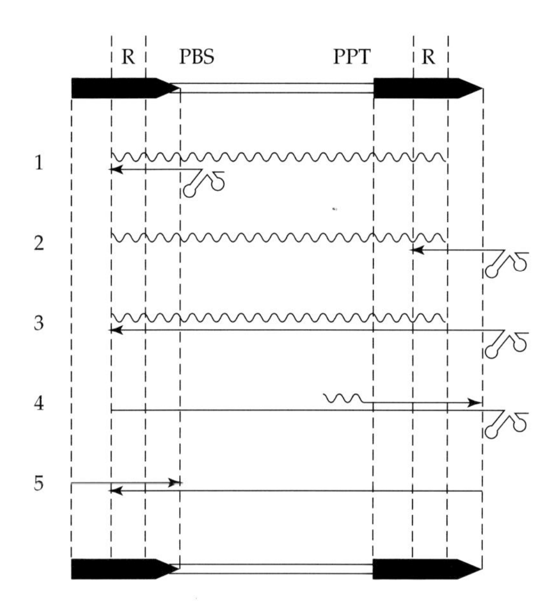

Figure 7.3 The steps by which a copy of an LTR element is produced by reverse transcription. Wavy lines denote RNA; solid lines are DNA; black terminal regions are the LTRs; and the stem–loop structure is the tRNA primer. R is the terminally redundant portion of the LTR contained within both ends of the mRNA; PBS is the tRNA binding site; and PPT is a polypurine tract, resistant to RNase activity. (1) A host tRNA with sequence complementary to the PBS primes reverse transcription of the lower strand, which proceeds to the end of R. (2) A strand switch occurs, with the previously synthesized DNA reannealing to the 3′ end of the mRNA. (3) Reverse transcription proceeds again to the 5′ end of the mRNA. (4) All but the PPT is digested by RNase activity, and the remaining fragment then serves as a primer for the production of a complementary portion of the top strand of DNA. (5) Finally, the two partial strands of DNA serve to prime second-strand synthesis in both directions, thereby reconstituting a complete element at the DNA level. A few intermediate steps are omitted. (Modified from Voytas and Boeke 2002.)

ment be reconstituted at the DNA level? The key resides in the fact that the portion of the LTR missing at the 5′ end of the transcript is present at the 3′ end, and vice versa, ensuring that each transcript contains all of the information necessary for the production of a full-length cDNA copy. After the initial priming of reverse transcription via a host tRNA, cDNA synthesis proceeds in the 5′ direction to make a partial copy of the 5′ LTR segment. Using the homology of LTR sequences at both ends of the transcript as a guide, a series of strand switches of primers followed by cDNA extension then leads to the reconstitution of a full-length element (see Figure 7.3). A few variants on this theme are known to exist (Levin 1995; Duncan et al. 2002).

This mechanism of cDNA production ensures that the complete right and left LTR sequences of newly produced elements are 100% identical, even if they had diverged in the parental molecule. Because LTRs are often hundreds of bases long, this unique feature of newborn LTR elements provides an exceptional opportunity for deciphering the evolutionary demography of a family of LTR elements within a host genome. With no mechanism to maintain their homogeneity after insertion (with the possible exception of gene conversion), the LTRs of individual elements are expected to diverge neutrally (i.e., at the mutation rate), allowing one to infer the age of an insertion from the magnitude of divergence between its two LTRs. Applying this logic to random samples of LTR elements within a genome, several studies have shed light on the average life spans of LTR element insertions.

The age distributions of such elements are often highly L-shaped, closely approximating an exponential distribution. Such distributions are expected under a steady-state birth–death model, and the half-life of an element can be estimated from the slope of the distribution on a log-arithmetic plot. For example, the age distribution of *hopi* family members in the rice genome implies that 50% of insertions are lost by the time their LTRs have diverged by just 1% at the nucleotide sequence level (Figure 7.4A). Similar results have been obtained for a broad range of other LTR families in rice (Ma et al. 2004), maize (San Miguel et al. 1998), wheat (San Miguel et al. 2002), peas

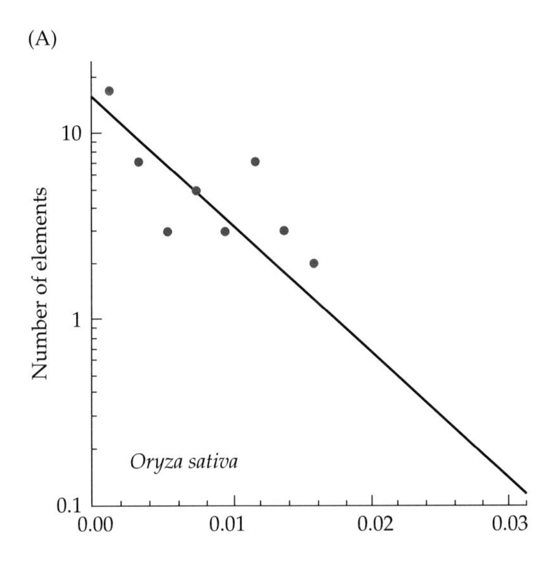

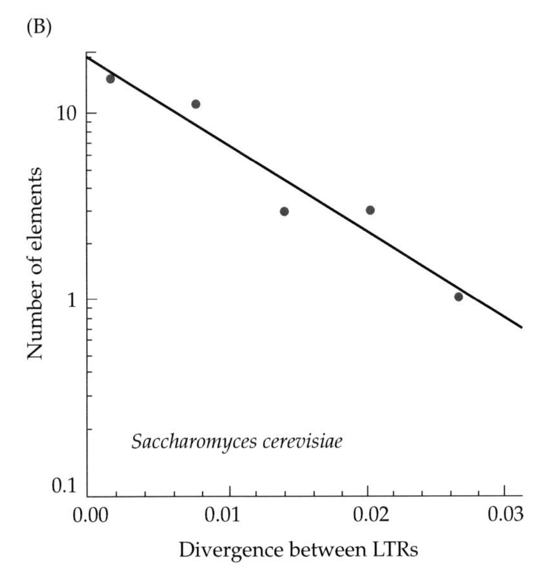

Figure 7.4 (A) Age distributions (in units of sequence divergence) of the *hopi* family of LTR elements in the rice genome (B) and of the *Ty1* and *Ty2* families in yeast. The negative slopes of the regressions yield half-life estimates of 0.010 and 0.014 (in terms of sequence divergence), respectively. (A, modified from Vitte and Panaud 2003; B, modified from Promislow et al. 1999.)

(Jing et al. 2005), and *Arabidopsis* (Devos et al. 2002; Pereira 2004). Because the average level of divergence of alleles at silent sites in land plants is generally on the order of 1% or less (see Chapter 4), this implies that a large fraction of LTR element insertions in plants are younger than segregating alleles, with most never reaching fixation at the population level.

These kinds of observations are not unique to plants. For example, in units of sequence divergence, the *Ty* elements of the yeast *S. cerevisiae* have an estimated half-life of 1.4% (Figure 7.4B). In addition, virtually all elements in 17 distinct LTR families in *D. melanogaster* are less than 1% divergent in their terminal regions, with 70% having identical intra-element LTRs (Bowen and McDonald 2001). Essentially the same observation has been made in *C. elegans* (Bowen and McDonald 1999). In striking contrast, LTR elements seem to survive much longer in the human genome, as revealed by average intra-element LTR divergences of 13.4% (Tristem 2000).

The distinct mobility mechanisms of non-LTR and LTR retrotransposons impose fundamental differences in their demographic properties. First, whereas non-LTR elements are often truncated (and hence dead on arrival) at the time of integration, new LTR insertions are generally of full length. Second, whereas a non-LTR element can be lost from the host genome only if it happens to be contained in a deletion event, LTR elements are subject to an internal mechanism of inactivation: intra-element recombination between the 5′ and 3′ LTRs results in a loss of intervening sequence, yielding a defective element in the form of a solo LTR. Third, LTR elements may gradually lose their ability to complete a replication cycle as their LTR sequences become too divergent to support primer strand switching. Thus, all other things being equal, LTR elements are expected to be more successful in producing autonomous progeny at the time of insertion than non-LTR elements, but less successful in maintaining them. Finally, because viruslike particles generally package more than one mRNA, the opportunity exists for template switching in the replication of heterologous mRNAs. Such events can generate entirely new families of elements (Temin 1991; Jordan and McDonald 1998; Kelly and Levin 2005).

Despite these differences, there are also similarities in the life history features of LTR and non-LTR elements. First, like non-LTR elements, LTR retrotransposons appear to depend almost entirely on vertical inheritance for survival. Few convincing cases of horizontal transmission have been uncovered, and these appear to involve rare hybridization events between congeneric species, as in the case of a *copia* element shared between *D. melanogaster* and *D. willistoni* (Jordan et al. 1999). Second, like non-LTR elements, autonomous LTR retrotransposons are subject to exploitation by nonautonomous derivatives (Jiang et al. 2002; Havecker et al. 2004). Third, a number of LTR elements have strong insertion-site preferences, often in regions containing arrays of genes. For example, of the five families of such elements in *S. cerevisiae*, four integrate primarily within the 750 bp region upstream of tRNA genes, while the fifth is targeted to telomeres (Voytas 1996; Zou et al. 1996). The *Tf1* element of *S. pombe* exhibits a strong preference for insertion in the promoter region of protein-coding genes (Kelly and Levin 2005).

Finally, it is worth noting that mobile elements that rely on the use of processed mRNA intermediates are not expected to contain introns, and this expectation is frequently exploited to formally demonstrate that an element qualifies as a retrotransposon: if a spliceosomal intron is introduced into an otherwise active retrotransposon, a new genomic insertion derived from such a construct is expected to lack the intron as a consequence of its removal during the intermediate stage of mRNA processing (Boeke et al. 1985; Kinsey 1993). This result has been seen with nearly all modified retrotransposons, although a few rare types of LTR elements do contain spliceosomal introns (Arkhipova et al. 2003; Goodwin and Poulter 2004; Evgen'ev and Arkhipova 2005). In these exceptional cases, only a subset of element transcripts may be spliced and translated into proteins in the cytoplasm, with unspliced forms being retained for packaging and replication.

Comparative analysis of reverse transcriptase sequences suggests that LTR elements can be subdivided into four major clades (Xiong and Eickbush 1990; Eickbush and Malik 2002). However, as in the case of non-LTR elements, most of the total diversity of such elements is found among deeply diverging lineages within each clade. Two of the four major clades , the *Ty1/copia* and *Ty3/gypsy* groups, named after elements described initially in *S. cerevisiae* and *D. melanogaster*, respectively, are very similar in most respects except for the relative position of the IN domain within the *pol* gene. Although the broad phylogenetic distribution of each group, combined with both groups' apparent absence from prokaryotes, suggests a presence in the stem eukaryote prior to the emergence of the major eukaryotic lineages, the origin of LTR retrotransposons is no less enigmatic than that of non-LTR elements. Malik and Eickbush (2001) have hypothesized that LTR elements arose via the fusion of a non-LTR element and a DNA-based transposon, with the former providing reverse transcriptase and the latter providing the integrase necessary for insertion into double-stranded DNA. In still a third step of eukaryotic retroelement evolution, retroviruses may arise when LTR retrotransposons acquire *env* genes from unrelated viruses, providing them with a capacity for extracellular transfer (Temin 1980; Malik et al. 2000; Frame et al. 2001; Pearson and Rohrmann 2006).

The *gypsy* element

Gypsy is one of about 30 LTR element families found in *D. melanogaster*. It is not particularly abundant, with just ten or so copies carried per individual (Rizzon et al. 2002), but the element is unusual in that it encodes an *env* gene, which may endow *gypsy* with infectious viruslike properties. *Gypsy* elements are found in many species of *Drosophila*, and although many of their *env* genes have acquired frameshift mutations, a comparison of substitutions at silent and replacement sites indicates that functional *env* genes are under strong purifying selection (Mejlumian et al. 2002; Pelisson et al.

2002), consistent with the idea that *gypsy* is a derived form of a retrovirus that has taken up residence in the *Drosophila* germ line (Terzian et al. 2001). Phylogenetic analyses of the elements in various *Drosophila* hosts support this view by pointing to the occurrence of horizontal transfer (Terzian et al. 2000; Herédia et al. 2004).

From the standpoint of mobilization, *gypsy* is clearly an unusual element. When lines of *D. melanogaster* with inactive elements are fed homogenized larvae containing viruslike particles from active elements, rates of incorporation can be as high as one per gamete (Kim et al. 1994; Pelisson et al. 2002). A role for *env* in transmission is implicated by the observation that such activity is inhibited by pretreating the homogenate with antibodies against the Env protein, but the mechanism by which Env is actually deployed in the movement of *gypsy* is less clear. Although heritable retrotranspositional events are generally thought to require germ line transcription of an element, *gypsy* is apparently expressed in female somatic cells and then somehow transferred to the oocytes, where integration into the germ line subsequently takes place (Chalvet et al. 1999). This is one of the only known examples of genetic transmission from the soma to the germ line.

There is significant genetic variation in the ability of *D. melanogaster* to resist the spread of *gypsy*. In particular, the *flamenco* gene, the products of which remain to be characterized, operates via a maternal effect. Mutant homozygous females at this locus provide a permissive environment for *gypsy* mobility in male and female progeny (Prud'homme et al. 1995), leading to a higher abundance of elements within individuals of such stocks (Lyubomirskaya et al. 2001). As in the case of the non-LTR *I* element of *D. melanogaster*, some form of RNA-mediated cosuppression appears to play a role in the posttranscriptional silencing of *gypsy*, as small RNAs homologous to *gypsy* (the hallmark of RNA interference, and potentially the products of the *flamenco* gene) are found in fly ovaries (Sarot et al. 2004).

Transposons

In contrast to retrotransposons, for which new insertions are derived from RNA intermediates, transposon movement relies exclusively on the excision of elements at the genomic (DNA) level. Transposition requires that an active element be transcribed to produce the transposase enzyme, but neither the element's mRNA nor a cDNA derived from it is involved in subsequent mobilization. Rather, after entering the nucleus, transposase molecules interact with random genomic copies of the element by binding to key recognition sequences contained within the 100–500 bp terminal inverted repeats (TIRs) flanking such elements. Once positioned at such sites, the transposase creates double-strand breaks and attempts to insert the liberated element elsewhere in the genome.

Despite the excision step, most transposon movement leads to an increase in copy number. Usually, the double-strand break inflicted at the source site is repaired by homologous recombination with the filled site on a sister chromatid (Engels et al. 1990; Gloor et al. 1991), restoring the original copy (Figure 7.5A). Duplicative transposition can also occur during germ cell division when a copy transposes from a replicated portion of a chromosome to a region that has not yet been replicated (Chen et al. 1992), ensuring that both descendent chromosomes inherit the new copy after replication (Figure 7.5B). As in the case of most retrotransposons, the relative simplicity of the insertion-site sequences of transposons generally ensures a wide, more or less random distribution of such locations over the host genome, although the physical mechanism of insertion may result in the localization of newborn elements near their parental sources. One exception is the *Pokey* element of the microcrustacean *Daphnia*, which is addressed to a specific site within the 18S ribosomal RNA repeat (Penton et al. 2002), similar to the situation for the *R1* and *R2* retrotransposons of arthropods.

Because gap repair is error prone, newborn transposons can sometimes be dead on arrival, and resident elements will inevitably succumb to inactivating mutations, so virtually all transposons are expected to eventually become nonautonomous. Moreover, because transposase is imported back into the nucleus with no memory of its genomic source, an autonomous element can mobilize any genomic copy of itself, provided the key recognition

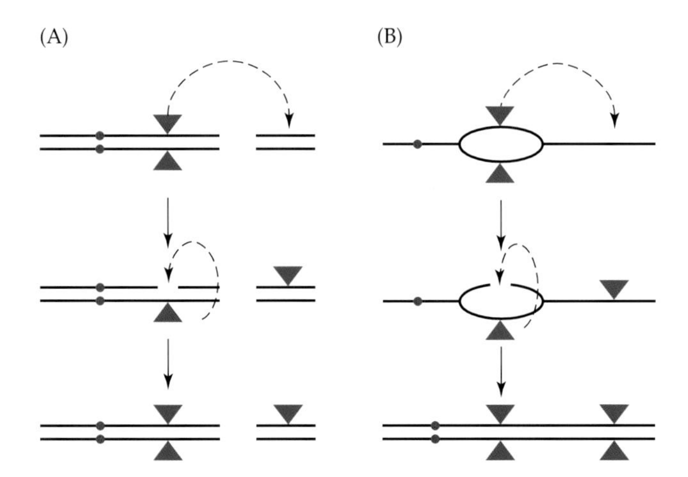

Figure 7.5 Duplicative transposition of a DNA-based transposon. In each case, the excised element is replaced by the intact copy by double-strand break repair using the sister chromatid. Centromeres of sister chromatids are denoted by solid circles; transposons are denoted by triangles; and a double-strand gap is denoted by a break in the chromatid. (A) The donor element moves to a replicated site on the same or a different chromosome, such that only a single daughter chromosome inherits a new element. (B) The donor element arises in a replicated region of the chromosome and moves to an unreplicated region, such that both daughter chromosomes inherit the new copy. Here, the solid bubble in the upper chromosome represents the region of replicated DNA, which will eventually expand to yield complete sister chromatids.

sites reside within the TIRs. Thus, as in the case of retrotransposons, incapacitated elements (in this case, lacking a functional transposase gene) can expand at the genomic level by exploiting the transposition machinery of active elements. Not surprisingly, of the many thousands of transposons that have been identified in various species, only a few have been formally demonstrated to be autonomous. A striking example of the proliferative ability of defective elements is provided by a superfamily of miniature inverted-repeat transposable elements (MITEs) that is widespread throughout animals and land plants, with up to tens of thousands of copies per genome in some cases (Bureau and Wessler 1994; X. Zhang et al. 2004). Completely devoid of protein-coding capacity, these large families of elements maintain themselves by taking advantage of autonomous members of the *mariner* family (Feschotte et al. 2003; Jiang et al. 2003).

A number of examples of horizontal transfers of transposons have been documented, perhaps the best case being that of the *mariner* elements in insects, other animals, and plants (Robertson 1993, 1997; Robertson and Zumpano 1997; Feschotte et al. 2002; Lampe et al. 2003). Although the exact mechanisms are unknown, transfers across species boundaries may be facilitated by the self-contained nature of transposon systems, which require no host factors for mobilization. As a consequence of this feature, transposons are capable of cleavage and integration when introduced to the genomes of even distantly related species (Rubin et al. 1999; Q. Zhang et al. 2000; Mamoun et al. 2000). Although transposons are widely distributed throughout the eukaryotic domain, this capacity for horizontal transfer greatly compromises the use of phylogenetic data to infer the times of origin of transposon families. Nevertheless, eubacteria harbor numerous lineages of insertion sequences (ISs) (Chandler and Mahillon 2002) that appear to be related to distinct lineages of eukaryotic transposons (Eisen et al. 1994; Kapitonov and Jurka 1999; Feschotte 2004) and to use essentially the same insertion mechanisms. Thus, there is again good reason to believe that DNA-based transposons have a very early origin.

The *P* element

One of the most intensely studied transposons is the *P* element of *Drosophila melanogaster*, originally discovered through its involvement in a hybrid dysgenesis syndrome similar to that involving the *I* retrotransposon. High rates of mutation, chromosomal breakage, male recombination (normally absent in *D. melanogaster*), and sterility (Kidwell 1983, 1994) are uniquely observed in the offspring of crosses between *P*-carrying males and *P*-free females, but not vice versa. Within the germ line, *P* element activity is regulated by a maternally transmitted condition known as *P* cytotype (Engels 1979), which appears to be specifically associated with *P* elements present in a telomere of the X chromosome (Niemi et al. 2004). Females lacking such elements transmit the *M*-cytotype condition, which is permissive to element activity.

Although *P* elements are now globally distributed in *D. melanogaster*, this appears to be a very recent condition. Laboratory strains of *D. melanogaster* isolated more than 50 years ago are entirely free of *P*. Moreover, no *P* elements have been found in any member of the *melanogaster* subgroup of *Drosophila* except *D. melanogaster* itself, which appears to have obtained it from *D. willistoni* (Daniels et al. 1990; Pinsker et al. 2001). Such a rapid and complete invasion is remarkably reminiscent of the *I* element system in the same species (noted above).

Like many mobile elements, *P* is active only in the germ line. In this case, tissue specificity is conferred by host-specific alternative splicing (Labrador and Corces 1997; Rio 2002). Unlike retrotransposons, transposons often harbor introns, and of the four carried by *P*, the third is retained in transcripts in somatic cells, resulting in a premature termination codon and the production of a nonfunctional transposase (Laski et al. 1986; Simmons et al. 2002). Such truncated transposases suppress transposition by competitively binding to the TIRs of inserted elements, restricting their access to functional transposase (Rio 2002). In some lines, a defective *P* element (*KP*) has a similar effect (Lee et al. 1998). This repressor mechanism is so powerful that it even somatically silences *P* elements in dysgenic crosses.

Rates of Insertion and Fitness Effects

Prior to considering the population biology of mobile elements, some basic information on the mobility rates and fitness effects of element insertions is essential, as both factors ultimately dictate whether a mobile element family can achieve long-term persistence in its host species. Because mobile elements are constantly subject to inactivating mutations, long-term persistence of an element family is absolutely dependent on the mobilization of existing autonomous copies, which by definition ensures the retention of functional replicative machinery in fresh insertions. In addition, the effects of new insertions on host fitness must not be too negative if they are to avoid selection at the host level long enough to produce still another generation of viable progeny elements.

Unfortunately, few attempts have been made to estimate the rate of element insertion, and almost all of these derive from mutation accumulation experiments in the fruit fly *D. melanogaster*. Such experiments involve the transmission of initially identical chromosomes through replicate full-sib mated lines for dozens of generations. With such small population sizes, random genetic drift overwhelms the ability of selection to oppose the accumulation of all but very highly deleterious insertions, and in accordance with the neutral theory (see Chapter 4), the increase in element numbers will be close to the rate of insertion. With *D. melanogaster*, new insertions can be identified by hybridizing element-specific DNA to polytene chromosomes and visualizing new positions under the microscope. Summarizing over a large number of element families, such studies lead to average insertion rate estimates of 1.2×10^{-4} per element per generation (Table 7.1). One

TABLE 7.1 Rates of mobile element insertion and loss (per element per generation) in *Drosophila melanogaster*

NUMBER OF FAMILIES	INSERTION	EXCISION	REFERENCE
11	1.2×10^{-4}	4.0×10^{-6}	Maside et al. 2000
9	1.0×10^{-4}	3.9×10^{-6}	Maside et al. 2001
17	1.8×10^{-4}	1.8×10^{-6}	Nuzhdin and Mackay 1995

Note: Rates are averaged over a large number of families, most of which are retrotransposons, from changes accumulated over 60–200 generations in parallel sets of initially identical, inbred lines.

limitation of these estimates is their inability to discriminate between autonomous and nonautonomous elements. Because not all potential parental elements are autonomous and not all insertions are complete, the rate of autonomous element insertion per autonomous element (which is the key to an element family's success) could be higher or lower than the estimates in Table 7.1. In contrast, the rate of element excision averages nearly two orders of magnitude less than the rate of insertion, just 3.2×10^{-6} per element per generation.

Other smaller studies with *D. melanogaster* have yielded qualitatively similar results (Eggleston et al. 1988; Harada et al. 1990; Pérez-González and Eickbush 2002), and the only studies for a non-fly species, involving the *Ty1* LTR element in yeast, suggest insertion rates on the order of 10^{-6}–10^{-4}, depending on the growth temperature, and excision rates about a hundred times lower (Curcio and Garfinkel 1991; Garfinkel et al. 2005). None of these estimates includes observations from highly active phases of hybrid dysgenesis, which can yield insertion rates approaching 0.01 per element per generation (Kidwell 1983; Busseau et al. 1994; Petrov et al. 1995). Extrapolating to the full set of mobile elements across the entire *Drosophila* genome, the total number of new insertions per gamete per generation is estimated to be 0.1–0.2 (Nuzhdin and Mackay 1995; Maside et al. 2000, 2001), so for this species there is little question that mobile element activity is a major force pushing the genome toward larger size.

Estimates of the average effects of mobile element insertions, again entirely confined to *D. melanogaster*, uniformly confirm the idea that insertions are typically deleterious to host fitness. On average, both homozygous and heterozygous insertions individually reduce fitness by 0.5%–1.5% (Eanes et al. 1988; Houle and Nuzhdin 2004; Pasyukova et al. 2004), although a study in one set of lines yielded average fitness effects as high as 13% (Mackay et al. 1992; Lyman et al. 1996), and approximately 10% of insertions have lethal effects (Engels 1989; Maside et al. 2000). In yeast, *Ty1* insertions, which target intergenic regions upstream of tRNA and rRNA genes, have small but significant negative fitness effects, averaging 0.6% per insertion (Blanc and Adams 2004). In the prokaryote *E. coli*, *Tn10* insertions cause an average 1%–3% reduction in fitness, depending on the environment (Cooper et al. 2005). Although most deleterious effects are likely to be direct consequences of insertions, it has been suggested that the relatively high heterozygous effects on viable progeny production may be indirect consequences of an elevated rate of ectopic recombination that results when insertions present on single chromosomes cause misalignment with homologs during meiosis (Pasyukova et al. 2004).

Regulation of Mobile Element Activity

Given the ubiquity of mobile elements and the general negative effects of new insertions, host species are expected to be under strong selection to restrict mobile element activity. In contrast, selection at the level of the mobile element family (effectively a parasite population) will encourage elevated insertion rates so long as these are not offset by excessive reductions in host fitness. Thus, a coevolutionary loop between various aspects of mobile element adaptation and host counteradaptation may be common, although the long-term stability of such situations remains unclear.

One aspect of mobile element evolution that will generally be mutually advantageous to both element and host is a restriction of element activity to the germ line, as damage inflicted on somatic cells will reduce the reliability of host carriers. This predicted pattern has been widely observed. For example, in *D. melanogaster*, the rDNA-inhabiting non-LTR elements *R1* and *R2* appear to move nearly exclusively in the male germ line (Pérez-González et al. 2003), whereas transposons such as *I* and *P* are largely silent except in the female germ line (Engels 1979; Bucheton et al. 2002; Rio 2002). The activity of the *ZAM* LTR element of *D. melanogaster* is also restricted to ovaries (LeBlanc et al. 2000), and the *Doc* non-LTR element exhibits its highest levels of transcription in ovaries and testes (Zhao and Bownes 1998). Similarly, *Tag1*, a transposon in *Arabidopsis*, moves almost exclusively in developing ovules and pollen (Galli et al. 2003). Yeast *Ty3* is about 50 times more active in meiotically than in mitotically dividing cells (Ribeiro-dos-Santos et al. 1997), and *Ty1* and *Ty5* are also more highly expressed during mating (Voytas and Boeke 2002; Sandmeyer et al. 2002). Although yeast is a single-celled organism, such a restriction in element activity will help ensure the survival of clonal lineages prior to mating. Finally, LINE mobility in humans appears to be largely restricted to male and female germ lines and suppressed in somatic tissues (Moran and Gilbert 2002). One known exception to this pattern is the *Tc1* non-LTR element in *C. elegans*, which is essentially silent in the germ line but highly mobile in mitotically dividing cells (Eide and Anderson 1985; Plasterk and van Luenen 2002).

Several additional observations indicate the existence of evolved constraints on mobile element activity. For example, the normally quiescent *C. elegans Tc1* has been shown to mobilize when placed in human cells (Schouten et al. 1998), *Tc3* rapidly mobilizes in zebrafish (Raz et al. 1998), and many other trans-species mobilizations have been induced in the laboratory with various members of the *Tc1/mariner* family, even in bacteria (Plasterk et al. 1999; Rubin et al. 1999). Mobile elements are also frequently activated by various kinds of stress to the host (Junakovic et al. 1986; Ratner et al. 1992; Engels 1989; Finnegan et al. 1989; Petrov et al. 1995; O'Neill, O'Neill et al. 1998; Labrador et al. 1999; Grandbastien et al. 1997, 2005; Grandbastien 1998; Capy et al. 2000), which may be a simple consequence of the compromised efficiency of host regulatory mechanisms in extreme environments.

The opportunity for self-regulation

Because the transmission of a mobile element from one generation to the next depends critically on host viability and reproductive potential, selection may oppose overly aggressive copies, and as just noted, substantial evidence does indeed support the idea that the physical activities of many mobile elements are strongly constrained. But is the regulation of mobile element activity an evolved feature of the element itself, of host-encoded immunity functions, or both? And if element-associated features are involved, is this a consequence of natural selection?

There is little question as to the involvement of the physical features of the elements themselves. Insertions of mobile elements may inadvertently alter local chromatin structure in ways that inhibit further element colonization (Ye et al. 2005), and in some cases the mere retention of inactivated element copies within a host genome provides a built-in mechanism for constraining the proliferation of autonomous copies. For example, as noted above, transcribed but otherwise defective elements can divert the mobility machinery from active copies, providing a buffer against the runaway proliferation of autonomous copies. However, such constrained behavior is probably nothing more than an indirect consequence of the basic life cycle of a mobile element. Because all mobile element copies must eventually succumb to mutational inactivation, the accumulation of nonautonomous elements is inevitable.

Consider the case of the highly abundant non-LTR *L1* elements of humans, which have extremely low levels of expression resulting from an inhibition of transcript elongation (Han et al. 2004). By simply engineering a series of silent-site substitutions within the *L1* open reading frame, Han and Boeke (2004) were able to increase the transposition rate 200-fold. There is no evidence that this constrained activity of endogenous *L1* elements is a consequence of direct selection (at the level of either the element or the host). Indeed, to date, there are no convincing examples of any autonomous eukaryotic mobile element having acquired a selectively promoted mechanism for regulating its own activity or that of other elements. Charlesworth and Langley (1986) present some compelling explanations for why this is so (Box 7.1).

Box 7.1 The Population Genetic Conditions for the Evolution of Mobile Element Self-Regulation

The selective advantage of a mutant element capable of self-regulation (i.e., of restricting the insertion rate of family members) is a function of the improvement in host fitness resulting from reduced element activity (which influences the self-regulating element's likelihood of being vertically transmitted to the next generation) and of the degree to which such an element is linked to the chromosomal region(s) in which it suppresses element activity (which determines the degree to which the advantage is transmitted to the next generation along with the element itself). Imagine an element that prevents the insertion of other elements in its immediate vicinity, assumed to be a fraction p of the entire genome. Such localized transposition immunity implies essentially complete linkage of the regulating element and the advantage that it produces.

The average reduction in the total insertion rate will be approximately $np\mu$, where μ is the insertion rate per element and n is the number of autonomous elements per genome (assumed to be randomly distributed, so that np is the expected number in the zone of regulation). For a genome in which an element family is in approximate equilibrium due to a balance between element insertion and selection at the host level, the average effect of an insertion on host fitness is approximately μ (see below), so the selective advantage associated with local transposition immunity is $s \approx np\mu^2$. Thus, with np unlikely to be much greater than 1, and μ^2 being on the order of 10^{-8} (see Table 7.1), $np\mu^2$ is expected to be no greater than 10^{-6}, and possibly much lower.

Another potential mechanism of self-regulation, called transposition repression by Charlesworth and Langley (1986), considers the hypothetical situation in which an element has a genome-wide ability to reduce the insertion rates of all other family members by some fraction f. In this case, given a total insertion rate of a family of $n\mu$, the maximum selective advantage of repression, approximately $fn\mu$, would be experienced if all insertions had dominant lethal or sterilizing effects. However, because the data cited in the previous section imply heterozygous fitness effects of about 1%, the selective advantage of a mutant element conferring genome-wide transposition repression should be more on the order of $s \approx 0.01 fn\mu$. Although some subtle issues regarding recombination rates have been glossed over here, the main point is that the selective advantage of transposition repression will generally be very small, again most likely less than 10^{-6}.

These qualitative results point to the extremely stringent conditions necessary for the adaptive evolution of self-regulatory mechanisms in mobile element systems. If selection is to promote a mutation conferring self-regulatory ability, the advantage of self-regulation (from the element's perspective) must exceed the power of random genetic drift at the element's insertion site, which must be essentially the same as that in the host species, approximately $1/N_g$, where N_g is approximately twice the effective population size of a diploid, outcrossing host species. Thus, if $s < 10^{-6}$, and perhaps substantially so, the host N_g must be considerably larger than 10^6 for self-regulation to evolve by positive selection. Given the data reported in Chapter 4, such conditions are not likely to be met in most multicellular eukaryotes, or even in many unicellular species. On the other hand, some prokaryotic species are likely to have a sufficiently large N_g to provide an adequate environment for the selective promotion of self-regulation. Thus, it is of interest that several prokaryotic mobile elements have evolved mechanisms for inhibiting transposition, often in a density-dependent fashion (Zerbib et al. 1990; Kleckner 1990; Craig 2002; Reznikoff 2002).

Host-encoded mechanisms for mobile element containment

Although the power of selection in eukaryotes may be generally insufficient for the emergence of element-encoded mechanisms of self-restraint, the benefits of reducing element activity can be much more substantial from the standpoint of the host. As noted above, from an element's perspective, the selective advantage of self-regulation will be roughly proportional to the number of active family members per host genome (n), but if a *host-encoded* mechanism for constraining element activity had broad utility, the advantage would be approximately proportional to the total number of autonomous elements across all element families. Thus, given the multiplicity of element families within hosts and their general negative effects on host fitness, it comes as no surprise that host mechanisms for regulating element activity have evolved. The more surprising observation is that few, if any, eukaryotic species have developed effective enough defensive mechanisms to eliminate mobile element activity entirely. Host-encoded mechanisms for reducing mobile element activity involve numerous aspects of cell biology, including transcription and RNA processing, chromatin dynamics, protein modification, and nuclear import from the cytoplasm (Lippman et al. 2003; Aye et al. 2004), but most of these mechanisms can be subdivided into four categories.

First, as discussed above in the context of the *I* and *gypsy* elements of *D. melanogaster*, homology-dependent suppression provides a powerful mechanism for silencing mobile elements. The involvement of an mRNA intermediate in both cases strongly implies the use of the posttranscriptional gene-silencing mechanism known as RNA interference (RNAi; see Chapter 3). However, it is not entirely clear how the double-stranded RNA-directed degradation associated with RNAi can control the activities of the single-stranded transcripts of mobile elements. One possibility is that as the number of copies of a mobile element increases within a host genome, there is an increasing likelihood that one or more copies will be erroneously transcribed from a noncoding strand, which would then lead to the spontaneous annealing of complementary copies to produce double-stranded RNA. The non-LTR *F* element of *D. melanogaster* is known to exhibit such behavior (Contursi et al. 1993). With transposons, an additional mechanism is possible: elements that are erroneously transcribed beyond the terminal inverted repeats will contain complementary end sequences at the mRNA level, which can then snap together by Watson-Crick base pairing (Sijen and Plasterk 2003).

Despite these uncertainties, several observations support the idea that RNAi plays a central role in constraining mobile element activity. For example, isolates of the nematode *C. elegans* that have lost the RNAi silencing pathway invariably have elevated rates of transposition (Ketting et al. 1999; Tabara et al. 1999). The RNAi pathway in this species also serves as an effective system for opposing the proliferation of double-stranded RNA viruses (Lu et al. 2005; Wilkins et al. 2005). Observations on the non-LTR *INGI* element in *Trypanosoma brucei* have yielded evidence of the production of small interfering molecules (Djiking et al. 2001), the hallmark of RNAi, suggesting that this pathway is at work in at least some unicellular species. (Posttranscriptional silencing also reduces the rate of *Ty1* retrotransposition in the yeast *S. cerevisiae*, although the absence of an RNAi pathway in this species implies the use of an alternative mechanism; Garfinkel et al. 2003.)

A second major mechanism of host control involves transcriptional silencing via the methylation of cytosines within genomic copies of mobile elements. In mammals, for example, a host-encoded DNA methyltransferase specifically active in spermatogonia and spermatocytes is responsible for the silencing of both LTR and non-LTR retrotransposons (Bourc'his and Bestor 2004), and related phenomena have been observed in the plant *Arabidopsis* (Miura et al. 2001) and the ascomycete *Ascobolus* (Faugeron et al. 1990; Rhounim et al. 1992; Goyon et al. 1996). Several key questions about this mode of transcriptional silencing remain only partially resolved. Is DNA methylation of mobile elements a direct cause or an indirect consequence of transcriptional silencing? How is DNA methylation specifically addressed to mobile elements, and how are methylated regions stably maintained across host generations?

Remarkably, at least in mammals and plants, the sites of DNA methylation appear to be guided by the short interfering RNAs produced by RNAi (Zilberman et al. 2003; Kawasaki and Taira 2004; Martienssen et al. 2004). This observation implies that transcriptional silencing is a by-product of posttranscriptional inactivation; that is, that the two major host defense mechanisms against mobile element activity are part of the same system. Such a dual mechanism could play a powerful role in reinforcing the silencing of mobile elements, as any demethylated elements that were transcribed would have the potential to yield some double-stranded RNAs, which would in turn elicit RNAi, returning the elements to a methylated state.

A third, highly unusual mechanism of mobile element control exists in the fungus *Neurospora crassa*, where the proliferation of all mobile elements is greatly curtailed, if not entirely prevented, by repeat-induced point mutation (RIP; Selker 1990). Although RIP occurs only during the sexual phase, it affects nearly all duplicated DNAs (not just mobile element insertions), and the alterations are permanent: both copies are generally riddled with mutations, and an unlinked duplicate has only about a 50% chance of going undetected. Not surprisingly, only a single active mobile element, the non-LTR *Tad* retrotransposon, has been found in a single *Neurospora* strain (Cambareri et al. 1994). However, most strains exhibit signs of prior *Tad* inactivation by RIP, and geographically adjacent clones can have different inactivated inserts (Kinsey et al. 1994; Cambareri et al. 1998; C. Anderson et al. 2001), so retrotransposition must be ongoing, albeit at a very low rate. Although it is remarkable that a mechanism of mobile element containment with this level of efficiency has been found in just a single species, RIP comes at a significant long-term cost: it essentially eliminates the possibility of any gene spawning a viable duplicate copy.

Finally, mammals appear to have an intrinsic intracellular immune defense against many retroviruses. Members of the multigene family of cytidine deaminases encourage the deamination of cytosines to uracils (see Chapter 6) during reverse transcription, thereby yielding mutagenized daughter viral genomes that are either destroyed by uracil DNA glycosylases (host proteins that eliminate such bases from DNA) or give rise to defective progeny upon replication (Bieniasz 2004). This defense system also appears to diminish the activity of mammalian retrotransposons by reducing cDNA production (Esnault et al. 2006; Muckenfuss et al. 2006).

The Population Biology of Mobile Elements

Now that we have outlined the natural history of mobile elements, we turn to their ability to persist and expand at the expense of their hosts. The one simple requirement for the long-term persistence of a mobile element family is identical to the viability criterion for any population: the average autonomous element must give rise to at least one autonomous daughter element before acquiring inactivating mutations or being eliminated by genomic deletion, ectopic recombination, or natural selection. A lower rate of daughter element production would imply a net replacement rate of less than one, a decline in the number of elements per host genome, and eventual element extinction. In contrast, because a fraction of element insertions will have deleterious consequences for host fitness, a highly aggressive element with excess daughter element production could eventually drive its host to extinction. Thus, the key to long-term mobile element success is an ability to stabilize at a copy number high enough to avoid stochastic loss but low enough to minimize the risk of host extinction. We will now outline some simple theory addressing these issues. Important technical reviews are provided by Charlesworth (1985) and Brookfield (1986).

Conditions for establishment

The influence of new insertions on host fitness is critical to an element's ability to colonize a host genome, as highly deleterious insertions will be eliminated by selection before they have a chance to replicate to additional, less deleterious sites. A key to mobile element proliferation is the class of insertions with the greatest propensity to increase: those with sufficiently mild deleterious effects on host fitness that they behave in a nearly neutral fashion (Box 7.2). Such insertions satisfy the condition $2N_g s < 1$, where N_g is the effective number of genes per locus in the host population (equivalent to the effective size of a haploid population and approximately twice that for a diploid population; see Chapter 4) and s is the fractional reduction in host fitness caused by a single insertion (assumed to have effect $2s$ in homozygotes).

The net replacement rate of a particular element (i.e., the number of new autonomous element insertions produced per element lifetime, R) is equal to the rate of insertion per generation (μ) times the number of generations the element is expected to survive. The latter quantity is simply equal to the reciprocal of the per-generation rate of element removal. If the parental copy is immune to selection, either because its effects are neutral or because drift overwhelms selection, its expected life span is $1/v$ generations, where v is the rate of removal by nonselective physical forces (e.g., genomic deletion, ectopic recombination, or degenerative mutations), but if the strength of selection is substantially greater than v and the population size is sufficiently large, the average life span is $1/s$ (the reciprocal of the rate of removal by selection). In these two extreme cases, $R = \mu/v$ and μ/s, respectively. Given the rate estimates for μ, v, and s provided above, μ/v is generally expected to be substantially greater than 1.0, implying a sufficient net replacement rate for establishment in the absence of efficient selection, but μ/s is expected to be substantially smaller than 1.0. Insertions are expected to have a continuous distribution of fitness effects, so many insertions will lie between these extremes. However, the case is made in Box 7.2 that the criterion for element establishment is approximately $\mu p(N_g)/v$, where $p(N_g)$ is the fraction of insertions that are effectively neutral given the host N_g.

This simple, qualitative relationship makes a fairly robust prediction about mobile element proliferation as a function of host population size. Regardless of the distribution of negative fitness effects of mutations, the function $p(N_g)$ must decrease with increasing N_g (Figure 7.6), but μ and v are defined by intracellular activities that should be independent of population size. At very small population sizes, $p(N_g) \approx 1$ because drift overwhelms selection,

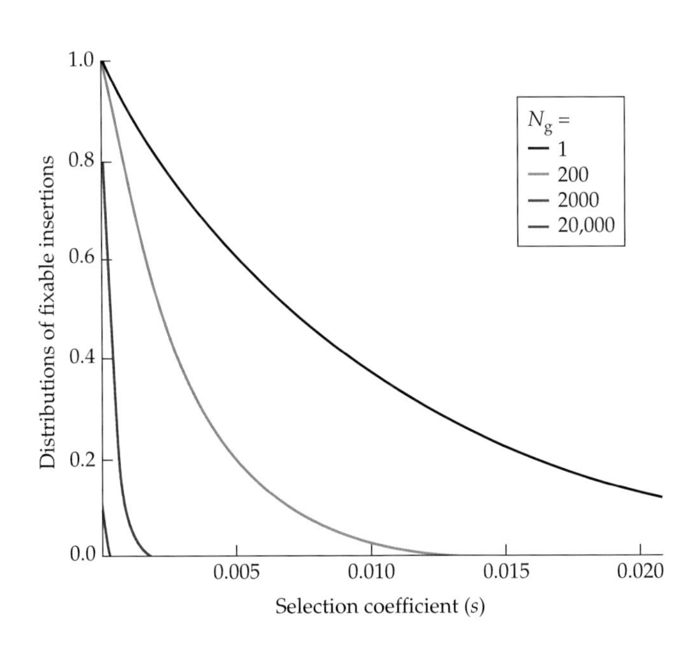

Figure 7.6 Hypothetical distributions of host fitness effects (s) of fixable mutations at various effective sizes for the host population (N_g). It is assumed that the input distribution of insertion effects is exponential with an average reduction in host fitness per insertion of 0.01 (black line). Distributions for larger N_g are obtained by multiplying the input distribution by the probabilities of fixation of deleterious mutations with effects s (see Chapter 4). The area under each curve relative to the input distribution is equivalent to the total fraction of inserts that are fixable at the host level, $p(N_g)$, which equals 1.000, 0.3167, 0.0375, and 0.0082 for the four increasing values of N_g. As N_g increases, selection removes all but the most mildly deleterious mutations.

Box 7.2 The Minimum Population Genetic Conditions for Establishment of a Mobile Element

If an autonomous mobile element is to have any hope of becoming established in a host population, it must give rise to at least one successful daughter element before it succumbs to inactivating physical mechanisms or is eliminated by natural selection. In a finite population, there is always a chance that an element will leave no descendants in the following generation as a consequence of drift alone, so the most meaningful analysis is to consider the reproductive capacity of an element on the basis of physical factors and selection alone.

For the simplest scenario in which the element has no deleterious effects and the host population is stable in size, every element is expected to contribute an average of one daughter copy by vertical descent to the next generation, but with a rate of element inactivation of v per generation, only a fraction $(1 - v)$ of these copies are functional. Thus, the average life span of a particular vertical line of descent is $1 + (1 - v) + (1 - v)^2 + (1 - v)^3 + \cdots = 1/v$ generations. With each autonomous copy giving rise to novel insertions at rate μ, the upper bound on the net replacement rate over the life span of a neutral element insertion is

$$R = \frac{\mu}{v} \tag{7.1}$$

The situation is more complicated for insertions that cause a reduction in host fitness because the efficiency of selection against new insertions will be a function of the effective population size of the host (see Chapter 4). The most stringent condition for element persistence arises if the host population is effectively infinite, so that selection at the host level prevents the fixation of all new insertions. In this case, each active element is expected to leave $(1 - s - v)$ copies by vertical descent in the following generation (Li and Nei 1972; García-Dorado et al. 2003), where s is the reduction in host fitness caused by a single insertion, so the average longevity in host generations is $1/(s + v)$. Following the preceding logic

$$R = \frac{\mu}{s+v} \tag{7.2A}$$

Thus, the condition for persistence, $R \geq 1$, or equivalently, $(\mu - s - v) > 0$, has a straightforward interpretation: the rate of insertion of new elements must exceed the rate of loss of old elements by selection and physical inactivation. Moreover, because the data presented

and element invasiveness is ensured provided $\mu > v$, although individual attempts at invasion may be thwarted by stochastic sampling loss while the element is initially at low frequency. However, for any element family, a threshold host N_g must exist above which invasiveness is not possible. The precise value of N_g at which this threshold occurs satisfies $p(N_g) \approx v/\mu$, which depends on the form of the distribution of s. For example, in Figure 7.6, where an exponential distribution of fitness effects of insertions with average $s = 0.01$ is assumed, $p(N_g) \approx 0.01$ when $N_g = 20{,}000$, so for this N_g, a positive net replacement rate for the element requires $v/\mu < 0.01$. Because the rates given in Table 7.1 imply $v/\mu \approx 0.01$, for such a distribution of insertion above indicate an average value of $s \approx 10^{-2}$ and $v \approx 10^{-6}$, the criterion for element family persistence in an infinite population is very close to

$$R = \frac{\mu}{s} \qquad (7.2B)$$

For finite populations, we resort to a more approximate argument, taking advantage of the preceding results and partitioning insertions into two fractions: those that are effectively neutral (and hence capable of going to fixation), $p(N_g)$, and those that are eliminated by selection, $[1 - p(N_g)]$, which implies

$$R \approx \mu\left\{\frac{p(N_g)}{v} + \frac{1 - p(N_g)}{s^*}\right\} \qquad (7.3A)$$

where $1/s^*$ denotes the average value of $1/s$ over all selectively perceived insertions. So long as s^* is substantially greater than both μ and v, as the data suggest, this expression reduces further to

$$R \approx \frac{\mu p(N_g)}{v} \qquad (7.3B)$$

In other words, the persistence of a mobile element family requires that the rate of insertion of fixable elements exceed the rate of physical loss.

In principle, with a large enough pool of insertions with deleterious effects substantial enough to prevent their fixation but mild enough to allow them to generate a successful daughter element before being eliminated by selection, persistence might be possible without any fixation (Charlesworth and Barton 2004). However, the conditions required for such a situation appear to be quite stringent. If selection is to prevent an insertion from going to fixation, s must be larger than $1/(2N_g)$, but if such an insertion is to have an appreciable chance of spawning an autonomous daughter element, s must be less than $(\mu - v)$. Thus, transient deleterious insertions must satisfy the approximate condition $1/(2N_g) < s < (\mu - v)$ if they are to ensure their initial establishment. Given the empirical results cited in the previous section, the upper limit to this range of s is approximately 10^{-4}, whereas the lower limit is 10^{-5} if $2N_g = 10^5$ and 10^{-8} if $2N_g = 10^8$. Thus, unless a large fraction of insertions have very tiny but effective selective effects, the subset of fixable insertions largely governs the invasiveness of an element family.

effects, any expansion of N_g beyond 20,000 would increase the efficiency of selection at the host level beyond the point that allows element persistence. One complication not considered here is that as mobile element families expand within genomes, $p(N_g)$ may increase even in the absence of a change in the host effective population size. For example, as more insertions accumulate within a host genome, the fraction of new insertions with functional effects on host fitness may decline simply because elements can begin to insert themselves into one another.

Empirical insight into these issues can be acquired by considering the results from fully sequenced genomes, which show a strong relationship between total genome size and the number of mobile elements (Kidwell 2002; Vieira et al. 2002; Lynch and Conery 2003b). The data suggest the presence of a threshold genome size below which mobile elements are unable to maintain themselves, an intermediate range in which only a fraction of species harbor them, and an upper threshold (~100 Mb) above which all species are infected with all three classes of elements (Figure 7.7). This threshold behavior with respect to genome size is qualitatively consistent with the theoreti-

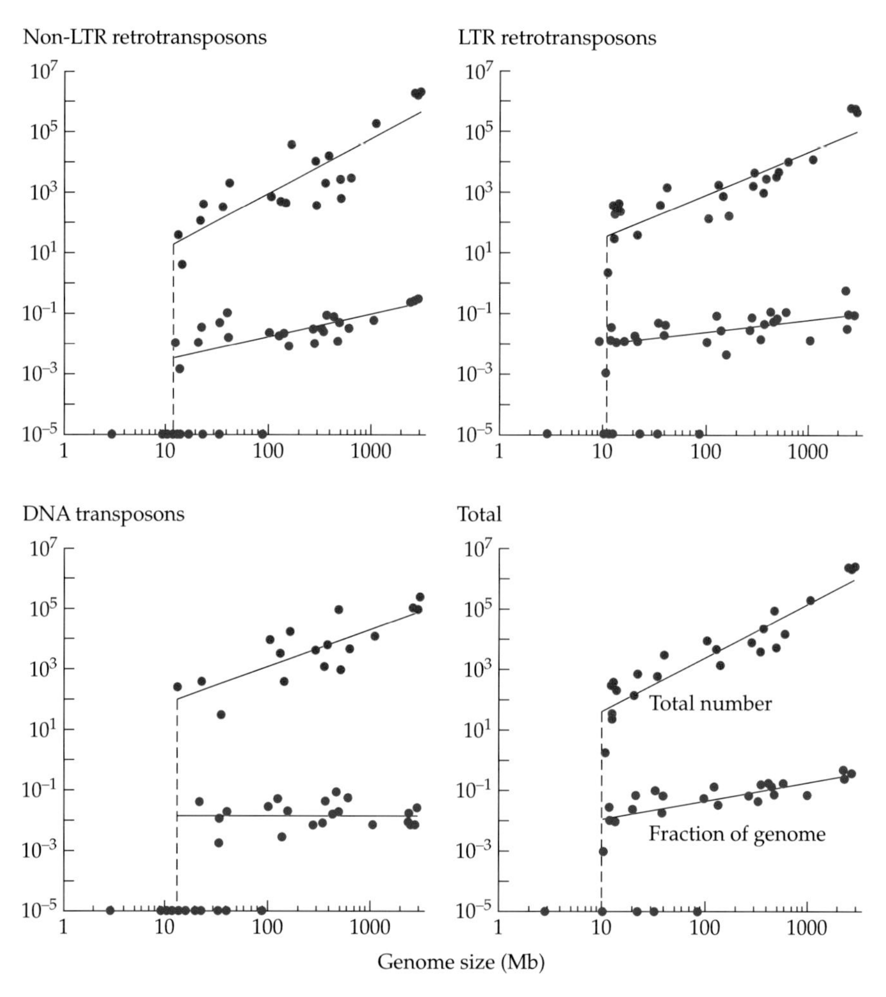

Figure 7.7 Expansion of the three major classes of mobile elements with genome size in species whose genomes have been fully sequenced. The scale on the *y* axis applies to both the numerical (red) and fractional (blue) contributions of the major classes. The regressions do not include species for which elements are entirely absent, although these are plotted as having *y*-axis values of 10^{-5}. (Updated from Lynch and Conery 2003b.)

cal expectations outlined above in that total genome size is inversely correlated with long-term effective population size (see Chapter 4).

Virtually all of the species that lie near or below the threshold for mobile element persistence are unicellular. The tiny microsporidian *Encephalitozoon cuniculi*, the kinetoplastid *Leishmania major*, the yeast *Ashbya gossypii*, the unicellular red alga *Cyanidioschyzon merolae*, and the apicomplexans *Cryptosporidium parvum*, *Plasmodium falciparum*, and *Toxoplasma gondii* appear to lack mobile elements entirely. Many yeasts, including *Saccharomyces cerevisiae* and *Schizosaccharomyces pombe*, harbor only a few hundred LTR retrotransposons, but no non-LTR retrotransposons and no transposons. Such conditions are not a consequence of immunity to mobile elements, as plant transposons are fully capable of mobilization when placed in a yeast genome (Weil and Kunze 2000), and even eukaryotic species with no known mobile elements are clearly susceptible to them (e.g., Beverley et al. 2002; Balu et al. 2005). Finally, insertion elements are extremely rare in prokaryotes (<10 elements per genome in almost all free-living species), with family members exhibiting shallow genealogical relationships, as expected for parasitic elements opposed by efficient selection at the host level (Wagner 2006).

Conditions for element number stabilization

Having evaluated the conditions for the establishment of mobile elements, we will now return to the central question from the host's point of view: what, if anything, prevents the runaway colonization of mobile elements? A key issue here is whether mobile element activity is regulated in a density-dependent manner. Negative density dependence, with insertion rates declining with increasing element number, would tend to stabilize the copy number per genome, whereas positive density dependence could lead to eventual host (and mobile element) extinction.

Models for copy number stabilization, formally developed by Charlesworth and Charlesworth (1983), Kaplan and Brookfield (1983), and Langley et al. (1983), have considered two alternative extremes: (1) elements whose dynamics are governed purely by the physical forces of insertion and excision (no selection), with some molecular mechanism regulating the insertion rate; and (2) elements controlled via negative selection on host carriers, with no genomic mechanism for regulation of element activities. The conditions for stability under the first model are straightforward: with an increasing number of elements per host (n), the insertion rate must decline, or the deletion rate must increase, or both (Figure 7.8B). If these functions are denoted as $\mu(n)$ and $v(n)$, respectively, then the equilibrium number of elements per host genome, $\tilde{n}$, is the value of n that satisfies $\mu(n) = v(n)$. Most evaluations of the relationship between element mobility rate and copy number (including those in the preceding examples) suggest a negative correlation, but the alternative is not out of the question. For example, insertion rates of the *copia* and *Doc* retrotransposons scale positively with copy number in *D. melanogaster* (Pasyukova et al. 1998).

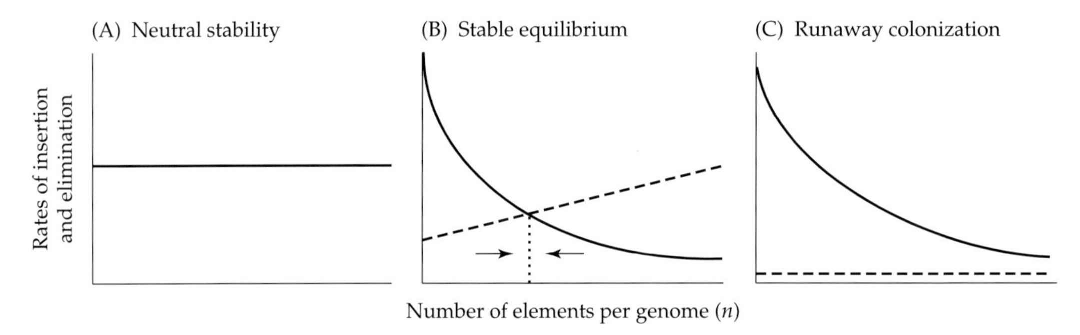

Figure 7.8 Conditions for stability of the average number of element copies per genome. Solid lines denote rates of insertion per element; dashed lines denote net rates of elimination by natural selection or physical processes (such as ectopic recombination). (A) The unlikely situation in which rates of insertion and elimination are identical for all values of n; in this case, all values of n denote neutrally stable equilibria. (B) A stable equilibrium exists at the point where the rates of insertion and elimination are identical. Below this point, element numbers increase, and above this point, element numbers decrease. (C) For every value of n, the rate of insertion exceeds the rate of elimination, so no equilibrium is attained.

Under the selection model, the logic developed in the previous section leads to the conclusion that $\tilde{n}$ is the value of n that satisfies $p(N_g,n)\mu = v$, such that the rate of colonization of effectively neutral elements is balanced by the rate of loss by physical factors. (Here we allow for the possibility that the fraction of neutral insertions is a function of both population size and element number per genome.) If s were independent of copy number, an equilibrium number of elements would exist only in the extremely improbable situation in which $p(N_g)\mu$ was exactly equal to v (Figure 7.8A), and even this equilibrium would be neutrally stable to perturbations. This is because for any new n, the rate of effective colonization, $p(N_g)\mu n$, would still be exactly equal to the rate of removal, vn, leading to a new transient equilibrium. Thus, under the pure selection model, copy number stabilization requires an increase in the average effect of an insertion with increasing element number, a phenomenon known as synergistic epistasis. Intermediate models in which both s and $\mu - v$ are density-dependent are possible, and multiple equilibrium points cannot be ruled out.

Before continuing with an empirical overview of the factors influencing the dynamics of mobile element numbers, it is essential to point out that elements capable of establishment need not always achieve an equilibrium state at which the rates of input and output balance. Indeed, they may only rarely do so. If a new element enters a naïve host genome that has no innate mechanisms for damping the element's activity, rapid proliferation throughout the entire host species may ensue. However, subsequent modifications of host and/or element might then prohibit further element proliferation, leading to a situation in which reentry or activation of a more aggressive element copy would be necessary to return the element family to a state of long-term viability. All of the detailed examples of mobile elements in *D. melanogaster* discussed above are consistent with the common occurrence of such boom-and-bust cycles, as are population surveys of nucleotide variation among elements.

Recall from Chapter 4 that the level of variation among silent sites for randomly sampled alleles in a population is expected to be $2N_g u$ under drift–mutation equilibrium, where u denotes the substitutional mutation rate per nucleotide. For a mobile element with an average n autonomous copies per haploid host genome, the expected equilibrium silent-site divergence among all elements (across all insertions sites) simply scales up to about $2nN_g u$ (Slatkin 1985; Brookfield 1986; Ohta 1986), as the effective number of element family members in the population is roughly n-fold greater than that for a single-locus gene. Thus, for a mobile element family that has achieved long-term equilibrium, the amount of silent-site variation among random copies (sampled over all insertion sites) is expected to be roughly n times that for randomly sampled host gene alleles. Quite contrary to this prediction, the average level of silent-site diversity for a variety of retrotransposons and transposons within *D. melanogaster* is roughly 100 times smaller than that for host genes, and a similar level of reduction is seen for the level of *interspecific* divergence among homologous elements (Sánchez-Gracia et al. 2005). Such results strongly suggest that mobile elements are periodically introduced into *D. melanogaster* via rare interspecific transfers (perhaps via inappropriate matings that do not lead to conventional gene exchange), with most families experiencing relatively rapid initial expansion and gradual contraction before ever reaching an insertion–loss equilibrium. Similar observations have been made in other *Drosophila* species (Silva and Kidwell 2004), leading to the suggestion that the long-term survival of many mobile element families is absolutely dependent on rare horizontal transfers to permissive environments in novel hosts (Robertson 1997; Hartl et al. 1997). Horizontal transfer need not be very frequent to allow persistence by such island-hopping; on average, an established element family need only make a single successful jump to a new host prior to local extinction.

Insights from population surveys

We will now consider empirical attempts to narrow down the list of alternative mechanisms for mobile element containment, again with the caveat that the vast majority of relevant data is restricted to *Drosophila*, leaving us with considerable uncertainty as to the generality of the results. Let us first dispense with the insertion–excision balance hypothesis, which predicts that the physical forces of element elimination or inactivation are sufficient to yield an equilibrium copy number without any assistance from natural selection. Such a situation could arise if the efficiency of host defense mechanisms such as RNAi and methylation increased with element number to the point at which $\mu(n) = v(n)$, or if an accumulation of nonautonomous elements led to a substantial reduction in the insertion capabilities of autonomous elements. However, the data from Table 7.1 clearly indicate that element numbers in *D. melanogaster* are not typically controlled by molecular regulatory mechanisms alone, as the rate of insertion is approximately 100 times higher than that of excision. If average mobile element family sizes in *D. melanogaster* are at equilibrium, these two rates should be identical under the insertion–excision balance model. Thus, alternative mechanisms for copy number stabilization must be sought, the most obvious of which is natural selection at the host level.

Although the average effects of mobile element insertions on host fitness are clearly negative, there has been some debate about the mechanisms underlying such effects. The most obvious consequence of an insertion is the direct disruption of the function of a neighboring or encompassing gene, but it is also possible that physiological effects of mobile elements compromise normal host cell functions (e.g., the attraction of host RNA polymerase to the promoters of retrotransposons might reduce transcription rates of normal host genes) (Nuzhdin 1999). However, if the latter effect were significant, incapacitated elements lacking 5′ promoters should be retained in a host population longer than autonomous elements. Because this is not the case (Yang and Nuzhdin 2003), it appears that the negative effects of mobile element insertions are primarily direct manifestations of structural aspects of gene disruption rather than indirect responses to element metabolism.

Chromosomal surveys of insertion-site occupancies in natural populations strongly support the hypothesis that these negative effects are of sufficient magnitude to be strongly resisted by natural selection. For example, if insertions were neutral with respect to host fitness, then specific insertions with high population-level frequencies should be just as common as those with low frequencies, except in the case of recent invasions (Charlesworth and Charlesworth 1983). However, surveys of *D. melanogaster* populations consistently show that most insertions are present in no more than a small percentage of sampled individuals (Charlesworth 1985; Pérez-González and Eickbush 2001, 2002), with only a small minority reaching high frequency or fixation (Hey 1989; Montchamp-Moreau et al. 1993; Petrov et al. 2003; Franchini et al. 2004). Similar observations have been made with natural isolates from *E. coli* (Sawyer et al. 1987), the yeast *Saccharomyces paradoxus* (Fingerman et al. 2003), peas (Jing et al. 2005), and *Arabidopsis thaliana* (Figure 7.9).

On the other hand, support for the idea that the direct deleterious effects of element insertions increase with the copy number per host genome in a way that might lead to element number stabilization is equivocal at best. Mutation accumulation experiments in which small inbred lines acquire mutations in an effectively neutral fashion have sometimes shown an accelerating rate of fitness loss through time, consistent with synergistic epistasis among consecutive mutations, but more often than not such studies have revealed no such pattern (Keightley and Eyre-Walker 1999; Lynch et al. 1999; Peters and Keightley 2000; Fry 2004; Maisnier-Patin et al. 2005). The inter-

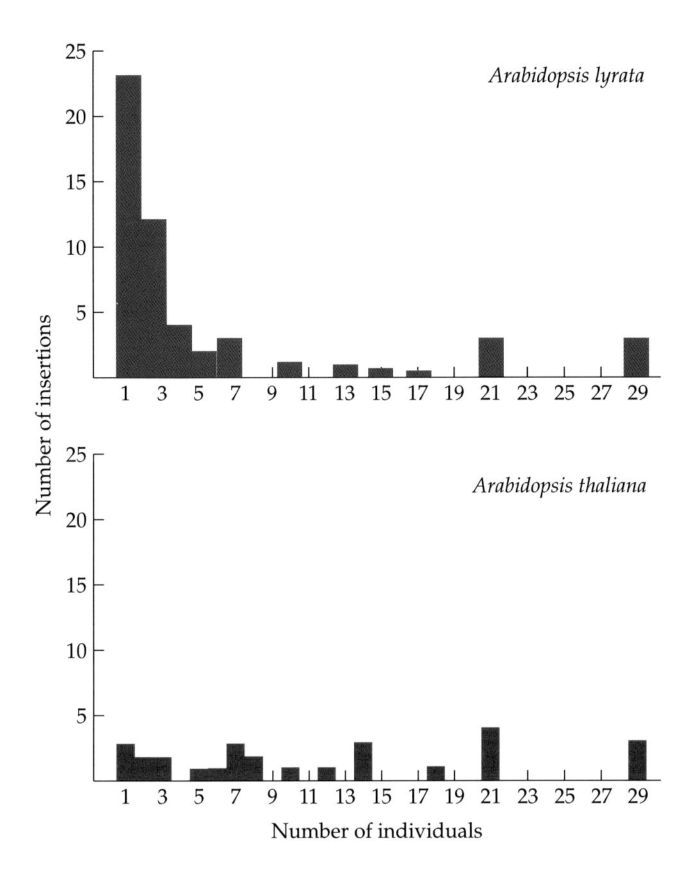

Figure 7.9 The distribution of frequencies for insertions of the *Ac*-like transposon in 29 randomly sampled individuals of outcrossing (*A. lyrata*) and self-fertilizing (*A. thaliana*) species of *Arabidopsis*. For example, 23 insertions at specific sites were found in only single individuals and 12 were present in two individuals of *A. lyrata*. (Modified from Wright et al. 2001.)

pretation of many of these studies is clouded by the fact that the assayed mutations come from all sources, not just mobile element activity, but specific studies of mobile element insertions have yielded no evidence for average synergistic effects of new insertions (Clark and Wang 1997; Elena and Lenski 1997; Cooper et al. 2005).

There is, however, an indirect physical mechanism by which element removal could effectively operate like synergistic epistasis. Ectopic recombination between similar mobile element sequences dispersed throughout a host genome can cause an elevated rate of production of aneuploid gametes and reduced fertility in element-laden individuals (Langley et al. 1988; Devos et al. 2002). Because nonhomologous recombination depends on pairs of elements, its frequency is expected to be a quadratic function of copy number, naturally creating a rapid increase in the magnitude of negative effects with an increase in element number. The establishment of convincing support for the ectopic recombination hypothesis has been difficult, however. For example, although the elevated abundance of mobile elements in some regions of low recombination in *D. melanogaster* is consistent with this model (Langley et al. 1988; Charlesworth and Langley 1989; Bartolomé et al. 2002; Kaminker et al. 2002; Rizzon et al. 2002; Dimitri et al. 2003; Casals et al. 2005), because the efficiency of natural selection is also reduced in low-recombination regions (see Chapter 4), the same pattern is expected in the absence of epistasis.

Given these conflicting observations, is there any justification for thinking that mobile elements ever achieve equilibrium densities? As noted above, element number stability requires that the power of selection be great enough that the net colonization rate of effectively neutral insertions can be balanced by the physical removal rate, such that $p(N_g,n)\mu(n) = v(n)$. Two observations suggest that this is not the case. First, with the insertion rate for established elements in *Drosophila* being roughly 100 times greater than the excision rate (see Table 7.1), $p(N_g,n)$ must be roughly 0.01 if such elements are being kept in check by selection. The *Drosophila* evidence reviewed above suggests that the average insertion reduces host fitness by about 1% and that the distribution of selection coefficients is approximately negative exponential, in which case $p(N_g,n) \approx 0.01$ requires that $N_g \approx 20{,}000$ (see Figure 7.6). As considerable evidence suggests that average $N_g \approx 10^6$ in *Drosophila* (see Chapter 4), actual $p(N_g,n)$ must be considerably smaller than 0.01, which implies that the current effective colonization rate of mobile elements in *Drosophila* is substantially smaller than the rate of their physical elimination, i.e., that element abundances are in excess of equilibrium expectations. Second, the distributions of insertion-site occupancies within populations take on diagnostic forms depending on the relative rates of insertion and excision and on the power of selection, but under the assumption of equilibrium, the observed low levels of occupancy of most insertion sites in *Drosophila* imply $2N_gs > 10$ (Charlesworth and Langley 1989; Biémont 1992; Biémont et al. 1994). This estimate is inconsistent with data derived from laboratory assays of the fitness effects of new insertions, as it suggests an average s on the order of 2.0×10^{-4} after factoring out the estimated N_g for this species (Charlesworth et al. 1992). Thus, insertion-site occupancies also suggest that mobile elements in *Drosophila* are generally more common than expected on the basis of laboratory estimates of insertion rates and effects.

Taken together, the data for *Drosophila* are explained most easily if, rather than attaining equilibrium, mobile elements are simply products of boom-and-bust cycles. As noted above, when elements initially enter a naïve host environment, insertion rates are often at least two orders of magnitude greater than those of well-established elements. Such a condition can be sufficient to initially overcome the rate of element loss via selection, rapidly seeding the population with numerous low-frequency elements, but eventually giving rise to an unsustainable situation when $\mu(n)$ declines to the point at which $p(N_g,n)\mu(n) < v(n)$. Theory suggests that such cycles are particularly likely when nonautonomous elements compete with their autonomous relatives (Kaplan et al. 1985).

Some final insight into the mechanisms opposing mobile element proliferation derives from observations on highly self-fertilizing species, which present a dramatically different environment for mobile element proliferation than outcrossing species. Because the entire genome of a selfer is essentially homozygous and inherited in an effectively clonal fashion, all genomic regions are expected to experience high and roughly equivalent amounts of selective interference due to linkage. In addition, homozygosity at insertion sites should minimize the incidence of ectopic exchange resulting from misalignment of chromosomes during meiosis (Montgomery et al. 1991). Thus, the genome-wide population of insertions in self-fertilizing species is expected to expand as a consequence of the reduced efficiency of selection, and the abundance of mobile elements is expected to be uncorrelated with the recombination rate along chromosomes, unless the density of functionally vulnerable DNA is associated with local levels of recombination. Consistent with this hypothesis, in two self-fertilizing species, the nematode *Caenorhabditis elegans* and the plant *Arabidopsis thaliana*, there is no association between retrotransposon abundance and the local recombination rate, although transposons are somewhat more abundant in regions of high recombination, which are also relatively gene-poor (Duret et al. 2000; Wright et al. 2003a). In addition, the frequency distribution of mobile element insertions in *A. thaliana* is nearly symmetrical, whereas that in the outcrossing *Arabidopsis lyrata* is highly biased toward low-frequency insertions (Figure 7.9). Although not ruling out a possible role for ectopic recombination in outcrossing species, these results provide additional compelling support for the hypothesis that gene disruption effects play a major role in the dynamics of mobile element numbers.

Mobile elements and species extinction

It is striking that not a single case has yet been uncovered that is reasonably consistent with the regulation of element numbers at a stable equilibrium. In principle, host lineages might still achieve relatively stable total numbers of elements (summed over all element families) while experiencing considerable turnover in the identities of the most active families over evolutionary time. However, given the evidence that individual element families are capable of dramatic expansions on time scales of decades in the appropriate population genetic environment, that such elements have average negative effects on host fitness, and that transient population bottlenecks driven by external ecological forces can occasionally reduce the efficiency of selection against deleterious element insertions, can we rule out the possibility that mobile element activity significantly influences the long-term survival of some host lineages? Our general ignorance of the factors driving the background extinction rates of phylogenetic lineages suggests that it is premature to do so.

In very large populations with efficient selection, the fitness load associated with mobile elements need not be great. For example, assuming that the mutational mechanisms of element removal are weak relative to the forces of insertion and selection, then for an effectively infinite population, the equilibrium frequency of insertions (q) for each of m potential insertion sites is simply the ratio of the rate of site-specific insertion, $\mu\tilde{n}/m$, to the rate of element removal by natural selection, s; that is, $q = \mu\tilde{n}/(ms)$. Under a multiplicative fitness model, with each insertion reducing host fitness by a fraction $(1 - s)$, the equilibrium fitness is $(1 - s)^{2qm}$, where $2qm$ is the expected number of insertions per diploid individual (Charlesworth and Charlesworth 1983), which, after substituting for q, reduces to roughly $e^{-2\mu\tilde{n}}$, or simply $(1 - 2\mu\tilde{n})$ if $2\mu\tilde{n}$ is small. In other words, in an effectively infinite population, the load associated with mobile elements is approximately equal to the total diploid insertion rate, $2\mu\tilde{n}$ per generation. Under such conditions, the selective effect of an individual insertion (s) has no influence on the equilibrium host fitness because the insertion frequency is an inverse function of s, which cancels out when the number of elements is multiplied by their individual effects. The same point was made long ago by Haldane (1937) in a more general consideration of the mutation load. Thus, for large populations, with μ on the order of 10^{-4}, the reduction in host fitness would be no greater than 1% for an element family of moderate size (50 copies per host genome).

This result relies on the assumption that the population size is so large that natural selection can prevent essentially any insertion from becoming fixed, an implausible condition for large multicellular species, as most clearly illustrated by the millions of LINE element fixations that have occurred in the human genome (see Chapter 3). Fixations of mildly deleterious insertions are problematic for a host species because, in the absence of compensatory mutations, they cause a cumulative decline in average fitness, eventually leading to the point at which the population can no longer sustain its current density (because the net replacement rate of individuals is less than one). By magnifying the power of genetic drift, and hence reducing the ability of natural selection to eliminate new insertions, the decline in population size will accelerate the fixation of future deleterious mutations, driving the population to a still smaller size and eventually resulting in extinction via a mutational meltdown (Lynch et al. 1993, 1995a,b). The extent to which such a condition will ever be reached depends on a number of factors, including those discussed above in the context of element number regulation: the degree to which an increase in element number results in molecular changes that reduce the level of element activity (e.g., via the erosion of element activity by the accumulation of nonautonomous elements) and the likelihood of genomic excision processes removing elements subsequent to their fixation.

A matter of special relevance to the mutational meltdown model is the distribution of fitness effects of insertions. Deleterious mutations for which $s \approx 1/N_g$ are most harmful to populations, as they constitute the class of elements with the greatest negative effects that are still highly vulnerable to fixation by random genetic drift (Gabriel et al. 1994; Lande 1994). Generously assuming that all insertions have effects of this magnitude and are fixed at the neutral rate, then with n autonomous elements per genome, the expected long-term fitness loss associated with each generation of insertions would be $n\mu/N_g$. If $n\mu$ were no greater than 1.0, then even a vertebrate or vascular plant with a small N_g of 10^4 (see Chapter 4) would have a cumulative rate of fitness loss of less than 10^{-4} per generation, so any extinction event driven by mobile element accumulation would have a fairly long prelude. Nevertheless, because the background extinction rate revealed by the fossil record is on the order of 10^{-7} to 10^{-6} per year, with higher rates applying to organisms with larger sizes (Stanley 1979), this crude calculation suggests that internal genomic processes driven by selfish elements can be quantitatively sufficient to play an influential role in species longevities. Such a view may seem implausible to those used to interpreting every aspect of the fossil record in terms of putative ecological forces, but there is no objective basis for the latter position. In both land plants and vertebrates, species with larger genomes (owing to larger numbers of mobile elements) do have higher risks of extinction (Vinogradov 2003, 2004), although it remains necessary to tease out cause versus effect.

All other things being equal, asexual lineages are expected to be especially vulnerable to extinction resulting from mobile element proliferation (Lynch et al. 1993; Gabriel et al. 1994). As noted above, the absence of outcrossing effectively converts an entire genome into a single linkage unit, thereby maximizing the amount of selective interference among linked sites and minimizing the ability of natural selection to eradicate mildly deleterious insertions. In addition, the absence of meiosis may largely render ectopic recombination inoperable, although some mitotic recombinational activity may remain. There is little question that asexual lineages of eukaryotes are relatively short-lived. Nearly every major group of animals harbors one or more parthenogenetic lineages, indicating a general absence of cytogenetic barriers to a transition to asexuality, yet nearly all parthenogenetic species appear to be less than a million years old (Judson and Normark 1996; Butlin 2002). Might these relatively short life spans be a consequence of deleterious mutation accumulation? The matter is not entirely settled, as numerous hypotheses have been suggested for the short life spans of asexual lineages (Kondrashov 1993), but the one study that has searched for mutational degeneration at the molecular level revealed a concentration of such changes in obligately asexual lineages of the microcrustacean *Daphnia* in comparison to its sexual relatives (Paland and Lynch 2006). Although this work focused on the sequences of host protein-coding genes, evidence that the general principles extend to mobile element accumulation derives from observations on *Candida albicans*, a highly clonal species of yeast, whose genome is quite enriched with mobile elements relative to the genomes in related species (Kim et al. 1998; Goodwin and Poulter 2000).

With so few data on the behavior of mobile elements in organisms with alternative reproductive modes, it is premature to conclude that asexual species will always be exceptionally vulnerable to mobile element accumulation. Some have even argued the opposite; that is, that the stochastic loss of autonomous elements from closed lineages should eventually lead to an alternative stable state of complete genomic sanitization (Arkhipova and Meselson 2000), and by extension, that sexual reproduction may have originated as a by-product of a selfish element that modified its host genome to overcome such problems (Hickey 1982). Theoretical studies indicate that the complete purging of mobile elements is indeed possible in asexual populations, but unless effective population sizes are in excess of the reciprocal of the excision rate ($N_g > 10^6$ from Table 7.1), mobile elements are more likely to drive asexual host species to extinction (Dolgin and Charlesworth 2006). Nevertheless, even the latter work ignores a key biological observation noted above: the confinement of the activities of many mobile elements to meiotically dividing cells. Thus, the key to understanding the consequences of host asexuality for mobile element proliferation comes down to a basic cell biological issue and a basic population genetic issue: the extent to which element activity is reduced by a transition to nonmeiotic reproduction, and the extent to which the stochastic accumulation of active elements within host lineages offsets the stochastic inactivation of elements within lineages.

One asexual species group that may be illuminating in this regard is the bdelloid rotifers, a family of morphologically diverse species probably exceeding 50 million years in age (a major outlier to the usual pattern of short asexual life spans). Is the unusual longevity of this group a consequence of an exceptional degree of immunity to mobile elements or of a lucky founder event involving an element-free ancestor? Although it was initially suggested that bdelloids are uniquely devoid of retrotransposons relative to other invertebrates (Arkhipova and Meselson 2000), subsequent work revealed a unique class of bdelloid retrotransposons (Arkhipova et al. 2003), and some DNA-based transposons that are abundant in bdelloids appear to be absent from some sexually reproducing rotifers (Arkhipova and Meselson 2000, 2005). Thus, it is not entirely clear whether bdelloid genomes are strikingly depleted in overall mobile element composition. In any event, it should also be recognized that members of this group are ubiquitously distributed and remarkably abundant (a glass of tap water left in the open air will usually acquire bdelloids within a week). In principle, such species may have large enough effective population sizes to compensate for the reduction in the efficiency of selection resulting from the absence of meiotic recombination. These matters will take on additional significance when we consider nonrecombining genomic regions within sexually reproducing species: organelles (see Chapter 11) and sex chromosomes (see Chapter 12).

Positive Effects of Mobile Element Insertions

Summarizing to this point, three independent sets of observations point to the net deleterious nature of mobile elements: direct observations on the average effects of insertions; population surveys of insertion frequencies and element age distributions; and the paucity of such elements in hosts with large effective population sizes. Nonetheless, a number of investigators prefer to emphasize the beneficial side of such elements (e.g., Kidwell and Lisch 2000, 2002; Wessler 2006), a view derived in part from ideas promoted by the discoverer of mobile elements, who suggested that element activation during times of stress enables host genomes to modify themselves in potentially favorable ways (McClintock 1984). Perhaps the simplest and most compelling argument opposing this view derives from the fact that transposition and retrotransposition factors are virtually always encoded by the elements themselves, not by the host genomes. Because numerous nonautonomous elements are known to be activated in *trans*, it is clear that host-encoded mobilization proteins could be relied on to regulate overall element activity if selection favored such a situation.

As we will see below, it is certainly true that numerous mobile element insertions have spawned beneficial host mutations. Given the extraordinary diversity of mutational types produced by such elements and their large contribution to the total mutation pool in multicellular species, it would be astonishing if this were not the case. However, the fact that mobile elements can generate mutations with positive effects need not imply that such elements are maintained to enhance the long-term adaptive potential of host lineages. Like chromosomal breakage resulting from X-ray irradiation and nucleotide damage resulting from free radicals, mobile element activity is largely a pathology resulting from the limitations of cell biology.

Adaptive evolution opportunistically takes advantage of the mutations that happen to present themselves during a period of selection, and it remains unclear whether there are many types of functional evolutionary changes that can be accomplished only with mobile elements. For example, although mobile element insertions can influence the expression of adjacent host genes in numerous ways, gene regulation can also be accomplished by alternative routes, such as modification of transcription factor binding sites by small-scale nucleotide substitution, insertions, and deletions (see Chapter 10). Nevertheless, a unique feature of mobile elements that encourages their indirect promotion is their tight linkage to the mutations they precipitate, which induces hitchhiking with any chance beneficial host mutations they create (Chao et al. 1983; Zeyl et al. 1996). In principle, such events can be sufficient to maintain an element family within a host genome, so long as the approximate requirement noted in the previous section is fulfilled: that a fixable autonomous insertion occurs within the life span of an average element. A few compelling examples of such hitchhiking events exist. For example, the expansion to near fixation of mutant alleles conferring insecticide resistance has been documented on several occasions in

Drosophila (Daborn et al. 2002; Schlenke and Begun 2004; Aminetzach et al. 2005), and in each case, the mutation has been a knockout induced by a mobile element insertion.

As first emphasized by McClintock (1956, 1984), mobile element insertions have an enormous potential to modify the activity of genes in their near vicinity in more subtle ways. Because mobile elements are equipped with their own promoters and regulatory elements, their insertion into host gene regulatory regions may result in modifications of transcription rates (Conte et al. 2002; Jordan et al. 2003; Han et al. 2004; Lesage and Todeschini 2005; Marino-Ramírez et al. 2005). Similar effects can result from insertions of 3′ retrotransposon sequences containing transcription termination sequences, polyadenylation signals, and/or micro RNA targets (Roy-Engel et al. 2005; Smalheiser and Torvik 2006), and some insertions may impose physical effects that insulate genes from prior local regulatory sites. Mobile element insertions can also have more indirect effects, as when an adjacent host gene in opposite orientation to a retrotransposon is silenced by RNAi effects generated by read-through antisense transcripts initiated from the element insertion site (Kashkush et al. 2003; Puig et al. 2004).

Mobile element insertions may also have structural effects on the products of adjacent host genes. For example, insertions into host gene UTRs can alter points of transcription initiation and/or termination, and insertions into introns can modify the locations of splice sites (see Chapter 9). In addition, as noted above, the frequent extension of non-LTR element transcripts into the territory of downstream genes provides a mechanism for duplicating host gene sequences. Complete duplication events result in increased levels of gene dosage, whereas partial duplications can lead to the production of chimeric genes with novel functions. Although the mechanism of acquisition is unknown, plant MULE elements are known to scavenge fragments of host coding DNA when they insert themselves, providing an additional route to generating novel gene functions when these fragments move to appropriate locations (Jiang et al. 2004).

As discussed in Chapter 5, mobile elements are involved in host chromosome stability in a number of eukaryotes. In a few species (most notably *Drosophila*), specific families of non-LTR elements are central to telomere maintenance, and the telomerase enzyme deployed by most other species appears to contain a reverse transcriptase domain from an ancient retrotransposon. In addition, the transcriptional activities of centromeric mobile elements appear to play a key role in defining the centromere via indirect epigenetic effects. Mobile elements also sometimes provide mechanisms for patching double-strand breaks of chromosomes (Moore and Haber 1996; Teng et al. 1996).

A few intriguing possibilities have been suggested for direct benefits derived from element-encoded proteins. An overview of several likely examples of co-option of viral proteins by mammalian genomes is presented by Britten (2004), but perhaps the most impressive instances of eukaryotic exploitation of retroviruses involve the *env* gene, which normally is deployed in viral fusion to host cell membranes. For example, syncytin, a protein specifically expressed in the mammalian placenta and thought to be involved in its formation through cell fusions, is closely related to the Env protein of mammalian retroviruses and appears to have arisen independently in primates and rodents (Mi et al. 2000; Dupressoir et al. 2005). Remarkably, *env* has also been independently domesticated in the *Drosophila* and mosquito lineages, possibly as a mechanism for shielding cells from retroviral infection (Malik and Henikoff 2005). These and many more types of observations demonstrate that although selection generally acts to eliminate mobile elements, vertical inheritance also provides an opportunity for the host to turn the tables by exploiting mutant elements that are fortuitously advantageous.

8 Genomic Expansion by Gene Duplication

Dating back to the pre-molecular era (Muller 1940; Haldane 1933; Ohno 1970; reviewed in Taylor and Raes 2004), substantial attention has been given to the idea that gene duplication is the major mechanism for the origin of new gene functions, and it is now firmly established that the refashioning of duplicate genes is a major contributor to the origin of adaptive evolutionary novelties (numerous examples are cataloged in Ganfornina and Sanchez 1999; Patthy 1999b; True and Carroll 2002; Hurley et al. 2005; Irish and Litt 2005; Nei and Rooney 2005). However, these attention-grabbing examples need not be representative of the fates of *average* gene duplicates, as there is a logical distinction between the processes responsible for the initial establishment of duplicate genes and their secondary modification by mutation and natural selection. Virtually all new genes must arise from accidental duplications of preexisting genes or parts thereof, which implies an initial state of a single copy in a single member of the population. Thus, understanding the processes that facilitate the expansion versus contraction of gene number requires an appreciation of the molecular processes that give rise to duplication events and of the population genetic forces that influence the dynamics of newly arisen genes.

If beneficial mutations resulting in new gene functions were the primary means for preserving duplicate genes, then because such mutations are rare, species with enormous population sizes would be expected to carry the largest numbers of genes. That this is not the case immediately suggests that other defining forces must be at work. A key point to be made below is that random genetic drift and degenerative mutations commonly play a central role in the growth of gene number in populations that are sufficiently small in size. Although the preservation of duplicate genes by degenerative mutations may seem counterintuitive, we will see that this process is a natural outcome of the structure of eukaryotic genes. Once this principle is understood, the conclusion that phylogenetic changes in gene number may be substantially influenced by nonadaptive processes becomes inescapable, although as noted above, this need not imply that gene duplication is a minor player in the origin of new gene functions.

Prior to addressing the population genetic mechanisms by which duplicate genes evolve, the context of the problem will first be established by considering the rates at which such genes arise and the time periods over which they typically survive. Although substantial evidence suggests that many key evolutionary lineages of multicellular eukaryotes have experienced one or more complete genome doublings (polyploidization) at some time in the distant past (Wolfe 2001), it will be seen that gene duplication by smaller-scale processes is an ongoing feature of all organisms. The genome is a dynamic playing field on which new genes are continually arising via duplication events, with most being eliminated fairly quickly by drift and/or natural selection, some simply replacing their parental copies, and a few experiencing functional changes that ensure their long-term preservation along with their ancestral family members. Even in the absence of any net growth in genome size, this continual turnover of genes has further evolutionary implications, as it passively promotes the origin of microchromosomal rearrangements. In this sense, the gene duplication process provides fuel for both of the major engines of evolution: adaptive phenotypic change within lineages and the creation of new lineages by speciation.

The Evolutionary Demography of Duplicate Genes

The power of gene duplication as an evolutionary force depends on the rate at which duplicate genes arise. Although there is currently no simple way to estimate this rate directly in laboratory experiments, information from complete genome sequences provides an indirect approach (Lynch and Conery 2000, 2003a). Through comparative sequence analysis, the total pool of duplicate genes within a genome can be identified, and under the assumption that silent sites accumulate nucleotide changes at a relatively constant rate, the relative age of each duplicate pair can be estimated from the silent-site divergence between pair members. The resultant age distribution of duplicate pairs can then be used to estimate the average rates of origin and elimination of duplicate genes, using the same principles that demographers use to estimate birth and death rates of individuals in natural populations.

If it can be assumed that birth and death rates of duplicate genes have remained roughly constant over the time scale of observation, a particularly powerful analysis becomes possible. Under a steady-state birth/death process, the expected frequency of duplicates declines exponentially with age, with the time-zero intercept of the age distribution providing information on the birth rate and the slope providing information on the death rate. To see this, recall the simple model introduced in Chapter 3,

$$n_t = n_{t-1} + B(1 + n_{t-1}) - Dn_{t-1}$$

where n_t denotes the number of genomic copies of a gene at generation t (in excess of the baseline single-copy requirement for viability), B denotes the rate of gene birth (applied to the baseline and extra copies), and D denotes the rate of gene death (applied only to the excess copies, and otherwise assumed to be independent of copy number). Both B and D are stochastic variables, but averaging over a large pool of genes (the entire genome), some specific patterns can be predicted. At equilibrium ($n_t = n_{t-1}$), the expected number of excess copies per gene is $n_{tot} = B/(D - B)$, which is just a function of the ratio B/D. If $D >> B$ (justified by data presented below), $n_{tot} \approx B/D$, which is much less than 1, so the total birth rate per gene family, $B(1 + n_{tot})$, is close to B each generation. Furthermore, because a constant fraction D is lost each generation, the steady-state age distribution is close to

$$n_i = B(1 - D)^i$$

as graphed in Figure 8.1 (Lynch and Conery 2003a). Under this model, a log-linear plot of n_i versus i is expected to yield a straight line with the expected form

$$\log(n_i) = \log B + i \cdot \log(1 - D)$$

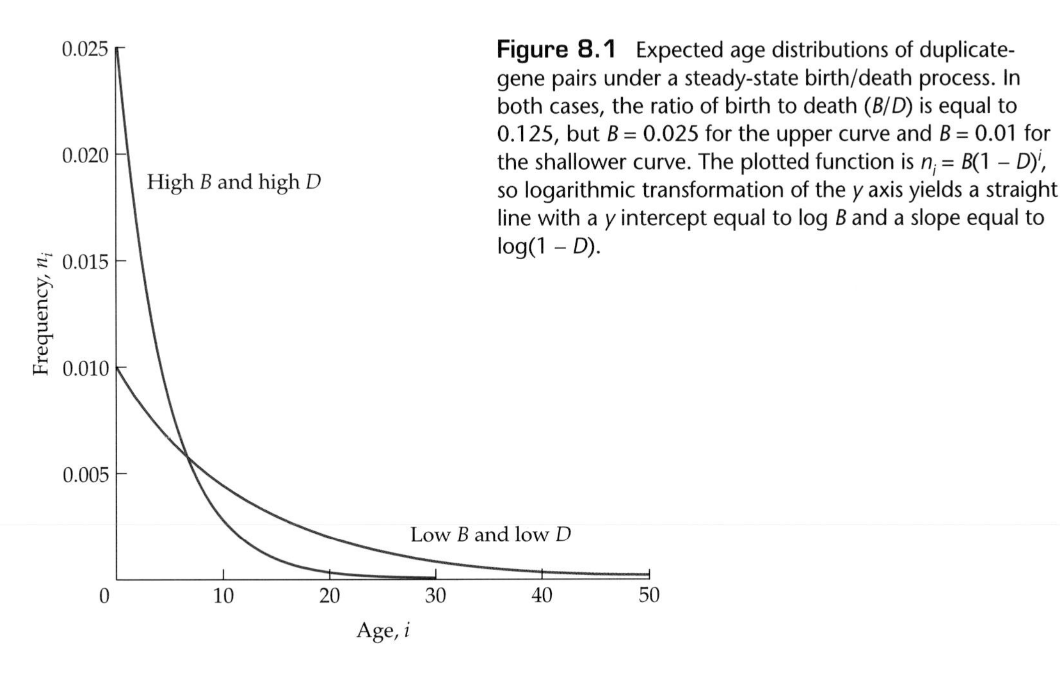

Figure 8.1 Expected age distributions of duplicate-gene pairs under a steady-state birth/death process. In both cases, the ratio of birth to death (B/D) is equal to 0.125, but $B = 0.025$ for the upper curve and $B = 0.01$ for the shallower curve. The plotted function is $n_i = B(1 - D)^i$, so logarithmic transformation of the y axis yields a straight line with a y intercept equal to log B and a slope equal to $\log(1 - D)$.

Thus, by fitting a linear regression to the logarithmic transformation of the age distribution, estimates of the birth and death rates of duplicate genes can be acquired by setting the intercept and slope equal to log B and log(1 – D), respectively. The age distribution (n_i) for such an analysis can be acquired by querying an entire genome for the number of gene pairs of each age i (binning in units of silent-site divergence) and dividing by the total number of genes (not including the duplicates themselves).

We have already encountered a duplicate-gene age distribution for the human genome (see Figure 3.1) that closely approximates the exponential form suggested by the preceding model, and another such distribution for the *Caenorhabditis elegans* genome appears in Figure 8.2A. Many additional eukaryotic genomes exhibit this pattern, at least as a first approximation

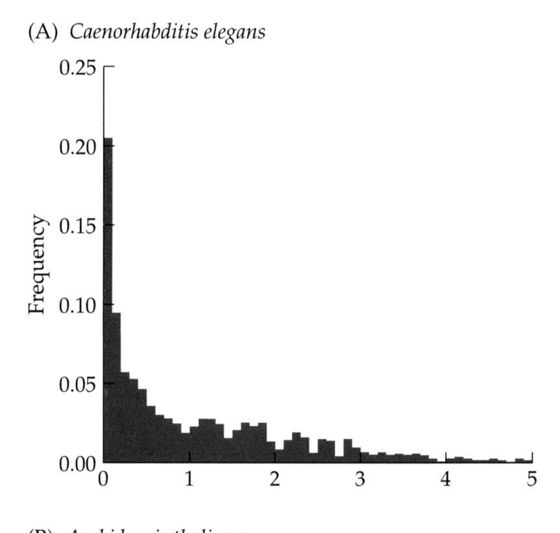

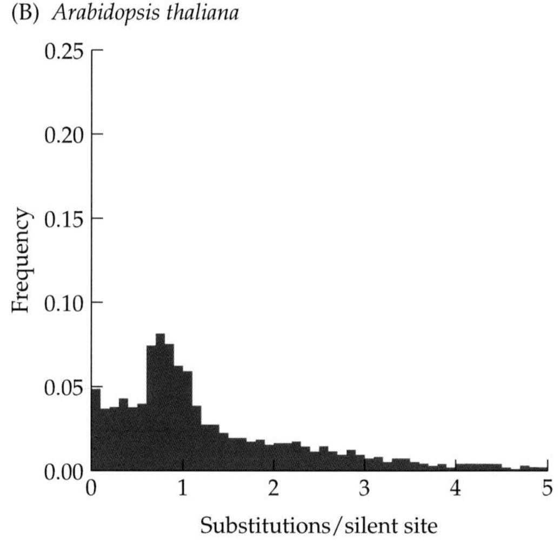

Figure 8.2 Age distributions of duplicate genes in the nematode *Caenorhabditis elegans* (A) and the plant *Arabidopsis thaliana* (B). The approximate sample sizes are 1100 and 3200, respectively. The large internal peak in the *Arabidopsis* age distribution is a reflection of an ancient polyploidization event. (From Lynch and Conery 2003a.)

(Lynch and Conery 2000, 2003a; Achaz et al. 2001; Blanc and Wolfe 2004). However, nonequilibrium situations, such as a "baby boom" resulting from a past polyploidization event, are sometimes inferred from an intermediate peak in the age distribution (Gu et al. 2002a; McLysaght et al. 2002; Jaillon et al. 2004; Vandepoele et al. 2004; Maere et al. 2005; Figure 8.2B). In principle, bulged distributions can also result from periods of reduced duplicate-gene loss.

Application of the steady-state solution to the genomes of diverse eukaryotic species yields birth rate estimates for duplicate genes that are generally in the range of 0.001 to 0.01 per gene on the time scale of 1% divergence (Table 8.1). There is no general phylogenetic pattern to these values, as both unicellular species and animals have average values of about 0.004, and the one land plant for which data are available, *Arabidopsis thaliana*, has an estimate of 0.003. It should be emphasized that these estimated rates of gene duplication apply to single genes in single individuals. They are not population-level rates of origin of new duplicates (which are equivalent to $2NB$, where N is the number of individuals in a diploid population), nor are they equivalent to fixation rates of new duplicates (which would be diminished by processes of duplicate-gene loss).

TABLE 8.1 Estimated rates of origin (B) and loss (D) of duplicate genes for eukaryotes

SPECIES	B	D	B/D
Unicellular species			
Plasmodium falciparum	0.0003	0.167	0.0018
Saccharomyces cerevisiae	0.0025	0.324	0.0077
Schizosaccharomyces pombe	0.0016	0.386	0.0042
Encephalitozoon cuniculi	0.0117	0.487	0.0240
Animals			
Homo sapiens	0.0049	0.081	0.0605
Mus musculus	0.0030	0.134	0.0224
Fugu rubripes	0.0043	0.189	0.0228
Caenorhabditis elegans	0.0028	0.136	0.0206
Drosophila melanogaster	0.0011	0.229	0.0048
Anopheles gambiae	0.0062	0.190	0.0326
Plants			
Arabidopsis thaliana	0.0032	0.033	0.0970

Source: Lynch and Conery 2003a.

Note: Both B and D are defined on a time scale of 1% divergence for silent sites. As noted in the text, the ratio B/D provides an estimate of the average number of excess copies per gene resulting from the stochastic birth/death process.

It is difficult to apply the age distribution approach to prokaryotic species because of their greatly reduced number of genes. However, a downwardly biased average estimate of *B* for 73 species of prokaryotes, which simply counts the number of duplicates with silent-site divergence of less than 1% and does not account for early gene loss, is about 0.002 (Lynch and Conery 2003a), well within the range of eukaryotic species. As many other indirect observations support the idea that rates of gene duplication (some involving horizontal transfer) are high in prokaryotes (Anderson and Roth 1977; Brenner et al. 1995; Lawrence and Ochman 1998; Bergthorsson and Ochman 1999; Lerat et al. 2005), the relatively small number of genes within prokaryotic genomes is a consequence of a relatively high rate of attrition of gene duplicates.

As a first approximation, these observations suggest that the rate of duplication of entire genes is only slightly less than the rate at which nucleotide substitutions occur at silent sites. Given this scaling and the fact that the amount of silent-site substitution per generation increases with generation time (see Chapter 4), it follows that the per-generation rate of gene duplication increases from unicellular to multicellular species. Recalling the estimates for silent-site divergence given in Figure 4.5, 1% silent-site divergence is roughly equivalent to 6×10^6, 8×10^5, and 2×10^5 generations in unicellular eukaryotes, invertebrates, and vertebrates, respectively, which in turn imply birth rate estimates of 0.06%, 0.5%, and 2% per gene per 10^6 generations for these three groups. Using an average rate of silent-site substitution for vascular plants derived from six independent studies (Gaut et al. 1996; Li 1997; Lynch 1997; Koch et al. 2000), 8.1 per site per billion years, and assuming two generations per year, the birth rate of gene duplicates in *Arabidopsis* is about 0.8% per gene per 10^6 generations.

There are a number of significant caveats with respect to these estimates. For example, most of them assume an equilibrium age distribution of duplicate genes and an approximately constant rate of silent-site divergence. Because duplicate genes generally have identical sequences at the time of origin and are also often in close spatial proximity, they are expected to be subject to stochastic homogenization by gene conversion. In principle, gene conversion between duplicate pairs could also be mediated by recombination with reverse-transcribed mRNAs, in which case highly expressed genes would be especially prone to rejuvenation (Pyne et al. 2005). Such events would result in a nonlinear relationship between the age of a pair and the rate of silent-site divergence until the level of divergence exceeded the point beyond which homology-dependent conversion is possible, potentially leading to overestimates of *B*.

Despite these concerns, the results reported above appear to be quite robust. Using rather different approaches, Cotton and Page (2005) obtained a human duplication rate estimate identical to that given above when a 20-year generation time was assumed. Furthermore, studies of segmental duplications in the human population, which reveal hundreds of presence/absence polymorphisms of duplication spans of up to several hundred kilobases in length (Iafrate et al. 2004; Sebat et al. 2004; Tuzun et al. 2005; Khaja et al. 2006; Redon et al. 2006), imply minimum duplication rates of about 2% per gene per million years (van Ommen 2005). In addition, Maere et al. (2005) arrived at a birth rate estimate for *Arabidopsis* identical to that given above for land plants. An exceedingly tiny estimate (~0.001%/gene/MY) reported for *Saccharomyces cerevisiae* (Gao and Innan 2004) is clearly an error, as the authors actually considered the rate of gene preservation (which is diminished by mechanisms of gene loss) rather than birth. Finally, because specific genes in *Drosophila* have been found to duplicate at rates as high as 1.0 to 100 per gene per 10^6 generations (Gelbart and Chovnick 1979; Shapira and Finnerty 1986), the indirect estimates suggested above are not biologically unrealistic and may be quite conservative.

Substantial differences in the loss rates of duplicate genes also exist between unicellular and multicellular species. On the time scale of 1% silent-site divergence, D averages 0.34 (SE = 0.07) for unicellular species, 0.18 (0.03) for invertebrates, 0.13 (0.03) for vertebrates, and 0.033 for *Arabidopsis*, but because 1% silent-site divergence requires many more generations in unicellular than in multicellular species, the per-generation loss rates are actually higher in multicellular species. On a time scale of 10^6 generations, estimated loss rates are ~0.04 for unicellular eukaryotes, 0.10 for invertebrates, and 0.28 for vertebrates and *Arabidopsis*, implying half-lives for duplicate genes in these three groups of 16.7, 7.0, and 2.5 millions of generations, respectively.

These results suggest that with respect to gene content, the eukaryotic genome is highly dynamic. As a consequence of a stochastic balance between gene birth and death rates, total genome size may remain approximately constant for long periods, but throughout such periods there is likely to be continual turnover with respect to the specific genes that are present in redundant copies. Given that the probability of a gene duplicating by the time a silent site experiences 0.01 substitutions is roughly 0.004, the rate of duplication per gene is about 40% of the rate of mutation per nucleotide site, raising the possibility that changes in gene content may often rival changes in gene sequence as a mechanism of phenotypic evolution. Indeed, on a time scale of roughly 250 million years, nearly every gene in a typical eukaryotic genome can be expected to duplicate at least once. On the other hand, as is the case for most amino acid replacement substitutions within genes, most duplicate genes appear to be evolutionarily short-lived, with typical half-lives of just a few million years.

Origins of segmental duplications

The molecular mechanisms by which duplicate genes arise are diverse, ranging from complete genome duplication (discussed in the following section) to more restricted duplications of smaller chromosomal regions. The latter,

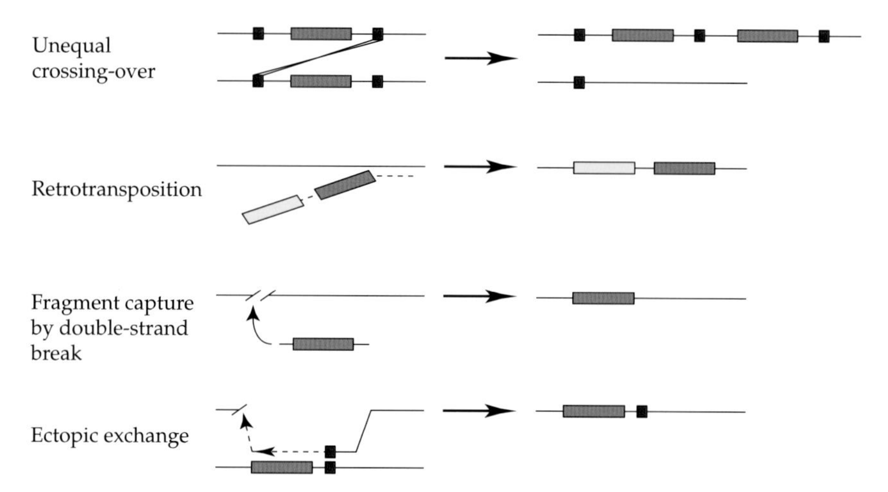

Figure 8.3 Four mechanisms for the origin of gene-sized duplications. Unequal crossing-over: A crossover occurs between two regions of sequence similarity (red) at nonhomologous sites, yielding one chromosome with a duplication and another with a deletion. Retrotransposition: A gene that has been transcribed along with an upstream retrotransposon (yellow) is inserted into a new site after reverse transcription of the mRNA intermediate (dashed line). Capture by double-strand break: An exogenous fragment containing a gene sequence (possibly originating by reverse transcription) is inserted into a chromosomal break point. Ectopic exchange: A double-strand break initiates a recombination event by invading a nonhomologous site, at which a gene copy is generated by strand extension prior to reannealing of the broken chromosome.

broadly defined as segmental duplications, arise by multiple pathways (Figure 8.3):

- Many newly arisen gene duplicates are tandemly associated with their parental copy, having arisen from local chromosomal events such as replication slippage or nonhomologous unequal crossing-over. Such local duplications can then be subsequently dispersed by various mechanisms of chromosomal rearrangement.
- Sloppy transcription of non-LTR retrotransposons occasionally leads to the replication of downstream genes and their reinsertion elsewhere in the genome, as emphasized in Chapter 3. The products of such events will often be nonfunctional (e.g., when the duplication span fails to incorporate key regulatory or coding domains), but the possibility also exists for the acquisition of entirely new expression patterns (e.g., when an insertion fortuitously incorporates regulatory sequences at the new site).
- Duplicates can originate via the capture of DNA inserts during the repair of double-strand breaks. Fragments of mitochondrial DNA and retrotransposon-derived cDNAs appear to be particularly common substrates

for this process (Ricchetti et al. 1999; Yu and Gabriel 1999; Lin and Waldman 2001a,b).

- Protruding ends of double-strand breaks may invade ectopic sites with short regions of homology, transiently using the invaded chromosome as a template for strand extension and then reattaching the two free ends (Gorbunova and Levy 1997). As in the case of LTR-derived duplications, whether insertions generated during double-strand break repair will contain one or more functional genes or simply be "dead on arrival" is entirely a matter of chance.

Structural analyses of the youngest cohorts of duplicate genes (<10% sequence divergence at silent sites) in the *C. elegans* genome provide some insight into these issues (Katju and Lynch 2003, 2006; Thomas 2006). In this species, the distribution of duplication-span lengths is highly L-shaped, with a mean of just 1.4 kb and only 30% of duplication events exceeding 2.5 kb in length (Figure 8.4). Thus, because the average length of a *C. elegans* gene from start codon to termination codon is about 1.9 kb, only 50% of newborn duplicates appear to be complete. Approximately 20% of newborn *C. elegans* genes are partial, in the sense that one member of the pair (presumably the parental copy) contains unique exonic DNA, and 30% are chimeric, with each member of the pair containing unique exons. The degree to which partial or chimeric duplicates are operable remains to be determined, but

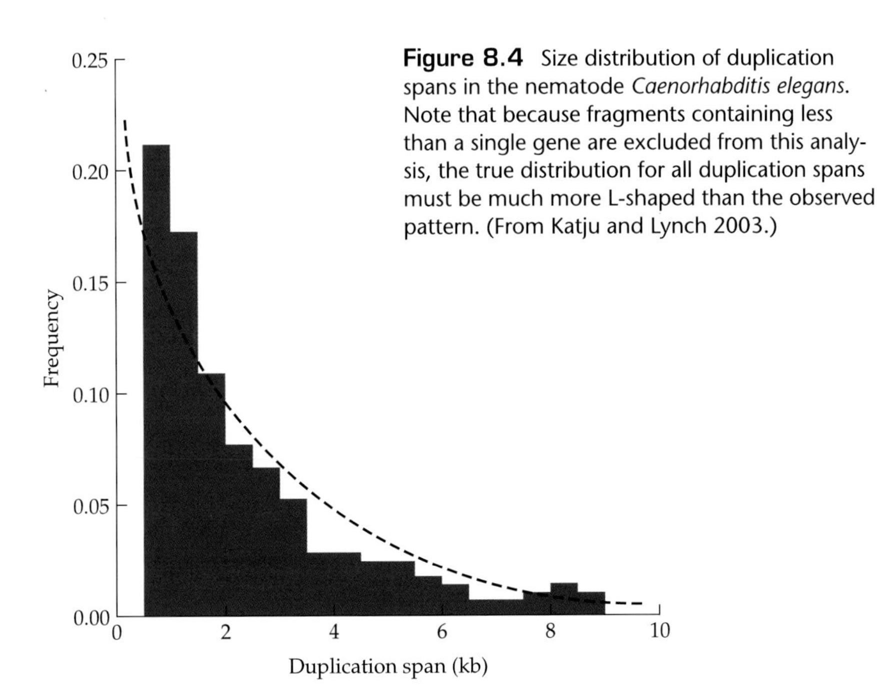

Figure 8.4 Size distribution of duplication spans in the nematode *Caenorhabditis elegans*. Note that because fragments containing less than a single gene are excluded from this analysis, the true distribution for all duplication spans must be much more L-shaped than the observed pattern. (From Katju and Lynch 2003.)

because the relative frequencies of such genes do not radically change in older cohorts, and because the expression of such genes is just as frequent as that of complete duplicates, some functional role seems likely.

Almost all (~90%) newborn duplicates in *C. elegans* are tandemly arranged. However, although tandem duplications are generally thought to arise via unequal crossing-over, which should lead to tail-to-head orientations of adjacent members, about 70% of newly arisen *C. elegans* duplicates are inverted and on opposite strands in tail-to-tail or head-to-head orientations. Tandem inversions can arise during replication if the DNA polymerase transiently switches strands, moving in the opposite direction on the complementary strand for a while, before switching back to the original strand (Bi and Liu 1996; Lin et al. 2001). Such inverted duplicates may be more stable than direct tandem repeats, which may be easily lost through the very same mechanisms that lead to their creation. L-shaped length distributions and inverted duplications are commonly observed in other species (Fischer et al. 2001; Bensasson et al. 2003; Thomas et al. 2004; Zhang et al. 2005).

Whole genome duplication

Although the previous analyses have focused on segmental duplications involving one to a few genes, entire genomes are sometimes duplicated via polyploidization events. Over the course of evolutionary history, whole-genome duplication events have been fairly common in plants, and although they are much rarer in animals (Ramsey and Schemske 1998; Otto and Whitton 2000), they nevertheless appear to occupy key positions in the animal phylogeny (Figure 8.5). Remarkably, as discussed below, the lineages of most of the model systems adopted by molecular, cell, and developmental biologists may have experienced at least one polyploidization event in the distant past.

The mechanisms of polyploidization are varied. Autopolyploids originate endogenously, with all alleles at a locus deriving from the same species, whereas allopolyploids arise via hybridization events. Both mechanisms often involve the participation of unreduced gametes, either at the outset in the case of autopolyploidy or subsequent to hybridization in the case of allopolyploidy. The initial stages of such events are made difficult by the fact that the incipient polyploid individual will generally be embedded within a population of diploids, and hence confronted with a high likelihood of yielding progeny with intermediate ploidy levels, which can inhibit the production of viable gametes. For example, the mating of a tetraploid with a diploid will produce triploid offspring that experience major problems during meiosis. Self-fertilization provides a simple mechanism for bypassing such problems in land plants, but most animals are obligate outcrossers.

Documenting an ancient polyploidization event can be difficult for several reasons. A clear signature of polyploidy is the presence of long colinear duplication spans of genes or, ideally, entire chromosomes. However, as

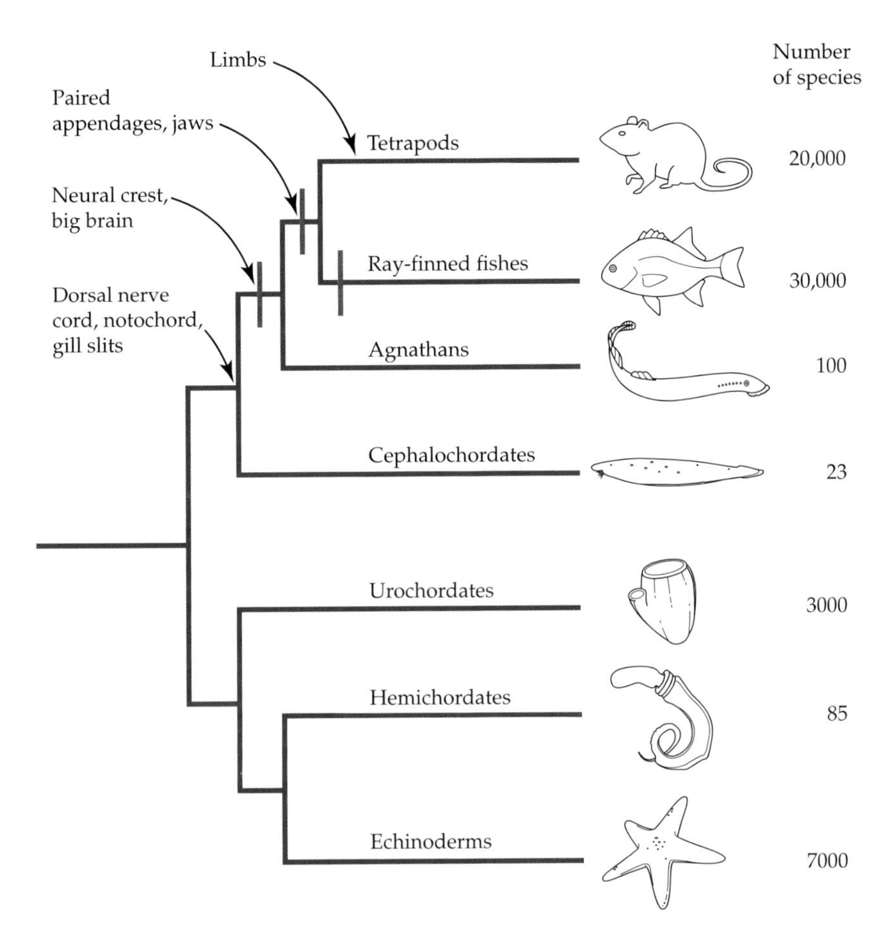

Figure 8.5 The chordate phylogeny. Lineage-specific morphological innovations are shown to the left, and numbers of species in each lineage are shown to the right. Blue bars denote periods of hypothesized genome duplication. The exact positions of the two duplications preceding the evolution of jawed vertebrates are uncertain, as is their polyploid origin; the duplication at the right denotes a polyploidization event deep in the ray-finned fish lineage, which has been followed by secondary polyploidization events in various fish lineages (not shown).

described above, many duplicate genes are eventually lost, leaving large gaps in what would otherwise be continuous stretches of genes. Secondary rearrangements of chromosomal segments (e.g., inversions and translocations) can further obscure the signature of polyploidization, as can background segmental duplications that increase the number of copies of genes beyond the expectation based on polyploidy alone. Bioinformaticians have attempted to grapple with these problems by factoring out apparently young gene duplicates and simultaneously analyzing several species, but acquiring strong evidence for ancient polyploidization events is still a formidable task. Here, we briefly highlight four putative eukaryotic genome duplication events, all involving lineages that are central to experimental biology.

Although polyploidy is often regarded as a much more common feature of multicellular than unicellular species, the first detailed genomic analysis of a polyploidization event derived from the discovery of a large number of duplicate chromosomal fragments in the yeast *S. cerevisiae* (Wolfe and Shields 1997). Only about 8% of the originally duplicated genes survive today, and there has been substantial secondary movement of chromosomal segments (Figure 8.6). Nevertheless, the fact that nearly all gene pairs within surviving fragments have the same orientation with respect to centromeres supports the idea that these segments originally arose via whole-genome duplication, a conclusion that is also consistent with phylogenetic analysis (Langkjaer et al. 2003). More recent work has revealed an even more striking case of polyploidization in the unicellular ciliate (*Paramecium tetraurelia*), where an apparent three rounds of whole-genome duplication has yielded ~40,000 genes (Aury et al. 2006).

The *Arabidopsis* genome project was initiated on the assumption that this species would have a simple genome relative to that of other plants, but it soon became apparent that the chromosomal contents of this species reflect a complex history of duplication events. The bulge in the age distribution of *Arabidopsis* gene duplicates appearing at an average level of silent-site divergence of 0.8 (see Figure 8.2) suggests an ancient phase of genome ampli-

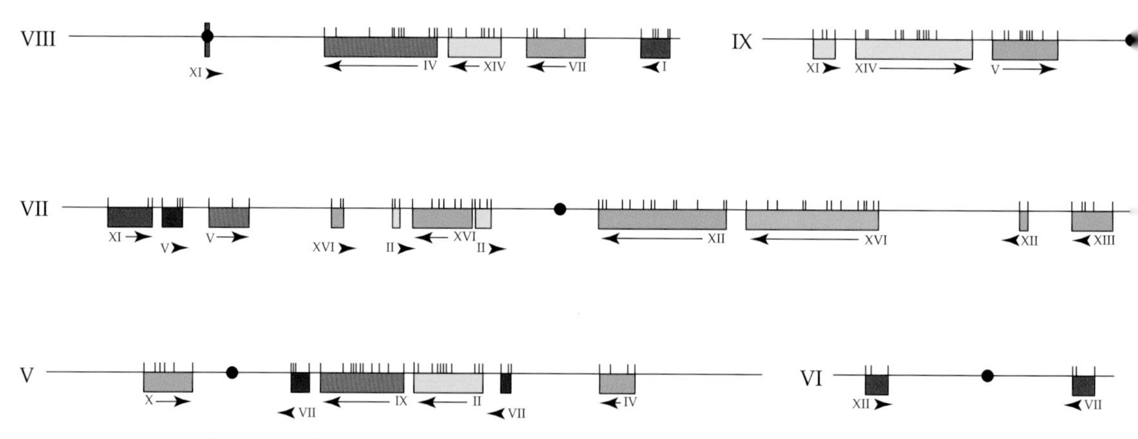

Figure 8.6 Examples of remnant chromosomal segments derived from an ancient whole-genome duplication in the yeast *S. cerevisiae*. Roman numerals denote chromosome numbers. Colored blocks of genes denote groups of genes (indicated by tick marks) for which duplicated spans exist on another chromosome (given by the Roman numeral underlying the block). Arrows denote the relative orientations of the full set of genes within a block. The solid circles demarcate centromeres. The mixture of blocks of different ancestries on individual chromosomes presumably results from chromosomal rearrangement subsequent to polyploidization. (From Wolfe and Shields 1997.)

fication. Because a very large fraction of duplicate genes of this age will have been lost, and because the silent sites of the remaining pairs are saturated with mutations, it is difficult to infer the mechanisms underlying such ancient bouts of gene amplification from age distributions alone. However, additional work that takes chromosomal positions into account suggests that the *Arabidopsis* genome is a product of at least two, and probably three, polyploidization events (Simillion et al. 2002; Bowers et al. 2003): the first prior to the monocot–dicot split; the second between 160 and 230 MYA (after the divergence of monocots and dicots); and the third between 20 and 85 MYA. An earlier suggestion of a fourth round of polyploidization (Vision et al. 2000) has not been upheld. Additional observations on other angiosperm groups suggest the occurrence of numerous independent polyploidization events in descendent sublineages (Wendel 2000; Paterson et al. 2004). Large-scale partial genome duplications have also occurred in plants. In rice, for example, a single-chromosome duplication appears to have occurred around 54 MYA, the approximate time of origin of the major grass lineages (Vandepoele et al. 2003).

Although polyploidization events are generally rare in animals, the ray-finned fish lineage appears to be exceptional. The discovery that the zebrafish has seven Hox clusters, compared with four in tetrapods, led to the hypothesis that it is an ancient polyploid (Amores et al. 1998). More in-depth analyses of Hox and other genes in medaka and pufferfish suggest that the polyploidization event took place prior to the radiation of the ray-finned fish lineage (Taylor et al. 2001a, 2003; Christoffels et al. 2004; Jaillon et al. 2004; Crow et al. 2006). Under this interpretation, the presence of seven Hox clusters in the zebrafish, rather than the predicted eight, implies the loss of one cluster following polyploidization. As in land plants, this basal polyploidization event was followed by numerous secondary genome duplications, including those at the bases of the lineages of salmonids (Johnson et al. 1987), suckers (Ferris and Whitt 1979), sturgeon (Ludwig et al. 2001), and carp (David et al. 2003).

Taking one step further back in the vertebrate phylogeny, the presence of four Hox clusters throughout the tetrapod lineage, as well as multiple copies of a number of other genes in tetrapods relative to invertebrates, suggests the occurrence of substantial gene duplication prior to the divergence of ray-finned fishes and tetrapods. Based on rather limited data and indirect inference, Ohno (1970) first launched the idea that two adjacent rounds of polyploidization preceded the emergence of the major lineages of jawed vertebrates. Subsequently christened the "2R" hypothesis (for "two rounds" of polyploidization), Ohno's conjecture implies that the ancestral jawed vertebrate was an octoploid. This idea has been extraordinarily difficult to test, largely because of the enormous time since the putative duplication events took place (450–550 MYA). Over such a long period, enough chromosomal rearrangements, gene removals, and secondary segmental duplications have taken place to all but obliterate most ancestral linkage groups, and the silent sites of protein-coding genes are so saturated with mutations that accurate dating of divergence times between candidate ancestral duplicates is essentially impossible.

Hughes and Friedman (2003) have questioned the validity of the 2R hypothesis on several grounds. First, based on the assumption that one duplication yields two copies of each gene and a second then yields four, they suggest that the 2R hypothesis implies a 4:1 ratio of gene family sizes between vertebrates and invertebrates. In contrast, the peak ratio of human gene family sizes to those of *Drosophila* (or *Caenorhabditis*) is close to 1:1, although ratios as large as 4:1 are found in many cases. Even in a comparison of vertebrates with the invertebrate cephalochordate *Branchiostoma*, a ratio of 2:1 is more common than 3:1, which in turn is more common than 4:1 (Furlong and Holland 2002). One weakness of using such observations to reject the 2R hypothesis, however, is that the 4:1 ratio prediction ignores the ubiquitous turnover of duplicated genes. Even if the half-life of vertebrate duplicate genes were as long as 100 million years (see below), almost all duplicates arising from an event about 500 MYA would have been silenced, and many smaller-scale secondary duplications would also have occurred. Thus, ratios of tetrapod:invertebrate gene family sizes are not particularly informative with respect to the 2R hypothesis, nor do they provide grounds for ruling out alternative hypotheses, such as ongoing segmental duplication.

Second, Hughes and Friedman (2003), as well as Martin (1999), argue that the 2R hypothesis predicts an (AB)(CD) topology for the genealogical trees of four-member families of vertebrate genes, with the AB and CD pairs of genes representing paralogs from the most recent genome duplication event and the (AB) versus (CD) clades representing the first duplication event (Figure 8.7). Although well over half of the estimated gene genealogies deviate from the (AB)(CD) expectation, there are again substantial reasons to question the informativeness of such an analysis. First, a large fraction of phylogenies with deep and contiguous internal nodes can be expected to deviate from the predicted pattern due to errors in phylogenetic construction alone. Second, with secondary duplications, rearrangements, and gene silencings overlaid on the original polyploidization events, a wide array of tree topologies is expected (recall, for example, Figure 1.5). Finally, if the original duplication events resulted from autopolyploidy, deviations from the (AB)(CD) pattern can arise naturally from allelic sorting prior to conversion of the octoploid state to functional diploidy (Furlong and Holland 2002). Thus, the failure to consistently recover an (AB)(CD) topology for vertebrate genes with four family members has little power for testing the 2R hypothesis.

Despite these uncertainties about the validity of the 2R hypothesis, there is little question that considerable gene duplication of some sort occurred deep in the chordate lineage. For example, after factoring out the large number of fairly recent duplications in the human genome, Panopoulou et al. (2003) dated a peak of ancient vertebrate duplicate genes at about 500 MYA. Similarly, by performing phylogenetic analyses on putative paralogous link-

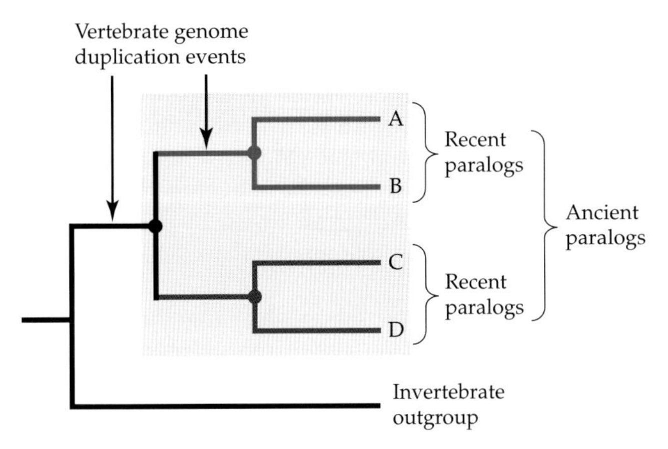

Figure 8.7 Idealized tree of relatedness of the members of a vertebrate gene family under the 2R hypothesis. The gray area denotes a set of genes within a single vertebrate genome resulting from two prior genome duplication events. A single copy of the gene (black dot) exists in the ancestral vertebrate as well as in the invertebrate outgroup. In the first round of polyploidization, this gene is duplicated to form the paralogous copies denoted by the green and blue dots. A second round of polyploidization then gives rise to four copies (A and B in the green lineage, and C and D in the blue lineage). As discussed in the text, this (AB)(CD) expectation requires that both copies survive the initial polyploidization event, and it also ignores the numerous ways in which secondary duplications and gene losses can lead to the same tree structure regardless of whether the 2R hypothesis is correct.

age groups using *Drosophila* and *Caenorhabditis* as outgroups, McLysaght et al. (2002) found a peak in the age distribution of duplicate genes prior to the emergence of vertebrates but subsequent to the protostome–deuterostome split. Additional analyses suggest that this peak postdates the split between cephalochordates and vertebrates (Figure 8.5) but predates the origin of jawed vertebrates (Abi-Rached et al. 2002; Escriva et al. 2002; Furlong and Holland 2002; Gu et al. 2002a; Robinson-Rechavi et al. 2004). Such timing is consistent with Ohno's conjecture that a massive phase of gene duplication provided the genetic substrate deployed in the subsequent emergence of the unique morphological innovations of jawed vertebrates (see Figure 8.5). Whether such extensive duplication activity involved one, two, or even three polyploidization events or was simply a period of substantial segmental duplication remains uncertain, but the most thorough study to date, a joint analysis of gene locations and family-member phylogenies, supports the 2R hypothesis (Dehal and Boore 2005). The unresolved matter of whether amplification of gene number had a causal role in phenotypic diversification is another issue entirely.

Finally, it is notable that genes arising via polyploidization generally exhibit enhanced longevity relative to those arising by segmental duplications. For example, based on the results cited above for yeast, the half-life of a duplicate gene arising by a localized duplication event is about 17×10^6 generations, whereas that for a duplicate arising via the ancient polyploidization event is about 27×10^6 years (and presumably many more generations) (Table 8.2). Similarly, vertebrate genes arising by segmental duplication have half-lives of about 2×10^6 years, whereas those arising by polyploidization in ray-finned fishes have average half-lives of about 45×10^6 years. Finally, the half-lives of genes arising by polyploidization in plants appear to be approximately tenfold greater than those arising by small-scale events.

One possible explanation for the greater longevity of duplicates arising via polyploidization involves dosage requirements. When a complete genome is duplicated, each gene's expression is expected to remain in the same stoichiometric relationship with all of its interacting partners, a balance that may be favored by selection. In contrast, when a single gene is duplicated, it is immediately out of balance with its partners, potentially leading to functional difficulties. This hypothesis is consistent with two observations in yeast: first, that duplicate genes whose products participate in protein complexes are relatively underrepresented among duplications with segmental origins; and second, that members of interacting pairs of genes tend to be co-duplicated (Papp et al. 2003a). The enhanced longevity of duplicate genes derived from polyploidization may also be facilitated by the complete conservation of surrounding regulatory sites and the ancestral pattern of gene expression at the time of duplication.

Table 8.2 Proportions of genes surviving a polyploidization event

PHYLOGENETIC GROUP	FRACTION SURVIVING	DATE OF EVENT (MYA)	HALF-LIFE (MY)	REFERENCE
Animals				
Catastomids	0.50	50	50	Ferris and Whitt 1979
Cyprinus carpio (carp)	0.60	12	16	David et al. 2003
Loaches	0.25	28	14	Ferris and Whitt 1977
Salmonids	0.50	100	100	Allendorf et al. 1975
Xenopus laevis (frog)	0.77	30	80	Hughes and Hughes 1993
Land plants				
Arabidopsis thaliana	0.33	50	31	Ermolaeva et al. 2003
Oryza sativa (rice)	0.21	70	31	Paterson et al. 2004
Zea mays	0.72	11	23	Gaut and Doebley 1997
Yeast				
Saccharomyces cerevisiae	0.08	100	27	Wolfe and Shields 1997

Mechanisms for the Preservation of Duplicate Genes

To be successful in the long term, a duplicate gene must first drift toward fixation, and then, once it has risen to a high frequency, the selective forces for its maintenance must be sufficiently large to prevent its subsequent loss by degenerative mutation. The preceding results indicate that the vast majority of gene duplicates arising by segmental duplication experience an early exit from the population, most probably failing to ever reach fixation. However, a minority of duplicates can be retained for long periods. The precise mechanisms by which duplicate genes are preserved, three of which are discussed below, have a fundamental bearing on genome evolution. For example, the reciprocal preservation of both members of a pair of duplicates leads to an expansion in genome size, while the preservation of a new unlinked duplicate combined with the loss of the ancestral copy has no effect on gene number, but does induce an alteration of the genetic map.

Neofunctionalization

One of the more notable mechanisms for the preservation of a pair of gene duplicates is the process of neofunctionalization, whereby one copy acquires a beneficial mutation that results in a new function. Models of neofunctionalization via gene duplication generally assume that new beneficial functions are acquired at the expense of essential ancestral functions, the unspoken reasoning being that selectively advantageous mutations with no negative pleiotropic effects on wild-type fitness should have had no barriers to fixation prior to duplication. Under this reasoning, the temporary phase of redundancy provided by gene duplication is thought to release one copy from prior selective constraints, thereby enabling it to take on a previously forbidden adaptive feature (e.g., Ohno 1970). Although it is frequently assumed that the newly arisen copy will be the recipient of a new function, natural selection makes no decision as to which copy to tinker with. Instead, the early trajectories of the members of a duplicate pair are defined largely by their allelic ancestry or by early chance mutational events that occur in one copy or the other. The simplest version of this model, which assumes the duplication to be initially fixed in the population, extends back at least to Haldane (1933) and was explored quantitatively by Walsh (1995). In most of the following discussion, however, we will consider the more realistic case in which a duplicate arises as a single copy in a single member of the population, as the early phase of establishment can be critical.

Although most theoretical treatments of neofunctionalization have focused on mutations arising after the duplication event, Spofford (1969) made the key observation that the process need not await the arrival of new mutations. Her reasoning was based on the simple fact that the spectrum of mutations arising after a duplication event must be the same as that of mutations arising prior to duplication. Under this view, prior to duplication, a mutant allele endowed with a beneficial function at the expense of an essential ancestral feature may be maintained at low frequency by balancing selection, provided that heterozygotes have a selective advantage (s) relative to wild-type homozygotes. Such ancestral polymorphisms can facilitate the route to neofunctionalization in two ways (Figure 8.8). First, if the duplicate locus is founded by a neofunctionalized allele, its fixation can be promoted by positive selection while the original locus retains the ancestral function. Second, if the duplicate locus is founded by a "wild-type" allele, which drifts by chance to a high enough frequency, the selective regime at the ancestral locus will be altered from balancing selection to positive selection for the previously low-frequency neofunctional allele. In either case, the final outcome is functionally equivalent to the fixation of heterozygosity, with one locus becoming essentially monomorphic for the wild-type allele and the other for the neofunctionalized allele.

A striking example of this process involves the evolution of insecticide resistance in the mosquito *Culex pipiens* (Lenormand et al. 1998). The acetyl-

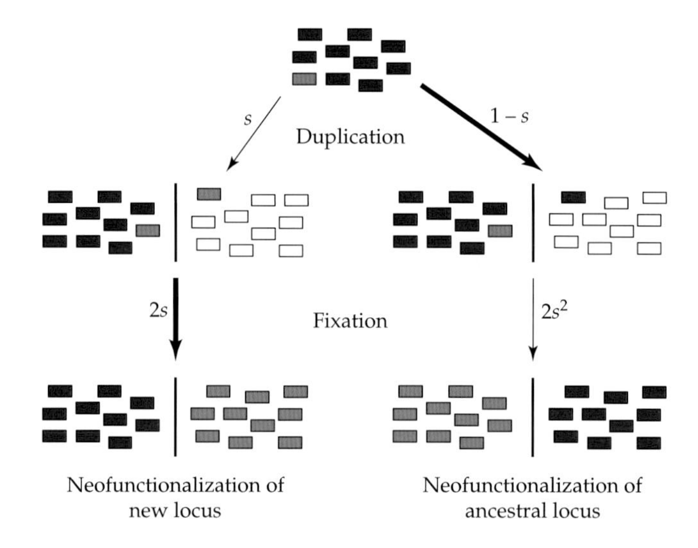

Figure 8.8 Neofunctionalization following duplication of an ancestral locus carrying an adaptive polymorphism. The individual boxes denote alleles in the gene pool. Prior to duplication, the green (neofunctional) allele is unable to go to fixation because, despite its advantage in heterozygotes, it is lethal in the homozygous state due to the absence of an essential wild-type function. The duplication event begins with either the wild-type (red) allele or the neofunctional (green) allele, with the relative frequencies of the starting states (denoted by arrows with different thicknesses and the algebraic expressions along the arrows) equal to the allele frequencies at the ancestral locus. Horizontal lines separate genes at the two loci; "absentee" alleles at the time of the duplication are denoted by hollow boxes. Although the path on the left is less likely at the duplication stage, once initiated, it is more likely to go to fixation because of the initial elevated advantage of the green allele at the new locus.

cholinesterase enzyme in this species normally plays an essential role in the central nervous system, but a mutant allele at the locus also confers resistance to organophosphate insecticides. Because the mutant allele reduces fitness in the absence of insecticides (Berticat et al. 2002), it is maintained at a very low frequency in insecticide-free environments. However, when mosquitoes were challenged with organophosphates, a linked combination of a wild-type and a resistant-type allele rose rapidly to a high frequency, presumably because the first member of the pair provided the essential ancestral gene functions while the second ameliorated the toxicity of the environment. The linked pair of duplicates may have been present in the absence of insecticides, but kept at a very low frequency by the negative effects of increased dosage. Similarly, trichromatic vision in some primates appears to have become established by the parallel fixation of preexisting opsin alleles following the duplication of an ancestral locus (Jacobs et al. 1996; Dulai et al. 1999).

The conditions necessary for the maintenance of neofunctional alleles at an ancestral locus by balancing selection have been worked out (Lynch et al. 2001). The key requirement is that the power of random genetic drift be sufficiently weak relative to the strength of balancing selection (approximately $1/N_g < s^2/8$, where N_g is the effective number of genes per locus—equivalent to the effective population size for a haploid and approximately twice that for an outcrossing diploid; see Chapter 4). Provided this condition is met, the neofunctional allele will be present at the ancestral locus with approximate expected frequency s (e.g., an allele that is lethal in the homozygous state but increases fitness by 1% in the heterozygous state would have an expected frequency of 1%, provided that $N_g > 80{,}000$). Thus, a mild heterozygote advantage combined with a moderately large effective population size is sufficient to poise a population for progression toward neofunctionalization following a duplication event, eliminating the waiting time for new mutations.

The probability of preservation of a pair of duplicate genes by neofunctionalization depends on several additional factors (Lynch et al. 2001; Walsh 2003). In large populations, the probability of neofunctionalization increases with s^2, provided the duplicate loci are unlinked. This scaling can be understood most easily by noting that the new locus will be founded by a wild-type allele with probability $(1 - s)$ and by a neofunctionalized allele with probability s (see Figure 8.8). In the former case, the founder allele will have a selective advantage defined by the frequency of neofunctional homozygotes at the ancestral locus (s^2), because "absentee" homozygotes at the new locus are lethal on this genetic background. In the latter case, the marginal selective advantage of the founder allele over the absentee allele is simply the selective advantage s. The probability of neofunctionalization by each of these paths is equal to the product of the probability of the starting condition and the fixation probability (which, in a large population, is approximately equal to twice the selective advantage; see Chapter 4). Assuming small s so that $(1 - s) \approx 1$, each path has a probability of approximately $2s^2$.

This overall neofunctionalization probability is actually an upper bound, as it assumes the duplicate loci to be unlinked. This point can most easily be understood by considering a pair of tandemly linked duplicates in which both members have the same initial allelic state. In the extreme case of complete linkage, a tandem pair of neofunctionalized alleles cannot proceed to fixation, as this would lead to the complete loss of the essential ancestral function.

In this latter case, as well as in the more general case in which $N_g < 8/s^2$ and the neofunctional allele cannot be maintained at the ancestral locus by selection, permanent neofunctionalization by gene duplication requires a starting point at which both loci harbor only wild-type alleles. Such a condition imposes a particularly precarious phase of initial establishment, making it unlikely that the probability of neofunctionalization will exceed $1/(2N)$, the initial frequency of the active allele at the new locus, unless there is an intrinsic selective premium on the duplication itself (e.g., a dosage advantage). If the duplication is initially neutral, it will drift to fixation with probability $1/(2N)$, and its subsequent fate will be determined by the relative rates of fixation of inactivating versus preservational mutations. Given an assumed initial state of neutrality, the rate of inactivation is simply equal to the degenerative mutation rate, whereas the rate of preservation by neofunctionalizing mutations is equal to the rate of origin of such mutations times the probability of fixation (necessarily less than 1). Thus, because we generally expect most mutations to be deleterious, the probability of neofunctionalization in the absence of neofunctional alleles in the base population is likely to be just a small fraction of $1/(2N)$ (Walsh 1995).

On the other hand, even in the absence of preexisting neofunctional alleles, if there is an immediate selective advantage for a duplication, as would be the case if a larger amount of the gene product were beneficial (Zhang 2003; Francino 2005; Kondrashov and Kondrashov 2006), then the probability of the first (preservational) step toward neofunctionalization can be greater than $1/(2N)$. As noted in Chapter 4, positive selection will be effective only if the intrinsic advantage of a duplicate (s_d) is sufficiently greater than the power of random genetic drift (i.e., $2N_g s_d > 1$). Moreover, after a duplicate has been preserved by this process, neofunctionalization will follow only if the origin of a new function at one locus provides a benefit (s) that significantly exceeds the initial advantage associated with gene dosage [i.e., $2N_g(s - s_d) > 1$]. All of these observations reinforce the general principle that the neofunctionalization of a duplicate gene is a large-population phenomenon.

Finally, under all of the theory discussed above, neofunctionalization evolves through the progressive modification of an active allele. Ohno (1970) suggested an alternative scenario by which a silenced gene duplicate (i.e., a transient pseudogene) might provide the substrate for the origin of a novel function (see also Marshall et al. 1994). By accumulating a series of molecular changes in a neutral fashion, an inactivated locus might eventually yield a beneficial product that would be impossible to acquire via natural selection alone. Imagine, for example, a silent allele at a duplicated locus separated from a beneficial state by intermediate mutational steps associated with low fitness (if expressed). If, after the fortuitous acquisition of mutations with jointly beneficial effects, such an allele were to be reactivated somehow, it would be strongly favored by natural selection, possibly displacing the ancestral locus (in the absence of negative pleiotropy) or coexisting permanently with it (in the presence of a trade-off). In principle, this mechanism for vaulting an adaptive valley might be facilitated by gene conversion between the silenced and expressed loci (Hansen et al. 2000). To date, however, there is no evidence that these sorts of Lazarus effects play an important role in evolution.

The masking effect of duplicate genes

Because all loci harbor suboptimal alleles due to the recurrent introduction of deleterious mutations, it is sometimes thought that duplicate genes have an intrinsic selective advantage associated with their ability to mask the effects of deleterious mutations at the ancestral locus. But is the magnitude of such an indirect advantage great enough to promote the permanent preservation of duplicate genes? Fisher (1935) realized that even in an effectively infinite population (in which the efficiency of selection is maximized), two genes with identical functions will not be mutually maintained by this process unless their mutation rates to defective alleles are identical. If this is not the case, the gene with the higher mutation rate will eventually be silenced by the differential accumulation of genetic load. In principle, permanent retention of duplicate genes might be achieved by a delicate balance between differential selective advantages and differential mutation rates of a pair of duplicate loci (e.g., with one gene operating more efficiently and the other having a lower mutation rate to null alleles) (Nowak et al. 1997), but the conditions required under this model are unrealistically stringent.

In finite populations, not even identical mutation rates are sufficient for the permanent retention of duplicate genes via the masking effect (Clark 1994; Lynch et al. 2001; O'Hely 2006). Consider, for example, a segregating recessive lethal allele, which in a large population has an equilibrium homozygote frequency at a single-copy locus equal to the null mutation rate, μ_c (Crow and Kimura 1970). Under the masking model, μ_c is equal to the initial selective advantage of an otherwise redundant functional duplicate at a new locus, as the duplicate modifies fitness only in individuals that have no functional gene at the ancestral locus. However, because μ_c is also the rate of silencing of genes at the duplicate locus, these two factors cancel exactly, rendering the duplicate effectively neutral and vulnerable to eventual loss by random genetic drift. If null alleles have fitness effects in heterozygotes, a duplicate gene can potentially mask the effects of suboptimal genotypes in a larger fraction of the population, but even in this case, the net advantage of the masking effect is on the order of the mutation rate.

A deleterious allele causing a reduction of fitness in heterozygotes of *hs* has an expected frequency $q = \mu_c/(hs)$ under selection–mutation balance, so with a fraction $2(1 - q)q \approx 2q$ of the population heterozygous at the original locus, the selective advantage of a newly arisen duplicate is just $2q \cdot hs = 2\mu_c$ (Otto and Yong 2002; Proulx and Phillips 2005). After subtracting out the rate of loss of the duplicate gene by mutation, and recalling the definition of effective neutrality (see Chapter 4), we see that the condition for the maintenance of a duplicate gene via the masking effect is $2N_g\mu_c > 1$. In other words, the effective gene number per locus must exceed $1/(2\mu_c)$.

Although the preceding arguments raise significant questions about the sufficiency of the power of the masking effect to maintain duplicate genes, they do not rule out the possibility that duplicates maintained by other processes still have a buffering potential. However, although genetic redundancy almost certainly masks some deleterious mutational effects, quantitative information on the matter is scant. In the yeast *S. cerevisiae* and the nematode *C. elegans*, the fitness consequences of knockouts of single members of pairs of duplicate genes are smaller than those of knockouts of single-copy genes, and the effects of knockouts increase with the magnitude of sequence divergence between paralogs (Gu et al. 2003; Conant and Wagner 2004). Taken at face value, such observations suggest a diminishing incidence of overlapping functions as a duplicate pair ages. However, studies of extant duplicates suffer from ascertainment bias. In yeast, for example, those genes that have the smallest knockout effects when present in a single copy are the most likely to exist as duplicates (He and Zhang 2005a; Prachumwat and Li 2006), so the small effects of deletions of single members of such pairs are not entirely due to a buffering effect, but at least in part a simple consequence of their nonessentiality.

Finally, it should be noted that the masking effects described above are concerned with compensation for mutationally silenced alleles, whereas duplicate genes might also provide a buffer against transient cellular mishaps causing localized absence of gene expression by normally active genes (Tautz 1992; Nowak et al. 1997). The potential sources of such developmental errors include somatic mutations and errors in transcription and translation. Following the logic outlined above, for this masking mechanism to maintain a duplicate gene by natural selection, the fitness consequences of developmental errors at a locus would have to exceed the rate of origin of null mutations at the duplicate locus, and this net difference, in turn, would have to exceed the power of random genetic drift.

In summary, despite their seductive nature, the various masking models for the preservation of duplicate genes require rather special sets of mutational conditions and very large effective population sizes to enable the very weak selective advantages of redundancy to come to prominence. Perhaps the most serious challenge to the idea that masking plays a prominent role in duplicate-gene retention is the general paucity of duplicate genes in prokaryotes despite their haploid nature and exceptionally high N_g.

Subfunctionalization

Given the difficulties with the various masking hypotheses, neofunctionalization has often been assumed to be the *only* mechanism that can permanently preserve duplicate genes. Under this view, however, the vast majority of new gene duplicates are expected to be lost within a relatively short time because of the rarity of neofunctionalizing relative to nonfunctionalizing mutations. In the absence of positive selection, a fraction $[1 - 1/(2N)]$ of newly arisen gene duplicates will be lost by random genetic drift in an average of just $2\ln(N_g)$ generations (Kimura and Ohta 1969), a flash on the evolutionary time scale. Moreover, the small remaining fraction, $1/(2N)$, that manages to drift to fixation is also expected to fall victim to silencing mutations relatively quickly. If $N_g\mu_c << 1$, the average time to silencing of a fixed duplicate is approximately equal to the mean waiting time for the appearance of a null mutation at one of the loci, $1/(2\mu_c)$ generations. On the other hand, when $N_g\mu_c >> 1$, null mutations are common, and the time to silencing depends largely on the time required for one of them to drift to fixation at one of the loci, about $2N_g$ generations (see Chapter 4), with the slight initial masking advantage of the new allele (described above) prolonging its survival up to $5N_g$ generations (Watterson 1983; Lynch and Force 2000b).

Although these predictions of a relatively rapid demise of the vast majority of duplicate genes are in rough accord with the evolutionary demographic analyses reported above for segmental duplications, they are inconsistent with the large levels of duplicate-gene retention in ancient polyploids (see Table 8.2). Moreover, the lack of evidence that polyploid lineages have evolved unusually large numbers of new gene functions suggests the involvement of preservational mechanisms other than neofunctionalization. As noted above, positive selection associated with dosage requirements may play a role in polyploid lineages, but a more general mechanism for duplicate-gene preservation, relevant even to segmental duplications, derives from a broader view of gene structure than assumed under the classic model.

Many eukaryotic genes, particularly those in multicellular species, have complex, modular regulatory regions, alternative splicing mechanisms, and/or functional domain structures. Such genes are naturally endowed with independently mutable subfunctions, in the sense that mutations that cause the loss of function in one particular tissue or developmental period do not necessarily affect other tissue- or timing-specific aspects of their expression. The widespread existence of such structural complexities leads to the prediction that duplicate-gene preservation will sometimes result from the partitioning of ancestral gene functions through complementary loss-of-function mutations in paralogous copies (Figure 8.9). Under the DDC (Duplication–Degeneration–Complementation) model of Force et al. (1999), subfunctionalization is driven entirely by degenerative mutations, which we know to be much more common than beneficial mutations (Lynch et al. 1999). Special cases of this model have also been discussed by Hughes (1994) and Stoltzfus (1999), and Taylor and Raes (2004) cite earlier relevant references.

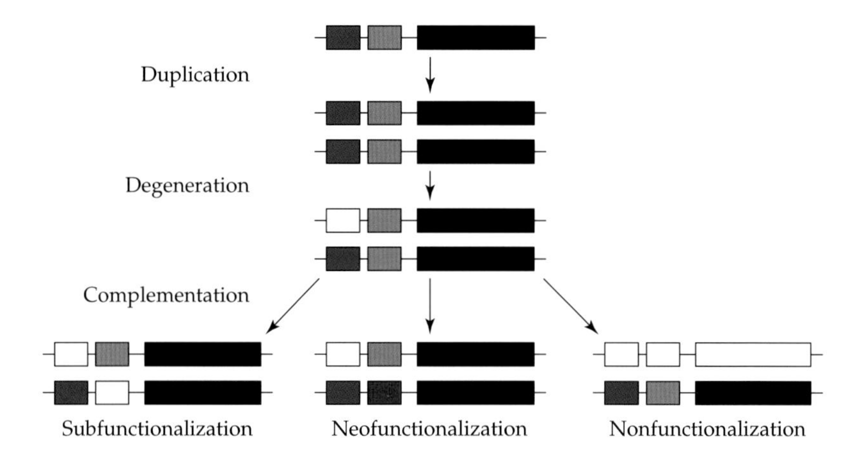

Figure 8.9 The DDC model for the alternative fates of duplicate genes. The ancestral gene is depicted as having two independently mutable regulatory regions (one blue and one green), each driving expression in a particular tissue or developmental period. Solid boxes denote fully functional regulatory and coding regions, whereas open boxes denote loss of function, and the red box denotes the gain of a new beneficial function. Each pair of genes reflects the fixed haploid state of the population. Following the duplication event, the first degenerative mutation eliminates a subfunction of one of the copies. The second mutational event then dictates the final fate of the pair—subfunctionalization, with the second copy acquiring a complementary loss-of-subfunction mutation; neofunctionalization, with the second copy acquiring a novel, beneficial expression pattern at the expense of an ancestral subfunction; or nonfunctionalization, with the first copy losing all functional ability. (From Force et al. 1999.)

Duplicate-gene preservation by subfunctionalization is a two-step process. First, one of the genes becomes fixed for a mutation that eliminates an essential subfunction, permanently preserving the second copy. Loss of an alternative subfunction by the second copy then reciprocally preserves the first copy. If the effective population size is sufficiently small that segregating null mutations are typically rare ($N_g \mu_c << 1$), the fate of a newly arisen duplicate gene under the DDC model depends almost entirely on the relative rates of origin of mutations that abrogate single subfunctions (μ_r) versus those that cause complete nonfunctionalization (μ_c), and the probability of subfunctionalization can be approximated with combinatorial logic. Consider, for example, a gene with two independently mutable subfunctions, with the rate of mutation to new beneficial functions being of negligible importance. The probability that a newborn duplicate will drift to fixation is its initial frequency, $1/(2N)$, and having arrived at that point, the probability that the first fixed mutation will eliminate a subfunction from one of the genes is simply the fraction of mutations that are of the subfunctionalizing type, $2\mu_r/(2\mu_r + \mu_c)$. After fixation of a subfunctionalized allele at one locus, the intact locus is no longer free to lose its now unique subfunction, so it accepts mutations at rate μ_r, whereas the partially debilitated locus is free to become completely silenced (at rate $\mu_r + \mu_c$). Thus, the fraction of permissible mutations that eliminate the complementary subfunction at the intact locus in the second step is $\mu_r/(2\mu_r + \mu_c)$. The probability of subfunctionalization (P_{sub}) is equal to the product of these three probabilities, $\alpha^2/(4N)$, where $\alpha = 2\mu_r/(2\mu_r + \mu_c)$ is the fraction of degenerative mutations that eliminate single subfunctions. Under this two-subfunction model, P_{sub} approaches a maximum of $1/(4N)$ as α approaches 1 (all mutations are of the subfunctionalizing type) because there is a 50% probability that the first two mutations will be incurred by the same copy (leading to nonfunctionalization) and a 50% probability that the two copies will incur complementary mutations.

A number of factors can increase the probability of subfunctionalization above $1/(4N)$, but none of them is likely to move the upper bound beyond $1/(2N)$, the probability of initial fixation of an entirely neutral duplication. For example, increasing the number of independently mutable subfunctions increases P_{sub} by increasing the number of paths by which complementary loss-of-function mutations can be acquired by the two copies (Lynch and Force 2000b). Mutations with partial effects on gene expression will also increase the probability of duplicate-gene preservation, even providing a retention mechanism for duplicates whose expression patterns cannot be subdivided spatially or temporally in development (Lynch and Force 2000b; Duarte et al. 2006). Such preservation, known as quantitative subfunctionalization, occurs whenever the total capacity of both loci is degraded to the extent that their joint presence is needed to fulfill the requirements of the single-copy ancestral gene (as in the case of duplicate enzyme genes that acquire activity-reducing mutations). Consider, for example, a gene with a single function, and let n be the number of mutations with partial effects necessary to completely eliminate its function. If μ_p is the total rate of origin of such mutations and $\rho = \mu_p/(\mu_p + \mu_c)$ is the fraction of the total pool of mutations with partial effects, then the upper limit to P_{sub} under this model, $\rho^2/(2N)$, is approached as the average effects of partially debilitating mutations decline to zero (i.e., as n approaches infinity). This limit approaches a maximum value of $1/(2N)$ as mutations with partial effects become more predominant (i.e., as ρ approaches 1). Finally, as noted above, newly arisen duplicates need not always be complete, and if critical regulatory elements are missing from the flanking regions at the time of the duplication event, the first step toward subfunctionalization may be fulfilled at the outset (Averof et al. 1996). For a gene with two subfunctions, the loss of one subfunction at the time of duplication would increase the probability of subfunctionalization from $\alpha^2/(4N)$ to $\alpha/(4N)$.

The subfunctionalization model makes unique predictions about the scaling of the probability of duplicate-gene preservation with population size (Box 8.1). Assuming that there is no initial selective consequence of subfunctionalization, duplicate genes should be preserved more often by this mechanism in small populations and somewhat more commonly when they are tandemly linked. These predictions contrast with those of the neofunctionalization model, which predicts that duplicate genes should be preserved more often in large populations and more commonly when they are unlinked. These divergent expectations have a potentially important bearing on our understanding of the mechanisms leading to genome expansion. If neofunctionalization is the predominant mechanism of duplicate-gene preservation, then larger populations, which harbor more targets for rare beneficial mutations, should experience a greater expansion of gene number (provided the duplication rate itself is comparable, and factoring out quantum changes due to polyploidization). Noting that the mouse lineage has

Box 8.1 Effective Population Size, Linkage, and the Probability of Subfunctionalization

The expressions derived in the text for the probability of subfunctionalization assume a sufficiently small effective population size that (1) each step in the process proceeds to completion before the next key mutational event occurs and (2) differences in the mutational vulnerabilities of alternative alleles have negligible selective consequences. These conditions, which are generally met if the effective population size is small enough that $N_g\mu_c << 1$ (where μ_c is the rate of nonfunctionalizing mutation per generation), result in a situation in which the rate of subfunctionalization at a locus is essentially independent of both the population size and the degree of linkage between duplicates. Assuming diploidy, the population-level rate of duplication is $2NB$, where N is the population size and B is the physical rate of gene duplication and, conditional on duplication, the probability of subfunctionalization is $\alpha^2/(4N)$, where α is the fraction of degenerative mutations that eliminate single subfunctions. The rate of subfunctionalization is the product of these two components, $B\alpha^2/2$ (Lynch and Force 2000b).

In larger populations, both population size and degree of linkage play important roles in determining the probability of subfunctionalization (Lynch et al. 2001). Specifically, for $N_g\mu_c > 1$ and unlinked duplicates, P_{sub} approaches zero at large N_g because during the first phase of the process (the ~$2N_g$ generations required for a newly arisen duplicate to drift to initial fixation), essentially all descendants of the initial duplicate will acquire silencing mutations.

The situation is slightly more complicated for linked duplicates. The probability of subfunctionalization again declines to zero at large N_g, but at a somewhat higher threshold value of N_g than for unlinked duplicates. A linked pair of duplicates initially has a weak selective advantage (approximately equal to μ_c) over a single-copy gene, resulting from the fact that complete inactivation of a "two-copy" allele requires the silencing of both members of the pair. This advantage gives a linked pair of duplicates a slight boost in the initial fixation process relative to the neutral expectation of $1/(2N)$. However, once such a pair becomes subfunctionalized, the tables are turned: because such a linked pair requires two coding regions to carry out the ancestral subfunctions, it is a larger mutational target than a fully functional single-copy allele, and therefore has a permanent selective disadvantage (again equal to μ_c). These types of insights would not have been possible without a formal population genetic analysis.

experienced roughly 60% more gains and 45% fewer losses of duplicate genes established prior to the rodent–primate divergence than has the great ape lineage (assumed to have a smaller long-term effective population size), Shiu et al. (2006) have argued that positive selection plays a more important role in duplicate-gene preservation than does degenerative mutation. However, because the generation length of rodents is substantially shorter than that of primates, on a per-generation basis (the relevant scale for genomic evolution), there are actually substantially more gains and losses of duplicates in the human lineage (approximately 6 and 19 times more, respectively, if the average generation time leading to humans is assumed to be ten times that for rodents). Thus, these comparative analyses of mammalian duplicates actually support a central role for subfunctionalizing and nonfunctionalizing mutations, as do numerous other observations to be discussed below.

By postulating a preservational process driven entirely by degenerative mutations, the subfunctionalization model provides a null hypothesis for the interpretation of patterns of survival of duplicate genes. Subfunctionalization, however, may be the beginning, not the end, of new evolutionary pathways. Consider, for example, a single-copy locus that is a victim of a "jack of all trades, master of none" syndrome, such that an adaptive conflict exists between its subfunctions. If such a gene is duplicated, complementary loss-of-subfunction mutations are expected to alter the selective landscape experienced by the two members of the duplicate pair, enabling each copy to become more refined to its specific subset of tasks (Piatigorsky and Wistow 1991; Hughes 1994) and perhaps opening up previously unavailable pathways to neofunctionalization. By this means, two of the most common forms of genomic upheaval, gene duplication and degenerative mutation, may provide a unique mechanism for the creation of novel evolutionary opportunities through the elimination of pleiotropic constraints.

The Fates of Duplicated Protein Sequences

The alternative models for the maintenance of duplicate genes motivate several questions. First, do duplicate genes experience an early phase of relaxed purifying selection? Second, do functional novelties in duplicate genes arise out of preexisting polymorphisms (the balancing-selection model), as refinements of preexisting subfunctions in multifunctional single-copy genes (the adaptive-conflict model), or as de novo modifications following duplication? Third, are the mutational events that drive the critical phase of early preservation typically beneficial (neofunctionalization model) or degenerative (subfunctionalization model), and how frequently do such mutations influence regulatory versus coding sequences? Fourth, are the rates of molecular evolution in the two copies equal or asymmetrical (the latter being the expectation if just one copy embarks on an exploratory evolutionary pathway)? Answering these questions has been a major challenge.

The presentation in the previous section indicates that the relative incidence of nonfunctionalizing, subfunctionalizing, and neofunctionalizing mutations should be a fundamental determinant of the fates of duplicate genes, but direct information on this matter is lacking. Thus, most attempts to understand the evolution of duplicate genes have resorted to a more indirect approach: comparative analysis of paralogous sequences in extant species. The cumulative nature of gene evolution substantially restricts the information that can be gleaned from such studies. In order to distinguish between the mutational events that lead to duplicate-gene establishment versus subsequent divergence, one would like to follow the historical record of substitutional changes incurred by both members of the pair from the time of origin to the time of preservation or elimination, but comparative analysis reveals only the end products of evolution. In addition, due to our poor understanding of the functional significance of noncoding regions of genes, almost all research in this area has been restricted to coding region evolution.

As noted in previous chapters, contrasts in the number of amino acid replacement substitutions per replacement site (R) and the number of silent substitutions per silent site (S) can provide crude information on the average form of selection operating on a pair of genes: an R/S ratio smaller than one implies that selection on replacement sites is predominantly of a purifying form, while a ratio greater than one implies directional selection for a change in function in one or both sequences. This criterion for distinguishing between directional and purifying selection is actually somewhat ambiguous in that an R/S ratio smaller than one averaged over the full length of a coding sequence does not preclude the possibility of strong adaptive divergence in one region of the protein on a more general background of purifying selection. In addition, an R/S close to one, which is consistent with neutrality, could also arise if one member of a pair were under purifying selection and the other under positive selection. On the other hand, a gene-wide estimate of R/S greater than one provides essentially unambiguous evidence of directional selection in one or both pair members. A deeper problem is that the estimates of R and S for any pair of extant duplicates are cumulative outcomes of the joint evolutionary pressures operating on both loci since the initial duplication event. In a simple two-gene analysis, a brief early phase of relaxed selection ($R/S \approx 1.0$) could easily be obscured by a subsequent prolonged phase of purifying or positive selection.

Patterns of molecular evolution

Some information on the temporal dynamics of selection on gene duplicates can be gleaned from observations on the joint distribution of R and S for the entire assemblage of gene pairs within a species, from the youngest newborn cohort to the less abundant ancient pairs. Although such an approach is unable to reveal the historical order of the mutational events associated with any particular gene pair, it may help reveal the average temporal pattern of selection experienced by all cohorts of duplicates. If, for example, the intensity of purifying selection operating on duplicate genes typically increases with the age of a pair, this pattern should be reflected in a reduction in R/S in cohorts of increasing age, whereas the opposite is expected if selection is progressively relaxed.

One way to achieve insight into this matter is to assume a particular pattern in the way in which the instantaneous ratio of replacement to silent mutation rates (dR/dS) changes through time to yield a cumulative relationship between observed R values in gene pairs with different values of S (Box 8.2; Figure 8.10). From the observed behavior of cumulative R and S over the entire set of duplicates within a particular genome, it is then possible to indirectly infer the historical changes in average cohort-specific values of dR/dS that best explain the cumulative data. Fits of such a model to observations on complete genomic sets of gene duplicates from eukaryotes reveal fairly consistent patterns (Lynch and Conery 2000, 2003a), most notably a clear tendency for dR/dS to decrease with increasing S (Figure 8.11). The asymptotic values of dR/dS at low S for animals, yeasts, and land plants range from 0.36 to 1.00, with an overall average of 0.77 (implying that only about 23% of replacement substitutions are removed from relatively young duplicates by selection). In contrast, estimates of dR/dS at high S fall in the narrow range of 0.02–0.09, with an overall average of 0.05. Thus, whereas a substantial fraction of young duplicate pairs experience a phase of highly relaxed selection, by the time a duplicate pair has diverged about 10% at silent sites, reestablishment of a purifying selection regime has generally increased the stringency of selection against amino acid changes more than tenfold, to the point at which only about 5% of replacement mutations are able to proceed to fixation. Only a small fraction of intermediate-aged duplicates exhibit rates of replacement-site substitution in excess of the neutral expectation, no more than expected by chance in a large sample of multiple comparisons.

The conclusion that newly arisen gene duplicates generally experience an initial phase of relaxed selection is supported by a number of other studies. For example, the average cumulative R/S values for paralogs within a variety of prokaryotic and eukaryotic species are higher than those for single-copy genes separated by speciation events, but still virtually always less than 1.0 (Li 1985; Van de Peer et al. 2001; Kondrashov et al. 2002; Nembaware et al. 2002; Seoighe et al. 2003). Despite the consistency of these observations, numerous issues remain unresolved. In particular, elevated levels of dR/dS in young duplicates could be a consequence of relaxed selection against degenerative mutations at some sites, positive directional selection at others, or both. In addition, there may be an intrinsic bias to whole-genome analyses that lump all functional categories of genes together if there is significant variation in the duplicability of different gene categories. Heterogeneous patterns of gene duplicability do appear to be common. For example, protein-coding genes that evolve slowly and are expressed broadly when present in a single copy are preserved as duplicates more frequently than rapidly evolving or narrowly expressed protein-coding genes (Yang et al. 2003; Davis and Petrov 2004; Jordan et al. 2004; Brunet et al. 2006; Chap-

Box 8.2 Indirect Inference of Historical Patterns of Molecular Evolution in Duplicate Protein-Coding Genes

To account for the possibility that the ratio of instantaneous rates of replacement and silent substitutions (dR/dS) changes through time, a mathematical function such as that outlined in Figure 8.10A can be employed. This particular function allows for two different phases of divergence as well as a gradual transition between them. Assuming positive m, dR/dS starts with an expected value of $1/(a - b)$ at $S = 0$ (newly arisen duplicates) and declines to $1/a$ as S approaches infinity (ancient duplicates). (It is assumed here that both a and b are positive; a negative value of b would cause dR/dS to increase with S, but such behavior has not been seen with empirical data.) Other mathematical functions with more complex behavior can be constructed, but the specific model in Figure 8.10A is especially useful because it can be integrated to yield a simple algebraic expression for the historical development of the cumulative ratio of substitutions per replacement site and per silent site, as shown in Figure 8.10B. It is this cumulative R/S ratio that is observed in empirical studies, as opposed to the instantaneous ratio, which can vary from cohort to cohort.

On a log–log plot of R versus S, points with equal R/S ratios fall on

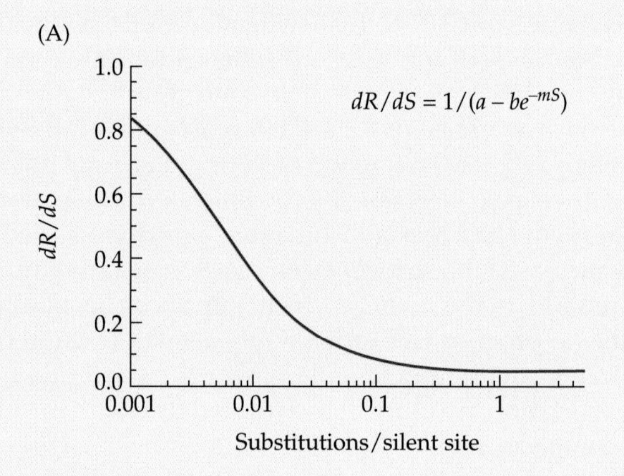

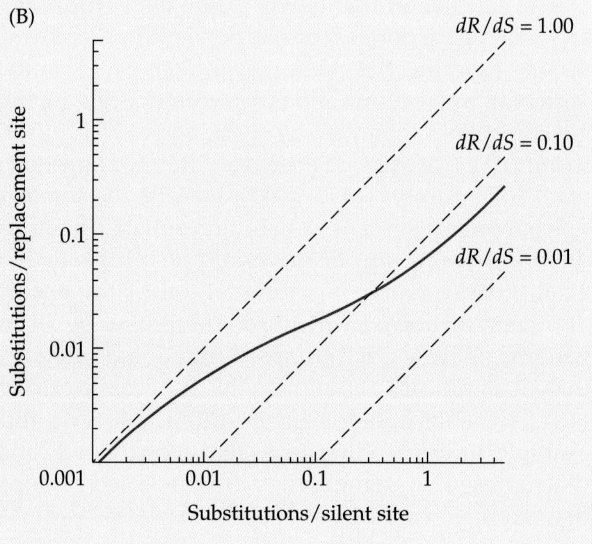

Figure 8.10 The change in the R/S ratio with increasing evolutionary time (measured in units of S). (A) The instantaneous ratio of replacement to silent substitutions (i.e., the ratio for mutations arising within the current cohort), defined by the inset equation with $a = 20$, $b = 19$, and $m = 10$. In this particular example, the ratio of replacement to silent substitutions initiates at 1.0 for newly arisen duplicates ($S = 0.0$) and gradually declines to a stable ratio of 0.05 as $S \to \infty$ (B) The cumulative behavior of R vs. S, obtained by integrating the equation in the top panel (see Equation 8 in Lynch and Conery 2003a). Here, the dashed lines represent points of equal R/S.

a diagonal line, with the height of the line being defined by the magnitude of R/S. Thus, with the preceding model, a linear relationship appears between log S and log R when S is sufficiently small that $dR/dS \approx 1/(a - b)$. The response of log R to log S (the solid curve) then becomes shallower as a transition is made to a constant and lower ratio ($dR/dS \approx 1/a$). During a period in which genes are evolving in a neutral fashion, the response will converge with the diagonal describing $dR/dS = 1.0$ (the main diagonal in Figure 8.10B).

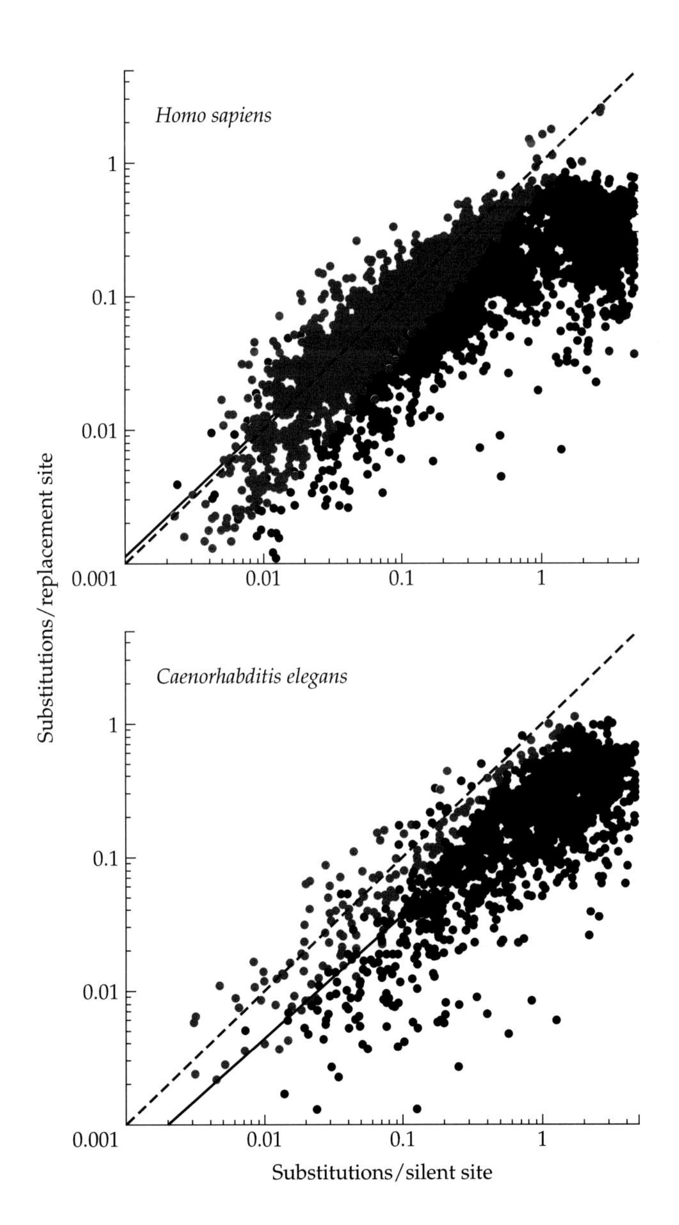

Figure 8.11 R versus S plots for duplicate gene pairs in two animal species. Red points denote pairs for which R is not significantly different from S (i.e., for which neutral evolution cannot be rejected). The diagonal dashed line denotes the neutral expectation, $R = S$, whereas the solid curve (mostly hidden behind the data points) is the fitted function described in the text. (From Lynch and Conery 2003a.)

man et al. 2006). The enhanced frequency of successful duplications of particular gene classes might be a consequence of their elevated susceptibility to subfunctionalization, a pattern that is consistent with large multidomain proteins having higher probabilities of preservation following polyploidization (Chapman et al. 2006). The elevated preservation of short genes with strong tissue-specific expression patterns following segmental duplication events (Urrutia and Hurst 2003) may be a simple consequence of an enhanced likelihood of being fully contained within a duplication span. Regardless of the underlying mechanism, the innate tendency of genes that are more prone to successful duplication to also have lower amino acid substitution rates implies that the decline in R/S in older duplicates seen in Figure 8.11 may be not only a consequence of an increase in the stringency of purifying selection on individual gene pairs over time, but also a reflection of the elevated extinction rate of duplicate pairs with high R.

Although the high values of R/S frequently observed in young cohorts of duplicate genes imply that mutations that never become fixed in single-copy genes are often able to accumulate in a nearly neutral fashion early after gene duplication, such observations do not address the common notion that just one member of a duplicate pair experiences an altered evolutionary trajectory after duplication. Resolution of this issue requires that both members of a paralogous pair be compared with an appropriate single-copy gene within an outgroup species; if both members of the pair are evolving under the same levels of constraint, they should be equally divergent from the outgroup sequence. In the first study of this kind, using mammalian outgroup sequences, Hughes and Hughes (1993) found no evidence for unequal rates of duplicate-gene evolution in the tetraploid frog *Xenopus laevis*, and similar results have been obtained with other taxa (Kondrashov et al. 2002; L. Zhang et al. 2002). Most of these analyses focused on relatively old duplicates, however, raising the possibility that an early phase of asymmetry was obscured by subsequent substitutional changes. Indeed, analysis of relatively young ($S < 0.3$) duplicate pairs from the human genome reveals a very high incidence (~60%) of asymmetrically evolving pairs (P. Zhang et al. 2003), and results from several other animals and yeasts also point to significant asymmetry in sequence divergence (Van de Peer et al. 2001; Conant and Wagner 2003; Chain and Evans 2006). The tendency for the more rapidly evolving members of human gene pairs to accumulate changes evenly across the molecule while the more slowly evolving copies exhibit spatially uneven substitution patterns is consistent with a relaxation of selection on the more rapidly evolving pair members (P. Zhang et al. 2003), as is the observation that the more slowly evolving members of yeast duplicate pairs tend to be more embedded in protein interaction networks and more critical to fitness (Kim and Yi 2006).

The elevated level of dR/dS, as well as the rate asymmetry, in young duplicate genes may simply be a consequence of chance events setting one copy on a trajectory of $dR/dS = 1.0$ until it is completely nonfunctionalized. However, the eventual return of surviving pairs of duplicate genes to a trajectory of low dR/dS is also consistent with the hypothesis that the early phase of relaxed selection frequently reflects the initial preservation of duplicates by subfunctionalizing degenerative mutations, as is the observed narrowing of gene-specific expression domains following duplication (described below). Single-copy vertebrate genes with more restricted tissue-specific patterns of expression are known to evolve more rapidly at the amino acid level, putatively because of a reduction in selective constraints (Hastings 1996; Duret and Mouchiroud 2000), so it is plausible that the early phase of high dR/dS in duplicate genes is a reflection of a narrowing of paralog expression patterns. In a broader evaluation of the possibility of subfunctionalization at the level of protein structure, Dermitzakis and Clark (2001) introduced a computational test for spatial variation in substitution patterns in the coding regions of paralogs and found that about half of mouse and human duplicate genes exhibit significant spatial variation among paralogous copies, some of which appears to be associated with functional domains.

Although these patterns are suggestive, statistical considerations raise significant caveats with respect to interpretations derived from observations on asymmetrical paralog divergence. As noted above, the fates of most duplicate genes are likely to be determined by the first few mutations incurred by one or both pair members. However, there is essentially no statistical power to detect significant asymmetry among paralogs until the average member has incurred a few dozen mutations (Lynch and Katju 2004). Moreover, from the standpoint of the masking model of duplicate-gene preservation, the question of interest is whether the two members of a pair have evolved in a more symmetrical pattern than expected by chance, not whether the evolutionary pattern is unbalanced. Testing for symmetry is even more difficult than testing for asymmetry. Even with an average of 100 to 300 substitutions per gene, it is virtually impossible to detect an exceptionally symmetrical rate of evolution across a pair of duplicates: a chance difference of just zero or one mutations per gene still has a cumulative probability of more than 5% under a model of equal underlying rates (Lynch and Katju 2004).

The Case for Subfunctionalization

Prior to the formal development of the DDC model, circumstantial evidence for duplicate-gene preservation via subfunctionalization of regulatory regions had been revealed through studies of polyploid fishes, which repeatedly demonstrated tissue specificity of expression of duplicated enzyme loci (Ferris and Whitt 1977, 1979). These observations have recently been supplemented by a substantial number of investigations in the zebrafish, a member of an ancient polyploid lineage that still retains roughly 25% of its original gene pairs in a functional state (Amores et al. 1998; Postlethwait et al. 2000). A key to such analyses has been the availability of orthologous single-copy genes in tetrapods. Comparison of zebrafish gene expression patterns with those in the homologous tissues of tetrapod outgroup species (usually mouse or chicken) provides insight into the mechanisms by which zebrafish paralogs may have been preserved. In virtually every well-characterized case, subfunctionalization has been implicated.

Consider, for example, the two zebrafish genes for microphthalmia-associated transcription factor, *mitfa* and *mitfb*. Expression of the first gene is restricted to neural crest, and that of the second is restricted to the epiphysis and olfactory bulb (Lister et al. 2001; Altschmied et al. 2002). The products of these two zebrafish paralogs appear to be homologous to the two alternatively spliced forms of the product of the single-copy locus found in tetrapods, with subfunctionalization resulting from deletions in both regulatory and coding regions (such that each copy has adopted a single splicing variant). As another example, consider the two zebrafish cytochrome P450 aromatase genes (Chiang, Yan et al. 2001). One of these is expressed in the ovary and the other in the brain, whereas the orthologous single-copy gene in tetrapods is expressed in both tissues. Similarly, the zebrafish has two *sox11* genes, one of which is expressed in the anterior and the other in the posterior somites, whereas the single *sox11* product in the mouse is expressed in all somites (de Martino et al. 2000). Many other zebrafish genes appear to have become subfunctionalized following gene duplication (Westin and Lardelli 1997; Nornes et al. 1998; Force et al. 1999; Chiang, Pai et al. 2001; Quint et al. 2000; Bruce et al. 2001; McClintock et al. 2002; de Souza et al. 2005; Liu et al. 2005). Nevertheless, it should be noted here that the DDC model originally postulated subfunctionalization as a *process* of duplicate-gene preservation by degenerative mutation, whereas it is formally possible that *patterns* of partitioned gene subfunctions between extant duplicate genes may have arisen secondarily, with the initial phase of preservation having been driven by other mechanisms.

Observed patterns of subfunctionalization are by no means a peculiarity of polyploid fishes, and just a few additional examples are cited here. First, genome-wide analyses of the mRNAs of duplicate genes provide strong support for the idea that the partitioning of ancestral alternatively spliced variants, as seen in the case of the zebrafish *mitf* genes, is a common fate of duplicate genes throughout the animal kingdom (Kopelman et al. 2005; Su et al. 2006). Second, the nematode *Caenorhabditis elegans* has two β-catenin genes, one of which plays a role in cell signaling and the other in cell adhesion, whereas a single gene fulfills both functions in most other animals (Grimson et al. 2000; Korswagen et al. 2000). The functional differences between the paralogous *C. elegans* genes appear to be due to alterations in the coding region. Third, in the barnacle *Sacculina carcini*, two *engrailed* duplicates are expressed late in development. The expression of one member of this pair is restricted to the nervous system and that of the other to the epidermis, whereas a single gene is responsible for both expression patterns in other arthropods (Gibert et al. 2000). Fourth, all vertebrates harbor two members of the Snail developmental gene family, *Snail* and *Slug*, the summed expression of which is conserved among all species (Locascio et al. 2002).

Remarkably, however, the three major expression domains of the paralogs appear to have shifted from one copy to the other in various vertebrate lineages, presumably because of ongoing regulatory region exchange between the duplicates. Fifth, experiments in the mouse have shown that the combination of a coding region from a single member of a paralogous Hox gene pair with the regulatory elements of both pair members is sufficient for normal development (Tvrdik and Capecchi 2006). Although such manipulations may be viewed as effectively reversing the subfunctionalization process and recreating the ancestral single-copy locus, precise interpretations are rendered difficult by the large numbers of accumulated mutations in the paralogous copies. Finally, in maize (*Zea mays*), two copies of the *p1 myb*-like transcriptional activator partition expression patterns in male and female reproductive structures and leaves, whereas the single-copy ortholog in closely related teosinte is responsible for all of these expression patterns (P. Zhang et al. 2000). Numerous cases of suspected subfunctionalization in maize are associated with the reciprocal loss of conserved noncoding regions (presumably regulatory elements) (Langham et al. 2004).

These observations, and many more, show that there are numerous potential paths to subfunctionalization, including the reciprocal silencing of tissue-specific promoters, the adoption of alternative splicing forms, and modification of the coding regions of multifunctional proteins. In addition, although theoretical considerations have focused on the role of degenerative mutations in initiating the subfunctionalization pathway, other poorly understood processes may facilitate the process (Rodin and Riggs 2003). For example, in numerous animals, including nematodes and mammals, the X chromosome is inactivated in the male germ line. This means that any autosomal gene with male germ line expression will immediately lose that subfunction if duplicated to the X. The autosomal copy is then free to lose expression in somatic tissues, but forced to retain male germ line expression. Several autosomal/X pairs of *C. elegans* gene duplicates exhibit this pattern (Maciejowski et al. 2005). In addition, following allopolyploidization (the joining of genomes from two different species), plants often experience rapid rates of genomic rearrangement and/or abrupt epigenetic changes in methylation patterns, which can spontaneously induce restrictions in the tissue-specific expression patterns of paralogs (Adams et al. 2003; Osborn et al. 2003). Complementary reciprocal epigenetic silencing can lead to essentially instantaneous subfunctionalization, with the epigenetically silenced subfunctions eventually being replaced by neutrally accumulating background degenerative mutations, providing one mechanism by which subfunctionalization might proceed to completion even in very large populations.

Although unicellular species have no opportunities for tissue-specific specialization of gene expression, gene function partitioning may occur at the subcellular level (e.g., via the modification of transit signals for localization of proteins to specific organelles; Silva-Filho 2003; Schmidt et al. 2003). As discussed above, however, even if such physical opportunities exist, the probability of subfunctionalization is greatly reduced in populations of unicellular species with sufficiently large genetic effective sizes. Nonetheless, substantial evidence suggests that duplicate genes in the yeast *S. cerevisiae* are often victims of subfunctionalization. For example, empirical studies in which *S. cerevisiae* paralogs were swapped with the single-copy gene from the outgroup species *S. kluyveri* have provided compelling evidence for subfunctionalization at the level of coding DNA (van Hoof 2005). A gradual loss of ancestral subfunctions by paralogous copies of *S. cerevisiae* genes is suggested by observed reductions in the similarity of expression patterns (measured across different environments), the number of shared regulatory motifs, the number of shared interacting protein partners, and the number of shared functional domains with increasing age of paralogs (Gu et al. 2002b; Papp et al. 2003b; He and Zhang 2005b; van Hoof 2005). In contrast, the total numbers of regulatory motifs and of interacting protein partners for each member of a pair appear to remain roughly constant or even increase over time, suggesting an approximate balance between gains and losses of such elements (Papp et al. 2003b; He and Zhang 2005b), a pattern also found in mammalian duplicate gene pairs by Huminiecki and Wolfe (2004). The recruitment of novel regulatory elements is not surprising, as small patches of DNA are even more likely to be duplicated than entire genes. However, answers to the many questions raised by these observations will require additional work, like that of van Hoof (2005), with outgroup *Saccharomyces* species containing single-copy orthologs of the *S. cerevisiae* paralogs.

Speciation via the Divergent Resolution of Duplicate Genes

Most studies of duplicate genes have focused on their potential role in the origin of evolutionary novelties through the establishment of new gene functions. However, given the high rate at which duplicate genes arise, move to unlinked positions, and become randomly silenced or subfunctionalized, gene duplication may be an equally important contributor to the other major engine of evolution: the origin of new species (Lynch and Force 2000a; Shpak 2005). Consider an unlinked pair of duplicate autosomal genes in a diploid ancestral species, which then experience divergent silencing or subfunctionalization in two descendent lineages, effectively resulting in a map change (Figure 8.12). Because the F_1 hybrids of such lineages will be "presence/absence" heterozygotes at the two independently segregating loci, 1/4 of the F_1 gametes will contain null (absentee) alleles at both loci. Thus, if the gene is critical to gamete function, this single divergently resolved duplication will result in an expected 25% reduction in fertility. For a zygotically acting gene, 1/16 of the F_2 offspring from the interspecific cross will lack functional alleles at both loci, and another 1/4 will carry only a single functional allele. Thus, if the gene is haploinsufficient, 5/16 of the F_2 zygotes of such a cross will be inviable (and/or sterile). With n divergently resolved

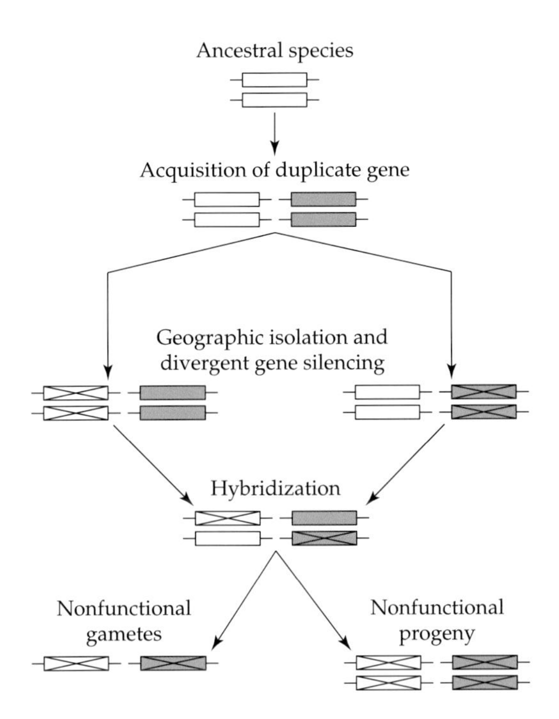

Figure 8.12 Divergent silencing of an ancestral duplicate gene in two geographically isolated lineages. One gene copy is denoted by a white box, the other by a green box, and an X denotes an inactivated gene. Gene pairs represent alleles at diploid loci. Progeny other than those indicated might have compromised fitness (e.g., individuals with just a single active gene).

duplicates, the expected fitness of hybrid progeny is $W = (1 - \delta)^n$, with δ denoting the reduction in hybrid fitness per map change. For example, with $\delta = 5/16$, $W = 0.024$ in the F_2 generation when $n = 10$, and $W = 5 \times 10^{-17}$ when $n = 100$. Observed rates of gene duplication indicate that this type of process is sufficiently powerful to yield nearly complete genomic incompatibility within a few million years of cessation of gene flow (Lynch and Force 2000a; Lynch 2002a; Shpak 2005), which is the approximate time scale over which postzygotic isolation generally occurs in animals (Parker et al. 1985; Coyne and Orr 1997; Sasa et al. 1998; Presgraves 2002; Price and Bouvier 2002).

Indirect evidence leaves little room for doubt that microchromosomal rearrangements resulting from gene duplication can be significant contributors to the establishment of postzygotic isolating barriers among species. Consider, for example, the mustards *Arabidopsis thaliana* and *Brassica oleraceae*, which are thought to have diverged from a common ancestor 10–20 MYA (Yang et al. 1999; Bancroft 2001). Although the two species contain many sets of orthologous chromosomal segments with long-range colinearity in gene content, alternative losses of orthologous gene copies in the two species are common (O'Neill and Bancroft 2000; Quiros et al. 2001). Complicating matters is the fact that *B. oleraceae* is a triploid derivative of the older lineage containing *A. thaliana*, which itself experienced two or three ancient polyploidization events, as described above. However, even this complexity is informative. For example, in an analysis of two triplicated paralogous chromosomal regions in *B. oleraceae*, O'Neill and Bancroft (2000) found that about two-thirds of the paralogous groups had at least one member silenced on at least one paralog. These results, as well as others (e.g., Ku et al. 2000; Bennetzen and Ramakrishna 2002), suggest that the primary mechanism of chromosomal repatterning in plants may be duplication of large chromosomal regions followed by random loss of component genes. Estimated rates of microchromosomal rearrangement in plants range from 0.3 to 6.0 per lineage per million years (Lagercrantz 1998; Burke et al. 2003), and most of these estimates are downwardly biased by imperfect resolution in the associated mapping projects. Thus, considering that many rearranged fragments may contain multiple genes, the number of map changes per million years may commonly exceed ten in land plants.

Such high rates of microchromosomal rearrangement are by no means unique to plants. For example, comparing the genomic sequence of the nematode *Caenorhabditis elegans* with that of its congener *C. briggsae*, Coghlan and Wolfe (2002) estimated that 4,030 rearrangements had occurred over a period of 80 million years, implying 25 rearrangements per lineage per million years. Reciprocal exchanges between two genomic locations, local inversions, and transpositions from one location to another contribute to the total pool of rearrangements at a ratio of 1:1:2, and the vast majority of the segments involved span five or fewer genes. Rates of microchromosomal rearrangements in vertebrates appear to be at least as high as those in nematodes, with roughly 1.5 large-scale (> 100 kb) rearrangements occurring per lineage per million years, usually as a consequence of duplicative transposition, and smaller-scale rearrangements arising at rates of at least 20 per lineage per million years (McLysaght et al. 2000; Locke et al. 2003; Pevzner and Tesler 2003). In contrast, the rates of microchromosomal rearrangement in both *Drosophila* (fruit flies) and *Anopheles* (mosquitoes) appear to be substantially lower, in both cases about 7 per lineage per million years, with the vast majority of rearrangements being within, rather than between, chromosomes (González et al. 2002; Ranz et al. 2001, 2003; Sharakhov et al. 2002). Finally, genomic comparisons of the yeasts *S. cerevisiae* and *Candida albicans* imply a rate of microchromosomal rearrangement of about 2.3 per lineage per million years (Seoighe et al. 2000).

Although microchromosomal rearrangements need not always originate via gene duplication events, a large fraction of such events almost certainly do, for the simple reason that the movement of a gene to a new unlinked location is much more easily accomplished if the transition is made gradually while the original locus is still intact. The magnitude of gene duplication activity discussed above appears to be fully compatible with this interpretation. Under a steady-state birth/death process, a population is expected to gain $2NB$ new duplications per locus per generation, each with a probability of fixation equal to $1/(2N)$ under the assumption of initial neutrality. Assuming that the fate of most duplicates is nonfunctionalization, and recalling that each fixed duplication has a 50% probability of being resolved in favor of the new locus, the steady-state rate of origin of map changes is simply ($GB/2$) per lineage per generation, where G is the total number of genes per genome. Using the estimates of B reported in Table 8.1 and the estimates of G from Table 3.2, the approximate expected number of map changes per million years is 2.5 for unicellular species, 15 for invertebrates, and 125 for vertebrates and plants. Not all such changes will necessarily induce hybrid incompatibilities, as some divergently resolved pairs of loci will be tandemly located and/or members of multigene families. Nevertheless, these collective results reinforce the idea that up to dozens of map changes per million years may passively accumulate in isolated lineages via small duplication events.

Duplication of autosomal genes is just one route by which map changes can be induced passively by divergent degenerative mutation, and to emphasize this point, we now consider three additional mechanisms. First, consider the situation in which an ancestral gene with a male-specific function is initially present on both sex chromosomes, with the copy on the X becoming silenced in one descendent lineage and the copy on the Y becoming silenced in a sister lineage. A cross between females of the first population (assumed to be XX) and males (XY) of the second would result in male progeny completely lacking in gene function, while the reciprocal cross would have active copies on both the X and the Y. Thus, duplication events involving genes on sex chromosomes have potential relevance to understanding the mechanisms underlying Haldane's rule, which states that interspecific genomic incompatibilities are most severe in heterogametic F_1 progeny (Orr 1997). A number of interesting reassignments of map locations of male-specific genes have been uncovered in mammals. For example, almost all extant mammals have autosomal *CDYL* and *CDYL2* genes, which carry out key housekeeping and testes-specific functions. However, in the lineage leading to humans, a copy of *CDYL* was duplicated to the Y chromosome, where it retained a function in spermatogenesis but lost the housekeeping function, while the autosomal loci experienced the opposite fate: loss of function in spermatogenesis but retention of the housekeeping role (Dorus et al. 2003).

Second, a remarkable set of examples of map changes induced by gene duplication in plants involves the movement of genes between organelle and nuclear genomes. Transfers of functional mitochondrial genes to the nuclear genome (accompanied by their subsequent loss from the mitochondrion) have occurred on many independent occasions within lineages of flowering plants, with the overall rate in some cases rivaling the rate of nucleotide substitution at silent sites (K. L. Adams et al. 2000, 2001, 2002). The details worked out for the mitochondrial respiratory protein gene *cox2* are particularly revealing. This gene was apparently duplicated to the nuclear genome of the ancestor of the Papilionoideae (a subfamily of legumes), transiently persisting as active copies in both genomes, with one or the other copy becoming randomly inactivated (in approximately equal frequencies, and by a variety of mechanisms) in almost all descendent lineages (Adams et al. 1999). Many closely related plant genera also exhibit complex patterns of nuclear transfer of mitochondrial ribosomal protein genes (Adams et al. 2002). Moreover, these kinds of intergenomic gene transfers are not restricted to plant mitochondria, as a study of the chloroplast *infA* gene also reveals large numbers of transfers to the nuclear genome (Millen et al. 2001).

Third, although the previous arguments focus entirely on the divergent resolution of duplicate genes driven by degenerative mutation, map changes can also be induced by neofunctionalization, provided the copies acquiring new functions do so at the expense of the old function (Figure 8.13). Such changes, which arise whenever the ancestral locus takes on the new function while the descendent locus retains the original function, are expected to arise in 25% to 50% of cases of duplicate-gene neofunctionalization, depending on the population size (Lynch et al. 2001). Because the probability of preservation of duplicate genes by neofunctionalization increases with population size, unlike the many genetic theories of speciation that rely on population bottlenecks, this version of the gene duplication model is also effective in very large populations.

Genetic theories of speciation have traditionally focused on two competing sets of hypotheses, each of which has numerous adherents and detractors (reviewed in Orr 1996; Rieseberg 2001; Coyne and Orr 2004). The Dobzhansky–Muller model postulates the accumulation of lineage-specific gene sequence changes that are mutually incompatible when brought together in a hybrid genome, whereas the chromosomal model invokes the accumulation of rearrangements that result in mis-segregation in hybrid backgrounds. Both models are based on rather stringent assumptions. For example, the Dobzhansky–Muller model invokes the evolution of coadapted complexes of epistatically interacting factors, none of which have yet been identified at the molecular level, whereas chromosomal models generally focus on major rearrangements, the fixation of which can be greatly inhibited by the reduction in fitness in chromosomal heterozygotes.

A notable feature of the gene duplication model of speciation described above is that it is consistent with *both* the Dobzhansky–Muller model and the chromosomal model while requiring fewer assumptions than either of them. The gene duplication model is effectively a chromosomal model of speciation, but because the rearrangements are microchromosomal, and hence unlikely to cause significant problems during meiosis, they accumulate passively without any alteration in within-species fitness. The gene duplication model also masquerades as a Dobzhansky–Muller model in that the map changes induced by divergent resolution result in pseudo-epistatic interactions without any changes at the gene level. Genomic incompatibilities arising from reassignments of genes to new locations appear superficially as epistatic interactions because the loss-of-function phenotype is determined by the number of active alleles at the two homologous loci in hybrid progeny. Thus, low-resolution analyses of species incompatibilities that fail to identify the specific underlying loci can lead to misinterpretations regarding the underlying genetic mechanism of postzygotic isolation.

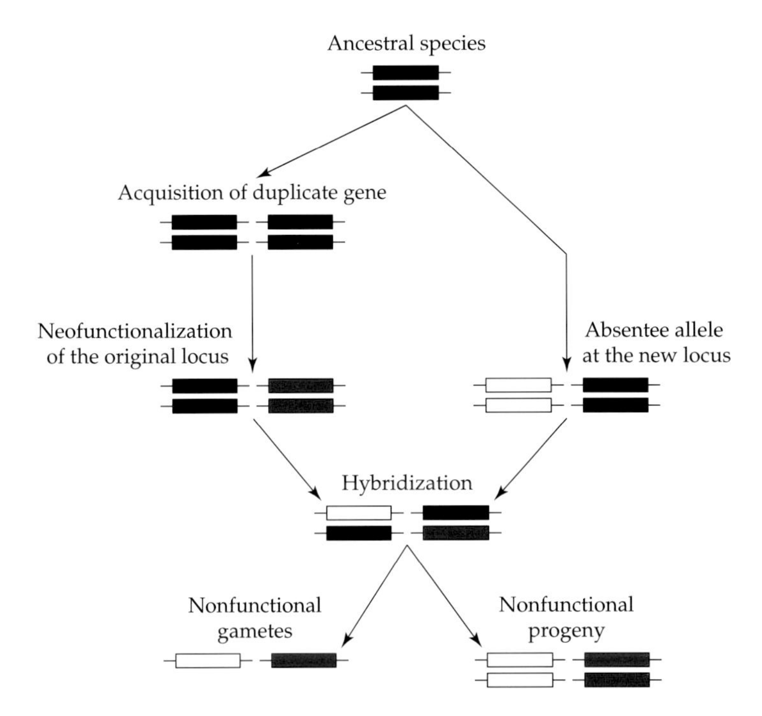

Figure 8.13 The origin of postzygotic incompatibility by neofunctionalization of an ancestral locus. A gene with the essential ancestral function is denoted by black, a neofunctionalized gene by red, and an empty locus by white. Neofunctionalization is assumed to have occurred at the expense of the original gene function. The species on the right either did not experience a duplication at the new locus or lost it prior to fixation.

The induction of map changes by neofunctionalization also blurs the distinction between species-isolating mechanisms based on adaptive genetic change and chromosomal rearrangements. For example, in a hybrid cross, a neofunctionalized ancestral locus would superficially appear to interact negatively with genetic factors residing in the single-copy species (see Figure 8.13). However, the incompatibility would not be a function of adaptive changes at the neofunctionalized locus, but simply an indirect consequence of the relocation of the ancestral gene function.

The vast majority of research on the genetic mechanisms of species-isolating barriers has focused on *Drosophila*, and most of this work, inspired by the classic Dobzhansky–Muller model, has concentrated on a search for "speciation" genes carrying the signature of local adaptation (Coyne and Orr 2004; Wu and Ting 2004). The genetic analysis of speciation is an extraordinarily difficult enterprise, as it is generally nearly impossible to determine whether a particular incompatibility factor played a formative role in species isolation or simply arose secondarily after interspecific gene flow had ceased.

Of even greater uncertainty, however, is whether a singular focus on adaptive mechanisms leads to a biased perspective on the origins of interspecific reproductive incompatibility. Given the generally high rate of microchromosomal rearrangement induced by gene duplication, an evaluation of the gene duplication model ought to be a priority of any study that seeks an unbiased perspective on the mechanisms of speciation, if for no other reason than to provide a null model against which hypotheses involving adaptation can be tested.

One remarkable observation that appears to be quite compatible with the gene duplication model of speciation involves the yeast *S. cerevisiae* and its close relatives, which exhibit numerous differences in gene order resulting from chromosomal rearrangements. Although the haploid offspring of crosses between such species are almost always sterile, after engineering the chromosomes to restore large-scale colinearity, Delneri et al. (2003) were able to increase fertility to 20%–30%. Some minor differences in gene order almost certainly went undetected in these constructs, so it is quite possible that restoration to complete colinearity would have had an even greater effect. Well-documented examples of the involvement of duplicate genes in the reproductive isolation of *Drosophila* species also exist. In two cases, a strong phase of positive selection operating on single paralogs has been implicated (Ting et al. 2004; Greenberg et al. 2006), but in some *D. melanogaster–D. simulans* hybrids, sterility appears to be a simple consequence of the movement of an essential gene to a new chromosomal location via an intermediate phase of gene duplication (and without a change in function), in full accordance with the model presented in Figure 8.12 (Masly et al. 2006). Other lines of evidence support the idea that isolating barriers are associated with small chromosomal rearrangements, although alternative mechanistic explanations (such as the capture of adaptive alleles by nonrecombining chromosomal inversions) have been suggested for such patterns (Noor et al. 2001; Rieseberg 2001; Navarro and Barton 2003).

Under the gene duplication model, certain groups of organisms are expected to be more prone to speciation than others, the most notable being lineages that experience a doubling in genome size. One potential example of such a key event was noted in Chapter 1: the colonization of ancestral eukaryotic genomes by endosymbiotic organelles. Considering the very large number of organelle-to-nucleus transfers that apparently occurred soon after the establishment of the mitochondrial progenitor (Martin et al. 1998), divergent resolution of duplicated organelle genes may have provoked the passive development of isolating barriers among a number of the basal eukaryotic lineages.

Polyploidization provides another enormous opportunity for the rapid proliferation of isolated lineages via the divergent resolution of duplicate genes. Following the first map changes induced by reciprocal silencing in sister polyploid taxa, the thousands of duplicate pairs still remaining in a functional state are free to become divergently resolved in subsequently isolated lineages, potentially yielding a large number of nested speciation events. The origin of species via polyploidy-associated map changes may be especially common in plants, almost all of which are descendents of one or more polyploidization events (Werth and Windham 1991). In addition, particularly striking support for the gene duplication model of species isolation is revealed by the genomic structures of lineages derived from the yeast polyploidization event noted above (Scannell et al. 2006). Pairwise comparisons of the three completely sequenced yeast genomes (*S. castellii*, *S. cerevisiae*, and *Candida glabrata*) identify 100 to 200 divergently resolved paralogous gene pairs (approximately 5% of the single-copy genes in different species are not orthologs), and phylogenetic analysis places much of the genomic repatterning during the period of species emergence. Finally, although the details remain to be worked out, one of the most striking examples of cryptic speciation, the reproductive isolation of 14 morphologically indistinguishable members of the *Paramecium aurelia* complex, also appears to have developed shortly after an ancestral polyploidization event (Aury et al. 2006).

Key genome doubling events may have facilitated the diversification of many major animal lineages as well. First, given the apparent twofold-to-threefold increase in gene content in basal animals relative to fungi, genome amplification may have been involved in the origin of the major animal phyla. Second, as described above, it also appears that a substantial amount of gene duplication occurred prior to the radiation of the major vertebrate lineages. Finally, it may be no coincidence that the most species-rich lineage of vertebrates, the ray-finned fishes (~30,000 species; see Figure 8.5), is a descendent of an ancient polyploidization event. At least some evidence points to divergent resolution in isolated lineages in this group (Cresko et al. 2003), and secondary polyploidization events within specific lineages are associated with enhanced rates of speciation (Ferris et al. 1979; Taylor et al. 2001b; Hoegg et al. 2004; Mank and Avise 2006; Volff 2005). In contrast, all of the fish lineages that branched off prior to the basal polyploidization event (e.g., bichirs, bowfin, gars, paddlefish, and sturgeons) contain just a handful of species. These kinds of observations suggest that adaptive radiations are associated with polyploidization events not only because gene duplication opens up novel evolutionary pathways for the origins of new gene functions, but also because polyploidization generates a population genetic environment that is highly conducive to the passive origin of reproductive barriers.

9 Genes in Pieces

The discovery that eukaryotic genes can contain intervening noncoding sequences (Berget et al. 1977; Chow et al. 1977; Evans et al. 1977; Goldberg et al. 1977) still stands as one of the most perplexing observations ever made in molecular genetics. Remarkably, these intragenic spacers, now known as introns (Gilbert 1978), are initially transcribed only to be eliminated during mRNA maturation. This processing is carried out by the spliceosome, a molecular machine rivaling the ribosome in complexity. Although completely absent from prokaryotes, spliceosomal introns appear to be present in all major eukaryotic lineages, as well as in some eukaryotic transposons (Stiller et al. 1998; Fast and Doolittle 1999; Archibald et al. 2002; Nixon et al. 2002; Simpson, Macquarrie, and Roger 2002). Thus, spliceosomal introns were almost certainly present in the stem eukaryote, and as we will see, there may have been two different spliceosomes at this time, each processing a different class of introns.

Explaining the origin of introns continues to be a major challenge for evolutionary genomics. Introns impose numerous burdens on their eukaryotic hosts, including additional investment in DNA and RNA; enhanced production of inappropriate transcripts by splicing errors; the need to selectively degrade excised introns; and an excess mutation rate to defective alleles associated with sites involved in splice-site recognition. A stark reminder of the nontrivial nature of these costs is the observation that at least a third of human genetic disorders are attributable to mutations that cause defective splice-site recognition (Culbertson 1999; Frischmeyer and Dietz 1999; Philips and Cooper 2000; Lopez-Bigas et al. 2005).

A puzzle of equal proportions is the phylogenetically uneven distribution of introns. Most multicellular species average five or more introns per protein-coding gene, while most unicellular eukaryotes average less than one.

Such disparities must ultimately be a consequence of differences in the molecular rates of intron origin and removal and/or the ability of natural selection to promote or prevent fixation of such modifications. Because the stem eukaryote appears to have been moderately intron rich, it is unlikely that the initial colonization of introns was promoted by their involvement in the adaptive evolution of the complex features of multicellular organisms. Rather, like other forms of nonessential DNA, introns may have more successfully colonized and been maintained in multicellular species simply because of the permissive nature of the population genetic environment (a lower effective population size allowing a larger fraction of mildly deleterious insertions to proceed to fixation). Nevertheless, despite their short-term intrinsic costs, once introns were established, they provided a potential substrate for the origin of novel mechanisms of gene regulation and transcript processing, opening up previously inaccessible paths of adaptive evolution.

Any rigorous theory about the intragenomic and phylogenetic distributions of introns must be consistent with known cellular mechanisms of intron birth, death, and processing. Thus, this chapter starts with an overview of the complex molecular machinery responsible for splicing. After considering the suspected molecular mechanisms of intron gain and loss and the costs of maintaining introns, the population genetic issues that bear on intron proliferation will be reviewed and put to the test by considering the incidence of introns in various phylogenetic lineages. Most of the discussion will focus on the spliceosomal introns of protein-coding genes, which constitute the vast majority of introns in nuclear genes, and these will hereafter be simply referred to as introns.

The Spliceosome(s)

Introns are processed out of mRNAs by the spliceosome (Figure 9.1), a large ribonucleoprotein (RNP) complex whose mechanisms have been studied most intensely in humans and two yeasts (*S. cerevisiae* and *S. pombe*) (reviewed in Burge et al. 1999; Hastings and Krainer 2001; Brow 2002; Kuhn and Käufer 2003). Eukaryotes invest very heavily in the spliceosome: about 1% of known human and yeast genes encode products deployed in splicing. Five small nuclear RNAs (the U1, U2, U4, U5, and U6 snRNAs) constitute the heart of the spliceosome in all eukaryotes (Collins and Penny 2005), and each of these is assembled into a snRNP (small nuclear RNP). Seven core proteins are shared by four of the snRNPs, U6 being exceptional in having its own set. Each snRNP incorporates an additional unique set of about ten proteins, and numerous non-snRNP proteins also contribute to the spliceosome. Although some have suggested that spliceosomes are preformed organelles (Stevens et al. 2002), recent data suggest that snRNPs assemble into spliceosomes in a stepwise fashion at each intron (Görnemann et al. 2005; Lacadie and Rosbash 2005). Moreover, the spliceosome appears to serve as a hub for additional mRNA development activities, including capping and preparation for nuclear export (Hirose and Manley 2000; Bent-

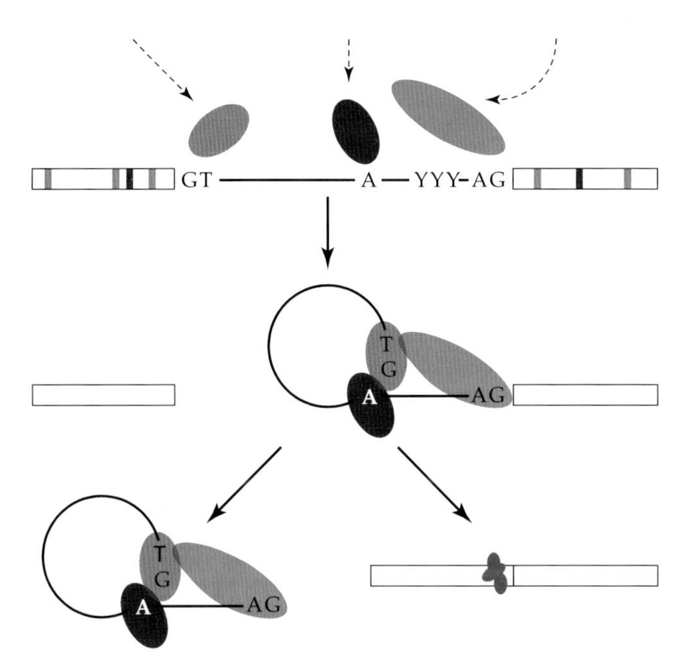

Figure 9.1 An approximate schematic of the anatomy of a spliceosomal intron and the steps in its removal from a precursor transcript. *Top*: The precursor mRNA (only part of which is shown) contains an intron, which usually starts with a GT and ends with an AG, always contains an internal branch point (A), and often has a downstream polypyrimidine (YYY) tract. The surrounding exons (open boxes) contain short motifs that serve as exon splicing enhancers and silencers (green and red bars). In the first step of splicing, the component snRNPs of the spliceosome (large colored ovals) begin to bind to appropriate locations on the pre-mRNA. *Middle*: The 5′ end of the intron is released and brought into contact with the branch point, forming a lariat. *Bottom*: The intron is completely released, and the two exons are spliced together. An exon junction complex (small orange ovals) is deposited just upstream of the splice site prior to export of the mRNA from the nucleus.

ley 2002; Maniatis and Reed 2002; Orphanides and Reinberg 2002; Proudfoot et al. 2002; Rappsilber et al. 2002; Zhou et al. 2002).

Despite its central significance to the biology of eukaryotes, the spliceosome is not an invariant structure. For example, several observations suggest that *S. cerevisiae*, a species that is extremely depauperate with respect to introns, has experienced an evolutionary simplification of its mRNA processing machinery. Altogether, about 200 unique proteins contribute to the human spliceosome (Zhou et al. 2002), but only 80 form the *S. cerevisiae* spliceosome (Brow 2002). The fission yeast *S. pombe* contains a number of spliceosomal components that are missing in *S. cerevisiae*, and of the components shared by all three species, those from *S. pombe* tend to be more similar to human than to *S. cerevisiae* orthologs (Abovich and Rosbash 1997;

Aravind et al. 2000; Käufer and Potashkin 2000; Kuhn and Käufer 2003). In addition, although mammals and *S. pombe* utilize exon splicing enhancers (ESEs, specific guide sequences encoded within exons) and SR (serine–arginine) proteins to orient the spliceosome to intron–exon junctions, neither have been found in *S. cerevisiae* (Nissim-Rafinia and Kerem 2002). A dramatic reduction in the complexity of the spliceosome has also been seen in the apicomplexan *Cryptosporidium*, which has a severalfold reduction in intron number compared with its relative *Plasmodium* (Templeton et al. 2004).

An unexpected layer of complexity in the evolution of the spliceosome was introduced with the discovery that many eukaryotes harbor not just one, but two, such molecular machines (Hall and Padgett 1994; Burge et al. 1998; Patel and Steitz 2003). The major spliceosome deals with the classic set of introns that usually start with GT and end with AG (see Figure 9.1), while the minor spliceosome processes another set. Although it was first thought that AT/AC ends were a defining feature of this second class of introns, earning them the initial name of ATAC introns, many have been found to have standard GT/AG ends, and they are now known as U12-dependent introns. In contrast, introns processed by the major spliceosome are referred to as U2-dependent, with U2 and U12 denoting homologous snRNAs used in the two spliceosomes. Of the five snRNAs employed in splicing, only U5 is used jointly by the two spliceosomes, although most (but not all) of the spliceosomal proteins appear to be shared (Hastings and Krainer 2001; Luo et al. 1999; Will et al. 1999, 2001, 2004; Schneider et al. 2002). Despite the overlapping components, there are fundamental differences between U2- and U12-dependent intron splicing, with the latter having a much more rigid set of splicing requirements (Burge et al. 1998; Patel and Steitz 2003). For example, the internal branch-point sequences and 5′ splice sites are much more highly conserved in U12-dependent introns, whereas the 3′ polypyrimidine tract typically found in U2-dependent introns is generally lacking (Burge et al. 1998). In addition, whereas the U1 and U2 snRNPs of the major spliceosome interact separately with 5′ splice sites and branch points, U11 (the U1 analog in the minor spliceosome) and U12 cooperatively bind both regions simultaneously (Frilander and Steitz 1999).

U12-dependent introns are present in amoebozoans, oomycetes, plants, *Drosophila*, vertebrates, and at least one fungus (Levine and Durbin 2001; Otake et al. 2002; Zhu and Brendel 2003; Schneider et al. 2004; Lorković et al. 2005). However, no evidence of a minor spliceosome has been found in nematodes or in any of a number of lineages of fungi and protists (Burge et al. 1998, 1999). Because the clades containing U12-dependent introns are distributed throughout the eukaryotic tree (see Chapter 1), and because most such introns utilize similar splice-site and branch-point sequences, the minor spliceosome appears to be ancient, perhaps even dating to the stem eukaryote, with many lineages subsequently losing all such introns.

How might the major and minor spliceosomes have arisen? One scenario, known as the fission–fusion model, assumes that the descendants of a single ancestral spliceosome diverged in two phylogenetically independent lineages that reunited by a subsequent hybridization event (Burge et al. 1998, 1999). There are, however, two significant difficulties with this hypothesis. First, during the period of isolation, the overall level of genomic divergence would have to be sufficiently small to allow reproductive compatibility between hybridizing lineages. Second, the magnitude of functional divergence between orthologous snRNAs would have to be substantial enough to allow their independent operation when reunited. Given the need for a spliceosome to service all of the introns within a host genome, it is difficult to see how this could occur unless the number of introns per genome were quite small. The ability of yeast to splice mammalian introns, despite the billion or so years separating these lineages, testifies to the evolutionary stability of the spliceosomal machinery (Trachtulec and Forejt 1999; Kunze et al. 2000).

An alternative model that avoids these limitations invokes an endogenous origin of two spliceosomes within a single species (Lynch and Richardson 2002). This might have been achieved by sequential seeding of the nucleus of the ancestral eukaryote with the descendants of two organelle-derived group II introns, the hypothesized original source of spliceosomal introns (see below). Such a scenario is conceptually appealing because the coevolution of the internal components of group II introns would provide a natural barrier to cross-talk between different elements (Toor et al. 2001). Under this hypothesis, the similar secondary structures of the snRNAs of the major and minor spliceosomes would have arisen by convergence, perhaps with the generalized pool of splicing proteins serving as a guiding force (Burge et al. 1999). If these ideas are correct, then other rare types of spliceosomes may await discovery. An intriguing possibility is presented by *Euglena gracilis*, which, in addition to having conventional U2-dependent introns, harbors a unique type of intron with long complementary ends that can be folded into a secondary structure (Tessier et al. 1995; Canaday et al. 2001).

How do we account for the apparent loss of the minor spliceosome from some lineages? In every species observed, U12-dependent introns are a tiny minority (<0.5%) of the total pool of spliceosomal introns. The highest count is 404 for the human genome (Levine and Durbin 2001), which contains over 150,000 U2-dependent introns. The rarity of U12-dependent introns is probably a consequence of several phenomena. First, the greater degree of conservation of branch-point and 5′ splice sites in U12 intron–containing alleles implies that they are more vulnerable to inactivating mutations than are U2 intron–containing alleles (Lynch 2002b). Second, some mutational changes convert U12-dependent introns into the more simply defined U2-dependent introns, whereas the reverse seems to occur rarely, if ever, probably because of the much more conserved recognition sequences of U12-dependent introns (Burge et al. 1998). Third, U12-dependent introns are spliced more slowly (Patel et al. 2002) and less accurately (Levine and Durbin 2001) than are U2-dependent introns. All three of these features are expected to result in shorter mean persistence times of U12-dependent introns over evolutionary time relative to their U2-dependent counterparts.

Should a genome then experience complete stochastic loss of U12-dependent introns, the functionless snRNAs of the minor spliceosome would eventually succumb to the passive accumulation of degenerative mutations, essentially immunizing the lineage against future recolonization by this class of introns. Under this stochastic-loss hypothesis, U12-dependent introns are confined to lineages that harbor large total numbers of introns (multicellular animals and plants) simply because the chance of complete loss is reduced when copy numbers are high.

The Introns Early–Introns Late Debate

Soon after the discovery of introns, an intense debate erupted over their time of origin and their role in protein evolution. One camp argued that introns played a critical role in the earliest stages of protein evolution by promoting recombinational shuffling of modular domains or "mini-genes" into combinations with new functions (Gilbert 1987). Various versions of this introns-early hypothesis exist, the most extreme being that introns were present in the first prokaryotic cells and were then lost from all descendant lineages except eukaryotes (Darnell 1978; Doolittle 1978). Under the contrasting introns-late view, spliceosomal introns were never present in prokaryotes, played no significant role in the formation of novel proteins, and were simply adventitious embellishments of genes with well-established functions (Orgel and Crick 1980; Cavalier-Smith 1985; Palmer and Logsdon 1991).

Given the complete absence of spliceosomal introns from prokaryotes, the introns-early school has relied on indirect lines of evidence to support its case. For example, one prediction of the introns-early hypothesis is that the majority of introns should reside between adjacent codons (phase 0 introns) rather than splitting codons (phase 1 and 2 introns), as the former arrangement preserves the reading frame after the shuffling of exons (Figure 9.2). A similar argument leads to the prediction that adjacent introns should tend to be in the same phase. Both kinds of phase bias do exist (Long et al. 1995), and the tendency for introns to lie between versus within protein architectural modules has also been cited as evidence for the involvement of introns in the early assembly of proteins (Roy et al. 1999).

These interpretations are compromised by the fact that qualitatively similar patterns of phase bias are expected under the introns-late hypothesis if introns tend to be inserted into particular dinucleotides and/or if postcolonization success depends on the intron phase. For example, as noted below, introns tend to be flanked on both sides by G nucleotides, and as a consequence, intron phase bias closely reflects the distribution of GG dinucleotides within and between codons (e.g., phase 1 introns commonly inhabit glycine codons, which always start with GG), and the sequences flanking introns tend to put them in amino acid contexts that fall between functional domains (Ruvinsky et al. 2005; Whamond and Thornton 2006). In addition, in the face of an error-prone spliceosome, as well as in the case

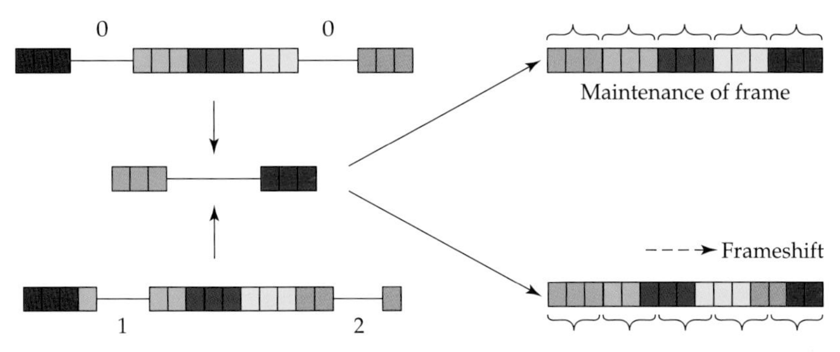

Figure 9.2 The consequences of exon swapping for protein sequence conservation. The color-coded codon boxes to the left denote the proper reading frames in ancestral sequences. The sets of triplets to the lower right consisting of mixed colors denote frameshifted codons. Intron phases are denoted by integers. *Top*: An exon flanked by two phase 0 introns is inserted into a phase 0 intron in another gene, resulting in the addition of three codons, but otherwise maintaining the reading frame. *Bottom*: Because the inserted exon is flanked by introns that split codons, the recipient gene experiences a frameshift downstream of the point of insertion.

of adaptive alternative splicing (see below), introns are expected to be more innocuous (less likely to encourage mRNAs with frameshifts and/or hybrid codons) if they are in phase 0 and in the same adjacent phases (Lynch 2002b; Figure 9.3). The similarity in the magnitude of phase biases in recent and ancient introns (Rogozin et al. 2003; Sverdlov et al. 2003; Coghlan and Wolfe 2004; Qiu et al. 2004; Nguyen et al. 2006) provides compelling evidence that these concerns are not misplaced.

In summary, although the fact that the number of introns in a few unicellular eukaryotes has dwindled to just a handful leaves open the remote possibility of complete spliceosomal intron loss from the main ancestral lines leading to all prokaryotes, the absence of any direct evidence for the early existence of introns in prokaryotes does not inspire confidence in the introns-early view. The population genetic principles to be discussed below provide a more formal challenge to an introns-early world, and even the strongest

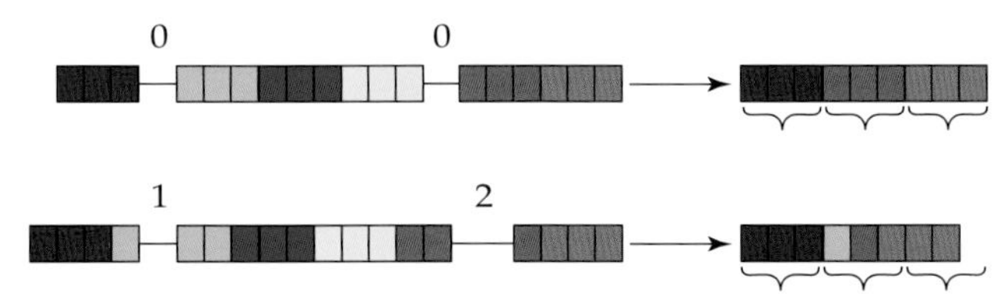

Figure 9.3 Skipping of an exon has different consequences depending on the phases of the surrounding introns. *Top*: The reading frame is maintained because the surrounding introns are in the same phase. *Bottom*: A frameshift results when the phases of the surrounding introns differ.

introns-early adherents now acknowledge that a large fraction of introns arose long after the origin of eukaryotes (de Souza et al. 1998; Roy et al. 1999; Fedorov, Cao et al. 2001). This conclusion does not rule out a creative role for exon shuffling in protein evolution in more recent times, however (Patthy 1999a). For example, about 10% of animal genes contain tandemly duplicated exons (Letunic et al. 2002), whose mutually exclusive splicing can achieve functional diversity once the exonic sequences have diverged sufficiently (Kondrashov and Koonin 2001).

A Group II Origin?

One of the most enduring mysteries about introns is the origin of the spliceosome, the only credible hypothesis to date invoking descent from a group II intron (Sharp 1985; Cech 1986; Jacquier 1990; Cavalier-Smith 1991; Lambowitz and Zimmerly 2004; Robart and Zimmerly 2005). Group II introns are often referred to as "self-splicing," which if taken literally, implies complete endowment with all the necessary machinery for their removal from host mRNAs. However, although most group II introns have a secondary RNA structure that facilitates splicing, and although many encode multifunctional proteins that assist in the splicing process, splicing is almost always facilitated by host-encoded factors. Group II introns have been found in the chromosomes and plasmids of numerous eubacteria (Dai and Zimmerly 2002) and archaea (Dai and Zimmerly 2003; Rest and Mindell 2003; Toro 2003). However, they are never very abundant in these contexts, and technically speaking, many of these prokaryotic elements are not introns at all, in that they lie between rather than within genes (Zimmerly et al. 2001; Dai and Zimmerly 2002). No group II intron has been found in a eukaryotic nuclear gene, but they are present in some protein-coding genes within the organelles of plants, fungi, and numerous protists (Bonen and Vogel 2001; see Chapter 11).

The ability to mobilize distinguishes group II introns from spliceosomal introns (Lambowitz et al. 1999; Hiller et al. 2000; Lambowitz and Zimmerly 2004). Group II intron integrations usually occur at a very specific insertion-site sequence (a process referred to as retrohoming), but insertions into nonspecific sites (retrotranspositions) occasionally occur (Cousineau et al. 2000; Dickson et al. 2001). There are at least three mobilization pathways (Figure 9.4). First, as with non-LTR retrotransposons (see Chapters 3 and 7), target-primed reverse transcription can occur at the site of insertion (Zimmerly et al. 1995; Yang et al. 1996). In this case, after the intron is removed from its host mRNA and subjected to translation, the insertion protein encoded within the intron binds directly back to the transcript, providing the machinery for reintegrating at a suitable genomic insertion site and cleaving the second strand to complete double-strand synthesis. Second, group II introns without endonuclease activity can sometimes insert themselves without second-strand cleavage by waiting for a DNA replication event to fill in the gap on the opposite strand (Zhong and Lambowitz 2003). Third, the intron RNA

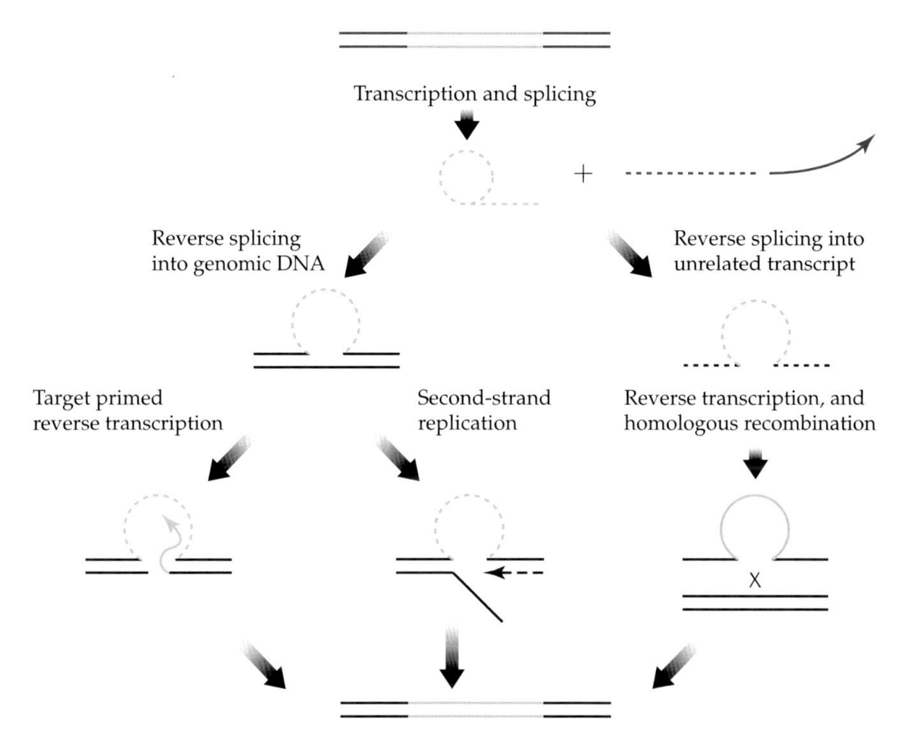

Figure 9.4 The three known pathways for the mobilization of a group II intron to a new site. DNA is denoted by solid lines and RNA by dotted lines. After transcription of the host gene, the intron is released as a lariat. In the two pathways on the left, the element is initially integrated as RNA, whereas in the pathway on the right, reverse transcription converts the element into DNA prior to its integration into a new genomic location via homologous recombination.

may be reverse-spliced into an unrelated transcript, which, if then reverse-transcribed, can introduce the intron into an ectopic site by homologous recombination (Cousineau et al. 2000).

This multiplicity of mobilization pathways, combined with phylogenetic evidence for the horizontal transfer of group II introns among eubacterial species and between microbes and eukaryotic organelles (Zimmerly et al. 2001), lends credence to the hypothesis that the nuclear genes of the stem eukaryote were initially populated by group II introns. It is plausible that the primordial mitochondrion was the source of such introns (Sheveleva and Hallick 2004), although other sources (including prior residence in the nuclear genome) cannot be ruled out. A list of striking structural and functional similarities between the excision mechanisms of group II introns and those of spliceosome-dependent introns is also consistent with the idea that the five spliceosomal snRNAs are direct descendants of the major subunits of the catalytic core of a group II intron (Michel and Ferat 1995; Hetzer et al.

1997; Burge et al. 1999; Sontheimer et al. 1999; Yean et al. 2000; Shukla and Padgett 2002; Valadkhan and Manley 2002), although the limited number of efficient biochemical mechanisms for splicing implies a nonnegligible probability of evolutionary convergence of two independent splicing pathways (Stahley and Strobel 2005).

Any transition from a self-splicing group II intron to a large population of eukaryotic spliceosome-dependent introns would necessarily involve a number of significant steps: (1) a cohesive self-splicing system would have to give rise to an elaborate fragmented spliceosomal system; (2) the fragmented components would have to retain their functionality while cooperatively operating in *trans*; (3) a set of splicing proteins would have to be recruited to form the full spliceosomal complex; and (4) the derived population of spliceosome-dependent introns would have to eventually displace the ancestral population of self-splicing introns. A variety of observations demonstrate the feasibility of each of these steps. For example, contrary to the usual situation in prokaryotes, plant organelles often harbor spatially isolated group II intron fragments (split introns), which cooperate to splice together host gene exons located in distant genomic regions (Malek and Knoop 1998; Dai and Zimmerly 2002; Robart and Zimmerly 2005). Although each fragment contains just a portion of the intronic sequence essential to splicing, when the products of separate transcripts are brought together in *trans*, a fully functional secondary RNA structure for splicing is assembled. Also contrary to the usual situation in prokaryotes, most organellar group II introns have lost their protein genes, relying instead on nuclear-encoded host proteins (Perron et al. 1999; Vogel et al. 1999; Bonen and Vogel 2001; Jenkins and Barkan 2001; Rivier et al. 2001; Mohr and Lambowitz 2003; Ostheimer et al. 2003). Thus, group II introns not only can fragment into mutually dependent RNA molecules (like the snRNAs of the spliceosome), but also can recruit various host proteins to facilitate the splicing process. The involvement of these proteins in the removal of group II introns from pre-mRNAs constitutes a sort of loosely defined spliceosome (Perron et al. 2004), albeit one that is quite different from that in the nucleus.

Given the tendency for group II introns to both fragment and mobilize, the spliceosome may have evolved largely by a series of neutral processes involving duplication and degeneration (Cavalier-Smith 1991; Stoltzfus 1999). Under this scenario, once a set of group II–derived, *trans*-acting fragments had become fully cooperative, related group II introns recognizable by this incipient spliceosome would have experienced relaxed selection and eventual loss of self-splicing abilities. Such a scenario is a special case of the subfunctionalization process (see Chapter 8), an evolutionary pathway that is unlikely to proceed in large populations. The key issue here is whether a newly arisen *trans*-acting fragment is likely to maintain functionality during the average $2N_g$ generations (where N_g is the effective number of genes per locus in the population) that it takes to drift to fixation. Letting u be the mutation rate per nucleotide site per generation and n be the number of nucleotide sites critical to fragment function, and following the logic laid out in Chapter 4, if $2N_gnu >> 1$, virtually all potential descendants of the founding fragment would be inactivated by mutations by the time fixation had occurred. Given that $2N_gu$ is generally greater than 0.1 in free-living prokaryotes (see Chapter 4), a fragment containing more than ten or so key nucleotides would have a difficult time becoming established by neutral processes in this type of population genetic background. Consistent with this view, all of the group II intron fragments known in eubacteria appear to be incapable of *trans* cooperation and may simply be incapacitated remnants of previously cohesive elements. On the other hand, because mutational inactivation is an insignificant barrier to subfunctionalization when $2N_gnu << 1$, the substantial reduction in $2N_gu$ that probably occurred during the origin of eukaryotes (see Chapters 1–4) would have been conducive to the emergence of the spliceosome by fragmentation.

Despite the attractiveness of the group II–seeding hypothesis for the origin of spliceosomal introns, several significant questions remain to be answered. First, how did mobile group II elements themselves originate? Did reverse transcriptase become associated with a self-splicing intron, endowing it with mobility? Or are group II introns non-LTR retrotransposons with acquired self-splicing ability? Whereas the joint use of target-primed reverse transcription by both classes of elements is an argument in favor of the latter, the absence of non-LTR elements from prokaryotes is not. Second, if the spliceosome arose via the subfunctionalization of a group II intron within the nuclear genome, then why have no nuclear group II introns yet been found? (A portion of one plant nuclear intron has been found to be homologous to a mitochondrial group II intron, but this may simply be a secondary transfer of an organelle fragment into a preexisting spliceosomal intron; Kudla et al. 2002.) Third, following subfunctionalization of the ancestral group II intron, how did the individual RNA fragments liberate themselves from their original host genes to become autonomous contributors to a more generalized splicing reaction? Given the antiquity of the events that gave rise to the spliceosome, definitive answers to these questions may be beyond reach, and we now return to more tractable matters.

Mechanisms of Origin and Loss

Although early comparative studies suggested that stochastic intron origin and loss are ongoing in all species (Hankeln et al. 1997; Frugoli et al. 1998; Logsdon et al. 1998; O'Neill, Brennan et al. 1998; Robertson 2000; Boudet et al. 2001; Watanabe and Ohama 2001; Feiber et al. 2002; Krzywinski and Besansky 2002; Wada et al. 2002), the physical processes leading to intron birth and death have remained enigmatic. One proposed mechanism of intron loss is reverse transcription of a processed mRNA (its introns already having been spliced out) followed by gene conversion of the parental gene (Fink 1986; Palmer and Logsdon 1991) (Figure 9.5). This mechanism does not account for the fact that most genes that lose one intron retain others (Krzywinski and Besansky 2002; Wada et al. 2002), unless only fractions of reverse tran-

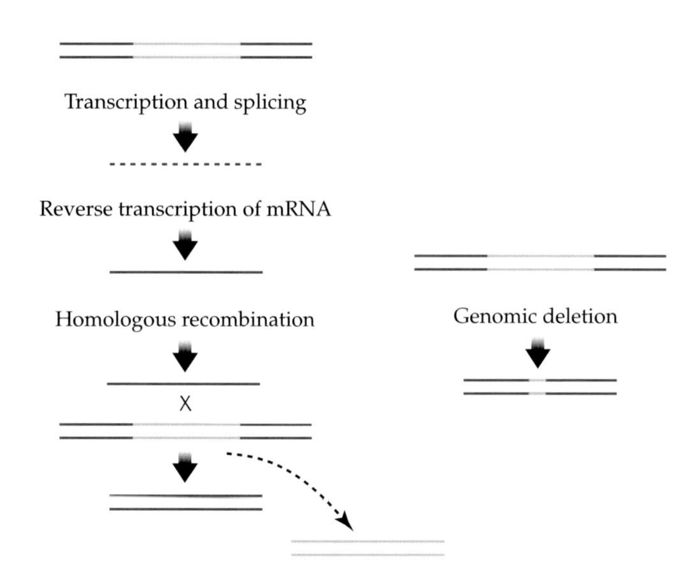

Figure 9.5 Two mechanisms of intron loss: reverse transcription followed by gene conversion (left), and genomic deletion (right). DNA is denoted by solid lines and RNA by dotted lines. Intronic DNA is shown in blue. Because the break points are arbitrary, genomic deletion may leave behind a small patch of intronic DNA (as shown) and/or eliminate flanking exonic DNA.

scripts are involved in such conversions. In principle, intron loss via reverse transcripts should exhibit a 3′ bias because reverse transcriptase moves in a 3′→ 5′ direction, sometimes failing to go to completion, but the entire process is expected to be rare in animals, as it must be restricted to genes whose transcripts are present in the germ line. Although some animals and a few fungi do exhibit a paucity of introns at the 3′ end, the general pattern is weak to nonexistent (Lynch and Kewalramani 2003; S. Cho et al. 2004; Nielsen et al. 2004; Niu et al. 2005; Roy and Gilbert 2005b; Jeffares et al. 2006).

An alternative mechanism of intron loss is simple genomic deletion. Such a modification need not always involve precise intronic removal, but if it yields a gain or loss of $3n$ coding nucleotides (where n is an arbitrary integer), the downstream reading frame will remain intact, resulting in codon addition or loss to the host gene. Provided that no in-frame termination codons are generated, and that the modified protein remains fully functional, such mutant alleles may sometimes spread through a population (Golding et al. 1994; Llopart et al. 2002).

The mechanisms of intron gain have been equally elusive. One suggested route involves the tandem duplication of an internal fragment of coding DNA containing an AGGT tetramer (Rogers 1990), a process that neatly creates an intervening sequence with conventional 5′ GT and 3′ AG splice sites as well as the G | G context noted above; ... AG | gt ... ag | GT ..., where the segment with lowercase letters denotes the new intron. Barriers to the de novo production of introns by this route include the need for fortuitous inclusions of a functional branch-point sequence, exon splicing enhancers, and so forth (see Figure 9.1), but circumstantial evidence for this mode of origin derives from sequence within novel introns in several fish genes that aligns well with surrounding exonic sequence (Venkatesh et al. 1999).

Transposable elements are a second potential source of introns (Purugganan and Wessler 1992; Kidwell and Lisch 2000). Retrotransposons cannot deliver introns, as introns are spliced out of the primary transcripts of such elements prior to mobilization, and there is no evidence that mobile element insertions typically carry the complete repertoire of recognition signals that would enable the spliceosome to treat them as de novo introns. However, mobile element insertions can sometimes alter the spatial configuration of cryptic splice sites within host genes so as to activate a new pattern of splicing that fortuitously excludes the element from the mature mRNA (Prêt et al. 1991; Giroux et al. 1994; Luehrsen et al. 1994; Nouaud et al. 1999; H. Lin et al. 2006). Such insertions usually induce some change in the coding sequence of the mature mRNA, and there is no guarantee that a functional product will remain, but cases are known in which transposons inserted into plant genes are spliced out precisely from mRNAs (Wessler 1989; Lal and Hannah 1999).

One of the most striking examples of an association between mobile elements and introns involves the *Tc1* and *Tc4* transposons of *C. elegans*, which are typically removed from host mRNAs prior to translation (Li and Shaw 1993; Rushforth and Anderson 1996). Although such pre-mRNA excisions are imprecise and stochastically leave behind small insertions, deletions, or substitutions, it appears that frameshifted transcripts are eliminated from cells by mRNA surveillance (described below), leaving only a subpopulation of mature mRNAs with continuous open reading frames. Even more remarkably, excision of *Tc1* elements *from the genome* can lead to de novo modifications of preexisting introns. As with most transposons, genomic excision of *Tc1* is imprecise, often leaving behind a TGTA insert in the host DNA (Carr and Anderson 1994), but in spite of the expected frameshift, many revertant alleles have no obvious phenotypic effects. This results from the *Tc1* remnant operating as a new 5′ splice site, T | gta, which is similar to the canonical 5′ splice site of *C. elegans* and alters the downstream reading frame in only a fraction of cases. These observations clearly suggest how mobile element insertion/excision cycles might give rise to expansions or contractions of intron–exon junctions (Carr and Anderson 1994), a phenomenon often referred to as intron sliding (Stoltzfus et al. 1997).

A third potential mechanism of intron birth, duplication of a released intron from an mRNA and its reintegration at an ectopic site (Sharp 1985; Lynch and Richardson 2002), requires three fortuitous steps. First, while remaining transiently combined with a recently excised intron, at least part of the spliceosome would have to reattach to a previously unoccupied but potentially useful splice site in a foreign mRNA. Second, the armed spliceosome would have to catalyze a reverse-splicing reaction to insert the intron. Third, the modified mRNA would need to be reverse-transcribed into a DNA capable of moving the intron into its new genomic site by homologous recombination with the host gene. Despite the intricacies involved, this model has a number of appealing features: (1) the inserted sequence would be immediately endowed with internal intron signature sequences; (2)

spliceosomal recruitment to the recipient mRNA by conventional processes provides a mechanism for preferential insertion into coding regions containing appropriate splice-site recognition sequences; and (3) insertion of the new intron need not occur at the expense of coding DNA. The proposed scenario is also mechanistically plausible. Immediately following splicing, the U2, U5, and U6 snRNPs (the spliceosomal components most likely to be catalytic) remain transiently attached to intron lariats (Burge et al. 1999), and the ~10–30-minute half-life of spliced introns (Clement et al. 2001) is sufficient to allow some reverse splicing. In principle, intron duplication might even proceed without an RNA intermediate for the recipient gene and without any involvement of the spliceosomal components. In almost all eukaryotes, retrotransposons produce reverse transcriptases and endonucleases for self-mobilization, and as noted in Chapter 7, this mobility apparatus occasionally switches from its parental transcript to alternative RNA templates, leading to novel insertions via target-primed reverse transcription. There are no obvious reasons why excised introns would be immune to this process.

Support for the reverse-splicing hypothesis derives from several observations. First, duplication of an intron into an ectopic site in its own host gene seems particularly likely given the close proximity of excised intron and host mRNA during normal splicing processes. A comparative analysis across *Drosophila* genomes has revealed three examples of intron duplication within the xanthine dehydrogenase locus (Tarrío et al. 1998), and five pairs of homologous introns have been found in the same host gene in *Caenorhabditis* (Coghlan and Wolfe 2004). Second, because intron gain by reverse splicing requires the donor and/or recipient gene to be transcriptionally active in the germ line of animals, it is noteworthy that intron gain is disproportionately common in *Caenorhabditis* germ line–expressed genes (Coghlan and Wolfe 2004). Third, the reverse-splicing hypothesis is consistent with the presence of spliceosomal introns in some of the tiny U1, U2, U5, and U6 snRNA genes of yeasts (Tani and Ohshima 1991; Takahashi et al. 1993, 1996). Spliceosomal introns are extremely uncommon in noncoding genes, but as the snRNAs are part of the catalytic core of the spliceosome, they reside in an optimal intracellular environment for acquiring introns by reverse splicing.

Do introns preferentially insert themselves into coding DNA with particular sequence motifs? In animals and land plants, introns are often surrounded by AG|...|(A/G), where |...| denotes the intron, although the bias is not great (Long et al. 1998). Dibb and Newman (1989) suggested that such patterns result from the preferential insertion of introns into fortuitous "proto-splice" sites in host genes, but it is difficult to rule out alternative hypotheses, including de novo production of such sites and random insertion followed by differential postcolonization success. For example, as noted above, the internal duplication of an exonic region containing an AGGT will yield an AG|gt ... ag|GT sequence, and flanking nucleotides involved in spliceosomal guidance can be promoted by natural selection (Zhuang and

Weiner 1989; Carmel et al. 2004). Nevertheless, the proto-splice site hypothesis draws support from an observed weak decline in the information content of flanking exonic regions with intron age in multicellular species (Kent and Zahler 2000; Sverdlov et al. 2003; Coghlan and Wolfe 2004), and from the observation that cryptic unoccupied splice sites in the exons of one species sometimes harbor introns in related species (Sadusky et al. 2004). One of the most striking examples of an insertion-site association is the unusually high incidence of phase 1 introns within the signal peptide cleavage sites commonly found at the glycine-rich N-terminal ends of mammalian cell-surface proteins, which, as noted above, serve as potential G|...|G insertion sites (Tordai and Patthy 2004). The observation that nuclear and mitochondrial ribosomal protein genes (which are believed to have separate stem eukaryotic and prokaryotic ancestries; see Chapter 1) contain numerous introns at identical nucleotide sites is also consistent with the hypothesis that intron colonization is promoted by local sequence contexts (Yoshihama et al. 2006).

The Intrinsic Cost of Introns

Introns impose a fundamental cost on their host genes at two levels. First, because introns are copied into precursor mRNAs, they decrease the efficiency and fidelity of transcript production. Cells produce dozens to hundreds of mRNAs per gene per cell cycle, so the investment in nucleotides at the RNA level can be considerable. Transcripts elongate at rates of ~20–40 bp per second in eukaryotic cells (Reines et al. 1996; Jackson et al. 2000), so an average human gene, with ~1.3 kb of total coding DNA interspersed among eight introns with average lengths of 4.8 kb, requires ~22 minutes for complete transcription (not including the untranslated end sequences), about thirty times the time required for an intron-free allele. Errors in splicing (discussed below) are also expected to yield unproductive transcripts, imposing a further wastage of nucleotides.

A second price paid by intron-containing alleles is an excess rate of mutation to nonfunctional alleles associated with nucleotide substitutions at sites required for proper spliceosomal recognition (Moore 2000; Lynch 2002b). The removal of each intron from a pre-mRNA depends on a diffuse pretranslational code (see Figure 9.1). For example, spliceosomal introns universally have specific short terminal sequence requirements for proper excision, almost always GT at the 5′ end and AG at the 3′ end in the case of U2-dependent introns, with longer tracts of conservation in the case of U12-dependent introns. Many animal, plant, and fungal introns also have polypyrimidine tracts of about ten nucleotides just upstream of the canonical AG, and in some species, such as *Caenorhabditis elegans*, this sequence is highly conserved (Blumenthal and Steward 1997; Simpson, Thow et al. 2002; Kupfer et al. 2004; Abril et al. 2005). In addition, an interior branch point A, located an appropriate distance from the downstream AG, is an essential feature of all spliceosomal introns. Although the sequence surrounding the branch point is generally quite variable, *S. cerevisiae* and many other yeasts exhibit essentially invariant seven-nucleotide branch points (TACTAAC) as well as invariant 5 bp sequences at 5′ splice sites (Burge et al. 1999; Bon et al. 2003). Remarkably, intense selection for splicing in mammalian cell cultures yields this same branch-point sequence (Lund et al. 2000), a finding that is consistent with the idea that the lack of conservation in multicellular species is a consequence of inefficient selection (to be discussed more fully below).

A number of other requirements for intron recognition and splicing are less specific. For example, land plant introns have elevated A/T contents relative to adjacent exons (Lorković et al. 2000), as do the introns of many other lineages (see Chapter 6), and in some cases the mere insertion of an A/T-rich region into a plant gene can result in the de novo production of an intron, with the spliceosome relying on the nearest cryptic splice signals for intron delimitation (Luehrsen and Walbot 1994). Vertebrate introns are commonly enriched with specific trinucleotides that serve as intron splicing enhancers (ISEs; Majewski and Ott 2002; Yeo et al. 2004), and the splicing efficiency of some mammalian introns is influenced further by intronic secondary structure in pre-mRNAs (Buratti and Baralle 2004). Either too low or too high a degree of sequence complementarity between the 5′ end of an intron and the U1 snRNA can inhibit splicing (Carmel et al. 2004), and circumstantial evidence also suggests that intron lengths can be under stabilizing selection (Schaeffer 2002; Parsch 2003; Ometto et al. 2005). Many more subtle aspects of splice-site recognition associated with intronic sequences probably remain to be discovered. For example, the introns of more than 95% of human genes contain latent (unused) 5′ splice sites that are more similar to the consensus splicing sequence than the ones that are actually used (Miriami et al. 2002). In most of these cases, use of the alternative downstream splice sites would introduce in-frame premature termination codons into the mature transcript. Remarkably, when such nonsense triplets are removed, the spliceosome switches to the more compliant site (B. Li et al. 2002), suggesting that such triplets are involved in splicing inhibition.

Despite its central importance, the information contained within introns is often insufficient for proper localization of the spliceosome, particularly in the case of very large introns containing numerous spurious recognition sites (Mount et al. 1992; Burge et al. 1999; Long and Deutsch 1999). Supplemental information is often contained within the surrounding exons in the form of exon splicing enhancers (ESEs) and silencers (ESSs). Typically four to ten nucleotides in length (Liu et al. 1998; Schaal and Maniatis 1999; Blencowe 2000), ESEs facilitate splicing by recruiting a variety of SR proteins, each of which contains one or more sequence-specific RNA recognition motifs and an arginine/serine-rich domain that helps secure the splicing apparatus (Graveley 2000). ESSs attract a separate set of proteins that inhibit local splicing, and the antagonism between these two types of activities, along with input from ISEs and ISSs, serves to coordinate the exact location of a splice site (Zhu et al. 2001; Z. Wang et al. 2004). The ancient use of ESSs is suggested by the demonstration that ESSs derived from vertebrate genes enhance the splicing of pre-mRNAs in the fission yeast *S. pombe* (C. J. Webb et al. 2005).

The critical importance of ESEs and ESSs is revealed by numerous observations on human genes. First, many human genetic disorders are caused by splicing errors resulting from point mutations in exons (Cooper and Mattox 1997; Nissim-Rafinia and Kerem 2002), with over 25% of nucleotide substitutions in silent sites resulting in a splicing defect (Pagani et al. 2005). Second, 2%–4% of exonic sequence matches the signatures of known ESSs, with about five such clumps per exon (Fairbrother et al. 2002). The significantly different frequencies of various nucleotide oligomers in intron-containing versus intron-free genes (Fedorov, Saxonov et al. 2001) is a likely reflection of selection for such sequences, as are the general tendency for silent-site sequences in exon termini to evolve unusually slowly (Majewski and Ott 2002; Halligan et al. 2004) and the reduced levels of sequence polymorphism in ESE motifs relative to other neighboring regions of the genome (Carlini and Genut 2006). This diverse set of observations leaves little doubt that the need for accurate splicing mechanisms has pronounced side effects on the coding sequence of exons, possibly even compromising the functional efficiency of the proteins encoded by some host genes (Marais et al. 2005).

The key point here is that mutations that alter the sequences of any nucleotide sites critical to intron processing can completely abrogate an intron's capacity to be spliced. Such intron-debilitating mutations will result in intron expansion or contraction, and the modified mRNA will yield a new protein with either a loss or a gain of amino acids, usually with an accompanying change in reading frame and loss of gene function. Summing over the conserved positions at the 5′ and 3′ splice sites, internal branch point, 3′ polypyrimidine tract, and exon splicing enhancers and silencers, the effective number of bases reserved for splicing appears to be on the order of 25 per intron, and perhaps somewhat higher for larger introns. Thus, letting the mutation rate per base per generation be u, each additional intron increases a gene's mutation rate to defective alleles by an amount in the neighborhood of $25u$, where $u \approx 10^{-9}$ in unicellular species and approaches 10^{-7} in large multicellular species (see Chapter 4).

Introns and Population Size

As noted in Chapter 3, whereas most multicellular species harbor an average of five to seven introns per protein-coding gene, many unicellular eukaryotes have only a handful in the entire genome. In principle, such disparities could reflect major differences in molecular mechanisms for intron gain or loss among eukaryotic lineages, but as yet there is no evidence that diversity in intron abundance is a consequence of variation in cell biological features. Thus, we must look to population-level processes that modify the success rate of newly derived intron-containing or intron-free alleles. One possibility is that the reduction in the efficiency of natural selection in species with smaller effective population sizes magnifies the probability of colonization and retention of innately deleterious introns (Lynch 2002b).

For newly arisen introns having no functional significance for the products of their host genes, the primary force opposing their ability to spread throughout a population is their excess mutation rate to defective alleles (s), and because this force is expected to be quite weak, selection will be ineffective in preventing intron colonization in populations experiencing substantial levels of random genetic drift. Letting N_g be the effective number of

Box 9.1 The Population Genetic Barrier to Intron Proliferation

As discussed in Chapter 4, differences among alleles in their rate of mutation to defective products have essentially the same effect on their fixation probabilities as weak selection. In the present context, this observation allows the use of standard diffusion theory to evaluate the probability that a newly arisen intron-containing allele will go to fixation. Letting s be the excess deleterious mutation rate for an intron-containing allele and N_g be the effective number of genes at the locus in the population (approximately twice the effective size of a diploid outcrossing population), the probability of fixation of a newly arisen intron relative to the neutral expectation (which is equivalent to the initial frequency of the mutation) is $2N_g s/(e^{2N_g s} - 1)$. The converse situation, the probability that a newly arisen intron-free allele will become fixed in an intron-containing background, can be evaluated by substituting $-s$ for s in this expression. The outcome depends entirely on the ratio of the power of differential mutation to the power of random genetic drift, $2N_g s$. For $2N_g s < 1$, the probabilities of fixation and loss are very close to the neutral expectation, but above $2N_g s = 1$, there is a precipitous decline in the fixation probability and an equally dramatic increase in the loss probability (Figure 9.6). Thus, $2N_g s > 1$ demarcates an approximate barrier to intron colonization and retention.

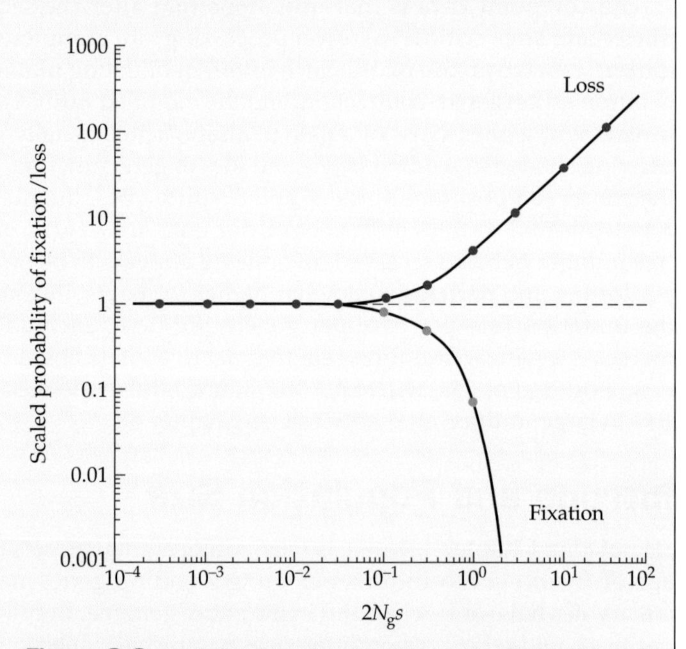

Figure 9.6 Probabilities of fixation of a newly arisen intron (lower curve) and of a newly arisen intron-free allele (upper curve) relative to the neutral expectation. Solid lines denote theoretical expectations, whereas points are derived from computer simulations. The dashed line is the neutral expectation. (From Lynch 2002b.)

genes at a locus in a population, the condition $2N_g s > 1$ defines the approximate population genetic barrier to intron colonization (Box 9.1; Figure 9.6). For situations in which $2N_g s << 1$, the power of random genetic drift, $1/N_g$, overwhelms the power of selection, enabling introns to proliferate in an effectively neutral manner, but when $2N_g s >> 1$, intron establishment is essentially prohibited. Because the preceding observations on splice-site recognition requirements indicate that $s \approx 25u$, where u is the mutation rate per base per generation, the barrier to intron colonization translates to $2N_g u > 0.04$. This benchmark is of practical utility because, as discussed in Chapter 4, observed levels of nucleotide diversity at silent sites in protein-coding genes (π_s) provide empirical estimates of $2N_g u$.

The limited data available are consistent with this hypothesized population genetic barrier to intron establishment. For almost all species with $\pi_s < 0.01$, most of which are multicellular, the mean number of introns per gene is nearly independent of π_s. Above this point, there is a drop-off in the incidence of introns, with the approximate threshold falling in the range of $0.02 < \pi_s < 0.06$ (Figure 9.7). Because of the short-term nature and potential bias of π_s estimates, and because of the stochastic nature of intron gain and loss, this relationship cannot be expected to be perfect, but it is compatible with the predicted $\pi_s = 0.04$ threshold. The most extreme cases of intron number reduction are the unicellular kinetoplastids *Leishmania major* and *Trypanosoma brucei*, neither of which has been found to contain spliceosomal introns, and the diplomonad *Giardia lamblia*, with only three introns detected genome-wide (Russell et al. 2005). However, not all unicellular species have

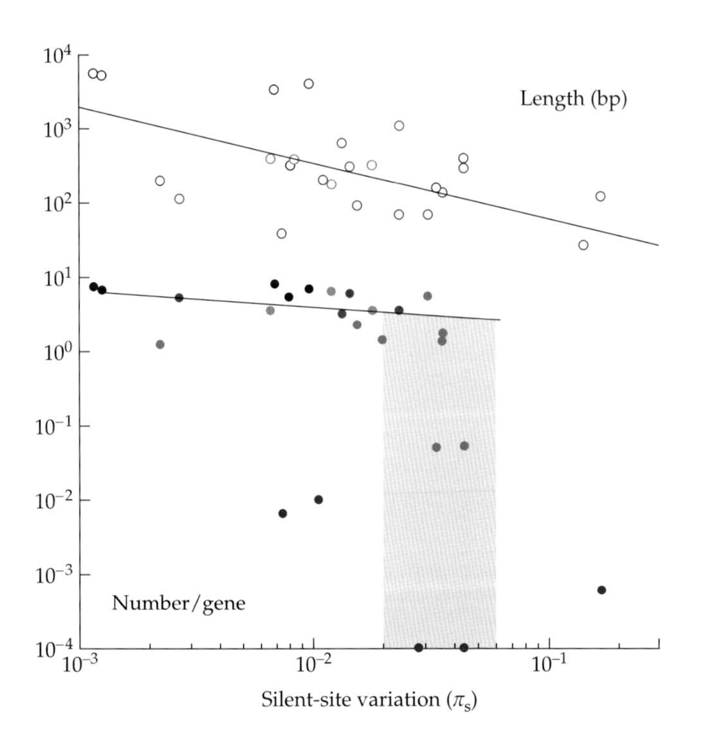

Figure 9.7 The relationship of the average number of introns per gene (solid points) and average intron length (bp, open points) to estimates of $2N_g u$ derived from levels of nucleotide variation at silent sites in protein-coding genes in eukaryotic species (see Chapter 4). All species-specific estimates of intron properties are based on surveys of the entire genome (black, vertebrates; red, invertebrates; green, land plants; purple, unicellular and oligocellular species). The shaded gray region denotes a rough transition between intron-rich and intron-poor taxa. (Data from Lynch 2006a and a few additional recent references.)

reduced numbers of introns. For example, *Plasmodium falciparum* (the malarial parasite), which has a much lower level of π_s than any other well-characterized unicellular eukaryote, has ~1.5 introns per gene.

Because the barrier to intron colonization is a function of the product of N_g and s, the preceding theory also predicts that introns with more stringent recognition requirements (larger s) will be less frequent. Quite striking in this regard is the substantial reduction in intron numbers in *S. cerevisiae* (which has only 250 in the entire genome) and other hemiascomycetes (Bon et al. 2003), which have a much higher degree of intronic sequence conservation than other well-studied eukaryotes. Fission yeast (*S. pombe*), for which intronic sequences are much less conserved, has roughly 20 times as many introns as *S. cerevisiae* (Wood et al. 2002). A high degree of conservation of splice-site sequences is also seen in the red alga *Cyanidioschyzon merolae*, whose entire genome contains only 27 introns (Matsuzaki et al. 2004), as well as within (and between) the deeply branching lineages containing *Giardia* and *Trichomonas* (Russell et al. 2005; Vaňáčova et al. 2005), both of which are highly depauperate with respect to introns. Similar reasoning helps explain the extreme rarity of U12- relative to U2-dependent introns, as the former have highly conserved 5′ splice sites (ATATCCTT) and branch-point sequences (TCCTTAAC), in contrast to the relaxed sequence requirements for U2-dependent introns described above.

Rates of Intron Gain and Loss

The average number of introns per protein-coding gene is ultimately a function of the relative rates of intron gain and loss by physical processes and of the ability of natural selection to eradicate or promote alleles with and without introns. Provided the effective population size and the physical rates of intron insertion and removal remain stable for a sufficiently long time, the average number of introns within a lineage is expected to approach a quasi-equilibrium level, although the precise locations of introns will still vary as a result of stochastic turnover. Sufficient data for estimating intron turnover rates exist for just a few species, but the results are quite consistent. For example, in a comparison of the genomes of the nematodes *C. elegans* and *C. briggsae*, 263 of ~2700 introns were found to be unique to a single species (Kent and Zahler 2000). Using 95 million years as the estimated divergence time between these two species (Stein et al. 2003), the rate of turnover per intron is inferred to be 0.51 per billion years per lineage. This estimate is a bit misleading, in that it is conditional on one member of a pair of species containing an intron at a site, but it can be standardized to account for the number of potential sites of occupancy by dividing by the average number of coding nucleotides per occupied site (assuming that introns are free to insert themselves at any position in a coding region). Application of this procedure to *Caenorhabditis* and several other pairs of multicellular species yields estimates in the narrow range of 0.001–0.003 turnovers per coding nucleotide per billion years (Table 9.1).

TABLE 9.1 Estimates of intron occupancy per coding nucleotide (P) and rates of intron turnover (Δ), birth (B), and death (D) for various pairs of multicellular species

SPECIES	CODING SITES/ GENE (BP)	INTRONS/ GENE	DIVER-GENCE TIME (MY)	P	Δ	B	D
Caenorhabditis	1250	5.0	95	0.0040	0.0020	0.0010	0.25
Drosophila	1660	4.7	50	0.0028	0.0022	0.0011	0.39
Mammals	1300	7.6	75	0.0058	0.0008	0.0004	0.07
Land plants	1800	6.2	150	0.0034	0.0033	0.0016	0.47
Cryptococcus	1620	5.3	18	0.0033	0.0004	0.0002	0.06
Plasmodium	1790	1.2	100	0.0007	0.00004	0.00002	0.03
Theileria	1506	2.2	82	0.0015	0.00003	0.00002	0.01

Note: All rates are per coding nucleotide per billion years. Average total numbers of coding nucleotides per gene are taken from Table 3.2. Data and times to common ancestors are taken from Kent and Zahler (2000) and Stein et al. (2003) for *C. elegans* and *C. briggsae;* Moriyama et al. (1998) and Russo et al. (1995) for *D. melanogaster* and *D. virilis;* Waterston et al. (2002) for *M. musculus* and *H. sapiens;* Ku et al. (2000) for *Arabidopsis* and *Lycopersicon;* Stajich and Dietrich (2006) for *Cryptococcus neoformans* var. *grubii* and var. *neoformans;* Castillo-Davis et al. (2004) and Roy and Hartl (2006), averaging, for *Plasmodium falciparum* and *P. yoelii;* and Pain et al. (2005) and Roy and Penny (2006) for *Theileria annulata* and *T. parva*. Intron numbers per gene are from Lynch (2006a).

Under the assumption of equilibrium numbers of introns per gene, joint information on intron turnover and intron occupancy can be used to estimate rates of intron birth and death (Box 9.2). Letting the number of introns per base pair of coding DNA be P, the molecular rates of intron gain per unoccupied site and loss per occupied site be B and D respectively, and $S = 2N_g s$, the equilibrium level of intron occupancy is $P \approx B/(De^S)$, provided the level of intron occupancy is less than 0.01, which is virtually always the case. For cases in which selection is ineffective ($S << 1$), the level of intron occupancy reduces further to a simple function of physical processes, $P \approx B/D$. The equilibrium rate of intron turnover within a lineage is $\Delta \approx (2SB)/(e^S - 1)$, which is very close to $\Delta \approx 2B$ provided that $S < 0.3$. Thus, as a first-order approximation, ignoring the terms involving selection, $\Delta/2$ provides a downwardly biased estimate of the intron birth rate B, and $\Delta/(2P)$ provides an upwardly biased estimate of the intron loss rate D. Because $S \approx 25\pi_s$, and π_s is generally in the range of 0.001–0.1 in eukaryotes (see Chapter 4), the bias of such estimates will be pronounced only for species at the high end of the π_s range (typically unicellular species), and even then, the estimates are unlikely to be off by much more than a factor of three.

Within the pairs of species in Table 9.1, the average numbers of introns per gene are highly similar, providing some justification for the use of these equilibrium expressions (although it is formally possible that two lineages could exhibit similar intron numbers because they are experiencing parallel patterns of expansion/contraction). For example, both species of *Caenorhab-*

Box 9.2 Intron Occupancy and Turnover

Here we examine the joint consequences of the molecular processes of intron gain and loss and the ability of natural selection to inhibit gains and promote losses. Specifically, we consider the situation in which these processes have remained constant for a long enough time for the level of intron occupancy to have reached an equilibrium between the net forces of gain and loss. Letting *P* denote the equilibrium occupancy, defined as the fraction of coding nucleotides that are immediately followed by an intron, the equilibrium point is the value of *P* at which the number of gains per unit time is equal to the number of losses,

$$P \cdot 2ND \cdot p_f^+ = (1-P) \cdot 2NB \cdot p_f^-$$

where *D* is the rate of intron loss per occupied site per gene, *B* is the rate of intron birth per unoccupied site per gene, and *N* is the size of the population (here assumed to be diploid, which has no influence on the following results). To understand this expression, note that for each occupied site, $2ND$ mutations to an intron-free state are experienced at the population level per generation, each of which goes to fixation with probability p_f^+; a similar explanation defines the rate of intron gains on the right side of the preceding equation.

From Chapter 4, the probabilities of fixation of beneficial (intron loss) and detrimental (intron gain) mutations are, respectively,

$$p_f^+ = \frac{S/(2N)}{1-e^{-S}}$$

$$p_f^- = -\frac{S/(2N)}{1-e^{S}}$$

where $S = 2N_g s$, with N_g being the effective number of genes at a locus and *s* being the selection coefficient (here equivalent to the excess mutation rate to defective alleles

ditis have ~5.0 introns per gene (Stein et al. 2003), and mice and humans have 7.4 and 7.7, respectively (Lynch and Conery 2003b). Applying the preceding formulae, the birth rates of introns in most lineages fall in the range of 0.0002–0.0016 per coding nucleotide site per billion years, whereas most estimated rates of intron loss are in the range of 0.06–0.57 per intron per billion years (see Table 9.1). (The estimates of *B* and *D* for the apicomplexans *Plasmodium* and *Theileria* fall substantially below these ranges, suggesting an overall reduction in the physical dynamics of introns in this lineage; Roy and Penny 2006.) Rates of the same order of magnitude have been inferred using other phylogenetic approaches that do not rely on an equilibrium assumption (S. Cho et al. 2004; Nielsen et al. 2004; Roy and Gilbert 2005c; Knowles and McLysaght 2006). Unless parallel gains of introns have caused a substantial misinterpretation of levels of intron sharing, these results indicate that introns within most lineages have been remarkably stable for the past 100 million years or so of evolutionary history, with average half-lives ($1/D$) on the order of several billion years, exceeding the age of eukaryotes.

caused by an individual intron). The use of these expressions assumes that the selective disadvantage of an intron gain is essentially equivalent to the selective advantage of an intron loss, which will not necessarily be the case for introns that have taken on key functions. Using the useful relationship $p_f^+/p_f^- = e^S$, the equilibrium occupancy is found to be

$$P = \frac{B/(De^S)}{B/(De^S)+1}$$

which reduces to

$$P = \frac{(B/D)}{(B/D)+1}$$

when selection is ineffective ($S << 1$).

The rate of intron turnover is the joint rate at which occupied sites lose introns and unoccupied sites gain them,

$$\Delta = [P \cdot 2ND \cdot p_f^+] + [(1-P) \cdot 2NB \cdot p_f^-]$$

This description is reasonable even under nonequilibrium conditions, but further simplification is possible at equilibrium, as substitution for P leads to

$$\Delta = \frac{2S/(e^S - 1)}{(1/B)+(1/De^S)}$$

which, when selection is ineffective, reduces to the harmonic mean of B and D,

$$\Delta = \frac{2}{(1/B)+(1/D)}$$

Some further details can be found in Lynch (2002).

Although crude, the birth rate estimates above provide some insight into the frequencies with which random insertions and deletions of DNA might give rise to new introns. From results discussed in Chapters 2 and 4, we know that the rate of insertion/deletion per nucleotide site (not including mobile element–induced events) is on the order of 10^{-10} to 10^{-8} per generation, whereas intron birth rates appear to be on the order of 10^{-16} to 10^{-11} per site per generation (after scaling the estimates in Table 9.1 by generation length), with the lower and higher ends of these ranges pertaining to unicellular and multicellular species, respectively. This means that no more than 0.0001%–0.1% of random insertion/deletion events need to give rise to new introns to account for the data. Thus, given that we have encountered several mechanisms that can sometimes spontaneously generate new introns—random internal duplications containing an AGGT tetramer, mobile element insertions, and AT-rich insertions—estimates of recent intron birth rates are not implausible in light of our knowledge of the molecular biology of today's organisms. Once the baseline recognition mechanisms of the spliceosome were established, eukaryotes must have become vulnerable to the passive inclusion of random coding region insertions that previously would have yielded nonfunctional alleles.

Open questions remain concerning the sources of the wide phylogenetic range of variation in intron numbers that exist among eukaryotic lineages. Did the ancestral eukaryote harbor large numbers of introns that were subsequently lost from some unicellular lineages? Do the large numbers of introns in multicellular species reflect massive intron gain in the ancestors of land plants and animals? Or did both types of events play a role? In an early phylogenetic analysis of the triose phosphate isomerase gene, Logsdon et al. (1995) found that 5 of 21 introns throughout the eukaryotic tree were shared by plants and animals. Given that these lineages appear to reside on opposite sides of the root of the eukaryotic tree (see Chapter 1), this observation suggested an abundance of introns in the stem eukaryote, substantial loss in most unicellular lineages, and extensive gain in the land plant and animal lineages. A similar conclusion arose from the observation that 8 of 56 known introns in actin genes are shared by species traversing the eukaryotic root (Bhattacharya and Weber 1997).

These matters were recently explored on a much larger scale with genome-wide phylogenetic analyses of introns in ancient genes conserved across deeply diverging lineages. For example, a parsimony analysis led Rogozin et al. (2003) to suggest that the basal bilaterian animal contained large numbers of introns, with significant losses occurring in the arthropod and nematode lineages and substantial gains occurring in vertebrates. Although an assumed absence of parallel intron losses may have yielded upwardly biased rates of intron gain and downwardly biased rates of intron loss in this study (Roy and Gilbert 2005a), independent support for an intron-rich ancestral animal derives from observations that most of the introns in sponges (basal animals) are also found in bilaterians (Müller et al. 2002) and that two-thirds of human introns are shared with annelids (Raible et al. 2005).

Going further back in time, Rogozin et al. (2003) also concluded that the stem eukaryote was intron rich. This hypothesis is consistent with several additional lines of evidence (Lynch and Richardson 2002; Roy and Gilbert 2005a; Vaňáčova et al. 2005; Lynch 2006a; Slamovits and Keeling 2006), although an array of statistical methodologies and assumptions have led to considerable debate over the relative rates of intron gain and loss on various branches of the eukaryotic tree (Qiu et al. 2004; Roy and Gilbert 2005c; Nguyen et al. 2005). To gain some insight into just how intron rich the stem eukaryote might have been, and into the difficulties associated with such an estimate, consider the observation that about 18% of the introns in animals occupy homologous insertion sites in land plants (Fedorov et al. 2002; Rogozin et al. 2003). If one is willing to ignore the possibility of parallel gains of introns, then, given that these two lineages are on opposite sides of the root of the eukaryotic tree (see Chapter 1) and that both have an average of approximately 5.0 introns per protein-coding gene, the 18% rate of intron sharing implies a minimum of one intron per protein-coding gene in the stem eukaryote. However, given the evidence in support of the proto-splice site hypothesis (described above) and the reliance of phylogenetic analyses on highly conserved genes (with stable insertion sites), it is by no means certain that parallel intron gains can be ruled out, especially on a time scale of a billion years. If parallel gains have occurred, the above estimate could be inflated substantially, but an even more intron-rich stem eukaryote cannot be ruled out either. Noting that animals and land plants diverged ~1.6 billion years ago (see Chapter 1), the rate of intron loss necessary to explain the current level of shared sites under the assumption of a steady-state average of five introns per gene is the solution to $e^{-3.2D} = 0.18$, which yields $D = 0.54$ per insertion per BY. Because this rate of loss is just slightly greater than the estimates in Table 9.1, extraordinary conditions do not have to be invoked to be compatible with a situation in which the stem eukaryote had as many introns as today's multicellular species. The substantial degree of uncertainty in these estimates highlights the need for an understanding of the molecular basis of intron gain and loss and the extent to which such events are likely to engender parallel changes.

Although it has been argued that an intron-rich stem eukaryote is inconsistent with the idea that introns are intrinsically deleterious (Roy and Gilbert 2006), that is not correct. Rather, such an inferred condition is entirely compatible with the hypothesis that the stem eukaryote experienced a long phase of relatively low N_g compared with many of today's unicellular eukaryotes (see Chapter 1). Such an episode would not only have facilitated the fixation of introns but, as noted above, could also have provided the conditions necessary for the emergence of the spliceosome itself. Moreover, the possibility that intron proliferation in the earliest stages of eukaryotic evolution was actively promoted by exceptional colonization mechanisms cannot be ruled out. Because the time between the origin of life and the origin of eukaryotes is ~1 BY (see Chapter 1), even in the complete absence of intron loss, a typical 1.5 kb coding region would be unlikely to acquire more than a single intron by the time of the early diversification of eukaryotes unless the early rate of intron insertion substantially exceeded the average estimate of B in Table 9.1 (~0.0006/nucleotide site/BY). Thus, a scenario in which the stem eukaryote was intron rich appears to require a phase in which the physical processes of insertion were much more active than in modern species and/or introns were favored by natural selection at the host level (yielding a rate of establishment in excess of the neutral expectation of B). Some possible selective mechanisms are discussed in the following section, but the fact that the stem eukaryote was almost certainly unicellular provides a compelling argument against the idea that the initial proliferation of introns was associated with the adaptive origin of features unique to multicellular species (e.g., tissue-specific alternative splicing).

Adaptive Exploitation of Introns

Although the initial establishment of spliceosomal introns may have been entirely a product of nonadaptive forces, once introns had spread throughout the eukaryotic genome, the new norm of genes in pieces would have provided a novel physical structure for adaptive exploitation by descent with modification. In today's multicellular eukaryotes, almost all of the major events in the production of mature mRNAs are tightly coupled with exon definition and/or intron splicing (Maniatis and Reed 2002). For example, interactions between various splicing factors and elongation factor promote transcript elongation (Ares et al. 1999; Fong and Zhou 2001), and splicing indirectly facilitates mRNA export to the cytoplasm via a protein complex specifically deposited on spliced mRNAs (Luo and Reed 1999; Reed and Hurt 2002; Rose 2004). In addition, splicing signals associated with the final intron can be essential for efficient transcription termination (Dye and Proudfoot 1999; McCracken et al. 2002) and polyadenylation (Niwa et al. 1990, 1992). Given this diverse array of services associated with introns and their processing, as well as the residence of functional noncoding sequences within some introns (Sironi et al. 2005) and the involvement of introns in alternative splicing and mRNA surveillance (described below), it is possible that few of today's eukaryotes could survive without introns. Even in the intron-depauperate yeast *S. cerevisiae*, experimental removal of introns from host genes results in reduced mRNA levels (Juneau et al. 2006). We now examine some of the ways in which introns have been suggested to be involved in adaptive evolution.

Modifiers of the recombination rate

Genome-wide analyses indicate that intron size and number increase in regions of low recombination in *D. melanogaster* (Carvalho and Clark 1999; Comeron and Kreitman 2000). Introns are also more abundant and longer in the G/C-poor regions of vertebrate genomes, which are relatively recombinationally quiescent (Duret et al. 1995). Although both trends are very weak, they have elicited both nonadaptive and adaptive explanations. Carvalho and Clark (1999) suggest that the inverse relationship between intron size and recombination rate is a simple reflection of the reduced efficiency of selection against insertion mutations in regions of low recombination. In contrast, Comeron and Kreitman (2000) argue that by magnifying the distance between some codons, intron expansion increases the intragenic recombination rate, thereby reducing the deleterious consequences of selective interference among tightly linked sites in coding exons. These two hypotheses could not be more different: the first invokes the expansion of introns as a symptom of the Hill–Robertson effect (see Chapter 4), and the second invokes it as a solution.

Are the population genetic conditions necessary for the selective promotion of intron expansion as a recombination frequency modifier likely to be met in many species? A simple argument suggests they are not. Because not all recombination events produce gametes with elevated fitness, and because the improvement in fitness associated with beneficial events is likely to be small, the selective advantage of a modifier of recombination is necessarily less than the fractional increase in the recombination rate, and probably substantially so. Yet a 100 bp expansion in the physical distance between selected sites (which exceeds the gradient of average intron size across the entire *Drosophila* genome) is expected to increase the recombination frequency per meiotic event by just 10^{-7} to 10^{-5} (see Figure 4.6). Thus, with effective population sizes in the range of 10^4 to 10^6 for animals (see Chapter 4) and the requirement for efficient selection being $2N_g s > 1$, it appears unlikely that anything other than an enormous increase in intron length could be promoted by selection for increased recombination frequency, and such alterations would probably have negative side effects associated with reduced splicing efficiency.

Carvalho and Clark's (1999) explanation for the association of intron abundance and size with recombination frequency is fully compatible with the population genetic theory presented above. If reduced intron size has a weak selective advantage (due, for example, to increased rates of transcription and/or increased accuracy of pre-mRNA processing), the efficiency of selection for intron-shortening mutations (and against intron-enlarging mutations) is expected to be weaker in regions of low recombination. This same logic provides an explanation for the hundredfold decline in average intron length with increasing $N_g u$ across eukaryotes (see Figure 9.7). Nevertheless, there are some remaining subtleties in intron size variation that remain to be explained, including a very weak positive association between intron size and recombination rate within the nematode *C. elegans* (Prachumwat et al. 2004) and differences in the dynamics of intron size in autosomal versus X-linked genes in *Drosophila* (Presgraves 2006). The extent to which the de novo length distributions of insertions and deletions vary with the level of recombinational activity (see Chapter 4) will need to be incorporated into future studies if the contribution from selection is to be factored out in an unbiased fashion. Even then, the very weak intron size gradients observed in previous intraspecific surveys indicate that the overall explanatory power of such work will be limited.

Alternative splicing

Introns provide multiple opportunities for enhancing protein diversity via the production of alternatively spliced forms of mRNAs, such as inclusion versus exclusion of an exon in the mature transcript (exon skipping), combinatorial mixing of various exons (exon swapping), and modifications of 5′ and/or 3′ splice sites (Mayeda et al. 1999; Smith and Valcárcel 2000; Graveley 2001; Ladd and Cooper 2002; Roberts and Smith 2002; Xing and Lee 2006) (Figure 9.8). At least 75% of human genes express alternatively spliced variants (Boue et al. 2003; Johnson et al. 2003), and high levels of alternative

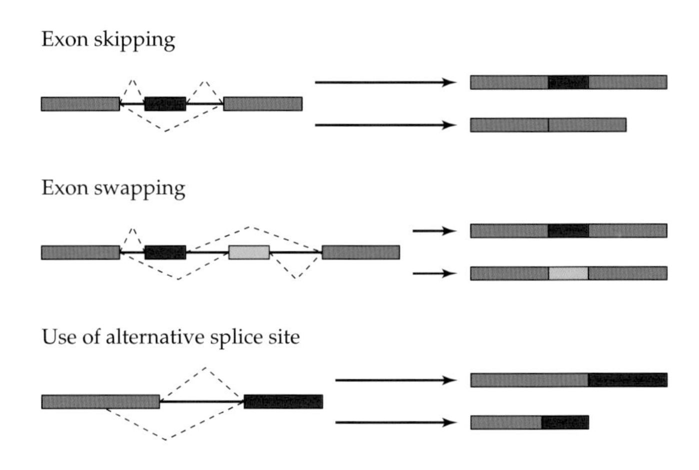

Figure 9.8 Some forms of alternative splicing. Colored bars denote exons, solid lines denote introns, and dashed lines demarcate regions that are spliced out to produce the alternative mRNAs on the right.

splicing have also been observed in other multicellular species (Brett et al. 2002; Okazaki et al. 2002; Stolc et al. 2004), with some genes producing dozens to hundreds of distinct mRNAs.

The mechanisms by which alternative splicing is accomplished are not entirely understood, but SR proteins, which bind exon splicing enhancers (as described above), are known to be involved. In humans, the SR family contains ten members with unique RNA-binding sequences (Graveley 2000), and other animals and land plants have multiple genes in this family as well (Lopato et al. 1996; Longman et al. 2000; Barbosa-Morais et al. 2006). Thus, variation in tissue-specific expression among SR family members provides a potential means for guiding the spliceosome toward alternative splice sites associated with different ESEs (Zahler et al. 1993), and further combinatorial possibilities are presented by ESSs, which are in turn bound by another variable family of proteins.

From the standpoint of adaptation, alternative splicing provides a potentially powerful mechanism for resolving conflicts between alternative allelic states. If two separate alleles encode unique products that exhibit a favorable interaction, the fitness-enhancing effect is experienced only by heterozygotes, and suboptimal homozygotes are produced anew each generation by segregation. In contrast, if alternative splicing can produce two (or more) products from the same allele in sufficient quantity, fixation of such an allele will ensure the expression of both products in all individuals, potentially in an advantageous tissue-specific manner.

Nevertheless, the degree to which alternatively spliced mRNAs are products of adaptive evolution, as opposed to being simple artifacts of an imperfect splicing process, remains unclear (Levine and Durbin 2001; Sorek et al. 2004; Chern et al. 2006). Only a few clear examples of adaptive alternative splicing have been uncovered, most notably the sex determination cascade in *Drosophila* (Tian and Maniatis 1993). Although numerous examples of alternative splicing in vertebrates superficially appear to be functionally critical (Stamm et al. 2005), the evidence is mostly circumstantial in nature. For example, about 77% of the alternatively spliced exons conserved between humans and mice are flanked by intronic sequence more highly conserved than that of constitutively spliced introns, suggesting important functional constraints (Sorek and Ast 2003; Xing and Lee 2006). Among alternatively spliced human genes, 10%–30% exhibit tissue-specific expression (Xu et al. 2002), and the alternatively spliced exons in these genes are very slightly biased toward lengths of $3n$, thereby preserving the reading frame (Xing and Lee 2005).

Despite these suggestive observations, some evidence, mostly from humans, suggests that a great deal of alternative splicing may be nonadaptive. First, alternative splicing commonly produces mRNAs with frameshifts that yield premature termination codons (Stamm et al. 2005; Sorek et al. 2004; Xing and Lee 2004), making them targets for immediate elimination by the nonsense-mediated decay (NMD) pathway (discussed in detail below). Although Mitrovich and Anderson (2000) and Lewis et al. (2003) have argued that the production of such transcripts provides an effective cellular mechanism for regulating gene expression, there are many more direct routes to silencing a gene than turning it on (see Chapter 10), and the fact that most alternatively spliced variants are not regulated by NMD in a tissue-specific manner further argues against this adaptive hypothesis (Pan et al. 2006). Second, ~75% of the alternatively spliced exons in humans are absent from mice, and these exons tend to be minor-form (numerically least abundant) variants with a substantial inflation in the incidence of frameshifts (Modrek and Lee 2003; Thanaraj et al. 2003; Sorek et al. 2004), consistent with the expectations for aberrant splicing products. Third, alternatively spliced exons exhibit higher rates of amino acid substitution (Chen et al. 2006) and higher levels of structural instability than constitutively spliced exons (Romero et al. 2006). Although both patterns could be a consequence of recent adaptive recruitment of new exons, they are also consistent with a relaxation of selection. Fourth, ~5% of human alternatively spliced exons are derived from insertions of primate-specific *Alu* retrotransposons (see Chapter 3) harboring latent splice sites. Nearly a third of such exons are just a single nucleotide substitution away from being constitutively expressed as exons that cause genetic disorders (Sorek et al. 2002; Lev-Maor et al. 2003; Kreahling and Graveley 2004).

Messenger RNA surveillance

One of the most significant services provided by introns is their indirect role in the cell's surveillance for transcripts harboring inappropriate premature termination codons (PTCs). PTC-containing mRNAs arise from the direct transcription of inherited mutant alleles as well as from errors in the transcription or splicing of otherwise functional alleles. Eukaryotes protect themselves from the accumulation of potentially deleterious truncated proteins resulting from such transcripts via the nonsense-mediated decay (NMD) pathway, which eliminates a substantial fraction of PTC-containing mRNAs during their first round of translation (Hentze and Kulozik 1999; Gonzalez et al. 2001; Lykke-Andersen 2001; Mango 2001; Maquat and Carmichael 2001; Wilusz et al. 2001; Maquat 2004, 2006). The need for mRNA surveillance appears to be substantial. The transcriptional error rate is on the order of 10^{-5} per nucleotide (Ninio 1991; R. J. Shaw et al. 2002), so with 3/64 of random codons denoting termination, base misincorporations are expected to yield 0.05%–0.5% PTC-containing transcripts for genes 10^3–10^4 coding nucleotides in length. Many additional inaccuracies undoubtedly arise from errors in transcription initiation and splicing (see Chapter 10).

Numerous human genetic disorders associated with PTC-containing alleles are rendered recessive in the heterozygous state via the silencing effects of NMD (Frischmeyer and Dietz 1999), and the power of NMD is illustrated by experiments in which the pathway has been silenced. Such studies show substantial increases in PTC-containing mRNAs, including those resulting from inappropriate alternative splicing and transcription from defective mobile elements and pseudogenes (Mendell et al. 2004; Mitrovich and Anderson 2005). Knockouts of the NMD pathway have small phenotypic effects in *S. cerevisiae* and *S. pombe* (Leeds et al. 1992; Dahlseid et al. 1998; Mendell et al. 2000), moderate fitness effects in *C. elegans* (Hodgkin et al. 1989), and lethal effects in mice (Medghalchi et al. 2001). The enhanced sensitivity of multicellular species to NMD inactivation may simply be a consequence of greater rates of production of erroneous transcripts in complex genomes with more opportunities for splicing errors, although this conclusion is clouded by evidence that some of the proteins in the NMD pathway have additional cellular functions (Maquat 2006).

The major requirement for a successful NMD pathway is a mechanism for distinguishing PTCs from correct termination codons. In mammals, guidance is usually provided by a multiprotein exon junction complex (the EJC) deposited 20–24 nucleotides upstream of exon–exon junctions at the time of splicing. During the first round of translation, these marked junctions allow the cell to infer the position of the true termination codon, which generally resides in the final exon (beyond the last EJC). If a termination codon is detected 50 or more nucleotides upstream of a marked exon–exon junction, the mRNA is targeted for selective degradation (Nagy and Maquat 1998). Although the central role of introns in this process is confirmed by the observation that intron-free mammalian genes are generally NMD insensitive (Maquat and Li 2001; Brocke et al. 2002), introns are not always required for NMD. For example, although the nuclear genome of *S. cerevisiae* contains only about 250 introns, this species is still broadly susceptible to NMD, with PTC recognition relying either on downstream sequence elements (DSEs) embedded within coding DNA (Ruiz-Echevarria et al. 1998) or on contextual information residing in the 3′ UTR (Amrani and Jacobson 2006).

It is unclear whether the mammalian EJC-based pathway or the *S. cerevisiae* DSE/UTR-based pathway more closely represents the ancestral mode of PTC recognition. However, because empirical work demonstrates that

NMD can operate on intron-free genes in *S. pombe* (Mendell et al. 2000) and on at least some intron-free genes in mammals (Cheng et al. 1994; Zhang et al. 1998; Rajavel and Neufeld 2001; Bühler et al. 2006), nematodes (Pulak and Anderson 1993), dipterans (Gatfield et al. 2003), and plants (van Hoof and Green 1996; Neu-Yilik et al. 2001; Kertész et al. 2006), it appears that two NMD pathways exist within some eukaryotic lineages. Mammals have come to rely predominantly on the EJC pathway, while *S. cerevisiae* and perhaps *S. pombe* predominantly utilize an intron-independent pathway. In *Drosophila*, PTC-containing intronless gene constructs introduced from other species are subject to NMD (Gatfield et al. 2003), suggesting that the mechanism of PTC detection is not a function of the coding sequence, but of some sort of mark, perhaps the poly(A) tail or a protein deposited on the 3′ UTR.

Although NMD is not known to occur in any prokaryote, homologs for all of the core components of the NMD pathway and the EJC are broadly distributed across the entire eukaryotic phylogeny, suggesting that both were present in the stem eukaryote (Lynch et al. 2006b). However, the EJC appears to have been independently lost from several species, including the microsporidian *Encephalitozoon*, the red alga *Cyanidioschyzon*, the kinetoplastid *Trypanosoma*, and the diplomonad *Giardia*. Remarkably, each loss of the EJC was accompanied by both a massive loss of introns and an apparent loss of the NMD pathway. Such a phylogenetic codistribution of events raises the possibility that NMD and the EJC may have cooperated in mRNA surveillance from a very early date. However, because introns (and therefore EJCs) are not a physical necessity for the operation of NMD in all genes or lineages, alternative explanations for the parallel loss of NMD and introns from specific lineages cannot be ruled out. For example, most of the species that have lost the NMD pathway have exceptionally compact genomes in all respects (see Chapter 3), being nearly devoid of mobile elements and having substantially diminished levels of intergenic DNA. By greatly reducing the production of erroneous transcripts, the absence of all of these features (including introns) is expected to relax the need for mRNA surveillance.

The establishment of an intron-based mechanism for NMD orientation may have dramatically altered the selective environment for further intron colonization (Lynch and Kewalramani 2003). At this point in eukaryotic evolution, not all newly arisen introns would have been unconditionally neutral or deleterious. Rather, appropriately located introns might have endowed their host genes with a selective advantage, potentially driving a positive feedback in genomic evolution by encouraging intron colonization at sites that maximize the efficiency of PTC detection. However, once sufficient spatial coverage for efficient NMD was achieved, the inherent disadvantages of introns would begin to outweigh their NMD-associated advantages, providing a natural barrier to runaway intron colonization.

Under this model, the selective consequences of new introns should depend on their insertion sites and on the geometric constraints associated with the mRNA surveillance process. For example, if the NMD machinery

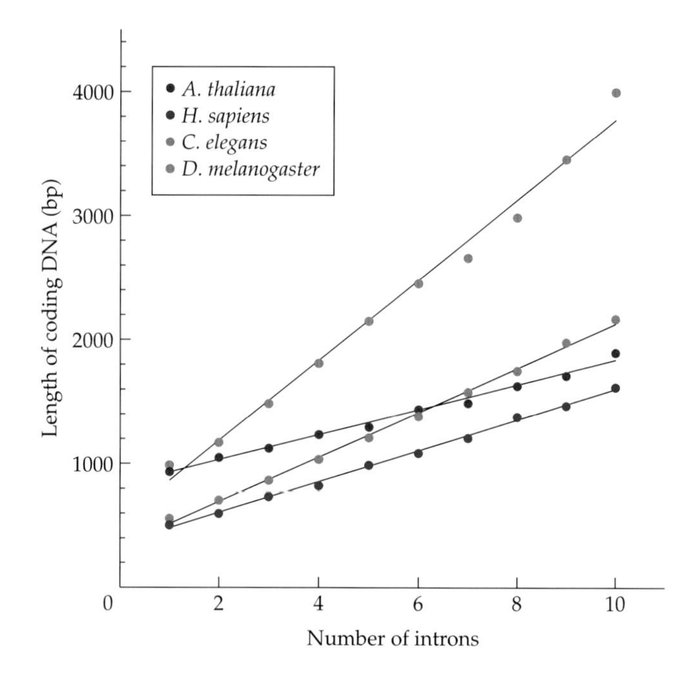

Figure 9.9 The average number of introns per protein-coding gene is linearly related to the length of the coding DNA in multicellular species. The results are based on the full set of annotated genes from four genomes. (From Lynch and Kewalramani 2003.)

were capable of scanning the entire length of a transcript (and no other selective factors influenced the spatial locations of introns), the optimal configuration would be a single intron at the 3′ end of every coding region. However, if the NMD machinery were somewhat nearsighted, and thus not fully capable of scanning hundreds to thousands of bases, a more uniform scattering of introns would be necessary for complete NMD coverage. Unfortunately, empirical information on the spatial requirements of NMD is quite limited. In humans, NMD is elicited by PTCs as far as 550 bp upstream of an intron in the triose phosphate isomerase gene (Zhang et al. 1998), 700 bp upstream in the β-globin gene (Neu-Yilik et al. 2001), and 1200 bp upstream in the HSP70 gene (Maquat and Li 2001). However, the relationship between NMD efficiency and the PTC–EJC distance is probably more gradual than a threshold function, and multiple introns may facilitate the process (Rajavel and Neufeld 2001). Some studies have suggested that more 5′-located introns elicit a higher level of NMD (van Hoof and Green 1996; Pulak and Anderson 1993), and the length of the surveillance tract may differ among species. For example, although no direct evidence has yet been forthcoming, it has been suggested that NMD requires two marked exon junctions in *C. elegans* (Mango 2001), and the first (rather than final) exon junction may be critical in rice (Isshiki et al. 2001). In *S. cerevisiae*, sensitivity to NMD is completely eliminated if the PTC is more than 300 bp upstream of a DSE (Ruiz-Echevarria et al. 1998).

Provided there are spatial limitations on the NMD scanning mechanism, one simple prediction of this model is that the number of introns per gene should increase linearly with the length of the gene; that is, with the amount of territory that needs to be scanned. A second prediction is that introns should be overdispersed within their host genes; that is, exon sizes should be more uniform than under a model of random insertion. Although the quantitative relationships differ among species, genome-wide surveys in multicellular organisms support both predictions. Intron number is consistently linearly related to gene size (Figure 9.9), and the average positions of consecutive introns tend to evenly partition the coding sequence (Figure 9.10; see also Sakurai et al. 2002; Mourier and Jeffares 2003). Although these patterns could arise from completely random colonization of introns, the low variance of exon size is statistically incompatible with such a "broken-stick" model (Lynch and Kewalramani 2003). That is, exon sizes are more even than expected by chance (Figure 9.11), consistent with the hypothesis of the operation of stabilizing selection on exon size.

It remains to be determined whether these patterns have been actively promoted by the requirements of NMD and/or by additional factors such as the existence of a minimal exon size for efficient splicing. Notably, *D. melanogaster*, the one species whose exon size distribution is very close to the random expectation, also reportedly has an intron-independent NMD process (Behm-Ansmant and Izaurralde 2006), although this pattern is less obvious in *Arabidopsis*, which also appears to have intron-independent NMD in at least some genes (van Hoof and Green 2006). Additional support for the operation of stabilizing selection on intron number and location derives from observations on the relative constancy of these features in the face of

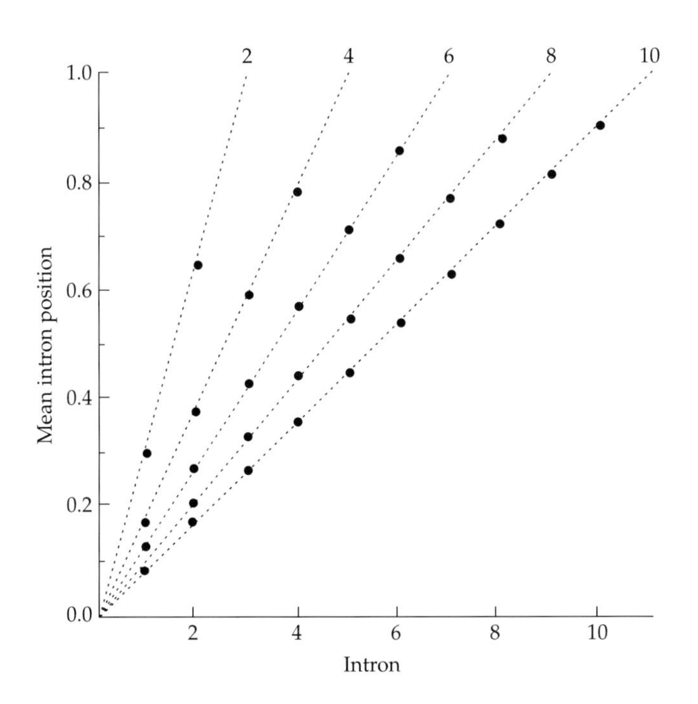

Figure 9.10 The average relative locations of introns in human genes with 2, 4, 6, 8, and 10 introns. The lengths of the coding regions of all genes are scaled to 1.0. The dotted lines denote the expected positions of consecutive introns in the absence of locational biases; for example, for genes with two introns, on average, the first intron is expected to be ~33% and the second ~67% down the length of the gene, which is very close to the actual data shown by the two solid points on the line labeled 2. (From Lynch and Kewalramani 2003.)

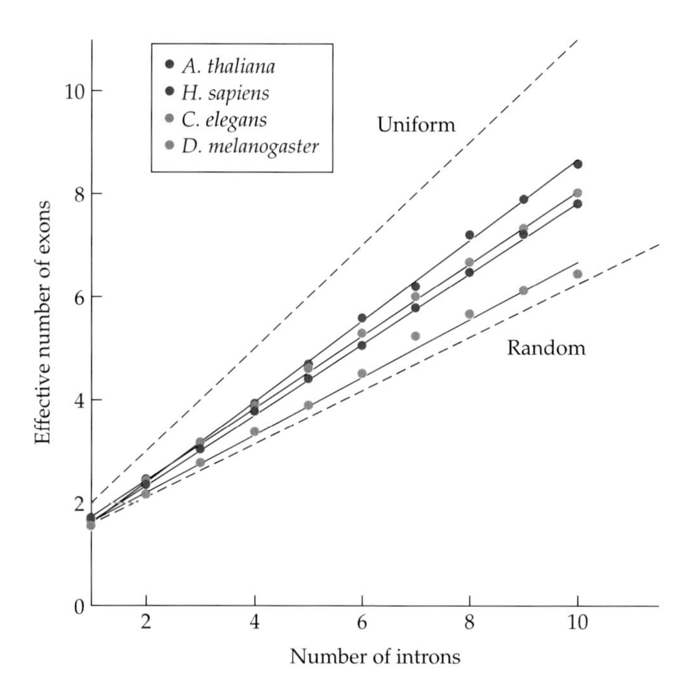

Figure 9.11 The average effective number of exons for genes in four multicellular species. Results are given for collections of genes with different numbers of introns. The effective number of exons is a function of the uniformity of exon sizes within genes, and where all exons are equal in size, has an expectation equal to the number of introns + 1 (upper dashed line). The lower dashed line gives the expectation under the "broken-stick" model, according to which introns are randomly distributed along the coding region. (From Lynch and Kewalramani 2003.)

significant change in intron size. In both nematodes and mammals, average intron size declines dramatically in genes with increasing levels of expression, while intron number remains constant (Castillo-Davis et al. 2002). In *Arabidopsis*, genes that are expressed in pollen, and hence exposed to stronger selection in the haploid state, have introns that are smaller than, but equally abundant as, those in genes expressed solely in diploid tissue (Seoighe et al. 2005), and curiously, the number of introns per gene in this species increases with the level of gene expression (Ren et al. 2006).

Interpretation of these observations needs to be tempered in light of our knowledge about rates of intron turnover. Even if an ancestral intron-based NMD mechanism was displaced by an intron-independent process in some lineages, the signature of past selection for intron location could remain for a very long time. If the average loss rate in Table 9.1 were to apply, about 83% of intron positions would be retained over a billion-year period in the absence of selection for subsequent intron movement.

10 Transcript Production

As noted in Chapter 1, the origin of eukaryotes coincided with numerous changes in the ways in which gene transcripts are processed into mature mRNAs and eventually translated into proteins. Eukaryotic gene expression is generally initiated by the recruitment of one or more transcription factors by multiple upstream regulatory elements, and these in turn help activate the transcription machinery in the vicinity of the correct transcription initiation site (Ptashne and Gann 2002). These key events must be sufficiently accurate to ensure that transcripts are initiated upstream of the translation initiation site, and elongation must proceed far enough to ensure the incorporation of the translation termination site. Both processes are generally guided by sequence information contained within noncoding regions of genes, the refinement of which imposes conflicting advantages and disadvantages. For example, as we have seen in the case of introns, any benefits of elaborate transcriptional control sequences must be weighed against their elevated mutation rate to defective alleles resulting from the increased size of the mutational target. A long region between a gene's transcription and translation initiation sites provides potential opportunities for finely tuning gene expression at the level of mRNA localization, stabilization, and/or translation, but increases the mutational rate of origin of premature translation initiation sites, landing sites for mobile element insertions, and so forth.

Although we now know a great deal about protein sequence evolution (Li 1997) and a fair amount about the temporal and spatial regulation of gene expression (Carroll et al. 2001; Davidson 2001), the evolutionary mechanisms sculpting the regions of eukaryotic genes that are essential to the development of productive transcripts are poorly understood (Wray et al. 2003). This chapter explores a number of central but seldom asked questions about the origins of the features of the terminal noncoding regions of genes. As with many other genomic features, it is unclear whether the complex architecture of the upstream and downstream regions of eukaryotic genes arose because the immediate selective advantages outweighed the negative mutational consequences. An alternative view is that nonadaptive processes played a key initial role in restructuring eukaryotic gene organization, yielding novel physical substrates for the subsequent evolutionary refinement of gene regulation mechanisms. The merit in exploring this view is embodied in the following question: if the origin of eukaryotic gene structure awaited the appearance of rare beneficial mutations, then why do all well-studied prokaryotes harbor much simpler forms of gene organization, despite their enormous population sizes (and large numbers of mutational targets)? In order to resolve these issues, we again need to consider the direction in which mutational processes are expected to drive the evolution of gene architectural features when the power of natural selection is overwhelmed by genetic drift.

Transcription and Translation Initiation

The total amount of transcriptional and translational activity within a cell is enormous. For example, in actively growing yeast (*S. cerevisiae*), individual genes are generally represented by 0.1–100 mRNAs per cell (with an average of 2–3), and each mRNA is responsible for the production of 10^2 to 10^6 proteins prior to decay and replacement (with an average of about 5000) (Ghaemmaghami et al. 2003). Approximately 70% of a yeast cell's mRNAs are engaged with one or more ribosomes at any time (Arava et al. 2003), although the number of ribosomes per mRNA is only about 20% of the maximum packing density, suggesting that translation is limited by the initiation process.

To ensure that transcription is initiated upstream of the translation initiation site (usually a methionine AUG), all organisms have core promoter sequences, usually immediately upstream of each transcription initiation site (or in some cases encompassing or following the latter). Generally consisting of no more than a dozen nucleotides, the core promoter is the site where information conveyed by specific transcription factors bound at more distant regulatory sites is integrated to signal the loading of the transcription pre-initiation complex (Smale and Kadonaga 2003). The position of the core promoter defines the point of transcription initiation, yielding a leader sequence between the sites of transcription and translation initiation. This sequence, the 5′ UTR (untranslated region), is retained as a permanent feature of the mature mRNA after the removal of introns during pre-mRNA processing (Figure 10.1).

Prokaryotes employ mechanisms that allow a very rapid transition from transcription to translation. In eubacteria, transcription initiation signals (hereafter, TISs) at core promoters are generally located close enough to the translation initiation codon to yield a 5′ UTR no more than a dozen

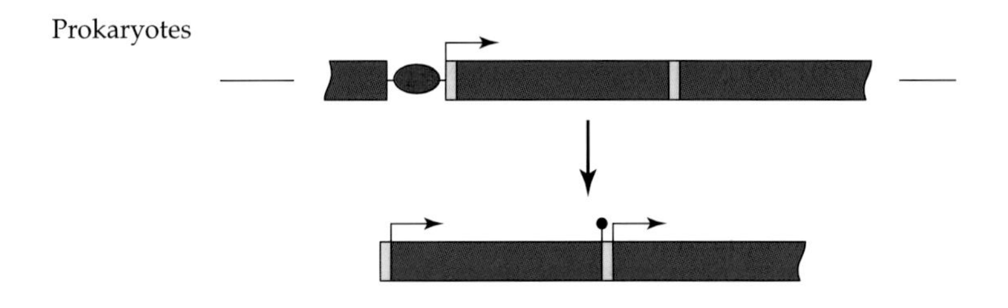

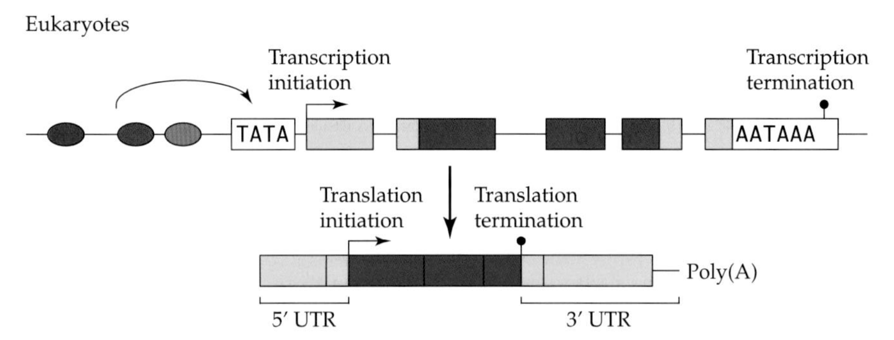

Figure 10.1 The comparative anatomy of gene and transcript structures in prokaryotes and eukaryotes. Blue bars denote coding DNA; yellow bars denote transcribed but untranslated DNA. Thin lines denote untranscribed intergenic DNA and introns within transcribed regions. Colored ovals denote transcription factor binding sites, which collectively convey information (curved arrow) to the core promoter (TATA being one of the possible eukaryotic motifs for the latter). AATAAA is a simplified depiction of the polyadenylation signal, and Poly(A) denotes the posttranscriptional addition of a poly(A) tail. Note that the lengths of various gene parts and numbers of transcription factor binding sites are not necessarily to scale, e.g., eukaryotic genes can harbor as many as several dozen introns, each of which can greatly exceed the length of its surrounding exons, and transcription factor binding sites can be much more numerous and widely distributed than illustrated.

nucleotides in length. Eubacterial genes are often aggregated into operons: cassettes of genes that are transcribed as continuous (polycistronic) mRNAs. Each gene in an operon is preceded by a Shine–Dalgarno sequence that binds the 16S small ribosomal RNA subunit, orienting it to the translation initiation codon (Shine and Dalgarno 1974). In the archaea, genes internal to operons often have Shine–Dalgarno sequences, whereas the leader genes often do not (Tolstrup et al. 2000; Slupska et al. 2001). In the archaeal species that has been investigated in the most detail, *Pyrobaculum aerophilum*, almost all mRNAs are leaderless; that is, the transcription and translation initiation sites are identical (Slupska et al. 2001), a feature also found in a few eubacteria (Weiner et al. 2000; Moll et al. 2002). Transcription of mitochondrial genomes is guided by simple ~9 bp core promoters and usually produces a small number of polycistronic transcripts that are cleaved into distinct leaderless mRNAs prior to translation (Gillham 1994; Taanman 1999).

Unlike the situation in prokaryotes, eukaryotic nuclear gene transcripts must be exported to the cytoplasm for translation, and this requirement necessitates several mRNA processing steps. Nevertheless, as in prokaryotes, eukaryotic nuclear genes employ relatively simple core promoters for transcription initiation. One of the most phylogenetically widespread eukaryotic TISs is the TATA box, which generally resides 20–40 nucleotides upstream of transcription initiation sites. The TATA box is employed in a substantial fraction of animal genes (about 42% in *Drosophila* and 32% in humans; Kutach and Kadonaga 2000; Suzuki, Tsunoda et al. 2001; Ohler et al. 2002) and in most yeast genes (Struhl 1989; Choi et al. 2002). Putative TATA boxes are also associated with the genes of *Paramecium* (Yamauchi et al. 1992), *Dictyostelium* (Hori and Firtel 1994), and *Entamoeba* (Singh et al. 1997). This wide phylogenetic distribution of the TATA box, as well as its involvement in archaeal transcription (Thomm 1996; Soppa 1999a,b; Bell et al. 2001; Slupska et al. 2001), suggests that the stem eukaryote utilized this sequence for transcription orientation, although not necessarily exclusively so.

Remarkably, only a few alternative core promoters have been found in eukaryotes, and only one of them has been found to be self-sufficient: the loosely defined initiator (Inr) sequence, which contains an internal transcription initiation site embedded within a semiconserved sequence of five to seven nucleotides. With a consensus sequence of Y-Y-A-N-(T/A)-Y-Y in animals (Lo and Smale 1996), where Y denotes a pyrimidine and N an arbitrary nucleotide, the complexity of Inr is even less than that of TATA. In *Drosophila* and humans, 65%–85% of all protein-coding genes are associated with Inr sequences (Suzuki, Taira et al. 2001; Suzuki, Tsunoda et al. 2001; Ohler et al. 2002). A self-sufficient Inr is not known in yeasts, although significant information can be encoded at the site of transcription initiation (Hahn et al. 1985; Nagawa and Fink 1985). However, Inr is used by a number of protists, including *Toxoplasma gondii* (Nakaar et al. 1998), *Trichomonas vaginalis* (Liston and Johnson 1999), and *Entamoeba histolytica* (Singh et al. 1997). This wide phylogenetic distribution of Inr elements, their rough sequence similarity in different eukaryotic lineages, and their utilization in archaea suggests that Inr, like TATA, was utilized by the stem eukaryote and then lost from some descendant lineages.

A substantial fraction of well-studied eukaryotic genes (including nearly half of human genes) do not appear to harbor a TATA box, an Inr sequence, or either of two other somewhat less well-defined TISs (Mitchell and Tijian 1989; Kadonaga 2002). However, such genes do generally harbor a sequence at the transcription initiation site that confers exceptional local physical flexibility on the DNA, as the TATA box and Inr do (Fukue et al. 2005). This observation suggests that TISs may be largely defined by mechanical properties, although proteins that specifically bind to TATA and Inr do exist.

Although the complexity of TISs does not appear to vary much across the major domains of life, eukaryotes are distinguished by a substantial expansion in the lengths of their 5′ UTRs relative to those of prokaryotes. Among the major eukaryotic groups, average 5′ UTR lengths are quite similar, ranging from just under 100 bp to just over 200 bp (Table 10.1). This approximate independence of 5′ UTR length and organismal complexity is in striking contrast to the order-of-magnitude interspecific differences in numbers and sizes of introns, numbers of mobile elements, and lengths of nontranscribed intergenic spacers (discussed in previous chapters). Equally striking is the relatively constant amount of dispersion about the mean for the full sets of 5′ UTR lengths within species, with the average coefficients of variation for vertebrates, invertebrates, plants, and unicellular species being 1.13, 1.15, 1.22, and 1.34 respectively (see Table 10.1).

TABLE 10.1 **The average lengths of 5′ UTRs (bp) in various eukaryotic lineages and their average coefficients of variation (ratio of standard deviation to mean) within species**

	NUMBER OF GENERA	MEAN	CV
Vertebrates			
Mammals	21	139	1.33
Birds	4	128	1.21
Frogs	4	103	1.14
Ray-finned fish	20	125	1.20
Agnathans	3	139	0.79
Invertebrates			
Invertebrate chordates	3	125	1.27
Echinodermata	4	206	1.09
Arthropoda	24	121	1.25
Mollusca	6	122	0.95
Annelida	2	76	0.77
Nematoda	8	67	1.15
Platyhelminthes	3	118	1.75
Porifera	3	85	0.95
Vascular plants	51	106	1.22
Unicellular species			
Fungi	17	149	1.22
Slime molds	2	107	1.17
Green algae	4	110	1.22
Apicomplexans	4	227	1.20
Kinetoplastids	2	163	1.43
Ciliates	2	126	1.80

Source: Lynch et al. 2005.

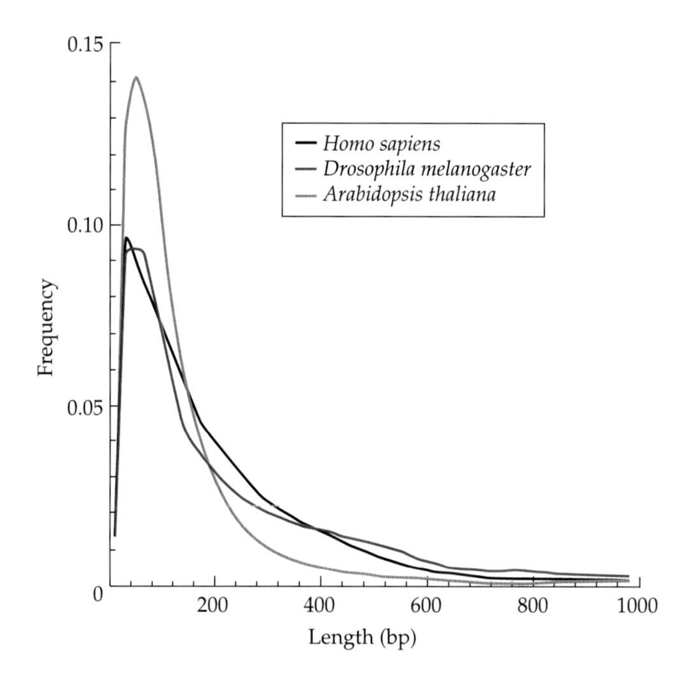

Figure 10.2 Frequency distributions of lengths of 5′ UTRs, derived from the database of Pesole et al. (2002). (From Lynch et al. 2005.)

Because the lengths of individual 5′ UTRs within species can range from tens to thousands of base pairs (Radford and Parish 1997; Pesole, Grillo et al. 2000; Pesole et al. 2001; Suzuki et al. 2000; Rogozin et al. 2001) (Figure 10.2), the relative invariance of the phylogenetic means and CVs does not appear to be due to a structural constraint, such as the need for a ribosomal landing pad. Although archaea initiate transcription by mechanisms similar to those in eukaryotes (Bell and Jackson 2001), their 5′ UTRs are very small, and 5′ UTRs in the diplomonad *Giardia*, a unicellular eukaryote, often consist of just a single nucleotide (Iwabe and Miyata 2001). A substantial fraction of 5′ UTRs in a variety of other unicellular eukaryotes have been found to be less than 25 bp in length, including those in the ciliate *Euplotes crassus* (Ghosh et al. 1994), the amoeba *Entamoeba histolytica* (Singh et al. 1997), and the trichomonad *Trichomonas vaginalis* (Liston and Johnson 1999). Experimental evidence suggests that such diminutive leader sequences are sufficient to support translation in mammals and yeasts, although the efficiency of translation can be reduced with UTRs shorter than about 30 bp (van den Heuvel et al. 1989; Maicas et al. 1990; Hughes and Andrews 1997).

The Premature Translation Initiation Problem

In contrast to the situation in prokaryotes, eukaryotic translation is usually preceded by the binding of the small 40S ribosomal subunit and multiple initiation factors, including the methionine initiator tRNA, to the 5′ cap structure of the mRNA (Kyrpides and Woese 1998; Saito and Tomita 1999b). This complex then scans downstream until it encounters an AUG (methionine) initiation codon, although the selection of a specific initiation site is sometimes influenced by a small surrounding set of nucleotides (Kozak 1987, 1994; Pesole, Gissi et al. 2000; Niimura et al. 2003). At this point, the full ribosome assembles and begins translation. No eukaryote is known to employ a Shine–Dalgarno sequence for translation initiation site localization, although some eukaryotic mRNAs do harbor sequences with complementarity to rRNAs that may function in translation control (Tranque et al. 1998; Mauro and Edelman 2002).

From the standpoint of allelic mutability, the scanning mechanism for translation initiation is potentially problematic because the acquisition of an AUG triplet in a 5′ UTR (a premature start codon, or PSC) will result in premature translation initiation. Such alterations will often result in nonfunctional alleles because two-thirds of random AUGs will be frameshifted with respect to the downstream coding sequence, and even when the triplet is in frame, the addition of N-terminal amino acids may lead to a defective protein. In a variety of species, 15%–55% of 5′ UTRs contain upstream AUGs (Suzuki et al. 2000; Peri and Pandey 2001; Rogozin et al. 2001), so it is clear that not all PSCs result in gene malfunction. Nevertheless, several human genetic disorders are known to result from the mutational origin of PSCs (Kozak 2002), and the fact that all species have a reduced incidence of PSCs relative to random expectations (Rogozin et al. 2001) implies that a substantial fraction of such sequences are opposed by selection. However, the deficit of PSCs within the 5′ UTRs of eukaryotes shows a strong gradient, with the bias being negligible beyond 500 bp upstream of the true translation initiation site (Figure 10.3). The observation that such deficits are much

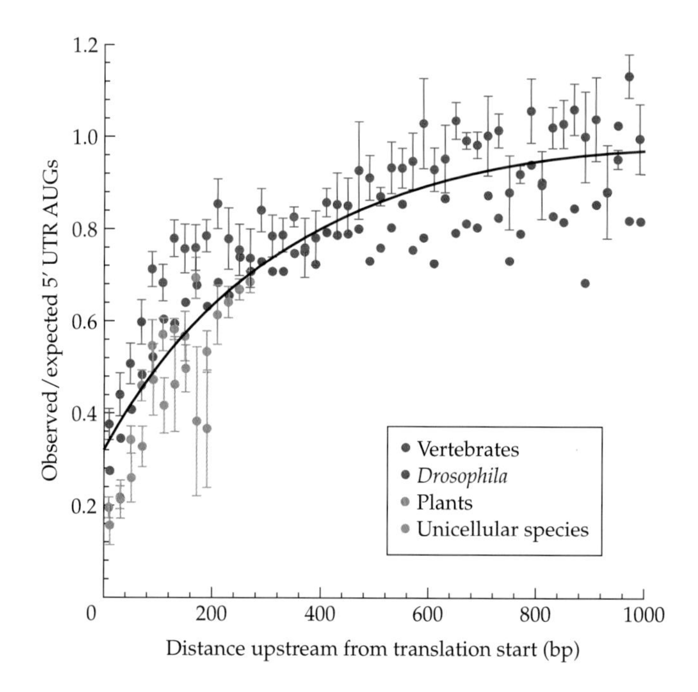

Figure 10.3 The incidence of observed AUG triplets, relative to the random expectation, at 20 bp intervals in 5′ UTR sequence increasingly upstream of the true translation initiation site. All data points are based on estimates from 50 or more genes using data from Pesole et al. (2002); standard errors are for the various genera within taxonomic groups. The solid line denotes the best fit to all of the data. (From Lynch et al. 2005.)

less pronounced within the introns in 5′ UTRs, regardless of position, strongly supports the idea that this pattern is driven by a translation-associated process (Lynch et al. 2005). Moreover, in prokaryotes, such deficits extend only 10–30 bp upstream of the true translation initiation point; that is, just up to the location of the Shine–Dalgarno sequence (Saito and Tomita 1999).

The near phylogenetic constancy in species-specific distributions of 5′ UTR lengths suggests the existence of a general (eukaryote-wide) mechanism of 5′ UTR evolution. One possibility is that such evolution is entirely driven by the stochastic mutational origin of alternative transcription initiation signals (TISs) and premature start codons (PSCs). Flexibility in the precise locations of TISs is supported by the empirical demonstration that the loss of a functional TIS can be compensated by fortuitous alternative matching sequences in the near vicinity (Hahn et al. 1985; Nagawa and Fink 1985). Such modifications are apparent within the yeast genus *Saccharomyces*, in which some 10% of TIS locations and/or motifs have changed among genes across species (Bazykin and Kondrashov 2006). Similarly, a comparison of mouse and human transcript sequences reveals that the vast majority of transcription initiation sites have moved at least 10 bp, with 20% moving more than 100 bp (Frith et al. 2006).

The simplest conceptual model for 5′ UTR evolution is one in which the transcription apparatus employs as a TIS the nearest appropriate sequence upstream of the proper translation initiation site (Lynch et al. 2005). Under this model, functional alleles require the absence of harmful PSCs within the UTR, but AUGs are free to accumulate in a neutral fashion upstream of the currently utilized TIS. Numerous types of functional and nonfunctional alleles can arise by mutation under this model (Figure 10.4). For example, all potential TISs are subject to mutational loss at rate nu, where n is the number of nucleotides in the TIS and u is the mutation rate per nucleotide. If such a mutation involves the currently utilized TIS, the UTR will expand out to the nearest upstream alternative TIS, and a nonfunctional allele will result if the extended region contains a harmful PSC. Alternatively, should a new TIS arise downstream of the previously employed TIS but in a location that still allows the inclusion of the translation initiation site in the transcript, the 5′ UTR will take on a shorter length. In parallel with these processes, PSCs are free to wink on and off neutrally upstream of the current TIS, whereas the subset of harmful PSCs that fall within the current UTR will result in nonfunctional alleles. Such behavior results in the evolution of a gradient of increasing density of AUGs 5′ to a gene's general TIS location (approaching random frequencies at positions far upstream), thereby producing a barrier to the upstream movement of a TIS by mutational processes. In the long run, the stochastic gain and loss of PSCs and TISs by these processes yields a steady-state distribution of viable 5′ UTR lengths that can be viewed as either the long-term history of UTR lengths of an individual gene or the expected snapshot genomic distribution of UTR lengths for all genes under similar mutational pressures.

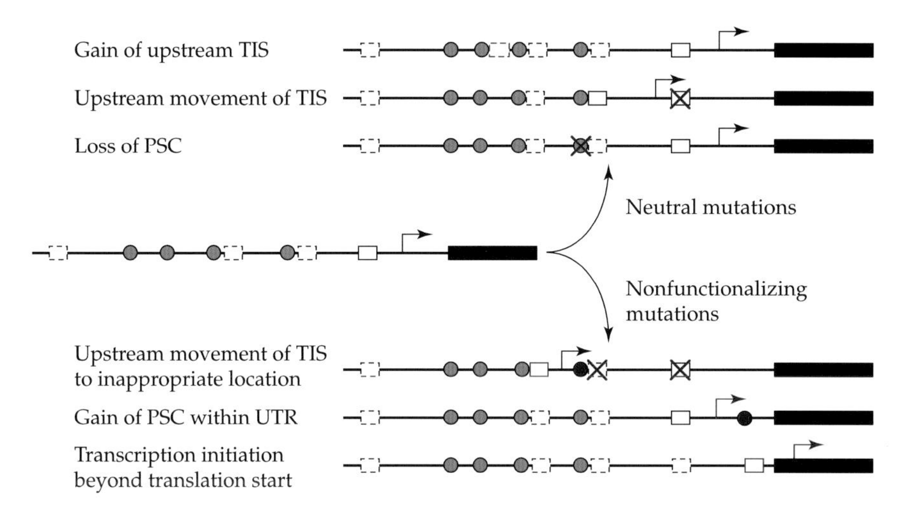

Figure 10.4 Schematic for some alternative allelic types associated with mutations in the 5′ UTR. Solid arrows denote active transcription initiation sites, and the leftmost edge of the solid box defines the true translation initiation codon; the span between these two points is the 5′ UTR. Small open boxes denote transcription initiation signals (TISs), here assumed to lie upstream of the transcription initiation site (dashed boxes indicate latent TISs, and solid boxes currently used TISs). Green circles denote harmless AUG triplets upstream of the UTR, whereas the red circle denotes a harmful PSC appearing within a UTR.

Because of the somewhat ill-defined nature of TISs, their mutational vulnerability (nu) can only be approximated. Given the widespread use of the TATA and Inr elements, n is at least 4, and the presence of putative conserved accessory elements in a variety of lineages implies that it could be as large as 10 (Singh et al. 1997; Soppa 1999a,b; Gourse et al. 2000; Ohler et al. 2002). However, the predictions of the preceding null model are quite robust to this uncertainty. For $n = 6$ and higher, the steady-state distribution of 5′ UTR length is nearly invariant, with an overall mean of 190 bp and a CV of 1.27 (Figure 10.5), both of which are roughly compatible with the observed data (see Table 10.1). Moreover, as can be seen by comparing Figures 10.2 and 10.5, the overall forms of the observed and expected 5′ UTR length distributions are very similar, except for the deficit of observed 5′ UTR lengths shorter than 50 bp.

At least two factors might contribute to this discrepancy. First, if transcription is initiated at variable positions relative to the TIS, some transcripts could inadvertently begin downstream of the translation initiation site when the TIS is close to the latter. Noisy transcription initiation sites, with ranges up to 10–100 bp, are known to occur in eubacteria (Nicolaides et al. 1995), mitochondria (Lizama et al. 1994; Carrodeguas and Vallejo 1997; Kühn et al. 2005), and nuclear genes (Hahn et al. 1985; Nagawa and Fink 1985; Bergsma

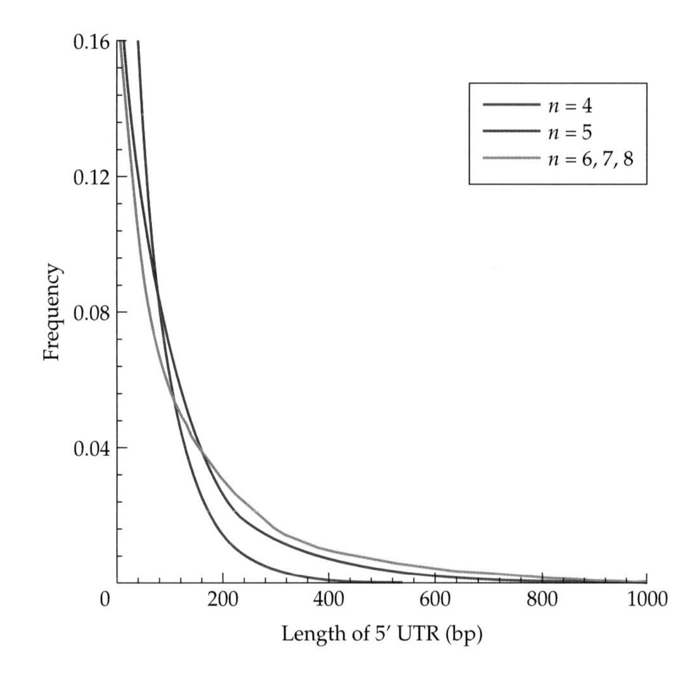

Figure 10.5 Expected steady-state frequency distributions for 5′ UTR lengths under a model in which TISs and PSCs randomly appear and disappear by mutational processes, as described in the text. Results are given for TISs of various lengths (in bp, denoted by *n*). Frequencies are given for bin widths of 20 bp. The results were obtained by computer simulation with $u = 10^{-7}$ per base pair per generation, using an empirical function for the probability that a PSC is potentially harmful (defined by the solid line in Figure 10.3). (From Lynch et al. 2005.)

et al. 1996; Yu et al. 1996; Kaji et al. 1998; Watanabe et al. 2002). In humans, for which the data are most extensive, many genes appear to have a continuum of transcription initiation sites (Kim et al. 2005), with an average range of 62 bp and an average standard deviation of 20 bp (Suzuki, Tsunoda et al. 2001), and similar observations have been made in the mouse (Carninci et al. 2006).

A second complication not directly accounted for by the preceding model concerns the presence of (external) introns in the 5′ UTR. By definition, a TIS lies upstream of all introns, but because introns are removed prior to translation, they are free to accumulate AUG triplets. In principle, a 5′ UTR intron is free to be lost, as it contributes nothing to the mature mRNA (although it may contain elements regulating transcription). However, the inclusion of AUGs within such introns can inhibit their loss in at least two ways. First, the appearance of a new TIS within an external intron will result in the incorporation of the downstream intronic sequence into the 5′ UTR of the mature mRNA, yielding a nonfunctional allele should the added region (previously intronic DNA) contain a harmful PSC. Second, any mutational elimination of either intronic splice site will result in the incorporation of the full intron into the 5′ UTR, yielding an even higher likelihood of PSC introduction. In this sense, external introns, once established, are expected to be exceptionally stable in location and resistant to size reductions. The elevated size of introns in 5′ UTRs relative to those in coding regions is consistent with this hypothesis (Hong et al. 2006). The presence of external introns can inhibit the evolution of overly short 5′ UTRs because a TIS must be located sufficiently far upstream of an intron to prevent transcription initiation within the intron (which would incorporate intronic sequence into the UTR).

We now return to the question of why the distributions of 5′ UTR lengths are so constant across diverse lineages of eukaryotes, focusing on the cost of long 5′ UTRs associated with the occasional mutational appearance of PSCs. Due to the stochastic expansion and contraction of UTRs (and hence alterations in mutational target size), the rate of origin of PSCs for individual genes will vary through time, but in the long run, an allelic lineage can be characterized by an average mutation rate to null alleles associated with PSCs. This excess mutation rate, which is equivalent to the average selective advantage (s) of maintaining a minimal 5′ UTR, turns out to be quite small (Lynch et al. 2005). Under the null model described above, s is a function of n, which defines the size of the TIS mutational target and also determines the vulnerability to PSCs (the larger the interval between alternative TISs, the higher the likelihood of incorporation of a PSC upon TIS loss). However, both the dependence on n and the overall cost of randomly maintained 5′ UTRs are weak, with $s \approx u$ when $n = 5$, $3u$ when $n = 6$, and $4u$ when $n = 8$.

Recalling the principle that weak differential mutation rates to null alleles effectively operate like selection with additive allelic effects on fitness, and the fact that selection is relatively ineffective if $2N_g s << 1$ (see Chapter 4), the preceding results suggest that $2N_g u$ must be on the order of 0.25 or greater if selection is to be capable of preventing the random expansion of 5′ UTRs. This limit is well above the upper bound on estimates of $2N_g u$ for multicellular species and even ~5 times greater than the mean for unicellular eukaryotes, but well within the range of possibility for prokaryotes (see Chapter 4). Thus, the effective population sizes of most eukaryotes are far too small for differential mutational vulnerability to significantly influence the evolution of 5′ UTR lengths, which in turn helps explain the approximate phylogenetic constancy of distributions of eukaryotic 5′ UTR lengths. In contrast, the long-term effective population sizes of most prokaryotes, and perhaps a few unicellular eukaryotes, may be adequate to allow the selective retention of alleles with 5′ UTRs of minimal length. As noted above, a few protist lineages may have average 5′ UTR lengths too small to be accommodated by the null model (Ghosh et al. 1994; Singh et al. 1997; Liston and Johnson 1999; Yee et al. 2000; Adam 2001).

These results support the idea that the reduction in N_g that accompanied the evolution of eukaryotes, particularly multicellular species, produced a population genetic environment conducive to the movement of TISs to random positions, subject only to the constraint imposed by the stochastic mutational production of PSCs. If this hypothesis is correct, then selection for gene-specific regulatory features need not be invoked to explain either the expansion of eukaryotic 5′ UTRs relative to those of prokaryotes or the thousandfold range of 5′ UTR lengths among genes within species. Nevertheless, once permanently established, expanded 5′ UTRs may have provided novel substrate for the evolution of mechanisms for posttranscriptional regulation of eukaryotic gene expression, providing still another example of how a reduction in population size can passively promote the emergence of forms of gene architecture that ultimately facilitate the evolution of organismal complexity.

Numerous structural features of 5′ UTRs (including their lengths) in today's eukaryotes influence rates of protein synthesis by modifying the efficiency of translation (Hughes 2006). For example, upstream open reading frames (uORFs) can slow the rate of translation by causing the ribosome to terminate and reinitiate, and secondary structure and/or internal ribosome entry sites may have similar indirect roles. However, although a number of authors have suggested that uORFs serve an adaptive function (Morris and Geballe 2000; Meijer and Thomas 2002; Vilela and McCarthy 2003), confirmation of the extent to which any features of 5′ UTRs are adaptive awaits empirical investigation. Because uORFs are generally on the order of 20 codons in length (approximately what is expected by chance), and because transcripts from some uORF-containing genes are degraded by the nonsense-mediated decay pathway (Ruiz-Echevarria and Peltz 2000), it is premature to reject the hypothesis that a substantial fraction of uORFs are either neutral or mildly deleterious. Although some uORFs appear to be evolutionarily conserved across lineages (Churbanov et al. 2005; Galagan et al. 2005), many of them may exist simply because downstream termination codons have neutralized the effects of PSCs, a view supported by the fact that upstream AUGs that are not followed by termination codons are much rarer than those that are (Iacono et al. 2005).

Transcription Termination

As important as the localization of proper transcription initiation sites is the need for an accurate mechanism to ensure that transcripts elongate beyond their translation termination sites. In eubacteria, a secondary structure involving one or more hairpins assembled beyond the translation termination codon in the developing mRNA combines with a downstream U-rich sequence to provide a sufficient signal for terminating transcription (Unniraman et al. 2002). Some archaeal genes also employ such structures (Clarens et al. 1995), but eukaryotic genes do not. Instead, prior to their export to the cytoplasm, eukaryotic mRNAs undergo two processing events at the 3′ end: removal of part of the trailing noncoding end from the immature transcript and addition of a poly(A) tail to the remaining mRNA. To accomplish these tasks, eukaryotic genes must contain sequence information specifying the proper points of both transcription termination and cleavage, as premature engagement in either process can yield a transcript devoid of a termination codon and/or a poly(A) tail. In principle, the same signal could serve both functions.

The mechanisms of eukaryotic transcription termination have been explored extensively only in mammals and yeasts, but in both cases a discrete polyadenylation signal defines the point of cleavage and poly(A) addition (Keller and Minvielle-Sebastia 1997; Graber et al. 1999; Wahle and Rüegsegger 1999; Zhao et al. 1999). Polyadenylation is essential to eukaryotic mRNA maturation: only polyadenylated messages are exported to the cytoplasm, and a poly(A) binding protein is involved in ribosome recruitment (Mendez and Richter 2001). In mammals, the poly(A) signal is almost always an AAUAAA sequence residing 10–30 nucleotides upstream of the cleavage site. This signal is generally sandwiched between a U-rich stimulatory upstream sequence and a U- or GU-rich downstream element (DSE) about 30 bp 3′ to the cleavage site. Cleavage occurs at an imprecise point between the poly(A) signal and the DSE (usually at an A in vertebrates). The transcription of many animal mRNAs is terminated prior to cleavage (Sadowski et al. 2003), but others appear to experience cleavage first (Osheim et al. 2002; Tollervey 2004). The exact position of transcription termination appears to be fairly random, often occurring at sites hundreds to thousands of base pairs downstream of the poly(A) signal (Proudfoot 1989; Dye and Proudfoot 2001; Carninci 2006).

Numerous variations in 3′ end processing exist. The upstream sequences and the poly(A) signals are more loosely defined in *S. cerevisiae* than in mammals, and there is no obvious DSE. However, the end sequences of *S. pombe* primary transcripts do contain DSEs and are much more similar to those in mammals than to those in *S. cerevisiae* in other ways as well (Humphrey et al. 1994). Some variants of the poly(A) signal have also been found in various protist lineages. In the trichomonad *Trichomonas vaginalis*, for example, UAAA serves as the signal for cleavage 10–30 nucleotides downstream (Espinosa et al. 2002), with the initial three nucleotides (UAA) also serving as the translation termination codon. The diplomonad *Giardia lamblia* uses AGURAAY, where R denotes a purine and Y a pyrimidine, about 10 nucleotides upstream of the cleavage site (Que et al. 1996). In prokaryotes, only a fraction of transcripts for a given gene become polyadenylated (often <50%), and poly(A) is not essential for growth (Sarkar 1997). Finally, downstream UTRs are entirely lacking in the transcripts of most mitochondrial genomes; instead, polycistronic transcripts are cleaved into individual components, and poly(A) tails are added directly to a terminal U or UA to form termination codons (Gillham 1994; Sarkar et al. 1997).

Average lineage-specific lengths of 3′ UTRs are about threefold longer than those of 5′ UTRs, falling in the narrow range of 200–500 bp with CVs generally close to 1.0 (Table 10.2). As in the case of 5′ UTRs, there is no obvious pattern in the data with respect to phylogeny or organismal complexity, and species-specific distributions are typically unimodal with a long tail to the right (Figure 10.6). However, a simple null model, like that presented above for the 5′ UTR, does not provide an immediate explanation for these patterns. For example, if the four nucleotides are present at equal frequencies, the expected distance from the translation termination codon to the first random consensus poly(A) signal, AAUAAA, would be about $1/(1/4)^6 = 4096$ bp, and the distribution of 3′ UTR lengths would be exponential (with a peak at 0). This rough calculation does not consider the additional con-

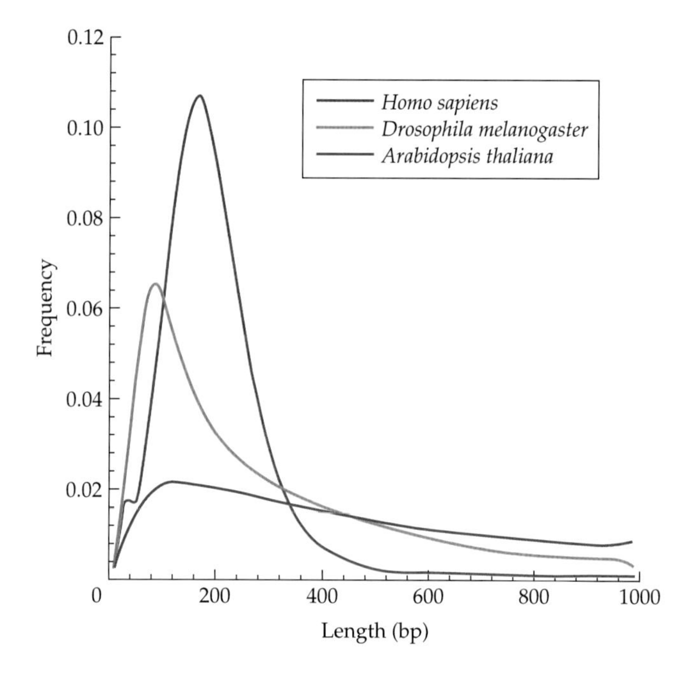

Figure 10.6 Frequency distributions of lengths of 3′ UTRs, derived from the database of Pesole et al. (2002).

straint imposed by the upstream and downstream U-rich patches that help define the poly(A) site, but additional sequence complexity would only increase the mean of the random null distribution of lengths.

How can the unexpectedly low means of 3′ UTR length distributions and the relative paucity of the shortest classes be explained? Overly short 3′ UTRs may be problematic because the maintenance of specific polyadenylation and cleavage signals upstream of the translation termination codon would compromise the coding sequence, but the rarity of long 3′ UTRs is less easily explained. Little is known about the mechanism by which the poly(A) signal is located during transcription, but it appears to occur prior to (or simultaneously with) the removal of introns (Zhao et al. 1999; Maniatis and Reed 2002; Proudfoot et al. 2002). However, because only about 1%–10% of the genes of animals and land plants have 3′ UTR introns (Pesole et al. 2001), intron removal does not account for the unexpectedly short mean lengths of 3′ UTRs. The proximity of adjacent genes must impose some selection for the avoidance of interference between the termination of transcription of one gene and initiation of transcription of another, but the space between the genes of multicellular species is generally much greater than the distributions in Figure 10.6. Genes with longer 3′ UTRs are transcribed more slowly (Chiaromonte et al. 2003), but their translation efficiency and mRNA stability are enhanced, although the effect is small beyond a length of about 27 bp (Tanguay and Gallie 1996a,b). The major problem imposed by overly long 3′ UTRs may be the appearance of spurious regulatory motifs (e.g., micro RNA binding sites; see Chapter 3).

TABLE 10.2 The average lengths of 3′ UTRs (bp) in various eukaryotic lineages and their average coefficients of variation (ratio of standard deviation to mean) within species

	NUMBER OF GENERA	MEAN	CV
Vertebrates			
Mammals	20	446	1.10
Birds	4	435	1.03
Frogs	4	479	1.13
Ray-finned fish	24	424	1.02
Agnathans	2	470	0.81
Invertebrates			
Invertebrate chordates	4	502	0.83
Echinoderms	4	850	1.10
Arthropoda	34	333	1.08
Mollusca	9	335	1.08
Nematoda	10	191	1.06
Platyhelminthes	5	214	1.02
Porifera	3	151	0.59
Vascular plants	76	232	0.56
Unicellular species			
Fungi	21	207	0.75
Slime molds	2	104	1.26
Green algae	3	571	0.64
Apicomplexans	5	327	1.21
Kinetoplastids	2	459	1.07
Ciliates	3	170	1.23

Source: Derived from the database of Pesole et al. 2002.

Contrary to the situation in 5′ UTRs, in which PSCs are selected against, one would not expect a bias in the incidence of translation initiation or termination codons within 3′ UTRs because sequences within them are expected to be translationally irrelevant. However, there is a 10%–30% excess of AUG triplets throughout the 3′ UTR relative to the random expectation, whereas termination codon triplets tend to be underrepresented by up to 25% in the first 100–200 bp (Figure 10.7). One simple mechanism that could drive the termination codon deficit in the terminal portion of genes is the stochastic gain and loss of termination codons: mutational gain of an in-frame termination codon upstream of the prior termination codon will yield a downstream UTR region relatively devoid of termination codons in the vicinity of the new one, assuming that alleles with 3′-truncated coding

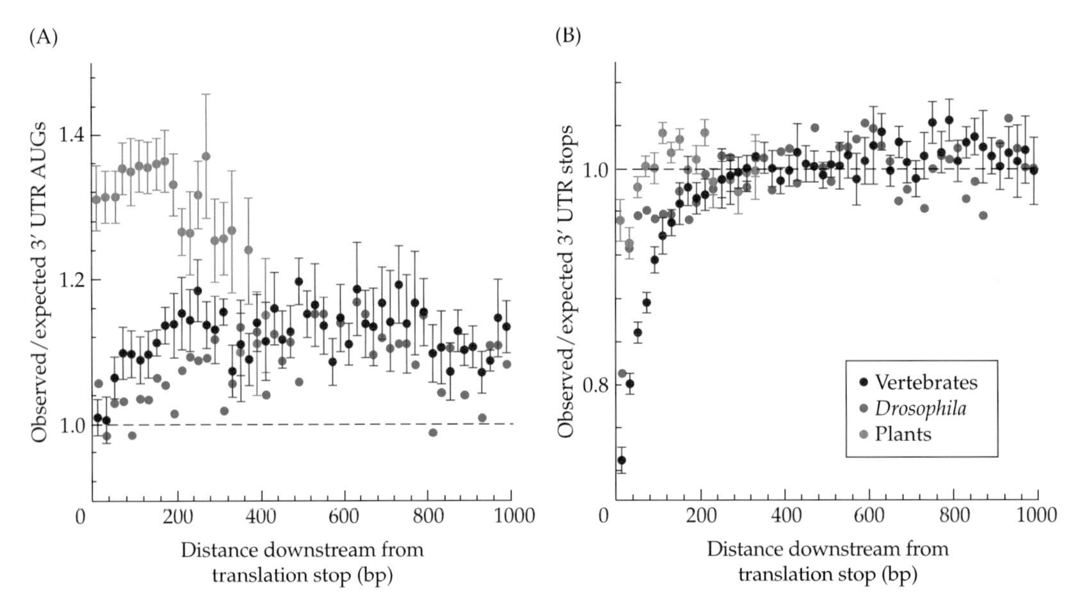

Figure 10.7 The fractions of observed initiation codon (AUG) triplets (A) and termination codon triplets (B) relative to the random expectations (dashed lines) at 20 bp intervals downstream of the true translation initiation site. (Data derived from the database of Pesole et al. 2002.)

regions are not always eliminated by selection. But how can an overabundance of initiation codon triplets in 3′ UTRs be explained? Rather than being simple linear molecules, the ends of eukaryotic mRNAs are bound together by a number of proteins to effectively form a circularized message (Preiss and Hentze 1999; Sachs and Varani 2000; Mendez and Richter 2001; Mazumder et al. 2003; Moore 2005). Such a structure appears to facilitate the return of the ribosomal subunits to the 5′ end of the message after each round of translation (Tanguay and Gallie 1996b). This process of recycling may also present the ribosomal subunits with the opportunity to interact with AUGs within the 3′ UTR prior to returning to the proper initiation codon, which although initially unproductive, could serve to keep the translational machinery localized. However, although the translation of uORFs (described above) formally demonstrates that reinitiation of scanning does occur within mRNAs, it remains to be seen whether such reengagement can also take place with downstream ORFs, as reinitiation and recycling are somewhat distinct processes (Rajkowitsch et al. 2004).

Finally, we return to the issue of the extreme rarity of introns within 3′ UTRs. Nagy and Maquat (1998) pointed out that the final introns within transcripts of human genes are generally no more than 50 bp downstream of the translation termination codon, and suggested that this is a necessary consequence of the reliance on a mRNA surveillance mechanism dependent on exon junction complexes (EJCs). Recall from Chapter 9 that EJC-dependent nonsense-mediated decay (NMD) relies on protein markers deposited upstream of each exon–exon junction; under this model, if the true termination codon resided more than 50 bp upstream of an intron, the mRNA would be identified as defective and subjected to decay. Nevertheless, although this model may be mechanistically correct for mammals, it appears that most eukaryotes, even those without EJC-dependent NMD, have very few 3′ UTR introns (Hong et al. 2006). A more general explanation for the paucity of 3′ UTR introns may simply be their evolutionary instability: in contrast to the situation in 5′ UTRs or coding regions, where the failure to splice out an intron sequence will generally lead to a nonfunctional allele, the mutation of a key splicing signal for a 3′ UTR intron (and subsequent integration of the intronic sequence into the UTR) may have negligible consequences for gene function, as the protein sequence remains unaltered.

In summary, as in the case of 5′ UTRs, the origin of eukaryotes was accompanied by an expansion in the size of 3′ UTRs. Eukaryotes are also unique in relying on a discrete polyadenylation signal downstream of the coding region. However, a fully satisfying explanation for the evolutionary origin of the features of 3′ UTRs remains to be developed. In today's eukaryotes, information encoded in 3′ UTRs plays a role in nuclear export and in the initiation, regulation, and subcellular localization of translation of many mRNAs (Jansen 2001; Macdonald 2001; Mendez and Richter 2001; Gebauer and Hentze 2004; Xie et al. 2005). But these novelties presumably arose secondarily, after the establishment of mRNA downstream sequence.

The Deployment of *trans* Splicing

The widespread process of removing introns and joining adjacent segments of exonic RNA, discussed in Chapter 9, is known as *cis* splicing (because the interacting cut sites are located on the same transcript). In some species, however, segments of different transcripts can be joined by *trans* splicing. The most common form of this process is spliced-leader (SL) *trans* splicing, whereby a short leader sequence, derived from a small nuclear RNA (snRNA), is added to the 5′ end of the transcript of a protein-coding gene.

The mechanism of *trans* splicing is remarkably similar to that of *cis* splicing: a small (15–45 nucleotide) leading sequence of the SL snRNA transcript (the spliced leader) is detached at a junction adjacent to a GU, analogous to a typical 5′ exon–intron junction, whereas the pre-mRNA is severed at a typical AG 3′ splice site (see Chapter 9). Thus, the discarded downstream portion of the SL snRNA functions as a 5′ end of an intron, whereas the discarded upstream portion of the pre-mRNA, often referred to as an outron, is equivalent to the 3′ end of an intron (Figure 10.8). Almost all of the spliceosomal components used in *cis* splicing are also employed in *trans* splicing (Furuyama and Bruzik 2002), although two unique proteins are associated with the SL transcript in nematodes (Denker et al. 2002). In addition, the SL snRNA of all characterized species has a modified nucleotide cap on the SL,

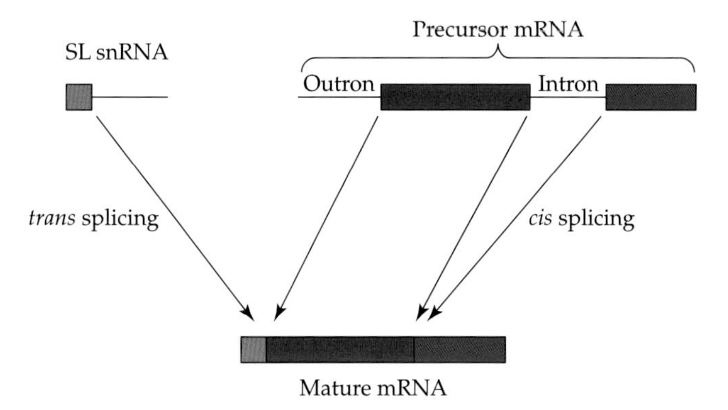

Figure 10.8 Spliced-leader (SL) *trans* splicing. Colored boxes denote sequences destined for mature mRNAs; horizontal lines denote intronic sequences. The spliced-leader sequence is transcribed from a different locus than the pre-mRNA, and its leading sequence (the spliced leader) is *trans*-spliced to the front end of the mRNA. *Cis* splicing denotes conventional splicing of an internal intron.

a conserved binding site for accessory proteins used in splicing (see Chapter 9), and a characteristic secondary structure. The similar mechanisms for SL *trans* splicing across distantly related lineages suggest that such processing was deployed in the stem eukaryote for at least some genes and probably has a common evolutionary origin with *cis* splicing.

Although SL *trans* splicing has not been observed in vertebrates or vascular plants, it appears to be phylogenetically widespread. Among animals, it has been found in some genes in cnidarians (Stover and Steele 2001), nematodes (Evans et al. 1997; Blumenthal and Gleason 2003; Lee and Sommer 2003), flatworms (Davis et al. 1995; Brehm et al. 2000; Zayas et al. 2005), rotifers (Pouchkina-Stantcheva and Tunnacliffe 2005), and urochordates (Vandenberghe et al. 2001; Ganot et al. 2004; Satou et al. 2006). It also occurs in euglenoids (Tessier et al. 1991; Frantz et al. 2000), and every mRNA is *trans*-spliced in trypanosomes (Campbell et al. 1984; Kooter et al. 1984; Milhausen et al. 1984). Because *trans* splicing is usually detected only after the discovery of identical upstream sequences on mature mRNAs of unrelated genes, it is likely that the phenomenon will be found in many additional eukaryotes following the characterization of full-length cDNA libraries.

Although not all eukaryotes engage in SL *trans* splicing, the ability to participate in other forms of *trans* splicing may be widely retained. For example, a few instances of *trans* splicing of internal exons from different transcripts have been discovered in *Drosophila* and other dipterans (Dorn et al. 2001; Horiuchi et al. 2003; Krauss and Dorn 2004), rodents (Takahara et al. 2002), and humans (Finta and Zaphiropoulos 2002). Such *trans* splicing can occur between the products of different alleles at the same locus or between transcripts from entirely different chromosomal locations. Remarkably, SL snRNAs from *C. elegans* and trypanosomes can function in mammalian cells to *trans*-splice an appropriate outron to transcripts (Bruzik and Maniatis 1992), suggesting the possibility that SL *trans*-splicing alleles could reemerge in various lineages if the SL snRNA genes were still present.

A clue to the functional significance of *trans* splicing is provided by the unusual genomic features of the species that employ the process. Most notably, although few nuclear genomes have operons (cassettes of cotranscribed genes), all species known to have them (kinetoplastids, nematodes, and some urochordates) process the component genes by SL *trans* splicing. Such an intermediate step in transcript individualization may be essential for eukaryotic operons because, unlike prokaryotic transcripts, eukaryotic mRNAs need to be processed prior to translation. The transcripts of genes within eukaryotic operons face an unusual problem at their 3′ ends. As noted above, the trailing tails of single-gene transcripts are generally rapidly degraded following cleavage from the upstream functional transcript. Thus, for genes internal to operons, some mechanism must exist to prevent destruction of a transcript following its cleavage from that of the upstream gene. The process has been studied most closely in *C. elegans*, in which a conventional poly(A) signal elicits cleavage and polyadenylation of the upstream transcript prior to processing of the downstream remnant, and an intervening U-rich intergenic region prevents downstream degradation and directs the attachment of the next spliced leader (Huang et al. 2001; Liu et al. 2003). Approximately 15% of *C. elegans* genes are contained within operons, with the first genes within operons utilizing one spliced-leader sequence and most internal genes utilizing another (Blumenthal and Steward 1997; Redmond and Knox 2001; Blumenthal et al. 2002).

Despite this association between operons and *trans* splicing, it remains unclear why these features evolved to such a significant extent in some lineages and not in others (Nilsen 2001). SL *trans* splicing requires a cellular investment that would otherwise be unnecessary: prior to their use in *trans* splicing, the SL snRNAs are subject to extensive processing (e.g., 5′ cap formation, binding of proteins necessary for splicing, and 3′ end trimming; Sturm et al. 1999; Zeiner et al. 2003). Nevertheless, other than the provisioning of the 5′ cap, which can be acquired by more conventional mechanisms for solo genes, the functions of the spliced leader have been elusive. In *Leishmania*, the SL sequence appears to contain determinants for efficient association with ribosomes (Zeiner et al. 2003), but such sequences have not been identified in other organisms. The 3′ end sequence of flatworm SLs is an AUG, which appears to be used as the translation initiation codon in at least some transcripts (Davis et al. 1995; Brehm et al. 2002; Cheng et al. 2006), but in other lineages the translation initiation codon is generally contained within the primary transcript.

The premature start codon (PSC) problem, discussed above in the context of 5′ UTR evolution, may provide a key to the mystery of SL *trans* splicing. With this method of transcript processing, most of the pre-mRNA sequence upstream of the translation initiation codon is eliminated via the removal of the outron. For example, in *C. elegans*, the distance between the *trans* splice site and the translation initiation codon averages only 10 bp and is seldom more than 30 bp (Blumenthal and Steward 1997). Moreover, in all known cases, the SL sequence added to the 5′ UTR is less than 50 bp (Table 10.3). Thus, SL *trans* splicing results in mature mRNAs with unusually short 5′ UTRs, thereby minimizing the likelihood of occurrence of a harmful PSC.

TABLE 10.3 Lengths (bp) of spliced-leader snRNAs for a variety of eukaryotes

	snRNA[a]	SL	REFERENCE
Euglenoids (six species)	95–107	25–27	Frantz et al. 2000
Trypanosomes (several species)	99	39	Gibson et al. 2000
Rotifers (two species)	106	23	Pouchkina-Stantcheva and Tunnacliffe 2005
Nematodes			
Caenorhabditis elegans			Blumenthal and Steward 1997
SL1	95	22	
SL2	110	22	
Haemonchus contortus	113	22	Redmond and Knox 2001
Flatworms			
Echinococcus multilocularis	104	36	Brehm et al. 2000
Taenia solium	96	33	Brehm et al. 2002
Schistosoma mansoni	92	33	Rajkovic et al. 1990
Hydra vulgaris			Stover and Steele 2001
SL-A	80	24	
SL-B	106	45	
Urochordates			
Ciona intestinalis	46	16	Vandenberghe et al. 2001
Oikopleura dioica	92	40	Ganot et al. 2004

[a]Refers to the entire length of the SL snRNA prior to the splicing of the leader sequence (SL) to the primary transcript.

Nevertheless, as discussed above, the extent to which this advantage translates into the evolution of SL *trans*-splicing alleles depends on the magnitude of the reduction in the mutation rate to PSCs relative to that for conventional transcripts. Experimental work with *C. elegans* shows that there are no significant barriers to the conversion of one type of allele to the other: the only difference between *trans*-spliced and conventional alleles is the presence of an outron (Conrad et al. 1991, 1993, 1995). Thus, without a significant enough selective advantage, a locus relying on SL *trans* is vulnerable to invasion by mutant non-*trans*-splicing alleles. Such conversions arise by two mutational pathways: (1) the outron may gain a 5′ splice site, which converts it to an upstream exon followed by a 5′ UTR intron; or (2) the outron may lose its 3′ splice site and be added to the leading end of the initial exon (Figure 10.9). Both paths would immediately yield null alleles if the recruited exon sequence contained a harmful PSC, and if a functional non-*trans*-splicing allele did initially emerge, an elevated mutation rate to nulls

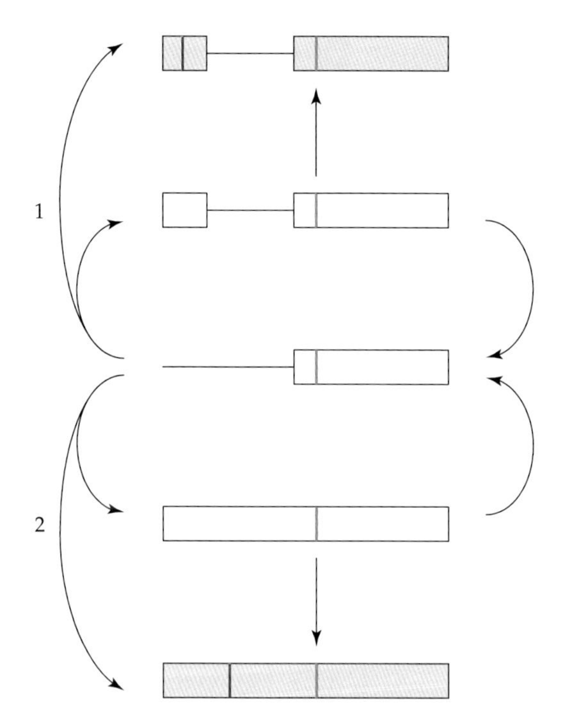

Figure 10.9 The paths by which *trans*-spliced and conventional alleles can be interconverted. For the *trans*-spliced allele in the center, the horizontal line on the left denotes the outron and the box on the right the first exon, with the vertical green line denoting the translation initiation codon. (1) Gain of a 5′ splice site in the outron. (2) Loss of the 3′ splice site from the outron. The shaded boxes denote null alleles resulting from the exposure of a premature start codon (red bar) upon the addition of new exonic DNA.

would result from the 5′ UTR expansion—all the more so for alleles losing the 3′ splice site.

Although the formal details of this model remain to be worked out, these simple arguments suggest that if the efficiency of selection were sufficiently weak (due, for example, to small effective population size), recurrent mutation to non-*trans*-splicing alleles could eventually result in the fixation of such derived alleles despite their weak mutational disadvantage. However, because the conversion of alleles from one functional type to another is a two-way street (i.e., through gain or loss of relevant splice sites, non-*trans*-splicing alleles can be converted back to the *trans*-splicing type), a quasi-equilibrium may be reached, with an approximately constant fraction of loci engaging in SL *trans* splicing. All lineages known to engage in SL *trans* splicing except trypanosomes apply it to only a moderate fraction of genes, although it remains to be seen whether these lineages represent cases of equilibria or transient phases en route to loss or fixation of SL dependence. In any event, should a point be reached at which SL *trans* splicing is lost from all loci, the SL locus would be released from selection and vulnerable to loss by inactivating mutations. Given the complexity of the SL locus, such a condition would probably lead to permanent loss of this mode of gene processing, a state that has apparently been reached by most eukaryotic lineages.

Evolution of Modular Gene Organization

In the preceding sections, we have considered a number of basic features of eukaryotic transcripts without regard to the mechanisms that dictate spatial and temporal patterns of gene expression. Yet the ability to regulate the transcription of various suites of genes in a tissue-specific, development-specific, and/or environment-specific manner is the hallmark of organisms with multiple cell types. To accomplish such tasks, the noncoding DNA of individual genes generally harbors numerous *cis*-acting elements that cooperatively interact with multiple *trans*-acting factors to finely tune levels of transcription. Collectively, the transcription factor binding sites of a gene are generally referred to as an enhancer. As noted above, the core promoter receives information from the enhancer and serves as the binding site for RNA polymerase and the basal transcriptional machinery.

As emphasized in Chapter 8, the regulatory elements of eukaryotic enhancers are often organized into modular units, so that individual mutations influence only a subset of a gene's total phenotypic effects. The presence of such gene subfunctions promotes the preservation of duplicate genes by degenerative mutations. But how do the modular subunits of gene expression on which subfunctionalization depends arise in the first place? Many evolutionary developmental biologists have argued that the emergence of this particulate form of gene regulation in multicellular species was a fundamental prerequisite for the origin of developmental modules capable of independent evolutionary trajectories, and some have suggested that such organization has been promoted by natural selection as a means for enhancing the evolvability of lineages (Raff 1996; Wagner 1996; Wagner and Altenberg 1996; Gerhart and Kirschner 1997; Hartwell et al. 1999; Niehrs and Pollet 1999; Carroll et al. 2001; Davidson 2001; Lipson et al. 2002; Cohen-Gihon et al. 2005). However, the causal link between genotypic and phenotypic modularity remains unclear, and prior to invoking natural selection as the driving force in the origin of modular gene structure, a formal theoretical framework incorporating known population genetic mechanisms needs to be considered. An essential first step is to evaluate the situation in which selection is a negligible force (Zuckerkandl 2001; Force et al. 2005). As we will see below, a model based entirely on drift and mutation yields a simple potential explanation for the emergence of the modular nature of gene expression, providing a null model against which to contrast the predictions of verbal adaptive arguments.

Before we explore the population genetic issues, a brief overview of eukaryotic transcriptional control is in order. Transcription factor genes are generally expressed at specific stages of development at a hierarchy of levels of organization. Some transcription factors are organ-specific, some are specific to individual tissues within an organ, some are expressed in specific cell types within a tissue, and still others have narrow subcellular patterns of localization. By virtue of their ability to bind DNA and/or to act as cofactors in binding one another together in functionally significant ways,

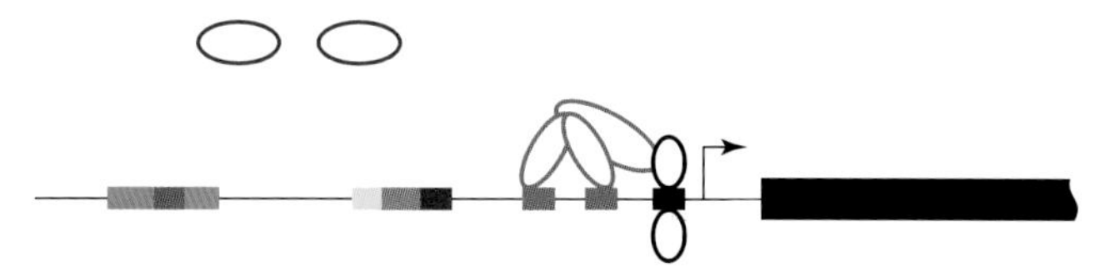

Figure 10.10 The anatomy of an idealized gene regulatory region. The long black bar at the right denotes the coding region, and the short black bar is the generic core promoter through which all transcriptional information passes to initiate transcription at the point of the arrow. The small colored bars denote transcription factor binding sites (TFBSs) that are specific to transcription factors of the same color (open ovals). As an example, each of the blue TFBSs is shown to be bound to a specific transcription factor, and the pair of factors then binds to a cofactor, producing a complex that transmits information to the core promoter. The green, purple, yellow, and red TFBSs would activate transcription within cells expressing their appropriate transcription factors. However, this gene would be silent in a cell expressing only the pink transcription factors, as the complementary TFBSs are absent from the gene's regulatory region.

the full suite of transcription factors and their combinatorial expression provide a heterogeneous informational environment within the organism. Eukaryotic genes exploit this information by incorporating arrays of transcription factor binding sites (hereafter, TFBSs) to finely tune their expression at various levels of morphological, physiological, and behavioral organization. For example, each of the color-coded TFBSs in the gene shown in Figure 10.10 can be viewed as a substrate for tissue-specific transcription factors denoted by the same color; the gene will be turned on only in cells expressing an appropriate mix of transcription factors. Further fine-tuning is achieved through the use of transcription factors that act as repressors of gene expression and through modifications in the spatial distribution of cooperatively acting TFBSs. The data suggest that the match between enhancer elements and transcription factors serves simply to modify the probability that a gene will be expressed (i.e., as a binary on/off switch) rather than to modify the actual rate of transcription once engaged (Walters et al. 1995; Fiering et al. 2000).

Although the entire regulatory domain of a gene may substantially exceed the length of the coding region, each of the binding sites associated with individual transcription factors is usually no more than a dozen base pairs in length (e.g., Wilkins 2002; Gasch et al. 2004). Individual genes can harbor dozens of such elements, some of which can be partially overlapping (e.g., the yellow and red elements that overlap to produce the orange region in Figure 10.10), entirely embedded (e.g., the purple element within the green element), on opposite strands, and/or in inverse orientation. Nonoverlapping elements, single or otherwise (e.g., the blue elements), fulfill our definition of a subfunction, as each nucleotide in their constituent binding sites can be modified without affecting expression in other tissues. Because small regulatory elements can arise by de novo mutation or transpositional insertion, there are many potential ways in which the numbers, locations, and types of TFBSs for a gene can become modified (Brosius 1999; von Dassow and Monro 1999; Edelman et al. 2000; Stone and Wray 2001; Rockman and Wray 2002; Jordan et al. 2003; Wray et al. 2003; Harbison et al. 2004; MacArthur and Brookfield 2004). By this means, entire suites of genes harboring functionally similar but evolutionarily independent regulatory elements can come to be turned on and off in a highly synchronous fashion. Likewise, modifications of the DNA-binding domains of individual transcription factors can result in cells, tissues, and/or organs with novel features.

The passive emergence of modularity

Redundant copies of binding sites for particular transcription factors are often present in gene regulatory regions, and various permutations of TFBSs can often yield functionally equivalent patterns of gene expression (Bonneton et al. 1997; Hancock et al. 1999; Ludwig et al. 2000, 2005; P. J. Shaw et al. 2002; Dermitzakis and Clark 2002; Arnosti 2003; Dermitzakis et al. 2003b). It then follows that the stochastic turnover of transcriptional control elements by nearly neutral processes can play a significant role in the evolution of regulatory region architecture. This is particularly true in multicellular species because, although some regulatory region structures will endow an allele with higher fitness than others, the range of effectively equivalent states will increase in populations with small effective sizes, in which the efficiency of selection is reduced. Provided these effectively neutral solutions are still sufficient for gene function, the intrusion of new TFBSs can proceed passively, in ways that may be prevented by selection in much larger populations.

The central idea to be explored here is that the same kinds of duplication and degeneration processes that lead to the subfunctionalization of duplicate genes promote the emergence of the subfunctions themselves. To simplify the discussion, we will assume that subfunctions are defined by transcription factor binding sites (or integrated regions of such sites) that are separable from other such sites, both mutationally and functionally (Yuh et al. 1998; Force et al. 1999; Arnosti 2003). However, as emphasized in Chapter 8, the same principles apply to subfunctions defined by functional motifs in coding regions or by alternative splice-site junctions, transcription initiation sites, polyadenylation signals, and so forth.

Before proceeding, it is essential to recall that we are not attempting to explain the expression of a gene in a new temporal or spatial context. Rather, we are trying to understand how a gene that is initially under the control of ubiquitously expressed transcription factors comes to be regulated by tissue-specific transcription factors while retaining the same overall expression pattern. The process envisioned here, subfunction fission, involves conventional processes of descent with modification: the progressive structural modification of a preexisting enhancer via consecutive processes of partial duplication and partial loss of regulatory information (Force et al. 2005). The first phase involves the accretion of new semi-redundant TFBSs combined with the degeneration of one or more ancestral sites, which yields a semi-independent enhancer (Figure 10.11, top). The second phase involves tan-

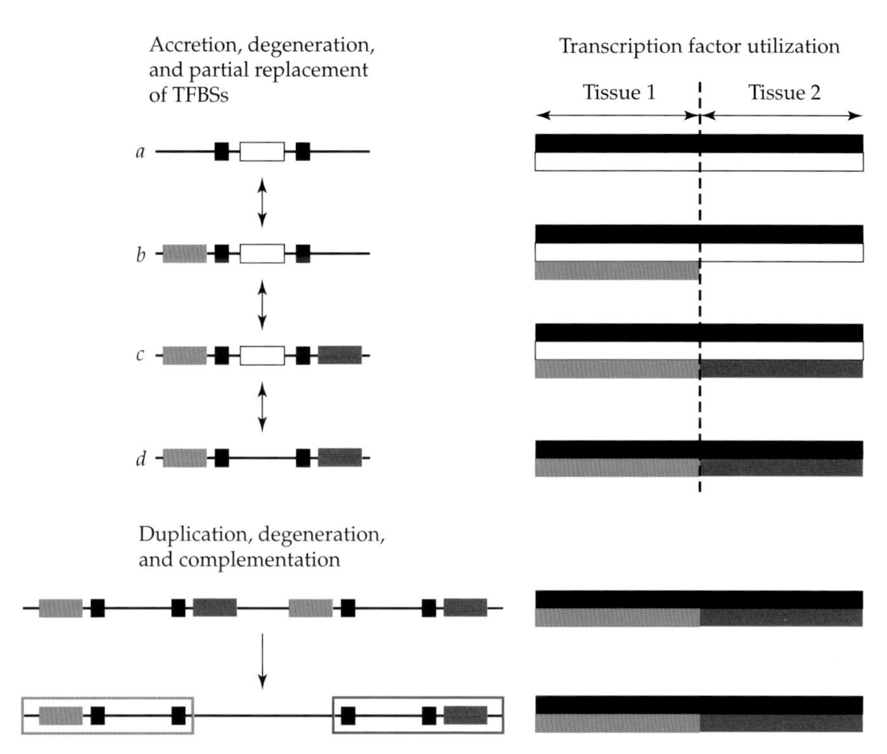

Figure 10.11 A hypothetical scenario in which an allele with two independently mutable subfunctions can arise from an allele with a single generalized expression mechanism. Regulatory regions are depicted on the left, with each regulatory element color-coded according to the transcription factor that binds to it. On the right, the spatial patterns of allele-specific utilization of transcription factors are color-coded. Transcription factors denoted by black and white are ubiquitously expressed, whereas those denoted by green and red are expressed in single, nonoverlapping tissues. Expression of the original allele requires three transcription factor binding sites (one white and two black). In the first phase of allelic evolution, the regulatory region undergoes the sequential accretion of green and red elements, which together are redundant with respect to the white element, which is then lost in a neutral fashion by degenerative mutation. At this point, the descendant allele has a semi-independent mode of expression, as the two black elements are still essential to expression in both tissues. In the second phase, the entire enhancer region is tandemly duplicated, with each component then losing a complementary (red or green) element. The resultant allele has two independent subfunctions, denoted by the green and red open boxes, as a mutation in either region has effects that are confined to a single tissue. Note that throughout all of these stages of allelic evolution, there has been no change in the spatial pattern of expression of the gene.

dem duplication of the semi-independent enhancer, followed by the formation of two entirely independent regulatory subfunctions by complementary degenerative mutations (Figure 10.11, bottom). Other than the fact that smaller regulatory elements are involved, the events during the second phase are conceptually identical to those underlying the subfunctionalization of duplicate genes (see Chapter 8).

We start by considering how informational accretion can transform an allele *a* with a single subfunction (with its expression in all tissues driven by the same transcription factors) into an allele with two overlapping regulatory subfunctions (Figure 10.11, top left). Here we assume a gene with essential functions in two tissues, which each express a unique mix of transcription factors (those expressed only in tissue 1 are shown in green, those expressed only in tissue 2 in red, and those expressed ubiquitously in black and in white). Under this model, the ancestral allele is subject to the stochastic accretion of TFBSs for the tissue-specific (green and red) transcription factors. However, because the green and red regulatory elements are together functionally redundant with respect to the white (but not the black) element, once allele *c* is established, the white element can be lost by degenerative mutation. At this point, the black elements, which represent other essential TFBSs (nonredundant with respect to the white, green, or red elements), must be retained and shared. The product at the end of the first phase is thus an allele *d* with partially overlapping subfunctions; mutations in either the green or red elements will alter expression in single tissues, while mutations in the black elements will alter expression in both tissues.

Because the binding sites for all three types of transcription factors are subject to both mutational gain and loss, there is not necessarily a permanent allelic state under this model, as the alternative classes of shared and semi-independently regulated alleles are free to mutate back and forth (hence, the two-way arrows at the top left of Figure 10.11). Nevertheless, it is informative to consider the circumstances under which the semi-independently regulated allele *d* is likely to rise to high frequency, as its presence is essential to completing the transition to an allele with two entirely independent subfunctions. Some of the technical details are outlined in Box 10.1, and further analysis is provided in Force et al. (2005). The key observation is that rates of mutational origin and loss of alternative TFBSs appear to be sufficiently small that the population genetic environments of most multicellular organisms enable alternative alleles like those at the upper left of Figure 10.11 to drift freely back and forth to transient states of fixation in an effectively neutral fashion. This observation implies that as soon as transcription factors with spatially and/or temporally restricted patterns of expression evolved in eukaryotes, large numbers of broadly expressed genes would have become vulnerable to the accumulation of alleles with semi-independent regulation even in the absence of any direct selection for such gene architecture.

If, at the close of the accretion/degeneration phase, no additional ubiquitously expressed transcription factors were critical to a gene's expression (e.g., if the black TFBSs in Figure 10.11 were absent), the descendant allele would have arrived at a state of modular architecture, with expression in the two tissues being independently regulated. However, the residual presence of one or more shared transcription factors would impose a requirement for a second phase of events for the emergence of two completely independent enhancers. In the example shown in Figure 10.11 (lower left), duplication of the entire regulatory domain of allele *d* opens up such an opportunity: the duplicate regulatory regions lose complementary tissue-specific TFBSs (red on the left and green on the right), yielding a modular pair of enhancers, each driving expression in a single tissue and each independently mutable with respect to the other. Because the rate of tandem duplication is inversely related to the length of a duplication span (Katju and Lynch 2003) and the rate of tandem duplication of gene-sized (>1 kb) fragments is already quite high (on the order of 1% per million years; see Chapter 8), the rate of spontaneous entry into this second phase (conditional on the presence of allele *d*) is likely to be at least as rapid as the completion of the first phase.

However, not all novel alleles with modular architectures will be free to become established throughout a population. Alleles with modular architectures will almost always have elevated total numbers of TFBSs (e.g., relative to allele *a*, the final modular allele in Figure 10.11 has twice the number of TFBSs), and hence will be larger targets for mutational degeneration. Following the general principles outlined in previous chapters, if the excess mutation rate to null alleles associated with the expansion of regulatory region complexity is μ_r, then the passive emergence of modular gene architecture will proceed in an effectively neutral fashion if $2N_g\mu_r << 1$ (see Box 10.1). However, in populations with large enough N_g to violate this condition, the probability of enhancer fission declines precipitously (Force et al. 2005).

These observations raise significant questions about both the necessity and the sufficiency of natural selection as a determining force in the emergence of complex patterns of gene regulation. Not only can modular forms of gene structure emerge in the absence of any direct selection for such architectural features, but in sufficiently large populations, such changes are actually opposed by selection (unless they are immediately accompanied by phenotypic advantages that substantially offset the mutational disadvantages). As we will see in the final chapter, this view of gene structure evolution has several implications for our understanding of the connections that may exist between the evolution of complexity at the genomic and organismal levels.

The demise of operons

Coregulation of the expression of genes mutually engaged in metabolic and developmental processes is critical to all organisms, particularly to multicellular species that rely on the cascading effects of gene networks to pro-

Box 10.1 Conversion to Independently Regulated Alleles by Effectively Neutral Processes

A key first step in the evolution of a modular regulatory region is the emergence of semi-independent enhancers from an ancestral allele that is activated by ubiquitously expressed transcription factors. Such a process initially requires the stochastic gain of tissue-specific regulatory elements, either by de novo mutation to an appropriate sequence or by the accretion of such elements via duplication from alternative genomic sites. The following discussion relies on the schematic in Figure 10.11, and for simplicity, it is assumed that the white, green, and red TFBSs are all gained and lost at the same rates, μ_b and μ_d respectively. These rates determine the rates of conversion from one allelic type to another, and a transition matrix approach provides a basis for estimating the average number of generations transpiring between an initial state fixed for allele *a* and a transient state of fixation for allele *d* (Force et al. 2005).

Assuming that two adjacent alleles in an allelic series are selectively equivalent, the expected transition time from one allele to the fixation of a novel derivative is a function of the arrival time of the requisite mutation and its time to drift to fixation. Letting N_g be the effective number of genes per locus in a population, provided that $2N_g\mu_b << 1$ and $2N_g\mu_d << 1$, the waiting times for mutations, $1/\mu_b$ and $1/\mu_d$, are much greater than the time for a neutral mutation to drift to fixation, $2N_g$. In this case, the transition time to the semi-independently regulated state is essentially independent of population size and entirely determined by the relative values of μ_b and μ_d (Figure 10.12). Although the most direct route from allele *a* to allele *d* involves two gains (green and red) and one loss (white) of regulatory elements, such elements can be repeatedly gained and lost, so the average time to fixation of allele *d* exceeds the sum of the arrival times of the requisite changes; that is, it is greater than $(2/\mu_b) + (1/\mu_d)$. In general, the transition time from *a* to *d* declines as the ratio of μ_b to μ_d increases, eventually reaching an asymptote at $1/\mu_d$ generations at high μ_b/μ_d, as the final degenerative step involving the loss of the generalized (white) element becomes the limiting factor.

The conditions $2N_g\mu_b << 1$ and $2N_g\mu_d << 1$ may be frequently met in eukaryotes. From Chapter 4, we know that the average values of $2N_gu$, where u is the substitutional mutation rate per nucleotide, are roughly 0.004, 0.015, 0.026, and 0.057 for vertebrates, land plants, invertebrates, and eukaryotic microbes, respectively. Thus, given that transcription factor binding sites typically contain just ten or so nucleotides, we can expect $2N_g\mu_d$ to be on the order of ten times $2N_gu$, and hence much less than 1 for most multicellular species. (Accounting for small-scale insertions and deletions would not alter this qualitative conclusion; see Chapter 3.) Obtaining an estimate of $2N_g\mu_b$ is less straightforward because a TFBS may appear in many places in the vicinity of a gene. However, if we assume that an appropriate TFBS can arise anywhere within a span of L nucleotides, then for a site containing n specific nucleotides, the spontaneous rate of emergence of appropriate TFBSs from a random DNA sequence with equal usage of all four nucleotides is $\mu_b \approx L(u/3)(0.75n)(0.25)^{n-1}$ (this equation takes into account the probability that a preexisting span of n nucleotides has the appropriate sequence at $n - 1$ sites and the rate at which new mutations at the remaining site give rise to the appropriate nucleotide for completing the TFBS). For a typical TFBS with $n = 10$, $\mu_b \approx 0.01u$ when $L = 10^3$, and $\mu_b \approx u$ when $L = 10^5$, implying that for the de novo origin of TFBSs by stochastic nucleotide substitution, $2N_g\mu_b$ will generally be on the order of $2N_gu$ or less. The insertion of preexisting TFBSs by duplication from ectopic locations will increase μ_b to an unknown, but probably nontrivial, degree.

As N_g increases beyond the point at which $2N_g\mu_b = 1$ and $2N_g\mu_d = 1$, the time required to make the transition to the semi-independently regulated state approaches a

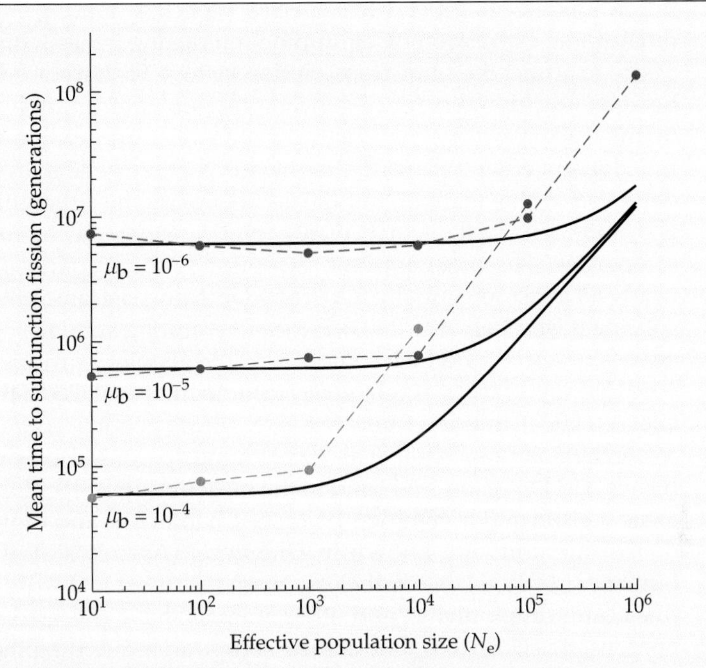

Figure 10.12 Time to transient fixation of a semi-independently regulated allele (*d* in Figure 10.11), starting from a state of fixation for an allele under unitary control (*a*). Solid lines denote the expectations under complete neutrality (i.e., assuming all alleles to be selectively equivalent) derived in Force et al. (2005). The data points were obtained by computer simulation, which accounted for differential allele-specific rates of nonfunctionalization by mutation, as described in the text. In this example, the rates of birth and death of transcription factor binding sites are assumed to be equal. Note that the effective population size is equal to N_g, the effective number of genes per locus, for a haploid population, and approximately $0.5N_g$ for an outcrossing diploid.

linear scaling with N_g (see Figure 10.12). This pattern is not simply a consequence of the greater times needed for mutant alleles to drift to fixation, however, as a purely neutral theory (which assumes alleles *a* to *d* to be selectively equivalent, as shown by the solid lines Figure 10.12) underestimates the transition time by nearly an order of magnitude. Rather, the reduced ability of the semi-independently regulated allele *d* to become established at large N_g is a result of *d* having an additional essential TFBS relative to allele *a*. This factor causes an increase in the mutation rate of allele *d* to nonfunctional (and hence nonfixable) alleles, imposing a weak selective disadvantage on the order of μ_d.

To evaluate the probability that a duplicated semi-independent enhancer (*d*) will subsequently acquire a completely modular structure (as in the lower part of Figure 10.11), recall the general theory outlined in Chapter 8. Such a fate requires the tandem duplication of the semi-independent enhancer, with each copy then incurring a complementary loss of a tissue-specific TFBS prior to one of the enhancers being entirely silenced. The probability of such an outcome (following enhancer duplication) depends on the relative rates of loss of tissue-specific (green or red) and shared (black) TFBSs, μ_d and μ_c respectively. Provided the population size is sufficiently small that $2N_g\mu_c << 1$, for the two-tissue case illustrated in Figure 10.11 and starting from a single copy of a mutant allele with a duplicated enhancer, the probability of fixation of enhancer fission is $2[2\mu_d/(2\mu_d + \mu_c)]^2/N_g$, which, if $\mu_d = \mu_c$, is $8/(9N_g)$, very close to the neutral expectation of $1/N_g$.

duce different cell types. Most species, especially among eukaryotes, coordinate gene expression by equipping interacting genes with similar sets of TFBSs to attract the same transcription factors. Prokaryotes, however, often have an additional layer of gene expression coordination made possible by the use of operons, polycistronic units of cotranscribed and oftentimes functionally related genes. By no means is every gene in every prokaryote deployed in this manner, but the contrast with the situation in eukaryotes is striking. The vast majority of eukaryotic genomes are completely devoid of operons. The most extreme exception is the trypanosomes, in which nearly all genes are initially transcribed into polycistronic units prior to *trans* splicing (Campbell et al. 2003). Exceptions among animals include the nematode *C. elegans*, for which some 15% of the genome is arranged into at least 1000 operons of two to eight genes (Blumenthal et al. 2002; Blumenthal and Gleason 2003), and the tunicate *Ciona*, whose genome contains at least 400 two-gene operons (Satou et al. 2006). Some operons also exist in flatworms (Davis and Hodgson 1997) and in the nucleomorphs of chlorarachniophyte algae (Gilson and McFadden 1996; see Chapter 11), and individual cases of dicistronic units are known in *Drosophila* (Andrews et al. 1996) and in humans (Lee 1991). Nevertheless, given that most well-studied eukaryotes (including fungi, plants, and other animals) have revealed no evidence of operons, it is clear that they are the exception rather than the rule.

Despite the rarity of operons, it is common for sets of colocalized genes in eukaryotes to be under the partial control of shared enhancer sequences. Microarray analyses of gene expression in yeasts (Cohen et al. 2000; Lee and Sonnhammer 2003), *Arabidopsis* (Williams and Bowles 2004), *Caenorhabditis* (Roy et al. 2002; Lercher, Blumenthal et al. 2003), *Drosophila* (Boutanaev et al. 2002; Spellman and Rubin 2002), and mammals (Lercher et al. 2002; Sémon and Duret 2006) all indicate the presence of correlated patterns of expression of small groups of (two to ten) adjacent genes. In yeasts, there is a weak tendency for adjacent genes to have similar functions (Cohen et al. 2000), but such associations appear to be absent in multicellular species. Human housekeeping genes are more clustered than expected by chance (Lercher et al. 2002), and mammalian globin and Hox gene clusters are under the partial control of shared regulatory elements, called locus control regions, while maintaining subsidiary gene-specific regulatory regions (Q. Li et al. 2002; Spitz et al. 2003), but these are exceptions. Moreover, colocalization of functionally related genes outside of operons is seen in many prokaryotes (Overbeek et al. 1999; Lathe et al. 2000), so this type of spatial patterning is by no means a unique feature of eukaryotes. Thus, from the standpoint of gene expression coordination, the most radical change associated with the emergence of eukaryotes was the widespread loss of operons.

Operons enable organisms to respond to environmental and/or developmental challenges in an efficient manner, and there is no compelling reason to think that they would be intrinsically disadvantageous in eukaryotes. Polycistronic transcripts could pose a potential structural problem for eukaryotes, given that the scanning mechanism of translation initiation requires ribosomal reentry at downstream genes. However, the processing of polycistronic transcripts into monocistronic units by *trans* splicing in some lineages (described above) clearly indicates that there are no insurmountable cytological barriers to operons in eukaryotes. Why, then, are most eukaryotic genomes devoid of operons? Does positive selection for single-gene control need to be invoked to explain such a transition?

The conversion of an operon containing genes under unitary control to a set of component genes under individual control is a form of modularization at the genomic level. That is, all genes within an operon are influenced equally by each mutation influencing transcription initiation, whereas single-gene transcripts are independently mutable. Thus, following the logic developed in the preceding section, it can be argued that the loss of operons is an expected passive response to a reduction in the efficiency of selection for weakly advantageous operon architectures in populations experiencing reduced effective sizes. As noted above, because small local duplications are common and because simple enhancer elements can arise at an appreciable rate by single-base substitutions, all genes are vulnerable to the chance invasion of TFBSs capable of driving independent expression. However, by increasing the size of the mutational target, the accumulation of regulatory elements imposes a weak selective disadvantage, equivalent to roughly ten times the per-nucleotide mutation rate for each control element (u). This implies that unless $2N_g u > 0.1$, which is rarely the case in eukaryotes (see Chapter 4), selection will be incapable of keeping a mutant self-regulated allele from going to fixation. In contrast, prokaryotic populations, for which $2N_g u > 0.1$ is common, will be relatively resistant to the breakup of operons by the mutational origin of single-gene control processes. Thus, as emphasized in many other contexts in previous chapters, direct selection for genomic streamlining need not be invoked to explain the elevated levels of operons in prokaryotes. A similar idea has been suggested by Price, Huang et al. (2005), although these authors put more emphasis on positive selection for the coordinated expression of genes with similar, complex regulatory domains than on the negative mutational consequences of maintaining such regulatory machinery for individual genes.

Both of these hypotheses depart radically from the "selfish-operon" model of Lawrence and Roth (1996), which postulates that operons evolve not because of their immediate benefits to the host genome, but because a spatial aggregation of functionally integrated genes facilitates the horizontal transfer of an operon to alternative host species. The selfish-operon model is motivated by two assumptions: (1) that the metabolic advantages of an operon structure are too weak to maintain such organization in the face of stochastic microchromosomal rearrangements and (2) that the stochastic origin of gene-specific promoters will eventually overwhelm the benefits of coregulation. However, both assumptions conflict with the observations made above. The fact that functionally related genes can remain aggregated to some extent even outside of operons appears to be inconsistent with the first assumption, whereas the mutational cost of independent gene regulation is inconsistent with the second point. The fact that essential genes are frequently contained within prokaryotic operons is also inconsistent with the selfish-operon model, as such genes are expected to be incapable of undergoing the periodic phases of loss and transfer postulated by the model (Price, Huang et al. 2005). Although these arguments need not imply that the selfish-operon hypothesis is devoid of explanatory power for all operons in all prokaryotes, the model certainly cannot explain the presence of operons in eukaryotes, in which there is no significant horizontal transfer.

A clear prediction of the theory presented here is that eukaryotic operons will most likely be found within species with exceptionally large effective population sizes. Thus, it is of interest that all eukaryotes known to contain numerous operons exhibit the hallmarks of genomic architecture expected under a long-term regime of large N_g. For example, among unicellular eukaryotes, trypanosomes are distinct in being nearly devoid of both introns and mobile elements, and the genome of the chlorarachniophytic algal nucleomorph contains the smallest known introns and is also lacking in mobile elements. Likewise, among animals, nematodes have the smallest known introns, are relatively depauperate with respect to mobile elements, and probably have relatively high ancestral N_g (Cutter et al. 2006). These are, of course, anecdotal observations, and a more formal survey remains to be done.

11 Expansion and Contraction of Organelle Genomes

As noted in Chapter 1, a defining event in the origin of eukaryotes was the internalization of a eubacterium, which prior to eukaryotic diversification became genetically and phenotypically modified to form the mitochondrion, locked into an obligate symbiosis with its host cell. A second major endosymbiotic event, the colonization of a descendant eukaryotic lineage by the cyanobacterial precursor of the chloroplast, brought the capacity for photosynthesis into the eukaryotic domain. Other early endosymbiotic relationships may have contributed to the establishment of the various organelles contained within eukaryotic cells, but only mitochondria and plastids (also known as chloroplasts in photosynthetic tissues) retain the history of such events in internal genomes.

As we will see in this chapter, organelle genomes have architectures that are radically different from those of eukaryotic nuclear genomes, and although they have some structural similarities to eubacterial genomes, the differences can be equally striking. Most notably, there has been a dramatic asymmetry in the evolution of host and endosymbiont genomes, with the latter becoming miniaturized with respect to gene content, in part by exporting genes to the nucleus. In addition, many organelle genomes have acquired changes in the genetic code, and in some cases, the sequences of their messenger RNAs are radically altered by posttranscriptional editing. The protein-coding genes in some organelle lineages harbor significant numbers of self-splicing introns, which are absent from nuclear coding genes, but organelle genomes are completely devoid of the spliceosomal introns that commonly inhabit nuclear genes.

Organelle genomes pose special opportunities and challenges for the general ideas proposed in previous chapters. Most of the patterns encountered up to now have drawn from the full domain of cellular life, leaving little room for independent evaluation of the contention that nonadaptive forces play a central role in the diversification of genome architecture. Because they represent independent branches on the tree of life, mitochondrial and plastid genomes provide such an option, although at first glance they also present a conundrum. Most notably, in stark contrast to the parallel patterns of nuclear genome evolution observed in the two major multicellular lineages of eukaryotes (animals and land plants), the organelle genomes in these two groups have evolved in radically different ways. The mitochondrial genomes of animals are extraordinarily streamlined, whereas plant organelle genomes are exceptionally bloated with noncoding DNA.

Given its reliance on the universal population genetic processes of mutation and random genetic drift, the theory presented in the preceding chapters should apply to both nuclear and organelle genomes. What, then, are the unique biological features of organelles that encourage evolutionary trajectories different from those experienced by nuclear genomes within the same cells? Although it has been argued in prior chapters that numerous aspects of the phylogenetic diversification of nuclear genomes are attributable to differences in the power of random genetic drift, it has also been repeatedly emphasized that the influence of the population genetic environment on genomic evolution is determined by the product of the effective number of genes per locus (N_g) and the mutation rate per nucleotide site (u). With this in mind, the case will be made here that disparities in mutation rates play a much more central role in organelle genome diversification than in the case of nuclear genomes.

Before we explore these issues, an overview of the putative ancestral states and basic biology of organelle genomes is in order. The data suggest that because of the haploid and uniparental inheritance of most organelle genomes, their N_g is lower than that of nuclear genomes in the same species, but not dramatically so. However, some organelle lineages (in particular, those in animals) have experienced a substantial increase in the mutation rate, whereas others (land plant mitochondria and chloroplasts) are unusually mutationally quiescent. These mutation rate differences provide a potentially unifying explanation for a wide range of previously disconnected observations on animal and land plant organelles, including differences in genome architecture, modifications in the genetic code, and the existence of mRNA editing. The chapter will then consider the susceptibility of nonrecombining organelle genomes to eventual extinction via deleterious mutation accumulation, and finally conclude with an overview of the features of more recent endosymbioses involving eubacterial inhabitants of a variety of arthropod species. In principle, these latter endosymbionts provide opportunities for still further independent tests of general hypotheses in evolutionary genomics.

Establishment and Stabilization of Organelles

Although the ultimate roots of the eukaryotic nuclear genome remain somewhat uncertain, one thing is clear: the common ancestor of most (and probably all) eukaryotes contained a mitochondrion-like organelle. The evidence for this assertion is straightforward and compelling: nearly all extant eukaryotes harbor mitochondria, and the few that do not nevertheless exhibit signs of their prior existence. Based on cytological evidence, mitochondria were long suspected of being bacterial in origin (Mereschkowsky 1905; Margulis 1981), and modern molecular phylogenetic evidence strongly supports their monophyletic origin, most likely from a member of the α-proteobacteria. Whereas some evidence supports an affinity with the Rickettsiales subgroup, which contains numerous intracellular parasites (Andersson et al. 1998; Gray et al. 1999; Gabaldón and Huynen 2003), a purple photosynthetic nonsulfur bacterial origin cannot be ruled out (Esser et al. 2004; Cavalier-Smith 2006b).

The early stages of mitochondrial evolution

Because the origin of the mitochondrion is a singular, ancient event, the factors underlying its initial establishment are largely a matter of conjecture. Virtually all such speculation has started with the assumption that the first cells to harbor mitochondria experienced significant benefits from doing so (Figure 11.1). However, given the numerous aspects of genomic complexity in eukaryotic cells that can arise by effectively neutral processes (e.g., preservation of duplicate genes, proliferation of introns or mobile elements, and modularization of regulatory regions), it is by no means certain that adaptive forces drove the initial establishment of the mitochondrion. For example, a simple capture of a key host gene by the primordial mitochondrion could have led to the permanent enslavement of the host species with little or no overall change in host fitness. In addition, even if adaptive processes did play a role in the initial stages of endosymbiosis, the clues to the underlying events may not reside in the known benefits of modern-day mitochondria, which may have arisen secondarily.

Some functions of modern-day mitochondria were probably entirely absent from the primordial endosymbiont (Gabaldón and Huynen 2004). One potential example is the role played by mitochondrion-derived ribosomes in the translation of nuclear-derived mRNAs in mammalian sperm (Gur and Breitbart 2006). The mitochondrion also plays a central role in programmed cell death during animal development (Kroemer et al. 1998; Joza et al. 2001), although here, the possibility exists that the capacity for inducing cell death is derived from an ancestral mitochondrion that killed cells

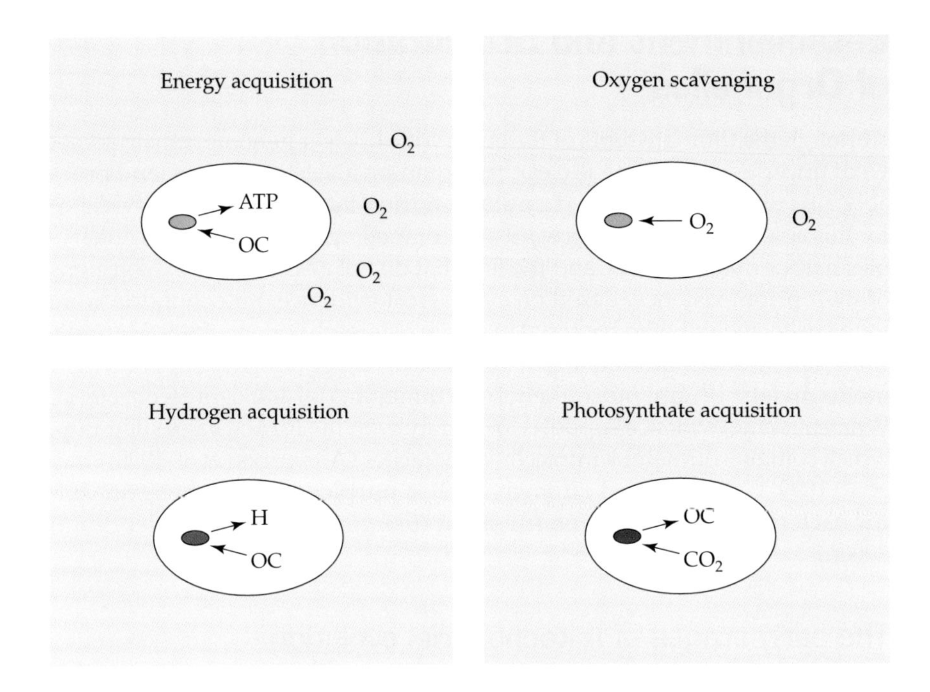

Figure 11.1 A simplified overview of four adaptive hypotheses for the initial establishment of the mitochondrion. The host cell is depicted as a white oval, and the small internal oval represents the primordial mitochondrion (gold denotes an oxygen-utilizing organelle, blue an anaerobic hydrogenosome, and purple an anaerobic purple photosynthetic bacterium). Arrows indicate the exchanges of materials between host cell and endosymbiont that are most relevant to the various hypotheses (with OC denoting organic carbon). The relative numbers of oxygen symbols indicate external environmental levels of oxygen (zero in the bottom two panels).

that did not contain it (Kobayashi 1998). Such an endosymbiont may sound far-fetched, but killer bacteria with exactly these properties exist in *Paramecium* (Preer et al. 1971), and another ciliate, *Tetrahymena thermophila*, deploys aggregates of mitochondria in the degeneration of old "macronuclei" (ciliates have dual nuclei, with all transcription occurring from the "somatic" macronucleus) (Kobayashi and Endoh 2005). The mere existence of such systems is relevant to any discourse on the origins of organelles, as it demonstrates the potential for some endosymbionts to promote their own existence without providing any metabolic advantages to their host cells. These substantial caveats should be kept in mind as we consider four proposed adaptive hypotheses for the origin of the mitochondrion (see Figure 11.1).

Many believe that the key contributor to the establishment of the mitochondrion was its provisioning of the host with energy. It is certainly true that the mitochondrion plays a vital role in energy production in nearly all of today's aerobic eukaryotes: within the cytoplasm, glycolysis converts glucose to pyruvate, which, upon entering the mitochondrion, is subjected to an oxygen-consuming process (aerobic respiration) that produces some 36 molecules of ATP per molecule of glucose. However, there is one obvious concern with this energy acquisition hypothesis: although the primordial endosymbiont must have been capable of ATP production, it need not have possessed a mechanism to transport ATP to its host (Amiri et al. 2003; Gabaldón and Huynen 2004), let alone an incentive to do so.

A second hypothesis for the emergence of the mitochondrion views the consumption of oxygen in a rather different way. Rising oxygen levels around the time of origin of eukaryotes (~2 BYA) may have posed a toxic threat to anaerobic species, thereby providing a selective advantage for mechanisms for removing intracellular oxygen. Under this scenario, an aerobic mitochondrion that scavenged oxygen from its host cell maintained the status quo for host cell metabolism (Kurland and Andersson 2000), and the host provided a sanctuary for its detoxifying inhabitant, but the reliance of the host on the mitochondrion for ATP production arose secondarily.

A third hypothesis postulates that the primordial mitochondrion emerged in a fully anaerobic environment in a methanogenic host that initially derived nothing more than a waste product from its endosymbiont: hydrogen produced from the fermentation of organic substrates (Martin and Müller 1998). Aside from the small demands placed on the endosymbiont, a key motivation for this hypothesis is the observation that the few existing amitochondriate eukaryotes generally harbor an alternative organelle, the hydrogenosome, which anaerobically processes glucose to yield CO_2, acetate, hydrogen, and a small amount of ATP. From this starting point, it is not too difficult to imagine the eventual conversion of an autotrophic methanogen, previously reliant on environmental hydrogen, to a heterotrophic lifestyle requiring host-derived substrates for energy and carbon.

Under this scenario, the ancestral endosymbiont might have been capable of both aerobic and anaerobic metabolism, in which case the mitochondrion and the hydrogenosome may represent alternative evolutionary outcomes in which one or the other pathway has been relinquished (Palmer 1997; Embley et al. 2003; Sutak et al. 2004). Unfortunately, unlike mitochondria, most hydrogenosomes do not contain a genome, rendering a molecular phylogenetic analysis difficult (Dyall et al. 2004; Hrdy et al. 2004; Tjaden et al. 2004). The analysis of proteins imported into the hydrogenosomes of trichomonads, most likely a very basal group (see Chapter 1), has led some to suggest that hydrogenosomes have an origin independent of the mitochondrion, but this is a demanding alternative, as hydrogenosomes are found in several distantly related lineages, including anaerobic ciliates (Akhmanova et al. 1998) and microsporidian fungi (van der Giezen et al. 2002; Williams et al. 2002). The issue is further clouded by the fact that the phylogenetically basal diplomonads (e.g., *Giardia*), once thought to have evolved prior to the origin of the mitochondrion, appear to harbor mitochondrion-derived genes in their nuclear genome (Roger et al. 1998) and also contain a highly reduced, genome-free, mitochondrion-like organelle (the mitosome) (Tovar et al. 2003). Phylogenetic analysis of the one hydrogenosome known to contain a genome, from an anaerobic ciliate inhabiting the hindguts of cockroaches, is entirely consistent with the hypothesis that mitochondria and hydrogenosomes have diverged from a common ancestor (Boxma et al. 2005).

Finally, although all of the above hypotheses assume that the primordial mitochondrion was a heterotroph dependent on host-derived organic substrates, Cavalier-Smith (2006b) suggests an autotrophic origin in the form of a purple nonsulfur bacterium. Under this hypothesis, the primordial endosymbiont relied on host-produced CO_2 as a carbon source while providing the host with some of the products of photosynthesis. Because the extant members of the purple nonsulfur bacteria are capable of photosynthesis only under anaerobic conditions, this hypothesis (like the preceding two) presumes a significant period of anoxia in the historical development of the stem eukaryote. If this assumption is correct, the eventual loss of photosynthetic capacity by the protomitochondrion may have occurred by simple relaxation of selection following the transition of the host to an aerobic environment.

Intergenomic transfer

Regardless of which, if any, of the preceding scenarios is correct, a key step in stabilizing the transition from an autonomous endosymbiont to a host-dependent mitochondrion must have been the transfer of at least a few genes central to endosymbiont metabolism and/or replication to the host's nuclear genome. Such transfers almost certainly involved an initial phase of gene duplication, with the gene being retained in both genomes until the nuclear copy had established full functionality, rendering the organelle copy expendable. Although the time scale over which such movement occurred has not been determined, the relatively small number of genes contained within all mitochondria suggests a fairly early and massive amount of gene transfer. In yeasts, in which these issues have been studied most closely, approximately 850 nuclear-encoded proteins are imported to the mitochondrion (Karlberg et al. 2000; Sickmann et al. 2003), and the numbers for invertebrates and vertebrates are approximately 2000 and 4000, respectively (Richly et al. 2003). Some of these genes were almost certainly derived from the primordial mitochondrion, although others were most likely present in the initial host genome, subsequently acquiring a new organelle-associated function (as described below).

Of the substantial barriers to the transfer of an essential organelle gene to the nuclear genome, the physical transport of organelle DNA is perhaps the easiest challenge to meet and is ongoing even in today's eukaryotes. Nearly all nuclear genomes harbor numerous insertions with high similarity to contemporary mitochondrial DNA (Thorsness and Fox 1990, 1993; J. V. Lopez et al. 1996; Stupar et al. 2001; Adams and Palmer 2003; Bensasson et al. 2003; Richly and Leister 2004; Leister 2005). Of course, to have much likelihood of spawning a productive transfer, any such insertions must completely encompass the organelle gene. A more considerable challenge is the necessity of either inserting into a transcriptionally active position in the nuclear genome or secondarily acquiring transcription factor binding sites prior to being pseudogenized by degenerative mutations. Finally, complete functional transfer to the nucleus requires the acquisition of a mechanism for orienting the nuclear-encoded protein back to the mitochondrion; either the transcript itself or the resultant protein must carry some sort of information for use in subcellular localization. This challenge has usually been resolved at the protein level, with most nuclear-encoded organelle proteins having a precursor stage with an N-terminal sequence that helps guide the protein to the organelle membrane. After being bound to receptors at the mitochondrial surface and threaded through membrane pores, these presequences are eliminated proteolytically. However, the mRNAs of nuclear-encoded mitochondrial proteins having eubacterial affinities (and therefore likely to have a mitochondrial origin) also tend to be differentially translated by ribosomes bound to mitochondrial surfaces (as opposed to ribosomes in the cytosol) (Marc et al. 2002).

Although the complete series of events outlined above may appear to be highly improbable, each step is supported by observations on recent mitochondrion-to-nucleus transfers in plants (Kadowaki et al. 1996; Adams et al. 2000; Kubo et al. 2000; Sandoval et al. 2004). Moreover, transfers made during the early establishment of the mitochondrion need not have involved the de novo origin of novel features. For example, a substantial number of bacterial proteins appear to have fairly simple N-terminal sequences that preadapt them to organelle import (Herrmann 2003; Lucattini et al. 2004; Ralph et al. 2004). Once such sequences were established in the nuclear genome via the first successful organelle-to-nucleus gene transfers, various duplication mechanisms may have led to their recruitment by additional mitochondrial proteins (Adams and Palmer 2003; Timmis et al. 2004). In addition, many of the nuclear-encoded proteins from which the entry sites to mitochondria are built appear to be derived from proteins used in transport across eubacterial membranes (Gabriel et al. 2001; Cavalier-Smith 2006b), and may have been involved in the earliest stages of mitochondrial assimilation.

Finally, it should be noted that the coalition between host and mitochondrial genomes was not simply a one-way street involving the relocation of organelle genes to the nucleus. Although organelles can import DNA from exogenous sources (Knoop 2004), there are no known cases in which a nuclear gene has been successfully incorporated into an organelle. Nevertheless, about 15% of the nuclear-encoded proteins deployed to the mitochondrion (including components of the protein import machinery itself) appear to be derived from the eukaryotic host genome, and some of these have no recognizable homologs in prokaryotes (Andersson et al. 2003; Herrmann 2003; Gabaldón and Huynen 2004; Dolezal et al. 2006). This observation suggests that the origin of the mitochondrion induced an environmental shift that promoted the neofunctionalization of host-encoded proteins. A number of genes of the primordial mitochondrion that were transferred to the nucleus now have functions that are carried out entirely in the cytoplasm, so they too have experienced neofunctionalization (Gabaldón and Huynen 2003).

The origins of plastids

The establishment of the mitochondrion is not the only endosymbiotic event that left a broad mark on eukaryotes. Colonization of the common ancestor of green plants, red algae, and glaucophytic algae by a cyanobacterium gave rise to the plastid (called the chloroplast in photosynthetic tissues) (Tomitani et al. 1999; Moreira et al. 2000), whose establishment appears to have involved events similar to those experienced by the mitochondrion. For example, approximately 20% of land plant nuclear genes appear to be derived from the primordial plastid genome, of which 1000–4000 produce proteins that are targeted to the plastid. Even larger numbers are addressed to other subcellular locations (Martin et al. 2002), having become neofunctionalized following duplication to the nucleus and loss from the organelle.

Unlike mitochondria, which appear to have been vertically inherited throughout the eukaryotic domain, plastids have been horizontally transferred across several distantly related lineages (Cavalier-Smith 2002a, 2003; Bhattacharya et al. 2004) (Figure 11.2). For example, the chlorarachniophytes, a group of amoeboid algae, appear to have obtained their photosynthetic capability by engulfing a green alga (Van de Peer et al. 1996). Although this secondary endosymbiont initially also contained a mitochondrion and a nucleus, only the latter was retained, and it now forms a highly reduced derivative called a nucleomorph, sandwiched between the inner and outer membranes of the plastid. Another secondary endosymbiotic event involving a green alga appears to have given rise to the euglenozoans (which include the euglenoids and the kinetoplastids). In this case, however, only the plastid was retained, and even that was lost from the nonphotosynthetic kinetoplastids, which nevertheless still contain some plantlike genes in the nuclear genome (Hannaert et al. 2003).

The engulfment of a red alga gave rise to still another secondary endosymbiotic event, in this case yielding the putative common ancestor of the diverse group of species sometimes referred to as the chromalveolates (Cavalier-Smith 1999; Keeling et al. 2004; Harper et al. 2005; Li et al. 2006). One sublineage of this group, the cryptomonads, retains both a plastid and a nucleomorph from this event (Douglas et al. 2001), much like the chlorarachniophytes described above. However, all other sublineages (including the diatoms and kelps) have lost the nucleomorph, and still other lineages of uncertain affinities (including the apicomplexans, ciliates, and oomycetes) have lost the plastid as well (Morden and Sherwood 2002). Remarkably, the nonphotosynthetic apicomplexans (including *Plasmodium*)

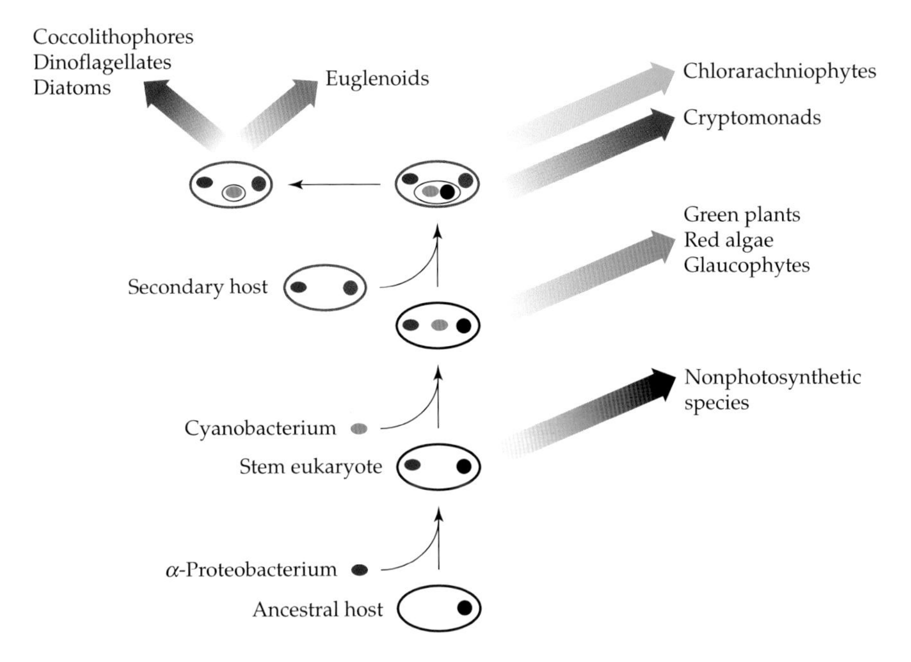

Figure 11.2 Hypothetical paths to the acquisition of photosynthesis by eukaryotes. Solid circles and ovals denote nuclear and organelle genomes, respectively. Broad arrows are color-coded to denote monophyletic events. The mitochondrion (gray arrow) and the primary chloroplast (dark green arrow) each have a single origin. Secondary chloroplasts have evolved at least three times (light green, blue, and brown arrows), with the nucleomorph having been lost independently from the two lineages at the upper left. The internal ovals surrounding secondarily derived chloroplasts denote additional membranes acquired from the donor species. Note that this schematic provides only an overview of the physical paths to organelle genome acquisition; it does not represent the phylogenetic relationships of the depicted lineages.

appear to contain a modified plastid (the apicoplast) derived from still another secondary event, the engulfment of a green alga (Funes et al. 2002; Cai et al. 2003). Although the function of the apicoplast is unknown, approximately 8% of the nuclear-encoded proteins in *Plasmodium* are imported to this organelle (Zuegge et al. 2001). Finally, most dinoflagellates appear to contain tertiary endosymbionts: their photosynthetic ability is a consequence of the engulfment of a member of the brown algal lineage, which (to complicate matters further) may have displaced a previously sequestered secondary plastid from the dinoflagellate host cell (Ishida and Green 2002; Yoon et al. 2002; Hackett et al. 2004). All of these observations have helped resolve the phylogenetic relationships of many of the major eukaryotic lineages outlined in Chapter 1.

One final issue with respect to the stabilization of organelle genomes needs to be addressed here. Given that most organelle genes can be physically transferred to the nucleus, at least over long evolutionary time spans, and given that the reverse seems not to occur, why do almost all lineages retain at least small organelle genomes? In a number of lineages, of animals in particular, the mutation rate is significantly lower in the nucleus than in the mitochondrion (as we will see below), a factor that might seem to encourage residence in the nucleus (Allen and Raven 1996; Blanchard and Lynch 2000). However, even if there is a long-term advantage to relocating an organelle gene to the nucleus, it appears unlikely that the short-term benefit is large enough to selectively advance such a change on the time scale of a typical fixation event (described below).

One hypothesis for the retention of organelle genomes is that their few remaining genes tend to produce hydrophobic proteins, many of which need to be integrated into the organelle membrane and are difficult to import (Von Heijne 1986; Adams and Palmer 2003). It has also been suggested that genes involved in electron transport need to be directly regulated by the local redox balance to avoid overproduction of toxic and mutagenic free oxygen radicals (Allen 2003). Although neither of these possibilities can be entirely ruled out, there are clearly additional factors at play. In particular, as we will see below, most organelle genomes have experienced at least one change in the genetic code, often involving termination codons. Once such a shift has been implemented, it may be essentially impossible to transfer an organelle gene to the nucleus without yielding mistranslated products. Although land plant organelles are exceptional in retaining the genetic code used by the nucleus, they also have remarkably low mutation rates, so from this standpoint, transfer to the nucleus would actually be disadvantageous.

Genome Content and Organization

Most organelles are thought to harbor single circular genomes like those of prokaryotes. However, this conclusion has usually been based on sequence analysis rather than direct observation, and many organelle genomes may exist as linear head-to-tail concatamers produced by rolling-circle replication (Williamson 2002; Bendich 2004). Linear mitochondrial genomes have been formally documented in some fungi (Fukuhara et al. 1993; Forget et al. 2002; Rycovska et al. 2004), other unicellular species (Burger et al. 2000; Conway et al. 2000; Fan and Lee 2002; Nosek and Tomáška 2003), and a small fraction of animals (Pont-Kingdon et al. 2000; Shao et al. 2006), and cannot be ruled out for plants (Oldenburg and Bendich 2004). There are also scattered deviations from the theme of a single organelle chromosome. One of the strangest genomes known is that of the kinetoplastid mitochondrion, to be discussed in more detail below, which consists of thousands of circular molecules containing cryptic messages that require posttranscriptional editing. The mitochondrial genome of *Amoebidium parasiticum* consists of hundreds of short linear fragments, each containing just one or two genes (Burger, Forget et al. 2003). Dinoflagellates (Zhang, Cavalier-Smith et al. 1999, 2002; Howe et al. 2003) and mesozoans (a group of simple 20–30-cell parasites) (Watanabe et al. 1999) also have single-gene chromosomes in the form of minicircles. Although these aberrations suggest numerous avenues for future research on chromosome structure, including the mechanisms by which fragmented organelle genomes avoid stochastic chromosomal loss during cell division, the remainder of this chapter will focus on genome content.

Despite the extreme reduction in organelle genome sizes, the general scaling of genome content with overall genome size in mitochondria (Figure 11.3) is similar to that in nuclear genomes (see Figure 2.1). Expansions of the largest mitochondrial genomes are only weakly associated with increases in gene number, with a near-threshold increase in investment in intronic

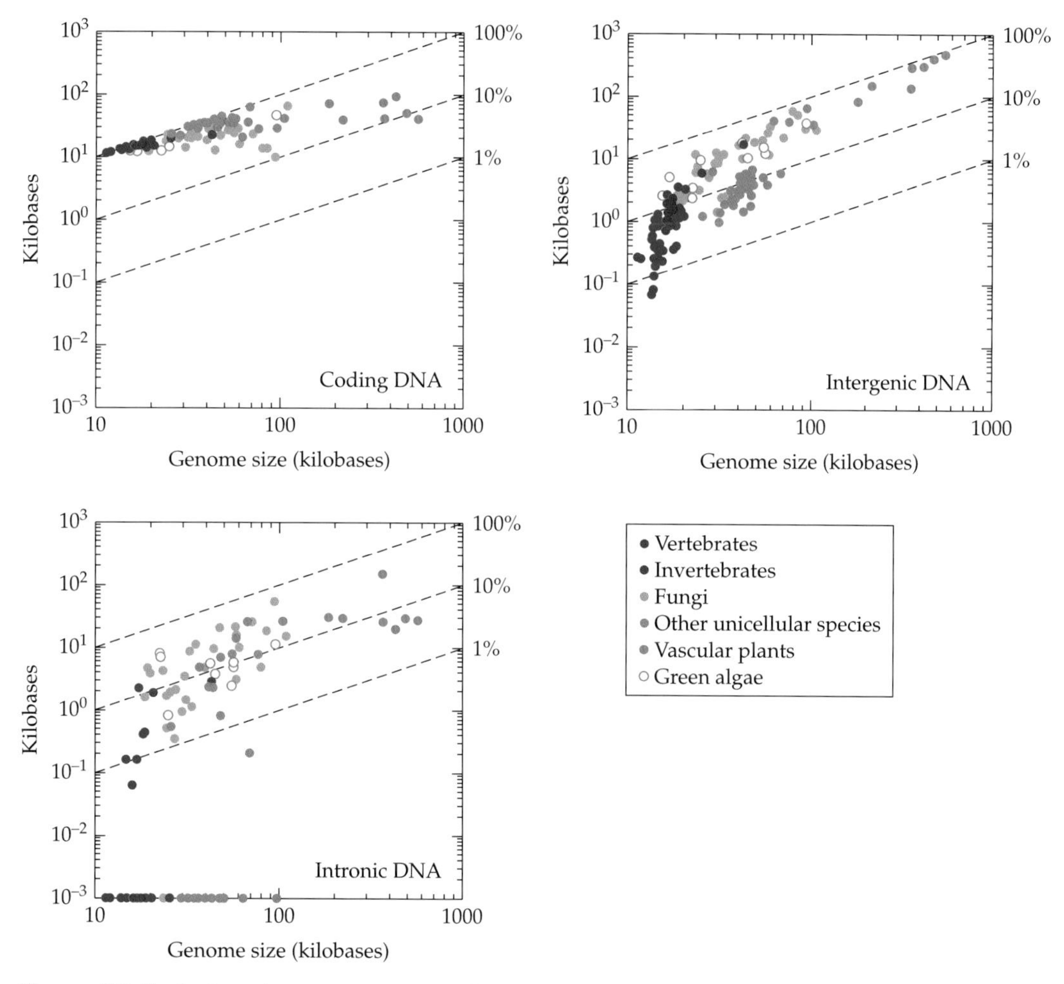

Figure 11.3 Scaling of genome content with genome size in mitochondrial genomes. The diagonal lines define points of equal proportional contribution to total genome size. (Updated from Lynch et al. 2006b.)

and intergenic DNA occurring at the low end of the genome size spectrum. In addition, there is considerable continuity of scaling within and between major groups of organisms, as in nuclear genomes.

In contrast, there are significant differences in the positions of various phylogenetic groups along the size gradient in mitochondrial versus nuclear genomes. As noted in preceding chapters, the independent excursions of animals and land plants into multicellularity were accompanied by very similar expansions in noncoding nuclear DNA. In contrast, the mitochondrial genomes of these two lineages followed radically different evolutionary trajectories; relative to the mitochondrial genomes of unicellular species, those in animals are miniaturized, while those in land plants are bloated. This contrasting pattern strongly supports the contention that major differences in genome architecture are driven by factors other than the evolution of organismal complexity.

The smallest mitochondrial genomes are found in animals. Generally in the range of 14–20 kb in size and nearly all sharing the same 13 protein-coding genes, 22 transfer RNA genes, and 2 ribosomal genes, these diminutive genomes are highly derived, having lost many of the genes that are broadly distributed across the various unicellular lineages (Gray et al. 1998). Land plant mitochondrial genomes reside at the opposite end of the scale, falling in the range of 180–1600 kb (Ward et al. 1981) and containing between 29 and 59 protein-coding genes, usually along with full sets of transfer and ribosomal RNAs. Most animal mitochondria consist of 90%–95% coding DNA, even exhibiting overlaps between the initiation and termination codons of some adjacent genes, whereas the vast majority (>90%) of land plant mitochondrial DNA is noncoding.

No mitochondria contain spliceosomal introns, and mobile elements are almost always absent, one possible exception being a small double-hairpin element distributed among the intergenic regions of the mitochondrial genome of *Allomyces* (a fungus) (Paquin et al. 1997). However, mitochondrial group I and II introns are found throughout the eukaryotic domain, albeit with a very uneven distribution (see Figure 11.3). Unicellular species are often completely devoid of such introns and rarely have more than ten (the main exception being an expansion in the number of group I introns in some fungal lineages). No green algae are known to have more than ten introns per mitochondrial genome, but almost all land plants have twenty to thirty, suggesting a uniquely large expansion in intron number in this lineage. In contrast, animal mitochondria lack introns, with just a few exceptions: the genomes of two cnidarians, a coral (*Acropera*; van Oppen et al. 2002) and a sea anemone (*Metridium*; Pont-Kingdon et al. 1998), contain one and two introns, respectively, whereas that of the placozoan *Trichoplax adhaerens* contains four (Dellaporta et al. 2006). As we will see below, these exceptions are potentially revealing with respect to the mechanisms promoting the maintenance of organelle introns.

Because of the diversity of ancestries associated with plastids and the limited number of whole-genome sequences, the interpretation of data on

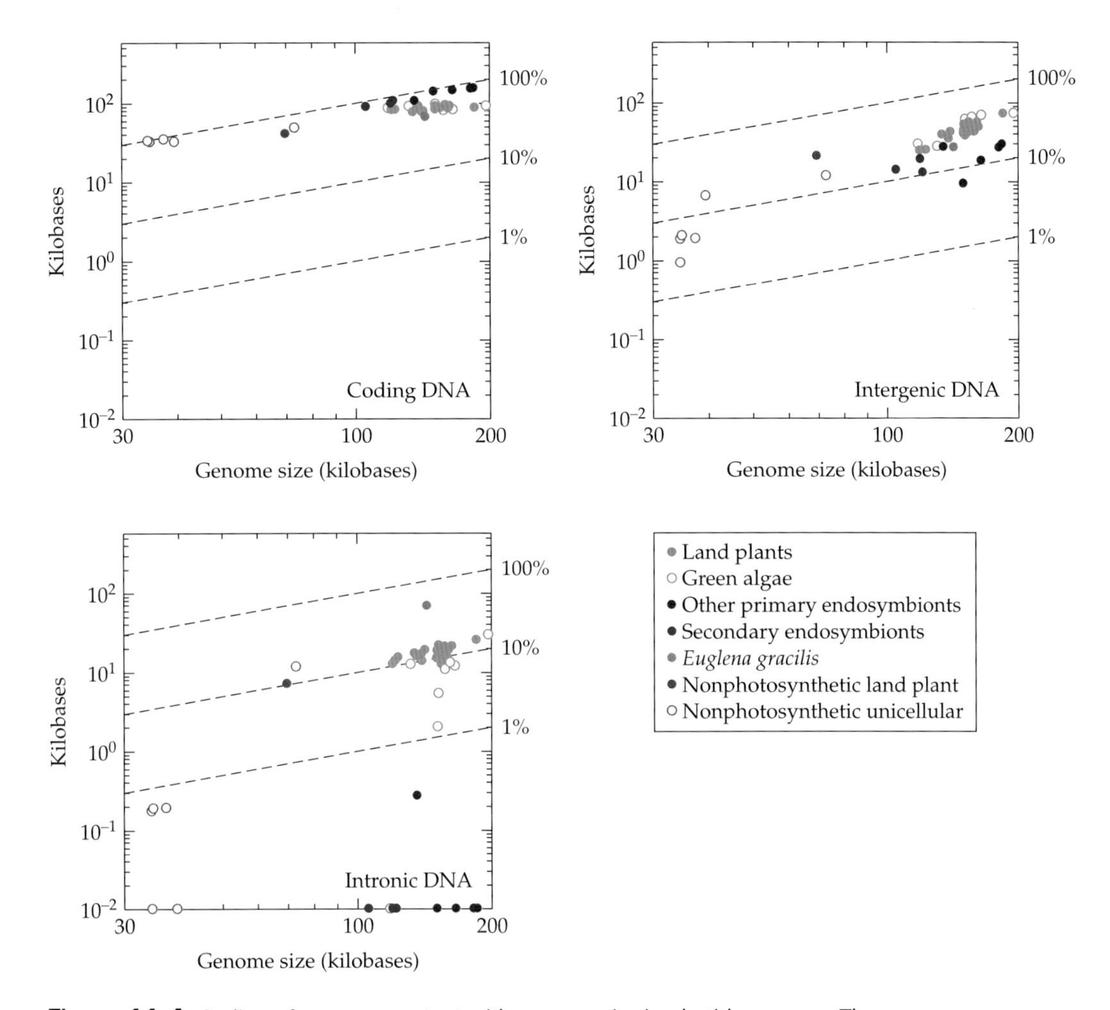

Figure 11.4 Scaling of genome content with genome size in plastid genomes. The diagonal lines define points of equal proportional contribution to total genome size. Algae with "other primary endosymbionts" include the red algae and a glaucocystophyte, whereas those with secondary endosymbionts include a diatom, a haptophyte, and a cryptophyte. The plastid of *Euglena gracilis*, another secondary endosymbiont, is depicted separately. Derived plastids from nonphotosynthetic organisms are from one land plant (*Epifagus*), two unicellular parasitic algae (*Helicosporidium* sp. and *Euglena longa*), and four apicomplexans. (Updated from Lynch et al. 2006b.)

plastid genomes is less straightforward than in the case of mitochondria, although three patterns are notable (Figure 11.4). First, as in the case of nuclear and mitochondrial genomes, expansion in size in the largest plastid genomes is largely a consequence of the proliferation of intergenic and intronic DNA. Of the lineages derived from the primary endosymbiotic

TABLE 11.1 Average content of plastid (cp) versus mitochondrial (mt) genomes

	GENOME SIZE (BP)	NUMBER OF GENES		INTRONS/ GENE	FRACTION NONCODING		
		PROTEIN	tRNA		INTRON	INTERGENIC	*N*
Land plants							
cp	150,290	87	33	0.16	0.12	0.30	35
mt	311,760	39	22	0.57	0.18	0.58	9
Green algae							
cp	167,210	92	31	0.07	0.05	0.38	8
mt	41,550	22	17	0.17	0.11	0.25	12
Red algae							
cp	169,950	206	32	0.00	0.00	0.12	4
mt	31,600	29	24	0.00	0.05	0.06	3

Source: Data are derived from annotated genomes deposited at the National Center for Biotechnology Information.

Note: Introns/gene refers to protein-coding loci, whereas the fraction of intronic DNA includes that within tRNA and rRNA genes. The number of genomes entering each analysis is denoted by *N*.

event, chlorophytes (green algae and land plants) have retained nearly 60% fewer protein-coding genes than their red algal relatives while harboring substantially more intergenic and intronic DNA (Table 11.1). Second, the plastid genomes of algae that have experienced secondary endosymbiotic events have lost numerous genes, with a dramatic proliferation of introns occurring in the euglenoid lineage. Finally, the expansion of noncoding DNA in land plant plastids is much less substantial than in mitochondria; although the latter contain less than half the number of genes in the former, the average genome size is twofold higher (see Table 11.1).

The Population Genetic Environment

Previous chapters have promoted the view that much of the diversity in gene and genomic architecture is a consequence of the relative power of two nonadaptive forces, mutation and random genetic drift. Before we attempt to extend this reasoning to organelle genomes, some fundamental population genetic issues must be revisited. As noted in Chapter 4, interactions between mutation, recombination, and numbers of reproductive individuals define the effective number of gene copies at a locus in a population (N_g). In the nuclear genome, all three of these factors work in the same direction to cause a reduction in N_g in multicellular species relative to that in unicellular taxa. However, given the unique aspects of inheritance and mutational susceptibility in organelles, these patterns need not hold in mitochondrial and plastid genomes.

Mutation

Several factors may promote unusual patterns of mutation in the mitochondrion. First, as sites of aerobic respiration, mitochondria generate high levels of free oxygen radicals, producing an internal environment with an exceptional potential for causing DNA damage, including the deamination of cytosine to uracil and the oxidative modification of guanine to 8-oxoguanine, which respectively cause C:G→T:A transitions and C:G→A:T transversions (see Chapter 6). Such mutation pressure is consistent with the nearly universal A/T bias that exists in mitochondrial genomes (Table 11.2). Second, in contrast to nuclear DNA, mitochondrial DNA is frequently replicated within nondividing cells, increasing the opportunities for replication errors per cell cycle. Third, mitochondria generally contain multiple genomic copies, enhancing the likelihood that newly arisen mutations will be involved in gene conversions. By promoting some types of mutations and eliminating others, biased gene conversion within an organelle can effectively alter the mutation profile, as mutations that are incapable of becoming fixed within an organelle are unable to become fixed at the population level (Birky and Walsh 1992). Unfortunately, almost nothing is known about the magnitude or direction of gene conversion forces in mitochondria, although the single study of the matter that has been done, on the plastid genome of tobacco, indicates a strong conversion bias toward A/T (Khakhlova and Bock 2006), the opposite of what is generally seen in nuclear genomes (see Chapter 6).

The mutation rate is a function of the accuracy of the replication machinery and the ability of repair enzymes to correct pre-replication damage. The base misincorporation rate of the mitochondrial DNA polymerase (pol γ),

TABLE 11.2 Mean A/T content of mitochondrial and plastid genomes as a fraction of the entire genome

	MEAN	SD	*N*
Mitochondrial genomes			
Vertebrates	0.59	0.04	33
Invertebrates	0.67	0.07	36
Land plants	0.58	0.04	9
Green algae	0.66	0.06	12
Fungi	0.74	0.07	36
Unicellular species	0.72	0.06	26
Plastid genomes			
Land plants	0.64	0.04	27
Green algae	0.67	0.05	9

Source: Data from records at the National Center for Biotechnology Information.

which is universally encoded in the nucleus, is several times lower than that of polymerases deployed in nuclear genome replication (Kunkel and Alexander 1986; Johnson and Johnson 2001). However, in contrast to nuclear genomes, almost no mitochondrial genome encodes DNA repair proteins, such functions having either been transferred to the nucleus or lost entirely in the early establishment of the primordial endosymbiont. Almost everything we know about mitochondrial DNA repair derives from studies of budding yeast and human cells, and even that knowledge is limited (Bohr et al. 2002; Kang and Hamasaki 2002; Mason and Lightowlers 2003). Of the three major pathways for eliminating DNA damage—base excision repair, mismatch repair, and nucleotide excision repair—the last appears to be completely absent from the mitochondrion of both species. In contrast, both groups deploy a base excision repair process that may be even more efficient than that in the nuclear genome (Bohr et al. 2002). Vertebrates appear to have more proteins allocated to mitochondrial base excision repair than yeasts (Foury et al. 2004), but this need not imply a higher degree of maintenance fidelity. Indeed, mitochondrial base excision repair appears to be mutagenic, in that damaged bases are often replaced by incorrect nucleotides (Phadnis et al. 2006). In yeast mitochondria, such errors can be eliminated by nuclear-encoded mismatch repair enzymes. Such proteins are not known to enter human mitochondria, however, and although some mechanism for mismatch repair is present in vertebrates, the absence of strand bias implies that correct and incorrect bases are equally likely to be altered (Mason et al. 2003). Unfortunately, these limited observations on mitochondrial DNA repair provide no quantitative information on mitochondrial mutation rates, which ultimately depend on both the amount of damage per cell cycle and the efficiency of repair.

The only attempt to directly estimate the germ line mutation rate of mitochondrial DNA involves a survey of long-term mutation accumulation lines of the nematode *C. elegans* (Denver et al. 2000). Because these lines were maintained by single-progeny descent from self-fertilizing individuals, natural selection was effectively eliminated, so the rate of mutation accumulation is expected to be very close to the mutation rate (in accordance with the neutral theory; see Chapter 4). At the level of nucleotide substitutions, this study revealed a mutation rate of 9.7×10^{-8} per site per generation, which corresponds to 8.9×10^{-6} per site per year, a rate about ten times greater than the directly observed nuclear rate for this species (Denver et al. 2004). (Strictly speaking, this estimate and all others discussed here should be viewed as net mutation rates after accounting for any biases associated with within-organelle gene conversion.) The ratio of mitochondrial transitions to transversions in this study was 4:1, consistent with observations on somatic mutations in human mitochondria (Khrapko et al. 1997) and with the commonly observed excess of transitions over transversions at silent sites in interspecies comparisons (Horai and Hayasaka 1990; Xia et al. 1996).

All other attempts to estimate the mitochondrial mutation rate are based on the enumeration of substitutions at silent sites in pairs of species with geologically based divergence time estimates, relying on the assumption that such sites are neutral (see Chapter 4). These studies reveal dramatic differences among major phylogenetic groups (Table 11.3). In mammals, the rate of silent-site substitution averages 34×10^{-9} per site per year and is fairly constant across species, so that the per-generation rate scales inversely with generation length. Qualitatively similar rates per absolute time have been found for other animals, with the exception of sharks and corals. In stark contrast, the absolute rate of silent-site substitution in plants averages just 0.36×10^{-9} per site per year, suggesting a hundredfold difference in mitochondrial mutation rates between animals and plants.

Uncertainties remain about the quantitative reliability of phylogenetically based mitochondrial mutation rate estimates, with some evidence suggesting they are substantially downwardly biased (as in the nuclear genome; see Chapter 4). For example, on an absolute time scale, the direct estimate for *C. elegans* noted above is about 150 times greater than the largest indirect rate obtained for any species (see Table 11.3), and the average estimate of the human mitochondrial mutation rate obtained from pedigree analysis exceeds that obtained from phylogenetic analysis by a factor of about 23 (Cavelier et al. 2000; Howell et al. 2003; Santos et al. 2005). There are at least two potential explanations for such discrepancies. First, as discussed in Chapter 6, selection associated with translation or other aspects of mRNA processing (codon bias) could cause the rate of divergence of silent sites to be less than the actual mutation rate. Second, the vast majority of human mitochondrial mutations appear to be concentrated at a few single-nucleotide hot spots (Khrapko et al. 1997; Zheng et al. 2006). Mutational hot spots can profoundly bias phylogenetic estimates of mutation rates, as pairs of species with low levels of average divergence may nevertheless have experienced multiple substitutions at a small number of hot spots.

These quantitative uncertainties aside, the data in Table 11.3 strongly suggest the presence of major phylogenetic differences in the mutagenicity of mitochondrial versus nuclear environments. In vertebrates and invertebrates (excluding corals), the average substitution rates at silent sites in the mitochondrion are 19 and 8 times those in the nuclear genome, respectively. In contrast, the land plant mitochondrial substitution rate is generally just 5% of that of the nuclear genome. Because the estimated ratio of mitochondrial to nuclear mutation rates averages 1.6 for unicellular species and is just 0.57 for corals (a basal lineage of animals), a high mitochondrial mutation rate appears to be largely a feature of bilaterian animals. The mechanisms responsible for the mutational quiescence of mitochondria in plants and corals are unclear, but differences in mismatch repair capabilities may be a factor. The only animal known to encode a mismatch repair enzyme in its mitochondrion is a coral (Pont-Kingdon et al. 1998), and land plants harbor a closely related variant of this enzyme (Abdelnoor et al. 2006). Nevertheless, whatever mechanism is responsible for the extraordinarily low mutation rate in plant mitochondria, it is not an entirely invariant feature. Two genera of plants (*Plantago* and *Pelargonium*) contain species with mitochondrial

TABLE 11.3 Rates of nucleotide substitution mutation per nucleotide site in mitochondria, derived from phylogenetic comparisons using silent sites in protein-coding genes (in units of 10^{-9}/site)

	RATE PER YEAR	RATE PER GENERATION	MT/NU[a]
Mammals			
Whales	28.80	648.00	13.09
Great apes	26.13	522.66	20.91
Lemurs	35.98	97.15	29.30
Big cats	67.68	473.76	17.92
Horses	36.50	109.50	26.96
Seals	20.30	62.93	17.65
Rodent	21.81	10.90	12.74
Birds			
Herons	19.20	192.00	10.91
Woodpeckers	24.71	74.12	14.04
Chicken/quail	8.10	8.10	4.59
Reptiles and amphibians			
Turtles	8.20	357.52	32.79
Frogs	20.37	20.37	16.60
Newts	17.72	115.20	
Fishes			
Centrarchidae	—	—	17.17
Characiformes	—	—	47.33
Sharks	0.70	11.90	4.38
Bilaterian invertebrates			
Drosophila	34.00	3.40	2.34
Mosquitoes	—	—	1.43
Crickets	—	—	18.03
Aphids	—	—	18.00
Butterflies	—	—	2.95
Tiger beetles	16.60	—	—
Snails	57.00	—	0.82
Sea urchins	81.00	—	7.54
Nematodes	—	—	10.26
Corals	0.50	—	0.57
Land plants			
Grasses	0.34	0.34	0.05
Palms	0.15	1.50	0.06
Brassicaceae	0.60	0.90	0.04

TABLE 11.3 (Continued)

	RATE PER YEAR	RATE PER GENERATION	MT/NU[a]
Fungi and unicellular species			
Aspergillus	—	—	1.84
Cryptococcus	—	—	0.73
Haemoproteus	—	—	0.47
Paramecium	—	—	1.40
Plasmodium	—	—	0.42
Trypanosoma	—	—	4.04
Yeasts	—	—	2.14

Source: Updated from Lynch et al. (2006b), with additional references for snails (Hellberg and Vacquier 1999), sea urchins (McCartney et al. 2000; Zigler et al. 2005), and coral (Hellberg 2006).

[a]Ratio of rates in mitochondrial (mt) versus nuclear (nu) genomes.

silent-site substitution rates as much as 4000 times greater than those reported above (Palmer et al. 2000; Y. Cho et al. 2004), and a switch from maternal to paternal inheritance in gymnosperms can increase organelle mutation rates by severalfold (Whittle and Johnston 2002).

Where does the plastid fit in with respect to mutation rates? Comparative studies of grasses (rice and maize) and palms imply ratios of silent-site substitution rates in the plastid versus nuclear genomes of 0.22 and 0.08, respectively (Gaut et al. 1996). Recalling the relative rates of mitochondrial silent-site substitution in Table 11.3, these observations suggest that the land plant plastid mutation rate is two to four times that of the land plant mitochondrion. A similar conclusion can be derived from the earlier work of Wolfe et al. (1987).

Transmission and recombination

In contrast to nuclear genomes, all organelle genomes are replicated ameiotically, and in the vast majority of animals and land plants, they are inherited uniparentally, usually through the mother. Although the numbers of organelles per oocyte can range up to 10^4 or so, the number of genome copies actually inherited by progeny appears to be substantially smaller. The issue of organelle transmission dynamics has been addressed on several occasions through the serendipitous discovery of heteroplasmic females (carrying two distinct mitochondrial types), generated either by mutation or by rare cases of paternal leakage. Letting p and $(1-p)$ denote the frequencies of two haplotypes in a mother, the variance in their frequencies among progeny follows from simple binomial sampling, $p(1-p)/n_o$, where n_o (the effective number of organelle genomes within an individual) can be ascertained from the degree of dispersion of the haplotype frequencies among progeny.

(This formula is identical in form to that for genetic drift of two autosomal alleles within a population; see Chapter 4.) Using this approach, or a close variant of it, the effective number of mitochondrial genomes per female transmission is estimated to be ten in mice and humans (Jenuth et al. 1996; Marchington et al. 1997) and two in cows (Ashley et al. 1989). Thus, in these mammals, heteroplasmic lineages become homoplasmic in just a few generations, although the mechanisms responsible for the underlying transmission bottlenecks are unclear (candidates include direct organelle destruction, differential replication, and random partitioning of the cytoplasm; for a review, see Burt and Trivers 2006). Heteroplasmy in the slime mold *Physarum* also tends to be resolved within just a couple of generations (Meland et al. 1991), and in *S. cerevisiae*, heteroplasmy is generally lost within 20 generations, suggesting an effective number of organelle genomes per individual of about five (Birky et al. 1978). In contrast, in *Drosophila* and crickets, the effective number of organelle genomes per female appears to be significantly larger, on the order of two to three hundred (Solignac et al. 1983; Rand and Harrison 1986).

Such rapid loss of heteroplasmy bears on the issue of organelle effective population sizes in two significant ways. First, the multiplicity of genomes within individuals is expected to have only a minor effect on the long-term effective number at the population level (N_g), which will be primarily dictated by the effective number of transmitting parents (Birky et al. 1983). This is because the depth of the entire genealogy of organelles within a population will be almost entirely determined by the time separating organelle lineages among (as opposed to within) hosts. Second, even though organelle genomes are capable of recombination (Thyagarajan et al. 1996), the opportunities for generating novel recombinant genotypes are quite restricted. Because recombination between molecules varying at zero or one sites yields identical offspring molecules, genetically effective recombination requires the participation of pairs of molecules with at least two polymorphic nucleotide sites. With within-individual divergence times shorter than a few hundred generations, such levels of variation will only rarely develop by mutation.

Thus, genetically effective recombination between organelles generally requires biparental transmission to bring divergent genomes into contact. The degree to which such situations arise has been debated considerably (Eyre-Walker and Awadalla 2001; McVean 2001). Given the thousandfold or greater difference in the numbers of mitochondria usually carried in eggs and in sperm, and the fact that sperm mitochondria are generally targeted for destruction (Satoh and Kuroiwa 1991), biparental encounters of animal mitochondria are clearly much less frequent than in the case of nuclear genes. Observations of recombinant events in the somatic cells of rare heteroplasmic humans provide direct evidence that recombination *can* occur in animals (Kraytsberg et al. 2004; Zsurka et al. 2005), but to have evolutionary consequences, recombination must occur in the germ line, and it is here that the picture becomes murkier.

Because of the extremely low likelihood of directly observing recombination events in an organelle, attempts have been made to address the problem indirectly by looking for the hallmark of recombination, a decline in linkage disequilibrium with increasing physical distance between polymorphic nucleotide sites in samples from natural populations (see Chapter 4). Some observations of this sort seem consistent with the occurrence of recombination in animal mitochondria (Awadalla et al. 1999; Piganeau et al. 2004), but a number of potential artifacts can cloud such interpretations (Jorde and Bamshad 2000; Kivisild and Villems 2000; Kumar et al. 2000; Parsons and Irwin 2000; Elson et al. 2001; Wiuf 2001; Innan and Nordborg 2002). Berlin and Ellegren (2001) tested the hypothesis of clonal inheritance of mitochondrial DNA in a bird by comparing the population-level genealogy of the mitochondrion with that of a molecular marker on the W sex chromosome, which is strictly maternally inherited without recombination. There was complete concordance between the two sets of markers, a finding that is consistent with the hypothesis that both mitochondrion and W chromosome are coinherited entirely through females. Thus, to date, there is no compelling evidence for the occurrence of significant recombination within animal mitochondria.

In contrast to the situation in animals, mitochondrial inheritance in fungi is often biparental, and recombination does occur (Wilkie and Thomas 1973; Silliker et al. 1996; MacAlpine et al. 1998; Saville et al. 1998; Ling et al. 2000; J. B. Anderson et al. 2001). The situation is less clear in plants, in which organelle inheritance is usually uniparental. Plant mitochondrial genomes commonly contain dispersed repetitive DNAs that promote the production of subgenomic-sized circles by recombination (Palmer and Shields 1984; Newton 1988; Gillham 1994), and plant nuclear genomes appear to encode a mitochondrion-targeted RecA protein for homologous recombination, which has been lost from the fungal and animal lineages (Z. Lin et al. 2006). However, although such intragenomic exchange can lead to the production of genomic rearrangements, its effect on sequence evolution may be negligible. Circumstantial evidence suggests the occurrence of some effective recombination in the mitochondria of the annual plant *Silene* (Städler and Delph 2002), which harbors an unusually high level of nucleotide polymorphism, but the extremely low levels of within-species variation in the organelle DNA of most plants (described below) imply negligible levels of genetically effective recombination.

Effective population size

It is commonly argued that genes contained within organelles have effective population sizes about one-fourth those of nuclear genes in diploid species, the rationale being that each sexual reproductive event involves the participation of four nuclear gene copies (two from each parent), but just one organelle gene copy (generally from the mother). This argument ignores two key issues, however. First, as emphasized in Chapter 4, N_g can be more strongly influenced by selective interference associated with linkage than by the number of transmitting units. The "one-fourth" rule implicitly assumes an identical level of selective interference in nuclear and organelle genes, but this assumption is almost certainly violated, perhaps even in different directions, in various phylogenetic lineages. Given the restricted recombination in organelle genomes, it is tempting to conclude that selective interference due to linkage is more substantial for organelle than for nuclear genes. However, even if recombination rates are greatly reduced in organelle genomes, nuclear chromosomes contain many more genes, and it is also unclear whether the distributions of the fitness effects of mutations that drive selective interference are equivalent among the two types of genomes. In plant organelle genomes, the mutation rate is so much lower than in the nuclear genome that selective interference may be a negligible force, but animal mitochondria may present the opposite condition. Second, the "one-fourth" rule implicitly assumes equal levels of variance in reproductive success among males and among females. However, in many species, animals in particular, a much smaller fraction of males participate in mating than females. The demographic events in the nontransmitting sex are irrelevant to the evolution of an organelle genome, but because the effective population size of a biparentally inherited gene is most strongly influenced by the number of breeders in the rarer sex, a male bottleneck could actually result in a lower effective number of nuclear genes than maternally inherited organelle genes (Birky et al. 1983).

Given this array of uncertainties, empirical observation is the only means of clarifying the relative effective numbers of organelle and nuclear genes. The problem will be addressed here in an indirect way, using neutral polymorphism and divergence data. Recall from Chapter 4 that the expected number of nucleotide substitutions between silent sites in randomly sampled nuclear genes within a species (π_{sn}) is $2N_{gn}u_n$ at mutation–drift equilibrium, where N_{gn} is the effective number of nuclear genes at the locus and u_n is the mutation rate per nuclear nucleotide per generation. Similarly, the expected silent-site diversity in a mitochondrial gene (π_{sm}) can be denoted as $2N_{gm}u_m$. If an estimate of the ratio of mitochondrial to nuclear divergence rates for silent sites (d_m/d_n) among closely related taxa is available, then, under the assumption of neutrality, it can be used as an estimate of u_m/u_n (see Table 11.3), thereby allowing the ratio of effective sizes (N_{gm}/N_{gn}) to be approximated by $(\pi_{sm}/\pi_{sn})/(d_m/d_n)$.

The data for these computations are not extensive, and as ratios based on limited data, the estimates for individual species are not uniformly reliable, but the group means are consistent enough to be informative (Table 11.4). The most significant observation is that although deviations from the "one-fourth" rule exist, they are generally fairly small. For mammals, the average N_{gm}/N_{gn} is 1.29 (SE = 0.43), which is significantly greater than the null expectation of 0.25, consistent with the fact that variance in male reproductive success often exceeds that for females in mammalian species (Clutton-Brock 1988). The average estimates of N_{gm}/N_{gn} for other vertebrates,

TABLE 11.4 Estimates of the effective number of genes per locus for mitochondrial versus nuclear genes

	N_{gm}/N_{gn}
Mammals	
Erinaceus	0.17
Gorilla	2.20
Homo	0.26
Loxodonta	1.94
Microunga	3.52
Mus	0.57
Ovis	0.09
Pan	1.54
Birds	
Aethia	0.02
Reptiles/amphibians	
Eleutherodactylus	2.55
Emys	1.81
Fishes	
Prochilodus	0.05
Arthropods	
Aedes	0.17
Anopheles	0.21
Drosophila	0.05
Gryllus	0.17
Lutzomyia	0.12
Nematodes	
Caenorhabditis	1.29
Haemonchus	0.16
Unicellular species	
Candida	0.32
Cryptococcus	0.76
Saccharomyces	0.26
Paramecium	0.27
Plasmodium	1.16
Trypanosoma	0.10

Source: Lynch et al. 2006b.

invertebrates, and unicellular species, 1.11 (0.64), 0.31 (0.16), and 0.48 (0.16), respectively, are uniformly greater than 0.25, although no individual deviation is significant. Polymorphism data are scant for land plant mitochondria, so only crude estimates of N_{gm}/N_{gn} are possible for them. For the four plant species for which data are available, three of the estimates of π_{sm} are 0.0000, and one is 0.0017 (Lynch et al. 2006a). Thus, recalling from Chapter 4 that average π_{sn} for land plants is 0.015 (0.003), and from Table 11.3 that average d_m/d_n for land plants is 0.05, it appears likely that average N_{gm}/N_{gn} for land plants is also in the vicinity of 0.5–1.0. Average silent-site heterozygosity for land plant chloroplasts is 0.0031 (0.0009) (Lynch et al. 2006a), and recalling from above that average $d_c/d_n \approx 0.015$ for land plants (where c denotes the chloroplast), N_{gc}/N_{gn} also appears to be in the vicinity of 1.0.

Mutation Pressure and the Diversification of Organelle Genomes

Despite the crude nature of the preceding calculations, they consistently point to the conclusion that effective numbers of genes per locus are generally of the same order of magnitude for organelle and nuclear genomes within species. Thus, the well-documented tendency for nuclear N_g to decline with increasing organism size (see Chapter 4) should generally extend to organelles, and given the similarity of nuclear N_g in animals and land plants, this further implies that the divergent patterns of organelle genome evolution in these two lineages are not products of differences in the power of random genetic drift. This leaves mutation as a likely determinant.

As noted above, under the assumption of neutrality, the average number of substitutions per silent site separating randomly sampled alleles within a species (π_s) provides an estimate of twice the ratio of the power of mutation to the power of random genetic drift ($2N_g u$). In the nuclear genome, this measure tends to be substantially smaller in vertebrates than in invertebrates as a consequence of the reduced N_g in the former (see Chapter 4), whereas both groups have comparable levels of silent-site variation in the mitochondrion (Table 11.5). Although Bazin et al. (2006) have suggested that this absence of pattern in π_{sm} among animal mitochondria is a consequence of a relatively constant effective population size for organelle genomes across species, the data in Tables 11.3 and 11.4 suggest otherwise: invertebrates have larger effective population sizes but lower per-generation mutation rates (u) than vertebrates, with these two opposing effects roughly canceling. In contrast, as alluded to in the previous section, π_s in plant organelle genomes is at least an order of magnitude lower than in those of bilaterian animals (see Table 11.5) as a consequence of depressed mutation rates in land plant organelles. In light of these fundamental differences in the relative power of mutation and random genetic drift, we now consider a potentially unifying explanation for a diverse set of differences between animal and land plant organelle genomes.

TABLE 11.5 Average estimates of silent-site nucleotide diversity for organelle genomes (π_s) in various taxonomic groups

	MEAN	SE	*N*
Mammals	0.0406	0.0087	12
Birds	0.0169	0.0053	4
Salamanders	0.0481	0.0148	3
Fishes	0.0362	0.0150	6
Echinoderms	0.0117	0.0036	2
Arthropods	0.0276	0.0056	17
Mollusks	0.0135	0.0068	6
Nematodes	0.0677	0.0084	8
Platyhelminthes	0.0399	0.0063	5
Fungi	0.0120	0.0046	3
Land plant mt	0.0004	0.0004	4
Land plant cp	0.0031	0.0009	17

Source: Lynch et al. 2006a.

Note: Results are given for groups for which estimates are available for two or more species. SE = standard error; N = number of species upon which mean estimate is based.

The proliferation of noncoding DNA

As noted in prior chapters, intronic and intergenic DNA increase the susceptibility of a gene to degenerative changes by increasing the size of the mutational target. The magnitude of this hazard depends on the number of nucleotide sites associated with an embellishment critical to gene function (n) and the per-nucleotide mutation rate (u), which together define the overall mutational disadvantage, $s = nu$. If $2N_g s << 1$, or equivalently, $2N_g u << 1/n$, the power of drift ($1/N_g$) is sufficiently large that an allele with an elevated deleterious mutation rate will behave in an effectively neutral fashion and hence be free to go to fixation. In contrast, sufficiently large $2N_g u$ serves as an effective barrier to the establishment of extragenic DNA.

To extend this principle to organelle genomes, we first revisit the problem of introns. In the case of the nuclear genome, $n \approx 25$ for introns serviced by the spliceosome, leading to an approximate threshold for intron colonization of $2N_g u < 0.04$ (see Chapter 9). Because organelle introns are more self-contained (i.e., they do not rely on an external spliceosome for removal) and assume specific secondary structures with many stems and loops, they must harbor a larger number of nucleotides critical to proper splicing than do typical spliceosomal introns. Thus, because $n > 25$, the threshold value of $2N_g u$ for organelle intron colonization must be considerably lower than 0.04. Several observations are compatible with this hypothesis.

First, the data in Table 11.5 indicate that average mitochondrial $2N_g u$ for land plants and bilaterian animals bracket the likely threshold value for organelle intron colonization, and whereas the mitochondrial genomes of bilaterian animals are completely devoid of introns, land plants contain between 0.3 and 0.8 group II introns per mitochondrial protein-coding gene (Lynch et al. 2006a). Although the mitochondrial mutation rate in green algae (the group most closely related to land plants) is unknown, the elevation of N_g in unicellular species suggests a higher $2N_g u$, and hence a less permissive environment for intron colonization, in such species. Thus, the observation that the average mitochondrial protein-coding gene in green algae contains fewer than 0.2 introns, and that most other unicellular eukaryotes contain none at all, suggests substantial intron colonization of land plant mitochondria subsequent to the reduction in $2N_g u$. Remarkably, although the mitochondria of both the liverwort *Marchantia* and angiosperms contain 25 group II introns, only one of these is in a conserved position (Knoop 2004). Such a condition is consistent with these two land plant lineages having reached an approximate equilibrium intron number determined by a stochastic birth–death process (see Chapter 9), although the possibility that both lineages are still experiencing parallel phases of intron gain cannot be ruled out.

Second, as noted above, the mutation rate in land plant plastids is a few times higher than in the mitochondrion. This observation implies that the plastid provides a somewhat less conducive environment for intron colonization than the plant mitochondrion, and consistent with this expectation, the intron density in the protein-coding genes of land plant plastids is less than a third that in the mitochondrion (see Table 11.1). The average number of introns in the protein-coding genes of green algal plastids is even lower, with several such species harboring no plastid introns at all. Most algae that have obtained their plastids by secondary endosymbiosis are also devoid of plastid introns, with one exception: the plastid of *Euglena gracilis* contains a remarkably high average of 2.2 introns per protein-coding gene (Hallick et al. 1993). Because euglenoids are thought to be obligately asexual, it is possible that the effective population sizes of such species are exceptionally low.

Third, although introns are universally absent from the mitochondria of bilaterian animals, the mitochondria of anemones, corals, and placozoans generally contain one to four group II introns. This pattern is consistent with the greatly reduced mutation rate observed in the mitochondrial genomes of anemones and corals relative to other animals (see Table 11.3), yielding a condition very much like that observed in plant organelles: the near-absence of mitochondrial nucleotide polymorphisms in global collections of such species (Shearer et al. 2002; Hellberg 2006). Although nothing is known about $2N_g u$ in placozoans, phylogenetic analysis suggests that they too may have low mitochondrial mutation rates (Dellaporta et al. 2006).

Finally, as noted in Figure 11.3, the scaling of the amount of intergenic DNA in organelle genomes parallels that for intronic DNA, with most bilaterian animal mitochondria containing 12% or less intergenic DNA and the largest plant mitochondrial genomes containing 80% or more. Land plant plastid genomes contain substantially lower amounts of intergenic DNA than do land plant mitochondria (see Table 11.1), and an exception among animal mitochondria is that of the placozoan *Trichoplax*, which contains some 42% intergenic DNA (Dellaporta et al. 2006). The mitochondria of almost all unicellular eukaryotes contain less than 10% intergenic DNA, although those of green algae and fungi contain averages of 25%–30% (less than half the level for land plants). Thus, as in the case of introns, it appears that prior to the emergence of multicellularity, most organelles contained small fractions of intergenic DNA (something on the order of 10%), with land plants experiencing a dramatic expansion in this component. One final exception that helps make this case is the wine yeast *Hanseniaspora uvarum*, which has the smallest known fungal mitochondrial genome (<19 kb, compared with a fungi-wide average of 43 kb), with only 5% intergenic DNA, as well as a substantial elevation in the mutation rate as inferred from phylogenetic analysis (Pramateftaki et al. 2006).

This diverse set of observations is consistent with the hypothesis that the dramatic divergence of organelle genome architectures by bilaterian animals and multicellular land plants from a more intermediate ancestral state is a simple indirect consequence of the hundredfold difference in mutation rates between these two lineages. As a consequence of their low mutation rates, plant organelle genomes provide population genetic environments (reflected in $2N_g u$) that are slightly more conducive to the accumulation of noncoding DNA than the nuclear genomes of multicellular species (see Chapter 4). In contrast, with their extremely high mutation rates, the mitochondria of bilaterian animals provide a population genetic environment that is comparable to that of the nuclear genomes of unicellular eukaryotes (and in some cases approaches that of prokaryotes), which encourages the elimination of mutationally disadvantageous noncoding DNA.

One alternative to the mutational-hazard hypothesis for the divergent evolution of organelle genome structure is the "race to replication" hypothesis, which postulates that a selective premium on high replication rates imposes strong selection for reduced genome sizes (Rand 2001; Selosse et al. 2001). Such a scenario is plausible in the sense that organelle genomes can multiply repeatedly within single cell cycles, setting up a competitive situation within cells, and it is well established that rapidly replicating but otherwise defective mitochondrial genomes can displace functional copies within the cells of yeasts and other species (Selosse et al. 2001; Taylor et al. 2002; Bernardi 2005). However, as in the case of replication rate arguments for the evolution of small microbial genome sizes (see Chapter 2), the "race to replication" hypothesis for organelle genome size evolution is confronted with several challenges. First, the wide range of variation in levels of extragenic DNA across the organelle genomes of animals, unicellular eukaryotes, and land plants (see Figure 11.3) remains to be explained. Why would selection for high replication rates be greater in animal mitochondria than in those of unicellular species, and why are land plants apparently immune to such effects? Second, regardless of whether a streamlined organelle genome can spread throughout a single organelle or even an entire host individual, establishment of such variants at the species level requires that the bearers of a reduced organelle genome have a selective advantage (or be neutral). Renegade organelle genomes with significant deletions from functional genes are expected to be purged at the population level, and with the nuclear genome being replicated just once per cell cycle, it is difficult to see how the generation time of an individual would be limited by the replication rate of its organelles. Third, if intra-organelle gene conversion occurs at a high enough rate, any replication advantage of a mutant genome might be overwhelmed by chance intracellular processes (Walsh 1992). Finally, unlike the "race to replication" hypothesis, the mutational-hazard hypothesis for organelle genome evolution has explanatory power that extends well beyond matters of noncoding DNA evolution, as will be seen in the following two sections.

Modifications of the genetic code

It has long been known that the "universal genetic code" is not actually invariant. Alterations of the code in nuclear genomes are rare, but a dozen or so protist lineages (mostly ciliates) and a few infectious bacterial species exhibit reassignments of single codons to different amino acids (Knight, Freeland, and Landweber 2001; Knight, Landweber, and Yarus 2001). The situation is dramatically different in the mitochondrion. Most mitochondrial lineages of unicellular eukaryotes have experienced one or two code changes, but more striking are the mitochondrial genomes of bilaterian animals, all of which have three to five changes in the code. At least twelve unique changes have occurred throughout the entire bilaterian phylogeny (and possibly many more; Abascal et al. 2006). In contrast, no reassignments are known in the organelle genomes of land plants, nor have any been found in the mitochondrial genomes of cnidarians. To understand the potential basis for this dichotomy in the incidence of organelle genetic code changes, some appreciation of the extreme conditions required for code modifications is necessary.

Messenger RNAs are decoded by a series of transfer RNAs (tRNAs), each of which reads one or more unique triplets via an anticodon and delivers a specific amino acid. Thus, a minimum requirement for a genetic code modification is that a tRNA be altered in such a way that it adds a novel amino acid to the growing polypeptide chain every time the old codon is encountered by the ribosome. This can happen if the anticodon of a tRNA is altered to recognize a novel codon, or if a tRNA modification results in charging by a different amino acid. Such tRNA mutants arise naturally, but the barriers to the permanent incorporation of a code change into a genome are enormous. In nuclear genomes containing thousands of genes, any reassignment of a codon is likely to alter the functions of hundreds of proteins, resulting in the rigid enforcement of the conventional nuclear genetic code.

This selective barrier is substantially relaxed in organelle genomes with no more than a few dozen protein-coding genes (just 13 in animals), in which the stochastic turnover of nucleotides at silent sites results in a small but appreciable chance that a codon will be entirely absent from a genome at any time. Indeed, about 70% of sequenced mitochondrial genomes completely lack one or more codons, and it is common for single codons to constitute less than 5% of the members of twofold or fourfold redundant sets for specific amino acids (Swire et al. 2005). As pointed out in the "codon-capture" model, the transient absence of a codon provides a simple first step in the alteration of a genetic code, as any unutilized codon is free to be reassigned once it reappears by mutation (Osawa et al. 1992; Jukes and Osawa 1993), but several additional conditions must be met (Figure 11.5). First, a tRNA must be available for recruitment, and in organelle genomes, just one tRNA often decodes all of the codon variants for a particular amino acid. Thus, tRNA reassignment will generally require the modification of the anticodon of a tRNA so as to expand its decoding capacity. Second, the tRNA that previously decoded the now missing codon must be lost by degenerative mutation or modified so as to no longer recognize the missing codon.

Codons	Transfer RNAs	
(1) AAG----AAA----AAG----AAG----AAG----AAG lys lys lys lys lys lys	Lys-AA(A/G)	Asn-AA(U/C)
(2) AAG----AAG----AAG----AAG----AAG----AAG lys lys lys lys lys lys	Lys-AA(A/G)	Asn-AA(U/C)
(3) AAG----AAG----AAG----AAG----AAG----AAG lys lys lys lys lys lys	Lys-AAG	Asn-AA(U/C)
(4) AAG----AAG----AAG----AAG----AAG----AAG lys lys lys lys lys lys	Lys-AAG	Asn-AA(A/U/C)
(5) AAG----AAG----AAG----AAG----AAG----AAA lys lys lys lys lys asn	Lys-AAG	Asn-AA(A/U/C)

Figure 11.5 A hypothetical example of a change in the genetic code by codon capture. (1) Initially, lysine is encoded by both AAA and AAG, both of which are recognized by the same tRNA. In contrast, the asparagine tRNA recognizes AAC and AAU codons. (2) The rare AAA codon is lost by mutation and random genetic drift. (3) The loss of AAA provides an environment that allows tRNA-Lys to acquire a change such that it now recognizes only AAG. (4) tRNA-Asn, which was previously selected to recognize C or U but to avoid A and G in the third position, is now partially released from selection, and it neutrally acquires an alteration that allows recognition of AAA, AAC, and AAU, but not AAG. (5) An AAA codon that arises by subsequent mutation is now read as Asn. This particular change in the genetic code actually occurred in the mitochondrial genome of echinoderms.

If this is to occur, the genome must not revert to use of the superfluous codon at sites where the original amino acid is functionally significant before a potential tRNA recruit acquires its new function by fortuitous mutation, as this would enforce the maintenance of the original tRNA for that codon. Third, once the new tRNA has become established by chance, the missing codon must reappear in the genome in such a way that the loading of a new amino acid will be beneficial (or at least not harmful) to individual fitness, so as to stabilize the newly evolved tRNA.

Although this informal description of the codon-capture model glosses over numerous details, the key point is that codon reassignments by semi-neutral processes require the fortuitous occurrence of several mutational events in the same linked genome. The small sizes and high degree of linkage of organelle genomes predispose them to modifications of the genetic code, and the exceptionally high mutation rates in the mitochondrial genomes of bilaterian animals may greatly elevate the probability of such changes simply by reducing the waiting time for the joint occurrence of multiple mutations. Under this view, the hundredfold reduction in the mutation rate of plant mitochondrial genomes (combined perhaps with their roughly threefold greater number of protein-coding genes) is the primary impediment to the origin of code changes, possibly only delaying their ultimate occurrence rather than preventing them entirely. In addition, the few microbial eukaryotes that have experienced a mitochondrial genetic code change may have achieved that state by the combined effects of an elevated mutation rate (relative to land plants) and an increase in the number of mutational targets (individuals per species). Finally, given their intrinsic rarity (just one per gene), it is not surprising that termination codons are exceptionally vulnerable to codon reassignments, constituting 10 of the 25 known cases of mitochondrial genetic code changes. In most cases, UGA is reassigned to tryptophan, which is ordinarily encoded only by UGG. Such modifications are readily explained by the codon-capture model, as tRNA-Trp must simply be modified to recognize UG(A/G).

Some questions have been raised about the universality of the effectively neutral codon-capture model, the main concern being that genetic code changes may not always have started with absent codons (Knight, Freeland, and Landweber 2001; Knight, Landweber, and Yarus 2001; Swire et al. 2005). An alternative adaptive hypothesis, the "ambiguous-codon" model of Schultz and Yarus (1994), postulates that the reassigned codons are actually present during the initiation of genetic code transitions, with selection somehow favoring a transformation of all such sites to an alternative amino acid (perhaps one less energetically expensive). Such an endpoint could be achieved if the tRNA for the disfavored codon were modified to recruit the preferred amino acid, or if the tRNA for the preferred amino acid were modified to recognize the disfavored codon. The obvious challenge to this hypothesis is its requirement for a setting in which the simultaneous replacement of one amino acid by another in multiple proteins is beneficial (or at least not harmful) with respect to fitness.

Messenger RNA editing

A long history of research in molecular biology leads us to expect that coding information at the DNA level will provide a reliable prediction of protein sequences, and with just a few exceptions, this has generally been found to be true for nuclear genes (Gerber and Keller 2001; Keegan et al. 2001; Reenan 2001). However, some organelles use posttranscriptional editing to modify mRNA, tRNA, and/or rRNA sequences. The most spectacular display of posttranscriptional editing occurs in the mitochondrial genes of kinetoplastids (e.g., *Trypanosoma*), where insertion and deletion of Us (uracils) affects about 90% of all codons (Simpson et al. 2000; Horton and Landweber 2002). With editing at such a massive scale, the underlying genomic sequences are literally nonsensical. The complex tasks of insertion/deletion editing in such species are accomplished via the unique structure of the kinetoplastid mitochondrial genome: a vast network of intertwined circular molecules, including several 20–40 kb maxicircles carrying the cryptic gene sequences as well as thousands of 0.5–3.0 kb minicircles, which carry guide RNA (gRNA) templates for the addition/removal of Us in the immature maxicircle-derived mRNAs (Koslowsky et al. 1992).

No other mitochondrial lineage engages in editing as extensively as kinetoplastids, although some slime molds insert nucleotides (usually C) approximately once every 25 nucleotides in mRNAs and once every 40 nucleotides in rRNAs and tRNAs (Horton and Landweber 2000; Cheng et al. 2001; Byrne and Gott 2004). In dinoflagellate mitochondria, editing modifies 2%–4% of the total sequence, mostly via three different types of nucleotide substitutions at amino acid replacement sites (Lin et al. 2002; Zhang and Lin 2005). This sporadic distribution of editing, along with the diverse mechanisms employed (Table 11.6), strongly suggests that such processes have evolved independently in different lineages.

Animals and land plants differ dramatically with respect to the incidence of editing in organelles. A few animals, such as centipedes (Lavrov et al. 2000), snails (Yokobori and Pääbo 1995), squids (Tomita et al. 1996), and marsupials (Janke and Pääbo 1993), use editing to repair base mismatches in the

TABLE 11.6 A summary of some aspects of editing of organelle mRNAs

PHYLOGENETIC GROUP	MODIFICATIONS	REFERENCE
Physarum (slime mold) mt	Primarily C insertions; but some A and U insertions, C→U substitutions, and a few dinucleotide insertions	Byrne and Gott 2004
Dinoflagellate mt	Primarily A→G, U→C, C→U	Lin et al. 2002
Kinetoplastid mt	U insertion and deletion	Landweber and Gilbert 1993
Plant mt	C→U, U→C	Gillham 1994
Plant cp	C→U, U→C	Tsudzuki et al. 2001

stems of single mitochondrial tRNAs, a process that has also been observed in a rhizopod amoeba (Lonergan and Gray 1993) and some basal fungi (Paquin et al. 1997; Laforest et al. 2004). However, whereas mRNA editing appears to be entirely absent from animal mitochondria, it is used extensively in land plant organelles. In *Arabidopsis* mitochondria, for example, 441 editing sites are present in coding regions along with smaller numbers in introns and intergenic DNA, nearly all of them changing C to U (Giegé and Brennicke 1999). Similar levels of C→U mRNA editing are found in the mitochondria of other land plants, including some liverworts (Malek et al. 1996; Freyer 1997). Although editing is much less extensive in land plant plastids, it is still prominent, with about 25–30 editing sites per genome in angiosperms (Tsudzuki et al. 2001) and up to several hundred sites in ferns and hornworts (Kugita et al. 2003; P. G. Wolf et al. 2004). Plastid mRNA editing shares many mechanistic features with editing in mitochondria (Tillich et al. 2006). These observations, combined with the absence of mRNA editing from the organelles of green algae, suggest a dramatic expansion of organelle editing with the origin of multicellular plants (Hiesel et al. 1994).

The vast majority of mRNA editing in land plant organelles occurs at amino acid replacement (rather than silent) sites, with the changes often ensuring the preservation of amino acids that are highly conserved across distantly related species (Maier et al. 1996; Tsudzuki et al. 2001). Although this type of observation motivates the idea that editing provides a genomic buffer against the accumulation of deleterious mutations (Cavalier-Smith 1997; Horton and Landweber 2002; Smith 2006), several observations raise significant doubts about this adaptive interpretation. First, if the buffering hypothesis were correct, we would expect editing to be most common in genomes with high mutation rates, exactly the opposite of what is observed. Second, the buffering hypothesis ignores the complexities of the editing process, which necessarily relies on the sequence stability of both the recognition sites in the organelle transcripts and the editing apparatus itself. Although the exact mechanism of mRNA editing is unknown in plants, *cis* recognition sites for *trans*-acting editing factors appear to constitute a minimal span of about 23 bp (Choury et al. 2004; Miyamoto et al. 2004), and contrary to the situation in kinetoplastids, the *trans*-acting factors appear to be editing-site-specific proteins encoded in the nucleus (Kotera et al. 2005; Shikanai 2006). It is difficult to imagine a net advantage for editing if the processing of each site depends on the integrity of a large number of *cis* and *trans* factors. Third, editing in plant organelles is quite noisy, resulting in the production of a heterogeneous pool of transcripts, some of which are incompletely edited and others containing erroneous editorial changes (Phreaner et al. 1996; Inada et al. 2004). In many cases, completely edited transcripts are the exception rather than the rule (Schuster et al. 1990), which again hardly seems advantageous to the individual.

Perhaps the most significant challenge to the hypothesis that mRNA editing is maintained by selection derives from the following observation. Recall that the vast majority of mRNA editing in plant organelles involves conversions of C to U. Because C→T mutations at these sites eliminate the need for editing, such mutations are expected to accumulate at the neutral rate under the buffering hypothesis, as an allele with an encoded T should be selectively equivalent to one that simply acquires a C→U replacement by editing. However, although editing sites do turn over frequently between species (only 45% of the editing sites in plastids and 58% of those in the mitochondrion are shared between monocots and dicots; Tsudzuki et al. 2001; Shields and Wolfe 1997), the conversion of C→U editing sites to unedited Ts appears to be a non-neutral process. Under the neutral hypothesis, the rates of nucleotide substitution at editing sites should equal those at silent sites. However, the former rate is four times larger than the latter (Shields and Wolfe 1997), and this discrepancy is likely to be an underestimate, as plant organelle editing sites tend to be in UNA contexts (where N is the edited site), which have lower than average mutation rates (Tillich et al. 2006). Thus, the loss of editing sites from land plant organelles appears to be promoted by positive selection, which implies that editing is selectively disadvantageous.

Aside from the problem of inaccuracies in editing, alleles that rely on editing for the development of productive mRNAs have a fundamental cost: an increased vulnerability to inactivating mutations. Following the same logic discussed above for introns, the minimal intrinsic mutational disadvantage of an organelle editing site will be approximately equal to the total mutation rate summed over the nucleotide sites reserved for editing-site recognition. Based on the size of such sites, this excess mutation rate appears to be approximately $20u$ (not including the cost of maintaining the editing apparatus). Because this disadvantage is expected to be approximately 100 times greater in animal than in land plant mitochondria due to the inflated mutation rate in the former, the absence of mRNA editing in animals is in accordance with the hypothesis that the mutation-associated disadvantage of editing is simply too great to allow its establishment (or maintenance) in that lineage. The reduced level of editing in land plant plastids versus mitochondria is also consistent with this hypothesis, as the former have higher mutation rates than the latter. Finally, if this hypothesis is correct, then one would expect to see a dramatic reduction in the incidence of editing in the mitochondrial genomes of plants such as *Plantago* and *Pelargonium*, which have experienced hundredfold increases in the mutation rate (Palmer et al. 2000), and this is indeed the case (Parkinson et al. 2005).

Although the preceding arguments lead to a reasonably satisfying neutral hypothesis for the phylogenetic distribution of editing, substantial questions remain as to how such processes initially became established. Stoltzfus (1999) and Horton and Landweber (2002) have suggested a scenario for the origin of guide RNAs in kinetoplastid mitochondria by effectively neutral mechanisms, although that is a rather specialized system. Our attention here will focus on the editing system in land plants, which is challenging enough. To account for such a system, a mechanism must exist for the establishment of multiple nuclear-encoded editing factors, each specialized for the alteration of a small number of organelle sites (perhaps even single sites), and this in turn presumably requires the existence of factors with editing potential prior to the origin of mRNA sites requiring editing. The long-term survival of an editing factor would be ensured so long as an essential editing site did not revert to the preferred nucleotide. Getting to this point, however, would be difficult with a nuclear-encoded factor, which would have to remain in tight linkage disequilibrium with the serviced organelle site while both are en route to fixation. Uniparental inheritance of organelles facilitates such associations, as half the gametes of the transmitting parent will contain the appropriate nuclear–cytoplasmic combination, but the probability of dissociation is still considerable.

Not all aspects of such a scenario are far-fetched. For example, the one nuclear factor that has been identified for a particular editing site in the plastid of *Arabidopsis* is a member of a large protein family involved in a wide range of organelle mRNA maturation activities (Shikanai 2006). Such proteins may be predisposed to recruitment to editing processes. The size of this family in *Arabidopsis* (about 450 genes), which greatly exceeds that in other well-characterized eukaryotes (including animals), closely approximates the number of mitochondrial editing sites in *Arabidopsis*, raising the possibility that the entire family diversified and proliferated in a stepwise fashion with the establishment of editing sites in land plants. Nevertheless, the mechanisms by which editing factors acquire their apparent site specificity remain mysterious. A really creative selfish editor might inflict the organelle genomic change necessary to ensure its own survival, and although no such element is known to exist, it is intriguing that some proteins involved in the editing of vertebrate nuclear mRNAs are capable of inducing site-specific mutations (Smith 2006).

In summary, although the origin of mRNA editing is one of more enigmatic aspects of genome evolution, there is no evidence that such processes have originated to buffer mutational damage. Nor is there any evidence that the occupation of dry land by plants led to an altered mutagenic environment that promoted the need for such buffering (Smith 2006). In addition, the hypothesis that editing promotes the generation of adaptive variation at the RNA level (Tillich et al. 2006) is entirely without support. In short, it is difficult to reject the hypothesis that mRNA editing in organelles arose by nearly neutral processes in mutationally permissive environments.

Muller's Ratchet and the Mitochondrion

Because of their reduced ability to shed mutations by recombination, organelle genomes are expected to be uniquely susceptible to degradation by a process known as Muller's ratchet (Muller 1964; Felsenstein 1974). In the absence of segregation and recombination, parents cannot produce offspring with a reduced number of deleterious mutations, except in the rare case of back or compensatory mutations. Thus, in an asexual population, if by chance the fittest class of individuals (those with the smallest deleterious mutation load) produces no surviving offspring, or only offspring that have acquired at least one new deleterious mutation, a nearly irreversible decline in fitness is experienced. There is an appreciable chance of such an event each generation because recurrent mutation pressure generally reduces the fittest class to a small fraction of the total population (Haigh 1978). Moreover, each time the currently fittest class is lost from the population, the previously second-best class becomes subject to the same stochastic process, eventually suffering an identical fate. Once this process has proceeded to the point at which the mutation load is so high that the average individual cannot replace itself, the population size must begin to decline. This decline further enhances the magnitude of random genetic drift, promoting increasingly higher rates of deleterious mutation accumulation and ultimately culminating in population extinction (Lynch and Gabriel 1990; Lynch et al. 1993, 1995a,b). Such mutational meltdowns are likely to be a primary cause of the typically short life spans of obligately asexual lineages (Lynch and Gabriel 1990; Lynch et al. 1993; Paland and Lynch 2006), except in cases in which population sizes are so enormous that back and/or compensatory mutations provide a remedy.

Sexual reproduction facilitates the avoidance of such a fate by enabling parents to produce progeny with reduced numbers of deleterious mutations via segregation and recombination (Maynard Smith 1978; Pamilo et al. 1987; Charlesworth et al. 1993; Lynch et al. 1995a,b). Indeed, this mechanism of shedding mutations appears to be so powerful that the nuclear genomes of large outcrossing populations are nearly invulnerable to mutational meltdowns on time scales shorter than hundreds of thousands of generations (Lande 1994; Lynch et al. 1995a,b; Schultz and Lynch 1997). Nevertheless, the possibility still exists that the viability of sexual species is compromised by the gradual mutational decay of their nonrecombining organelle genomes (Takahata and Slatkin 1983; Hastings 1992; Gabriel et al. 1993; Reboud and Zeyl 1994; Loewe 2006).

Resolution of this issue will ultimately depend on empirical observations on the rates and effects of organelle mutations. The fact that the effective number of genes residing at mitochondrial loci is generally no greater, and often considerably smaller, than the number at nuclear loci (see Table 11.4) is consistent with the idea that the efficiency of selection against deleterious mutations is reduced in organelle genomes. However, several features peculiar to organelles might reduce the rate of deleterious mutation accumulation. For example, replication of the mitochondrion independently of the cell cycle induces a within-generation drift process in the germ line, which might enhance the efficiency of selection by increasing the variance in mutant organelle content among maternal progeny (Takahata and Slatkin 1983; Bergstrom and Pritchard 1998). In mammals, there is also considerable loss of the female germ cell population via apoptosis during development, leading Krakauer and Mira (1999) to suggest that selective cell mortality acts as a sieve against deleterious mutation–containing organelles, a hypothesis that has some empirical support (Perez et al. 2000).

Despite these concerns, several indirect lines of evidence point to the exceptional vulnerability of organelle genomes to deleterious mutation accumulation. For example, numerous of human genetic disorders resulting from mitochondrial mutations are associated with deleterious alleles with population frequencies as high as 5% (Brown and Wallace 1994; Wallace 1994; Simon and Johns 1999; Ruiz-Pesini et al. 2000; Kivisild et al. 2006). In addition, population surveys of nucleotide sequence variation in organelles consistently reveal that ratios of nonsynonymous (amino acid replacement) to synonymous (silent-site) polymorphisms segregating within populations exceed those for divergence between closely related species. The most reasonable interpretation of this pattern is that some mildly deleterious replacement mutations are able to expand to frequencies high enough to be observed in population surveys, but not so high as to go to fixation. For the great apes, the within-species ratio of polymorphism at replacement sites to polymorphism at silent sites is 10 times greater than that for fixed differences among species (Nachman et al. 1996; Templeton 1996; Hasegawa et al. 1998; Wise et al. 1998). There is similar (13-fold) inflation in mice (Nachman et al. 1994), and although the disparity is smaller, it is nevertheless significant in *Drosophila* (1.5-fold; Ballard and Kreitman 1994; Rand et al. 1994), birds (1.9-fold; Fry 1999), and the annual plant *Silene* (2.5-fold; Städler and Delph 2002). These patterns are in complete opposition to those seen in nuclear genomes, in which there is generally an excess of replacement substitution at the level of divergence, a presumed reflection of fixation of adaptive mutations (see Chapter 4). Because mildly deleterious mutations almost certainly have a wide range of selective effects (Lynch and Walsh 1998), if such mutations are more capable of rising to moderate frequencies in organelle than in nuclear genomes, the fixation probabilities of subclasses with the mildest deleterious effects must be elevated as well.

A more direct approach to obtaining insight into the matter of deleterious mutation accumulation would be to examine the evolutionary fates of genes with identical functions in the mitochondrial and nuclear genomes of the same species. Unfortunately, this is not possible with protein-coding genes, as there are no nuclear/mitochondrial pairs with homologous functions. However, transfer RNA (tRNA) and ribosomal RNA (rRNA) genes provide an ideal venue for such analysis, as most organelle genomes contain full sets of both, which like their nuclear counterparts, play central roles in mRNA translation. The extreme conservation of the primary, secondary, and tertiary structure of prokaryotic and eukaryotic nuclear tRNAs (Kimura 1983; Söll and RajBhandary 1995) and rRNAs (Smith 2001) implies that natural selection is exceptionally successful at maintaining the optimal molecular architecture of these genes. For example, in all prokaryotes and all eukaryotic nuclear genomes, tRNAs have a standard cloverleaf secondary structure, with their roughly seventy bases mostly contained in three loops and four stem structures. Thirteen of these bases (not including the three nucleotides in the anticodon sites) are essentially invariant across the tRNAs for all amino acids in all prokaryotes and all nuclear genomes, and the average rate of nucleotide substitution across the molecule is just 1%–3% of the neutral expectation (Lynch 1997, 1998; Lynch and Blanchard 1998). Because this degree of constancy must have been present in the primordial organelle genomes, any deviations of organelle-encoded tRNAs from the canonical prokaryotic/nuclear architecture are likely to reflect a reduction in the efficiency of selection imposed by the population genetic environment of organelles. The evidence for such a shift is compelling.

First, contrary to the situation in the nuclear genome, there are no invariant sites in organelle tRNAs, and every region of the animal mitochondrial tRNA molecule evolves at a higher rate than the homologous region in the nuclear tRNA. These differences are not simply consequences of elevated mitochondrial mutation rates, because the ratio of the observed substitution rate to the neutral expectation (the width of selective sieve, which estimates the fraction of mutations that avoid elimination by selection; see Box 3.1) is elevated in mitochondrial tRNAs of all phylogenetic groups by factors of 1.3–13.1 relative to the same ratio in nuclear tRNAs (Table 11.7). A similar picture emerges from rRNA analyses, which yield increases in the width of the selective sieve ranging from 2.2 to 10.0. Second, animal mitochondrial tRNAs exhibit a wide array of structural deviations from the canonical form of prokaryotic/nuclear tRNAs, including losses of entire arms in some cases (Wolstenholme 1992). The width of the selective sieve for insertion/deletion mutations in the loops of mitochondrial tRNAs is five times that for nuclear tRNAs (see Table 11.7). Third, for animals, plants, and fungi, the average binding strength of mitochondrial tRNA stems is 40%–90% that of nuclear tRNA stems, largely due to the higher incidence of A:T versus G:C bonds (two vs. three hydrogen bonds) in the former (Lynch 1997, 1998). Such modifications lead to conditions in which the stems of mammalian mitochondrial tRNAs initiate melting well below normal physiological temperatures (Ueda et al. 1985; Kumazawa et al. 1989).

Although compensatory mutations in tRNA molecules, such as the restoration of Watson–Crick base pairs, may eventually mitigate some of the negative effects of single-base changes (Steinberg and Cedergren 1994; Steinberg et al. 1994; Watanabe et al. 1994; Wolstenholme et al. 1994; Kern and Kondrashov 2004), experimental evidence suggests that the net effects of the structural modifications noted above compromise the functional efficiency of protein synthesis. For example, bovine mitochondrial tRNA-Phe has been shown to have an unusually low rate of amino acid loading (Kumazawa et al. 1989), and an in vitro translation system containing bovine mitochondrial ribosomes and elongation factors revealed that the rate of phenylalanine loading increases approximately tenfold when *E. coli* tRNA-Phe is substituted for the bovine mitochondrial tRNA (Kumazawa et al. 1991). The bovine mitochondrial tRNA for serine, which lacks an entire arm, has also been shown to be functionally compromised (Hanada et al. 2001).

Could the increased width of the selective sieve for organelle genes be a simple consequence of the relaxation of selection in organelles (i.e., smaller selection coefficients), as suggested by Brown et al. (1982) and Kumazawa

TABLE 11.7 Estimates of the width of the selective sieve (measured as the ratio of the rate of evolution of RNA genes to the rate of synonymous substitution in protein-coding genes in the same genome) for organelle (or) vs. nuclear (nu) genes

	WIDTH OF SIEVE (ω)[a]		sN_g[b]	
	or	nu	or	nu
tRNA substitutions				
Mammal mt[c]	0.18	0.07	1.40	2.05
Invertebrate mt	0.38	0.10	0.85	1.80
Yeast mt	0.50	0.20	0.63	1.33
Plant mt	1.18	0.09	0.00	1.87
Plant cp	0.12	0.09	1.68	1.87
rRNA substitutions				
Mammal mt	0.14	0.02	1.58	2.82
Invertebrate mt	0.44	0.20	0.73	1.33
Yeast mt	0.80	0.08	0.22	1.95
Plant mt	0.14	0.02	1.58	2.82
Plant cp	0.07	0.02	2.05	2.82
tRNA insertion/deletions				
Mammal mt	0.26	0.05	1.14	2.26

[a]The results for substitutions are from Lynch (1997) and Lynch and Blanchard (1998), whereas those for insertion/deletions, which apply to tRNA loops, are from Lynch 1996. The results for tRNAs are averages over all 20 amino acid groups, while those for rRNAs are averages over the small and large subunits.

[b]The quantity sN_g is equal to the product of the deleterious selection coefficient and the effective number of genes per locus.

[c]Mitochondrial and chloroplast genes are denoted by mt and cp.

and Nishida (1993), rather than an outcome of a reduction in the efficiency of selection (smaller N_g)? Insight into this matter can be acquired by recalling from Box 3.1 that if most mutations are either neutral or deleterious, the width of the selective sieve can be expressed as $\omega \approx 2sN_g/(e^{2sN_g} - 1)$, where s is the average selection coefficient opposing new mutations and N_g is the effective number of genes at a locus in the population. Solving this equation for organelle (o) and nuclear (n) genes using the data for ω in Table 11.7 provides estimates of s_oN_{go} and s_nN_{gn}, and taking the ratio of these estimates and applying the N_{go}/N_{gn} ratios given above (see Table 11.4) then yields estimates of the ratio of average selective intensities in the two cellular environments, s_o/s_n. For mammals, the average value of N_{go}/N_{gn} is about 1.3, so from Table 11.7 the estimated ratio of selective intensities is (1.40/2.05)/1.3

= 0.5 for tRNAs and (1.58/2.82)/1.3 = 0.4 for rRNAs. For invertebrates, the average value of N_{go}/N_{gn} is about 0.3, which implies ratios of selective intensities of 1.6 for tRNAs and 1.8 for rRNAs. Similar calculations suggest s_o/s_n of 1.1 and 0.3 for yeast tRNAs and rRNAs, respectively, and 0.9 and 0.7 for land plant plastid tRNAs and rRNAs, respectively. If selectively driven codon bias is more significant in nuclear than in mitochondrial genes, as seems likely to be the case, these estimates of s_o/s_n will be downwardly biased by inflated estimates of the width of the selective sieve for nuclear genes. Thus, although crude, these calculations suggest that the absolute strength of selection opposing mutations in organelle genes is similar to that in nuclear genes, if not slightly higher, implying that the greater width of the selective sieve in organelle genomes is indeed a consequence of a reduction in the efficiency of selection. There is one potential outlier to this conclusion: for land plant mitochondrial tRNAs, the estimate of ω is approximately equal to the neutral expectation of 1.0 (see Table 11.7). However, this estimate may be upwardly biased by the inadvertent inclusion of some laterally transferred genes from the plastid (Joyce and Gray 1989; Marechal-Drouard et al. 1990; Nakazono and Hirai 1993).

These results provide the basis for a rough assessment of the degree to which the long-term fitness of eukaryotes might be imperiled by their nonrecombining organelle genomes. Consider the situation in animal mitochondria, in which the mutation rate is quite high, averaging 3×10^{-8} per site per year (see Table 11.3), with ~12,000 bases encoding tRNAs, rRNAs, amino acid replacement sites, and origins of replication. If it is assumed that nuclear tRNA and rRNA mutations escaping the selective sieve are essentially neutral with respect to fitness, then the difference in the sieve widths between organelle and nuclear genes, which averages 0.20 in animals, represents the fraction of newly arisen mutations that are both deleterious and capable of fixation in organelles. Multiplying these three factors together, the average genome-wide rate of deleterious mutation in mitochondria is roughly 10^{-4} per year. The average effect of a new mitochondrial mutation on fitness is roughly 0.01 in animals, which is much too large to allow fixation unless effective population sizes are on the order of 100 or smaller (which is not the case), but because the distribution of mutational effects is quite L-shaped (Lynch and Walsh 1998), plenty of very mildly deleterious mutations are available for fixation. As discussed in Chapter 7, the mutations that do the greatest long-term damage to a population have selection coefficients on the order of the reciprocal of the effective population size, as these have the largest fitness effects that are subject to substantial random genetic drift (Lynch and Gabriel 1990; Lande 1994). Using the results from Chapter 4 and Table 11.4, a typical animal species has an organelle effective population size (N_g) of approximately 10^5, so the maximum rate of fitness decline associated with mitochondrial mutations in a natural population is on the order of $10^{-4} \times 10^{-5} = 10^{-9}$ per year (ignoring the fixations of mutations of larger effects that might occur during population bottlenecks). Assuming multiplicative fitness effects of mutations, this level of mutation accumulation would yield just a 50% decline in fitness since the origin of animals about 700 million years ago (see Chapter 1), during which time compensatory and/or back mutations could reduce the rate of fitness loss to an even lower level.

Thus, the observed levels of mutation accumulation in organelle genomes do not necessarily impose a substantial threat to species survival on time scales less than 10^8 million years or so. However, because the preceding estimates are derived from surviving lineages, and because ecological processes may cause N_g to be substantially less than 10^5 in some lineages for extended periods, it is premature to dismiss the possibility of a role for buildups of mitochondrial mutation loads in the differential proliferation of some eukaryotic lineages. One potentially important complication that has been omitted from the previous analysis concerns a unique evolutionary consequence of uniparental inheritance: a maternally inherited genome can be selected only on the basis of its influence on female fitness. This means that sexually antagonistic organelle mutations, with positive effects on females but negative effects on male performance, can be driven to fixation despite their potential threat to long-term population viability (Gemmell et al. 2004).

Insights from More Recent Endosymbioses

Although mitochondria and plastids represent the most widespread cases of microbial endosymbiosis, many more phylogenetically restricted endosymbionts are known. The eubacterial inhabitants of various insect lineages have received the most attention. Because most such species are strictly vertically inherited like mitochondria, their N_g is again expected to be comparable to that of the host species, although the polymorphism and divergence data necessary for testing this hypothesis are limited. For the one system for which data are available, the γ-proteobacterium *Buchnera aphidicola* residing within the aphid *Uroleucon ambrosiae*, the endosymbiont's N_g is estimated to be about 1.9 times that of the host mitochondrion (Lynch 2006b), which itself is quite low (Funk et al. 2002), and its $2N_g u$ is 25 times smaller than the average for free-living and/or pathogenic prokaryotes (Lynch 2006b). Thus, although surveys of other species will be necessary to evaluate the generality of this conclusion, it is reasonable to expect the genomes of arthropod endosymbionts to exhibit the symptoms of a reduction in the efficiency of natural selection associated with reduced N_g.

This expectation appears to be upheld (Moran and Plague 2004). For example, four sequenced *B. aphidicola* genomes, each from a different aphid lineage, have an extraordinarily high A/T content (~74%), consistent with a strongly mutation-driven process, and comparative analyses suggest elevated long-term rates of molecular evolution and the accumulation of mildly deleterious mutations in protein-coding and ribosomal RNA genes (Moran 1996; Lambert and Moran 1998; Shigenobu et al. 2000; Tamas et al. 2002; Herbeck et al. 2003; Bastolla et al. 2004; van Ham et al. 2003). Containing just 362–630 protein-coding genes, *Buchnera* genomes are among the smallest known in prokaryotes (Latorre et al. 2005; Pérez-Brocal et al. 2006), and many hundreds of genes (including many involved in DNA replication and repair) were probably lost early in the establishment of the lineage (Klasson and Andersson 2006), putatively in part because of the reduced efficiency of selection for gene maintenance in a stable host cell environment. A similar series of events is recorded in other eubacterial endosymbionts, such as *Blochmannia* (an inhabitant of ants), *Wigglesworthia* (an inhabitant of tsetse flies), and *Carsonella* (an inhabitant of sap-feeding psyllid insects) (Akman et al. 2002; Gil et al. 2003; Canbäck et al. 2004; Degnan et al. 2005; Fry and Wernegreen 2005; Nakabachi et al. 2006).

Given the likely reduction in $2N_g u$, the genomes of these microbes might be expected to evolve elevated levels of noncoding DNA, as the nuclear and mitochondrial genomes of eukaryotes have done, but they are often nearly devoid of mobile elements and contain approximately the same proportion of noncoding DNA as other prokaryotes (Moran and Plague 2004). Taken at face value, such observations might be viewed as a major challenge to the central premise of the mutational-hazard hypothesis: that low $2N_g u$ encourages the accumulation of nonfunctional DNA. However, the genomic response to reduced $2N_g u$ depends not only on the mutational cost of excess DNA, but also on the relative rates of insertion and deletion of extragenic DNA. If deletion mutations were to arise more frequently than insertions, relaxed selection would lead to a reduction, rather than an increase, in genome size. Thus, contrary to the claims of some authors (Gregory and DeSalle 2005), the rarity of genomic obesity in arthropod endosymbionts need not be incompatible with the hypothesis that genome evolution is driven by basic population genetic processes.

There are at least two reasons why the spectrum of mutations in some microbial endosymbionts may evolve in the direction of a deletion bias. First, as noted in Chapter 7, species with a strictly clonal mode of inheritance provide an unstable habitat for mobile elements (a major source of insertions in most genomes): overly aggressive mobile elements may cause the extinction of asexual hosts, whereas overly quiescent elements will eventually become completely inactivated by mutations (Arkhipova and Meselson 2000; Dolgin and Charlesworth 2006). Such a filtering process may explain why prokaryotes in putatively early stages of host associations (described below) and *Wolbachia*, a broadly distributed endosymbiont with a capacity for horizontal transfer and recombination, contain large numbers of insertion elements (Jiggins 2002; Moran and Plague 2004; Baldo et al. 2006), whereas more ancient endosymbionts with strict vertical transmission are devoid of them. Second, although it has not been determined directly whether deletion mutations associated with replication and/or chromosome maintenance outnumber insertions in *Buchnera* and allies, this is likely to be the case. Most ancient arthropod endosymbionts share the unique attribute of having lost conventional mismatch repair pathways, and it is known that defective mismatch repair greatly magnifies the deletion rate by promoting recombination between nonorthologous sequences (Nilsson et al. 2005).

Observed accumulations of deletions in pseudogenes in *Rickettsia* are consistent with such behavior (Andersson and Andersson 1999), although it is not possible to exclude a role for selection in this process.

Three specific examples clearly demonstrate that genome size reduction is by no means universal in arthropod endosymbionts. First, *Sodalis glossinidius*, an endosymbiont of tsetse flies, has only 51% coding DNA (Toh et al. 2005), by far the lowest value for any well-characterized prokaryote except for the intracellular parasite *Mycoplasma leprae* (48%). Perhaps it is no coincidence that unlike most other arthropod endosymbionts, *S. glossinidius* has retained its DNA repair genes. Second, the genomes of three of the four extracellular symbionts associated with the marine oligochaete *Olavius algarvensis* (which lacks a mouth, gut, and anus) contain 300–400 mobile elements (Woyke et al. 2006), an extraordinarily high number for a prokaryote. Third, thousands of parasitoid wasp species use endosymbiotic viruses (from the Polydnaviridae family) to facilitate progeny development within parasitized lepidopteran larvae. Although most viruses are pathogens with substantial capacity for horizontal transmission and presumably high N_g, as reflected in their extremely streamlined genomes (Lynch 2006b), the chromosomes of polydnaviruses are integrated directly into the host nuclear genome, subjecting them to the same drift and mutation processes as the wasps within which they reside. The genomes of polydnaviruses are enormous (~200–600 kb), with 71%–83% noncoding DNA, including numerous mobile elements, and 10%–20% of the protein-coding genes harbor spliceosomal introns (all highly unusual features for DNA viruses) (Espagne et al. 2004; B. A. Webb et al. 2005). Thus, when placed in a permissive population genetic environment, even viruses can succumb to the syndrome of genomic changes seen in eukaryotes with low $2N_gu$.

One other unique form of endosymbiosis merits consideration here. As noted above, a number of eukaryotes have acquired the capacity for photosynthesis via secondary endosymbiosis; that is, by engulfing members of the primary plastid-carrying lineage. The ancestors of the cryptomonad and chlorarachniophytic algae were the recipients of red and green algal endosymbionts, respectively, and in both cases, the nucleus of the endosymbiont still survives as a nucleomorph. One complete nucleomorph genome has been analyzed for each lineage (the respective host cells being *Guillardia theta* and *Bigelowiella natans*). The two nucleomorphs exhibit strikingly similar patterns of evolution, including massive genomic streamlining, with one notable exception. The *B. natans* nucleomorph genome is just 0.37 Mb in length and contains only 331 protein-coding genes (Gilson et al. 2006), whereas that of *G. theta* is 0.55 Mb in length and contains 464 protein-coding genes (Douglas et al. 2001). In both cases, the intergenic spacers are extraordinarily tiny (often as small as 1 bp, with some genes overlapping by up to 100 bp), and many genes are cotranscribed, even when on opposite strands (Williams et al. 2005). Both genomes are entirely devoid of mobile elements, but the incidence of introns differs substantially. Spliceosomal introns are common in the *B. natans* nucleomorph, averaging 2.6 per gene, although they have the smallest known average size of any eukaryotic introns (19 bp). In contrast, the *G. theta* nucleomorph is depauperate with respect to spliceosomal introns, containing only 17 within the entire set of protein-coding genes, although they are more substantial in length (42–52 bp). However, the tRNA genes of the *G. theta* nucleomorph harbor an unusually large number of introns (nearly half of them contain at least one, and these are in locations not previously observed in any eukaryote), although many of these introns are extremely tiny (as small as 3 bp) (Kawach et al. 2005). This difference in the incidence of introns between the two lineages might be a simple consequence of different ancestral conditions in the colonizing endosymbiont, as a substantial range of variation in intron density exists among unicellular eukaryotes (see Chapter 9). However, a deeper understanding of nucleomorph evolution will require information on aspects of the population genetic environment associated with such genomes, including effective population sizes and the rates of nucleotide substitutions, insertions, and deletions.

12 Sex Chromosomes

The overview of organelle genomic evolution in the previous chapter provided a supplement to our previous focus on autosomal genes, demonstrating further how a variety of DNA-level features can depend on the population genetic properties of the chromosomes on which they reside. Our final overview of genomic evolution addresses this matter in still another way by considering the features of sex chromosomes. Although sex chromosomes reside side by side with their autosomal counterparts in the nuclear environment, they have unique recombinational and mutational features, which can, in principle, alter their vulnerability to nonadaptive forces of evolution. Moreover, because sex chromosomes generally evolve from autosomes, most alterations in the architectural features of their genes can confidently be interpreted as derived traits, as opposed to being potential artifacts of ancient ancestry. As we will see below, such alterations arise remarkably rapidly, providing a powerful indictment of mutation, recombination, and random genetic drift as major sculptors of genomic evolution.

One nomenclatural issue needs to be addressed at the outset. In most species with morphologically distinguishable sex chromosomes, the male is the heterogametic (XY) sex, whereas females are homogametic (XX). However, in a few groups, most notably birds and lepidopterans, the situation is reversed, with females being heterogametic; such systems are often referred to as ZW/ZZ. To simplify the presentation, this chapter will for the most part focus on XY/XX systems, although it will be seen that many of the peculiar features of the Y in XY/XX systems extend to the W in ZW/ZZ systems. It should also be noted that the definition of a sex chromosome is not always so clear-cut. In principle, this special notation may simply apply to any pair of chromosomes on which the sex determination locus resides, regardless of the size of the differentiated region, although we will generally be focusing on the special cases in which the two chromosomes have become morphologically distinguishable from each other and non-recombining.

The evolutionary biology of sex chromosomes encompasses a diversity of subjects. Some of the central questions concern the matters of why sex chromosomes originate in the first place and why they are largely confined to multicellular animals and land plants (Ohno 1967; Bull 1983). Because the Y chromosome resides only in males, it provides a unique target for the refinement of the functions of genes with male-specific features, so sex chromosomes are also of special relevance to the field of sexual selection and the evolution of sexual dimorphism. Sex chromosomes also raise the challenge of understanding how gene expression is regulated so as to equalize the total activity of X-linked genes with generalized functions in males and in females, which contain one and two X chromosomes, respectively. Finally, nonrecombining sex chromosomes are especially vulnerable to the invasion of selfish elements promoting sex chromosome segregation distortion in the heterogametic sex; that is, the differential transmission of the X versus the Y into successful male gametes (Jaenike 2001; Burt and Trivers 2006). Although each of these issues could easily constitute an entire book or book chapter (as the preceding references testify), the focus here will be on the evolutionary consequences of residence on a sex chromosome for the basic architectural features of a gene, and details on the above matters will be covered only to the extent necessary to accomplish this end.

There are several basic asymmetries with respect to the population genetics of X and Y chromosome evolution. First, with the exception of a small pseudoautosomal region, recombination is entirely suppressed across the Y chromosome, whereas the X is able to fully recombine in the homogametic sex. On this basis, the Y is expected to be much more vulnerable to selective sweeps and selective interference between simultaneously segregating mutations. Second, in a population with an even sex ratio, there are three times as many X as Y chromosomes, so on the basis of numbers alone, the Y is expected to be more vulnerable to random genetic drift. Third, at least in animals, the mutation rate appears to be substantially elevated in the male relative to the female germ line, so that the Y chromosome is a much more mutagenic environment than the X, which spends only a third of its time in males. Fourth, because of the haploid nature of both sex chromosomes in the heterogametic sex, recessive mutations are expected to be exceptionally sensitive to selective forces. Fifth, as noted above, the asymmetry in the sexual background of the X and the Y imposes different patterns of selection for female-specific and male-specific gene functions.

In the following pages, it will be seen that the joint operation of these factors generally leads to a greatly elevated vulnerability of the Y chromosome to mutational degradation as well as to a repatterning of the gene contents of both sex chromosomes. Although natural selection plays a central role in the latter aspect of sex chromosome differentiation, it will become clear that nonadaptive forces have a substantial influence on many aspects of sex chromosome evolution, including the expansion of extragenic DNA, the establishment of certain multigene families, and possibly even the origin of sex chromosomes themselves.

The Origins of Sex Chromosomes

One of the defining features of fully developed sex chromosomes is the extreme suppression of recombination on the chromosome confined to the heterogametic sex. Although the mechanisms by which such a state first becomes established are not entirely clear (Charlesworth 1996b; Charlesworth and Charlesworth 2000), comparative genome mapping indicates that nonrecombining sex chromosomes have emerged on multiple occasions. For example, the mammalian XY/XX (male/female) system appears to be nonhomologous to the avian ZW/ZZ (female/male) system (Nanda, Haaf et al. 2002), although a potential link has been revealed by observations from a complex chromosomal chain in the platypus (Grutzner et al. 2004). In addition, as will be noted below, nonrecombining sex determination regions have arisen in numerous ray-finned fish lineages. In the fly *Megaselia scalaris*, the sex determination factor appears to be carried by a mobile element that has been transposed to different chromosomal locations in different populations (Traut and Wollert 1998). The specific loci involved in sex determination appear to differ dramatically among species, and the protein sequences of such loci evolve at exceptionally rapid rates (O'Neil and Belote 1992; Whitfield et al. 1993; Ferris et al. 1997; Haag and Doty 2005; Hill et al. 2006).

The initial steps in the origin of sex chromosomes most likely involve suppression of recombination (perhaps entirely fortuitous) around a sex determination locus, as recombinants of such genes would probably generate reproductively inferior progeny (Nei 1969). But how does such a locus arise in the first place? One possibility involves an initial hermaphroditic stage, followed by the emergence of two mutations at a locus influencing sex organ differentiation: one incurred by an allele on the proto-X, which suppresses the production of male gametes, and another arising on the proto-Y, which causes female sterility (Ohno 1967; Bull 1983; Charlesworth 1991). If the proto-Y-linked allele were dominant to the proto-X-linked mutation, two XY individuals would be incapable of a successful mating (eliminating the possibility of YY genotypes): XX individuals would be female, and XY individuals would be male. The transition to this end state would not be a simple matter, however (Figure 12.1). As the instantaneous coestablishment of both mutations would be extremely unlikely, an intermediate state involving hermaphrodites and pure males or females would be expected under this model, the likelihood of which would depend in turn on the selective advantage of separate sexes (Charlesworth and Charlesworth 1978). In the case of the medaka fish, the initiation of a sex

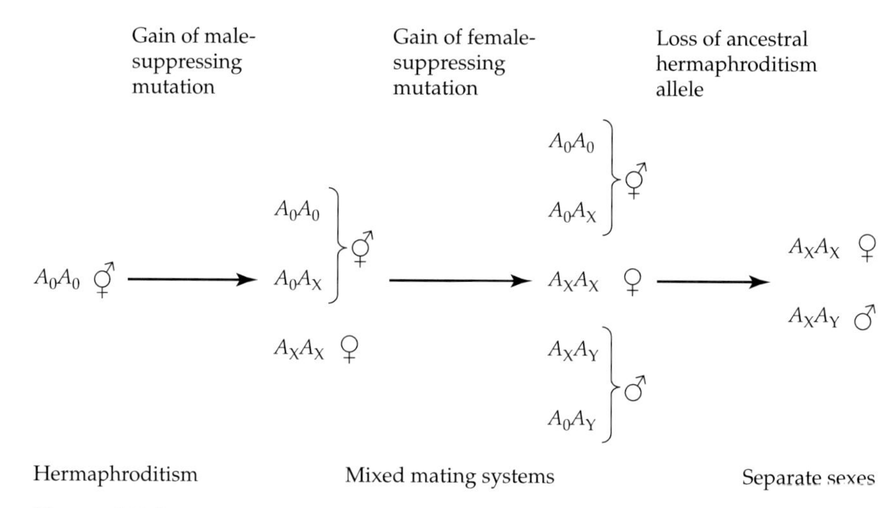

Figure 12.1 A potential two-step pathway to the evolution of a chromosomal region specifying separate sexes from an ancestral condition of hermaphroditism. Here the proto-X mutation is assumed to be recessive and the proto-Y mutation to be dominant. The final loss of the ancestral allele (A_0) could be facilitated by natural selection if hermaphroditism had a significant cost. In principle, the order of the two steps could be reversed, leading to the initial evolution of males, and the intermediate steps could be maintained as balanced polymorphisms for prolonged periods.

determination region appears to have been precipitated by the duplication of a gene involved in the development of the undifferentiated gonad, which when transferred to a new chromosomal location, became specialized for expression in the testes (Kondo et al. 2004, 2006).

Complete differentiation of sex chromosomes requires an outward expansion of recombination suppression around the sex determination locus, although in species such as *Drosophila*, in which there is no male recombination, a Y chromosome can be nonrecombining from the time of origin. Because recombination is homology-dependent (see Chapters 2, 4, and 6), a nonrecombining region might simply grow passively once a critical level of sequence differentiation is reached in the core region. Such a process might also be nucleated in regions of intrinsically low recombination (e.g., centromeric regions), and might in some cases be facilitated by the accumulation of chromosomal rearrangements. Whatever the mechanism, studies of both the human X and the chicken Z have revealed a series of strata of different ages, thereby implicating a gradual emergence of the nearly fully nonrecombining sex chromosomes (Lahn and Page 1999; Handley et al. 2004; Ross et al. 2005). Such differentiation appears to be molded by the stochastic recruitment of duplicated autosomal fragments by the X, followed by their recombination onto and subsequent degeneration on the Y (Graves 1995).

The Population Genetic Environment

As emphasized in Chapters 4 and 11, because new mutations cannot be freed from their original genetic background, nonrecombining genomes are subject to a relaxation in the efficiency of natural selection. The net result is a reduction of the chromosomal effective population size (N_g), which jointly reduces the probability of fixation of beneficial mutations and increases the probability of fixation of deleterious mutations. However, although this effect alone might be sufficient to lead eventually to the mutational degradation of a nonrecombining Y (Charlesworth and Charlesworth 2000), several other distinguishing features of Y chromosomes are expected to exacerbate the situation.

First, the haploid and sex-limited nature of the Y will cause a further reduction in its effective population size in several ways. Each mating involves four copies of each autosome and three X chromosomes, but only a single Y. Moreover, because the Y spends all of its time in males, while the X spends two-thirds of its time in females, any difference in the effective numbers of males and females in a population will further influence the disparity of N_g associated with these two chromosomes. In animals, it is often the case that only a fraction of males are successful in fertilizing females, further suppressing the relative effective number of Y chromosomes.

Second, Y-linked genes are only subject to selection for male-specific functions, whereas those on the X and autosomes are selected for their joint effects in the two sexes. Not surprisingly, Y chromosomes tend to be enriched with genes with testes-specific functions, many of which are present in multiple copies (Skaletsky et al. 2003). Because a gene that moves onto the Y will be released from any prior selective constraints associated with its effects on female performance, residence on the Y can transform a locus previously kept static by opposing selective pressures in the two sexes into one under positive selection for male function, potentially enhancing the frequency of selective sweeps on the Y. Rice (1998) was able to demonstrate this effect experimentally by taking advantage of the absence of recombination in male *Drosophila*. The use of chromosomal markers enabled him to treat a full set of autosomes and the true Y like a single giant Y chromosome, transmitting the entire artificial linkage group only through male progeny (without recombination). Within just 41 generations, the autosomes associated with these synthetic Y chromosomes acquired changes that improved male fitness at the expense of female fitness (as revealed when they were inserted back into female backgrounds).

Third, Y chromosomes typically experience higher average mutation rates than chromosomes that cycle through females. To see this, we rely on a simple method of estimating the relative mutagenicity of males by comparing levels of interspecific divergence at silent sites in Y-linked versus X-linked (or autosomal) genes. For example, assuming neutrality, the divergence of Y-linked silent sites proceeds at a per-generation rate equal to the male-specific mutation rate, $u_{♂}$, whereas that for X-linked sites proceeds at rate ($2u_{♀}$

TABLE 12.1 Relative mutation rates in males versus females based on silent-site divergence estimates for genes on sex chromosomes

Oryzias (medaka fish)	8.2	Zhang 2004
Oncorhynchus (salmonids)	6.0	Ellegren and Fridolfsson 2003
Hominids	5.2	Makova and Li 2002
Felidae (cats)	4.4	Slattery and O'Brien 1998
Perissodactyls (horses, rhinos)	3.9	Goetting-Minesky and Makova 2006
Passeriform birds	3.6	Axelsson et al. 2004
Sheep/goats	3.4	Lawson and Hewitt 2002
Gallus (chickens, quail, turkeys)	2.2	Berlin et al. 2006
Mus/Rattus	1.8	Chang et al. 1994
Drosophila	1.3	Bauer and Aquadro 1997
Silene (plant)	0.6	Laporte et al. 2005

$+ u_{♂})/3$ because X chromosomes spend two-thirds of their time in XX females. Letting d_X and d_Y denote observed levels of silent-site divergence between two species for X- and Y-linked genes, the ratio of male to female mutation rates $(u_{♂}/u_{♀})$ can be estimated as $2(d_Y/d_X)/[3-(d_Y/d_X)]$. This ratio exceeds 1.0 in every animal species examined (Table 12.1), and virtually identical results are obtained with insertions/deletions as with substitutions (Makova et al. 2004). Because many mutations arise during periods of DNA replication, this pattern is expected in animals in which the number of germ line cell divisions in males generally exceeds that in females, increasing with male age (Miyata et al. 1987; Crow 1993; Chang et al. 1994), an idea first suggested by Haldane (1935). This hypothesis can be tested by restricting the analysis to highly mutable CG dinucleotide sites in mammals, which are thought to mutate in a largely replication-independent manner (see Chapter 6). In such analyses, the male bias is reduced by more than 50% (Mikkelsen et al. 2005; Taylor et al. 2006), consistent with the hypothesis that the bulk of mutational events at the remaining sites are driven by replication-associated processes.

Further insight into the degree to which the population genetic environment is altered for genes residing on sex chromosomes can be acquired by considering a now familiar metric: the amount of segregating variation within natural populations at silent sites in protein-coding genes, which provides an estimate of the product of twice the effective number of genes per locus and the per-generation mutation rate per nucleotide ($2N_g u$). In all cases for which data are available, levels of variation on the Y are substantially less than those on both the X and the autosomes, in some cases being undetectable (Table 12.2). That this feature is associated with the lack of recombination, and is not simply a peculiarity of male-specific chromosomes, can be seen from the results for chickens (*Gallus gallus*). As noted above, the

TABLE 12.2 Nucleotide diversity at silent sites for population samples of randomly sampled alleles for protein-coding genes, which provide estimates of $2N_g u$

	NUCLEOTIDE DIVERSITY		RATIOS			
	Y	X	Y/A	X/A	Y/X	
Mammals:						
Homo sapiens	0.00013	0.00040	0.134	0.438	0.319	Sachidanandam et al. 2001; Shen et al. 2000; Stone et al. 2002
Pan troglodytes	0.00067	0.00131	0.500	0.978	0.511	Stone et al. 2002
Ovis aries	0.00009	—	0.024	—	—	Meadows et al. 2004
Lynx lynx	0.00000	0.00016	—	—	0.000	Hellborg and Ellegren 2004
Canis lupus	0.00004	0.00038	—	—	0.105	Hellborg and Ellegren 2004
Rangifer tarandus	0.00000	0.00050	—	—	0.000	Hellborg and Ellegren 2004
Bos taurus	0.00000	0.00025	—	—	0.000	Hellborg and Ellegren 2004
Microtus agrestis	0.00017	0.00080	—	—	0.212	Hellborg and Ellegren 2004
Mus domesticus	—	0.00078	—	0.067	—	Nachman 1997
***Drosophila*:**						
D. melanogaster	—	0.01900	—	1.210	—	Andolfatto 2001
D. miranda	0.00012	0.00711	0.007	0.435	0.016	Bachtrog and Charlesworth 2002
D. simulans	—	0.02340	—	0.850	—	Andolfatto 2001
Plants:						
Silene dioica	0.00330	0.02576	0.203	1.585	0.128	Laporte et al. 2005
Silene latifolia	0.00000	0.03744	0.000	1.617	0.000	Laporte et al. 2005
	W	Z	W/A	Z/A	W/Z	
Birds:						
Gallus gallus	0.00007	0.00200	0.011	0.308	0.035	Berlin and Ellegren 2004
Ficedula albicollis	—	0.00150	—	0.416	—	Borge et al. 2005
Ficedula hypoleuca	—	0.00100	—	0.370	—	Borge et al. 2005

Source: The autosomal (A) data are either taken directly from the cited reference or from the survey in Lynch 2006a.

Note: In birds, females (ZW) are the heterogametic sex, with W being analogous to Y (i.e., W is the non-recombining chromosome, found only in females).

TABLE 12.3 Ratios of effective population sizes (N_g) for genes on Y chromosomes, X chromosomes, and autosomes, derived from the information in Tables 12.1 and 12.2

	Y/A	X/A	Y/X
Homo sapiens	0.080	0.567	0.147
Pan troglodytes	0.298	1.264	0.236
Ovis aries	0.015	—	—
Microtus agrestis	—	—	0.149
Mus domesticus	—	0.067	—
D. melanogaster	—	1.260	—
D. miranda	0.006	0.455	0.014
D. simulans	—	0.889	—
Silene dioica	0.271	1.468	0.186
Silene latifolia	0.000	1.497	0.000
	W/A	Z/A	W/Z
Gallus gallus	0.018	0.271	0.064

female is the heterogametic sex in birds, and at least for this one species, the nonrecombining female-specific (W) chromosome exhibits just 4% of the silent-site diversity on the recombining Z.

Using the relative mutation rates given in Table 12.1 to yield ratios of chromosome-specific mutation rates (u_Y/u_X) and dividing these into the average diversity ratios (e.g., $2N_{gY}u_Y/2N_{gX}u_X$) in Table 12.2 yields the estimated ratios of effective population sizes in Table 12.3. These results demonstrate that the effective number of Y chromosomes is always less than 30% of that of autosomes and X chromosomes, and often substantially so. Thus, there is little question that the Y chromosome is exceptionally vulnerable to the accumulation of mildly deleterious mutations.

Degeneration of the Y Chromosome

The evidence for Y chromosome degeneration is overwhelming. For example, the human Y chromosome harbors only 78 protein-coding genes and encodes only 27 distinct proteins (Skaletsky et al. 2003), less than 1% of the number on other chromosomes, but has an elevated abundance of mobile element insertions (Erlandsson et al. 2000). About 30% of the euchromatic region of the human Y consists of degenerated sequence with homology to recognizable sequences on the X, while about 10% consists of recently (<3 MYA) transposed X sequences that have acquired twice the density of mobile elements found on the X. Approximately 25% of the euchromatic Y in humans consists of eight palindromes ranging in length from 9 kb to 1.5 Mb. Each of these regions consists of duplicate spans of DNA in tail-to-tail orientations on opposite strands, kept in a nearly identical state by intrapalindrome gene conversion (Rozen et al. 2003). Although they too contain many pseudogenes, these palindromes harbor the bulk of the active genes on the Y, most of which have been recruited from autosomes and have testes-specific expression (Noordam and Repping 2006). Many of these Y-linked genes exhibit levels of divergence at amino acid replacement sites that approach those at silent sites (Kuroki et al. 2006), a condition consistent with a scenario of reduced efficiency of selection. The mutational inactivation of several human Y-linked genes on the chimpanzee Y provides further support for this contention (Hughes et al. 2005).

This general syndrome of a largely decomposing Y chromosome has been observed in numerous other species. For example, *Drosophila melanogaster* is a member of a lineage containing an ancient X/Y pair. There is no recognizable homology between the two, and the Y contains only a dozen or so genes, many of which harbor enormous introns (Kurek et al. 2000; Carvalho et al. 2001). As in humans, most of the Y-linked genes in this species have been recruited from autosomes and have male-specific functions, consistent with a pattern of duplication and subfunctionalization (Carvalho 2002). In *D. pseudoobscura*, the genes from the original Y chromosome were translocated back to an autosome, after which their introns and intergenic regions declined dramatically in size (Carvalho and Clark 2005). Given the complete lack of recombination on the Y, this particular observation provides compelling evidence that extragenic DNA is a consequence of inefficient selection, not an adaptation to magnify the recombination rate (see Chapter 9). In contrast to the situation in *D. pseudoobscura*, the Y chromosome of *D. miranda* has fused to an autosome, enforcing a nonrecombining environment on the latter since about 1 MYA (Figure 12.2). The autosome-derived genes on this neo-Y show many signs of degeneration, including mobile element insertions, elevated levels of amino acid replacement substitutions, and reduced expression, with about one-third having been completely silenced by frameshift and/or nonsense mutations (and their X-linked homologs retaining function) (Steinemann and Steinemann 1997, 2005a,b; Bachtrog 2003, 2005, 2006).

Increased rates of replacement substitutions (relative to silent substitutions) as well as accumulations of mobile element insertions and other repetitive DNAs have also been noted on the Y chromosomes of mice (Tucker et al. 2003), members of the cat (Felidae) family (Slattery and O'Brien 1998), the liverwort *Marchantia* (Ishizaki et al. 2002), and the plant *Silene* (Filatov 2005; Nicolas et al. 2005; Hobza et al. 2006). Likewise, the nonrecombining female-specific W chromosome of the silkworm *Bombyx mori* is enriched with mobile elements (Abe et al. 2005). Moreover, although stickleback and medaka fish do not yet have fully differentiated Ys, the localized sex determination regions are littered with mobile element insertions, repeat ele-

Figure 12.2 The evolution of a neo-X and neo-Y pair of chromosomes, resulting from the fusion of the Y and an autosome. The X and the free autosome will cosegregate during meiosis in male *Drosophila*, in which there is no recombination, but will experience recombination and independent segregation in females. The autosome that becomes physically attached to the Y experiences no recombination and is expected to accumulate degenerative mutations (cross-hatches).

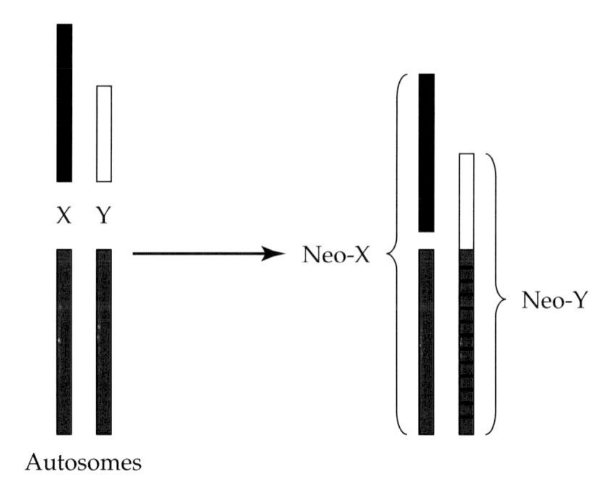

ments, and pseudogenes (Nanda, Kondo et al. 2002; Peichel et al. 2004; Kondo et al. 2006). The same is true in papaya (Liu et al. 2004), and a substantial enrichment of mobile element insertions exists in the region of the mating-type locus in the fungus *Microbotryum violaceum* (Hood et al. 2004). After the X and Y cease communication, degradation of the Y can be quite rapid, as Rice (1994) demonstrated with manipulated Y-chromosomal constructs in laboratory populations of *D. melanogaster* in just a few dozen generations.

Once sufficient degradation of Y-linked genes with general functions has occurred, modifier loci that down-regulate the Y and/or promote higher X expression in males may be promoted by selection (Charlesworth 1996b). By relaxing selection on the Y even further, such dosage compensation is expected to exacerbate Y deterioration, eventually leading to the complete demise of all Y-linked genes other than those critical to male function. As noted above, a common feature of Y chromosomes is an amplification of the copy number of testes-specific genes, and it is tempting to interpret such change as a simple response to positive selection for increased male function of genes restricted to the male sex. At least one example of positive selection on the protein-coding sequences of Y-linked mammalian genes appears to support this idea (Gerrard and Filatov 2005), but this study also noted that the efficiency of selection against mildly deleterious mutations was compromised at such loci (relative to that on the X). As noted in Chapter 8, tight linkage elevates the probability of preservation of duplicate genes by processes driven by degenerative mutations, such as quantitative subfunctionalization (Lynch et al. 2001). This is because a pair of completely linked duplicates transiently endows a chromosome with a weak mutational advantage, as each member of the pair serves as a backup for the other. Thus, the tendency for Y-linked genes to be present in multiple copies may simply reflect the same nonadaptive conditions that drive all of the other features of Y chromosome evolution, rather than being solely a consequence of positive selection for increased dosage of male-specific proteins.

Is the deterioration of the Y solely a consequence of its increased vulnerability to deleterious mutation accumulation? A rather different hypothesis for Y degeneration (as well as for the evolution of dosage compensation) postulates a process driven not so much by degenerative mutations as by the more rapid incorporation of adaptive mutations into X-linked than into Y-linked genes (Orr and Kim 1998). Adaptive selective sweeps are not expected to result in a net loss of fitness for any chromosome, as the deleterious load swept along will always be smaller than the advantage of the selected mutation. However, as noted above, the likelihood of a beneficial mutation being swept to fixation is reduced on a nonrecombining chromosome because of the elevated burden of interference from linked background mutations. Thus, a lower rate of accumulation of positively selected mutations might lead to a situation in which the fitness of Y-linked alleles lags far enough behind that of their X-linked counterparts to encourage the selective advancement of mechanisms that increase the expression of X-linked genes at the expense of those on the Y. If this adaptive hypothesis for the degeneration of the Y is correct, then degenerated sex chromosomes should evolve more readily in populations with large effective sizes (in which the fixation probabilities of beneficial mutations are elevated). This prediction is completely contrary to the observed pattern. Despite the widespread distribution of genetic sex determination, fully differentiated sex chromosomes are unknown outside of the animal and land plant lineages. Several fungal species have moderately sized chromosomal regions associated with sex determination (in some cases, on the order of 100 to 500 kb in length, with greatly suppressed recombination) (Fraser and Heitman 2003, 2004), and *Chlamydomonas* has such a region extending over 1 Mb (Ferris et al. 2002), but these exceptions are dwarfed by the numerous times that full-fledged sex chromosomes have independently emerged in multicellular species.

It is often argued that these somewhat expanded sex determination regions of fungi and algae, as well as the small nonrecombining sex determination regions in some fishes, represent "incipient sex chromosomes" in early stages of an evolutionary progression toward full-fledged sex chromosomes. However, there is no obvious reason why sex chromosome evolution should have been delayed in such species, nor is there any evidence to support the idea that organisms with separate sexes are destined to eventually acquire such chromosomal constitutions. An alternative view is that many eukaryotic lineages simply do not provide the appropriate population genetic environment for the emergence of sex chromosomes, regardless of the available time span. The differential phylogenetic distribution of sex chromosomes may instead simply reflect the natural predisposition of populations with small effective sizes to the evolution of sex chromosomes by degenerative mutations. It could be countered that sex chromosomes are unlikely to evolve in unicellular species without pronounced sexual dimorphism, but a case can also be made for the opposite causal connection; that is, that the origin of sex chromosomes facilitates the evolution of sexual differentiation at the phenotypic level (Rice 1984).

Retailoring the X Chromosome

Although the X chromosome is nowhere near as inhospitable a place for gene residence as the Y, it still presents a distinct environment relative to autosomes, raising the question of whether its contents also experience unique evolutionary trajectories. First, because the X chromosome does not recombine in males, it will generally experience two-thirds the level of recombination on an average autosome, provided the level of recombination is otherwise equivalent in the two sexes. However, in species such as *Drosophila*, in which even the autosomes are nonrecombining in males, the average rate of recombination on the X (which spends two-thirds of its time in females) will be elevated to four-thirds that on autosomes. Second, for each mating pair, there are only three X chromosomes for every four autosomes. This difference contributes to an expected reduction in the effective population size for X-linked genes, although the effect may be offset if the effective number of females is sufficiently greater than that of males. Third, because of the greater amount of time they spend in females, X chromosomes will experience lower average mutation rates than autosomal genes, although this effect is not large: with the largest ratio of male to female mutation rates given in Table 12.1, the ratio of mutation rates on the X versus autosomes is 0.74. Fourth, because the X is haploid in males, the exposure of recessive mutations to selection is magnified.

Given the numerous uncertainties about the relative contributions of these factors to the population genetic environment of the X, we again resort to empirical data on silent-site variation (π_s) to indirectly infer the magnitude of $2N_g u$ for X-linked genes. With just one exception (*Mus domesticus*), the ratios of π_s for X-linked versus autosomal genes are within the range of 0.4 to 1.6 (see Table 12.2), and after factoring out differences in mutation rates, the ratios of effective numbers of genes (N_g) are in the range of 0.6 to 1.5 (see Table 12.3). These results provisionally suggest that, contrary to the situation on the Y, neither the magnitude of random genetic drift nor the mutation rate on the X chromosome deviates greatly from that on autosomes. This overall similarity in N_g and u appears to be reflected by the absence of any major deviations between the architectural features of X chromosomes and autosomes. For example, the human X chromosome contains about 1000 genes, a number similar to that on a typical autosome, and its density of intergenic DNA, mobile elements, and pseudogenes is only slightly elevated relative to autosomes (Kohn et al. 2004; Ross et al. 2005; Drouin 2006).

On the other hand, the gene contents of X chromosomes are often qualitatively different than those of autosomes. In *D. melanogaster*, for example, the X chromosome is demasculinized, in the sense that only 5% of its genes have male-biased expression, which is about half the level for such genes on autosomes (Parisi et al. 2003). In *C. elegans*, testes-specific and germ line–intrinsic genes are almost exclusively present on autosomes, in striking contrast to genes with oocyte-biased expression, which are randomly distributed throughout the genome (Reinke et al. 2000). In contrast, genes with male-biased expression are randomly distributed across the X and autosomes in mosquitoes (Hahn and Lanzaro 2005), whereas in mammals the X contains an excess of genes associated with male sexual reproduction, brain function, and musculature (Saifi and Chandra 1999; Wang et al. 2001; Lercher, Urrutia et al. 2003; Vallender and Lahn 2004). In both *D. melanogaster* and mammals, gene duplications arising via the transposition of reverse transcripts of mRNAs appear to move more frequently from the X to autosomes than vice versa, and an exceptionally high fraction of these novel genes have testes-specific expression (Betrán et al. 2002; Emerson et al. 2004). It remains unclear whether such differential distributions of male-specific genes are driven by physical factors causing a directional bias in movement and establishment, by in situ molecular evolutionary change promoted by selection, or both.

One feature that may encourage the adaptive repatterning of the contents of the X chromosome relates to its haploid versus diploid state in males versus females. It has been argued that the hemizygosity (the presence of a single copy) of the X chromosome in males sets up a selective asymmetry that not only facilitates the fixation of recessive beneficial mutations, but also specifically promotes the evolution of male-advantageous alleles with antagonistic effects on female fitness (Rice 1984; Charlesworth et al. 1987). All male-expressed X-linked mutations are exposed to immediate selection in males, but if such mutations are recessive, they will be sheltered from selection in females until they rise to adequate frequencies to be exposed in homozygotes (and even then, they will be conditionally neutral if they are silent in females). Under this reasoning, mutations that are specifically beneficial to male fitness are expected to have a higher probability of fixation if they reside on the X than on an autosome. The degree to which the fixation probability will be magnified will depend on the degree to which such mutations influence female fitness. However, even if such recessive mutations have strong negative pleiotropic effects in homozygous females, they should still be capable of rising to high frequencies and should potentially still be vulnerable to fixation, especially if females are capable of selecting mates based on X-linked traits (Albert and Otto 2005). Under this reasoning, because autosomal genes are no more sheltered in females than in males, an enrichment of male-specific X-linked genes is always expected unless other factors are involved. Thus, the observed paucity of male-specific X-linked genes in *Drosophila* and *Caenorhabditis* implicates a significant involvement of something other than the hemizygous effect.

One complicating factor with X-linked genes in animals is their frequent exposure to dosage compensation: a modification of allele-specific expression that approximately equalizes the total locus-specific expression in males and females. Might dosage compensation in some way alter the expectations of the preceding theory to make it more concordant with observations on male-specific gene location? Two types of observations suggest that it does not. First, invoking dosage compensation does not help explain the demasculinization of the X chromosome in *Drosophila* and *Caenorhabditis*, as dosage compensation in these species operates to equalize the summed expression of both X chromosomes in females with that of the single X in males (Lucchesi 1998; Nusinow and Panning 2005), satisfying the assumption that recessive mutations will be sheltered in heterozygous females.

Second, although the enrichment of male-specific genes on mammalian X chromosomes is superficially in accord with the logic outlined above, the way in which the expression of X-linked genes is regulated in mammals is inconsistent with the assumptions of the theory. In most mammals, dosage compensation is accomplished by the inactivation of a random X chromosome in each female cell (Lucchesi 1998; Nusinow and Panning 2005). Thus, there is no masking of recessive X-linked alleles at dosage-compensated loci at the cellular level in females, making it unclear whether the temporary sheltering of sexually antagonistic features can provide an explanation for the association of male-specific genes with the mammalian X chromosome. About 15% of the genes on the mammalian X are partially to entirely insensitive to dosage compensation in females (Carrel and Willard 2005), although it remains to be seen whether these genes are peculiar to male/female function. Although male-specific X-linked genes of placental mammals exhibit higher rates of protein sequence evolution than genes on autosomes (Torgerson and Singh 2003; Wang and Zhang 2004; Lu and Wu 2005), future work will be necessary to ascertain whether this is a consequence of adaptive evolution or reduced efficiency of selection against mildly detrimental mutations.

Marsupial mammals may provide an ideal substrate for clarifying these matters, as in these species it is always the paternal X that is inactivated in females (Graves 1995). In this case, paternally derived X chromosomes are neutral in daughters and absent in sons, whereas maternally derived X chromosomes will be selected for their full set of functions in daughters as well as in sons. That is, in marsupials all active X-linked genes are fully expressed, regardless of the sex, completely eliminating the complications of any masking effects of dosage compensation.

As a possible explanation for the paucity of male-specific X-linked genes in invertebrates, we revisit the assumption of recessivity. X-linked genes spend only a third of their time in males, whereas autosomal genes are equally distributed between the sexes. Thus, if male-beneficial mutations are not recessive but rather at least partially dominant, or if individuals are highly inbred (as in the case of *Caenorhabditis*), selection for such mutations might be more effective if they reside on autosomes. This idea may have merit in that, contrary to the situation in mammals, average rates of protein sequence evolution of genes on the X and on autosomes are not significantly different in *Drosophila* (Bartolomé et al. 2005; Thornton et al. 2006). Such equality is inconsistent with the idea that advantageous mutations are typically recessive, which would promote higher rates of adaptive divergence for X-linked genes.

Finally, although the preceding arguments focus on aspects of gene expression in somatic cells, the degree to which expression of the X chromosome is modified in the male and female germ lines may be a more central determinant of the location of sex-specific genes (Wu and Xu 2003). In *C. elegans*, for example, the X chromosome appears to be inactivated in the male germ line during mitosis and early meiosis (Reinke et al. 2000; Fong et al. 2002), a process that is distinct from dosage compensation in female cells. Thus, in this species, there is essentially a complete relaxation of selection on the functions of X-linked genes in male gonadal tissue, leaving only functions at the level of the soma as potential male targets for selection on the X. Similarly, in mammals, despite the overall enrichment of the X with male-specific genes, there is a paucity of X-linked genes involved in late stages of spermatogenesis, in accordance with X inactivation in the male germ line at the onset of meiosis (Khil et al. 2004).

In summary, with data bearing on all of the above X chromosome issues being derived from only three species, all of which are animals, it is premature to promote a general interpretation of the existing data in this area. However, it can be stated fairly confidently that, unlike the Y, the X generally does not provide a sufficiently distinct environment with respect to the forces of mutation, recombination, and/or random genetic drift to drive substantial changes in the architecture of individual genes. Rather, with its differential distribution between the sexes, the X provides a compelling substrate for potentially deciphering the roles that the degree of dominance and sexual antagonism play in the molecular evolution of individual gene products.

13 Genomfart

As we move into the next phase of evolutionary biology, we can be confident of two things: the basic theoretical machinery for understanding the evolutionary process is well established, and we will soon be effectively unlimited by the availability of information at the DNA level. The main question now is how to use this information to productively advance the development of a more mature science of evolution. What follows is a departure from the somewhat impersonal style of the previous chapters, in which an attempt was made to integrate the key findings of molecular biology and population genetics into a cohesive evolutionary genomics framework. First, I will give a very brief overview of the historical development of evolutionary theory, as this helps clarify the roots of some of the controversies that still exist today. Second, I will comment on the current state of affairs in evolutionary biology, particularly the perception of softness in the field that has been encouraged by the propagation of evolutionary ideas by those with few intentions of being confined by the constraints of prior knowledge. Third, as considerable territory has been covered in prior chapters, I will summarize the central findings, place them in the context of prior conflicts, and highlight the major challenges that lie ahead.

To the extent that the ideas in this book have merit, there are two fundamental ways to move the field of evolutionary genomics forward. First, the overview presented in the preceding chapters provides only the seeds for a comprehensive understanding of genomic evolution. Substantial additional research, with more statistical rigor than was used on the previous pages, will be necessary to determine the full scope of the notion that non-adaptive forces play a central role in the tailoring of genomic architecture. In any such efforts it will be essential to remember that today's products of evolution are remnants from the past, and that the current evolutionary forces operating on a genomic segment may have nothing to do with those involved in its origin. Second, if nonadaptive factors have influenced the evolution of functionally important components of the genome, we must ask whether such processes have had cascading effects on evolution at higher levels of organization; that is, at the cellular and whole-organism levels. Because evolution proceeds with the resources that it has in hand (i.e., via descent with modification), and because phenotypes are functions of genes with particular structural and functional attributes, such a possibility cannot be dismissed lightly.

Some readers may be curious about the title of this chapter. Several years ago, following an international conference on bioinformatics where many of us pondered the potential meaning of the newly emerging genomic data, we were treated to an evening dinner on a ferry navigating through the Stockholm archipelago. Sitting at a table with Eugene Koonin, I happened to glance through the window, where, rising out of the sea, was a sign proclaiming "Genomfart." At the time, this proclamation seemed a rather fitting description of the day's activities, but *genomfart* is Swedish for "place of passage," or "the way forward." At the close of this chapter, I will suggest some ways in which knowledge about the evolution of gene and genomic architecture and the further development of population genetic theory might contribute to our understanding of the fundamental issues that concern most evolutionary biologists: the mechanisms that encourage versus constrain phenotypic evolution.

The Origins of Modern Evolutionary Theory

Since the 1859 publication of Darwin's major treatise, *On the Origin of Species by Means of Natural Selection*, evolutionary biology has undergone a succession of transformations. For nearly 60 years, the field wallowed in uncertainty regarding mechanisms, even after Mendel's observations on segregation and independent inheritance had become widely known (Provine 1971; Stoltzfus 2006b). Around the turn of the twentieth century, the Mendelian school was convinced that the particulate nature of inheritance satisfactorily eliminated most of the lingering doubts about the mechanics of the evolutionary process, whereas the biometrics school argued just as vehemently that evolutionarily significant variation was a function of ill-defined factors with very small effects and that traits with observable Mendelian features were mere oddities, irrelevant to evolution.

A series of empirical observations from plant breeding began to raise doubts about some of the extreme views of the biometricians (reviewed in Lynch and Walsh 1998), but it was Fisher's landmark 1918 paper (drawing in part on earlier work by Pearson and Yule; see Hill 1984 for a review) that demolished the barriers between these two schools of thought. Here, Fisher formally demonstrated that the focal traits of the biometricians—continuously distributed phenotypes associated with complex traits—are entirely consistent with Mendel's laws. At the age of 28, he had simultaneously removed the major barriers to our understanding of the sources of natural variation, clarified the distinction between genetic and environmental effects, laid the foundation for the development of a formal theory connecting the fields of genetics and evolution, and established the modern science of statistics. Although this paper is one of the most important articles written in the life sciences in the twentieth century, it is by no means an easy read, and remarkably few biologists are even aware of its existence.

This early era of confusion in evolutionary biology might never have occurred had we known the molecular basis of genetic variation, but that would take many more years. Miescher first described a peculiar component of pus (nuclein, later to be called DNA) in 1871 (for a review, see Dahm 2005), which a long series of studies eventually revealed to be the source of hereditary variation. But it was not until 1953 that Watson and Crick clarified the physical and chemical nature of DNA, and even then, we had only the vaguest notion of the nature of the genetic material and the mechanisms by which it is mutationally modified. Nevertheless, in the brief 35 years between Fisher's paper and the elucidation of the structure of DNA, our understanding of the population genetic basis of evolutionary change made substantial advances. Although this progress involved a good deal of mathematical expertise, most of the theory was made possible by several well-accepted genetic observations: the Mendelian nature of inheritance, the ubiquity of recombination in sexual species, and the mutability of the (still undefined) gene.

Early practical successes came in the field of quantitative genetics, which largely deals with complex, multilocus traits. For example, Wright's (1934a,b) seminal work on threshold characters made significant contributions to our understanding of major genetic disorders, many of which appear superficially to have a non-Mendelian basis. Wright elegantly showed that the sudden appearance and disappearance of major phenotypes is fully consistent with expectations for a dichotomous trait with an underlying biochemical or cellular determinant having a multilocus basis. Like Fisher's paper, most of Wright's contributions are invisible to the vast majority of biologists outside of evolutionary biology, even to those whose careers are devoted to understanding the disparity between major quantitative traits in related taxa.

Virtually all of science is derivative: new ideas almost always grow out of old concepts, some prior interpretations are shown to be wrong, and the field moves on. What is remarkable about the field of population genetics is that nearly all of the mathematical foundations developed prior to the molecular era not only still stand today, but are making major contributions to our understanding of the molecular basis of evolutionary change, to the development of rational programs in plant and animal breeding, and to the localization and characterization of the genetic defects underlying human genetic disorders. Shortly after the genetic code was cracked, a rapid succession of technological breakthroughs led to major advances in our ability to reveal variation at the molecular level: protein sequencing in the 1950s, rapid population-level surveys of protein variants in the 1960s, ribosomal

RNA sequencing in the 1970s, gene sequencing at the DNA level in the 1980s, and whole-genome sequencing in the 1990s. Each of these brief episodes brought the need for new methods of data analysis and interpretation, and in each case the fundamental framework for such study was drawn largely from preexisting population genetic machinery. Many embellishments have been added to the theory, and views have changed on the relative power of alternative evolutionary forces, but no keystone principle of population genetics has been overturned by an observation in molecular, cellular, or developmental biology. Thus, although we do not yet fully understand the connections between variation at the molecular and phenotypic levels, we can be confident that the basic theoretical machinery for understanding the evolutionary process is in place.

Evolution as a Science in the Post-Darwinian World

Despite the tremendous theoretical and physical resources now available, the field of evolutionary biology continues to be widely perceived as a soft science. Here I am referring not to the problems associated with those pushing the view that life was created by an intelligent designer, but to a more significant internal issue: a subset of academics who consider themselves strong advocates of evolution but who see no compelling reason to probe the substantial knowledge base of the field. Although this is a heavy charge, it is easy to document. For example, in his 2001 presidential address to the Society for the Study of Evolution, Nick Barton presented a survey that demonstrated that about half of the recent literature devoted to evolutionary issues is far removed from mainstream evolutionary biology.

With the possible exception of behavior, evolutionary biology is treated unlike any other science. Philosophers, sociologists, and ethicists expound on the central role of evolutionary theory in understanding our place in the world. Physicists excited about biocomplexity and computer scientists enamored with genetic algorithms promise a bold new understanding of evolution, and similar claims are made in the emerging field of evolutionary psychology (and its derivatives in political science, economics, and even the humanities). Numerous popularizers of evolution, some with careers focused on defending the teaching of evolution in public schools, are entirely satisfied that a blind adherence to the Darwinian concept of natural selection is a license for such activities. A commonality among all of these groups is the near-absence of an appreciation of the most fundamental principles of evolution. Unfortunately, this list extends deep within the life sciences.

Consider the steady stream of recent books by authors striving to define a new field called evolutionary developmental biology (e.g., Arthur 1997; Gerhart and Kirschner 1997; Davidson 2001, 2006; Carroll et al. 2001; West-Eberhard 2003; Carroll 2005a; Kirschner and Gerhart 2005). The plots of all these books are similar: first, it is claimed that observations from developmental biology demonstrate major inadequacies in current evolutionary theory, and then a new view of evolution that eliminates many of the central shortcomings of the field is promised. Developmental biologists are correct in pointing out that evolutionary theory has not yet specifically connected genotypes to phenotypes in a molecular/cell biological sense. However, extraordinary claims call for extraordinary evidence, and none of these treatises provides any formal examples of the fundamental inability of evolutionary theory to explain patterns of morphological diversity. Those who argue that microevolutionary theory has made no contributions to our understanding of the evolution of form may wish to consult the substantial body of quantitative genetic literature on multivariate evolution. Such work is by no means fully satisfactory, as it is couched in terms of statistics (variances and covariances) rather than the molecular features of individual genes, but a more precise evolutionary framework for linking genes and morphology will not be possible until a critical mass of generalities on the matter has emerged at the molecular, cellular, and developmental levels.

For the vast majority of biologists, evolution is nothing more than natural selection. This view reduces the study of evolution to the simple documentation of differences between species, proclamation of a belief in Darwin, and concoction of a superficially reasonable tale of adaptive divergence (see Table 13.1 for a sampling of such ideas from the genomic biology literature). A common stance in cellular and developmental biology is that the elucidation of differences in molecular genetic pathways between two species (usually very distant species) completes the evolutionary story. No need to dig any deeper—because natural selection surely produced the end products, the population genetic details do not matter. In individual cases, this type of informal thinking may do little harm, but in the long run it undermines the very scientific basis of evolutionary biology.

There are two fundamental issues here. First, the notion that interspecific differences at the molecular level reveal the mechanism of evolution ignores the fundamental distinction between the outcome of evolution and the events that lead to such changes. For example, although most animal developmental biologists argue that it was shocking to discover that the development of all animals is based on modifications of the same sets of ancient genes, many evolutionary biologists regard this view with some surprise. It is, of course, easy to criticize based on 20/20 hindsight, but we have known for decades that all eukaryotes share most of the same genes for transcription, translation, replication, nutrient uptake, core metabolism, cytoskeletal structure, and so forth. Why would we expect anything different for development? Although knowing that Hox genes play a central role in the development of all animals provides insight into the genetic scaffold from which body plans are built, it does not advance our knowledge of the evolutionary *process* much beyond knowing that all vertebrates share a heritage of calcified skeletons. It need not even tell us that such genes were involved in the initial stages of differentiation (Alonso and Wilkins 2005).

TABLE 13.1 A sampling of unsubstantiated claims from molecular, cellular, and developmental biology regarding adaptive evolution, with some specific examples from genomics

Complexity: The notion that all additional layers of complexity at the genomic, molecular, subcellular, cellular, and developmental levels are entirely the products of adaptive processes.

Genome size is modulated so as to optimize cell size and cell division rates.

Duplicated genes are always actively preserved by positive selection for new functions.

Genes encode upstream open reading frames in their 5′ untranslated regions to slow down the rate of transcription.

The incorporation of nonoptimal codons into genes serves as a mechanism for modulating the translation rate.

Genes are alternatively spliced to yield aberrant messenger RNAs so as to reduce the level of gene expression via the nonsense-mediated decay pathway.

Modularity: The notion that the evolution of independently operating regulatory elements and networks of interacting genes owe their origins to the direct effects of natural selection.

Gene network modules serve as "plug-ins" that enable saltatory evolution.

Operons represent selfish groups of genes whose coordinated expression facilitates horizontal transfer.

Robustness: The notion that genomes are constructed so as to minimize the damage resulting from mutation and/or physiological perturbations.

Duplicate genes serve as a buffer for inactivated paralogs.

Biased gene conversion toward G+C composition has evolved to offset mutation pressure toward A+T content.

G+C-rich vertebrate isochores have evolved to stabilize double-stranded DNA.

Editing of messenger RNAs serves as a mutational buffer.

Evolvability: The notion that the genetic architecture and cellular/developmental machinery of a species evolves so as to magnify the ability to evolve.

Introns exist to enhance the production of new genes by exon shuffling.

Mobile elements are activated during times of stress to generate beneficial mutations.

Microsatellites and mononucleotide runs are maintained to enhance levels of mutability.

Prions and heat-shock proteins serve as cellular mechanisms for revealing potentially adaptive phenotypes during times of stress.

Note: Most of these topics have been touched on in earlier chapters. Although an adaptive basis for some of these features may eventually be shown to exist, simple nonadaptive mechanisms can already explain the origins of many such patterns.

A vast chasm of stepwise (and partially overlapping) changes may separate today's products of evolution, and understanding those steps is what distinguishes evolutionary biology from comparative biology.

Second, the uncritical acceptance of natural selection as an explanatory force for all aspects of biodiversity (without any direct evidence) is not much different than invoking an intelligent designer (without any direct evidence).

True, we have actually seen natural selection in action in a number of well-documented cases of phenotypic evolution (Endler 1986; Kingsolver et al. 2001), but it is a leap to assume that selection accounts for all evolutionary change, particularly at the molecular and cellular levels. The blind worship of natural selection is not evolutionary biology. It is arguably not even science. Natural selection is just one of several evolutionary mechanisms, and the failure to realize this is probably the most significant impediment to a fruitful integration of evolutionary theory with molecular, cellular, and developmental biology.

It should be emphasized here that the sins of panselectionism are by no means restricted to developmental biology, but simply follow the tradition embraced by many areas of evolutionary biology itself, including paleontology and evolutionary ecology (as cogently articulated by Gould and Lewontin 1979). The vast majority of evolutionary biologists studying morphological, physiological, and/or behavioral traits almost always interpret their results in terms of adaptive mechanisms, and they are so convinced of the validity of this approach that virtually no attention is given to the null hypothesis of neutral evolution, despite the availability of methods to do so (Lande 1976; Lynch and Hill 1986; Lynch 1994). For example, in a substantial series of books addressed to the general public, Dawkins (e.g., 1976, 1986, 1996, 2004) has deftly explained a bewildering array of observations in terms of hypothetical selection scenarios. Dawkins's effort to spread the gospel of the awesome power of natural selection has been quite successful, but it has come at the expense of reference to any other mechanisms, and because more people have probably read Dawkins than Darwin, his words have in some ways been profoundly misleading. To his credit, Gould, who is also widely read by the general public, frequently railed against adaptive storytelling, but it can be difficult to understand what alternative mechanisms of evolution Gould had in mind.

What would Darwin think of this? Presumably, he would be amazed at how much progress has been made in understanding the genetic basis of evolution, as well as satisfied to see that natural selection has been rigorously validated as a major evolutionary force. But Darwin brought a lot more to the table than natural selection. His greatest contribution was in the promotion of integrative thinking in biology—not the integrative biology sometimes incorporated in the labels of academic departments that exclude entire areas of molecular and cell biology, but a concerted effort to solve problems by bringing together observations from all relevant areas. Darwin did not view all of evolution as being a product of natural selection, and even had a crude idea of random genetic drift: "Variations neither useful nor injurious would not be affected by natural selection, and would be left either a fluctuating element, as perhaps we see in certain polymorphic species, or would ultimately become fixed, owing to the nature of the organism and the nature of the conditions" (1859).

There is, of course, a substantial difference between the kind of popular evolutionary literature produced by Dawkins and Gould and the more technical knowledge base that has grown from the past century of theoretical and empirical study of the mechanisms of evolution, but this distinction is often missed by nonevolutionary biologists. With interest in evolution now emerging among scientists working at all levels of biological organization, it is time to stop the trivialization of the field of evolutionary biology. A strong belief in cells does not make one a cell biologist, and a strong belief in Darwin's principle of natural selection is not a sufficient condition for understanding evolution. The standards for research in evolutionary biology should be set no lower than those in any other area of scientific inquiry.

Nothing in Evolution Makes Sense Except in the Light of Population Genetics

In a few decades, we have gone from a state of no molecular data at all to one in which hundreds of genomes are available to the general public, and in the very near future, every moderately well-funded laboratory will be capable of acquiring multiple whole-genome sequences overnight. It would be easy to sit back and assume that the goals of evolutionary biology will soon be met. However, in some sense we are at a new beginning. Although cataloguing the products of evolution is a tangible goal, understanding how these endpoints were reached is quite another matter.

Evolution is a population genetic process governed by four fundamental forces, which jointly dictate the relative abilities of genotypic variants to expand throughout a species. Darwin articulated a clear but informal description of one of those forces, selection (including natural and sexual selection), whose central role in the evolution of complex phenotypic traits is universally accepted, and for which an elaborate formal theory in terms of change in genotypic frequencies now exists (Crow and Kimura 1970; Bürger 2000). The remaining three evolutionary forces, however, are nonadaptive in the sense that they are not a function of the fitness properties of individuals: mutation (broadly including insertions, deletions, and duplications) is the fundamental source of variation on which natural selection acts; recombination (including crossing-over and gene conversion) assorts variation within and among chromosomes; and random genetic drift ensures that gene frequencies will deviate a bit from generation to generation independently of other forces. Given the century of theoretical and empirical work devoted to the study of evolution, the only logical conclusion is that these four broad classes of mechanisms are, in fact, the only fundamental forces of evolution. Their relative intensity, directionality, and variation over time define the way in which evolution proceeds in a particular context.

As emphasized in previous chapters, an understanding of molecular biology combined with studies of molecular variation within species yields reasonably reliable estimates of the relative strengths of the forces of evolution operating at the genomic level, and such work leaves little doubt that all four major forces are of central significance. Not only is it impossible to understand the diversification of genomes purely in terms of natural selection, but many aspects of genomic evolution can only be understood by invoking a negligible level of adaptive involvement. This conclusion raises significant technical challenges because all three nonadaptive forces of evolution—mutation, recombination, and random genetic drift—are stochastic in nature and can generally only be understood in probabilistic terms. It is well known that most biologists abhor all things mathematical, but the quantitative details really do matter in evolutionary biology.

Some have attempted to marginalize the contributions of population genetics to our understanding of evolution by pointing to the "beanbag" genetics debate that occurred in the middle of the twentieth century (for a historical overview, see Felsenstein 1975). Such criticism is a product of a misunderstanding, however, as the tensions during this period were not about the population genetic nature of the evolutionary process, but about the need to incorporate interaction (epistatic) effects into the existing framework, something that population geneticists have invested heavily in over the past few decades (e.g., Wolf et al. 2000; Carter et al. 2005). Nevertheless, numerous biologists, particularly in the area of development, have expressed reservations about the entire population genetic enterprise. Consider this quote from Carroll (2005a): "Since the Modern Synthesis, most expositions of the evolutionary process have focused on microevolutionary mechanisms. Millions of biology students have been taught the view (from population genetics) that 'evolution is change in gene frequencies.' Isn't that an inspiring theme? This view forces the explanation towards mathematics and abstract descriptions of genes, and away from butterflies and zebras. … The evolution of form is the main drama of life's story, both as found in the fossil record and in the diversity of living species. So, let's teach that story. Instead of 'change in gene frequencies,' let's try 'evolution of form is change in development.'" Many similar statements could be quoted from other authors.

Even ignoring the fact that the vast majority of species are unicellular, differentiated mainly by metabolic rather than developmental features, this type of statement paints an inaccurate portrait of the current field of evolutionary biology. Evolution is much more than a storytelling exercise, and the goal of population genetics is not to be inspiring, but to be explanatory. From the standpoint of its phenotypic products, evolution is indeed more than a change in gene frequencies: organisms are far more than the sum of their parts, just as genes are more than the sum of their functional components. But if we are concerned with the process of evolutionary change, then evolution is indeed a change in gene frequencies (or more accurately, a change in genotype frequencies). Population genetics provides an essential framework for understanding how such changes come about, and more importantly, grounds us in reality by clarifying the pathways that are open or closed to evolutionary exploitation in various contexts. As Carroll (2005a) states in a somewhat different context, "Simplification may indeed be necessary for news articles, but it can distort the more complex and subtle realities of evolutionary patterns and mechanisms."

Population geneticists do not claim to have solved every problem in evolution—far from it. But as far as we know, the basic theoretical machinery to do so is largely in place. Population geneticists do not aspire to exclusive ownership of the field of evolutionary biology. Indeed, one of the central points of the preceding chapters is that a full understanding of the evolutionary process will be impossible without substantial input from molecular, cellular, and developmental biologists. However, such integration needs to be a two-way street. Evolutionary biologists have thought quite a lot about evolution, and individuals from outside the field who claim to have solved a major evolutionary enigma might want to consider why their ideas have not previously come to the forefront. Have such ideas been ignored, or have they faded into the background because their feasibility is known to be marginal? The population genetic basis of evolutionary change is now so well established that those who claim its inadequacy should certainly bear the burden of explanation.

The Passive Emergence of Genomic Complexity by Nonadaptive Processes

Throughout this book, we have adhered to the principle of uniformitarianism, the well-accepted view that the natural forces that operate today also did so in the past (at least back to the point at which cells, and therefore individuals, had evolved). Regardless of the structural and metabolic features of cells, the four major forces of evolution apply to all organisms, including those that have gone extinct. No genome can replicate without error, and mutation creates genetic variation in fitness, which by definition ensures the operation of selection, even in a constant environment. All organisms, even prokaryotes, also experience some recombination. Finally, all populations are finite in size and therefore subject to chance fluctuation in genotype frequencies, the magnitude of which depends on the absolute population size, the breeding system, and the chromosomal features of the genome.

Even among those who acknowledge the joint operation of the four major forces in all contexts, statements are commonly made that natural selection is the only guiding force of evolution, with mutation creating variation but never controlling its ultimate direction (for a review, see Stoltzfus 2006b). This view, especially cherished by those studying continuously distributed (e.g., morphological, life history, and physiological) traits, derives from two types of arguments. First, artificial selection experiments with diverse organisms are generally capable of generating changes in mean phenotypes well beyond anything seen in the base population in just a few dozen generations (Falconer and Mackay 1996). Such observations inspired the development of the infinitesimal model, which assumes that standing genetic variation reflects a balance between opposing forces of selection and mutation distributed over an effectively infinite number of loci, each with minuscule effects (Kimura 1965; Lande 1975; Bulmer 1980). Under one extreme form of this model, directional selection changes the mean phenotype without altering the genetic variance, so variation is always available. Second, much of the earliest work in theoretical population genetics downplayed the role of mutation in a different way. In particular, Fisher (1930) and Haldane (1932) argued that mutation operating in a direction contrary to selection could never overcome the force of selection.

Neither of these arguments for a marginal role of mutation in directional evolution represents a situation in which the theory was structurally incorrect. Rather, both are simply based on questionable a priori assumptions. The infinitesimal model assumes a symmetrical distribution of mutational effects independent of the prior state; the Fisher–Haldane argument assumes that mutation is always a weak force relative to selection; and both ignore the complications that arise in finite populations. In the context of genomic evolution, we now know that the idea that mutation is a nondirectional force is substantially off the mark. Nucleotide composition is influenced considerably by biases in mutation and gene conversion, and many other aspects of genomic architecture, including mobile element proliferation, arise via internal drivelike mechanisms. With some vestiges tracing back to Darwin (1859, 1866) and Morgan (1925), the notion that internal mutation pressures can be a directional force in evolution is not new (Dover 1982; Nei 1987, 2005; Cavalier-Smith 1997) and has recently been revisited in a more formal way by Yampolsky and Stoltzfus (2001) and Stoltzfus (2006a). But it is fair to say that most biologists view evolution as a process dominated by ecological forces external to the organism.

Given the centrality of this issue, the conditions that must be fulfilled if mutation is to impose a directional bias on evolution are worth reiterating. Suppose there are two alternative alleles at a locus, with the mutation rate of $a \rightarrow A$ being m times that of $A \rightarrow a$, but with type A having a selective advantage s over type a. Letting the effective number of genes per locus be N_g, the fixation probability of a mutation to type A is e^S times that for a mutation to type a, where $S = 2N_g s$ (see Box 6.2). Thus, because the rate of transition from one allelic type to another is equal to the product of the mutation rate and the fixation rate, the ratio of probabilities of being in state A versus state a is simply me^S. Because it jointly incorporates the effects of mutation, selection, and random genetic drift (even the local effects of recombination are contained within N_g), this simple expression makes a powerful statement.

If there is no mutational bias ($m = 1$) and no selection ($S = 0$), both states are equally probable (Figure 13.1). However, if the force of selection is weak ($e^S \approx 1$), because either the absolute value of s or the effective population size is small ($N_g << |1/s|$), the pattern of genomic evolution will be strongly governed by mutational bias (m). Only if S is much greater than one (the

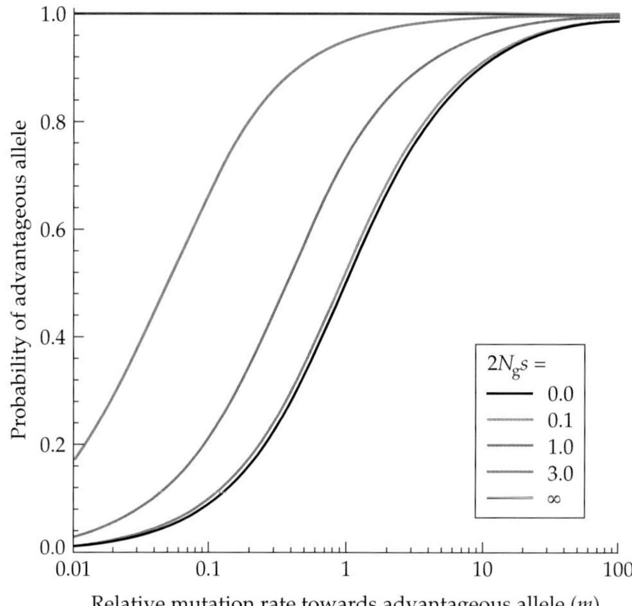

Figure 13.1 The long-term probability that an allele residing at a biallelic locus will be of the selectively advantageous type, given a selective advantage s, an effective number of gene copies at the locus in the population of N_g, and a mutation rate to the beneficial allele m times that in the reverse direction. The curved lines plot the function $me^S/(1 + me^S)$, as described in the text. The situation in which $S = 2N_g s = 0.0$ denotes neutrality, which leads to an expected frequency of the advantageous allele equal to $m/(1 + m)$. The situation in which $2N_g s = \infty$ denotes an effectively infinite population size, such that random genetic drift is a negligible force, and the expected allele frequency is very close to 1.0. When $2N_g s \ll 1.0$ (effective neutrality), allelic states closely resemble the expectations under mutation pressure alone.

worldview of Fisher and Haldane) will the population be in state A with a probability approaching 1.0. Moreover, these conclusions hold regardless of the absolute power of mutation relative to selection. Numerous results presented in previous chapters indicate that for all forms of mutation, whether advantageous or deleterious, the condition $N_g \ll |1/s|$ is frequently met. That is, the conditions on this planet are such that effective population sizes are often too small to prevent internal mutation pressures from driving patterns of genomic evolution.

Thus, at least at the genomic level, there can no longer be any doubt that directional aspects of evolution are governed by all four major forces. Furthermore, the data from comparative genomics summarized in the preceding chapters support the more specific contention that many of the basic embellishments of the eukaryotic genome originated by nonadaptive processes operating contrary to the expected direction of natural selection.

The central conclusions, some of which constitute a rather substantial departure from many of the traditional beliefs in evolutionary biology, and others a reinterpretation of those beliefs, can be summarized as follows:

- The same population genetic processes govern patterns of evolution in prokaryotes and eukaryotes, although differences in the relative power of various nonadaptive forces result in fundamentally different evolutionary trajectories at the genomic level.
- Internal mutation pressure plays a central role in genomic evolution, often completely overwhelming the external forces of natural selection and driving change in a directional manner.
- Although genetic drift is a random process, population size, by dictating the efficiency of selection, modulates the evolution of genomic complexity by nonadaptive mechanisms, inhibiting certain kinds of evolutionary change and encouraging others.
- Because each additional layer of gene structural complexity increases the mutation rate to defective alleles, mutation operates as a weak selective force, serving as a barrier to the emergence of gene and genomic complexity in populations of sufficiently large size.
- By enhancing the permissiveness of the population genetic environment for the passive emergence of gene architectural complexity, the nonadaptive force of random genetic drift sets the stage for future paths of adaptive evolution in novel ways that would not otherwise be possible.

Although this broad set of claims may appear to be overly grandiose, inviting the same sort of criticism that I have launched in the preceding paragraphs, mutation and random genetic drift are universal evolutionary forces (even more universal than natural selection), applying to all organisms and all chromosomal regions. Thus, provided that the power of one or both of these forces is often high relative to that of natural selection operating at the genomic level, as the data suggest, it should not be too surprising that a wide range of heretofore disconnected patterns in genomic evolution are potentially encompassed by a single theoretical framework.

If the explanatory power of this theory is judged to be overly simplistic, it should be relatively easy to demonstrate that the aspects of genomic evolution outlined in Table 13.2 cannot be the products of nonadaptive forces. Simply claiming that natural selection is all-powerful will not do, for under that worldview, natural selection can explain anything, and therefore explains nothing. If a more specific adaptive counterargument is to be formally defensible, then simpler nonadaptive models must be shown to be inadequate, and to accomplish that, something must be known about the expected pattern of evolution in the absence of selection. If nothing else, the ideas presented above serve as a null hypothesis about the expected pattern of genomic evolution in the absence of selection.

In further pursuing these ideas, care must be taken to avoid descending into a world of nonadaptive storytelling. For example, as we learn more

TABLE 13.2 Aspects of gene and genomic architecture that appear to be explainable only after accounting for nonadaptive evolutionary forces

Transition from the RNA world to the DNA world
Genomic streamlining in microbial species
Nucleotide composition variation within and among genomes: A/T composition, strand asymmetry, isochores, synonymous codon usage
Centromere expansion in multicellular lineages
Differential proliferation of mobile elements in unicellular versus multicellular species
Gene number: preservation of duplicate genes by degenerative mutations (subfunctionalization)
Origin of the spliceosome by subfunctionalization and proliferation of introns in lineages of multicellular species
Expansion of UTRs (untranslated regions) of the messenger RNAs of eukaryotes
Origin of modular regulatory regions in eukaryotic genes
Demise of operons in eukaryotes
Variation in organelle genomic architecture: lean in animals, bloated in land plants
Messenger RNA editing in plant organelle genomes
Restriction of sex chromosomes to multicellular lineages

about the physical processes of genomic mutability (e.g., the rates and distribution of nucleotide substitutions, gene duplications, transpositions, and other forms of insertion and deletion), it may be found that the null models presented above are inadequate in some respects. But that will not alter the basic challenge. If evolutionary genomics is to be more than a hand-waving exercise, it is not so important that the ideas herein withstand the test of time, so long as they can be formally shown to be false in a hypothesis-testing framework.

Finally, it is worth emphasizing that the views presented above bear on the long-standing debate in evolutionary biology regarding the predictability of evolutionary trajectories. Because of the vagaries of the external environment and the stochastic nature of mutation, recombination, and genetic drift, it is generally believed that evolution can never be a predictive science, any more than meteorology will ever be able to predict the weather weeks in advance. Nevertheless, despite our inability to forecast the precise changes that natural selection will elicit at the molecular or phenotypic levels, it now seems clear that the nonadaptive forces of random genetic drift and mutation do leave strong and consistent signatures at the genomic level in independent phylogenetic lineages, regardless of the exact nature of external ecological factors. One of the most dramatic examples of such parallel evolutionary patterns concerns the nuclear genomes of animals and land plants. Although many lineage-specific peculiarities exist at the genomic level (e.g., scrambled genes and dual nuclei in ciliates, massive mitochondrial mRNA editing in trypanosomes), changes in the mutation rate (u) and/or the effective population size (N_g) (or more precisely, the product $N_g u$, the ratio of the power of mutation to that of drift) generally leave predictable footprints at the level of gene and genomic architecture when viewed in the context of known molecular biological constraints.

This being said, it must be emphasized that genomic evolution operates on very long time scales (on the order of the reciprocal of the mutation rate). In contrast, estimates of $N_g u$ derived from silent-site polymorphism data (π_S; see Chapter 4) have short-term applicability (reflecting events over just the past $2N_g$ generations) and can suffer from considerable statistical inaccuracies and biases, all of which reduces their utility as indicators of the conditions under which particular genomic architectures have evolved. This problem represents a fundamental limitation on our ability to precisely define the population genetic conditions that have driven the evolution of genomic complexity versus simplicity. Because no method exists for the estimation of longer-term average $N_g u$, to minimize the impact of stochastic species-specific events, most of the preceding analyses have relied on lineage-wide average estimates of π_S to approximate the long-term population genetic environment for major lineages.

Do the Roots of Organismal Complexity Also Reside in Nonadaptive Processes?

Some of the deepest mysteries in evolutionary biology concern the mechanisms that gave rise to the transitions between unicellular and multicellular species. Multicellularity is widely viewed as a unique attribute of eukaryotes, somehow made possible by the origin of a more complex cellular architecture and, without question, with the assistance of natural selection. However, it is difficult to defend this assertion in any formal way. One central problem is that only two lineages of complex multicellular organisms exist: animals and vascular plants. True, both are eukaryotic, and one might add fungi to the list, although the number of fungal cell types is not large and there is some question as to whether multicellularity was ancestral to the major phylogenetic group containing animals, fungi, and slime molds (see Chapter 1). In any event, the probability that two or three origins of multicellularity simply arose by chance within eukaryotes as opposed to prokaryotes is somewhere on the order of 1/8 to 1/4, well below the general standards of statistical validity. Of course, many other eukaryotes are capable of producing a few different cell types, but the same is true of prokaryotes, some species of which produce radically different cell morphologies.

Nevertheless, King (2004) states that "this historical predisposition of eukaryotes to the unicellular lifestyle begs the question of what selective advantages might have been conferred by the transition to multicellularity;" and Jacob (1977) argues that "it is natural selection that gives direction to changes, orients chance, and slowly, progressively produces more complex structures, new organs, and new species." The vast majority of biologists almost certainly agree with such statements. Where, then, is the direct supportive evidence for the assumption that complexity is entirely rooted in adaptive processes? No existing observations support such a claim, and given the massive global dominance of unicellular species over multicellular eukaryotes, in terms of both species richness and numbers of individuals, if there is an advantage of organismal complexity, one can only marvel at the inability of natural selection to promote it. As noted in Chapter 4, multicellular species experience reduced population sizes and reduced recombination rates, both of which diminish the efficiency of selection, thereby encouraging the accumulation of hazardous extragenic DNA. It may be no coincidence that multicellular species also have substantially higher extinction rates than unicellular species (Raup 1978; Stanley 1985).

Although some aspects of the roots of the cellular interactions that constitute development must reside in the resolution of adaptive conflicts between the advantages of cell–cell cooperation and the individual benefits of going it alone (e.g., Maynard Smith and Szathmáry 1995; Michod 1999; Michod and Roze 2001), it need not follow that natural selection is a sufficient force for the exit from the unicellular world. The concept of a "great chain of being," with all lineages gradually evolving toward some advanced grade, has long been abandoned, and earlier chapters have provided numerous examples of simplification at the genomic level, the case of the yeast *S. cerevisiae* being particularly well documented. Many developmental genes previously thought to have originated in the vertebrate lineage, owing to their absence in arthropods and nematodes, are now known to be present in basal lineages of animals lacking mesoderm (the cnidarians) and hence must have been lost from triploblastic lineages (Technau et al. 2005). Numerous examples of morphological simplification exist in animals, such as limb loss in various tetrapods, coelom loss in nematodes (assuming the validity of the Ecdysozoa grouping; see Chapter 1), and mouth and anus loss in hydrothermal vent worms. A plausible, albeit controversial, argument has even been made that prokaryotic cell architecture is a simplified derivative of that of eukaryotes (Kurland et al. 2006). One could argue that multicellularity arises after a species evolves new proteins that enable cells to stick together (Kirschner and Gerhart 2005), but large numbers of prokaryotes are capable of that, and one could just as easily argue that multicellularity arises when a unicellular species loses a reliable mechanism of cell separation.

Is there any justification for thinking that nonadaptive processes could have played a role in the evolution of something as intricate as cell architectural or developmental features? Although there is no need to abandon the idea that many of the external morphological or behavioral manifestations of cellular complexity in today's organisms are adaptive, understanding the roots of the hallmarks of multicellularity (e.g., multiple cell types, complex patterns of gene expression, and mechanisms of cell signaling) requires knowledge of the processes that facilitate the prior establishment of the genomic material that makes their evolution possible. The challenge here is that although numerous embellishments of eukaryotic genes are exploited in functionally adaptive ways in today's multicellular species, most such modifications are mutationally disadvantageous, and there is no evidence that they were of *immediate* adaptive value to the first unicellular species to harbor them. There are two broad mechanisms by which nonadaptive evolution at the genomic level might precipitate a cascade of events at higher levels of organization, and in both cases, the salient point is that certain types of cellular/developmental architectures are more likely to emerge evolutionarily in some population genetic environments than in others, independent of external selection pressures.

First, intrinsically deleterious genome-level changes, such as those resulting from intron and mobile element proliferation, may impose selection pressure for intracellular defense mechanisms. Remarkably little thought has been given to the evolution of cellular architecture except in the context of hypothetical external selective factors (e.g., Cavalier-Smith 2006a). However, it has been argued that the origin of spliceosomal introns, by imposing a need to process mRNAs prior to their exposure to ribosomes, provided the evolutionary pressure leading to the origin of the nuclear membrane (Martin and Koonin 2006; Lopez-Garcia and Moreira 2006), and Koonin (2006) has gone further in suggesting that the nonsense-mediated decay (NMD; see Chapter 9) and ubiquitin signaling pathways evolved as downstream mechanisms for minimizing the accumulation of aberrant transcripts and proteins resulting from splicing errors. In principle, this line of thinking could be taken in a number of additional directions. For example, the assembly of the spliceosomal subunits occurs in the Cajal body, a subnuclear organelle (Stanìk and Neugebauer 2006), and aberrant transcripts flagged by the NMD pathway are degraded in cytoplasmic P bodies (Sheth and Parker 2006). Nevertheless, resolving matters of causality is not so straight-forward here. All three hypothetical lines of defense against intron-associated problems appear to be distributed throughout the entire eukaryotic domain and hence were almost certainly present in the stem eukaryote. Thus, the possibility exists that the colonization of nuclear genes by introns followed the origin of permissive cellular features, rather than the other way around, although in either case, the idea that internal constraints played a role in cellular evolution would be secure.

Second, because cellular and developmental features reflect the transformation of gene-level information into gene expression, the potential directions of phenotypic evolution must ultimately be defined by the physical materials existing at the genomic level, the nature of which appears to be strongly influenced by aspects of the population genetic environment. Reductions in effective population sizes are expected to lead to passive increases in intron number and size, expansions in UTR lengths, losses of operons, the modularization of regulatory regions, and the preservation of duplicate genes by subfunctionalization, among other things (see Table 13.2).

Thus, as will be discussed more fully in the following section, to the extent that an increase in genomic and gene architectural complexity is a precondition for the emergence of greater complexity at the organismal level, a long-term synergism may exist between the mechanisms of nonadaptive evolution at the DNA level and adaptive evolution at the phenotypic level. If this view is correct, then the relatively simple phenotypes of the Earth's smallest organisms are neither an inevitable outcome of the cell biological features of such species nor a reflection of selection for metabolic efficiency, but an indirect consequence of enormous effective population sizes, which, by reducing the power of random genetic drift, impose a barrier on the passive emergence of genomic complexity.

Unfortunately, the emerging field of evolutionary developmental biology is based almost entirely on a paradigm of natural selection, and the near-absence of the concept of nonadaptive processes from the lexicon of those concerned with cellular and developmental evolution does not follow from any formal demonstration of the negligible contribution of such mechanisms, but simply reflects the failure to consider them. There is no fundamental reason why cellular and developmental features should be uniquely immune to nonadaptive evolutionary forces. One could even argue that the stringency of natural selection is reduced in complex organisms with behavioral and/or growth form flexibilities that allow individuals to match their phenotypic capabilities to the local environment.

An Entrée to Understanding the Origins of Genomic Determinants of Organismal Complexity

The tone of dissent in the preceding paragraphs should not be misinterpreted as disrespect for the more reductionistic areas of biology, all of which are essential for resolving the functional underpinnings of the *products* of evolution. However, cataloging patterns of biodiversity is not equivalent to ascertaining the evolutionary mechanisms by which such patterns emerged. To date, cellular and developmental studies have contributed minimally to our understanding of such *processes*, and indeed, since their inception, have been strangely distant from mainstream evolutionary thinking (for a historical overview, see Wilkins 2002). With the advent of new methods for measuring gene expression and protein interactions, the opportunities for advancement of knowledge in the areas of cellular and developmental evolution are enormous, but real progress will require the integration of well-established principles from evolutionary genetics.

Almost all molecular, cellular, and developmental biologists eschew intraspecific variation, focusing instead on typological characterizations of a few model species, for obvious practical reasons. However, because virtually every complex trait in every species exhibits significant genetic variation (Lynch and Walsh 1998), it is likely that many textbook examples of genetic/developmental pathways derived from single clones or inbred lines are unrepresentative of patterns in nature, and some may be positively misleading. Will the field of comparative development learn to utilize variation and incorporate population-level thinking to enhance interpretation? Or will the field remain shrouded in a kind of fifteenth-century mysticism, packaged in a gallery of beautiful embryo paintings? Although there is still a very long way to go, it is encouraging that a scaffold connecting evolutionary genetics, genomics, and developmental biology is beginning to emerge. Interested readers are encouraged to start with Johnson and Porter (2000, 2001), Stern (2000), Delattre and Felix (2001), True and Haag (2001), Rockman and Wray (2002), Wray et al. (2003), Force et al. (2005), and Haag and Molla (2005).

Today's field of comparative development is heavily invested in the idea that modularity and repatterning of regulatory element utilization are the central determinants of the evolution of organismal complexity (Carroll 2005b; Davidson 2006), a view inspired in part by King and Wilson's (1975) suggestion that much of phenotypic evolution is driven by regulatory rather than structural mutations. This is not a universally accepted view, even among developmental biologists (e.g., Alonso and Wilkins 2005), and the dismissal of gene duplication as a major force in the origin of evolutionary novelties (Behe and Snoke 2004; Carroll 2005b) ignores a vast amount of data that suggest otherwise (see Chapter 8). Nonetheless, because development always involves cross-talk between multiple gene products, a logical starting point is to consider the mechanisms of origin of the gene structural features that allow such transactions to take place.

Despite the seductive nature of the idea that the apparently well-designed regulatory modules with functional significance in today's organisms could have arisen only via natural selection, it remains to be determined how the alterations necessary for the construction of genetic pathways come about. There is no evidence that gene regulatory modules associated with complex (multigene) functions arise as de novo integrated units, although some developmental biologists seem to believe otherwise (e.g., Davidson and Erwin 2006). Rather, like all aspects of evolution, changes in genetic pathways must be products of descent with modification, with mutant alleles arising at individual loci independently of processes occurring at other interacting loci. The following is a brief sketch of how stepwise mutations, operating in population genetic environments common to multicellular species, might encourage the emergence of the kinds of gene structures and genetic pathway organizations that are thought to be essential for the building of complex organisms, without any direct involvement of selection.

First, as noted in Chapter 10, in populations of sufficiently small size, the joint processes of regulatory region duplication and subsequent degenerative mutation can lead to the passive emergence of modular gene regulatory structures (with unique transcription factors governing expression in different spatial/temporal contexts) from an initial state in which the entire expression breadth of the gene is under unified control (i.e., regulated via the same transcription factors in all tissues or stages of development). During such a transition, a new gene architecture emerges beneath a constant spatial and/or temporal pattern of gene expression, with no direct selection for modularity per se. However, because of the mutational cost of allelic complexity, the completion of such repatterning is less likely to occur in situations in which the excess degenerative mutation rate of a modified allele exceeds the power of drift. Thus, contrary to popular belief, natural selection not only may be an insufficient mechanism for the origin of genetic modularity, but may actually promote the opposite result: alleles under unified transcriptional control as well as operons (spatially contiguous genes under unified control). Under this view, the reductions in effective population size that probably accompanied the origin of eukaryotes and the subsequent emergence of the animal and land plant lineages may have played pivotal roles in the emergence of modular gene architecture. Moreover, such neutral transitions may help explain numerous cases of "developmental system drift," whereby seemingly similar morphological structures in closely related species have come to be achieved by substantially different developmental/regulatory mechanisms (Sommer 1997; Ludwig et al. 2000; Johnson and Porter 2001; True and Haag 2001; Ruvinsky and Ruvkun 2003; Tsong et al. 2006).

Second, the transition to a modular gene regulatory structure following population size reduction may not always await the de novo origin of new regulatory elements following a population bottleneck. Rather, due to the recurrent mutational production of alternative transcription factor binding sites, large populations can be expected to harbor low frequencies of semi-modular alleles with redundant modes of gene activation (see Figure 10.11), which, owing to the masking effects of redundancy, have reduced rates of mutation to defective alleles. Thus, while the *fixation* of modularized alleles may be inhibited in large populations, their immediate precursors may be maintained at moderately high frequencies in such settings (Wagner 1999, 2001). This implies that a population that has experienced a prolonged phase of large effective size can actually be poised to make a transition to a state of modularity following a population size reduction. This same point was made in Chapter 8 with respect to the neofunctionalization of duplicate genes: alleles with novel beneficial mutations that simultaneously eliminate key ancestral functions cannot go to fixation prior to duplication, but in populations of sufficiently large size, they will be maintained at low frequencies, which greatly increases the probability of neofunctionalization following a duplication event. Wright's shifting balance theory, developed in the pre-molecular era, provides an interesting historical precedent for the idea that fluctuations in population size and structure may greatly alter the paths open to evolution (reviewed in Provine 1986).

Third, the emergence of independently mutable subfunctions in a modularized allele may have second-order effects that contribute to adaptive evolution in significant ways. For example, if the ancestral allele under unified control was subject to pleiotropic constraints associated with shared regulatory regions, modularization may open up previously inaccessible evolutionary pathways by eliminating a "jack of all trades, master of none" syndrome. Such genomic repatterning may be even further facilitated following the duplication of the entire gene, as complementary degenerative mutations can then result in the partitioning of cellular tasks among paralogous copies, which can in turn lead to further refinements in the ability to carry out individual subfunctions (Force et al. 1999). Indeed, it is relevant that both processes—the emergence of gene subfunctions and their subsequent partitioning among paralogous genes—occur most readily in the same small to medium-sized populations. Operating together, these processes provide a powerful but passive mechanism for the gradual remodeling of entire genetic pathways (Figure 13.2).

Finally, a striking aspect of developmental pathways that has yet to be explained in adaptive terms concerns their oftentimes baroque structure (Wilkins 2002, 2005). It is common for linear pathways to consist of a series of genes whose products are essential to the activation/deactivation of the next downstream member, with only the expression of the final component in the series having an immediate phenotypic effect. For example, the product of gene D may be necessary to turn on gene C, whose product is necessary to turn on gene B, which product finally turns on gene A. Pathways involving only inhibitory steps also exist, and these lead to an alternating series of high and low expression, depending on the state of the first gene in the pathway. For example, gene D may produce a product that inhibits the expression of gene C, whose silence allows gene B to be turned on, whose product inhibits the expression of gene A. It is unclear that such complexity has any advantages over the simple constitutive expression or self-regulation of the final member of the pathway.

In principle, pathway augmentation may be driven entirely by the nonadaptive processes of duplication, degeneration, and random genetic drift. Consider the series of events in Figure 13.3. Initially, a single gene A carries out some function in a constitutive fashion, but in a series of steps, it becomes completely reliant on an upstream activation factor B. A scenario like this could unfold in the following way. First, gene A becomes sensitive to activation by gene B, either because gene A has acquired a *cis* modification that causes activation by B, or because some transcription factor B acquires a mutation that causes it to serve as a *trans* activator of A. At this point, gene A has redundant activation pathways and is therefore vulnerable to loss of one of them. Should a degenerative mutation cause gene A to lose the ability to self-regulate, gene B will have been established as an essential activator of gene A. This process can be repeated anew as gene B acquires sensitivity to a further upstream transcription factor and loses the ability to be constitutively expressed.

As in the case of subfunction fission and duplicate-gene subfunctionalization, the probability of establishment of these types of changes will depend on the effective population size. A redundantly regulated allele has a weak mutational advantage equal to the rate of loss of a regulatory site

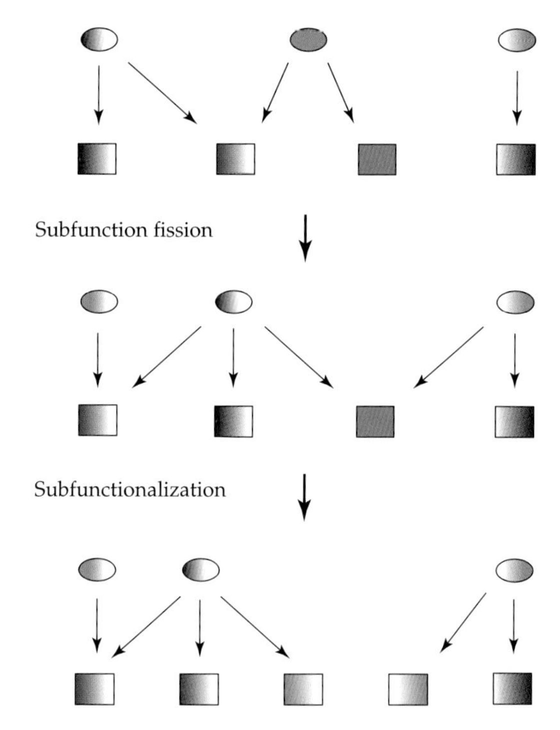

Figure 13.2 Reorganization of a gene regulatory network by subfunction fission and duplicate-gene subfunctionalization. Initially, three transcription factors (TFs; ovals) drive the expression of four genes (squares) such that there are two independent networks (resulting in the final red/blue/green products on the left and the maroon product on the right). The gradient of shading within ovals and boxes denotes the regional expression of genes (e.g., left = anterior, right = posterior, and solid = ubiquitous). Initially, the green gene experiences subfunction fission in such a way that its expression is no longer driven by a ubiquitous TF (olive), but instead by two more localized TFs (black and orange), which together cover all relevant tissues. The functional aspects of the olive TF are then engaged only in anterior expression (yielding a potential release from pleiotropic constraints), as no posterior genes are any longer dependent on its expression. Now exploiting two tissue-specific TFs, the green gene may also experience a reduction in pleiotropic constraints. In addition, the new modular regulatory architecture of the green gene sets the stage for subfunctionalization following gene duplication, with the black and orange TFs each coming to drive the expression of a different paralog. The anterior and posterior gene networks are now entirely independent.

(u_l): one such mutation will result in the nonfunctionalization of either a self-regulated or an upstream-dependent allele, but will leave the function of a redundantly regulated allele unaltered. If the effective size of a population is sufficiently small that $1/N_g >> u_l$, such an advantage will be impervious to selection, and the population will evolve to an allelic state that sim-

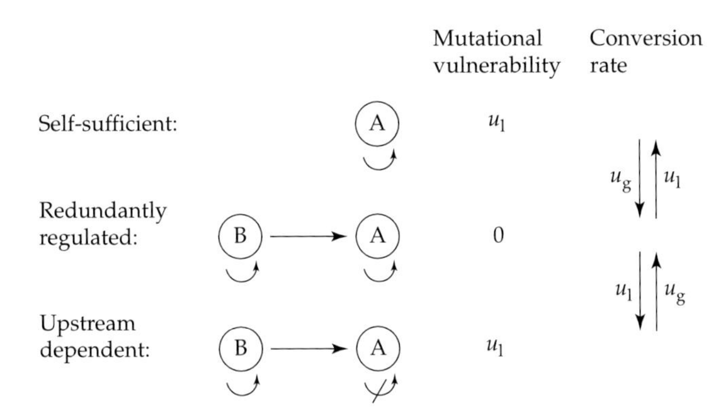

Figure 13.3 An example of the growth of a genetic pathway by augmentation with an upstream transcription factor and loss of self-regulatory ability. Mutational rates of gain and loss of regulatory abilities are denoted by u_g and u_l, here for simplicity assumed to be the same for self-activation and upstream activation. The redundantly regulated gene remains active in the face of single-loss mutations.

ply depends on the relative rates of gain and loss of regulatory sites (u_g and u_l in Figure 13.3), eventually leading to the establishment of an obligatory pathway. In contrast, if $1/N_g << u_l$, the accumulation of upstream-dependent alleles will be inhibited by their weak mutational burden, and in recombining species, by their inactivation when they appear in genetic backgrounds that fail to support A–B cross-talk. Thus, whereas small population sizes may promote the passive elongation of genetic pathways, large population sizes have the opposite effect.

Although the preceding arguments demonstrate that small population size provides a permissive environment for the emergence of complex gene-level transactions without any direct selection for complexity, this does not mean that such alterations cannot occur in very large populations. However, if such changes are to occur in a large N_g context, they must have substantial immediate selective advantages to offset the mutational burden of gene structural complexity. Imagine a new mutational alteration of gene structure that confers a change in development (or any other organismal feature) with a selective advantage in terms of survival and offspring production equal to s_a. The net advantage is then $s_a - u_d$, where u_d is the mutational burden of the modified allele (i.e., the excess mutation rate to defective alleles, owing to the larger mutational target). Following the arguments presented above, a reasonable probability of fixation by natural selection requires that $2N_g(s_a - u_d) > 1$, such that the net selective advantage $(s_a - u_d)$ exceeds the power of random genetic drift, $1/N_g$. If $2N_g(s_a - u_d) < -1$, owing to a sufficiently large mutational disadvantage, fixation will be thwarted regardless of the allele's selective advantage.

Evolvability

The emergence of systems biology and the various -omics fields (genomics, proteomics, metabolomics, transciptomics, etc.) and their focus on "discovery-based science" (e.g., Ideker et al. 2001) has led to the promotion of four of the major buzzwords in biology today: complexity, modularity, robustness, and evolvability (see Table 13.1). Many view such attributes to be emergent properties of organisms that cannot be explained by the sum of their parts, arguing that adaptive mechanisms not previously appreciated by evolutionary biologists are required to explain their existence. Those who promote the concept of the adaptive evolution of the above features are by no means members of the intelligent design movement, which invokes unknown mechanisms to explain biodiversity. However, as emphasized above, the burden of evidence for those who embrace an all-powerful guiding hand of natural selection should be no less stringent than what one would demand of a creationist.

Interest in the emergent properties of organisms is derived in part from the massive amount of data now available on genome sequences and gene expression and a desire to reduce this information down to a few generalized descriptors. Such a philosophy has historical precedence in the field of systems ecology, which developed in the 1960s in response to the growing amount of information on ecosystem properties and a desire to reduce the data down to indices of energy flow, connectivity, diversity, and stability. Today, this field is essentially extinct, having been displaced by population and community ecology as it became clear that the detailed interactions of individual species do matter. Will the field of systems biology suffer the same fate, or is there really something special about the emergent properties of organisms? The preceding paragraphs have already raised significant questions about the adaptive nature of complexity, modularity, and robustness, but one final matter of tension over views of the evolutionary process merits consideration.

All replicating populations are capable of evolution, but it has recently been argued that some species (multicellular organisms, in particular) have evolved to be better at it than others, with natural selection directly advancing features of genomic architecture, genetic networks, and developmental pathways so as to enhance their future ability to generate heritable adaptive variation. Such speculation is almost entirely restricted to molecular and cell biologists, engineers, physicists, and those who study digital organisms (e.g., Depew and Weber 1985; Gerhart and Kirschner 1997; Kirschner and Gerhart 1998, 2005; Rutherford and Lindquist 1998; True and Lindquist 2000; Caporale 2003; Earl and Deem 2004; Bloom et al. 2006; Federici and Downing 2006), and there are good reasons to question its validity (e.g., Williams 1966; Dickinson and Seger 1999; Partridge and Barton 2000; Brookfield 2001; Sniegowski and Murphy 2006). However, the issues apparently warrant reiteration, and it is worth doing so in the context of our knowledge from comparative genomics.

Before proceeding, it must be emphasized that the word "evolvability" is deployed in very different ways by evolutionary biologists and by those outside the field. In quantitative genetics, the term "evolvability" has long had a definition closely related to the concept of heritability, i.e., the relative amount of standing variation that is subject to a response to natural selection (Houle 1992; Lynch and Walsh 1998). The above-mentioned authors, however, use the word in a rather different way, loosely defining "evolvability" as the ability of a lineage to generate useful adaptive variation via mutational flexibility. Under neither view is there any question regarding the existence of variation in evolvability, as it has been known for decades that different organisms and classes of traits have different propensities to respond to natural selection (Falconer and Mackay 1996). Less clear, however, is whether the ability to evolve is itself actively promoted by directional selection. Four reasons for skepticism follow.

First, evolution is a population-level feature: individuals are subject to selection, whereas populations evolve. Thus, in order for an aspect of evolvability to be advanced directly by adaptive mechanisms, selection must operate efficiently at a higher level of organization than the individual. Not only must populations with different levels of evolvability exist, but evolvability must be positively correlated with population longevity and/or productivity. Because populations survive longer than individuals, such group selection is necessarily a weaker force than individual selection (in general, very much weaker) and necessarily operates on longer time scales. Thus, if evolvability is to be subject to selective advancement, at least three stringent conditions must be met: (1) the group-level advantages of any genomic attribute that enhances evolvability must exceed any conflicting pressures operating at the individual level; (2) the enhanced capacity for rapid evolutionary change must persist over the time scale of group-level selection; and (3) while en route to fixation at the population level, the alleles that promote evolvability must remain tightly linked to the loci underlying the phenotypic traits whose evolution is advantageous. The evidence for individual-level selection on phenotypes is overwhelming (Endler 1986; Kingsolver et al. 2001). The evidence for the operation of group-level selection in nature, aside from kin selection in the evolution of social behavior (Hamilton 1964a,b; Wilson 1975), is weak at best, although some investigators remain more optimistic than others [see the exchange between Coyne et al. (2000) and Goodnight and Wade (2000)].

Second, there is no evidence that a long-term elevation in the level of evolvability is generally advantageous. Surely, a complete inability to evolve would not be a good thing, but inaccuracies in replication prevent such a scenario from ever occuring. The dynamics of the heritabilities of quantitative traits are extraordinarily complex, and selection operating at the individual level can modify the frequencies of epistatically interacting genes in ways that can either indirectly increase or decrease heritabilities, regardless of the advantage of the traits under selection (Carter et al. 2005). In addition, one can just as easily point to a long list of pathologies that can arise from an overly rapid proliferation of a new phenotype, including the overexploitation of resources resulting from the evolution of specialization, the reduced population size resulting from selective mortality, and the increase in mutational load associated with variance-generating mechanisms. Indeed, such scenarios have motivated a completely different, and equally speculative, view: that selection can favor mechanisms that suppress evolvability (Altenberg 2005). Moreover, theoretical studies have also shown that the kinds of complexities that are often the focus of those enamored with evolvability (e.g., increased dimensionality and modularity) can actually inhibit the rate of adaptive evolution, rather than promote it (Orr 2000a; Welch and Waxman 2003; Haygood 2006). Although the arguments are technical, they are no more abstract than the verbal reasoning of the evolvability school.

Third, there is no evidence that differences in short-term evolvability are anything more than simple by-products of phylogenetic diversity in the variation-generating and variation-maintaining factors that exist among species for purely physical reasons. For example, recombination and mutation rates vary by orders of magnitude among species, with no sudden discontinuities in the lineages imagined to be most evolvable (animals and land plants), and such variation appears to be largely a consequence of variation in chromosome length and number of germ line cell divisions per generation (see Chapter 4). The idea that alterations in the numbers and functions of prions and heat shock proteins serve to release new forms of cryptic variation on which natural selection can operate (e.g., Rutherford and Lindquist 1998; True and Lindquist 2000) is not very different from the now discredited idea that mobile elements are maintained to maximize the ability to evolve through a time of stress (see Chapter 7). To draw an analogy, free oxygen radicals, arising from intracellular metabolism, are a major source of mutation, but it is exceedingly unlikely that oxidative metabolism arose to promote evolutionary change.

Fourth, no observation from the several hundred genomes that have been sequenced can be taken as support for the idea that genomic architectural changes have been promoted in multicellular lineages so as to enhance their ability to evolve. Indeed, other than the appearance of spliceosomal introns, some forms of mobile elements, and organelles in the stem eukaryote (all of which arose prior to the origin of multicellularity), there is no discontinuity in the basic features of genomes across the entire domain of cellular life. Yes, on average, the genomic attributes of multicellular eukaryotes are more complex than those of unicellular lineages, but such differences did not arise for the reasons imagined by the evolvability school. Many unicellular species are excluded from certain evolutionary pathways that are open to multicellular species, and vice versa, but this is simply an indirect consequence of the altered power of random genetic drift in these different contexts, not a direct outcome of natural selection for the ability to engage in particular evolutionary pursuits.

In summary, if there is any validity to the notion that the ability to evolve is molded by natural selection, answers to several fundamental questions will need to be provided. What would serve as an objective measure of evolvability? Given such a definition, where are the direct examples of evolvability of traits being advanced by natural selection? What are the expected phylogenetic patterns of evolvability in the absence of selection? What would constitute a test of an evolvability hypothesis in a real living system, and what types of evidence would be acceptable as a falsification of the hypothesis?

Closing Comments

Because it deals with observations on historical outcomes, often in the face of incomplete information on intermediate steps (especially at the molecular level), the field of evolution attracts significantly more speculation than the average area of science. Nevertheless, the substantial body of well-tested theory established over the past century lays the groundwork for understanding the pathways that are open to evolutionary exploration in various population genetic contexts, providing guidance as to the likely reality of alternative evolutionary hypotheses. Because the nonadaptive forces of mutation, recombination, and random genetic drift are now readily estimated in multiple species using molecular data, there is no longer any justification for rejecting the practical utility of population genetic theory based on its quantitative unreliability. And because of the enormous knowledge base that has grown out of molecular biology, there is no longer any compelling reason for blindly launching adaptive stories about genomic evolution without a reasonably precise evaluation of the likelihood of nonadaptive alternatives.

It is now clear that many (and probably most) aspects of genomic biology that superficially appear to have adaptive roots—including the numerous features that contribute to complexity, modularity, robustness, and evolvability—are almost certainly also products of nonadaptive processes, whereas others are readily understood only by invoking a near-complete relaxation of natural selection. It would have been difficult to reach such conclusions in the absence of a formal population genetic framework, but they are equally reliant on observations from molecular and cellular biology. If the view that nonadaptive processes have played a central role in driving genomic evolution is correct, biological complexity should be viewed not as an extraordinarily low-probability outcome of unobservable adaptive challenges, but as an expected product of the special population genetic features of DNA-based genomes. A rather similar point was made previously by Kauffman (1993), although his conclusions were derived from models far removed from the mainstream of population genetics. The next step will be to evaluate whether nonadaptive evolutionary forces are largely confined to the level of DNA or whether the more visible aspects of biodiversity (at the cellular and organismal levels) can be understood only by extending our view beyond an idealized world of Darwinian adaptation.

Glossary

acrocentric a chromosome whose centromere is located near a tip.

allele a particular variant at a specific genetic locus.

allopolyploidy the form of **polyploidy** in which the full set of chromosomes is derived from a single ancestral species.

alpha-proteobacteria a group of eubacteria containing the purple photosynthetic bacteria, the pathogenic rickettsia bacteria, and a number of other free-living species.

alternative splicing the use of different intron–exon junctions to obtain alternative mRNAs from a single gene.

alveolates a monophyletic lineage of eukaryotes, united by the presence of alveoli (a system of sacs underlying the cell surface); contains the ciliated protozoa the **apicomplexans**.

anticodon the nucleotide triplet in a **transfer RNA** used in the recognition of **codons** in **messenger RNAs**.

antisense DNA the DNA on the strand complementary to the one from which transcription of a protein-coding gene proceeds.

AP endonuclease an enzyme used in DNA repair that recognizes sites missing a base and makes an incision that allows the insertion of an appropriate nucleotide.

apicomplexans a group of obligately parasitic, unicellular eukaryotes; includes the malarial parasite *Plasmodium*.

apicoplast an organelle within the apicomplexans that is derived from a chloroplast but no longer functional as such; thought to be involved in lipid metabolism.

archaea one of the two major **monophyletic** lines of descent among **prokaryotes**.

autopolyploidy the form of **polyploidy** in which the full set of chromosomes is derived from two ancestral species.

autonomous element a mobile genetic element fully equipped with coding regions for the proteins involved in mobilization and insertion.

autotroph an organism that uses carbon dioxide as a carbon source, and light or the oxidation of inorganic compounds as a source of energy.

bacteriophage a prokaryotic virus.

balancing selection a selection scenario in which an allelic polymorphism is maintained by heterozygote advantage (owing to the fact that the most fit genotype carries both alleles).

base pair (bp) a nucleotide position in double-stranded DNA.

B chromosome small and generally nonessential chromosomes found in some eukaryotes, which are capable of long-term proliferation by virtue of containing a centromere.

bikonts the eukaryotic group in which all members have cells with two flagella at some stage of the life cycle, or are thought to be derived from ancestors with such features.

bilaterians the group of **triploblast** animals that are ancestrally bilaterially symmetrical.

branchpoint sequence a central adenosine nucleotide and/or small motif surrounding it within an intron, used in establishing a lariat in the first phase of splicing.

BYA billion years ago.

Cajal bodies organelles within the nuclei of eukaryotic cells, within which some assembly of the subunits of the spliceosome occurs.

cDNA complementary DNA; a DNA copy of an mRNA derived by the use of **reverse transcriptase**.

centromere the region of a chromosome where sister chromatids join during mitosis and **kinetochores** assemble to provide anchor points for spindle formation during cell division.

chromalveolates a diverse and putatively monophyletic assemblage of eukaryotes, which includes the dinoflagellates, diatoms, coccolithophores, cryptomonads, oomycetes, and apicomplexans, all of which are united by an ancestral event that gave rise to a chlorophyll c producing plastid via the preservation of an engulfed red alga (which was subsequently lost in some lineages).

chromatin the complex of DNA and histone proteins that make up eukaryotic chromosomes.

***cis*-acting** refers to a factor that affects the expression of a gene at a nearby chromosomal location; e.g., a **transcription factor** binding site or a **core promoter**.

***cis*-splicing** a mechanism used to attach two segments of RNA derived from the same transcript, after eliminating an intervening region.

chlorarachniophytes a group of unicellular marine phytoplankton with amoeboid features and a capacity for acquiring energy by both photosynthesis and ingestion of other small microbes.

cnidarians a basal group of animals lacking mesoderm; includes corals, hydras, jellyfishes, and sea anemones.

coalescence time the number of generations in the past to which a group of alleles within a population trace to a single common ancestor.

codon a nucleotide triplet that corresponds to one of the 64 possibilities under a genetic code.

codon bias the degree to which the usage of alternative codons for a particular amino acid deviates from some null model, usually the expectation in the absence of selection.

consensus sequence a sequence that is viewed as representative of a broader set of related sequences.

core promoter a short motif that binds the transcription initiation machinery, thereby promoting the initiating transcription of the associated gene.

cosuppression the post-transcriptional silencing of a gene via a double-stranded RNA intermediate; analogous to **RNA interference**.

crossover a reciprocal exchange between two homologous chromosomes.

cryptomonads a group of photosynthetic, flagellated unicellular eukaryotes, generally inhabiting planktonic environments.

cytidine deaminase an enzyme that removes an $-NH_2$ group from cytosine, converting it to uracil.

cytoskeleton the dynamic, internal structural support system in eukaryotic cells, comprised of several filamentous proteins.

C value the total amount of DNA in a haploid genome.

deme a subpopulation.

dicer an enzyme used in the processing of miRNAs and siRNAs.

diffusion theory a branch of mathematics that allows the prediction of the probability distribution of allele frequencies at a locus, resulting from the joint operation of directional forces (e.g., mutation and/or selection) and the stochastic force of random genetic drift.

dinoflagellates a diverse group of photosynthetic eukaryotic flagellates, mostly unicellular; inhabit both marine and freshwater ecosystems.

diploblast basal lineages of metazoans, notable for their absence of mesoderm (e.g., the cnidarians, which include corals, anemones, and jellyfishes).

diplomonads a group of unicellular eukaryotes, lacking mitochondria; includes the parasite *Giardia*.

diploid having two gene copies at a genetic locus; as in virtually all animals and land plants.

DNA glycosylase an enzyme that removes erroneous uracil residues from DNA, leaving empty sites.

DNA polymerase an enzyme used to replicate DNA.

ecdysozoa the phylogenetic grouping that contains all molting invertebrate animals; not necessarily monophyletic.

ectopic recombination recombination between nonhomologous sites, which necessarily leads to a chromosomal rearrangement.

effective neutrality refers to the situation in which the intensity of selection operating on an allele is small enough that the impact of selection is overwhelmed by the stochastic changes in allele frequencies caused by random genetic drift, i.e., $1/(4N_e) > |s|$, or equivalently $|4N_e s| > 1$.

effective number of genes a per-locus measure, equal to the **effective population size** for a haploid species and double that for a diploid species.

effective population size the size of an ideal **Wright–Fisher population** that most closely reflects the evolutionary behavior of a nonideal natural population; generally denoted as N_e.

endonuclease an enzyme capable of cutting a double-strand of DNA.

endopolyploid a state in which a somatic cell has had one or more genomic duplications.

endosymbiont an organism residing within the cells of a host organism, and generally providing a benefit to the latter; includes the mitochondria and chloroplasts of eukaryotes.

enhancer a region of DNA containing one or more **transcription factor binding sites** responsible for the regulation of gene expression.

epigenetic refers to a heritable change on a chromosome at a level other than the DNA itself, e.g., methylation or histone modifications.

error catastrophe the point at which the damage introduced by mutation is so great that the most highly fit genotype is unable to maintain itself in the population.

eubacteria one of the two major **monophyletic** lines of descent among **prokaryotes**.

euchromatin the relatively open, gene-rich regions of eukaryotic chromosomes.

euglenozoa a group of unicellular eukaryotes, which includes the euglenoids and the kinetoplastids.

eukaryotes cells with membrane-bound organelles, including a nucleus.

exon a portion of a transcription unit that is retained in a mature messenger RNA after splicing.

exon junction complex a protein complex deposited just upstream of each exon-exon junction of the mature messenger RNAs of most eukaryotes.

exon splicing enhancer (ESE) a sequence motif residing in an exon that positively influences the utilization of a neighboring site as a splicing substrate.

exon splicing silencer (ESS) a sequence motif residing in an exon that negatively influences the utilization of a neighboring site as a splicing substrate.

exonuclease an enzyme that catalyzes the removal of nucleotides from DNA or RNA.

fatty acid esters lipids used in the cell membranes of the **eubacteria**.

5′ cap a modified base at the 5′ end of eukaryotic messenger RNAs, essential for the initiation of translation.

5′ untranslated region (5′ UTR) the leader sequence on a messenger RNA (prior to the translation initiation codon), which is not translated.

fixation the increase of an allele to the point at which it has achieved a frequency of 1.0.

functional domain a region of a protein with specific functional significance.

gene conversion the process by which one participant in a recombination event is converted to the sequence of the partner participant; occurs during almost all recombination events, but is not necessarily associated with crossing-over.

genetic code the complete set of 64 nucleotide triplets (codons) and their assignments to the 20 amino acids used in translation as well as to termination codons.

genetic draft see **hitchhiking effect**.

genetic map ordered arrays of molecular markers along chromosomes with the distances between markers defined by the expected number of **crossovers** per meiosis.

germ line the specific cell lineage within a multicellular organism that gives rise to gametes.

group II intron a form of self-splicing intron largely restricted to prokaryotes and eukaryotic organelles.

haploid having a single gene copy at a genetic locus; as in all prokaryotes and some unicellular eukaryotes.

haploinsufficient implies that a single-gene dose (in a diploid species), caused by the presence of a null allele at a locus, compromises organism fitness.

haplotype a particular version of a chromosomal region defined by a unique set of molecular markers.

helicase an enzyme involved in the separation of the two strands of a DNA molecule.

hemizygous refers to a locus in a diploid individual for which there is only a single copy, e.g., a Y-linked gene in an XY male.

heteroduplex a double-stranded form of DNA whose two strands are derived from different sources; as in the recombination intermediate between two parental chromosomes.

heterogametic sex the sex containing two nonrecombining sex chromosomes, e.g., males in an X/Y system.

heteroplasmy refers to the presence of two distinct organelle genotypes within a single cell.

heterotroph an organism that relies on organic compounds (generally derived from **autotrophs** or other heterotrophs) as carbon sources.

Hill-Robertson effect see **selective interference**.

hitchhiking effect the change in allele frequency at a locus caused by the stronger effects of selection on a linked locus.

homeothermy the maintenance of stable body temperatures by internal physiological mechanisms, primarily in birds and mammals.

homogametic sex the sex containing a pair of morphologically similar sex chromosomes, e.g., XX females in an X/Y system.

homologous recombination recombination involving regions of two partner chromosomes that are ancestrally related, e.g., two alleles at a locus.

horizontal gene transfer gene acquisition from a distantly related lineage, as opposed to vertical transmission by direct inheritance from a parent.

Hox cluster a linear array of a specific set of developmental genes that is central to body formation in animals; the spatial arrangement of such genes is generally collinear with anterior–posterior patterns of expression.

hybrid dysgenesis an elevated rate of mobile genetic element activity that occurs when species are hybridized or when permissive and nonpermissive strains are mated.

hydrogenosome a membrane-bound organelle of anaerobic eukaryotes, believed to be related to the mitochondrion, which yields hydrogen and ATP.

index of subdivision a measure of the degree to which the gene pool of a species varies among population segments; generally denoted by *F*, which falls in the range of zero (panmixia) to one (complete isolation of subpopulations).

integrase an enzyme used by transposons to insert copies into the host genome.

intron a portion of a transcription unit that is removed from a mature messenger RNA.

intron splicing enhancer (ISE) a sequence motif residing in an intron that positively influences the utilization of a neighboring site as a splicing substrate.

intron splicing silencer (ISS) a sequence motif residing in an intron that negatively influences the utilization of a neighboring site as a splicing substrate.

isochore a chromosomal region, often tens to hundreds of kb in length, of fairly homogeneous nucleotide composition, differing from adjacent regions.

isoprene ethers lipids used in the cell membranes of the **archaea**.

karyotype a broad description of the chromosomal set in a eukaryotic species, including chromosome number and centromere locations.

kilobase (kb) one thousand nucleotide bases.

kinetochore a protein structure that assembles on the **centromere**, linking the chromosome to the spindle fibers during cell division.

kinetoplastids a group of unicellular eukaryotes, many of which are parasitic (e.g., *Trypanosoma* and *Leishmania*).

linkage disequilibrium the degree of nonrandom association among markers at two chromosomal locations, often measured as the difference between the observed frequency of a gamete type and the expectation based on random assortment at the individual marker positions.

LINEs *l*ong *i*nterspersed *e*lements; a large family of retrotransposons found in the primate lineage.

locus usually refers to the chromosomal location of a genetic factor, although sometimes used synonymously with gene.

locus control region a region containing regulatory elements shared by two or more genes.

LTR element *l*ong *t*erminal *r*epeat element; one of the two major categories of retrotransposons;

noted for the presence of direct terminal repeats, which are identical at the time of insertion.

macroevolution arbitrarily refers to long-term evolutionary change (on time scales greater than a few millions of years); but also is often used to describe the view that long-term evolution is a process based on processes operating beyond the population level (such processes are rarely explicitly outlined, and arguably do not exist).

macronucleus one of the two nuclei contained within the cells of all ciliated protozoans; the micronucleus is transcriptionally silent and serves as the germline, while the macronucleus (a processed version of the former) is the source of all transcription and is destroyed at the time of sexual reproduction.

major-form exon an alternatively-spliced exon that is included in transcripts most of the time.

megabase (Mb) one million nucleotide bases.

messenger RNA (mRNA) a mature gene transcript, obtained after introns have been removed from an mRNA precursor.

metacentric a chromosome whose centromere is centrally located.

metapopulation the complete network of subpopulations comprising the gene pool of an entire species.

methanogens obligately anaerobic members of the archaea that produce methane as a by-product of the metabolism of organic matter.

microarray a fabricated array of spotted DNAs used for evaluating relative levels of transcripts or fragments of genomic DNA.

microevolution arbitrarily refers to short-term evolutionary change (on time scales of millions of years or less); but also is often used to describe the view that long-term evolution is a process based on principles of population genetics that apply to all known living organisms.

micro RNA (miRNA) a small (generally 20 to 30 bp) RNA processed from a transcribed "hairpin" precursor RNA; generally used in the regulation of gene expression by complementary binding to a nearly identical motif in the 3′ UTRs of transcripts.

microsatellite a small repetitive DNA involving repeats of 1 to ~6 nucleotides, e.g., a dinucleotide repeat.

microsporidians a group of animal parasites with reduced genomes; belong to the fungal lineage.

microtubules protein filaments that serve a number of structural functions with the eukaryotic cell, including formation of the cytoskeleton.

minor-form exon an alternatively-spliced exon that is included in transcripts only infrequently.

mismatch repair the process by which noncomplementary bases paired in double-stranded DNA are corrected by replacing one of the noncomplementary bases, which returns the site to a standard Watson–Crick (A:T or G:C) pair.

mobile-genetic element a segment of DNA capable of mobilizing and inserting into additional genome regions; includes **transposons** and **retrotransposons**.

mononucleotide run a sequential run of one particular nucleotide at the DNA or RNA level; e.g., AAAAA.

monophyletic refers to a natural evolutionary lineage that contains a number of related species and includes their common ancestor; for example, the vertebrate group is monophyletic because all of its members derive from a single common ancestor, but an "algal" grouping would not be monophyletic because there are numerous lineages of photosynthetic microbes distantly distributed over the tree of life.

Muller's ratchet the process by which the best-fit genome in a population stochastically gains deleterious mutations by random genetic drift and/or mutation pressure.

mutational spectrum the full description of the frequency distribution of all forms of mutations arising in an organism, e.g., the 12 forms of base substitutions, and the insertion/deletion size spectrum.

mutational meltdown a phase of rapid extinction that results when a population experiences a situation in which the rate of fitness loss induced by recurrent spontaneous mutations exceeds the ability of natural selection to purge the mutational load.

MYA million years ago.

neofunctionalization the origin of a new gene function by mutation.

neutral model a population genetic model that excludes selective forces; among other things, serves as an essential null hypothesis for testing for natural selection.

neutral mutation a newly arisen allele that has no advantages or disadvantages relative to the alleles currently existing at the locus.

nonautonomous element a mobile genetic element lacking in one or more protein coding regions that are essential for mobilization and/or insertion, which nevertheless can be activated by parasitizing the mobilization machinery of related autonomous elements.

noncoding RNA (ncRNA) loosely defined as any transcript that does not encode for protein.

nonfunctionalization the loss of gene function by mutation.

nonhomologous end joining the repair of a double-strand chromosome break by the direct ligation of the two ends back together; mutations may be introduced in the process.

nonLTR element *non-l*ong *t*erminal *r*epeat element; one of the two major categories of retrotransposons; noted for the absence of direct terminal repeats.

nonsense-mediated decay a surveillance pathway in eukaryotes that identifies and eliminates aberrant messenger RNAs containing premature termination codons.

nonsynonymous site a nucleotide within a coding region for which substitutions give rise to an alteration of the amino-acid sequence; also known as a replacement site.

N terminus the end of a protein bearing a free $-NH_2$ group; conventionally designated as the front end of a protein when the sequence is written out.

nucleomorph a remnant nuclear genome contained within a chloroplast-containing inclusion in two lineages of algae (the cryptomonads and the chlorararchniophytes), which was acquired via a secondary endosymbiotic event (with the host species engulfing a photosynthetic eukaryote).

nucleosome a repeating subunit of the chromosome, consisting of the DNA coiled around a set of histone proteins.

oligocellular refers to a species with a small number of cell types, e.g., fungi.

oligomer a specific, short nucleotide sequence.

operon a cassette of cotranscribed and sometimes functionally related genes; generally but not entirely restricted to prokaryotes.

open reading frame a stretch of DNA that translates into a continuous tract of amino acids without intervening stop codons.

opisthokonts the eukaryotic assemblage containing animals, fungi, and slime molds; believed to be monophyletic.

origin of replication (ORI) a location on a chromosome from which strand replication is initiated, usually bidirectionally.

orphan gene a generic term, used to refer to a gene with no known orthologous gene in another species.

ortholog a gene that is specifically related to a copy in another species by vertical descent.

outgroup a lineage that is clearly outside of the phylogenetic lineage under consideration, e.g., a bird for a mammalian phylogeny.

overdominant refers to a form of gene action in which the joint effects of two alleles in a heterozygote at a diploid locus are outside the range of that for the alternative homozygotes.

paralog a gene that is related to another within the same genome, having arisen by a duplication event.

parthenogenesis reproduction via an unfertilized egg.

P bodies "processing bodies"; organelles in the cytoplasm of eukaryotic cells, within which degradation of mRNAs takes place.

phagocytosis cellular ingestion involving the enveloping of the food particle by the cell membrane, which is then internalized as a food vacuole.

placozoans a basal group of marine invertebrates having only four cell types.

polyadenylation the addition of a string of adenines to the trailing end of a messenger RNA.

poly(A) tail a homopolymeric run of adenines added to the end of transcripts.

polycistronic transcription refers to the processing of several genes from an **operon** into a single transcriptional unit.

polymerase chain reaction (PCR) a laboratory procedure used to amplify a small template into large quantities of DNA for further molecular analysis.

polymorphism the presence of alternative alleles at a locus, or nucleotide bases at a nucleotide site.

polyphyletic refers to an unnatural evolutionary grouping that contains distantly related lineages, but does not include their common ancestor (see **monophyletic**).

polyploidy a condition in which the genome has been derived from one or more complete genome duplication events (relative to the ancestral diploid state).

polytene chromosome a giant chromosome resulting from repeated rounds of chromosome replication without cell division, most notably found in the salivary-gland cells of *D. melanogaster*.

post-transcriptional silencing any mechanism by which a messenger RNA is rendered unable to progress towards translation; e.g., the binding of a **micro RNA** to a **3′ UTR**.

precursor messenger RNA the initial transcript of a gene, prior to processing events such as capping and intron removal.

premature start codon (PSC) a translation initiation codon upstream of the proper start site, i.e., in the 5′ UTR.

primase an enzyme that synthesizes a short RNA primer used in the initial step of DNA replication.

processed pseudogenes a putatively nonfunctional gene derived from the reverse transcription of an mRNA and reinsertion into the genome.

prokaryotes cells lacking membrane-bound organelles, often referred to as bacteria; includes two unique, divergent clades, the eubacteria, and the archaea.

promoter a DNA sequence that helps localize the point of transcription initiation.

protists a general term used to describe unicellular eukaryotes.

pseudoautosomal region the small, recombining region of an otherwise nonrecombining sex chromosome, typically at the tip.

pseudogene a nonfunctional copy of a previously normally functioning gene.

purifying selection a form of natural selection that favors the maintenance of a particular allelic state, thereby encouraging the removal of most mutant alleles.

purines the nucleotides adenine (A) and guanine (G).

pyrimidines the nucleotides cytosine (C) and thymine (T).

random genetic drift fluctuations in allele frequencies across generations resulting from the derivation of gametes and offspring from a finite number of parents.

replacement site a nucleotide within a coding region for which substitutions give rise to an alteration of the amino-acid sequence; also known as a nonsynonymous site.

replicon a region of the genome whose replication is associated with a single **origin of replication**.

retrohoming site fidelity characteristic of the insertion of numerous group II mobile elements.

retrotransposons a broad class of mobile-genetic elements, which move by use of a messenger RNA intermediate.

retrovirus a virus containing an RNA genome.

reverse transcriptase an enzyme that produces a complementary strand of DNA from an RNA template.

rhodophytes the "red algae," a monophyletic lineage of unicellular/oligocellular photosynthetic eukaryotes.

ribonucleoprotein a generic term for a functional composite containing one or more proteins and RNAs.

ribonucleotide reductases proteins used in the production of the nucleotide precursors: dAMPs, dCMPs, and dGMPs.

ribosomal RNA (rRNA) an RNA subunit of the ribosome.

ribosome the cellular organelle at which mRNAs are translated into proteins.

rooted phylogenetic tree the root on a phylogenetic tree denotes the lineage containing the most recent common ancestor from which all species in the tree descend.

RNA interference (RNAi) a cellular response by which the recognition of double-stranded RNA leads to the post-transcriptional silencing of a transcribed region.

RNase H an enzyme that degrades the RNA in RNA:DNA hybrid molecules.

RNA polymerase an enzyme involved in the copying of a segment of nucleic acid (usually genomic DNA) into a complementary RNA (usually a transcript).

segmental duplication a duplicated region of sub-chromosomal length.

selection coefficient a measure of the selective advantage or disadvantage of an allele relative to the standard index of 1.0; i.e., a selective disadvantage of 0.01 implies a 1% reduction in fitness for an individual carrying a single copy of the gene (assuming additive effects at a diploid locus); generally denoted as *s*.

selective interference the reduction in the efficiency of selection operating on a locus as a consequence of simultaneous selection operating on linked loci.

selective sweep the rapid passage of a particular variant within a genomic region through a population, promoted by the force of directional selection.

Shine-Delgarno sequence a short sequence just upstream of the translation initiation point in prokaryotic messenger RNAs; complementary to a key sequence in the ribosome.

silent site a nucleotide within a coding region for which, due to the redundancy of the genetic code, substitutions do not give rise to an alteration of the amino-acid sequence; also known as a synonymous site.

SINEs *s*mall *i*nterspersed *e*lements; a large family of retrotransposons found in the primate lineage; dependent on LINEs for mobility.

single-nucleotide polymorphism (SNP) refers to a segregating polymorphism at a single nucleotide site.

small RNA (sRNA) a generic term that encompasses miRNAs and siRNAs.

small interfering RNA (siRNA) a small (generally 20 to 30 bp) RNA processed from a longer double-stranded RNA by the **RNA interference** pathway; deployed in posttranscriptional gene silencing.

small nuclear RNA (snRNA) refers to a member of a heterogeneous group of small RNAs whose functions are confined to the nucleus, including those involved in splicing introns out of precursor mRNAs and in telomere maintenance.

small nuclear ribonucleoprotein (snRNP) a complex of RNA and protein deployed in the nucleus; most notably, in the construction of the spliceosome.

small nucleolar RNA (snoRNA) an RNA molecule involved in the chemical modifications made in the construction of ribosomes; often encoded within an intron of a ribosomal protein gene.

spliced-leader sequence a separately transcribed external exon that is spliced onto the front end of the primary transcripts in trypanosomes and several other species.

spliceosome the complex molecular machine responsible for the removal of introns from precursor messenger RNAs for eukaryotic nuclear genes.

stem eukaryote the direct lineage containing the most recent common ancestor of all eukaryotes.

stochastic variable.

stramenopiles a diverse group of unicellular/oligocellular eukaryotes; includes the diatoms, brown algae, and oomycetes.

subfunctionalization the partial loss of one or more functions of a gene resulting from degenerative mutation.

synergistic epistasis the situation in which the joint effect of multiple mutations is greater than the sum of their individual effects, usually considered in the context of deleterious mutations, in which case fitness would decline at an accelerating rate with increasing numbers of mutations.

synonymous site a nucleotide within a coding region for which, due to the redundancy of the genetic code, substitutions do not give rise to an alteration of the amino acid sequence; also known as a silent site.

target-primed reverse transcription the mechanism by which nonLTR retrotransposons are inserted into DNA.

telomeres the ends of linear chromosomes.

terminal inverted repeats (TIRs) small inverted sequences, generally found at the two ends of a transposon.

tetrad the array of four haploid products resulting from meiosis.

3′ poly(A) tail a string of adenines added to the end of a **3′ UTR** of a mature messenger RNA, essential for translation initiation.

3′ untranslated region (3′ UTR) the trailing sequence on a messenger RNA (beyond the translation-termination codon), which is not translated.

thymidylate synthase catalyzes the production of dTMP nucleotide precursors by the methylation of dUMPs.

***trans*-acting** refers to a factor that affects the expression of a gene at another chromosomal location; e.g., a transcription factor.

transcription factor a protein whose function is to regulate the expression of another gene, usually by binding to a small binding site in the target gene's vicinity.

transcription factor binding site a small DNA motif responsible for binding a specific protein used in the regulation of a gene's transcription.

transcriptome the complete set of transcripts in a cell or organism.

transfer RNA (tRNA) a small clover-leaf shaped molecule that serves as a vehicle for delivering a specific amino acid during the translation of mRNA.

transgene a gene derived from an exogenous source, generally inserted into a foreign lab organism by various molecular techniques.

transition a mutational change from **purine** to purine or **pyrimidine** to pyrimidine.

transposase an enzyme used by a transposon during the "cut-and-paste" process used in the integration of a new copy.

transposons a broad class of mobile-genetic elements, which move by use of "cut-and-paste" mechanisms.

***trans*-splicing** a mechanism used to attach two segments of RNA derived from separate transcripts.

transversion a mutational change from **purine** to **pyrimidine** or vice versa.

trichomonads a group of unicellular eukaryotes, lacking mitochondria; includes the human sexual parasite *Trichomonas*.

triploblasts animals with three cell layers (ectoderm, mesoderm, and endoderm).

tubulin one of the classes of filamentous proteins involved in cytoskeletal support in eukaryotic cells.

tunicate a phylogenetic group of invertebrate chordate, which includes the sea squirts.

unikonts the eukaryotic group for which all members have cells with single flagella at some stage of the life cycle, or are thought to be derived from ancestors with such features.

urochordates a subphylum of chordates, which includes sea squirts and larvaceans.

upstream open reading frame (uORF) a short open reading frame in the 5′ UTR of a transcript, followed by a termination codon, and preceding the true translation initiation site.

Wright–Fisher population an idealized random-mating population with discrete generations, with each member contributing equally and synchronously to an effectively infinite gamete pool, from which pairs of gametes are drawn randomly to produce the next generation.

Literature Cited

Abad, J. P., B. De Pablos, K. Osoegawa, P. J. De Jong, A. Martin-Gallardo and A. Villasante. 2004. Genomic analysis of *Drosophila melanogaster* telomeres: Full-length copies of HeT-A and TART elements at telomeres. *Mol. Biol. Evol.* 21: 1613–1619.

Abascal, F., D. Posada, R. D. Knight and R. Zardoya. 2006. Parallel evolution of the genetic code in arthropod mitochondrial genomes. *PLoS Biol.* 4: 711–718.

Abdelnoor, R. V., A. C. Christensen, S. Mohammed, B. Munoz-Castillo, H. Moriyama and S. A. Mackenzie. 2006. Mitochondrial genome dynamics in plants and animals: Convergent gene fusions of a *MutS* homologue. *J. Mol. Evol.* 63: 165–173.

Abe, H., K. Mita, Y. Yasukochi, T. Oshiki and T. Shimada. 2005. Retrotransposable elements on the W chromosome of the silkworm, *Bombyx mori*. *Cytogenet. Genome Res.* 110: 144–151.

Abi-Rached, L., A. Gilles, T. Shiina, P. Pontarotti and H. Inoko. 2002. Evidence of en bloc duplication in vertebrate genomes. *Nature Genet.* 31: 100–105.

Abovich, N. and M. Rosbash. 1997. Cross-intron bridging interactions in the yeast commitment complex are conserved in mammals. *Cell* 89: 403–412.

Abril, J. F., R. Castelo and R. Guigo. 2005. Comparison of splice sites in mammals and chicken. *Genome Res.* 15: 111–119.

Achaz, G., P. Netter and E. Coissac. 2001. Study of intrachromosomal duplications among the eukaryotic genomes. *Mol. Biol. Evol.* 18: 2280–2288.

Ackermann, M. and L. Chao. 2006. DNA sequences shaped by selection for stability. *PLoS Genet.* 2: 224–230.

Ackermann, R. R. and J. M. Cheverud. 2004. Detecting genetic drift versus selection in human evolution. *Proc. Natl. Acad. Sci. USA* 101: 17946–17951.

Adam, R. D. 2001. Biology of *Giardia lamblia*. *Clin. Microbiol. Rev.* 14: 447–475.

Adams, K. L. and J. D. Palmer. 2003. Evolution of mitochondrial gene content: Gene loss and transfer to the nucleus. *Mol. Phylogenet. Evol.* 29: 380–395.

Adams, K. L., K. Song, P. G. Roessler, J. M. Nugent, J. L. Doyle, J. J. Doyle and J. D. Palmer. 1999. Intracellular gene transfer in action: Dual transcription and multiple silencings of nuclear and mitochondrial *cox2* genes in legumes. *Proc. Natl. Acad. Sci. USA* 96: 13863–13868.

Adams, K. L., D. O. Daley, Y. L. Qiu, J. Whelan and J. D. Palmer. 2000. Repeated, recent and diverse transfers of a mitochondrial gene to the nucleus in flowering plants. *Nature* 408: 354–357.

Adams, K. L., M. Rosenblueth, Y. L. Qiu and J. D. Palmer. 2001. Multiple losses and transfers to the nucleus of two mitochondrial succinate dehydrogenase genes during angiosperm evolution. *Genetics* 158: 1289–1300.

Adams, K. L., Y. L. Qiu, M. Stoutemyer and J. D. Palmer. 2002. Punctuated evolution of mitochondrial gene content: High and variable rates of mitochondrial gene loss and transfer to the nucleus during angiosperm evolution. *Proc. Natl. Acad. Sci. USA* 99: 9905–9912.

Adams, K. L., R. Cronn, R. Percifield and J. F. Wendel. 2003. Genes duplicated by polyploidy show unequal contributions to the transcriptome and organ-specific reciprocal silencing. *Proc. Natl. Acad. Sci. USA* 100: 4649–4654.

Adams, S. P., T. P. Hartman, K. Y. Lim, M. W. Chase, M. D. Bennett, I. J. Leitch and A. R. Leitch. 2001. Loss and recovery of *Arabidopsis*-type telomere repeat sequences 5′-(TTTAGGG)(n)-3′ in the evolution of a major radiation of flowering plants. *Proc. Biol. Sci.* 268: 1541–1546.

Aggarwal, B. D. and B. R. Calvi. 2004. Chromatin regulates origin activity in *Drosophila* follicle cells. *Nature* 430: 372–376.

Agrawal, N., P. V. Dasaradhi, A. Mohmmed, P. Malhotra, R. K. Bhatnagar and S. K. Mukherjee. 2003. RNA interference: Biology, mechanism, and applications. *Microbiol. Mol. Biol. Rev.* 67: 657–685.

Aguinaldo, A. M., J. M. Turbeville, L. S. Linford, M. C. Rivera, J. R. Garey, R. A. Raff and J. A. Lake. 1997. Evidence for a clade of nematodes, arthropods and other moulting animals. *Nature* 387: 489–493.

Ahn, B. Y. and D. M. Livingston. 1986. Mitotic gene conversion lengths, coconversion patterns, and the incidence of reciprocal recombination in a *Saccharomyces cerevisiae* plasmid system. *Mol. Cell. Biol.* 6: 3685–3693.

Ahn, B. Y., K. J. Dornfeld, T. J. Fagrelius and D. M. Livingston. 1988. Effect of limited homology on gene conversion in a *Saccharomyces cerevisiae* plasmid recombination system. *Mol. Cell. Biol.* 8: 2442–2448.

Akashi, H. 1994. Synonymous codon usage in *Drosophila melanogaster*: Natural selection and translational accuracy. *Genetics* 136: 927–935.

Akashi, H. 1995. Inferring weak selection from patterns of polymorphism and divergence at "silent" sites in *Drosophila* DNA. *Genetics* 139: 1067–1076.

Akashi, H. 2003. Translational selection and yeast proteome evolution. *Genetics* 164: 1291–303.

Akashi, H. and S. W. Schaeffer. 1997. Natural selection and the frequency distributions of "silent" DNA polymorphism in *Drosophila*. *Genetics* 146: 295–307.

Akashi, H., R. M. Kliman and A. Eyre-Walker. 1998. Mutation pressure, natural selection, and the evolution of base composition in *Drosophila*. *Genetica* 102/103: 49–60.

Akhmanova, A., F. Voncken, T. van Alen, A. van Hoek, B. Boxma, G. Vogels, M. Veenhuis and J. H. Hackstein. 1998. A hydrogenosome with a genome. *Nature* 396: 527–528.

Akman, L., A. Yamashita, H. Watanabe, K. Oshima, T. Shiba, M. Hattori and S. Aksoy. 2002. Genome sequence of the endocellular obligate symbiont of tsetse flies, *Wigglesworthia glossinidia*. *Nature Genet.* 32: 402–407.

Aksoy, S., S. Williams, S. Chang and F. F. Richards. 1990. *SLACS* retrotransposon from *Trypanosoma brucei gambiense* is similar to mammalian *LINEs*. Nucleic Acids Res. 18: 785–792.

Albert, A. Y. and S. P. Otto. 2005. Sexual selection can resolve sex-linked sexual antagonism. *Science* 310: 119–121.

Allen, J. F. 2003. The function of genomes in bioenergetic organelles. *Phil. Trans. Roy. Soc. Lond.* B *Biol. Sci.* 358: 19–37.

Allen, J. F. and J. A. Raven. 1996. Free-radical-induced mutation vs. redox regulation: Costs and benefits of genes in organelles. *J. Mol. Evol.* 42: 482–492.

Allendorf, F. W., F. M. Utter and B. P. May. 1975. Gene duplication within the family *Salmonidae*. II. Detection and determination of the genetic control of duplicate loci through inheritance studies and the examination of populations. In C. L. Markert, ed., *Isozymes*, vol. 4, *Genetics and Evolution*, 414–432. Academic Press, New York.

Alonso, C. R. and A. S. Wilkins. 2005. The molecular elements that underlie developmental evolution. *Nature Rev. Genet.* 6: 709–715.

Altenberg, L. 2005. Evolvability suppression to stabilize far-sighted adaptations. *Artif. Life* 11: 427–443.

Altman, A. L. and E. Fanning. 2004. Defined sequence modules and an architectural element cooperate to promote initiation at an ectopic mammalian chromosomal replication origin. *Mol. Cell. Biol.* 24: 4138–4150.

Altschmied, J., J. Delfgaauw, B. Wilde, J. Duschl, L. Bouneau, J. N. Volff and M. Schartl. 2002. Subfunctionalization of duplicate *mitf* genes associated with differential degeneration of alternative exons in fish. *Genetics* 161: 259–267.

Ames, B. N., M. K. Shigenaga and T. M. Hagen. 1995. Mitochondrial decay in aging. *Biochim. Biophys. Acta* 1271: 165–170.

Aminetzach, Y. T., J. M. Macpherson and D. A. Petrov. 2005. Pesticide resistance via transposition-mediated adaptive gene truncation in *Drosophila*. *Science* 309: 764–767.

Amiri, H., O. Karlberg and S. G. Andersson. 2003. Deep origin of plastid/parasite ATP/ADP translocases. *J. Mol. Evol.* 56: 137–150.

Amor, D. J., K. Bentley, J. Ryan, J. Perry, L. Wong, H. Slater and K. H. Choo. 2004. Human centromere repositioning "in progress." *Proc. Natl. Acad. Sci. USA* 101: 6542–6547.

Amores, A., A. Force, Y. L. Yan, L. Joly, C. Amemiya, A. Fritz, R. K. Ho, J. Langeland, V. Prince, Y. L. Wang et al. 1998. Zebrafish hox clusters and vertebrate genome evolution. *Science* 282: 1711–1714.

Amrani, N. and A. Jacobson. 2006. All termination events are not equal: Premature termination in yeast is aberrant and triggers NMD. In L. E. Maquat, ed., *Nonsense-Mediated mRNA Decay*, 15–26. Landes Bioscience, Georgetown, TX.

Ananiev, E. V., L. G. Polukarova and Y. B. Yurov. 1977. Replication of chromosomal DNA in diploid *Drosophila melanogaster* cells cultured in vitro. *Chromosoma* 59: 259–272.

Anderson, C., Q. Tang and J. A. Kinsey. 2001. Elimination of active *tad* elements during the sexual phase of the *Neurospora crassa* life cycle. *Fungal Genet. Biol.* 33: 49–57.

Anderson, J. B., C. Wickens, M. Khan, L. E. Cowen, N. Federspiel, T. Jones and L. M. Kohn. 2001. Infrequent genetic exchange and recombination in the mitochondrial genome of *Candida albicans*. *J. Bacteriol.* 183: 865–872.

Anderson, R. P. and J. R. Roth. 1977. Tandem genetic duplications in phage and bacteria. *Annu. Rev. Microbiol.* 31: 473–505.

Andersson, J. O. and S. G. Andersson. 1999. Genome degradation is an ongoing process in *Rickettsia*. *Mol. Biol. Evol.* 16: 1178–1191.

Andersson, S. G., A. Zomorodipour, J. O. Andersson, T. Sicheritz-Ponten, U. C. Alsmark, R. M. Podowski, A. K. Naslund, A. S. Eriksson, H. H. Winkler and C. G. Kurland. 1998. The genome sequence of *Rickettsia prowazekii* and the origin of mitochondria. *Nature* 396: 133–140.

Andersson, S. G., O. Karlberg, B. Canback, and C. G. Kurland. 2003. On the origin of mitochondria: A genomics perspective. *Phil. Trans. Roy. Soc. Lond.* B *Biol. Sci.* 358: 165–177.

Andolfatto, P. 2001. Contrasting patterns of X-linked and autosomal nucleotide variation in *Drosophila melanogaster* and *Drosophila simulans*. *Mol. Biol. Evol.* 18: 279–290.

Andolfatto, P. 2005. Adaptive evolution of non-coding DNA in *Drosophila*. *Nature* 437: 1149–1152.

Andolfatto, P. and M. Przeworski. 2001. Regions of lower crossing over harbor more rare variants in African populations of *Drosophila melanogaster*. *Genetics* 158: 657–665.

Andrews, J., M. Smith, J. Merakovsky, M. Coulson, F. Hannan and L. E. Kelly. 1996. The *stoned* locus of *Drosophila melanogaster* produces a dicistronic transcript and encodes two distinct polypeptides. *Genetics* 143: 1699–1711.

Anglana, M., F. Apiou, A. Bensimon and M. Debatisse. 2003. Dynamics of DNA replication in mammalian somatic cells: Nucleotide pool modulates origin choice and interorigin spacing. *Cell* 114: 3853–3894.

Arava, Y., Y. Wang, J. D. Storey, C. L. Liu, P. O. Brown and D. Herschlag. 2003. Genome-wide analysis of mRNA translation profiles in *Saccharomyces cerevisiae*. *Proc. Natl. Acad. Sci. USA* 100: 3889–3894.

Aravind, L., H. Watanabe, D. J. Lipman and E. V. Koonin. 2000. Lineage-specific loss and divergence of functionally linked genes in eukaryotes. *Proc. Natl. Acad. Sci. USA* 97: 11319–11324.

Archibald, J. M., C. J. O'Kelly and W. F. Doolittle. 2002. The chaperonin genes of jakobid and jakobid-like flagellates: Implications for eukaryotic evolution. *Mol. Biol. Evol.* 19: 422–431.

Ares, M., Jr., L. Grate and M. H. Pauling. 1999. A handful of intron-containing genes produces the lion's share of yeast mRNA. *RNA* 5: 1138–1139.

Aris-Brosou, S. and Z. Yang. 2003. Bayesian models of episodic evolution support a late Precambrian explosive diversification of the Metazoa. *Mol. Biol. Evol.* 20: 1947–1954.

Arisue, N., M. Hasegawa and T. Hashimoto. 2005. Root of the Eukaryota tree as inferred from combined maximum likelihood analyses of multiple molecular sequence data. *Mol. Biol. Evol.* 22: 409–420.

Arkhipova, I. R. and M. Meselson. 2000. Transposable elements in sexual and ancient asexual taxa. *Proc. Natl. Acad. Sci. USA* 97: 14473–14477.

Arkhipova, I. R. and M. Meselson. 2005. Diverse DNA transposons in rotifers of the class Bdelloidea. *Proc. Natl. Acad. Sci. USA* 102: 11781–11786.

Arkhipova, I. R. and H. G. Morrison. 2001. Three retrotransposon families in the genome of *Giardia lamblia*: Two telomeric, one dead. *Proc. Natl. Acad. Sci. USA* 98: 14497–14502.

Arkhipova, I. R., K. I. Pyatkov, M. Meselson and M. B. Evgen'ev. 2003. Retroelements containing introns in diverse invertebrate taxa. *Nature Genet.* 33: 123–124.

Armengol, L., M. A. Pujana, J. Cheung, S. W. Scherer and X. Estivill. 2003. Enrichment of segmental duplications in regions of breaks of synteny between the human and mouse genomes suggest their involvement in evolutionary rearrangements. *Hum. Mol. Genet.* 12: 2201–2208.

Arndt, P. F., D. A. Petrov and T. Hwa. 2003. Distinct changes of genomic biases in nucleotide substitution at the time of mammalian radiation. *Mol. Biol. Evol.* 20: 1887–1896.

Arndt, P. F., T. Hwa and D. A. Petrov. 2005. Substantial regional variation in substitution rates in the human genome: Importance of GC content, gene density, and telomere-specific effects. *J. Mol. Evol.* 60: 748–763.

Arnosti, D. N. 2003. Analysis and function of transcriptional regulatory elements: Insights from *Drosophila*. *Annu. Rev. Entomol.* 48: 579–602.

Arthur, W. 1997. *The Origin of Animal Body Plans*. Cambridge University Press, Cambridge.

Ashley, M. V., P. J. Laipis and W. W. Hauswirth. 1989. Rapid segregation of heteroplasmic bovine mitochondria. *Nucleic Acids Res.* 17: 7325–7331.

Aury, J. M., O. Jaillon, L. Duret, B. Noel, C. Jubin, B. M. Porcel, B. Segurens, V. Daubin, V. Anthouard, N. Aiach et al. 2006. Global trends of whole-genome duplications revealed by the ciliate *Paramecium tetraurelia*. *Nature* 444: 171–178.

Averof, M., R. Dawes and D. Ferrier. 1996. Diversification of arthropod Hox genes as a paradigm for the evolution of gene functions. *Sem. Cell. Dev. Biol.* 7: 539–551.

Awadalla, P., A. Eyre-Walker and J. Maynard Smith. 1999. Linkage disequilibrium and recombination in hominid mitochondrial DNA. *Science* 286: 2524–2525.

Axelsson, E., N. G. Smith, H. Sundstrom, S. Berlin and H. Ellegren. 2004. Male-biased mutation rate and divergence in autosomal, Z-linked and W-linked introns of chicken and turkey. *Mol. Biol. Evol.* 21: 1538–1547.

Ayala, F. J., A. Rzhetsky and F. J. Ayala. 1998. Origin of the metazoan phyla: Molecular clocks confirm paleontological estimates. *Proc. Natl. Acad. Sci. USA* 95: 606–611.

Aye, M., B. Irwin, N. Beliakova-Bethell, E. Chen, J. Garrus and S. Sandmeyer. 2004. Host factors that affect *Ty3* retrotransposition in *Saccharomyces cerevisiae*. *Genetics* 168: 1159–1176.

Bachtrog, D. 2003. Adaptation shapes patterns of genome evolution on sexual and asexual chromosomes in *Drosophila*. *Nature Genet.* 34: 215–219.

Bachtrog, D. 2005. Sex chromosome evolution: Molecular aspects of Y-chromosome degeneration in *Drosophila*. *Genome Res.* 15: 1393–1401.

Bachtrog, D. 2006. Expression profile of a degenerating neo-Y chromosome in *Drosophila*. *Curr. Biol.* 16: 1694–1699.

Bachtrog, D. and B. Charlesworth. 2002. Reduced adaptation of a non-recombining neo-Y chromosome. *Nature* 416: 323–326.

Bailey, J. A., Z. Gu, R. A. Clark, K. Reinert, R. V. Samonte, S. Schwartz, M. D. Adams, E. W. Myers, P. W. Li and E. E. Eichler. 2002. Recent segmental duplications in the human genome. *Science* 297: 1003–1007.

Balakirev, E. S. and F. J. Ayala. 2003. Pseudogenes: Are they "junk" or functional DNA? *Annu. Rev. Genet.* 37: 123–151.

Baldauf, S. L., J. D. Palmer and W. F. Doolittle. 1996. The root of the universal tree and the origin of eukaryotes based on elongation factor phylogeny. *Proc. Natl. Acad. Sci. USA* 93: 7749–7754.

Baldauf, S. L., A. J. Roger, I. Wenk-Siefert and W. F. Doolittle. 2000. A kingdom-level phylogeny of eukaryotes based on combined protein data. *Science* 290: 972–977.

Baldo, L., S. Bordenstein, J. J. Wernegreen and J. H. Werren. 2006. Widespread recombination throughout *Wolbachia* genomes. *Mol. Biol. Evol.* 23: 437–449.

Ball, E. V., P. D. Stenson, S. S. Abeysinghe, M. Krawczak, D. N. Cooper and N. A. Chuzhanova. 2005. Microdeletions and microinsertions causing human genetic disease: Common mechanisms of mutagenesis and the role of local DNA sequence complexity. *Hum. Mutat.* 26: 205–213.

Ballard, J. W. and M. Kreitman. 1994. Unraveling selection in the mitochondrial genome of *Drosophila*. *Genetics* 138: 757–772.

Balu, B., D. A. Shoue, M. J. Fraser, Jr. and J. H. Adams. 2005. High-efficiency transformation of *Plasmodium falciparum* by the lepidopteran transposable element piggyBac. *Proc. Natl. Acad. Sci. USA* 102: 16391–16396.

Bamshad, M. and S. P. Wooding. 2003. Signatures of natural selection in the human genome. *Nature Rev. Genet.* 4: 99–111.

Bancroft, I. 2001. Duplicate and diverge: The evolution of plant genome microstructure. *Trends Genet.* 17: 89–93.

Bapteste, E. and H. Philippe. 2002. The potential value of indels as phylogenetic markers: Position of trichomonads as a case study. *Mol. Biol. Evol.* 19: 972–977.

Bapteste, E., H. Brinkmann, J. A. Lee, D. V. Moore, C. W. Sensen, P. Gordon, L. Durufle, T. Gaasterland, P. Lopez, M. Muller and H. Philippe. 2002. The analysis of 100 genes supports the grouping of three highly divergent amoebae: *Dictyostelium, Entamoeba,* and *Mastigamoeba*. *Proc. Natl. Acad. Sci. USA* 99: 1414–1419.

Barbosa-Morais, N. L., M. Carmo-Fonseca and S. Aparicio. 2006. Systematic genome-wide annotation of spliceosomal proteins reveals differential gene family expansion. *Genome Res.* 16: 66–77.

Bartel, B. and D. P. Bartel. 2003. MicroRNAs: At the root of plant development? *Plant Physiol.* 132: 709–717.

Bartolomé, C., X. Maside and B. Charlesworth. 2002. On the abundance and distribution of transposable elements in the genome of *Drosophila melanogaster*. *Mol. Biol. Evol.* 19: 926–937.

Bartolomé, C., X. Maside, S. Yi, A. L. Grant and B. Charlesworth. 2005. Patterns of selection on synonymous and nonsynonymous variants in *Drosophila miranda*. *Genetics* 169: 1495–1507.

Barton, N. H. and B. Charlesworth. 1998. Why sex and recombination? *Science* 281: 1986–1990.

Barton, N. H. and S. P. Otto. 2005. Evolution of recombination due to random drift. *Genetics* 169: 2353–2370.

Baserga, R. 1985. *The Biology of Cell Reproduction*. Harvard University Press, Cambridge, MA.

Bastolla, U., A. Moya, E. Viguera and R. C. van Ham. 2004. Genomic determinants of protein folding thermodynamics in prokaryotic organisms. *J. Mol. Biol.* 343: 1451–1466.

Bauer, V. L. and C. F. Aquadro. 1997. Rates of DNA sequence evolution are not sex-biased in *Drosophila melanogaster* and *D. simulans*. *Mol. Biol. Evol.* 14: 1252–1257.

Baumann, P. and T. R. Cech. 2001. Pot1, the putative telomere end-binding protein in fission yeast and humans. *Science* 292: 1171–1175.

Bazin, E., S. Glemin and N. Galtier. 2006. Population size does not influence mitochondrial genetic diversity in animals. *Science* 312: 570–572.

Bazykin, G. A. and A. S. Kondrashov. 2006. Rate of promoter class turn over in yeast evolution. *BMC Evol. Biol.* 6: 14.

Begun, D. J. and C. F. Aquadro. 1992. Levels of naturally occurring DNA polymorphism correlate with recombination rates in *D. melanogaster*. *Nature* 356: 519–520.

Begun, D. J., H. A. Lindfors, M. E. Thompson and A. K. Holloway. 2006. Recently evolved genes identified from *Drosophila yakuba* and *D. erecta* accessory gland expressed sequence tags. *Genetics* 172: 1675–1681.

Behe, M. J. and D. W. Snoke. 2004. Simulating evolution by gene duplication of protein features that require multiple amino acid residues. *Protein Sci.* 13: 2651–2664.

Behm-Ansmant, I. and E. Izaurralde. 2006. NMD in *Drosophila*: A snapshot into the evolution of a conserved mRNA surveillance pathway. In L. E. Maquat, ed., *Nonsense-Mediated mRNA Decay*, 151–164. Landes Bioscience, Georgetown, TX.

Bejerano, G., M. Pheasant, I. Makunin, S. Stephen, W. J. Kent, J. S. Mattick and D. Haussler. 2004. Ultraconserved elements in the human genome. *Science* 304: 1321–1325.

Beletskii, A. and A. S. Bhagwat. 1996. Transcription-induced mutations: Increase in C to T mutations in the nontranscribed strand during transcription in *Escherichia coli*. *Proc. Natl. Acad. Sci. USA* 93: 13919–13924.

Belfort, M., M. E. Reaban, T. Coetzee and J. Z. Dalgaard. 1995. Prokaryotic introns and inteins: A panoply of form and function. *J. Bacteriol.* 177: 3897–3903.

Belfort, M., V. Derbyshire, M. M. Parker, B. Cousineau and A. M. Lambowitz. 2002. Mobile introns: Pathways and proteins. In N. L. Craig, R. Craigie, M. Gellert and A. M. Lambowitz, eds., *Mobile DNA II*, 761–783. ASM Press, Washington, DC.

Bell, G. and A. O. Mooers. 1997. Size and complexity among multicellular organisms. *Biol. J. Linn. Soc.* 60: 345–363.

Bell, J. S. and R. McCulloch. 2003. Mismatch repair regulates homologous recombination, but has little influence on antigenic variation, in *Trypanosoma brucei*. *J. Biol. Chem.* 278: 45182–45188.

Bell, P. J. L. 2001. Viral eukaryogenesis: Was the ancestor of the nucleus a complex DNA virus? *J. Mol. Evol.* 53: 251–256.

Bell, S. D. and S. P. Jackson. 2001. Mechanism and regulation of transcription in archaea. *Curr. Opin. Microbiol.* 4: 208–213.

Bell, S. D., C. P. Magill and S. P. Jackson. 2001. Basal and regulated transcription in Archaea. *Biochem. Soc. Trans.* 29: 392–395.

Belle, E. M., L. Duret, N. Galtier and A. Eyre-Walker. 2004. The decline of isochores in mammals: An assessment of the GC content variation along the mammalian phylogeny. *J. Mol. Evol.* 58: 653–660.

Belshaw, R., V. Pereira, A. Katzourakis, G. Talbot, J. Paces, A. Burt and M. Tristem. 2004. Long-term reinfection of the human genome by endogenous retroviruses. *Proc. Natl. Acad. Sci. USA* 101: 4894–4899.

Bendich, A. J. 2004. Circular chloroplast chromosomes: The grand illusion. *Plant Cell* 16: 1661–1666.

Bennett, E. A., L. E. Coleman, C. Tsui, W. S. Pittard and S. E. Devine. 2004. Natural genetic variation caused by transposable elements in humans. *Genetics* 168: 933–951.

Bennett, M. D. 1972. Nuclear DNA content and minimum generation time in herbaceous plants. *Proc. Roy. Soc. Lond.* B *Biol. Sci.* 181: 109–135.

Bennetzen, J. L. and B. D. Hall. 1982. Codon selection in yeast. *J. Biol. Chem.* 257: 3026–3031.

Bennetzen, J. L. and E. A. Kellogg. 1997. Do plants have a one-way ticket to genomic obesity? *Plant Cell* 9: 1509–1514.

Bennetzen, J. L. and W. Ramakrishna. 2002. Numerous small rearrangements of gene content, order and orientation differentiate grass genomes. *Plant Mol. Biol.* 48: 821–827.

Bensasson, D., M. W. Feldman and D. A. Petrov. 2003. Rates of DNA duplication and mitochondrial DNA insertion in the human genome. *J. Mol. Evol.* 57: 343–354.

Bentley, D. 2002. The mRNA assembly line: Transcription and processing machines in the same factory. *Curr. Opin. Cell. Biol.* 14: 336–342.

Bentley, S. D., K. F. Chater, A. M. Cerdeno-Tarraga, G. L. Challis, N. R. Thomson, K. D. James, D. E. Harris, M. A. Quail, H. Kieser, D. Harper et al. 2002. Complete genome sequence of the model actinomycete *Streptomyces coelicolor* A3(2). *Nature* 417: 141–147.

Bentwich, I., A. Avniel, Y. Karov, R. Aharonov, S. Gilad, O. Barad, A. Barzilai, P. Einat, U. Einav, E. Meiri et al. 2005. Identification of hundreds of conserved and nonconserved human microRNAs. *Nature Genet.* 37: 766–770.

Berget, S. M., C. Moore and P. A. Sharp. 1977. Spliced segments at the 5′ terminus of adenovirus 2 late mRNA. *Proc. Natl. Acad. Sci. USA* 74: 3171–3175.

Bergman, C. M. and M. Kreitman. 2001. Analysis of conserved noncoding DNA in *Drosophila* reveals similar constraints in intergenic and intronic sequences. *Genome Res.* 11: 1335–1345.

Bergsma, D. J., Y. Ai, W. R. Skach, K. Nesburn, E. Anoia, S. Van Horn and D. Stambolian. 1996. Fine structure of the human galactokinase *GALK1* gene. *Genome Res.* 6: 980–985.

Bergstrom, C. T. and J. Pritchard. 1998. Germline bottlenecks and the evolutionary maintenance of mitochondrial genomes. *Genetics* 149: 2135–2146.

Bergthorsson, U. and H. Ochman. 1998. Distribution of chromosome length variation in natural isolates of *Escherichia coli*. *Mol. Biol. Evol.* 15: 6–16.

Bergthorsson, U. and H. Ochman. 1999. Chromosomal changes during experimental evolution in laboratory populations of *Escherichia coli*. *J. Bacteriol.* 181: 1360–1363.

Berlin, S. and H. Ellegren. 2001. Clonal inheritance of avian mitochondrial DNA. *Nature* 413: 37–38.

Berlin, S. and H. Ellegren. 2004. Chicken W: A genetically uniform chromosome in a highly variable genome. *Proc. Natl. Acad. Sci. USA* 101: 15967–15969.

Berlin, S., M. Brandstrom, N. Backstrom, E. Axelsson, N. G. Smith and H. Ellegren. 2006. Substitution rate heterogeneity and the male mutation bias. *J. Mol. Evol.* 62: 226–233.

Bernardi, G. 1993. The isochore organization of the human genome and its evolutionary history—a review. *Gene* 135: 57–66.

Bernardi, G. 2000. Isochores and the evolutionary genomics of vertebrates. *Gene* 241: 3–17.

Bernardi, G. 2005. Lessons from a small, dispensable genome: The mitochondrial genome of yeast. *Gene* 354: 189–200.

Bernardi, G., B. Olofsson, J. Filipski, M. Zerial, J. Salinas, G. Cuny, M. Meunier-Rotival and F. Rodier. 1985. The mosaic genome of warm-blooded vertebrates. *Science* 228: 953–958.

Bernardi, G., S. Hughes and D. Mouchiroud. 1997. The major compositional transitions in the vertebrate genome. *J. Mol. Evol.* 44(suppl.): S44–S51.

Berticat, C., G. Boquien, M. Raymond and C. Chevillon. 2002. Insecticide resistance genes induce a mating competition cost in *Culex pipiens* mosquitoes. *Genet. Res.* 79: 41–47.

Bertone, P., V. Stolc, T. E. Royce, J. S. Rozowsky, A. E. Urban, X. Zhu, J. L. Rinn, W. Tongprasit, M. Samanta, S. Weissman et al. 2004. Global identification of human transcribed sequences with genome tiling arrays. *Science* 306: 2242–2246.

Betancourt, A. J. and D. C. Presgraves. 2002. Linkage limits the power of natural selection in *Drosophila*. *Proc. Natl. Acad. Sci. USA* 99: 13616–13620.

Betrán, E., K. Thornton and M. Long. 2002. Retroposed new genes out of the X in *Drosophila*. *Genome Res.* 12: 1854–1859.

Beverley, S. M., N. S. Akopyants, S. Goyard, R. S. Matlib, J. L. Gordon, B. H. Brownstein, G. D. Stormo, E. N. Bukanova, C. T. Hott, F. Li et al. 2002. Putting the *Leishmania* genome to work: Functional genomics by transposon trapping and expression profiling. *Phil. Trans. Roy. Soc. Lond.* B *Biol. Sci.* 357: 47–53.

Bhattacharya, D. and K. Weber. 1997. The actin gene of the glaucocystophyte *Cyanophora paradoxa*: Analysis of the coding region and introns, and an actin phylogeny of eukaryotes. *Curr. Genet.* 31: 439–446.

Bhattacharya, D., H. S. Yoon and J. D. Hackett. 2004. Photosynthetic eukaryotes unite: Endosymbiosis connects the dots. *BioEssays* 26: 50–60.

Bhattacharyya, S. and M. A. Griep. 2000. DnaB helicase affects the initiation specificity of *Escherichia coli* primase on single-stranded DNA templates. *Biochemistry* 39: 745–752.

Bi, X. and L. F. Liu. 1996. DNA rearrangement mediated by inverted repeats. *Proc. Natl. Acad. Sci. USA* 93: 819–823.

Biémont, C. 1992. Population genetics of transposable DNA elements: A *Drosophila* point of view. *Genetica* 86: 67–84.

Biémont, C., F. Lemeunier, M. P. Garcia Guerreiro, J. F. Brookfield, C. Gautier, S. Aulard and E. G. Pasyukova. 1994. Population dynamics of the *copia, mdg1, mdg3, gypsy,* and *P* transposable elements in a natural population of *Drosophila melanogaster*. *Genet. Res.* 63: 197–212.

Bieniasz, P. D. 2004. Intrinsic immunity: A front-line defense against viral attack. *Nature Immunol.* 5: 1109–1115.

Bierne, N. and A. Eyre-Walker. 2006. Variation in synonymous codon use and DNA polymorphism within the *Drosophila* genome. *J. Evol. Biol.* 19: 1–11.

Bill, C. A., W. A. Duran, N. R. Miselis and J. A. Nickoloff. 1998. Efficient repair of all types of single-base mismatches in recombination intermediates in Chinese Hamster ovary cells: Competition between long-patch and G-T glycosylase-mediated repair of G-T mismatches. *Genetics* 149: 1935–1943.

Bill, C. A., D. G. Taghian, W. A. Duran and J. A. Nickoloff. 2001. Repair bias of large loop mismatches during recombination in mammalian cells depends on loop length and structure. *Mutat. Res.* 485: 255–265.

Birchler, J. A., R. K. Dawe and J. F. Doebley. 2003. Marcus Rhoades, preferential segregation and meiotic drive. *Genetics* 164: 835–841.

Birdsell, J. A. 2002. Integrating genomics, bioinformatics, and classical genetics to study the effects of recombination on genome evolution. *Mol. Biol. Evol.* 19: 1181–1197.

Birky, C. W., Jr. and J. B. Walsh. 1988. Effects of linkage on rates of molecular evolution. *Proc. Natl. Acad. Sci. USA* 85: 6414–6418.

Birky, C. W., Jr. and J. B. Walsh. 1992. Biased gene conversion, copy number, and apparent mutation rate differences within chloroplast and bacterial genomes. *Genetics* 130: 677–683.

Birky, C. W., Jr., C. A. Demko, P. S. Perlman and R. Strausberg. 1978. Uniparental inheritance of mitochondrial genes in yeast: Dependence on input bias of mitochondrial DNA and preliminary investigations of the mechanism. *Genetics* 89: 615–651.

Birky, C. W., Jr., T. Maruyama and P. Fuerst. 1983. An approach to population and evolutionary genetic theory for genes in mitochondria and chloroplasts, and some results. *Genetics* 103: 513–527.

Blair, J. E., K. Ikeo, T. Gojobori and S. B. Hedges. 2002. The evolutionary position of nematodes. *BMC Evol. Biol.* 2: 7.

Blanc, G. and K. H. Wolfe. 2004. Widespread paleopolyploidy in model plant species inferred from age distributions of duplicate genes. *Plant Cell* 16: 1667–1678.

Blanc, V. M. and J. Adams. 2004. *Ty1* insertions in intergenic regions of the genome of *Saccharomyces cerevisiae* transcribed by RNA polymerase III have no detectable selective effect. *FEMS Yeast Res.* 4: 487–491.

Blanchard, J. L. and M. Lynch. 2000. Organellar genes: Why do they end up in the nucleus? *Trends Genet.* 16: 315–320.

Blencowe, B. J. 2000. Exonic splicing enhancers: Mechanism of action, diversity and role in human genetic diseases. *Trends Biochem. Sci.* 25: 106–110.

Bloom, J. D., S. T. Labthavikul, C. R. Otey and F. H. Arnold. 2006. Protein stability promotes evolvability. *Proc. Natl. Acad. Sci. USA* 103: 5869–5874.

Blumenstiel, J. P., D. L. Hartl and E. R. Lozovsky. 2002. Patterns of insertion and deletion in contrasting chromatin domains. *Mol. Biol. Evol.* 19: 2211–2225.

Blumenthal, T. and K. S. Gleason. 2003. *Caenorhabditis elegans* operons: Form and function. *Nature Rev. Genet.* 4: 112–120.

Blumenthal, T. and K. Steward. 1997. RNA processing and gene structure. In D. L. Riddle, T. Blumenthal, B. J. Meyer and J. R. Preis, eds., *C. elegans II*, 117–145. Cold Spring Harbor Laboratory Press, Cold Spring Harbor, NY.

Blumenthal, T., D. Evans, C. D. Link, A. Guffanti, D. Lawson, J. Thierry-Mieg, D. Thierry-Mieg, W. L. Chiu, K. Duke, M. Kiraly and S. K. Kim. 2002. A global analysis of *Caenorhabditis elegans* operons. *Nature* 417: 851–854.

Boeke, J. D., D. J. Garfinkel, C. A. Styles and G. R. Fink. 1985. *Ty* elements transpose through an RNA intermediate. *Cell* 40: 491–500.

Bohr, V. A., T. Stevnsner and N. C. de Souza-Pinto. 2002. Mitochondrial DNA repair of oxidative damage in mammalian cells. *Gene* 286: 127–134.

Boissinot, S., P. Chevret and A. V. Furano. 2000. *L1* (LINE-1) retrotransposon evolution and amplification in recent human history. *Mol. Biol. Evol.* 17: 915–928.

Boissinot, S., A. Entezam, L. Young, P. J. Munson and A. V. Furano. 2004. The insertional history of an active family of *L1* retrotransposons in humans. *Genome Res.* 14: 1221–1231.

Bon, E., S. Casaregola, G. Blandin, B. Llorente, C. Neuvéglise, M. Munsterkotter, U. Guldener, H.-W. Mewes, J. Van Helden, B. Dujon and C. Gaillardin. 2003. Molecular evolution of eukaryotic genomes: Hemiascomycetous yeast spliceosomal introns. *Nucleic Acids Res.* 31: 1121–1135.

Bonen, L. and J. Vogel. 2001. The ins and outs of group II introns. *Trends Genet.* 17: 322–331.

Bonner, J. T. 1988. *The Evolution of Complexity*. Princeton University Press, Princeton, NJ.

Bonner, J. T. 2004. Perspective: The size-complexity rule. *Evolution* 58: 1883–1890.

Bonneton, F., P. J. Shaw, C. Fazakerley, M. Shi and G. A. Dover. 1997. Comparison of bicoid-dependent regulation of hunchback between *Musca domestica* and *Drosophila melanogaster*. *Mechs. Dev.* 66: 143–156.

Borge, T., M. T. Webster, G. Andersson and G. P. Saetre. 2005. Contrasting patterns of polymorphism and divergence on the Z chromosome and autosomes in two *Ficedula* flycatcher species. *Genetics* 171: 1861–1873.

Borkovich, K. A., L. A. Alex, O. Yarden, M. Freitag, G. E. Turner, N. D. Read, S. Seiler, D. Bell-Pedersen, J. Paietta, N. Plesofsky et al. 2004. Lessons from the genome sequence of *Neurospora crassa*: Tracing the path from genomic blueprint to multicellular organism. *Microbiol. Mol. Biol. Rev.* 68: 1–108.

Borts, R. H. and D. T. Kirkpatrick. 2006. The role of the genome in meiotic recombination. In L. H. Caporale, ed., *The Implicit Genome*, 208–224. Oxford University Press, Oxford.

Boucher, Y., M. Kamekura and W. F. Doolittle. 2004. Origins and evolution of isoprenoid lipid biosynthesis in archaea. *Mol. Microbiol.* 52: 515–527.

Boudet, N., S. Aubourg, C. Toffano-Nioche, M. Kreis and A. Lecharny. 2001. Evolution of intron/exon structure of DEAD helicase family genes in *Arabidopsis, Caenorhabditis*, and *Drosophila*. *Genome Res.* 11: 2101–2114.

Boue, S., I. Letunic and P. Bork. 2003. Alternative splicing and evolution. *BioEssays* 25: 1031–1034.

Bourc'his, D. and T. H. Bestor. 2004. Meiotic catastrophe and retrotransposon reactivation in male germ cells lacking Dnmt3L. *Nature* 431: 96–99.

Boutabout, M., M. Wilhelm and F. X. Wilhelm. 2001. DNA synthesis fidelity by the reverse transcriptase of the yeast retrotransposon *Ty1*. *Nucleic Acids Res.* 29: 2217–2222.

Boutanaev, A. M., A. I. Kalmykova, Y. Y. Shevelyov and D. I. Nurminsky. 2002. Large clusters of co-expressed genes in the *Drosophila* genome. *Nature* 420: 666–669.

Bowen, N. J. and J. F. McDonald. 1999. Genomic analysis of *Caenorhabditis elegans* reveals ancient families of retroviral-like elements. *Genome Res.* 9: 924–935.

Bowen, N. J. and J. F. McDonald. 2001. *Drosophila* euchromatic LTR retrotransposons are much younger than the host species in which they reside. *Genome Res.* 11: 1527–1540.

Bowers, J. E., B. A. Chapman, J. Rong and A. H. Paterson. 2003. Unravelling angiosperm genome evolution by phylogenetic analysis of chromosomal duplication events. *Nature* 422: 433–438.

Bowman, S., D. Lawson, D. Basham, D. Brown, T. Chillingworth, C. M., Churcher, A. Craig, R. M. Davies, K. Devlin, T. Feltwell et al. 1999. The complete nucleotide sequence of chromosome 3 of *Plasmodium falciparum*. *Nature* 400: 532–538.

Boxma, B., R. M. de Graaf, G. W. van der Staay, T. A. van Alen, G. Ricard, T. Gabaldón, A. H. van Hoek, S. Y. Moon-van der Staay, W. J. Koopman, J. J. van Hellemond et al. 2005. An anaerobic mitochondrion that produces hydrogen. *Nature* 434: 74–79.

Brasier, M. D., O. R. Green, A. P. Jephcoat, A. K. Kleppe, M. J. Van Kranendonk, J. F. Lindsay, A. Steele and N. V. Grassineau. 2002. Questioning the evidence for Earth's oldest fossils. *Nature* 416: 76–81.

Brehm, K., K. Jensen and M. Frosch. 2000. mRNA *trans*-splicing in the human parasitic cestode *Echinococcus multilocularis*. *J. Biol. Chem.* 275: 38311–38318.

Brehm, K., K. Hubert, E. Sciutto, T. Garate and M. Frosch. 2002. Characterization of a spliced leader gene and of *trans*-spliced mRNAs from *Taenia solium*. *Mol. Biochem. Parasitol.* 122: 105–110.

Bremer, H. and P. P. Dennis. 1996. Modulation of chemical composition and other parameters of the cell by growth rate. In F. C. Neidhardt, R. Curtiss III, J. L. Ingraham, E. C. C. Lin, K. B. Low, B. Magasanik, W. S. Reznikoff, M. Riley, M. Schaechter and H. E. Umbarger, eds., *Escherichia coli and Salmonella*: *Cellular and Molecular Biology*, 2nd ed., 1553–1569. ASM Press, Washington, DC.

Brenner, S. E., T. Hubbard, A. Murzin and C. Chothia. 1995. Gene duplications in *H. influenzae*. *Nature* 378: 140.

Brett, D., H. Pospisil, J. Valcárcel, J. Reich and P. Bork. 2002. Alternative splicing and genome complexity. *Nature Genet.* 30: 29–30.

Britten, R. J. 2004. Coding sequences of functioning human genes derived entirely from mobile element sequences. *Proc. Natl. Acad. Sci. USA* 101: 16825–16830.

Brocke, K. S., G. Neu-Yilik, N. H. Gehring, M. W. Hentze and A. E. Kulozik. 2002. The human intronless melanocortin 4-receptor gene is NMD insensitive. *Hum. Mol. Genet.* 11: 331–335.

Brocks, J. J., G. A. Logan, R. Buick and R. E. Summons. 1999. Archean molecular fossils and the early rise of eukaryotes. *Science* 285: 1033–1036.

Bromham, L. D. and M. D. Hendy. 2000. Can fast early rates reconcile molecular dates with the Cambrian explosion? *Proc. Roy. Soc. Lond.* B *Biol. Sci.* 267: 1041–1047.

Brookfield, J. F. 1986. The population biology of transposable elements. *Phil. Trans. Roy. Soc. Lond.* B *Biol. Sci.* 312: 217–226.

Brookfield, J. F. 2001. Evolution: The evolvability enigma. *Curr. Biol.* 11: R106–R108.

Brosius, J. 1999. Genomes were forged by massive bombardments with retroelements and retrosequences. *Genetica* 107: 209–238.

Brow, D. A. 2002. Allosteric cascade of spliceosome activation. *Annu. Rev. Genet.* 36: 333–360.

Brown, E. H., M. A. Iqbal, S. Stuart, K. S. Hatton, J. Valinsky and C. L. Schildkraut. 1987. Rate of replication of the murine immunoglobulin heavy-chain locus: Evidence that the region is part of a single replicon. *Mol. Cell. Biol.* 7: 450–457.

Brown, G. R., G. P. Gill, R. J. Kuntz, C. H. Langley and D. B. Neale. 2004. Nucleotide diversity and linkage disequilibrium in loblolly pine. *Proc. Natl. Acad. Sci. USA* 101: 15255–15260.

Brown, J. R. and W. F. Doolittle. 1995. Root of the universal tree of life based on ancient aminoacyl-tRNA synthetase gene duplications. *Proc. Natl. Acad. Sci. USA* 92: 2441–2445.

Brown, J. R. and W. F. Doolittle. 1997. Archaea and the prokaryote-to-eukaryote transition. *Microbiol. Mol. Biol. Rev.* 61: 456–502.

Brown, J. R., C. J. Douady, M. J. Italia, W. E. Marshall and M. J. Stanhope. 2001. Universal trees based on

large combined protein sequence data sets. *Nature Genet.* 28: 281–285.
Brown, M. D. and D. C. Wallace. 1994. Molecular basis of mitochondrial DNA disease. *J. Bioenerg. Biomembr.* 26: 273–289.
Brown, T. A., C. Cecconi, A. N. Tkachuk, C. Bustamante and D. A. Clayton. 2005. Replication of mitochondrial DNA occurs by strand displacement with alternative light-strand origins, not via a strand-coupled mechanism. *Genes Dev.* 19: 2466–2476.
Brown, T. C. and J. Jiricny. 1988. Different base/base mispairs are corrected with different efficiencies and specificities in monkey kidney cells. *Cell* 54: 705–711.
Brown, W. M., E. M. Prager, A. Wang and A. C. Wilson. 1982. Mitochondrial DNA sequences of primates: Tempo and mode of evolution. *J. Mol. Evol.* 18: 225–239.
Bruce, A. E., A. C. Oates, V. E. Prince and R. K. Ho. 2001. Additional *hox* clusters in the zebrafish: Divergent expression patterns belie equivalent activities of duplicate *hoxB5* genes. *Evol. Dev.* 3: 127–144.
Brunet, F. G., H. R. Crollius, M. Paris, J.-M. Aury, P. Gibert, O. Jaillon, V. Laudet and M. Robinson-Rechavi. 2006. Gene loss and evolutionary rates following whole-genome duplication in teleost fishes. *Mol. Biol. Evol.* 23: 1808–1816.
Bruzik, J. P. and T. Maniatis. 1992. Spliced leader RNAs from lower eukaryotes are trans-spliced in mammalian cells. *Nature* 360: 692–695.
Bucheton, A., I. Busseau and D. Teninges. 2002. *I* elements in *Drosophila melanogaster*. In N. L. Craig, R. Craigie, M. Gellert and A. M. Lambowitz, eds., *Mobile DNA II*, 796–812. ASM Press, Washington, DC.
Buckler, E. S. IV, T. L. Phelps-Durr, C. S. Buckler, R. K. Dawe, J. F. Doebley and T. P. Holtsford. 1999. Meiotic drive of chromosomal knobs reshaped the maize genome. *Genetics* 153: 415–426.
Bühler, M., S. Steiner, F. Mohn, A. Paillusson, and O. Mühlemann. 2006. EJC-independent degradation of nonsense immunoglobulin-μ mRNA depends on 3′ UTR length. *Nature Struct. Mol. Biol.* 13: 462–464.
Bull, J. J. 1983. *Evolution of Sex Determining Mechanisms*. Benjamin Cummings, Menlo Park, CA.
Bullerwell, C. E., L. Forget and B. F. Lang. 2003. Evolution of monoblepharidalean fungi based on complete mitochondrial genome sequences. *Nucleic Acids Res.* 31: 1614–1623.
Bulmer, M. 1987. Coevolution of codon usage and transfer RNA abundance. *Nature* 325: 728–730.
Bulmer, M. 1988. Codon usage and intragenic position. *J. Theor. Biol.* 133: 67–71.
Bulmer, M. 1991. The selection-mutation-drift theory of synonymous codon usage. *Genetics* 129: 897–907.
Bulmer, M. G. 1980. *The Mathematical Theory of Quantitative Genetics*. Oxford University Press, Oxford.
Buratti, E. and F. E. Baralle. 2004. Influence of RNA secondary structure on the pre-mRNA splicing process. *Mol. Cell. Biol.* 24: 10505–10514.
Bureau, T. E. and S. R. Wessler. 1994. *Stowaway*: A new family of inverted repeat elements associated with the genes of both monocotyledonous and dicotyledonous plants. *Plant Cell.* 6: 907–916.
Burge, C. B., R. A. Padgett and P. A. Sharp. 1998. Evolutionary fates and origins of U12-type introns. *Mol. Cell* 2: 773–785.
Burge, C. B., T. Tuschl and P. A. Sharp. 1999. Splicing of precursors to mRNAs by the spliceosomes. In R. F. Gesteland, T. R. Cech and J. F. Atkins, eds., *The RNA World*, 2nd ed., 525–560. Cold Spring Harbor Laboratory Press, Cold Spring Harbor, NY.
Burger, G., D. Saint-Louis, M. W. Gray and B. F. Lang. 1999. Complete sequence of the mitochondrial DNA of the red alga *Porphyra purpurea*: Cyanobacterial introns and shared ancestry of red and green algae. *Plant Cell* 11: 1675–1694.
Burger, G., Y. Zhu, T. G. Littlejohn, S. J. Greenwood, M. N. Schnare, B. F. Lang and M. W. Gray. 2000. Complete sequence of the mitochondrial genome of *Tetrahymena pyriformis* and comparison with *Paramecium aurelia* mitochondrial DNA. *J. Mol. Biol.* 297: 365–380.
Burger, G., L. Forget, Y. Zhu, M. W. Gray and B. F. Lang. 2003. Unique mitochondrial genome architecture in unicellular relatives of animals. *Proc. Natl. Acad. Sci. USA* 100: 892–897.
Burger, G., M. W. Gray and B. F. Lang. 2003. Mitochondrial genomes: Anything goes. *Trends Genet.* 19: 709–716.
Bürger, R. 2000. *The Mathematical Theory of Selection, Recombination, and Mutation*. John Wiley & Sons, New York.
Bürger, R. and W. J. Ewens. 1995. Fixation probabilities of additive alleles in diploid populations. *J. Math. Biol.* 33: 557–575.
Burke, J. M., L. Zhao, M. Salmaso, T. Nakazato, S. Tang, A. Heesacker, S. J. Knapp and L. H. Rieseberg. 2003. Comparative mapping and rapid karyotypic evolution in the genus *Helianthus*. *Genetics* 167: 449–457.
Burke, W. D., H. S. Malik, S. M. Rich and T. H. Eickbush. 2002. Ancient lineages of non-LTR retrotransposons in the primitive eukaryote, *Giardia lamblia*. *Mol. Biol. Evol.* 19: 619–630.
Burke, W. D., D. Singh and T. H. Eickbush. 2003. *R5* retrotransposons insert into a family of infrequently transcribed 28S rRNA genes of planaria. *Mol. Biol. Evol.* 20: 1260–1270.
Burt, A. and R. Trivers. 2006. *Genes in Conflict*. Harvard University Press, Cambridge, MA.
Bush, E. C. and B. T. Lahn. 2005. Selective constraint on noncoding regions of hominid genomes. *PLoS Comput Biol.* 1: 593–598.
Busseau, I., M. C. Chaboissier, A. Pelisson and A. Bucheton. 1994. *I* factors in *Drosophila melanogaster*: Transposition under control. *Genetica* 93: 101–116.
Bustamante, C. D., R. Nielsen and D. L. Hartl. 2002. A maximum likelihood method for analyzing pseudogene evolution: Implications for silent site evolution in humans and rodents. *Mol. Biol. Evol.* 19: 110–117.
Bustamante, C. D., A. Fledel-Alon, S. Williamson, R. Nielsen, M. T. Hubisz, S. Glanowski, D. M. Tanenbaum, T. J. White, J. J. Sninsky, R. D. Hernandez et

al. 2005. Natural selection on protein-coding genes in the human genome. *Nature* 437: 1153–1157.
Butlin, R. 2002. Evolution of sex: The costs and benefits of sex: New insights from old asexual lineages. *Nature Rev. Genet.* 3: 311–317.
Byrne, E. M. and J. M. Gott. 2004. Unexpectedly complex editing patterns at dinucleotide insertion sites in *Physarum* mitochondria. *Mol. Cell. Biol.* 24: 7821–7828.

Caballero, A. 1994. Developments in the prediction of effective population size. *Heredity* 73: 657–679.
Cai, X., A. L. Fuller, L. R. McDougald and G. Zhu. 2003. Apicoplast genome of the coccidian *Eimeria tenella*. *Gene* 321: 39–46.
Calboli, F. C., G. W. Gilchrist and L. Partridge. 2003. Different cell size and cell number contribution in two newly established and one ancient body size cline of *Drosophila subobscura*. *Evolution* 57: 566–573.
Camacho, J. P. M. 2005. B chromosomes. In T. R. Gregory, ed., *The Evolution of the Genome*, 224–289. Elsevier Academic Press, Burlington, MA.
Camacho, J. P., T. F. Sharbel and L. W. Beukeboom. 2000. B-chromosome evolution. *Phil. Trans. Roy. Soc. Lond.* B *Biol. Sci.* 355: 163–178.
Cambareri, E. B., J. Helber and J. A. Kinsey. 1994. *Tad1*–1, an active *LINE*-like element of *Neurospora crassa*. *Mol. Gen. Genet.* 242: 658–665.
Cambareri, E. B., R. Aisner and J. Carbon. 1998. Structure of the chromosome VII centromere region in *Neurospora crassa:* Degenerate transposons and simple repeats. *Mol. Cell. Biol.* 18: 5465–5477.
Campbell, D. A., D. A. Thornton and J. C. Boothroyd. 1984. Apparent discontinuous transcription of *Trypanosoma brucei* variant surface antigen genes. *Nature* 311: 350–355.
Campbell, D. A., S. Thomas and N. R. Sturm. 2003. Transcription in kinetoplastid protozoa: Why be normal? *Microbes Infect.* 5: 1231–1240.
Canaday, J., L. H. Tessier, P. Imbault and F. Paulus. 2001. Analysis of *Euglena gracilis* alpha-, beta- and gamma-tubulin genes: Introns and pre-mRNA maturation. *Mol. Genet. Genomics* 265: 153–160.
Canbäck, B., I. Tamas and S. G. Andersson. 2004. A phylogenomic study of endosymbiotic bacteria. *Mol. Biol. Evol.* 21: 1110–1122.
Caporale, L. H. 2003. Foresight in genome evolution. *Amer. Sci.* 91: 234–241.
Cappello, J., K. Handelsman and H. F. Lodish. 1985. Sequence of *Dictyostelium DIRS-1:* An apparent retrotransposon with inverted terminal repeats and an internal circle junction sequence. *Cell* 43: 105–115.
Capy, P., G. Gasperi, C. Biémont and C. Bazin. 2000. Stress and transposable elements: Co-evolution or useful parasites? *Heredity* 85: 101–106.
Carbone, C. and J. L. Gittleman. 2002. A common rule for the scaling of carnivore density. *Science* 295: 2273–2276.
Carbone, L., S. G. Nergadze, E. Magnani, D. Misceo, M. Francesca Cardone, R. Roberto, L. Bertoni, C. Attolini, M. Francesca Piras, P. de Jong et al. 2006. Evolutionary movement of centromeres in horse, donkey, and zebra. *Genomics* 87: 777–782.
Cargill, M., D. Altshuler, J. Ireland, P. Sklar, K. Ardlie, N. Patil, N. Shaw, C. R. Lane, E. P. Lim, N. Kalyanaraman et al. 1999. Characterization of single-nucleotide polymorphisms in coding regions of human genes. *Nature Genet.* 22: 231–238.
Carlini, D. B. 2005. Context-dependent codon bias and messenger RNA longevity in the yeast transcriptome. *Mol. Biol. Evol.* 22: 1403–1411.
Carlini, D. B. and J. E. Genut. 2006. Synonymous SNPs provide evidence for selective constraint on human exonic splicing enhancers. *J. Mol. Evol.* 62: 89–98.
Carlini, D. B., Y. Chen and W. Stephan. 2001. The relationship between third-codon position nucleotide content, codon bias, mRNA secondary structure and gene expression in the drosophilid alcohol dehydrogenase genes Adh and Adhr. *Genetics* 159: 623–633.
Carlton, J. M., S. V. Angiuoli, B. B. Suh, T. W. Kooij, M. Pertea, J. C. Silva, M. D. Ermolaeva, J. E. Allen, J. D. Selengut, H. L. Koo et al. 2002. Genome sequence and comparative analysis of the model rodent malaria parasite *Plasmodium yoelii yoelii*. *Nature* 419: 512–519.
Carmel, I., S. Tal, I. Vig and G. Ast. 2004. Comparative analysis detects dependencies among the 5′ splice-site positions. *RNA* 10: 828–840.
Carninci, P. 2006. Tagging mammalian transcription complexity. *Trends Genet.* 22: 501–510.
Carninci, P., A. Sandelin, B. Lenhard, S. Katayama, K. Shimokawa, J. Ponjavic, C. A. Semple, M. S. Taylor, P. G. Engstrom, M. C. Frith et al. 2006. Genome-wide analysis of mammalian promoter architecture and evolution. *Nature Genet.* 38: 626–635.
Carr, B. and P. Anderson. 1994. Imprecise excision of the *Caenorhabditis elegans* transposon Tc1 creates functional 5′ splice sites. *Mol. Cell. Biol.* 14: 3426–3433.
Carrel, L. and H. F. Willard. 2005. X-inactivation profile reveals extensive variability in X-linked gene expression in females. *Nature* 434: 400–404.
Carrodeguas, J. A. and C. G. Vallejo. 1997. Mitochondrial transcription initiation in the crustacean *Artemia franciscana*. *Eur. J. Biochem.* 250: 514–523.
Carroll, M. L., A. M. Roy-Engel, S. V. Nguyen, A. H. Salem, E. Vogel, B. Vincent, J. Myers, Z. Ahmad, L. Nguyen, M. Sammarco et al. 2001. Large-scale analysis of the *Alu* Ya5 and Yb8 subfamilies and their contribution to human genomic diversity. *J. Mol. Biol.* 311: 17–40.
Carroll, S. B. 2005a. *Endless Forms Most Beautiful*. W. W. Norton, New York.
Carroll, S. B. 2005b. Evolution at two levels: On genes and form. *PLoS Biol.* 3: 1160–1166.
Carroll, S. B., J. K. Grenier and S. D. Weatherbee. 2001. *From DNA to Diversity*. Blackwell Science, Malden, MA.
Carter, A. J. and G. P. Wagner. 2002. Evolution of functionally conserved enhancers can be accelerated in large populations: A population-genetic model. *Proc. Biol. Sci.* 269: 953–960.
Carter, A. J., J. Hermisson and T. F. Hansen. 2005. The role of epistatic gene interactions in the response to selection and the evolution of evolvability. *Theor. Pop. Biol.* 68: 179–196.

Carvalho, A. B. 2002. Origin and evolution of the *Drosophila* Y chromosome. *Curr. Opin. Genet. Dev.* 12: 664–668.

Carvalho, A. B. and A. G. Clark. 1999. Intron size and natural selection. *Nature* 401: 344.

Carvalho, A. B. and A. G. Clark. 2005. Y chromosome of *D. pseudoobscura* is not homologous to the ancestral *Drosophila* Y. *Science* 307: 108–110.

Carvalho, A. B., B. A. Dobo, M. D. Vibranovski and A. G. Clark. 2001. Identification of five new genes on the Y chromosome of *Drosophila melanogaster*. *Proc. Natl. Acad. Sci. USA* 98: 13225–13230.

Casals, F., M. Cáceres, M. H. Manfrin, J. González and A. Ruiz. 2005. Molecular characterization and chromosomal distribution of *Galileo, Kepler* and *Newton*, three foldback transposable elements of the *Drosophila buzzatii* species complex. *Genetics* 169: 2047–2059.

Casjens, S. 1998. The diverse and dynamic structure of bacterial genomes. *Annu. Rev. Genet.* 32: 339–377.

Castillo-Davis, C. I. and D. L. Hartl. 2002. Genome evolution and developmental constraint in *Caenorhabditis elegans*. *Mol. Biol. Evol.* 19: 728–735.

Castillo-Davis, C. I., S. L. Mekhedov, D. L. Hartl, E. V. Koonin and F. A. Kondrashov. 2002. Selection for short introns in highly expressed genes. *Nature Genet.* 31: 415–418.

Castillo-Davis, C. I., T. B. Bedford and D. L. Hartl. 2004. Accelerated rates of intron gain/loss and protein evolution in duplicate genes in human and mouse malaria parasites. *Mol. Biol. Evol.* 21: 1422–1427.

Cavelier, L., A. Johannisson and U. Gyllensten. 2000. Analysis of mtDNA copy number and composition of single mitochondrial particles using flow cytometry and PCR. *Exp. Cell Res.* 259: 79–85.

Cavalier-Smith, T. 1978. Nuclear volume control by nucleoskeletal DNA, selection for cell volume and cell growth rate, and the solution of the DNA C-value paradox. *J. Cell. Sci.* 34: 247–278.

Cavalier-Smith, T. 1985. Selfish DNA and the origin of introns. *Nature* 315: 283–284.

Cavalier-Smith, T. 1991. Intron phylogeny: A new hypothesis. *Trends Genet.* 7: 145–148.

Cavalier-Smith, T. 1997. Cell and genome coevolution: Facultative anaerobiosis, glycosomes and kinetoplastan RNA editing. *Trends Genet.* 13: 6–9.

Cavalier-Smith, T. 1998. A revised six-kingdom system of life. *Biol. Rev.* 73: 203–266.

Cavalier-Smith, T. 1999. Principles of protein and lipid targeting in secondary symbiogenesis: Euglenoid, dinoflagellate, and sporozoan plastid origins and the eukaryote family tree. *J. Eukaryot. Microbiol.* 46: 347–366.

Cavalier-Smith, T. 2002a. Chloroplast evolution: Secondary symbiogenesis and multiple losses. *Curr. Biol.* 12: R62–R64.

Cavalier-Smith, T. 2002b. The neomuran origin of archaebacteria, the negibacterial root of the universal tree and bacterial megaclassification. *Internatl. J. Syst. Evol. Microbiol.* 52: 7–76.

Cavalier-Smith, T. 2002c. The phagotrophic origin of eukaryotes and phylogenetic classification of Protozoa. *Internatl. J. Syst. Evol. Microbiol.* 52: 297–354.

Cavalier-Smith, T. 2003. Genomic reduction and evolution of novel genetic membranes and protein-targeting machinery in eukaryote-eukaryote chimaeras (meta-algae). *Phil. Trans. Roy. Soc. Lond.* B *Biol. Sci.* 358: 109–133.

Cavalier-Smith, T. 2005. Economy, speed and size matter: Evolutionary forces driving nuclear genome miniaturization and expansion. *Ann. Bot.* 95: 147–175.

Cavalier-Smith, T. 2006a. Cell evolution and Earth history: Stasis and revolution. *Phil. Trans. Roy. Soc. Lond.* B *Biol. Sci.* 361: 969–1006.

Cavalier-Smith, T. 2006b. Origin of mitochondria by intracellular enslavement of a photosynthetic purple bacterium. *Proc. Roy. Soc. Lond.* B *Biol. Sci.* 273: 1943–1952.

Cawley, S., S. Bekiranov, H. H. Ng, P. Kapranov, E. A. Sekinger, D. Kampa, A. Piccolboni, V. Sementchenko, J. Cheng, A. J. Williams et al. 2004. Unbiased mapping of transcription factor binding sites along human chromosomes 21 and 22 points to widespread regulation of noncoding RNAs. *Cell* 116: 499–509.

Cech, T. R. 1986. The generality of self-splicing RNA: Relationship to nuclear mRNA splicing. *Cell* 44: 207–210.

Chaboissier, M. C., A. Bucheton and D. J. Finnegan. 1998. Copy number control of a transposable element, the *I* factor, a *LINE*-like element in *Drosophila*. *Proc. Natl. Acad. Sci. USA* 95: 11781–11785.

Chain, F. J. and B. J. Evans. 2006. Multiple mechanisms promote the retained expression of gene duplicates in the tetraploid frog *Xenopus laevis*. *PLoS Genet.* 2: 478–490.

Chalvet, F., L. Teysset, C. Terzian, N. Prud'homme, P. Santamaria, A. Bucheton and A. Pelisson. 1999. Proviral amplification of the *Gypsy* endogenous retrovirus of *Drosophila melanogaster* involves *env*-independent invasion of the female germline. *EMBO J.* 18: 2659–2669.

Chamary, J. V. and L. D. Hurst. 2004. Similar rates but different modes of sequence evolution in introns and at exonic silent sites in rodents: Evidence for selectively driven codon usage. *Mol. Biol. Evol.* 21: 1014–1023.

Chamary, J. V. and L. D. Hurst. 2005. Evidence for selection on synonymous mutations affecting stability of mRNA secondary structure in mammals. *Genome Biol.* 6: R75.

Chambers, S. R., N. Hunter, E. J. Louis and R. H. Borts. 1996. The mismatch repair system reduces meiotic homeologous recombination and stimulates recombination-dependent chromosome loss. *Mol. Cell. Biol.* 16: 6110–6120.

Chambeyron, S. and A. Bucheton. 2005. *I* elements in *Drosophila:* In vivo retrotransposition and regulation. *Cytogenet. Genome Res.* 110: 215–222.

Chan, S. R. and E. H. Blackburn. 2004. Telomeres and telomerase. *Phil. Trans. Roy. Soc. Lond.* B *Biol. Sci.* 359: 109–121.

Chandler, M. and J. Mahillon. 2002. Insertion sequences revisited. In N. L. Craig, R. Craigie, M. Gellert and A. M. Lambowitz, eds., *Mobile DNA II*, 305–366. ASM Press, Washington, DC.

Chang, B. H., L. C. Shimmin, S. K. Shyue, D. Hewett-Emmett and W.-H. Li. 1994. Weak male-driven molecular evolution in rodents. *Proc. Natl. Acad. Sci. USA* 91: 827–831.

Chao, L., C. Vargas, B. B. Spear and E. C. Cox. 1983. Transposable elements as mutator genes in evolution. *Nature* 303: 633–635.

Chapman, B. A., J. E. Bowers, F. A. Feltus and A. H. Paterson. 2006. Buffering of crucial functions by paleologous duplicated genes may contribute cyclicality to angiosperm genome duplication. *Proc. Natl. Acad. Sci. USA* 103: 2730–2735.

Charlesworth, B. 1985. The population genetics of transposable elements. In T. Ohta and K. Aoki, eds., *Population Genetics and Molecular Evolution*, 213–232. Springer-Verlag, New York.

Charlesworth, B. 1991. The evolution of sex chromosomes. *Science* 251: 1030–1033.

Charlesworth, B. 1996a. The changing sizes of genes. *Nature* 384: 315–316.

Charlesworth, B. 1996b. The evolution of chromosomal sex determination and dosage compensation. *Curr. Biol.* 6: 149–162.

Charlesworth, B. and N. Barton. 2004. Genome size: Does bigger mean worse? *Curr. Biol.* 14: R233–R235.

Charlesworth, B. and D. Charlesworth. 1978. A model for the evolution of dioecy and gynodioecy. *Amer. Nat.* 112: 975–997.

Charlesworth, B. and D. Charlesworth. 1983. The population dynamics of transposable elements. *Genet. Res.* 42: 1–27.

Charlesworth, B. and D. Charlesworth. 2000. The degeneration of Y chromosomes. *Phil. Trans. Roy. Soc. Lond.* B *Biol. Sci.* 355: 1563–1572.

Charlesworth, B. and C. H. Langley. 1986. The evolution of self-regulated transposition of transposable elements. *Genetics* 112: 359–383.

Charlesworth, B. and C. H. Langley. 1989. The population genetics of *Drosophila* transposable elements. *Annu. Rev. Genet.* 23: 251–87.

Charlesworth, B., R. Lande and M. Slatkin. 1980. A neo-Darwinian commentary on macroevolution. *Evolution* 36: 474–498.

Charlesworth, B., J. A. Coyne and N. H. Barton. 1987. The relative rates of evolution of sex chromosomes and autosomes. *Amer. Nat.* 130: 113–146.

Charlesworth, B., A. Lapid and D. Canada. 1992. The distribution of transposable elements within and between chromosomes in a population of *Drosophila melanogaster*. I. Element frequencies and distribution. *Genet. Res.* 60: 103–114.

Charlesworth, B., M. T. Morgan and D. Charlesworth. 1993. Mutation accumulation in finite outbreeding and inbreeding populations. *Genet. Res.* 61: 39–56.

Chen, F. C., S. S. Wang, C. J. Chen, W.-H. Li and T. J. Chuang. 2006. Alternatively and constitutively spliced exons are subject to different evolutionary forces. *Mol. Biol. Evol.* 23: 675–682.

Chen, J., I. M. Greenblatt and S. L. Dellaporta. 1992. Molecular analysis of *Ac* transposition and DNA replication. *Genetics* 130: 665–676.

Chen, J. M., N. Chuzhanova, P. D. Stenson, C. Ferec and D. N. Cooper. 2005. Meta-analysis of gross insertions causing human genetic disease: Novel mutational mechanisms and the role of replication slippage. *Hum. Mutat.* 25: 207–221.

Chen, J. Y., P. Oliveri, F. Gao, S. Q. Dornbos, C. W. Li, D. J. Bottjer and E. H. Davidson. 2002. Precambrian animal life: Probable developmental and adult cnidarian forms from Southwest China. *Dev. Biol.* 248: 182–196.

Chen, S. L., W. Lee, A. K. Hottes, L. Shapiro and H. H. McAdams. 2004. Codon usage between genomes is constrained by genome-wide mutational processes. *Proc. Natl. Acad. Sci. USA* 101: 3480–3485.

Cheng, G., L. Cohen, D. Ndegwa and R. E. Davis. 2006. The flatworm spliced leader 3′-terminal AUG as a translation initiator methionine. *J. Biol. Chem.* 281: 733–743.

Cheng, J., P. Belgrader, X. Zhou and L. E. Maquat. 1994. Introns are *cis*-effectors of the nonsense-codon-mediated reduction in nuclear mRNA abundance. *Mol. Cell. Biol.* 14: 6317–6325.

Cheng, Y. W., L. M. Visomirski-Robic and J. M. Gott. 2001. Non-templated addition of nucleotides to the 3′ end of nascent RNA during RNA editing in *Physarum*. *EMBO J.* 20: 1405–1414.

Cheng, Z., M. Ventura, X. She, P. Khaitovich, T. Graves, K. Osoegawa, D. Church, P. DeJong, R. K. Wilson, S. Pääbo et al. 2005. A genome-wide comparison of recent chimpanzee and human segmental duplications. *Nature* 437: 88–93.

Chern, T. M., E. van Nimwegen, C. Kai, J. Kawai, P. Carninci, Y. Hayashizaki and M. Zavolan. 2006. A simple physical model predicts small exon length variations. *PLoS Genetics* 2: 606–613.

Cherry, L. M., S. M. Case and A. C. Wilson. 1978. Frog perspective on the morphological difference between humans and chimpanzees. *Science* 200: 209–211.

Cherry, L. M., S. M. Case, J. G. Kunkel, J. S. Wyles and A. C. Wilson. 1982. Body shape metrics and organismal evolution. *Evolution* 36: 914–933.

Chiang, E. F.-L., C.-I. Pai, M. Wyatt, Y.-L. Yan, J. Postlethwait and B.-C. Chung. 2001. Two *sox9* genes on duplicated zebrafish chromosomes: Expression of similar transcription activators in distinct sites. *Dev. Biol.* 231: 149–163.

Chiang, E. F.-L., Y.-L. Yan, Y. Guiguen, J. Postlethwait and B.-C. Chung. 2001. Two *Cyp19* (P450 aromatase) genes on duplicated zebrafish chromosomes are expressed in ovary or brain. *Mol. Biol. Evol.* 18: 542–550.

Chiaromonte, F., W. Miller and E. E. Bouhassira. 2003. Gene length and proximity to neighbors affect genome-wide expression levels. *Genome Res.* 13: 2602–2608.

Chikashige, Y., N. Kinoshita, Y. Nakaseko, T. Matsumoto, S. Murakami, O. Niwa and M. Yanagida. 1989. Composite motifs and repeat symmetry in *S. pombe* centromeres: Direct analysis by integration of NotI restriction sites. *Cell* 57: 739–751.

Chin, C. S., J. H. Chuang and H. Li. 2005. Genome-wide regulatory complexity in yeast promoters: Separation of functionally conserved and neutral sequence. *Genome Res.* 15: 205–213.

Cho, S., S. W. Jin, A. Cohen and R. E. Ellis. 2004. A phylogeny of *Caenorhabditis* reveals frequent loss of introns during nematode evolution. *Genome Res.* 14: 1207–1220.

Cho, Y., J. P. Mower, Y. L. Qiu and J. D. Palmer. 2004. Mitochondrial substitution rates are extraordinarily elevated and variable in a genus of flowering plants. *Proc. Natl. Acad. Sci. USA* 101: 17741–17746.

Choi, W. S., M. Yan, D. Nusinow and J. D. Gralla. 2002. In vitro transcription and start site selection in *Schizosaccharomyces pombe*. *J. Mol. Biol.* 319: 1005–1013.

Choury, D., J. C. Farre, X. Jordana and A. Araya. 2004. Different patterns in the recognition of editing sites in plant mitochondria. *Nucleic Acids Res.* 32: 6397–6406.

Chow, L. T., R. E. Gelinas, T. R. Broker and R. J. Roberts. 1977. An amazing sequence arrangement at the 5′ ends of adenovirus 2 messenger RNA. *Cell* 12: 1–8.

Christoffels, A., E. G. Koh, J. M. Chia, S. Brenner, S. Aparicio and B. Venkatesh. 2004. *Fugu* genome analysis provides evidence for a whole-genome duplication early during the evolution of ray-finned fishes. *Mol. Biol. Evol.* 21: 1146–1151.

Churbanov, A., I. B. Rogozin, V. N. Babenko, H. Ali and E. V. Koonin. 2005. Evolutionary conservation suggests a regulatory function of AUG triplets in 5′-UTRs of eukaryotic genes. *Nucleic Acids Res.* 33: 5512–5520.

Ciccarelli, F. D., T. Doerks, C. von Mering, C. J. Creevey, B. Snel and P. Bork. 2006. Toward automatic reconstruction of a highly resolved tree of life. *Science* 311: 1283–1287.

Clarens, M., A. J. Macario and E. C. de Macario. 1995. The archaeal *dnaK-dnaJ* gene cluster: Organization and expression in the methanogen *Methanosarcina mazei*. *J. Mol. Biol.* 250: 191–201.

Clark, A. G. 1994. Invasion and maintenance of a gene duplication. *Proc. Natl. Acad. Sci. USA* 91: 2950–2954.

Clark, A. G. and L. Wang. 1997. Epistasis in measured genotypes: *Drosophila* P-element insertions. *Genetics* 147: 157–163.

Claverie, J. M., H. Ogata, S. Audic, C. Abergel, K. Suhre and P. E. Fournier. 2006. Mimivirus and the emerging concept of "giant" virus. *Virus Res.* 117: 133–144.

Claycomb, J. M., M. Benasutti, G. Bosco, D. D. Fenger and T. L. Orr-Weaver. 2004. Gene amplification as a developmental strategy: Isolation of two developmental amplicons in *Drosophila*. *Dev. Cell* 6: 145–155.

Clement, J., S. Maiti and M. F. Wilkinson. 2001. Localization and stability of introns spliced from the *Pem* homeobox gene. *J. Biol. Chem.* 20: 16919–16930.

Clutton-Brock, T. (ed.). 1988. *Reproductive Success: Studies of Individual Variation in Contrasting Breeding Systems*. University of Chicago Press, Chicago.

Coghlan, A. and K. H. Wolfe. 2002. Fourfold faster rate of genome rearrangement in nematodes than in *Drosophila*. *Genome Res.* 12: 857–867.

Coghlan, A. and K. H. Wolfe. 2004. Origins of recently gained introns in *Caenorhabditis*. *Proc. Natl. Acad. Sci. USA* 101: 11362–11367.

Cohan, F. M. 1995. Does recombination constrain neutral divergence among bacterial taxa? *Evolution* 149: 164–175.

Cohen, B. A., R. D. Mitra, J. D. Hughes and G. M. Church. 2000. A computational analysis of whole-genome expression data reveals chromosomal domains of gene expression. *Nature Genet.* 26: 183–186.

Cohen-Gihon, I., D. Lancet and I. Yanai. 2005. Modular genes with metazoan-specific domains have increased tissue specificity. *Trends Genet.* 21: 210–213.

Cohn, M. and J. E. Edstrom. 1992. Chromosome ends in *Chironomus pallidivittatus* contain different subfamilies of telomere-associated repeats. *Chromosoma* 101: 634–640.

Collins, L. and D. Penny. 2005. Complex spliceosomal organization ancestral to extant eukaryotes. *Mol. Biol. Evol.* 22: 1053–1066.

Comeron, J. M. 2004. Selective and mutational patterns associated with gene expression in humans: Influences on synonymous composition and intron presence. *Genetics* 167: 1293–1304.

Comeron, J. M. 2006. Weak selection and recent mutational changes influence polymorphic synonymous mutations in humans. *Proc. Natl. Acad. Sci. USA* 103: 6940–6945.

Comeron, J. M. and T. B. Guthrie. 2005. Intragenic Hill-Robertson interference influences selection intensity on synonymous mutations in *Drosophila*. *Mol. Biol. Evol.* 22: 2519–2530.

Comeron, J. M. and M. Kreitman. 2000. The correlation between intron length and recombination in *Drosophila*: Dynamic equilibrium between mutational and selective forces. *Genetics* 156: 1175–1190.

Comeron, J. M. and M. Kreitman. 2002. Population, evolutionary and genomic consequences of interference selection. *Genetics* 161: 389–410.

Comeron, J. M., M. Kreitman and M. Aguade. 1999. Natural selection on synonymous sites is correlated with gene length and recombination in *Drosophila*. *Genetics* 151: 239–249.

Commoner, B. 1964. Roles of deoxyribonucleic acid in inheritance. *Nature* 202: 960–968.

Conant, G. C. and A. Wagner. 2003. Asymmetric sequence divergence of duplicate genes. *Genome Res.* 13: 2052–2058.

Conant, G. C. and A. Wagner. 2004. Duplicate genes and robustness to transient gene knock-downs in *Caenorhabditis elegans*. *Proc. Roy. Soc. Lond.* B *Biol. Sci.* 271: 89–96.

Conrad, R., J. Thomas, J. Spieth and T. Blumenthal. 1991. Insertion of part of an intron into the 5′ untranslated region of a *Caenorhabditis elegans* gene converts it into a *trans*-spliced gene. *Mol. Cell. Biol.* 11: 1921–1926.

Conrad, R., R. F. Liou and T. Blumenthal. 1993. Conversion of a *trans*-spliced *C. elegans* gene into a conventional gene by introduction of a splice donor site. *EMBO J.* 12: 1249–1255.

Conrad, R., K. Lea and T. Blumenthal. 1995. SL1 *trans*-splicing specified by AU-rich synthetic RNA inserted at the 5′ end of *Caenorhabditis elegans* pre-mRNA. *RNA* 1: 164–170.

Conte, C., B. Dastugue and C. Vaury. 2002. Promoter competition as a mechanism of transcriptional interference mediated by retrotransposons. *EMBO J.* 21: 3908–3916.

Contursi, C., G. Minchiotti and P. P. Di Nocera. 1993. Functional dissection of two promoters that control sense and antisense transcription of *Drosophila melanogaster* F elements. *J. Mol. Biol.* 234: 988–997.

Conway, D. J., C. Fanello, J. M. Lloyd, B. M. Al-Joubori, A. H. Baloch, S. D. Somanath, C. Roper, A. M. Oduola, B. Mulder, M. M. Povoa et al. 2000. Origin of *Plasmodium falciparum* malaria is traced by mitochondrial DNA. *Mol. Biochem. Parasitol.* 111: 163–171.

Conway Morris, S. 2000. The Cambrian "explosion": Slow-fuse or megatonnage? *Proc. Natl. Acad. Sci. USA* 97: 4426–4429.

Cook, P. R. 2001. *Principles of Nuclear Structure and Function*. John Wiley & Sons, New York.

Cooper, J. L. and S. Henikoff. 2004. Adaptive evolution of the histone fold domain in centromeric histones. *Mol. Biol. Evol.* 21: 1712–1718.

Cooper, T. A. and W. Mattox. 1997. The regulation of splice-site selection, and its role in human disease. *Amer. J. Hum. Genet.* 61: 259–266.

Cooper, T. F., R. E. Lenski and S. F. Elena. 2005. Parasites and mutational load: An experimental test of a pluralistic theory for the evolution of sex. *Proc. Roy. Soc. Lond.* B *Biol. Sci.* 272: 311–317.

Copenhaver, G. P., K. Nickel, T. Kuromori, M.-I. Benito, S. Kaul, X. Lin, M. Bevan, G. Murphy, B. Harris, L. D. Parnell et al. 1999. Genetic definition and sequence analysis of *Arabidopsis* centromeres. *Science* 286: 2468–2474.

Copenhaver, G. P., E. A. Housworth and F. W. Stahl. 2002. Crossover interference in *Arabidopsis*. *Genetics* 160: 1631–1639.

Cordaux, R., D. J. Hedges, S. W. Herke and M. A. Batzer. 2006. Estimating the retrotransposition rate of human *Alu* elements. *Gene* 373: 134–137.

Corsetti, F. A., S. M. Awramik and D. Pierce. 2003. A complex microbiota from snowball Earth times: Microfossils from the Neoproterozoic Kingston Peak Formation, Death Valley, USA. *Proc. Natl. Acad. Sci. USA* 100: 4399–4404.

Costantini, M., O. Clay, F. Auletta and G. Bernardi. 2006. An isochore map of human chromosomes. *Genome Res.* 16: 536–541.

Cotton, J. A. and R. D. Page. 2005. Rates and patterns of gene duplication and loss in the human genome. *Proc. Biol. Sci.* 272: 277–283.

Cousineau, B., S. Lawrence, D. Smith and M. Belfort. 2000. Retrotransposition of a bacterial group II intron. *Nature* 404: 1018–1021.

Cox, R. A. 2003. Correlation of the rate of protein synthesis and the third power of the RNA:protein ratio in *Escherichia coli* and *Mycobacterium tuberculosis*. *Microbiology* 149: 729–737.

Cox, R. A. 2004. Quantitative relationships for specific growth rates and macromolecular compositions of *Mycobacterium tuberculosis, Streptomyces coelicolor* A3(2) and *Escherichia coli* B/r: An integrative theoretical approach. *Microbiology* 150: 1413–1426.

Coyne, J. A. and H. A. Orr. 1997. "Patterns of speciation in *Drosophila*" revisited. *Evolution* 51: 295–303.

Coyne, J. A. and H. A. Orr. 2004. *Speciation*. Sinauer Associates, Sunderland, MA.

Coyne, J. A., N. H. Barton and M. Turelli. 2000. Is Wright's shifting balance process important in evolution? *Evolution* 54: 306–317.

Craig, N. L. 2002. Tn7. In N. L. Craig, R. Craigie, M. Gellert and A. M. Lambowitz, eds., *Mobile DNA II*, 423–456. ASM Press, Washington, DC.

Cresko, W. A., Y. L. Yan, D. A. Baltrus, A. Amores, A. Singer, A. Rodriguez-Mari and J. H. Postlethwait. 2003. Genome duplication, subfunction partitioning, and lineage divergence: Sox9 in stickleback and zebrafish. *Dev. Dyn.* 228: 480–489.

Crow, J. F. 1993. How much do we know about spontaneous human mutation rates? *Environ. Mol. Mutagen.* 21: 122–129.

Crow, J. F. 2000. The origins, patterns and implications of human spontaneous mutation. *Nature Rev. Genet.* 1: 40–47.

Crow, J. F. and M. Kimura. 1970. *An Introduction to Population Genetics Theory*. Harper and Row, New York.

Crow, K. D., P. F. Stadler, V. J. Lynch, C. Amemiya and G. P. Wagner. 2006. The "fish-specific" Hox cluster duplication is coincident with the origin of teleosts. *Mol. Biol. Evol.* 23: 121–136.

Culbertson, M. R. 1999. RNA surveillance: Unforeseen consequences for gene expression, inherited genetic disorders and cancer. *Trends Genet.* 15: 74–80.

Curcio, M. J. and K. M. Derbyshire. 2003. The outs and ins of transposition: From mu to kangaroo. *Nature Rev. Mol. Cell. Biol.* 4: 865–877.

Curcio, M. J. and D. J. Garfinkel. 1991. Regulation of retrotransposition in *Saccharomyces cerevisiae*. *Mol. Microbiol.* 5: 1823–1829.

Cutter, A. D. and B. A. Payseur. 2003. Selection at linked sites in the partial selfer *Caenorhabditis elegans*. *Mol. Biol. Evol.* 20: 665–673.

Cutter, A. D., S. E. Baird and D. Charlesworth. 2006. High nucleotide polymorphism and rapid decay of linkage disequilibrium in wild populations of *Caenorhabditis remanei*. *Genetics* 174: 901–913.

Daborn, P. J., J. L. Yen, M. R. Bogwitz, G. Le Goff, E. Feil, S. Jeffers, N. Tijet, T. Perry, D. Heckel, P. Batterham et al. 2002. A single p450 allele associated with insecticide resistance in *Drosophila*. *Science* 297: 2253–2256.

Dacks, J. B. and A. J. Roger. 1999. The first sexual lineage and the relevance of facultative sex. *J. Mol. Evol.* 48: 779–783.

Dacks, J. B., A. Marinets, W. F. Doolittle, T. Cavalier-Smith and J. M. Logsdon, Jr. 2002. Analyses of RNA polymerase II genes from free-living protists: Phylogeny, long branch attraction, and the eukaryotic big bang. *Mol. Biol. Evol.* 19: 830–840.

Dahlseid, J. N., J. Puziss, R. L. Shirley, A. L. Atkin, P. Hieter and M. R. Culbertson. 1998. Accumulation of mRNA coding for the ctf13p kinetochore subunit of *Saccharomyces cerevisiae* depends on the same factors that promote rapid decay of nonsense mRNAs. *Genetics* 150: 1019–1035.

Dahm, R. 2005. Friedrich Miescher and the discovery of DNA. *Dev. Biol.* 278: 274–288.

Dai, J., R. Y. Chuang and T. J. Kelly. 2005. DNA replication origins in the *Schizosaccharomyces pombe* genome. *Proc. Natl. Acad. Sci. USA* 102: 337–342.

Dai, L. and S. Zimmerly. 2002. Compilation and analysis of group II intron insertions in bacterial genomes: Evidence for retroelement behavior. *Nucleic Acids Res.* 30: 1091–1102.

Dai, L. and S. Zimmerly. 2003. ORF-less and reverse-transcriptase-encoding group II introns in archaebacteria, with a pattern of homing into related group II intron ORFs. *RNA* 9: 14–19.

Daley, J. M., P. L. Palmbos, D. Wu and T. E. Wilson. 2005. Nonhomologous end joining in yeast. *Annu. Rev. Genet.* 39: 431–451.

Daly, M. J., J. D. Rioux, S. F. Schaffner, T. J. Hudson and E. S. Lander. 2001. High-resolution haplotype structure in the human genome. *Nature Genet.* 29: 229–232.

Damuth, J. 1981. Population density and body size in mammals. *Nature* 290: 699–700.

Daniels, S. B., K. R. Peterson, L. D. Strausbaugh, M. G. Kidwell and A. Chovnick. 1990. Evidence for horizontal transmission of the *P* transposable element between *Drosophila* species. *Genetics* 124: 339–355.

Danis, E., K. Brodolin, S. Menut, D. Maiorano, C. Girard-Reydet and M. Mechali. 2004. Specification of a DNA replication origin by a transcription complex. *Nature Cell Biol.* 6: 721–730.

Darnell, J. E., Jr. 1978. Implications of RNA-RNA splicing in evolution of eukaryotic cells. *Science* 202: 1257–1260.

Darwin, C. 1859. *On the Origin of Species by Means of Natural Selection, or the Preservation of Favoured Races in the Struggle for Life*. John Murray, London.

Darwin, C. 1866. *The Variation of Animals and Plants under Domestication*. John Murray, London.

Datta, A. and S. Jinks-Robertson. 1995. Association of increased spontaneous mutation rates with high levels of transcription in yeast. *Science* 268: 1616–1619.

Datta, A., M. Hendrix, M. Lipsitch and S. Jinks-Robertson. 1997. Dual roles for DNA sequence identity and the mismatch repair system in the regulation of mitotic crossing-over in yeast. *Proc. Natl. Acad. Sci. USA* 94: 9757–9762.

David, L., S. Blum, M. W. Feldman, U. Lavi and J. Hillel. 2003. Recent duplication of the common carp (*Cyprinus carpio* L.) genome as revealed by analyses of microsatellite loci. *Mol. Biol. Evol.* 20: 1425–1434.

Davidson, E. H. 2001. *Genomic Regulatory Systems: Development and Evolution*. Academic Press, San Diego, CA.

Davidson, E. H. 2006. *The Regulatory Genome: Gene Regulatory Networks in Development and Evolution*. Academic Press, New York.

Davidson, E. H. and D. H. Erwin. 2006. Gene regulatory networks and the evolution of animal body plans. *Science* 311: 796–800.

Davis, J. C. and D. A. Petrov. 2004. Preferential duplication of conserved proteins in eukaryotic genomes. *PLoS Biol.* 2: 318–326.

Davis, R. E. and S. Hodgson. 1997. Gene linkage and steady state RNAs suggest *trans*-splicing may be associated with a polycistronic transcript in *Schistosoma mansoni*. *Mol. Biochem. Parasitol.* 89: 25–39.

Davis, R. E., C. Hardwick, P. Tavernier, S. Hodgson and H. Singh. 1995. RNA *trans*-splicing in flatworms: Analysis of *trans*-spliced mRNAs and genes in the human parasite, *Schistosoma mansoni*. *J. Biol. Chem.* 270: 21813–21819.

Dawkins, R. 1976. *The Selfish Gene*. Oxford University Press, New York.

Dawkins, R. 1986. *The Blind Watchmaker*. W. W. Norton, New York.

Dawkins, R. 1996. *Climbing Mount Improbable*. W. W. Norton, New York.

Dawkins, R. 2004. *The Ancestor's Tale: A Pilgrimage to the Dawn of Evolution*. Houghton Mifflin, Boston, MA.

Dawson, E., G. R. Abecasis, S. Bumpstead, Y. Chen, S. Hunt, D. M. Beare, J. Pabial, T. Dibling, E. Tinsley, S. Kirby et al. 2002. A first-generation linkage disequilibrium map of human chromosome 22. *Nature* 418: 544–548.

Dawson, S. C. and N. R. Pace. 2002. Novel kingdom-level eukaryotic diversity in anoxic environments. *Proc. Natl. Acad. Sci. USA* 99: 8324–8329.

Decottignies, A. 2005. Capture of extranuclear DNA at fission yeast double-strand breaks. *Genetics* 171: 1535–1548.

Degnan, P. H., A. B. Lazarus and J. J. Wernegreen. 2005. Genome sequence of *Blochmannia pennsylvanicus* indicates parallel evolutionary trends among bacterial mutualists of insects. *Genome Res.* 15: 1023–1033.

De Gobbi, M., V. Viprakasit, J. R. Hughes, C. Fisher, V. J. Buckle, H. Ayyub, R. J. Gibbons, D. Vernimmen, Y. Yoshinaga, P. de Jong et al. 2006. A regulatory SNP causes a human genetic disease by creating a new transcriptional promoter. *Science* 312: 1215–1217.

Dehal, P. and J. L. Boore. 2005. Two rounds of whole genome duplication in the ancestral vertebrate. *PLoS Biol.* 3: 1700–1708.

Dehal, P., P. Predki, A. S. Olsen, A. Kobayashi, P. Folta, S. Lucas, M. Land, A. Terry, C. L. Ecale Zhou, S. Rash et al. 2001. Human chromosome 19 and related regions in mouse: Conservative and lineage-specific evolution. *Science* 293: 104–111.

Deininger, P. L. and M. A. Batzer. 2002. Mammalian retroelements. *Genome Res.* 12: 1455–1465.

de Jong, P. J., A. J. Grosovsky and B. W. Glickman. 1988. Spectrum of spontaneous mutation at the *APRT* locus of Chinese hamster ovary cells: An analysis at the DNA sequence level. *Proc. Natl. Acad. Sci. USA* 85: 3499–3503.

de Lange, T. 2004. T-loops and the origin of telomeres. *Nature Rev. Mol. Cell. Biol.* 5: 323–329.

Delattre, M. and M. A. Felix. 2001. Microevolutionary studies in nematodes: A beginning. *BioEssays* 23: 807–819.

Dellaporta, S. L., A. Xu, S. Sagasser, W. Jakob, M. A. Moreno, L. W. Buss and B. Schierwater. 2006. Mito-

chondrial genome of *Trichoplax adhaerens* supports placozoa as the basal lower metazoan phylum. *Proc. Natl. Acad. Sci. USA* 103: 8751–8756.
Delneri, D., I. Colson, S. Grammenoudi, I. N. Roberts, E. J. Louis and S. G. Oliver. 2003. Engineering evolution to study speciation in yeasts. *Nature* 422: 68–72.
Delwiche, C. F. 1999. Tracing the thread of plastid diversity through the tapestry of life. *Amer. Natur.* 154: S164–S177.
de Martino, S., Y.-L. Yan, T. Jowett, J. H. Postlethwait, Z. M. Varga, A. Ashworth and C. A. Austin. 2000. Expression of *sox11* gene duplicates in zebrafish suggests the reciprocal loss of ancestral gene expression patterns in development. *Dev. Dyn.* 217: 279–92.
de Massy, B. 2003. Distribution of meiotic recombination sites. *Trends Genet.* 19: 514–522.
Denker, J. A., D. M. Zuckerman, P. A. Maroney and T. W. Nilsen. 2002. New components of the spliced leader RNP required for nematode *trans*-splicing. *Nature* 417: 667–670.
Denver, D. R., K. Morris, M. Lynch, L. L. Vassilieva and W. K. Thomas. 2000. High direct estimate of the mutation rate in the mitochondrial genome of *Caenorhabditis elegans*. *Science* 289: 2342–2344.
Denver, D. R., K. Morris, M. Lynch and W. K. Thomas. 2004. High mutation rate and predominance of insertions in the *Caenorhabditis elegans* nuclear genome. *Nature* 430: 679–682.
Denver, D. R., K. Morris, J. T. Streelman, S. K. Kim, M. Lynch and W. K. Thomas. 2005. The transcriptional consequences of mutation and natural selection in *Caenorhabditis elegans*. *Nature Genet.* 37: 544–548.
DePamphilis, M. L. (ed.). 1996. *DNA Replication in Eukaryotic Cells*. Cold Spring Harbor Laboratory Press, Plainview, NY.
Depew, D. J. and D. H. Weber (eds.). 1985. *Evolution at a Crossroads: The New Biology and the New Philosophy of Science*. MIT Press, Cambridge, MA.
Dermitzakis, E. T. and A. G. Clark. 2001. Differential selection after duplication in mammalian developmental genes. *Mol. Biol. Evol.* 18: 557–562.
Dermitzakis, E. T. and A. G. Clark. 2002. Evolution of transcription factor binding sites in mammalian gene regulatory regions: Conservation and turnover. *Mol. Biol. Evol.* 19: 1114–1121.
Dermitzakis, E. T., A. Reymond, R. Lyle, N. Scamuffa, C. Ucla, S. Deutsch, B. J. Stevenson, V. Flegel, P. Bucher, C. V. Jongeneel and S. E. Antonarakis. 2002. Numerous potentially functional but non-genic conserved sequences on human chromosome 21. *Nature* 420: 578–582.
Dermitzakis, E. T., A. Reymond, N. Scamuffa, C. Ucla, E. Kirkness, C. Rossier and S. E. Antonarakis. 2003a. Evolutionary discrimination of mammalian conserved non-genic sequences (CNGs). *Science* 302: 1033–1035.
Dermitzakis, E. T., C. M. Bergman and A. G. Clark. 2003b. Tracing the evolutionary history of *Drosophila* regulatory regions with models that identify transcription factor binding sites. *Mol. Biol. Evol.* 20: 703–714.
Dermitzakis, E. T., E. Kirkness, S. Schwarz, E. Birney, A. Reymond and S. E. Antonarakis. 2004. Comparison of human chromosome 21 conserved nongenic sequences (CNGs) with the mouse and dog genomes shows that their selective constraint is independent of their genic environment. *Genome Res.* 14: 852–859.
Des Marais, D. J. 2000. When did photosynthesis emerge on Earth? *Science* 289: 1703–1705.
de Souza, F. S., V. F. Bumaschny, M. J. Low and M. Rubinstein. 2005. Subfunctionalization of expression and peptide domains following the ancient duplication of the proopiomelanocortin gene in teleost fishes. *Mol. Biol. Evol.* 22: 2417–2427.
de Souza, S. J., M. Long, R. J. Klein, S. Roy, S. Lin and W. Gilbert. 1998. Toward a resolution of the introns early/late debate: Only phase zero introns are correlated with the structure of ancient proteins. *Proc. Natl. Acad. Sci. USA* 95: 5094–5099.
Devos, K. M., J. K. Brown and J. L. Bennetzen. 2002. Genome size reduction through illegitimate recombination counteracts genome expansion in *Arabidopsis*. *Genome Res.* 12: 1075–1079.
Dibb, N. J. and A. J. Newman. 1989. Evidence that introns arose at proto-splice sites. *EMBO J.* 8: 2015–2021.
Dickinson, W. J. and J. Seger. 1999. Cause and effect in evolution. *Nature* 399: 30.
Dickson, L., H. R. Huang, L. Liu, M. Matsuura, A. M. Lambowitz and P. S. Perlman. 2001. Retrotransposition of a yeast group II intron occurs by reverse splicing directly into ectopic DNA sites. *Proc. Natl. Acad. Sci. USA* 98: 13207–13212.
Dietrich, F. S., S. Voegeli, S. Brachat, A. Lerch, K. Gates, S. Steiner, C. Mohr, R. Pohlmann, P. Luedi, S. Choi et al. 2004. The *Ashbya gossypii* genome as a tool for mapping the ancient *Saccharomyces cerevisiae* genome. *Science* 304: 304–307.
Dimitri, P., N. Junakovic and B. Arca. 2003. Colonization of heterochromatic genes by transposable elements in *Drosophila*. *Mol. Biol. Evol.* 20: 503–512.
Diniz-Filho, J. A. F. and N. M. Torres. 2002. Phylogenetic comparative methods and the geographic range size-body size relationship in New World terrestrial Carnivora. *Evol. Ecol.* 16: 351–367.
Djikeng, A., H. Shi, C. Tschudi and E. Ullu. 2001. RNA interference in *Trypanosoma brucei*: Cloning of small interfering RNAs provides evidence for retroposon-derived 24–26-nucleotide RNAs. *RNA* 7: 1522–1530.
Dolezal, P., O. Smid, P. Rada, Z. Zubacova, D. Bursac, R. Sutak, J. Nebesarova, T. Lithgow and J. Tachezy. 2005. *Giardia* mitosomes and trichomonad hydrogenosomes share a common mode of protein targeting. *Proc. Natl. Acad. Sci. USA* 102: 10924–10929.
Dolezal, P., V. Likic, J. Tachezy and T. Lithgow. 2006. Evolution of the molecular machines for protein import into mitochondria. *Science* 313: 314–318.
Dolgin, E. S. and B. Charlesworth. 2006. The fate of transposable elements in asexual populations. *Genetics* 174: 817–827.

D'Onofrio, G., D. Mouchiroud, B. Aissani, C. Gautier and G. Bernardi. 1991. Correlations between the compositional properties of human genes, codon usage, and amino acid composition of proteins. *J. Mol. Evol.* 32: 504–510.

D'Onofrio, G., K. Jabbari, H. Musto and G. Bernardi. 1999. The correlation of protein hydropathy with the base composition of coding sequences. *Gene* 238: 3–14.

Doolittle, W. F. 1978. Genes in pieces: Were they ever together? *Nature* 272: 581–582.

Doolittle, W. F. 1998. You are what you eat: A gene transfer ratchet could account for bacterial genes in eukaryotic nuclear genomes. *Trends Genet.* 14: 307–311.

Doolittle, W. F. and C. Sapienza. 1980. Selfish genes, the phenotype paradigm and genome evolution. *Nature* 284: 601–603.

Doolittle, W. F., Y. Boucher, C. L. Nesbo, C. J. Douady, J. O. Andersson and A. J. Roger. 2003. How big is the iceberg of which organellar genes in nuclear genomes are but the tip? *Phil. Trans. Roy. Soc. Lond. B Biol. Sci.* 358: 39–57.

Dorn, R., G. Reuter and A. Loewendorf. 2001. Transgene analysis proves mRNA *trans*-splicing at the complex *mod(mdg4)* locus in *Drosophila*. *Proc. Natl. Acad. Sci. USA* 98: 9724–9729.

Dorus, S., S. L. Gilbert, M. L. Forster, R. J. Barndt and B. T. Lahn. 2003. The CDY-related gene family: Coordinated evolution in copy number, expression profile and protein sequence. *Hum. Mol. Genet.* 12: 1643–1650.

Dorus, S., E. J. Vallender, P. D. Evans, J. R. Anderson, S. L. Gilbert, M. Mahowald, G. J. Wyckoff, C. M. Malcom, and B. T. Lahn. 2004. Accelerated evolution of nervous system genes in the origin of *Homo sapiens*. *Cell* 119: 1027–1040.

dos Reis, M., R. Savva and L. Wernisch. 2004. Solving the riddle of codon usage preferences: A test for translational selection. *Nucleic Acids Res.* 32: 5036–5044.

Douglas, S., S. Zauner, M. Fraunholz, M. Beaton, S. Penny, L. T. Deng, X. Wu, M. Reith, T. Cavalier-Smith and U. G. Maier. 2001. The highly reduced genome of an enslaved algal nucleus. *Nature* 410: 1091–1096.

Dover, G. 1982. Molecular drive: A cohesive mode of species evolution. *Nature* 299: 111–117.

Drake, J. W. 1991. A constant rate of spontaneous mutation in DNA-based microbes. *Proc. Natl. Acad. Sci. USA* 88: 7160–7164.

Drake, J. W., B. Charlesworth, D. Charlesworth and J. F. Crow. 1998. Rates of spontaneous mutation. *Genetics* 148: 1667–1686.

Drouin, G. 2006. Processed pseudogenes are more abundant in human and mouse X chromosomes than in autosomes. *Mol. Biol. Evol.* 23: 1652–1655.

Drummond, D. A., J. D. Bloom, C. Adami, C. O. Wilke and F. H. Arnold. 2005. Why highly expressed proteins evolve slowly. *Proc. Natl. Acad. Sci. USA* 102: 14338–14343.

Duan, J. and M. A. Antezana. 2003. Mammalian mutation pressure, synonymous codon choice, and mRNA degradation. *J. Mol. Evol.* 57: 694–701.

Duarte, J. M., L. Cui, P. K. Wall, Q. Zhang, X. Zhang, J. Leebens-Mack, H. Ma, N. Altman and C. W. dePamphilis. 2006. Expression pattern shifts following duplication indicative of subfunctionalization and neofunctionalization in regulatory genes of *Arabidopsis*. *Mol. Biol. Evol.* 23: 469–478.

Dubey, D. D., S. M. Kim, I. T. Todorov and J. A. Huberman. 1996. Large, complex modular structure of a fission yeast DNA replication origin. *Curr. Biol.* 6: 467–473.

Dulai, K. S., M. von Dornum, J. D. Mollon and D. M. Hunt. 1999. The evolution of trichromatic color vision by opsin gene duplication in New World and Old World primates. *Genome Res.* 9: 629–638.

Duncan, L., K. Bouckaert, F. Yeh and D. L. Kirk. 2002. *Kangaroo*, a mobile element from *Volvox carteri*, is a member of a newly recognized third class of retrotransposons. *Genetics* 162: 1617–1630.

Dupressoir, A., G. Marceau, C. Vernochet, L. Benit, C. Kanellopoulos, V. Sapin and T. Heidmann. 2005. Syncytin-A and syncytin-B, two fusogenic placenta-specific murine envelope genes of retroviral origin conserved in Muridae. *Proc. Natl. Acad. Sci. USA* 102: 725–730.

Duret, L. 2000. tRNA gene number and codon usage in the *C. elegans* genome are co-adapted for optimal translation of highly expressed genes. *Trends Genet.* 16: 287–289.

Duret, L. and D. Mouchiroud. 1999. Expression pattern and, surprisingly, gene length shape codon usage in *Caenorhabditis*, *Drosophila*, and *Arabidopsis*. *Proc. Natl. Acad. Sci. USA* 96: 4482–4487.

Duret, L. and D. Mouchiroud. 2000. Determinants of substitution rates in mammalian genes: Expression pattern affects selection intensity but not mutation rate. *Mol. Biol. Evol.* 17: 68–74.

Duret, L., D. Mouchiroud and C. Gautier. 1995. Statistical analysis of vertebrate sequences reveals that long genes are scarce in GC-rich isochores. *J. Mol. Evol.* 40: 308–317.

Duret, L., G. Marais and C. Biémont. 2000. Transposons but not retrotransposons are located preferentially in regions of high recombination rate in *Caenorhabditis elegans*. *Genetics* 156: 1661–1669.

Duret, L., M. Semon, G. Piganeau, D. Mouchiroud and N. Galtier. 2002. Vanishing GC-rich isochores in mammalian genomes. *Genetics* 162: 1837–1847.

Dvořák, J., M. C. Luo and Z. L. Yang. 1998. Restriction fragment length polymorphism and divergence in the genomic regions of high and low recombination in self-fertilizing and cross-fertilizing *Aegilops* species. *Genetics* 148: 423–434.

Dyall, S. D. and P. J. Johnson. 2000. Origins of hydrogenosomes and mitochondria: Evolution and organelle biogenesis. *Curr. Opin. Microbiol.* 3: 404–411.

Dyall, S. D., W. Yan, M. G. Delgadillo-Correa, A. Lunceford, J. A. Loo, C. F. Clarke and P. J. Johnson. 2004. Non-mitochondrial complex I proteins in a hydrogenosomal oxidoreductase complex. *Nature* 431: 1103–1107.

Dye, M. J. and N. J. Proudfoot. 1999. Terminal exon definition occurs cotranscriptionally and promotes termination of RNA polymerase II. *Mol. Cell* 3: 371–378.

Dye, M. J. and N. J. Proudfoot. 2001. Multiple transcript cleavage precedes polymerase release in termination by RNA polymerase II. *Cell* 105: 669–681.

Eanes, W. F., C. Wesley, J. Hey and D. Houle. 1988. The fitness consequences of *P* element insertions in *Drosophila melanogaster*. *Genet. Res.* 52: 17–26.

Earl, D. J. and M. W. Deem. 2004. Evolvability is a selectable trait. *Proc. Natl. Acad. Sci. USA* 101: 11531–11536.

Echols, N., P. Harrison, S. Balasubramanian, N. M. Luscombe, P. Bertone, Z. Zhang and M. Gerstein. 2002. Comprehensive analysis of amino acid and nucleotide composition in eukaryotic genomes, comparing genes and pseudogenes. *Nucleic Acids Res.* 30: 2515–2523.

Eddy, S. R. 2001. Non-coding RNA genes and the modern RNA world. *Nature Rev. Genet.* 2: 919–929.

Edelman, G. M., R. Meech, G. C. Owens and F. S. Jones. 2000. Synthetic promoter elements obtained by nucleotide sequence variation and selection for activity. *Proc. Natl. Acad. Sci. USA* 97: 3038–3043.

Eder, V., M. Ventura, M. Ianigro, M. Teti, M. Rocchi and N. Archidiacono. 2003. Chromosome 6 phylogeny in primates and centromere repositioning. *Mol. Biol. Evol.* 20: 1506–1512.

Edgell, D. R. and W. F. Doolittle. 1997. Archaea and the origin(s) of DNA replication proteins. *Cell* 89: 995–998.

Edwards, N. S. and A. W. Murray. 2005. Identification of *Xenopus* CENP-A and an associated centromeric DNA repeat. *Mol. Biol. Cell* 16: 1800–1810.

Eggleston, W. B., D. M. Johnson-Schlitz and W. R. Engels. 1988. *P-M* hybrid dysgenesis does not mobilize other transposable element families in *D. melanogaster*. *Nature* 331: 368–370.

Ehrlich, M. and R. Y. Wang. 1981. 5-Methylcytosine in eukaryotic DNA. *Science* 212: 1350–1357.

Eichinger, L., J. A. Pachebat, G. Glockner, M. A. Rajandream, R. Sucgang, M. Berriman, J. Song, R. Olsen, K. Szafranski, Q. Xu et al. 2005. The genome of the social amoeba *Dictyostelium discoideum*. *Nature* 435: 43–57.

Eichler, E. E., N. Archidiacono and M. Rocchi. 1999. CAGGG repeats and the pericentromeric duplication of the hominoid genome. *Genome Res.* 9: 1048–1058.

Eickbush, T. H. 1997. Telomerase and retrotransposons: Which came first? *Science* 277: 911–912.

Eickbush, T. H. 2002. *R2* and related site-specific non-long-terminal repeat retrotransposons. In N. L. Craig, R. Craigie, M. Gellert and A. M. Lambowitz, eds., *Mobile DNA II*, 813–835. ASM Press, Washington, DC.

Eickbush, T. H. and H. S. Malik. 2002. Origins and evolution of retrotransposons. In N. L. Craig, R. Craigie, M. Gellert and A. M. Lambowitz, eds., *Mobile DNA II*, 1111–1147. ASM Press, Washington, DC.

Eide, D. and P. Anderson. 1985. Transposition of *Tc1* in the nematode *Caenorhabditis elegans*. *Proc. Natl. Acad. Sci. USA* 82: 1756–1760.

Eigen, M. and P. Schuster. 1977. The hypercycle: A principle of natural self-organization. Part A: Emergence of the hypercycle. *Naturwissenschaften* 64: 541–565.

Eisen, J. A. and P. C. Hanawalt. 1999. A phylogenomic study of DNA repair genes, proteins, and processes. *Mutat. Res.* 435: 171–213.

Eisen, J. A., M. I. Benito and V. Walbot. 1994. Sequence similarity of putative transposases links the maize *Mutator* autonomous element and a group of bacterial insertion sequences. *Nucleic Acids Res.* 22: 2634–2636.

Elena, S. F. and R. E. Lenski. 1997. Test of synergistic interactions among deleterious mutations in bacteria. *Nature* 390: 395–398.

Ellegren, H. and A. K. Fridolfsson. 2003. Sex-specific mutation rates in salmonid fish. *J. Mol. Evol.* 56: 458–463.

Elson, J. L., R. M. Andrews, P. F. Chinnery, R. N. Lightowlers, D. M. Turnbull and N. Howell. 2001. Analysis of European mtDNAs for recombination. *Amer. J. Hum. Genet.* 68: 145–153.

Embley, T. M., M. van der Giezen, D. S. Horner, P. L. Dyal and P. Foster. 2003. Mitochondria and hydrogenosomes are two forms of the same fundamental organelle. *Phil. Trans. Roy. Soc. Lond.* B *Biol. Sci.* 358: 191–201.

Emerson, J. J., H. Kaessmann, E. Betrán and M. Long. 2004. Extensive gene traffic on the mammalian X chromosome. *Science* 303: 537–540.

Emmanuel, E., E. Yehuda, C. Melamed-Bessudo, N. Avivi-Ragolsky and A. A. Levy. 2006. The role of AtMSH2 in homologous recombination in *Arabidopsis thaliana*. *EMBO Rep.* 7: 100–105.

Enard, W., M. Przeworski, S. E. Fisher, C. S. Lai, V. Wiebe, T. Kitano, A. P. Monaco and S. Pääbo. 2002. Molecular evolution of *FOXP2*, a gene involved in speech and language. *Nature* 418: 869–872.

Endler, J. A. 1986. *Natural Selection in the Wild*. Princeton University Press, Princeton, NJ.

Engels, W. R. 1979. Extrachromosomal control of mutability in *Drosophila melanogaster*. *Proc. Natl. Acad. Sci. USA* 76: 4011–4015.

Engels, W. R. 1989. *P* elements in *Drosophila*. In M. M. Howe and D. E. Berg, eds., *Mobile DNA*, 437–484. ASM Press, Washington, DC.

Engels, W. R., D. M. Johnson-Schlitz, W. B. Eggleston and J. Sved. 1990. High-frequency *P* element loss in *Drosophila* is homolog dependent. *Cell* 62: 515–525.

Enquist, B. J. and K. J. Niklas. 2001. Invariant scaling relations across tree-dominated communities. *Nature* 410: 655–660.

Erlandsson, R., J. F. Wilson and S. Pääbo. 2000. Sex chromosomal transposable element accumulation and male-driven substitutional evolution in humans. *Mol. Biol. Evol.* 17: 804–812.

Ermolaeva, M. D., M. Wu, J. A. Eisen and S. L. Salzberg. 2003. The age of the *Arabidopsis thaliana* genome duplication. *Plant Mol. Biol.* 51: 859–866.

Escriva, H., L. Manzon, J. Youson and V. Laudet. 2002. Analysis of lamprey and hagfish genes reveals a complex history of gene duplications during early vertebrate evolution. *Mol. Biol. Evol.* 19: 1440–1450.

Eskesen, S. T., F. N. Eskesen and A. Ruvinsky. 2004. Natural selection affects frequencies of AG and GT dinucleotides at the 5′ and 3′ ends of exons. *Genetics* 167: 543–550.

Esnault, C., J. Maestre and T. Heidmann. 2000. Human LINE retrotransposons generate processed pseudogenes. *Nature Genet.* 24: 363–367.

Esnault, C., J. Millet, O. Schwartz and T. Heidmann. 2006. Dual inhibitory effects of APOBEC family proteins on retrotransposition of mammalian endogenous retroviruses. *Nucleic Acids Res.* 34: 1522–1531.

Espagne, E., C. Dupuy, E. Huguet, L. Cattolico, B. Provost, N. Martins, M. Poirie, G. Periquet and J. M. Drezen. 2004. Genome sequence of a polydnavirus: Insights into symbiotic virus evolution. *Science* 306: 286–289.

Espinosa, N., R. Hernandez, L. Lopez-Griego and I. Lopez-Villasenor. 2002. Separable putative polyadenylation and cleavage motifs in *Trichomonas vaginalis* mRNAs. *Gene* 289: 81–86.

Esser, C., N. Ahmadinejad, C. Wiegand, C. Rotte, F. Sebastiani, G. Gelius-Dietrich, K. Henze, E. Kretschmann, E. Richly, D. Leister et al. 2004. A genome phylogeny for mitochondria among α-proteobacteria and a predominantly eubacterial ancestry of yeast nuclear genes. *Mol. Biol. Evol.* 21: 1643–1660.

Evans, D., D. Zorio, M. MacMorris, C. E. Winter, K. Lea and T. Blumenthal. 1997. Operons and SL2 *trans*-splicing exist in nematodes outside the genus *Caenorhabditis*. *Proc. Natl. Acad. Sci. USA* 94: 9751–9756.

Evans, P. D., S. L. Gilbert, N. Mekel-Bobrov, E. J. Vallender, J. R. Anderson, L. M. Vaez-Azizi, S. A. Tishkoff, R. R. Hudson and B. T. Lahn. 2005. Microcephalin, a gene regulating brain size, continues to evolve adaptively in humans. *Science* 309: 1717–1720.

Evans, R. M., N. Fraser, E. Ziff, J. Weber, M. Wilson and J. E. Darnell. 1977. The initiation sites for RNA transcription in Ad2 DNA. *Cell* 12: 733–739.

Evgen'ev, M. B. and I. R. Arkhipova. 2005. *Penelope*-like elements—a new class of retroelements: Distribution, function and possible evolutionary significance. *Cytogenet. Genome Res.* 110: 510–521.

Eyre-Walker, A. 1996. The close proximity of *Escherichia coli* genes: Consequences for stop codon and synonymous codon use. *J. Mol. Evol.* 42: 73–78.

Eyre-Walker, A. and P. Awadalla. 2001. Does human mtDNA recombine? *J. Mol. Evol.* 53: 430–435.

Eyre-Walker, A. and M. Bulmer. 1993. Reduced synonymous substitution rate at the start of enterobacterial genes. *Nucleic Acids Res.* 21: 4599–4603.

Eyre-Walker, A. and M. Bulmer. 1995. Synonymous substitution rates in enterobacteria. *Genetics* 140: 1407–1412.

Eyre-Walker, A. and L. D. Hurst. 2001. The evolution of isochores. *Nature Rev. Genet.* 2: 549–555.

Fairbrother, W. G., R. F. Yeh, P. A. Sharp and C. B. Burge. 2002. Predictive identification of exonic splicing enhancers in human genes. *Science* 297: 1007–1013.

Faith, J. J. and D. D. Pollock. 2003. Likelihood analysis of asymmetrical mutation bias gradients in vertebrate mitochondrial genomes. *Genetics* 165: 735–745.

Fajkus, J., E. Sykorova and A. R. Leitch. 2005. Telomeres in evolution and evolution of telomeres. *Chromosome Res.* 13: 469–479.

Falconer, D. S. and T. F. C. Mackay. 1996. *Introduction to Quantitative Genetics*, 4th ed. Longman, New York.

Falconer, D. S., I. K. Gauld and R. C. Roberts. 1978. Cell numbers and cell sizes in organs of mice selected for large and small body size. *Genet. Res.* 31: 287–301.

Fan, J. and R. W. Lee. 2002. Mitochondrial genome of the colorless green alga *Polytomella parva*: Two linear DNA molecules with homologous inverted repeat termini. *Mol. Biol. Evol.* 19: 999–1007.

Farh, K. K., A. Grimson, C. Jan, B. P. Lewis, W. K. Johnston, L. P. Lim, C. B. Burge and D. P. Bartel. 2005. The widespread impact of mammalian microRNAs on mRNA repression and evolution. *Science* 310: 1817–1821.

Fast, N. M. and W. F. Doolittle. 1999. *Trichomonas vaginalis* possesses a gene encoding the essential spliceosomal component, PRP8. *Mol. Biochem. Parasitol.* 99: 275–278.

Fast, N. M., L. Xue, S. Bingham and P. J. Keeling. 2002. Re-examining alveolate evolution using multiple protein molecular phylogenies. *J. Eukaryot. Microbiol.* 49: 30–37.

Faugeron, G., L. Rhounim and J. L. Rossignol. 1990. How does the cell count the number of ectopic copies of a gene in the premeiotic inactivation process acting in *Ascobolus immersus*? *Genetics* 124: 585–591.

Fay, J. C., G. J. Wyckoff and C.-I. Wu. 2001. Positive and negative selection on the human genome. *Genetics* 158: 1227–1234.

Federici, D. and K. Downing. 2006. Evolution and development of a multicellular organism: Scalability, resilience, and neutral complexification. *Artif. Life* 12: 381–409.

Fedorov, A., X. Cao, S. Saxonov, S. J. de Souza, S. W. Roy and W. Gilbert. 2001. Intron distribution difference for 276 ancient and 131 modern genes suggests the existence of ancient introns. *Proc. Natl. Acad. Sci. USA* 98: 13177–13182.

Fedorov, A., S. Saxonov, L. Fedorova and I. Daizadeh. 2001. Comparison of intron-containing and intron-lacking human genes elucidates putative exonic splicing enhancers. *Nucleic Acids Res.* 29: 1464–1469.

Fedorov, A., A. F. Merican and W. Gilbert. 2002. Large-scale comparison of intron positions among animal, plant, and fungal genes. *Proc. Natl. Acad. Sci. USA* 99: 16128–16133.

Feiber, A. L., J. Rangarajan and J. C. Vaughn. 2002. The evolution of single-copy *Drosophila* nuclear *4f-rnp* genes: Spliceosomal intron losses create polymorphic alleles. *J. Mol. Evol.* 55: 401–413.

Feil, E. J., E. C. Holmes, D. E. Bessen, M. S. Chan, N. P. Day, M. C. Enright, R. Goldstein, D. W. Hood, A. Kalia, C. E. Moore et al. 2001. Recombination within natural populations of pathogenic bacteria:

Short-term empirical estimates and long-term phylogenetic consequences. *Proc. Natl. Acad. Sci. USA* 98: 182–187.
Feldman, M. W. and S. P. Otto. 1991. A comparative approach to the population-genetics theory of segregation distortion. *Amer. Natur.* 137: 443–456.
Felsenstein, J. 1974. The evolutionary advantage of recombination. *Genetics* 78: 737–756.
Felsenstein, J. 1975. Review of *The Genetic Basis of Evolutionary Change*. *Evolution* 29: 587–590.
Felsenstein, J. 1988. Phylogenies from molecular sequences: Inference and reliability. *Annu. Rev. Genet.* 22: 521–565.
Felsenstein, J. 2003. *Inferring Phylogenies*. Sinauer Associates, Sunderland, MA.
Feng, D. F., G. Cho and R. F. Doolittle. 1997. Determining divergence times with a protein clock: Update and reevaluation. *Proc. Natl. Acad. Sci. USA* 94: 13028–13033.
Ferris, P. J., C. Pavlovic, S. Fabry and U. W. Goodenough. 1997. Rapid evolution of sex-related genes in *Chlamydomonas*. *Proc. Natl. Acad. Sci. USA* 94: 8634–8639.
Ferris, P. J., E. V. Armbrust and U. W. Goodenough. 2002. Genetic structure of the mating-type locus of *Chlamydomonas reinhardtii*. *Genetics* 160: 181–200.
Ferris, S. D. and G. S. Whitt. 1977. Duplicate gene expression in diploid and tetraploid loaches (Cypriniformes, Cobitidae). *Biochem. Genet.* 15: 1097–1112.
Ferris, S. D. and G. S. Whitt. 1979. Evolution of the differential regulation of duplicate genes after polyploidization. *J. Mol. Evol.* 12: 267–317.
Ferris, S. D., S. L. Portnoy and G. S. Whitt. 1979. The roles of speciation and divergence time in the loss of duplicate gene expression. *Theor. Pop. Biol.* 15: 114–139.
Feschotte, C. 2004. *Merlin*, a new superfamily of DNA transposons identified in diverse animal genomes and related to bacterial IS1016 insertion sequences. *Mol. Biol. Evol.* 21: 1769–1780.
Feschotte, C., N. Jiang and S. R. Wessler. 2002. Plant transposable elements: Where genetics meets genomics. *Nature Rev. Genet.* 3: 329–341.
Feschotte, C., L. Swamy and S. R. Wessler. 2003. Genome-wide analysis of *mariner*-like transposable elements in rice reveals complex relationships with *stowaway* miniature inverted repeat transposable elements (*MITEs*). *Genetics* 163: 747–758.
Fiering, S., E. Whitelaw and D. I. Martin. 2000. To be or not to be active: The stochastic nature of enhancer action. *BioEssays* 22: 381–37.
Filatov, D. A. 2005. Substitution rates in a new *Silene latifolia* sex-linked gene, *SlssX/Y*. *Mol. Biol. Evol.* 22: 402–408.
Filatov, D. A. and D. T. Gerrard. 2003. High mutation rates in human and ape pseudoautosomal genes. *Gene* 317: 67–77.
Filipski, J., J. P. Thiery and G. Bernardi. 1973. An analysis of the bovine genome by Cs_2SO_4-Ag density gradient centrifugation. *J. Mol. Biol.* 80: 177–197.
Fingerman, E. G., P. G. Dombrowski, C. A. Francis and P. D. Sniegowski. 2003. Distribution and sequence analysis of a novel *Ty3*-like element in natural *Saccharomyces paradoxus* isolates. *Yeast* 20: 761–770.
Fink, G. R. 1986. Pseudogenes in yeast? *Cell* 49: 5–6.
Finlay, B. J. 2002. Global dispersal of free-living microbial eukaryote species. *Science* 296: 1061–1063.
Finlay, B. J., G. F. Esteban, K. J. Clarke and J. L. Olmo. 2001. Biodiversity of terrestrial protozoa appears homogeneous across local and global spatial scales. *Protist* 152: 355–366.
Finlay, B. J., E. B. Monaghan and S. C. Maberly. 2002. Hypothesis: The rate and scale of dispersal of freshwater diatom species is a function of their global abundance. *Protist* 153: 261–273.
Finnegan, E. J., B. H. Taylor, S. Craig and E. S. Dennis. 1989. Transposable elements can be used to study cell lineages in transgenic plants. *Plant Cell* 1: 757–764.
Finta, C. and P. G. Zaphiropoulos. 2002. Intergenic mRNA molecules resulting from *trans*-splicing. *J. Biol. Chem.* 277: 5882–5890.
Fischer, G., C. Neuveglise, P. Durrens, C. Gaillardin and B. Dujon. 2001. Evolution of gene order in the genomes of two related yeast species. *Genome Res.* 11: 2009–2019.
Fisher, R. A. 1918. The correlation between relatives on the supposition of Mendelian inheritance. *Trans. Royal Soc. Edinburgh* 52: 399–433.
Fisher, R. A. 1930. *The Genetical Theory of Natural Selection*. Clarendon Press, Oxford.
Fisher, R. A. 1935. The sheltering of lethals. *Amer. Natur.* 69: 446–455.
Floyd, S. K. and J. L. Bowman. 2004. Gene regulation: Ancient microRNA target sequences in plants. *Nature* 428: 485–486.
Fong, Y. W. and Q. Zhou. 2001. Stimulatory effect of splicing factors on transcriptional elongation. *Nature* 414: 929–933.
Fong, Y., L. Bender, W. Wang and S. Strome. 2002. Regulation of the different chromatin states of autosomes and X chromosomes in the germ line of *C. elegans*. *Science* 296: 2235–2238.
Force, A., M. Lynch, B. Pickett, A. Amores, Y.-L. Yan and J. Postlethwait. 1999. Preservation of duplicate genes by complementary, degenerative mutations. *Genetics* 151: 1531–1545.
Force, A., W. A. Cresko, F. B. Pickett, S. Proulx, C. Amemiya and M. Lynch. 2005. The origin of subfunctions and modular gene regulation. *Genetics* 170: 433–446.
Forget, L., J. Ustinova, Z. Wang, V. A. Huss and B. F. Lang. 2002. *Hyaloraphidium curvatum*: A linear mitochondrial genome, tRNA editing, and an evolutionary link to lower fungi. *Mol. Biol. Evol.* 19: 310–319.
Forterre, P. 2005. The two ages of the RNA world, and the transition to the DNA world: A story of viruses and cells. *Biochimie* 87: 793–803.
Forterre, P. 2006a. The origin of viruses and their possible roles in major evolutionary transitions. *Virus Res.* 117: 5–16.
Forterre, P. 2006b. Three RNA cells for ribosomal lineages and three DNA viruses to replicate their

genomes: A hypothesis for the origin of cellular domain. *Proc. Natl. Acad. Sci. USA* 103: 3669–3674.
Fortna, A., Y. Kim, E. MacLaren, K. Marshall, G. Hahn, L. Meltesen, M. Brenton, R. Hink, S. Burgers, T. Hernandez-Boussard et al. 2004. Lineage-specific gene duplication and loss in human and great ape evolution. *PLoS Biol.* 2: 937–954.
Foury, F., J. Hu and S. Vanderstraeten. 2004. Mitochondrial DNA mutators. *Cell. Mol. Life Sci.* 61: 2799–2811.
Frame, I. G., J. F. Cutfield and R. T. Poulter. 2001. New BEL-like LTR-retrotransposons in *Fugu rubripes, Caenorhabditis elegans,* and *Drosophila melanogaster*. *Gene* 263: 219–230.
Franchini, L. F., E. W. Ganko and J. F. McDonald. 2004. Retrotransposon-gene associations are widespread among *D. melanogaster* populations. *Mol. Biol. Evol.* 21: 1323–1331.
Francino, M. P. 2005. An adaptive radiation model for the origin of new gene functions. *Nature Genet.* 37: 573–577.
Francino, M. P. and H. Ochman. 1997. Strand asymmetries in DNA evolution. *Trends Genet.* 13: 240–245.
Francis, D., N. D. Davies, J. A. Bryant, S. G. Hughes, D. R. Sibson and P. N. Fitchett. 1985. Effects of psoralen on replicon size and mean rate of DNA synthesis in partially synchronized cells of *Pisum sativum* L. *Exp. Cell Res.* 158: 500–508.
Francis, L. 2004. Microscaling: Why larger anemones have longer cnidae. *Biol. Bull.* 207: 116–129.
Frank, A. C. and J. R. Lobry. 1999. Asymmetric substitution patterns: A review of possible underlying mutational or selective mechanisms. *Gene* 238: 65–77.
Frank, A. C. and J. R. Lobry. 2000. Oriloc: Prediction of replication boundaries in unannotated bacterial chromosomes. *Bioinformatics* 16: 560–561.
Frankham, R. 1995. Effective population size / adult population size ratios in wildlife: A review. *Genet. Res.* 66: 95–107.
Frantz, C., C. Ebel, F. Paulus and P. Imbault. 2000. Characterization of *trans*-splicing in euglenoids. *Curr. Genet.* 37: 349–355.
Fraser, C. M., S. Casjens, W. M. Huang, G. G. Sutton, R. Clayton, R. Lathigra, O. White, K. A. Ketchum, R. Dodson, E. K. Hickey et al. 1997. Genomic sequence of a Lyme disease spirochaete, *Borrelia burgdorferi*. *Nature* 390: 580–586.
Fraser, J. A. and J. Heitman. 2003. Fungal mating-type loci. *Curr. Biol.* 13: R792–R795.
Fraser, J. A. and J. Heitman. 2004. Evolution of fungal sex chromosomes. *Mol. Microbiol.* 51: 299–306.
French, S. 1992. Consequences of replication fork movement through transcription units in vivo. *Science* 258: 1362–1365.
Freyer, R., M. C. Kiefer-Meyer and H. Kossel. 1997. Occurrence of plastid RNA editing in all major lineages of land plants. *Proc. Natl. Acad. Sci. USA* 94: 6285–6290.
Frilander, M. J. and J. A. Steitz. 1999. Initial recognition of U12-dependent introns requires both U11/5′ splice-site and U12/branchpoint interactions. *Genes Dev.* 13: 851–863.
Frischmeyer, P. A. and H. C. Dietz. 1999. Nonsense-mediated mRNA decay in health and disease. *Hum. Mol. Genet.* 8: 1893–1900.
Frisse, L., R. R. Hudson, A. Bartoszewicz, J. D. Wall, J. Donfack and A. Di Rienzo. 2001. Gene conversion and different population histories may explain the contrast between polymorphism and linkage disequilibrium levels. *Amer. J. Hum. Genet.* 69: 831–843.
Frith, M. C., J. Ponjavic, D. Fredman, C. Kai, J. Kawai, P. Carninci, Y. Hayashizaki and A. Sandelin. 2006. Evolutionary turnover of mammalian transcription start sites. *Genome Res.* 16: 713–722.
Frugoli, J. A., M. A. McPeek, T. L. Thomas and C. R. McClung. 1998. Intron loss and gain during evolution of the catalase gene family in angiosperms. *Genetics* 149: 355–365.
Fry, A. J. 1999. Mildly deleterious mutations in avian mitochondrial DNA: Evidence from neutrality tests. *Evolution* 53: 1617–1620.
Fry, A. J. and J. J. Wernegreen. 2005. The roles of positive and negative selection in the molecular evolution of insect endosymbionts. *Gene* 355: 1–10.
Fry, J. D. 2004. On the rate and linearity of viability declines in *Drosophila* mutation-accumulation experiments: Genomic mutation rates and synergistic epistasis revisited. *Genetics* 166: 797–806.
Fryxell, K. J. and W. J. Moon. 2005. CpG mutation rates in the human genome are highly dependent on local GC content. *Mol. Biol. Evol.* 22: 650–658.
Fryxell, K. J. and E. Zuckerkandl. 2000. Cytosine deamination plays a primary role in the evolution of mammalian isochores. *Mol. Biol. Evol.* 17: 1371–1383.
Fuerst, J. A. 2005. Intracellular compartmentation in planctomycetes. *Annu. Rev. Microbiol.* 59: 299–328.
Fukue, Y., N. Sumida, J. Tanase and T. Ohyama. 2005. A highly distinctive mechanical property found in the majority of human promoters and its transcriptional relevance. *Nucleic Acids Res.* 33: 3821–3827.
Fukuhara, H., F. Sor, R. Drissi, N. Dinouel, I. Miyakawa, S. Rousset and A. M. Viola. 1993. Linear mitochondrial DNAs of yeasts: Frequency of occurrence and general features. *Mol. Cell. Biol.* 13: 2309–2314.
Fullerton, S. M., A. B. Carvalho and A. G. Clark. 2001. Local rates of recombination are positively correlated with GC content in the human genome. *Mol. Biol. Evol.* 18: 1139–1142.
Funes, S., E. Davidson, A. Reyes-Prieto, S. Magallon, P. Herion, M. P. King and D. Gonzalez-Halphen. 2002. A green algal apicoplast ancestor. *Science* 298: 2155.
Funk, D. J., J. J. Wernegreen and N. A. Moran. 2002. Intraspecific variation in symbiont genomes: Bottlenecks and the aphid-*Buchnera* association. *Genetics* 157: 477–489.
Furlong, R. F. and P. W. Holland. 2002. Were vertebrates octoploid? *Phil. Trans. Roy. Soc. Lond.* B *Biol. Sci.* 357: 531–544.
Furnes, H., N. R. Banerjee, K. Muehlenbachs, H. Staudigel and M. de Wit. 2004. Early life recorded in archean pillow lavas. *Science* 304: 578–581.

Furuyama, S. and J. P. Bruzik. 2002. Multiple roles for SR proteins in trans splicing. *Mol. Cell. Biol.* 22: 5337–5346.

Gabaldón, T. and M. A. Huynen. 2003. Reconstruction of the proto-mitochondrial metabolism. *Science* 301: 609.

Gabaldón, T. and M. A. Huynen. 2004. Shaping the mitochondrial proteome. *Biochim. Biophys. Acta* 1659: 212–220.

Gabriel, K., S. K. Buchanan and T. Lithgow. 2001. The alpha and the beta: Protein translocation across mitochondrial and plastid outer membranes. *Trends Biochem. Sci.* 26: 36–40.

Gabriel, S. B., S. F. Schaffner, H. Nguyen, J. M. Moore, J. Roy, B. Blumenstiel, J. Higgins, M. DeFelice, A. Lochner, M. Faggart et al. 2002. The structure of haplotype blocks in the human genome. *Science* 296: 2225–2229.

Gabriel, W., M. Lynch and R. Bürger. 1994. Muller's ratchet and mutational meltdowns. *Evolution* 47: 1744–1757.

Galagan, J. E., C. Nusbaum, A. Roy, M. G. Endrizzi, P. Macdonald, W. FitzHugh, S. Calvo, R. Engels, S. Smirnov, D. Atnoor et al. 2002. The genome of *M. acetivorans* reveals extensive metabolic and physiological diversity. *Genome Res.* 12: 532–542.

Galagan, J. E., S. E. Calvo, C. Cuomo, L. J. Ma, J. R. Wortman, S. Batzoglou, S. I. Lee, M. Basturkmen, C. C. Spevak, J. Clutterbuck et al. 2005. Sequencing of *Aspergillus nidulans* and comparative analysis with *A. fumigatus* and *A. oryzae*. *Nature* 438: 1105–1115.

Galli, M., A. Theriault, D. Liu and N. M. Crawford. 2003. Expression of the *Arabidopsis* transposable element *Tag1* is targeted to developing gametophytes. *Genetics* 165: 2093–2105.

Galtier, N. and J. R. Lobry. 1997. Relationships between genomic G+C content, RNA secondary structures, and optimal growth temperature in prokaryotes. *J. Mol. Evol.* 44: 632–636.

Galtier, N., G. Piganeau, D. Mouchiroud and L. Duret. 2001. GC-content evolution in mammalian genomes: The biased gene conversion hypothesis. *Genetics* 159: 907–911.

Galtier, N., E. Bazin and N. Bierne. 2006. GC-biased segregation of noncoding polymorphisms in *Drosophila*. *Genetics* 172: 221–228.

Ganfornina, M. D. and D. Sanchez. 1999. Generation of evolutionary novelty by functional shift. *BioEssays* 21: 432–439.

Ganot, P., T. Kallesoe, R. Reinhardt, D. Chourrout and E. M. Thompson. 2004. Spliced-leader RNA trans splicing in a chordate, *Oikopleura dioica*, with a compact genome. *Mol. Cell. Biol.* 24: 7795–7805.

Gao, L. Z. and H. Innan. 2004. Very low gene duplication rate in the yeast genome. *Science* 306: 1367–1370.

García-Dorado, A., A. Caballero and J. F. Crow. 2003. On the persistence and pervasiveness of a new mutation. *Evolution* 57: 2644–2646.

Garfinkel, D. J., K. Nyswaner, J. Wang and J. Y. Cho. 2003. Post-transcriptional cosuppression of *Ty1* retrotransposition. *Genetics* 165: 83–99.

Garfinkel, D. J., K. M. Nyswaner, K. M. Stefanisko, C. Chang and S. P. Moore. 2005. *Ty1* copy number dynamics in *Saccharomyces*. *Genetics* 169: 1845–1857.

Garnier, O., V. Serrano and S. Duharcourt and E. Meyer. 2004. RNA-mediated programming of developmental genome rearrangements in *Paramecium tetraurelia*. *Mol. Cell. Biol.* 24: 7370–7379.

Gasch, A. P., A. M. Moses, D. Y. Chiang, H. B. Fraser, M. Berardini and M. B. Eisen. 2004. Conservation and evolution of *cis*-regulatory systems in ascomycete fungi. *PLoS Biol.* 2: 2202–2219.

Gaston, K. J. and T. M. Blackburn. 1996. Range size-body size relationships: Evidence of scale dependence. *Oikos* 75: 479–485.

Gatfield, D., L. Unterholzner, F. D. Ciccarelli, P. Bork and E. Izaurralde. 2003. Nonsense-mediated mRNA decay in *Drosophila*: At the intersection of the yeast and mammalian pathways. *EMBO J.* 22: 3960–3970.

Gaut, B. S. and J. F. Doebley. 1997. DNA sequence evidence for the segmental allotetraploid origin of maize. *Proc. Natl. Acad. Sci. USA* 94: 6809–6814.

Gaut, B. S., B. R. Morton, B. C. McCaig and M. T. Clegg. 1996. Substitution rate comparisons between grasses and palms: Synonymous rate differences at the nuclear gene *Adh* parallel rate differences at the plastid gene *rbcL*. *Proc. Natl. Acad. Sci. USA* 93: 10274–10279.

Gebauer, F. and M. W. Hentze. 2004. Molecular mechanisms of translational control. *Nature Rev. Mol. Cell. Biol.* 5: 827–835.

Gelbart, W. M. and A. Chovnick. 1979. Spontaneous unequal exchange in the rosy region of *Drosophila melanogaster*. *Genetics* 92: 849–859.

Gemmell, N. J., V. J. Metcalf and F. W. Allendorf. 2004. Mother's curse: The effect of mtDNA on individual fitness and population viability. *Trends Ecol. Evol.* 19: 238–244.

Gerber, A. P. and W. Keller. 2001. RNA editing by base deamination: More enzymes, more targets, new mysteries. *Trends Biochem. Sci.* 26: 376–384.

Gerhart, J. and M. Kirschner. 1997. *Cells, Embryos and Evolution*. Blackwell Science, Malden, MA.

Gerrard, D. T. and D. A. Filatov. 2005. Positive and negative selection on mammalian Y chromosomes. *Mol. Biol. Evol.* 22: 1423–1432.

Gesteland, R. F., T. R. Cech and J. F. Atkins. 1999. *The RNA World*. 2nd ed. Cold Spring Harbor Laboratory Press, Cold Spring Harbor, NY.

Ghaemmaghami, S., W. K. Huh, K. Bower, R. W. Howson, A. Belle, N. Dephoure, E. K. O'Shea and J. S. Weissman. 2003. Global analysis of protein expression in yeast. *Nature* 425: 737–741.

Ghosh, S., J. W. Jaraczewski, L. A. Klobutcher and C. L. Jahn. 1994. Characterization of transcription initiation, translation initiation, and poly(A) addition sites in the gene-sized macronuclear DNA molecules of *Euplotes*. *Nucleic Acids Res.* 22: 214–221.

Gibert, J. M., E. Mouchel-Vielh, E. Queinnec and J. S. Deutsch. 2000. Barnacle duplicate *engrailed* genes: Divergent expression patterns and evidence for a vestigial abdomen. *Evol. Dev.* 2: 194–202.

Gibson, W., L. Bingle, W. Blendeman, J. Brown, J. Wood and J. Stevens. 2000. Structure and sequence varia-

tion of the trypanosome spliced leader transcript. *Mol. Biochem. Parasitol.* 107: 269–277.
Giegé, P. and A. Brennicke. 1999. RNA editing in *Arabidopsis* mitochondria effects 441 C to U changes in ORFs. *Proc. Natl. Acad. Sci. USA* 96: 15324–15329.
Gierlik, A., M. Kowalczuk, P. Mackiewicz, M. R. Dudek and S. Cebrat. 2000. Is there replication-associated mutational pressure in the *Saccharomyces cerevisiae* genome? *J. Theor. Biol.* 202: 305–314.
Gil, R., F. J. Silva, E. Zientz, F. Delmotte, F. Gonzalez-Candelas, A. Latorre, C. Rausell, J. Kamerbeek, J. Gadau, B. Hölldobler et al. 2003. The genome sequence of *Blochmannia floridanus*: Comparative analysis of reduced genomes. *Proc. Natl. Acad. Sci. USA* 100: 9388–9393.
Gilad, Y., O. Man, S. Pääbo and D. Lancet. 2003. Human specific loss of olfactory receptor genes. *Proc. Natl. Acad. Sci. USA* 100: 3324–3327.
Gilad, Y., A. Oshlack, G. K. Smyth, T. P. Speed and K. P. White. 2006. Expression profiling in primates reveals a rapid evolution of human transcription factors. *Nature* 440: 242–245.
Gilbert, N., S. Lutz-Prigge and J. V. Moran. 2002. Genomic deletions created upon LINE-1 retrotransposition. *Cell* 110: 315–325.
Gilbert, W. 1978. Why genes in pieces? *Nature* 271: 501.
Gilbert, W. 1986. The RNA world. *Nature* 319: 618.
Gilbert, W. 1987. The exon theory of genes. *Cold Spring Harbor Symp. Quant. Biol.* 52: 901–905.
Gillespie, J. H. 2000. Genetic drift in an infinite population: The pseudohitchhiking model. *Genetics* 155: 909–919.
Gillham, N. W. 1994. *Organelle Genes and Genomes*. Oxford University Press, New York.
Gillooly, J. F., A. P. Allen, G. B. West and J. H. Brown. 2005. The rate of DNA evolution: Effects of body size and temperature on the molecular clock. *Proc. Natl. Acad. Sci. USA* 102: 140–145.
Gilson, P. R. and G. I. McFadden. 1996. The miniaturized nuclear genome of eukaryotic endosymbiont contains genes that overlap, genes that are cotranscribed, and the smallest known spliceosomal introns. *Proc. Natl. Acad. Sci. USA* 93: 7737–7742.
Gilson, P. R., V. Su, C. H. Slamovits, M. E. Reith, P. J. Keeling and G. I. McFadden. 2006. Complete nucleotide sequence of the chlorarachniophyte nucleomorph: Nature's smallest nucleus. *Proc. Natl. Acad. Sci. USA* 103: 9566–9571.
Gindullis, F., C. Desel, I. Galasso and T. Schmidt. 2001. The large-scale organization of the centromeric region in *Beta* species. *Genome Res.* 11: 253–265.
Giovannoni, S. J., H. J. Tripp, S. Givan, M. Podar, K. L. Vergin, D. Baptista, L. Bibbs, J. Eads, T. H. Richardson, M. Noordewier et al. 2005. Genome streamlining in a cosmopolitan oceanic bacterium. *Science* 309: 1242–1245.
Giroux, M. J., M. Clancy, J. Baier, L. Ingham, D. McCarty and L. C. Hannah. 1994. *De novo* synthesis of an intron by the maize transposable element Dissociation. *Proc. Natl. Acad. Sci. USA* 91: 12150–12154.
Gloor, G. B. 2002. The role of sequence homology in the repair of DNA double-strand breaks in *Drosophila*. *Adv. Genet.* 46: 91–117.
Gloor, G. B., N. A. Nassif, D. M. Johnson-Schlitz, C. R. Preston and W. R. Engels. 1991. Targeted gene replacement in *Drosophila* via *P* element-induced gap repair. *Science* 253: 1110–1117.
Goetting-Minesky, M. P. and K. D. Makova. 2006. Mammalian male mutation bias: Impacts of generation time and regional variation in substitution rates. *J. Mol. Evol.* 63: 537–544.
Gogarten, J. P., H. Kibak, P. Dittrich, L. Taiz, E. J. Bowman, B. J. Bowman, M. F. Manolson, R. J. Poole, T. Date, T. Oshima et al. 1989. Evolution of the vacuolar H+-ATPase: Implications for the origin of eukaryotes. *Proc. Natl. Acad. Sci. USA* 86: 6661–6665.
Gogarten, J. P., W. F. Doolittle and J. G. Lawrence. 2002. Prokaryotic evolution in light of gene transfer. *Mol. Biol. Evol.* 19: 2226–2238.
Goldberg, S., H. Schwartz and J. E. Darnell, Jr. 1977. Evidence from UV transcription mapping in HeLa cells that heterogeneous nuclear RNA is the messenger RNA precursor. *Proc. Natl. Acad. Sci. USA* 74: 4520–4523.
Golding, G. B. and R. S. Gupta. 1995. Protein-based phylogenies support a chimeric origin for the eukaryotic genome. *Mol. Biol. Evol.* 12: 1–6.
Golding, G. B., N. Tsao and R. E. Pearlman. 1994. Evidence for intron capture: An unusual path for the evolution of proteins. *Proc. Natl. Acad. Sci. USA* 91: 7506–7509.
Goldman, B. S., W. C. Nierman, D. Kaiser, S. C. Slater, A. S. Durkin, J. Eisen, C. M. Ronning, W. B. Barbazuk, M. Blanchard, C. Field et al. 2006. Evolution of sensory complexity recorded in a myxobacterial genome. *Proc. Natl. Acad. Sci. USA* 103: 15200–15205.
Gonzalez, C. I., A. Bhattacharya, W. Wang and S. W. Peltz. 2001. Nonsense-mediated mRNA decay in *Saccharomyces cerevisiae*. *Gene* 274: 15–25.
González, J., J. M. Ranz and A. Ruiz. 2002. Chromosomal elements evolve at different rates in the *Drosophila* genome. *Genetics* 161: 1137–1154.
Goodner, B., G. Hinkle, S. Gattung, N. Miller, M. Blanchard, B. Qurollo, B. S. Goldman, Y. Cao, M. Askenazi, C. Halling et al. 2001. Genome sequence of the plant pathogen and biotechnology agent *Agrobacterium tumefaciens* C58. *Science* 294: 2323–2328.
Goodnight, C. J. and M. J. Wade. 2000. The ongoing synthesis: A reply to Coyne, Barton, and Turelli. *Evolution* 54: 317–324.
Goodwin, T. J. and R. T. Poulter. 2000. Multiple LTR-retrotransposon families in the asexual yeast *Candida albicans*. *Genome Res.* 10: 174–191.
Goodwin, T. J. and R. T. Poulter. 2001. The *DIRS1* group of retrotransposons. *Mol. Biol. Evol.* 18: 2067–2082.
Goodwin, T. J. and R. T. Poulter. 2004. A new group of tyrosine recombinase-encoding retrotransposons. *Mol. Biol. Evol.* 21: 746–759.
Gorbunova, V. and A. A. Levy. 1997. Non-homologous DNA end joining in plant cells is associated with deletions and filler DNA insertions. *Nucleic Acids Res.* 25: 4650–4657.
Görnemann, J., K. M. Kotovic, K. Hujer and K. M. Neugebauer. 2005. Cotranscriptional spliceosome

assembly occurs in a stepwise fashion and requires the cap binding complex. *Mol. Cell* 19: 53–63.
Gottesman, S. 2005. Micros for microbes: Non-coding regulatory RNAs in bacteria. *Trends Genet.* 21: 399–404.
Gould, S. J. and R. C. Lewontin. 1979. The spandrels of San Marco and the Panglossian paradigm: A critique of the adaptationist programme. *Proc. Roy. Soc. Lond.* B *Biol. Sci.* 205: 581–598.
Gourse, R. L., W. Ross and T. Gaal. 2000. UPs and downs in bacterial transcription initiation: The role of the alpha subunit of RNA polymerase in promoter recognition. *Mol. Microbiol.* 37: 687–695.
Gouy, M. and C. Gautier. 1982. Codon usage in bacteria: Correlation with gene expressivity. *Nucleic Acids Res.* 10: 7055–7074.
Goyon, C., C. Barry, A. Gregoire, G. Faugeron and J. L. Rossignol. 1996. Methylation of DNA repeats of decreasing sizes in *Ascobolus immersus*. *Mol. Cell. Biol.* 16: 3054–3065.
Graber, J. H., C. R. Cantor, S. C. Mohr and T. F. Smith. 1999. *In silico* detection of control signals: mRNA 3′-end-processing sequences in diverse species. *Proc. Natl. Acad. Sci. USA* 96: 14055–14060.
Grandbastien, M. A. 1998. Activation of plant retrotransposons under stress conditions. *Trends Plant Sci.* 3: 181–187.
Grandbastien, M. A., H. Lucas, J. B. Morel, C. Mhiri, S. Vernhettes and J. M. Casacuberta. 1997. The expression of the tobacco *Tnt1* retrotransposon is linked to plant defense responses. *Genetica* 100: 241–252.
Grandbastien, M. A., C. Audeon, E. Bonnivard, J. M. Casacuberta, B. Chalhoub, A. P. Costa, Q. H. Le, D. Melayah, M. Petit, C. Poncet et al. 2005. Stress activation and genomic impact of *Tnt1* retrotransposons in Solanaceae. *Cytogenet. Genome Res.* 110: 229–241.
Graur, D. and W.-H. Li. 2000. *Fundamentals of Molecular Evolution*. Sinauer Associates, Sunderland, MA.
Graur, D. and W. Martin. 2004. Reading the entrails of chickens: Molecular timescales of evolution and the illusion of precision. *Trends Genet.* 20: 80–86.
Graveley, B. R. 2000. Sorting out the complexity of SR protein functions. *RNA* 6: 1197–1211.
Graveley, B. R. 2001. Alternative splicing: Increasing diversity in the proteomic world. *Trends Genet.* 17: 100–107.
Graves, J. A. 1995. The origin and function of the mammalian Y chromosome and Y-borne genes—an evolving understanding. *BioEssays* 17: 311–320.
Gray, M. W., B. F. Lang, R. Cedergren, G. B. Golding, C. Lemieux, D. Sankoff, M. Turmel, N. Brossard, E. Delage, T. G. Littlejohn et al. 1998. Genome structure and gene content in protist mitochondrial DNAs. *Nucleic Acids Res.* 26: 865–878.
Gray, M. W., G. Burger and F. Lang. 1999. Mitochondrial evolution. *Science* 283: 1476–1481.
Green, J. L., A. J. Holmes, M. Westoby, I. Oliver, D. Briscoe, M. Dangerfield, M. Gillings and A. J. Beattie. 2004. Spatial scaling of microbial eukaryote diversity. *Nature* 432: 747–750.
Green, P., B. Ewing, W. Miller, P. J. Thomas, E. D. Green and NISC Comparative Sequencing Program. 2003. Transcription-associated mutational asymmetry in mammalian evolution. *Nature Genet.* 33: 514–517.
Greenberg, A. J., J. R. Moran, S. Fang and C.-I. Wu. 2006. Adaptive loss of an old duplicated gene during incipient speciation. *Mol. Biol. Evol.* 23: 401–410.
Greenwood, T. A., B. K. Rana and N. J. Schork. 2004. Human haplotype block sizes are negatively correlated with recombination rates. *Genome Res.* 14: 1358–1361.
Gregory, T. R. 2001. The bigger the C-value, the larger the cell: Genome size and red blood cell size in vertebrates. *Blood Cells Mol. Dis.* 27: 830–843.
Gregory, T. R. 2002a. Genome size and developmental parameters in the homeothermic vertebrates. Genome 45: 833–838.
Gregory, T. R. 2002b. A bird's-eye view of the C-value enigma: Genome size, cell size, and metabolic rate in the class aves. *Evolution* 56: 121–130.
Gregory, T. R. (ed.). 2005a. *The Evolution of the Genome*. Elsevier Academic Press, Boston, MA.
Gregory, T. R. 2005b. Synergy between sequence and size in large-scale genomics. *Nature Rev. Genet.* 6: 699–708.
Gregory, T. R. 2005c. Macroevolution and the genome. In T. R. Gregory, ed., *The Evolution of the Genome*, 679–729. Elsevier Academic Press, Boston, MA.
Gregory, T. R. and R. DeSalle. 2005. Comparative genomics in prokaryotes. In T. R. Gregory, ed., *The Evolution of the Genome*, 585–675. Elsevier Academic Press, Boston, MA.
Gribaldo, S. and P. Cammarano. 1998. The root of the universal tree of life inferred from anciently duplicated genes encoding components of the protein-targeting machinery. *J. Mol. Evol.* 47: 508–516.
Grigoriev, A. 1998. Analyzing genomes with cumulative skew diagrams. *Nucleic Acids Res.* 26: 2286–2290.
Grimson, M. J., J. C. Coates, J. P. Reynolds, M. Shipman, R. L. Blanton, A. J. Harwood. 2000. Adherens junctions and β-catenin-mediated cell signaling in a non-metazoan organism. *Nature* 408: 727–731.
Grollman, A. P. and M. Moriya. 1993. Mutagenesis by 8-oxoguanine: An enemy within. *Trends Genet.* 9: 246–249.
Grosjean, H. and W. Fiers. 1982. Preferential codon usage in prokaryotic genes: The optimal codon-anticodon interaction energy and the selective codon usage in efficiently expressed genes. *Gene* 18: 199–209.
Grutzner, F., W. Rens, E. Tsend-Ayush, N. El-Mogharbel, P. C. O'Brien, R. C. Jones, M. A. Ferguson-Smith and J. A. Marshall Graves. 2004. In the platypus a meiotic chain of ten sex chromosomes shares genes with the bird Z and mammal X chromosomes. *Nature* 432: 913–917.
Gu, X. 1998. Early metazoan divergence was about 830 million years ago. *J. Mol. Evol.* 47: 369–371.
Gu, X., D. Hewett-Emmett and W.-H. Li. 1998. Directional mutational pressure affects the amino acid composition and hydrophobicity of proteins in bacteria. *Genetica* 102/103: 383–391.
Gu, X., Y. Wang and J. Gu. 2002a. Age distribution of human gene families shows significant roles of

both large- and small-scale duplications in vertebrate evolution. *Nature Genet.* 31: 205–209.

Gu, Z., D. Nicolae, H. H. Lu and W.-H. Li. 2002b. Rapid divergence in expression between duplicate genes inferred from microarray data. *Trends Genet.* 18: 609–613.

Gu, Z., L. M. Steinmetz, X. Gu, C. Scharfe, R. W. Davis and W.-H. Li. 2003. Role of duplicate genes in genetic robustness against null mutations. *Nature* 421: 63–66.

Gunasekera, A. M., S. Patankar, J. Schug, G. Eisen, J. Kissinger, D. Roos and D. F. Wirth. 2004. Widespread distribution of antisense transcripts in the *Plasmodium falciparum* genome. *Mol. Biochem. Parasitol.* 136: 35–42.

Gupta, R. S. 1998a. Life's third domain (Archaea): An established fact or an endangered paradigm? *Theor. Pop. Biol.* 54: 91–104.

Gupta, R. S. 1998b. Protein phylogenies and signature sequences: A reappraisal of evolutionary relationships among archaebacteria, eubacteria, and eukaryotes. *Microbiol. Mol. Biol. Rev.* 62: 1435–1491.

Gur, Y. and H. Breitbart. 2006. Mammalian sperm translate nuclear-encoded proteins by mitochondrial-type ribosomes. *Genes Dev.* 20: 411–416.

Haag, E. S. and A. V. Doty. 2005. Sex determination across evolution: Connecting the dots. *PLoS Biol.* 3: 21–24.

Haag, E. S. and M. N. Molla. 2005. Compensatory evolution of interacting gene products through multifunctional intermediates. *Evolution* 59: 1620–1632.

Haber, J. E. 2000. Lucky breaks: Analysis of recombination in *Saccharomyces*. *Mutat. Res.* 451: 53–69.

Hackett, J. D., H. S. Yoon, M. B. Soares, M. F. Bonaldo, T. L. Casavant, T. E. Scheetz, T. Nosenko and D. Bhattacharya. 2004. Migration of the plastid genome to the nucleus in a peridinin dinoflagellate. *Curr. Biol.* 14: 213–218.

Hahn, M. W. and G. C. Lanzaro. 2005. Female-biased gene expression in the malaria mosquito *Anopheles gambiae*. *Curr. Biol.* 15: R192–R193.

Hahn, S., E. T. Hoar and L. Guarente. 1985. Each of three "TATA elements" specifies a subset of the transcription initiation sites at the CYC-1 promoter of *Saccharomyces cerevisiae*. *Proc. Natl. Acad. Sci. USA* 82: 8562–8566.

Hahn, M. W., J. E. Stajich and G. A. Wray. 2003. The effects of selection against spurious transcription factor binding sites. *Mol. Biol. Evol.* 20: 901–906.

Hahn, M. W., T. De Bie, J. E. Stajich, C. Nguyen and N. Cristianini. 2005. Estimating the tempo and mode of gene family evolution from comparative genomic data. *Genome Res.* 15: 1153–1160.

Haigh, J. 1978. The accumulation of deleterious genes in a population—Muller's ratchet. *Theor. Pop. Biol.* 14: 251–267.

Hails, T., O. Huttner and A. Day. 1995. Isolation of a *Chlamydomonas reinhardtii* telomere by functional complementation in yeast. *Curr. Genet.* 28: 437–440.

Hajra, S., S. K. Ghosh and M. Jayaram. 2006. The centromere-specific histone variant Cse4p (CENP-A) is essential for functional chromatin architecture at the yeast 2-μm circle partitioning locus and promotes equal plasmid segregation. *J. Cell Biol.* 174: 779–790.

Haldane, J. B. S. 1932. *The Causes of Evolution*. Longmans, Green, and Co., New York.

Haldane, J. B. S. 1933. The part played by recurrent mutation in evolution. *Amer. Natur.* 67: 5–9.

Haldane, J. B. S. 1935. The rate of spontaneous mutation of a human gene. *J. Genet.* 31: 317–326.

Haldane, J. B. S. 1937. The effect of variation on fitness. *Amer. Natur.* 71: 337–349.

Haldane, J. B. S. 1957. The cost of natural selection. *J. Genetics* 55: 511–524.

Hall, S. E., G. Kettler and D. Preuss. 2003. Centromere satellites from *Arabidopsis* populations: Maintenance of conserved and variable domains. *Genome Res.* 13: 195–205.

Hall, S. L. and R. A. Padgett. 1994. Conserved sequences in a class of rare eukaryotic nuclear introns with non-consensus splice sites. *J. Mol. Biol.* 239: 357–365.

Hallick, R. B., L. Hong, R. G. Drager, M. R. Favreau, A. Monfort, B. Orsat, A. Spielmann and E. Stutz. 1993. Complete sequence of *Euglena gracilis* chloroplast DNA. *Nucleic Acids Res.* 21: 3537–3544.

Halligan, D. L., A. Eyre-Walker, P. Andolfatto and P. D. Keightley. 2004. Patterns of evolutionary constraints in intronic and intergenic DNA of *Drosophila*. *Genome Res.* 14: 273–279.

Halushka, M. K., J. B. Fan, K. Bentley, L. Hsie, N. Shen, A. Weder, R. Cooper, R. Lipshutz and A. Chakravarti. 1999. Patterns of single-nucleotide polymorphisms in candidate genes for blood-pressure homeostasis. *Nature Genet.* 22: 239–247.

Hamada, K., T. Horiike, H. Ota, K. Mizuno and T. Shinozawa. 2003. Presence of isochore structures in reptile genomes suggested by the relationship between GC contents of intron regions and those of coding regions. *Genes Genet. Syst.* 78: 195–198.

Hamilton, M. L., Z. Guo, C. D. Fuller, H. Van Remmen, W. F. Ward, S. N. Austad, D. A. Troyer, I. Thompson and A. Richardson. 2001. A reliable assessment of 8-oxo-2-deoxyguanosine levels in nuclear and mitochondrial DNA using the sodium iodide method to isolate DNA. *Nucleic Acids Res.* 29: 2117–2126.

Hamilton, W. D. 1964a. The genetical evolution of social behaviour. I. *J. Theor. Biol.* 7: 1–16.

Hamilton, W. D. 1964b. The genetical evolution of social behaviour. II. *J. Theor. Biol.* 7: 17–52.

Hammond, P. M. 1995. Described and estimated species numbers: An objective assessment of current knowledge. In D. Allsopp, R. R. Colwell and D. L. Hawksworth, eds., *Microbial Diversity and Ecosystem Function*, 29–71. CAB International, Wallingford, UK.

Han, J. S. and J. D. Boeke. 2004. A highly active synthetic mammalian retrotransposon. *Nature* 429: 314–318.

Han, J. S., S. T. Szak and J. D. Boeke. 2004. Transcriptional disruption by the *L1* retrotransposon and implications for mammalian transcriptomes. *Nature* 429: 268–274.

Han, T. M. and B. Runnegar. 1992. Megascopic eukaryotic algae from the 2.1-billion-year-old negaunee iron-formation, Michigan. *Science* 257: 232–235.

Hanada, T., T. Suzuki, T. Yokogawa, C. Takemoto-Hori, M. Sprinzl and K. Watanabe. 2001. Translation ability of mitochondrial tRNAsSer with unusual secondary structures in an in vitro translation system of bovine mitochondria. *Genes Cells* 6: 1019–1030.

Hanawalt, P. C. 1994. Transcription-coupled repair and human disease. *Science* 266: 1957–1958.

Hancock, J. M., P. J. Shaw, F. Bonneton and G. A. Dover. 1999. High sequence turnover in the regulatory regions of the developmental gene hunchback in insects. *Mol. Biol. Evol.* 16: 253–265.

Handley, L. J., H. Ceplitis and H. Ellegren. 2004. Evolutionary strata on the chicken Z chromosome: Implications for sex chromosome evolution. *Genetics* 167: 367–376.

Hankeln, T., H. Friedl, I. Ebersberger, J. Martin and E. R. Schmidt. 1997. A variable intron distribution in globin genes of *Chironomus*: Evidence for recent intron gain. *Gene* 205: 151–160.

Hannaert, V., E. Saavedra, F. Duffieux, J. P. Szikora, D. J. Rigden, P. A. Michels and F. R. Opperdoes. 2003. Plant-like traits associated with metabolism of *Trypanosoma* parasites. *Proc. Natl. Acad. Sci. USA* 100: 1067–1071.

Hansen, T. F., A. J. Carter and C.-H. Chiu. 2000. Gene conversion may aid adaptive peak shifts. *J. Theor. Biol.* 207: 495–511.

Harada, K., K. Yukuhiro and T. Mukai. 1990. Transposition rates of movable genetic elements in *Drosophila melanogaster*. *Proc. Natl. Acad. Sci. USA* 87: 3248–3252.

Harbison, C. T., D. B. Gordon, T. I. Lee, N. J. Rinaldi, K. D. Macisaac, T. W. Danford, N. M. Hannett, J. B. Tagne, D. B. Reynolds, J. Yoo et al. 2004. Transcriptional regulatory code of a eukaryotic genome. *Nature* 431: 99–104.

Harper, J. T., E. Waanders and P. J. Keeling. 2005. On the monophyly of chromalveolates using a six-protein phylogeny of eukaryotes. *Internatl. J. Syst. Evol. Microbiol.* 55: 487–496.

Harrison, P. M. and M. Gerstein. 2002. Studying genomes through the aeons: Protein families, pseudogenes and proteome evolution. *J. Mol. Biol.* 318: 1155–1174.

Harrison, P. M., H. Hegyi, S. Balasubramanian, N. M. Luscombe, P. Bertone, N. Echols, T. Johnson and M. Gerstein. 2002. Molecular fossils in the human genome: Identification and analysis of the pseudogenes in chromosomes 21 and 22. *Genome Res.* 12: 272–280.

Harrison, P. M., D. Milburn, Z. Zhang, P. Bertone and M. Gerstein. 2003. Identification of pseudogenes in the *Drosophila melanogaster* genome. *Nucleic Acids Res.* 31: 1033–1037.

Hartl, D. L. and C. H. Taubes. 1996. Compensatory nearly neutral mutations: Selection without adaptation. *J. Theor. Biol.* 182: 303–309.

Hartl, D. L., E. N. Moriyama and S. A. Sawyer. 1994. Selection intensity for codon bias. *Genetics* 138: 227–234.

Hartl, D. L., A. R. Lohe and E. R. Lozovskaya. 1997. Modern thoughts on an ancyent marinere: Function, evolution, regulation. *Annu. Rev. Genet.* 31: 337–358.

Hartl, D. L., S. K. Volkman, K. M. Nielsen, A. E. Barry, K. P. Day, D. F. Wirth and E. A. Winzeler. 2002. The paradoxical population genetics of *Plasmodium falciparum*. *Trends Parasitol.* 18: 266–272.

Hartman, H. and A. Fedorov. 2002. The origin of the eukaryotic cell: A genomic investigation. *Proc. Natl. Acad. Sci. USA* 99: 1420–1425.

Hartwell, L. H., J. J. Hopfield, S. Leibler and A. W. Murray. 1999. From molecular to modular cell biology. *Nature* 402: C47–C52.

Hasegawa, M., Y. Cao and Z. Yang. 1998. Preponderance of slightly deleterious polymorphism in mitochondrial DNA: Nonsynonymous/synonymous rate ratio is much higher within species than between species. *Mol. Biol. Evol.* 15: 1499–1505.

Hastings, I. M. 1992. Population genetic aspects of deleterious cytoplasmic genomes and their effect on the evolution of sexual reproduction. *Genet. Res.* 59: 215–225.

Hastings, K. E. M. 1996. Strong evolutionary conservation of broadly expressed protein isoforms in the troponin I gene family and other vertebrate gene families. *J. Mol. Evol.* 42: 631–640.

Hastings, M. L. and A. R. Krainer. 2001. Pre-mRNA splicing in the new millennium. *Curr. Opin. Cell. Biol.* 13: 302–309.

Haubold, B., J. Kroymann, A. Ratzka, T. Mitchell-Olds and T. Wiehe. 2002. Recombination and gene conversion in a 170-kb genomic region of *Arabidopsis thaliana*. *Genetics* 161: 1269–1278.

Havecker, E. R., X. Gao and D. F. Voytas. 2004. The diversity of LTR retrotransposons. *Genome Biol.* 5: 225.

Havilio, M., E. Y. Levanon, G. Lerman, M. Kupiec and E. Eisenberg. 2005. Evidence for abundant transcription of non-coding regions in the *Saccharomyces cerevisiae* genome. *BMC Genomics* 6: 93.

Hawk, J. D., L. Stefanovic, J. C. Boyer, T. D. Petes and R. A. Farber. 2005. Variation in efficiency of DNA mismatch repair at different sites in the yeast genome. *Proc. Natl. Acad. Sci. USA* 102: 8639–8643.

Haygood, R. 2006. Mutation rate and the cost of complexity. *Mol. Biol. Evol.* 23: 957–963.

Haywood-Farmer, E. and S. P. Otto. 2003. The evolution of genomic base composition in bacteria. *Evolution* 57: 1783–1792.

He, L. and G. J. Hannon. 2004. MicroRNAs: Small RNAs with a big role in gene regulation. *Nature Rev. Genet.* 5: 522–531.

He, X. and J. Zhang. 2005a. Higher duplicability of less important genes in yeast genomes. *Mol. Biol. Evol.* 23: 144–151.

He, X. and J. Zhang. 2005b. Rapid subfunctionalization accompanied by prolonged and substantial neofunctionalization in duplicate gene evolution. *Genetics* 169: 1157–1164.

Hedges, D. J., P. A. Callinan, R. Cordaux, J. Xing, E. Barnes and M. A. Batzer. 2004. Differential *Alu* mobilization and polymorphism among the

human and chimpanzee lineages. *Genome Res.* 14: 1068–1075.
Hedges, S. B. and S. Kumar. 2004. Precision of molecular time estimates. *Trends Genet.* 20: 242–247.
Hedges, S. B., H. Chen, S. Kumar, D. Y. Wang, A. S. Thompson and H. Watanabe. 2001. A genomic timescale for the origin of eukaryotes. *BMC Evol. Biol.* 1(1): 4.
Hedrick, P. 2005. Large variance in reproductive success and the N_e/N ratio. *Evolution* 59: 1596–1599.
Heidenreich, E., R. Novotny, B. Kneidinger, V. Holzmann and U. Wintersberger. 2003. Non-homologous end joining as an important mutagenic process in cell cycle-arrested cells. *EMBO J.* 22: 2274–2283.
Heilig, R., R. Eckenberg, J. L. Petit, N. Fonknechten, C. Da Silva, L. Cattolico, M. Levy, V. Barbe, V. de Berardinis, A. Ureta-Vidal et al. 2003. The DNA sequence and analysis of human chromosome 14. *Nature* 421: 601–607.
Hein, J. J., M. H. Schierup and C. H. Wiuf. 2004. *Gene Genealogies, Variation and Evolution*. Oxford University Press, Oxford.
Helbock, H. J., K. B. Beckman, M. K. Shigenaga, P. B. Walter, A. A. Woodall, H. C. Yeo and B. N. Ames. 1998. DNA oxidation matters: The HPLC-electrochemical detection assay of 8-oxo-deoxyguanosine and 8-oxo-guanine. *Proc. Natl. Acad. Sci. USA* 95: 288–293.
Hellberg, M. E. 2006. No variation and low synonymous substitution rates in coral mtDNA despite high nuclear variation. *BMC Evol. Biol.* 6: 24.
Hellberg, M. E. and V. D. Vacquier. 1999. Rapid evolution of fertilization selectivity and lysin cDNA sequences in teguline gastropods. *Mol. Biol. Evol.* 16: 839–848.
Hellborg, L. and H. Ellegren. 2004. Low levels of nucleotide diversity in mammalian Y chromosomes. *Mol. Biol. Evol.* 21: 158–163.
Hellmann, I., I. Ebersberger, S. E. Ptak, S. Pääbo and M. Przeworski. 2003. A neutral explanation for the correlation of diversity with recombination rates in humans. *Amer. J. Hum. Genet.* 72: 1527–1535.
Henikoff, S. and Y. Dalal. 2005. Centromeric chromatin: What makes it unique? *Curr. Opin. Genet. Dev.* 15: 177–184.
Henikoff, S., K. Ahmad and H. S. Malik. 2001. The centromere paradox: Stable inheritance with rapidly evolving DNA. *Science* 293: 1098–1102.
Hentze, M. W. and A. E. Kulozik. 1999. A perfect message: RNA surveillance and nonsense-mediated decay. *Cell* 96: 307–310.
Herbeck, J. T., D. J. Funk, P. H. Degnan and J. J. Wernegreen. 2003. A conservative test of genetic drift in the endosymbiotic bacterium *Buchnera*: Slightly deleterious mutations in the chaperonin groEL. *Genetics* 165: 1651–1660.
Herédia, F., E. L. Loreto and V. L. Valente. 2004. Complex evolution of *gypsy* in drosophilid species. *Mol. Biol. Evol.* 21: 1831–1842.
Herrmann, J. M. 2003. Converting bacteria to organelles: Evolution of mitochondrial protein sorting. *Trends Microbiol.* 11: 74–79.
Hetzer, M., G. Wurzer, R. J. Schweyen and M. W. Mueller. 1997. Trans-activation of group II intron splicing by nuclear U5 snRNA. *Nature* 386: 417–420.
Heun, P., S. Erhardt, M. D. Blower, S. Weiss, A. D. Skora and G. H. Karpen. 2006. Mislocalization of the *Drosophila* centromere-specific histone CID promotes formation of functional ectopic kinetochores. *Dev. Cell* 10: 303–315.
Hey, J. 1989. The transposable portion of the genome of *Drosophila algonquin* is very different from that in *D. melanogaster*. *Mol. Biol. Evol.* 6: 66–79.
Hey, J. and R. M. Kliman. 2002. Interactions between natural selection, recombination and gene density in the genes of *Drosophila*. *Genetics* 160: 595–608.
Hey, J. and J. Wakeley. 1997. A coalescent estimator of the population recombination rate. *Genetics* 145: 833–846.
Hickey, D. A. 1982. Selfish DNA: A sexually-transmitted nuclear parasite. *Genetics* 101: 519–531.
Hiesel, R., B. Combettes and A. Brennicke. 1994. Evidence for RNA editing in mitochondria of all major groups of land plants except the Bryophyta. *Proc. Natl. Acad. Sci. USA* 91: 629–633.
Higgs, P. G. 1998. Compensatory neutral mutations and the evolution of RNA. *Genetica* 102/103: 91–101.
Hill, R. C., C. E. de Carvalho, J. Salogiannis, B. Schlager, D. Pilgrim and E. S. Haag. 2006. Genetic flexibility in the convergent evolution of hermaphroditism in *Caenorhabditis* nematodes. *Dev. Cell* 10: 531–538.
Hill, W. G. 1975. Linkage disequilibrium among multiple neutral alleles produced by mutation in finite population. *Theor. Pop. Biol.* 8: 117–126.
Hill, W. G. (ed.). 1984. *Quantitative Genetics*. Part I. *Explanation and Analysis of Continuous Variation*. Van Nostrand Reinhold, New York.
Hill, W. G. and A. Robertson. 1966. The effect of linkage on limits to artificial selection. *Genet. Res.* 8: 269–294.
Hiller, R., M. Hetzer, R. J. Schweyen and M. W. Mueller. 2000. Transposition and exon shuffling by group II intron RNA molecules in pieces. *J. Mol. Biol.* 297: 301–308.
Hillier, L. W., R. S. Fulton, L. A. Fulton, T. A. Graves, K. H. Pepin, C. Wagner-McPherson, D. Layman, J. Maas, S. Jaeger, R. Walker et al. 2003. The DNA sequence of human chromosome 7. *Nature* 424: 157–164.
Hilliker, A. J., S. H. Clark and A. Chovnick. 1991. The effect of DNA sequence polymorphisms on intragenic recombination in the rosy locus of *Drosophila melanogaster*. *Genetics* 129: 779–781.
Hilliker, A. J., G. Harauz, A. G. Reaume, M. Gray, S. H. Clark and A. Chovnick. 1994. Meiotic gene conversion tract length distribution within the rosy locus of *Drosophila melanogaster*. *Genetics* 137: 1019–1026.
Hiriyanna, K. T. and T. Ramakrishnan. 1986. Deoxyribonucleic acid replication time in *Mycobacterium tuberculosis* H37 Rv. *Arch. Microbiol.* 144: 105–109.
Hirose, Y. and J. L. Manley. 2000. RNA polymerase II and the integration of nuclear events. *Genes Dev.* 14: 1415–1429.

Hirotsune, S., N. Yoshida, A. Chen, L. Garrett, F. Sugiyama, S. Takahashi, K. Yagami, A. Wynshaw-Boris and A. Yoshiki. 2003. An expressed pseudogene regulates the messenger-RNA stability of its homologous coding gene. *Nature* 423: 91–96.

Ho, S. Y. and G. Larson. 2006. Molecular clocks: When times are a-changin'. *Trends Genet.* 22: 79–83.

Hobza, R., M. Lengerova, J. Svoboda, H. Kubekova, E. Kejnovsky and B. Vyskot. 2006. An accumulation of tandem DNA repeats on the Y chromosome in *Silene latifolia* during early stages of sex chromosome evolution. *Chromosoma* 115: 376–382.

Hodgkin, J., A. Papp, R. Pulak, V. Ambros and P. Anderson. 1989. A new kind of informational suppression in the nematode *Caenorhabditis elegans*. *Genetics* 123: 301–313.

Hoegg, S., H. Brinkmann, J. S. Taylor and A. Meyer. 2004. Phylogenetic timing of the fish-specific genome duplication correlates with the diversification of teleost fish. *J. Mol. Evol.* 59: 190–203.

Hoffman, P. F. and D. P. Schrag. 2002. The snowball Earth hypothesis: Testing the limits of global change. *Terra Nova* 14: 129–155.

Hoffman, P. F., A. J. Kaufman, G. P. Halverson and D. P. Schrag. 1998. A neoproterozoic snowball earth. *Science* 281: 1342–1346.

Holmquist, G. P. 1992. Chromosome bands, their chromatin flavors, and their functional features. *Amer. J. Hum. Genet.* 51: 17–37.

Hong, X., D. G. Scofield and M. Lynch. 2006. Intron size, abundance and distribution within untranslated regions of genes. *Mol. Biol. Evol.* 23: 2392–2404.

Hood, M. E., J. Antonovics and B. Koskella. 2004. Shared forces of sex chromosome evolution in haploid-mating and diploid-mating organisms: *Microbotryum violaceum* and other model organisms. *Genetics* 168: 141–146.

Horai, S. and K. Hayasaka. 1990. Intraspecific nucleotide sequence differences in the major noncoding region of human mitochondrial DNA. *Amer. J. Hum. Genet.* 46: 828–842.

Hori, R. and R. A. Firtel. 1994. Identification and characterization of multiple A/T-rich cis-acting elements that control expression from *Dictyostelium* actin promoters: The *Dictyostelium* actin upstream activating sequence confers growth phase expression and has enhancer-like properties. *Nucleic Acids Res.* 22: 5099–5011.

Horiike, T., K. Hamada, S. Kanaya and T. Shinozawa. 2001. Origin of eukaryotic cell nuclei by symbiosis of Archaea in Bacteria is revealed by homology-hit analysis. *Nature Cell Biol.* 3: 210–214.

Horiuchi, T., E. Giniger and T. Aigaki. 2003. Alternative *trans*-splicing of constant and variable exons of a *Drosophila* axon guidance gene, *lola*. *Genes Dev.* 17: 2496–2501.

Horner-Devine, M. C., M. Lage, J. B. Hughes and B. J. Bohannan. 2004. A taxa-area relationship for bacteria. *Nature* 432: 750–753.

Horton, T. L. and L. F. Landweber. 2000. Evolution of four types of RNA editing in myxomycetes. *RNA* 6: 1339–1346.

Horton, T. L. and L. F. Landweber. 2002. Rewriting the information in DNA: RNA editing in kinetoplastids and myxomycetes. *Curr. Opin. Microbiol.* 5: 620–626.

Houle, D. 1992. Comparing evolvability and variability of quantitative traits. *Genetics* 130: 195–204.

Houle, D. and S. V. Nuzhdin. 2004. Mutation accumulation and the effect of *copia* insertions in *Drosophila melanogaster*. *Genet. Res.* 83: 7–18.

Housworth, E. A., E. P. Martins and M. Lynch. 2003. The phylogenetic mixed model. *Amer. Natur.* 163: 84–96.

Howe, C. J., A. C. Barbrook, V. L. Koumandou, R. E. Nisbet, H. A. Symington and T. F. Wightman. 2003. Evolution of the chloroplast genome. *Phil. Trans. Roy. Soc. Lond.* B *Biol. Sci.* 358: 99–106.

Howell, N., C. B. Smejkal, D. A. Mackey, P. F. Chinnery, D. M. Turnbull and C. Herrnstadt. 2003. The pedigree rate of sequence divergence in the human mitochondrial genome: There is a difference between phylogenetic and pedigree rates. *Amer. J. Hum. Genet.* 72: 659–670.

Howland, J. L. 2000. *The Surprising Archaea*. Oxford University Press, Oxford.

Hrdy, I., R. P. Hirt, P. Dolezal, L. Bardonova, P. G. Foster, J. Tachezy and T. M. Embley. 2004. *Trichomonas* hydrogenosomes contain the NADH dehydrogenase module of mitochondrial complex I. *Nature* 432: 618–622.

Huang, T., S. Kuersten, A. M. Deshpande, J. Spieth, M. MacMorris and T. Blumenthal. 2001. Intercistronic region required for polycistronic pre-mRNA processing in *Caenorhabditis elegans*. *Mol. Cell. Biol.* 21: 1111–1120.

Huang, S. W., R. Friedman, N. Yu, A. Yu and W.-H. Li. 2005. How strong is the mutagenicity of recombination in mammals? *Mol. Biol. Evol.* 22: 426–431.

Hudson, R. E., U. Bergthorsson and H. Ochman. 2003. Transcription increases multiple spontaneous point mutations in *Salmonella enterica*. *Nucleic Acids Res.* 31: 4517–4522.

Hughes, A. L. 1994. The evolution of functionally novel proteins after gene duplication. *Proc. Roy. Soc. Lond.* B *Biol. Sci.* 256: 119–124.

Hughes, A. L. and R. Friedman. 2003. 2R or not 2R: Testing hypotheses of genome duplication in early vertebrates. *J. Struc. Func. Genomics* 3: 85–93.

Hughes, A. L. and R. Friedman. 2005. Nucleotide substitution and recombination at orthologous loci in *Staphylococcus aureus*. *J. Bacteriol.* 187: 2698–2704.

Hughes, J. F. and J. M. Coffin. 2001. Evidence for genomic rearrangements mediated by human endogenous retroviruses during primate evolution. *Nature Genet.* 29: 487–489.

Hughes, J. F., H. Skaletsky, T. Pyntikova, P. J. Minx, T. Graves, S. Rozen, R. K. Wilson and D. C. Page. 2005. Conservation of Y-linked genes during human evolution revealed by comparative sequencing in chimpanzee. *Nature* 437: 100–103.

Hughes, M. J. and D. W. Andrews. 1997. A single nucleotide is a sufficient 5′ untranslated region for translation in an eukaryotic in vitro system. *FEBS Lett.* 414: 19–22.

Hughes, M. K. and A. L. Hughes. 1993. Evolution of duplicate genes in a tetraploid animal, *Xenopus laevis*. *Mol. Biol. Evol.* 10: 1360–1369.

Hughes, S., D. Zelus and D. Mouchiroud. 1999. Warm-blooded isochore structure in Nile crocodile and turtle. *Mol. Biol. Evol.* 16: 1521–1527.

Hughes, T. A. 2006. Regulation of gene expression by alternative untranslated regions. *Trends Genet.* 22: 119–122.

Huminiecki, L. and K. H. Wolfe. 2004. Divergence of spatial gene expression profiles following species-specific gene duplications in human and mouse. *Genome Res.* 14: 1870–1879.

Humphrey, T., C. E. Birse and N. J. Proudfoot. 1994. RNA 3′ end signals of the *S. pombe ura4* gene comprise a site determining and efficiency element. *EMBO J.* 13: 2441–2451.

Hunter, N., S. R. Chambers, E. J. Louis and R. H. Borts. 1996. The mismatch repair system contributes to meiotic sterility in an interspecific yeast hybrid. *EMBO J.* 15: 1726–1733.

Hurley, I., M. E. Hale and V. E. Prince. 2005. Duplication events and the evolution of segmental identity. *Evol. Dev.* 7: 556–567.

Hurst, L. D. and A. R. Merchant. 2001. High guanine-cytosine content is not an adaptation to high temperature: A comparative analysis amongst prokaryotes. *Proc. Biol. Sci.* 268: 493–497.

Huynen, M. A. and E. van Nimwegen. 1998. The frequency distribution of gene family sizes in complete genomes. *Mol. Biol. Evol.* 15: 583–589.

Hwang, D. G. and P. Green. 2004. Bayesian Markov chain Monte Carlo sequence analysis reveals varying neutral substitution patterns in mammalian evolution. *Proc. Natl. Acad. Sci. USA* 101: 13994–14001.

Hyrien, O., K. Marheineke and A. Goldar. 2003. Paradoxes of eukaryotic DNA replication: MCM proteins and the random completion problem. *BioEssays* 25: 116–125.

Iacono, M., F. Mignone and G. Pesole. 2005. uAUG and uORFs in human and rodent 5′ untranslated mRNAs. *Gene* 349: 97–105.

Iafrate, A. J., L. Feuk, M. N. Rivera, M. L. Listewnik, P. K. Donahoe, Y. Qi, S. W. Scherer and C. Lee. 2004. Detection of large-scale variation in the human genome. *Nature Genet.* 36: 949–951.

Ibrahim, K. M., S. J. Cooper and G. M. Hewitt. 2002. Testing for recombination in a short nuclear DNA sequence of the European meadow grasshopper, *Chorthippus parallelus*. *Mol. Ecol.* 11: 583–590.

Ideker, T., T. Galitski and L. Hood. 2001. A new approach to decoding life: Systems biology. *Annu. Rev. Genomics Hum. Genet.* 2: 343–372.

Ikemura, T. 1981. Correlation between the abundance of *Escherichia coli* transfer RNAs and the occurrence of the respective codons in its protein genes. *J. Mol. Biol.* 146: 1–21.

Ikemura, T. 1985. Codon usage and tRNA content in unicellular and multicellular organisms. *Mol. Biol. Evol.* 2: 13–34.

Ina, S., T. Sasaki, Y. Yokota and T. Shinomiya. 2001. A broad replication origin of *Drosophila melanogaster*, *oriDα*, consists of AT-rich multiple discrete initiation sites. *Chromosoma* 109: 551–564.

Inada, M., T. Sasaki, M. Yukawa, T. Tsudzuki and M. Sugiura. 2004. A systematic search for RNA editing sites in pea chloroplasts: An editing event causes diversification from the evolutionarily conserved amino acid sequence. *Plant Cell Physiol.* 45: 1615–1622.

Ingraham, J. L., O. Maaløe and F. C. Neidhardt. 1983. *Growth of the Bacterial Cell*. Sinauer Associates, Sunderland, MA.

Innan, H. and M. Nordborg. 2002. Recombination or mutational hot spots in human mtDNA? *Mol. Biol. Evol.* 19: 1122–1127.

Irish, V. F. and A. Litt. 2005. Flower development and evolution: Gene duplication, diversification and redeployment. *Curr. Opin. Genet. Dev.* 15: 454–460.

Ishida, K. and B. R. Green. 2002. Second- and third-hand chloroplasts in dinoflagellates: Phylogeny of oxygen-evolving enhancer 1 (*PsbO*) protein reveals replacement of a nuclear-encoded plastid gene by that of a haptophyte tertiary endosymbiont. *Proc. Natl. Acad. Sci. USA* 99: 9294–9299.

Ishizaki, K., Y. Shimizu-Ueda, S. Okada, M. Yamamoto, M. Fujisawa, K. T. Yamato, H. Fukuzawa and K. Ohyama. 2002. Multicopy genes uniquely amplified in the Y chromosome-specific repeats of the liverwort *Marchantia polymorpha*. *Nucleic Acids Res.* 30: 4675–4681.

Isshiki, M., Y. Yamamoto, H. Satoh and K. Shimamoto. 2001. Nonsense-mediated decay of mutant *waxy* mRNA in rice. *Plant Physiol.* 125: 1388–1395.

Iwabe, N. and T. Miyata. 2001. Overlapping genes in parasitic protist *Giardia lamblia*. *Gene* 280: 163–167.

Iwabe, N., K. Kuma, M. Hasegawa, S. Osawa and T. Miyata. 1989. Evolutionary relationship of archaebacteria, eubacteria, and eukaryotes inferred from phylogenetic trees of duplicated genes. *Proc. Natl. Acad. Sci. USA* 86: 9355–9359.

Jackson, D. A. and A. Pombo. 1998. Replicon clusters are stable units of chromosome structure: Evidence that nuclear organization contributes to the efficient activation and propagation of S phase in human cells. *J. Cell Biol.* 140: 1285–1295.

Jackson, D. A., A. Pombo and F. Iborra. 2000. The balance sheet for transcription: An analysis of nuclear RNA metabolism in mammalian cells. *FASEB J.* 14: 242–254.

Jacob, F. 1977. Evolution and tinkering. *Science* 196: 1161–1166.

Jacob, F., S. Brenner and F. Cuzin. 1963. On the regulation of DNA replication in bacteria. *Cold Spring Harbor Symp. Quant. Biol.* 28: 329–348.

Jacobs, G. H., M. Neitz, J. F. Deegan II and J. Neitz. 1996. Trichromatic colour vision in New World monkeys. *Nature* 382: 156–158.

Jacquier, A. 1990. Self-splicing group II and nuclear pre-mRNA introns: How similar are they? *Trends Biochem. Sci.* 15: 351–354.

Jaenike, J. 2001. Sex chromosome meiotic drive. *Annu. Rev. Ecol. Syst.* 32: 25–49.

Jaillon, O., J. M. Aury, F. Brunet, J. L. Petit, N. Stange-Thomann, E. Mauceli, L. Bouneau, C. Fischer, C.

Ozouf-Costaz, A. Bernot et al. 2004. Genome duplication in the teleost fish *Tetraodon nigroviridis* reveals the early vertebrate proto-karyotype. *Nature* 431: 946–957.
Jain, R., M. C. Rivera and J. A. Lake. 1999. Horizontal gene transfer among genomes: The complexity hypothesis. *Proc. Natl. Acad. Sci. USA* 96: 3801–3806.
Jain, R., M. Rivera, J. Moore and J. Lake. 2002. Horizontal gene transfer in microbial genome evolution. *Theor. Pop. Biol.* 61: 489–495.
Janke, A. and S. Pääbo. 1993. Editing of a tRNA anticodon in marsupial mitochondria changes its codon recognition. *Nucleic Acids Res.* 21: 1523–1525.
Jansen, R. P. 2001. mRNA localization: Message on the move. *Nature Rev. Mol. Cell. Biol.* 2: 247–256.
Javaux, E. J., A. H. Knoll and M. R. Walter. 2001. Morphological and ecological complexity in early eukaryotic ecosystems. *Nature* 412: 66–69.
Jeffares, D. C., T. Mourier and D. Penny. 2006. The biology of intron gain and loss. *Trends Genet.* 22: 16–22.
Jeffreys, A. J., J. K. Holloway, L. Kauppi, C. A. May, R. Neumann, M. T. Slingsby and A. J. Webb. 2004. Meiotic recombination hot spots and human DNA diversity. *Phil. Trans. Roy. Soc. Lond.* B *Biol. Sci.* 359: 141–152.
Jenkins, B. D. and A. Barkan. 2001. Recruitment of a peptidyl-tRNA hydrolase as a facilitator of group II intron splicing in chloroplasts. *EMBO J.* 20: 872–879.
Jensen, S., M. P. Gassama and T. Heidmann. 1999a. Cosuppression of *I* transposon activity in *Drosophila* by *I*-containing sense and antisense transgenes. *Genetics* 153: 1767–1774.
Jensen, S., M. P. Gassama and T. Heidmann. 1999b. Taming of transposable elements by homology-dependent gene silencing. *Nature Genet.* 21: 209–212.
Jensen, S., M. P. Gassama, X. Dramard and T. Heidmann. 2002. Regulation of *I*-transposon activity in *Drosophila*: Evidence for cosuppression of nonhomologous transgenes and possible role of ancestral *I*-related pericentromeric elements. *Genetics* 162: 1197–1209.
Jensen-Seaman, M. I., T. S. Furey, B. A. Payseur, Y. Lu, K. M. Roskin, C. F. Chen, M. A. Thomas, D. Haussler and H. J. Jacob. 2004. Comparative recombination rates in the rat, mouse, and human genomes. *Genome Res.* 14: 528–538.
Jenuth, J. P., A. C. Peterson, K. Fu and E. A. Shoubridge. 1996. Random genetic drift in the female germline explains the rapid segregation of mammalian mitochondrial DNA. *Nature Genet.* 14: 146–151.
Jiang, N., Z. Bao, S. Temnykh, Z. Cheng, J. Jiang, R. A. Wing, S. R. McCouch and S. R. Wessler. 2002. *Dasheng*: A recently amplified nonautonomous long terminal repeat element that is a major component of pericentromeric regions in rice. *Genetics* 161: 1293–1305.
Jiang, N., Z. Bao, X. Zhang, H. Hirochika, S. R. Eddy, S. R. McCouch and S. R. Wessler. 2003. An active DNA transposon family in rice. *Nature* 421: 163–167.
Jiang, N., Z. Bao, X. Zhang, S. R. Eddy and S. R. Wessler. 2004. Pack-MULE transposable elements mediate gene evolution in plants. *Nature* 431: 569–573.
Jiggins, F. M. 2002. The rate of recombination in *Wolbachia* bacteria. *Mol. Biol. Evol.* 19: 1640–1643.
Jin, W., J. R. Melo, K. Nagaki, P. B. Talbert, S. Henikoff, R. K. Dawe and J. Jiang. 2004. Maize centromeres: Organization and functional adaptation in the genetic background of oat. *Plant Cell* 16: 571–581.
Jing, R., M. R. Knox, J. M. Lee, A. V. Vershinin, M. Ambrose, T. H. Ellis and A. J. Flavell. 2005. Insertional polymorphism and antiquity of *PDR1* retrotransposon insertions in *Pisum* species. *Genetics* 171: 741–752.
Jinks-Robertson, S., M. Michelitch and S. Ramcharan. 1993. Substrate length requirements for efficient mitotic recombination in *Saccharomyces cerevisiae*. *Mol. Cell. Biol.* 13: 3937–50.
Johnson, A. A. and K. A. Johnson. 2001. Fidelity of nucleotide incorporation by human mitochondrial DNA polymerase. *J. Biol. Chem.* 276: 38090–38096.
Johnson, J. M., J. Castle, P. Garrett-Engele, Z. Kan, P. M. Loerch, C. D. Armour, R. Santos, E. E. Schadt, R. Stoughton and D. D. Shoemaker. 2003. Genome-wide survey of human alternative pre-mRNA splicing with exon junction microarrays. *Science* 302: 2141–2144.
Johnson, K. P. 2004. Deletion bias in avian introns over evolutionary timescales. *Mol. Biol. Evol.* 21: 599–602.
Johnson, K. R., J. E. Wright, Jr. and B. May. 1987. Linkage relationships reflecting ancestral tetraploidy in salmonid fish. *Genetics* 116: 579–591.
Johnson, N. A. and A. H. Porter. 2000. Rapid speciation via parallel, directional selection on regulatory genetic pathways. *J. Theor. Biol.* 205: 527–542.
Johnson, N. A. and A. H. Porter. 2001. Toward a new synthesis: Population genetics and evolutionary developmental biology. *Genetica* 112/113: 45–58.
Johnson, T. and N. H. Barton. 2002. The effect of deleterious alleles on adaptation in asexual populations. *Genetics* 162: 395–411.
Jordan, I. K. and J. F. McDonald. 1998. Evidence for the role of recombination in the regulatory evolution of *Saccharomyces cerevisiae Ty* elements. *J. Mol. Evol.* 47: 14–20.
Jordan, I. K., L. V. Matyunina and J. F. McDonald. 1999. Evidence for the recent horizontal transfer of long terminal repeat retrotransposon. *Proc. Natl. Acad. Sci. USA* 96: 12621–12625.
Jordan, I. K., I. B. Rogozin, G. V. Glazko and E. V. Koonin. 2003. Origin of a substantial fraction of human regulatory sequences from transposable elements. *Trends Genet.* 19: 68–72.
Jordan, I. K., Y. I. Wolf and E. V. Koonin. 2004. Duplicated genes evolve slower than singletons despite the initial rate increase. *BMC Evol. Biol.* 4: 22.
Jorde, L. B. and M. Bamshad. 2000. Questioning evidence for recombination in human mitochondrial DNA. *Science* 288: 1931.
Joyce, G. F. 2004. Directed evolution of nucleic acid enzymes. *Annu. Rev. Biochem.* 73: 791–836.

Joyce, P. B. and M. W. Gray. 1989. Chloroplast-like transfer RNA genes expressed in wheat mitochondria. *Nucleic Acids Res.* 17: 5461–5476.

Joza, N., S. A. Susin, E. Daugas, W. L. Stanford, S. K. Cho, C. Y. Li, T. Sasaki, A. J. Elia, H. Y. Cheng, L. Ravagnan et al. 2001. Essential role of the mitochondrial apoptosis-inducing factor in programmed cell death. *Nature* 410: 549–554.

Judd, S. R. and T. D. Petes. 1988. Physical lengths of meiotic and mitotic gene conversion tracts in *Saccharomyces cerevisiae*. *Genetics* 118: 401–410.

Judson, O. P. and B. B. Normark. 1996. Ancient asexual scandals. *Trends Ecol. Evol.* 11: 41–46.

Jukes, T. H. and S. Osawa. 1993. Evolutionary changes in the genetic code. *Comp. Biochem. Physiol. B* 106: 489–494.

Junakovic, N., C. DiFranco, P. Barsanti and G. Palumbo. 1986. Transposition of *copia*-like nomadic elements can be induced by heat shock. *J. Mol. Evol.* 24: 89–93.

Juneau, K., M. Miranda, M. E. Hillenmeyer, C. Nislow and R. W. Davis. 2006. Introns regulate RNA and protein abundance in yeast. *Genetics* 174: 511–518.

Kado, T., H. Yoshimaru, Y. Tsumura and H. Tachida. 2003. DNA variation in a conifer, *Cryptomeria japonica* (Cupressaceae *sensu lato*). *Genetics* 164: 1547–1559.

Kadonaga, J. T. 2002. The DPE, a core promoter element for transcription by RNA polymerase II. *Exp. Mol. Med.* 34: 259–264.

Kadowaki, K., N. Kubo, K. Ozawa and A. Hirai. 1996. Targeting presequence acquisition after mitochondrial gene transfer to the nucleus occurs by duplication of existing targeting signals. *EMBO J.* 15: 6652–6661.

Kaji, H., S. Tai, Y. Okimura, G. Iguchi, Y. Takahashi, H. Abe and K. Chihara. 1998. Cloning and characterization of the 5′-flanking region of the human growth hormone secretagogue receptor gene. *J. Biol. Chem.* 273: 33885–33888.

Kamal, M., X. Xie and E. S. Lander. 2006. A large family of ancient repeat elements in the human genome is under strong selection. *Proc. Natl. Acad. Sci. USA* 103: 2740–2745.

Kaminker, J. S., C. M. Bergman, B. Kronmiller, J. Carlson, R. Svirskas, S. Patel, E. Frise, D. A. Wheeler, S. E. Lewis, G. M. Rubin et al. 2002. The transposable elements of the *Drosophila melanogaster* euchromatin: A genomics perspective. *Genome Biol.* 3(12): RESEARCH0084.

Kampa, D., J. Cheng, P. Kapranov, M. Yamanaka, S. Brubaker, S. Cawley, J. Drenkow, A. Piccolboni, S. Bekiranov, G. Helt et al. 2004. Novel RNAs identified from an in-depth analysis of the transcriptome of human chromosomes 21 and 22. *Genome Res.* 14: 331–342.

Kanaya, S., Y. Yamada, Y. Kudo and T. Ikemura. 1999. Studies of codon usage and tRNA genes of 18 unicellular organisms and quantification of *Bacillus subtilis* tRNAs: Gene expression level and species-specific diversity of codon usage based on multivariate analysis. *Gene* 238: 143–155.

Kanaya, S., Y. Yamada, M. Kinouchi, Y. Kudo and T. Ikemura. 2001. Codon usage and tRNA genes in eukaryotes: Correlation of codon usage diversity with translation efficiency and with CG-dinucleotide usage as assessed by multivariate analysis. *J. Mol. Evol.* 53: 290–298.

Kang, D. and N. Hamasaki. 2002. Maintenance of mitochondrial DNA integrity: Repair and degradation. *Curr. Genet.* 41: 311–322.

Kapitonov, V. V. and J. Jurka. 1999. Molecular paleontology of transposable elements from *Arabidopsis thaliana*. *Genetica* 107: 27–37.

Kaplan, N. L. and J. F. Brookfield. 1983. The effect of homozygosity of selective differences between sites of transposable elements. *Theor. Pop. Biol.* 23: 273–280.

Kaplan, N. L., T. Darden and C. H. Langley. 1985. Evolution and extinction of transposable elements in Mendelian populations. *Genetics* 109: 459–480.

Kaplan, N. L., R. R. Hudson and C. H. Langley. 1989. The "hitchhiking effect" revisited. *Genetics* 123: 887–899.

Kapranov, P., S. E. Cawley, J. Drenkow, S. Bekiranov, R. L. Strausberg, S. P. Fodor and T. R. Gingeras. 2002. Large-scale transcriptional activity in chromosomes 21 and 22. *Science* 296: 916–919.

Karev, G. P., Y. I. Wolf, A. Y. Rzhetsky, F. S. Berezovskaya and E. V. Koonin. 2002. Birth and death of protein domains: A simple model of evolution explains power law behavior. *BMC Evol. Biol.* 2: 18.

Karev, G. P., F. S. Berezovskaya and E. V. Koonin. 2005. Modeling genome evolution with a diffusion approximation of a birth-and-death process. *Bioinformatics* 21 Suppl.: iii12–iii19.

Karlberg, O., B. Canback, C. G. Kurland and S. G. Andersson. 2000. The dual origin of the yeast mitochondrial proteome. *Yeast* 17: 170–187.

Karlin, S., J. Mrazek and A. M. Campbell. 1998. Codon usages in different gene classes of the *Escherichia coli* genome. *Mol. Microbiol.* 29: 1341–1355.

Kashkush, K., M. Feldman and A. A. Levy. 2003. Transcriptional activation of retrotransposons alters the expression of adjacent genes in wheat. *Nature Genet.* 33: 102–106.

Kaszas, E. and J. A. Birchler. 1998. Meiotic transmission rates correlate with physical features of rearranged centromeres in maize. *Genetics* 150: 1683–1692.

Katinka, M. D., S. Duprat, E. Cornillot, G. Metenier, F. Thomarat, G. Prensier, V. Barbe, E. Peyretaillade, P. Brottier, P. Wincker et al. 2001. Genome sequence and gene compaction of the eukaryote parasite *Encephalitozoon cuniculi*. *Nature* 414: 450–453.

Katju, V. and M. Lynch. 2003. The structure and early evolution of recently arisen gene duplicates in the *Caenorhabditis elegans* genome. *Genetics* 165: 1793–1803.

Katju, V. and M. Lynch. 2006. On the formation of novel genes by duplication in the *Caenorhabditis elegans* genome. *Mol. Biol. Evol.* 23: 1056–1067.

Käufer, N. F. and J. Potashkin. 2000. Analysis of the splicing machinery in fission yeast: A comparison with budding yeast and mammals. *Nucleic Acids Res.* 28: 3003–3010.

Kauffman, S. A. 1993. *The Origins of Order: Self Organization and Selection in Evolution.* Oxford Univ. Press, Oxford, UK.

Kawabe, A. and S. Nasuda. 2005. Structure and genomic organization of centromeric repeats in *Arabidopsis* species *Mol. Genet. Genomics* 272: 593–602.

Kawach, O., C. Voss, J. Wolff, K. Hadfi, U. G. Maier and S. Zauner. 2005. Unique tRNA introns of an enslaved algal cell. *Mol. Biol. Evol.* 22: 1694–1701.

Kawasaki, H. and K. Taira. 2004. Induction of DNA methylation and gene silencing by short interfering RNAs in human cells. *Nature* 431: 211–217.

Keegan, L. P., A. Gallo and M. A. O'Connell. 2001. The many roles of an RNA editor. *Nature Rev. Genet.* 2: 869–878.

Keeling, P. J. and J. D. Palmer. 2000. Parabasalian flagellates are ancient eukaryotes. *Nature* 405: 635–637.

Keeling, P. J., J. M. Archibald, N. M. Fast and J. D. Palmer. 2004. Comment on "The evolution of modern eukaryotic phytoplankton." *Science* 306: 2191.

Keightley, P. D. and A. Eyre-Walker. 1999. Terumi Mukai and the riddle of deleterious mutation rates. *Genetics* 153: 515–523.

Keightley, P. D., M. J. Lercher and A. Eyre-Walker. 2005. Evidence for widespread degradation of gene control regions in hominid genomes. *PLoS Biol.* 3: 282–288.

Keller, W. and L. Minvielle-Sebastia. 1997. A comparison of mammalian and yeast pre-mRNA 3′-end processing. *Curr. Opin. Cell Biol.* 9: 329–336.

Kellis, M., N. Patterson, M. Endrizzi, B. Birren and E. S. Lander. 2003. Sequencing and comparison of yeast species to identify genes and regulatory elements. *Nature* 423: 241–254.

Kelly, F. D. and H. L. Levin. 2005. The evolution of transposons in *Schizosaccharomyces pombe. Cytogenet. Genome Res.* 110: 566–574.

Kelman, L. M. and Z. Kelman. 2004. Multiple origins of replication in archaea. *Trends Microbiol.* 12: 399–401.

Kent, W. J. and A. M. Zahler. 2000. Conservation, regulation, synteny, and introns in a large-scale *C. briggsae-C. elegans* genomic alignment. *Genome Res.* 10: 1115–1125.

Kern, A. D. and F. A. Kondrashov. 2004. Mechanisms and convergence of compensatory evolution in mammalian mitochondrial tRNAs. *Nature Genet.* 36: 1207–1212.

Kertész, S., Z. Kerényi, Z. Mérai, I. Bartos, T. Pálfy, E. Barta and D. Silhavy. 2006. Both introns and long 3′-UTRs operate as cis-acting elements to trigger nonsense-mediated decay in plants. *Nucleic Acids Res.* 34: 6147–6157.

Ketting, R. F., T. H. Haverkamp, H. G. van Luenen and R. H. Plasterk. 1999. *Mut-7* of *C. elegans*, required for transposon silencing and RNA interference, is a homolog of Werner syndrome helicase and RNaseD. *Cell* 99: 133–141.

Khaitovich, P., G. Weiss, M. Lachmann, I. Hellmann, W. Enard, B. Muetzel, U. Wirkner, W. Ansorge and S. Pääbo. 2004. A neutral model of transcriptome evolution. *PLoS Biol.* 2: 682–689.

Khaitovich, P., S. Pääbo and G. Weiss. 2005. Toward a neutral evolutionary model of gene expression. *Genetics* 170: 929–939.

Khaja, R., J. Zhang, J. R. Macdonald, Y. He, A. M. Joseph-George, J. Wei, M. A. Rafiq, C. Qian, M. Shago, L. Pantano et al. 2006. Genome assembly comparison identifies structural variants in the human genome. *Nature Genet.* 38: 1413–1418.

Khakhlova, O. and R. Bock. 2006. Elimination of deleterious mutations in plastid genomes by gene conversion. *Plant J.* 46: 85–94.

Khil, P. P., N. A. Smirnova, P. J. Romanienko and R. D. Camerini-Otero. 2004. The mouse X chromosome is enriched for sex-biased genes not subject to selection by meiotic sex chromosome inactivation. *Nature Genet.* 36: 642–646.

Khrapko, K., H. A. Coller, P. C. Andre, X. C. Li, J. S. Hanekamp and W. G. Thilly. 1997. Mitochondrial mutational spectra in human cells and tissues. *Proc. Natl. Acad. Sci. USA* 94: 13798–13803.

Kidwell, M. G. 1983. Evolution of hybrid dysgenesis determinants in *Drosophila melanogaster*. *Proc. Natl. Acad. Sci. USA* 80: 1655–1659.

Kidwell, M. G. 1994. The evolutionary history of the *P* family of transposable elements. *J. Hered.* 85: 339–346.

Kidwell, M. G. 2002. Transposable elements and the evolution of genome size in eukaryotes. *Genetica* 115: 49–63.

Kidwell, M. G. and D. R. Lisch. 2000. Transposable elements and host genome evolution. *Trends Ecol. Evol.* 15: 95–99.

Kidwell, M. G. and D. R. Lisch. 2002. Transposable elements as sources of genomic variation. In N. L. Craig, R. Craigie, M. Gellert and A. M. Lambowitz, eds., *Mobile DNA II*, 59–92. ASM Press, Washington, DC.

Kim, A., C. Terzian, P. Santamaria, A. Pelisson, N. Prud'homme and A. Bucheton. 1994. Retroviruses in invertebrates: The *gypsy* retrotransposon is apparently an infectious retrovirus of *Drosophila melanogaster*. *Proc. Natl. Acad. Sci. USA* 91: 1285–1289.

Kim, J. M., S. Vanguri, J. D. Boeke, A. Gabriel and D. F. Voytas. 1998. Transposable elements and genome organization: A comprehensive survey of retrotransposons revealed by the complete *Saccharomyces cerevisiae* genome sequence. *Genome Res.* 8: 464–478.

Kim, S. H. and S. V. Yi. 2006. Correlated asymmetry of sequence and functional divergence between duplicate proteins of *Saccharomyces cerevisiae*. *Mol. Biol. Evol.* 23: 1068–1075.

Kim, T. H., L. O. Barrera, M. Zheng, C. Qu, M. A. Singer, T. A. Richmond, Y. Wu, R. D. Green, B. Ren. 2005. A high-resolution map of active promoters in the human genome. *Nature* 436: 876–880.

Kim, Y. and H. A. Orr. 2005. Adaptation in sexuals vs. asexuals: Clonal interference and the Fisher-Muller model. *Genetics* 171: 1377–1386.

Kimura, H., T. Takada, Y. Shingu, Y. Kato, H. Iyehara-Ogawa and T. Terado. 1998. Neocarzinostatin-induced mutations at the *hprt* locus in exponentially growing CHO cells, compared with spontaneous mutations. *Carcinogenesis* 19: 791–796.

Kimura, M. 1962. On the probability of fixation of mutant genes in populations. *Genetics* 47: 713–719.

Kimura, M. 1965. A stochastic model concerning the maintenance of genetic variability in quantitative characters. *Proc. Natl. Acad. Sci. USA* 54: 731–736.

Kimura, M. 1983. *The Neutral Theory of Molecular Evolution*. Cambridge University Press, Cambridge.

Kimura, M. 1985. The role of compensatory neutral mutations in molecular evolution. *J. Genet.* 64: 7–19.

Kimura, M. and T. Ohta. 1969. The average number of generations until fixation of a mutant gene in a finite population. *Genetics* 61: 763–771.

King, M. C. and A. C. Wilson. 1975. Evolution at two levels in humans and chimpanzees. *Science* 188: 107–116.

King, N. 2004. The unicellular ancestry of animal development. *Dev. Cell* 7: 313–325.

Kingsolver, J. G., H. E. Hoekstra, J. M. Hoekstra, D. Berrigan, S. N. Vignieri, C. E. Hill, A. Hoang, P. Gibert and P. Beerli. 2001. The strength of phenotypic selection in natural populations. *Amer. Natur.* 157: 245–261.

Kinsey, J. A. 1993. Transnuclear retrotransposition of the *Tad* element of *Neurospora*. *Proc. Natl. Acad. Sci. USA* 90: 9384–9387.

Kinsey, J. A., P. W. Garrett-Engele, E. B. Cambareri and E. U. Selker. 1994. The *Neurospora* transposon *Tad* is sensitive to repeat-induced point mutation (RIP). *Genetics* 138: 657–664.

Kipling, D. and P. E. Warburton. 1997. Centromeres, CENP-B and Tigger too. *Trends Genet.* 13: 141–145.

Kirschner, M. and J. Gerhart. 1998. Evolvability. *Proc. Natl. Acad. Sci. USA* 95: 8420–8427.

Kirschner, M. and J. Gerhart. 2005. *The Plausibility of Life*. Yale University Press, New Haven, CT.

Kirschvink, J. L., E. J. Gaidos, L. E. Bertani, N. J. Beukes, J. Gutzmer, L. N. Maepa and R. E. Steinberger. 2000. Paleoproterozoic snowball earth: Extreme climatic and geochemical global change and its biological consequences. *Proc. Natl. Acad. Sci. USA* 97: 1400–1405.

Kivisild, T. and R. Villems. 2000. Questioning evidence for recombination in human mitochondrial DNA. *Science* 288: 1931.

Kivisild, T., P. Shen, D. P. Wall, B. Do, R. Sung, K. Davis, G. Passarino, P. A. Underhill, C. Scharfe, A. Torroni et al. 2006. The role of selection in the evolution of human mitochondrial genomes. *Genetics* 172: 373–387.

Klahre, U., P. Crete, S. A. Leuenberger, V. A. Iglesias and F. Meins, Jr. 2002. High molecular weight RNAs and small interfering RNAs induce systemic post-transcriptional gene silencing in plants. *Proc. Natl. Acad. Sci. USA* 99: 11981–11986.

Klasson, L. and S. G. Andersson. 2006. Strong asymmetric mutation bias in endosymbiont genomes coincide with loss of genes for replication restart pathways. *Mol. Biol. Evol.* 23: 1031–1039.

Kleckner, N. 1990. Regulation of transposition in bacteria. *Annu. Rev. Cell Biol.* 6: 297–327.

Kliman, R. M. and J. Hey. 1994. The effects of mutation and natural selection on codon bias in the genes of *Drosophila*. *Genetics* 137: 1049–1056.

Kliman, R. M. and J. Hey. 2003. Hill-Robertson interference in *Drosophila melanogaster*: Reply to Marais, Mouchiroud and Duret. *Genet. Res.* 81: 89–90.

Knight, R. D., L. F. Landweber and M. Yarus. 2001. How mitochondria redefine the code. *J. Mol. Evol.* 53: 299–313.

Knight, R. D., S. J. Freeland and L. F. Landweber. 2001. Rewiring the keyboard: Evolvability of the genetic code. *Nature Rev. Genet.* 2: 49–58.

Knoll, A. H. 1992. The early evolution of eukaryotes: A geological perspective. *Science* 256: 622–627.

Knoll, A. H. 2003. *Life on a Young Planet*. Princeton University Press, Princeton, NJ.

Knoll, A. H. and S. B. Carroll. 1999. Early animal evolution: Emerging views from comparative biology and geology. *Science* 284: 2129–2137.

Knoop, V. 2004. The mitochondrial DNA of land plants: Peculiarities in phylogenetic perspective. *Curr. Genet.* 46: 123–139.

Knowles, D. G. and A. McLysaght. 2006. High rate of recent intron gain and loss in simultaneously duplicated *Arabidopsis* genes. *Mol. Biol. Evol.* 23: 1548–1557.

Kobayashi, I. 1998. Selfishness and death: Raison d'etre of restriction, recombination and mitochondria. *Trends Genet.* 14: 368–374.

Kobayashi, T. and H. Endoh. 2005. A possible role of mitochondria in the apoptotic-like programmed nuclear death of *Tetrahymena thermophila*. *FEBS J.* 272: 5378–5387.

Kobryn, K. and G. Chaconas. 2001. The circle is broken: Telomere resolution in linear replicons. *Curr. Opin. Microbiol.* 4: 558–564.

Koch, M. A., B. Haubold and T. Mitchell-Olds. 2000. Comparative evolutionary analysis of chalcone synthase and alcohol dehydrogenase loci in *Arabidopsis, Arabis,* and related genera (Brassicaceae). *Mol. Biol. Evol.* 7: 1483–1498.

Kohn, M., H. Kehrer-Sawatzki, W. Vogel, J. A. Graves and H. Hameister. 2004. Wide genome comparisons reveal the origins of the human X chromosome. *Trends Genet.* 20: 598–603.

Kohzaki, H. and Y. Murakami. 2005. Transcription factors and DNA replication origin selection. *BioEssays* 27: 1107–1116.

Kojima, K. K. and H. Fujiwara. 2003. Evolution of target specificity in *R1* clade non-LTR retrotransposons. *Mol. Biol. Evol.* 20: 351–361.

Kojima, K. K. and H. Fujiwara. 2004. Cross-genome screening of novel sequence-specific non-LTR retrotransposons: Various multicopy RNA genes and microsatellites are selected as targets. *Mol. Biol. Evol.* 21: 207–217.

Kojima, K. K. and H. Fujiwara. 2005. An extraordinary retrotransposon family encoding dual endonucleases. *Genome Res.* 15: 1106–1117.

Komaki, K. and H. Ishikawa. 2000. Genomic copy number of intracellular bacterial symbionts of aphids varies in response to developmental stage and morph of their host. *Insect Biochem. Mol. Biol.* 30: 253–258.

Kondo, M., I. Nanda, U. Hornung, M. Schmid and M. Schartl. 2004. Evolutionary origin of the medaka Y chromosome. *Curr. Biol.* 14: 1664–1669.
Kondo, M., U. Hornung, I. Nanda, S. Imai, T. Sasaki, A. Shimizu, S. Asakawa, H. Hori, M. Schmid, N. Shimizu and M. Schartl. 2006. Genomic organization of the sex-determining and adjacent regions of the sex chromosomes of medaka. *Genome Res.* 16: 815–826.
Kondrashov, A. S. 1993. Classification of hypotheses on the advantage of amphimixis. *J. Hered.* 84: 372–387.
Kondrashov, A. S. 1995. Contamination of the genome by very slightly deleterious mutations: Why have we not died 100 times over? *J. Theor. Biol.* 175: 583–594.
Kondrashov, A. S. 2003. Direct estimates of human per nucleotide mutation rates at 20 loci causing Mendelian diseases. *Hum. Mutat.* 21: 12–27.
Kondrashov, F. A. and A. S. Kondrashov. 2006. Role of selection in fixation of gene duplications. *J. Theor. Biol.* 239: 141–151.
Kondrashov, F. A. and E. V. Koonin. 2001. Origin of alternative splicing by tandem exon duplication. *Hum. Mol. Genet.* 10: 2661–2669.
Kondrashov, F. A., I. B. Rogozin, Y. I. Wolf and E. V. Koonin. 2002. Selection in the evolution of gene duplications. *Genome Biol.* 3(2): RESEARCH0008.
Kondrashov, F. A., A. Y. Ogurtsov and A. S. Kondrashov. 2006. Selection in favor of nucleotides G and C diversifies evolution rates and levels of polymorphism at mammalian synonymous sites. *J. Theor. Biol.* 240: 616–626.
Kong, A., D. F. Gudbjartsson, J. Sainz, G. M. Jonsdottir, S. A. Gudjonsson, B. Richardsson, S. Sigurdardottir, J. Barnard, B. Hallbeck, G. Masson et al. 2002. A high-resolution recombination map of the human genome. *Nature Genet.* 31: 241–247.
Koonin, E. V. 2006. The origin of introns and their role in eukaryogenesis: A compromise solution to the introns-early versus introns-late debate? *Biol. Direct* 1: 22.
Koonin, E. V. and W. Martin. 2005. On the origin of genomes and cells within inorganic compartments. *Trends Genet.* 21: 647–654.
Koonin, E. V., A. R. Mushegian, M. Y. Galperin and D. R. Walker. 1997. Comparison of archaeal and bacterial genomes: Computer analysis of protein sequences predicts novel functions and suggests a chimeric origin for the archaea. *Mol. Microbiol.* 25: 619–637.
Koonin, E. V., Y. I. Wolf and G. P. Karev. 2002. The structure of the protein universe and genome evolution. *Nature* 420: 218–223.
Kooter, J. M., T. De Lange and P. Borst. 1984. Discontinuous synthesis of mRNA in trypanosomes. *EMBO J.* 3: 2387–2392.
Kopelman, N. M., D. Lancet and I. Yanai. 2005. Alternative splicing and gene duplication are inversely correlated evolutionary mechanisms. *Nature Genet.* 37: 588–589.
Korswagen, H. C., M. A. Herman and H. C. Clevers. 2000. Distinct β-catenins mediate adhesion and signalling functions in *C. elegans*. *Nature* 406: 527–532.
Koslowsky, D. J., H. U. Goringer, T. H. Morales and K. Stuart. 1992. In vitro guide RNA/mRNA chimaera formation in *Trypanosoma brucei* RNA editing. *Nature* 356: 807–809.
Kotera, E., M. Tasaka and T. Shikanai. 2005. A pentatricopeptide repeat protein is essential for RNA editing in chloroplasts. *Nature* 433: 326–330.
Kozak, M. 1987. At least six nucleotides preceding the AUG initiator codon enhance translation in mammalian cells. *J. Mol. Biol.* 196: 947–950.
Kozak, M. 1994. Determinants of translational fidelity and efficiency in vertebrate mRNAs. *Biochimie* 76: 815–821.
Kozak, M. 2002. Pushing the limits of the scanning mechanism for initiation of translation. *Gene* 299: 1–34.
Kozlowski, J., M. Konarzewski and A. T. Gawelczyk. 2003. Cell size as a link between noncoding DNA and metabolic rate scaling. *Proc. Natl. Acad. Sci. USA* 100: 14080–14085.
Krakauer, D. C. and A. Mira. 1999. Mitochondria and germ-cell death. *Nature* 400: 125–126.
Krauss, V. and R. Dorn. 2004. Evolution of the *trans*-splicing *Drosophila* locus *mod(mdg4)* in several species of Diptera and Lepidoptera. *Gene* 331: 165–176.
Krauss, V., M. Pecyna, K. Kurz and H. Sass. 2005. Phylogenetic mapping of intron positions: A case study of translation initiation factor eIF2γ. *Mol. Biol. Evol.* 22: 74–84.
Krawczak, M., E. V. Ball and D. N. Cooper. 1998. Neighboring-nucleotide effects on the rates of germ-line single-base-pair substitution in human genes. *Amer. J. Hum. Genet.* 63: 474–488.
Kraytsberg, Y., M. Schwartz, T. A. Brown, K. Ebralidse, W. S. Kunz, D. A. Clayton, J. Vissing and K. Khrapko. 2004. Recombination of human mitochondrial DNA. *Science* 304: 981.
Kreahling, J. and B. R. Graveley. 2004. The origins and implications of *Alu*ternative splicing. *Trends Genet.* 20: 1–4.
Kriventseva, E. V., I. Koch, R. Apweiler, M. Vingron, P. Bork, M. S. Gelfand and S. Sunyaev. 2003. Increase of functional diversity by alternative splicing. *Trends Genet.* 19: 124–128.
Kroemer, G., B. Dallaporta and M. Resche-Rigon. 1998. The mitochondrial death/life regulator in apoptosis and necrosis. *Annu. Rev. Physiol.* 60: 619–642.
Krzywinski, J. and N. J. Besansky. 2002. Frequent intron loss in the *White* gene: A cautionary tale for phylogeneticists. *Mol. Biol. Evol.* 19: 362–366.
Ku, H. M., T. Vision, J. Liu and S. D. Tanksley. 2000. Comparing sequenced segments of the tomato and *Arabidopsis* genomes: Large-scale duplication followed by selective gene loss creates a network of synteny. *Proc. Natl. Acad. Sci. USA* 97: 9121–9126.
Kubo, N., X. Jordana, K. Ozawa, S. Zanlungo, K. Harada, T. Sasaki and K. Kadowaki. 2000. Transfer of the mitochondrial *rps10* gene to the nucleus in rice: Acquisition of the 5′ untranslated region followed by gene duplication. *Mol. Gen. Genet.* 263: 733–739.
Kudla, J., F. J. Albertazzi, D. Blazevic, M. Hermann and R. Bock. 2002. Loss of the mitochondrial cox2 intron 1 in a family of monocotyledonous plants

and utilization of mitochondrial intron sequences for the construction of a nuclear intron. *Mol. Genet. Genomics* 267: 223–230.

Kugita, M., Y. Yamamoto, T. Fujikawa, T. Matsumoto and K. Yoshinaga. 2003. RNA editing in hornwort chloroplasts makes more than half the genes functional. *Nucleic Acids Res.* 31: 2417–2423.

Kuhn, A. N. and N. F. Käufer. 2003. Pre-mRNA splicing in *Schizosaccharomyces pombe:* Regulatory role of a kinase conserved from fission yeast to mammals. *Curr. Genet.* 42: 241–251.

Kühn, K., A. Weihe and T. Börner. 2005. Multiple promoters are a common feature of mitochondrial genes in *Arabidopsis*. *Nucleic Acids Res.* 33: 337–346.

Kumar, S. and S. B. Hedges. 1998. A molecular timescale for vertebrate evolution. *Nature* 392: 917–920.

Kumar, S., P. Hedrick, T. Dowling and M. Stoneking. 2000. Questioning evidence for recombination in human mitochondrial DNA. *Science* 288: 1931.

Kumazawa, Y. and M. Nishida. 1993. Sequence evolution of mitochondrial tRNA genes and deep-branch animal phylogenetics. *J. Mol. Evol.* 37: 380–398.

Kumazawa, Y., T. Yokogawa, E. Hasegawa, K. Miura and K. Watanabe. 1989. The aminoacylation of structurally variant phenylalanine tRNAs from mitochondria and various nonmitochondrial sources by bovine mitochondrial phenylalanyl-tRNA synthetase. *J. Biol. Chem.* 264: 13005–13011.

Kumazawa, Y., C. J. Schwartzbach, H. X. Liao, K. Mizumoto, Y. Kaziro, K. Miura, K. Watanabe and L. L. Spremulli. 1991. Interactions of bovine mitochondrial phenylalanyl-tRNA with ribosomes and elongation factors from mitochondria and bacteria. *Biochim. Biophys. Acta* 1090: 167–172.

Kunkel, T. A. and P. S. Alexander. 1986. The base substitution fidelity of eucaryotic DNA polymerases. Mispairing frequencies, site preferences, insertion preferences, and base substitution by dislocation. *J. Biol. Chem.* 261: 160–166.

Kunz, B. A., K. Ramachandran and E. J. Vonarx. 1998. DNA sequence analysis of spontaneous mutagenesis in *Saccharomyces cerevisiae*. *Genetics* 148: 1491–1505.

Kunze, B., T. Hellwig-Burgel, D. Weichenhan and W. Traut. 2000. Transcription and proper splicing of a mammalian gene in yeast. *Gene* 246: 93–102.

Kupfer, D. M., S. D. Drabenstot, K. L. Buchanan, H. Lai, H. Zhu, D. W. Dyer, B. A. Roe and J. W. Murphy. 2004. Introns and splicing elements of five diverse fungi. *Eukaryot. Cell* 3: 1088–1100.

Kurek, R., A. M. Reugels, U. Lammermann and H. Bunemann. 2000. Molecular aspects of intron evolution in dynein encoding mega-genes on the heterochromatic Y chromosome of *Drosophila* sp. *Genetica* 109: 113–123.

Kurland, C. G. and S. G. Andersson. 2000. Origin and evolution of the mitochondrial proteome. *Microbiol. Mol. Biol. Rev.* 64: 786–820.

Kurland, C. G., L. J. Collins and D. Penny. 2006. Genomics and the irreducible nature of eukaryote cells. *Science* 312: 1011–1014.

Kuroki, Y., A. Toyoda, H. Noguchi, T. D. Taylor, T. Itoh, D. S. Kim, D. W. Kim, S. H. Choi, I. C. Kim, H. H. Choi et al. 2006. Comparative analysis of chimpanzee and human Y chromosomes unveils complex evolutionary pathway. *Nature Genet.* 38: 158–167.

Kutach, A. K. and J. T. Kadonaga. 2000. The downstream promoter element DPE appears to be as widely used as the TATA box in *Drosophila* core promoters. *Mol. Cell. Biol.* 20: 4754–4764.

Kyrpides, N. C. and C. R. Woese. 1998. Universally conserved translation initiation factors. *Proc. Natl. Acad. Sci. USA* 95: 224–228.

Labrador, M. and V. G. Corces. 1997. Transposable element-host interactions: Regulation of insertion and excision. *Annu. Rev. Genet.* 31: 381–404.

Labrador, M., M. Farre, F. Utzet and A. Fontdevila. 1999. Interspecific hybridization increases transposition rates of *Osvaldo*. *Mol. Biol. Evol.* 16: 931–937.

Lacadie, S. A. and M. Rosbash. 2005. Cotranscriptional spliceosome assembly dynamics and the role of U1 snRNA: 5′ss base pairing in yeast. *Mol. Cell* 19: 65–75.

Ladd, A. N. and T. A. Cooper. 2002. Finding signals that regulate alternative splicing in the post-genomic era. *Genome Biol.* 3: Reviews0008.

Lafay, B. and P. M. Sharp. 1999. Synonymous codon usage variation among *Giardia lamblia* genes and isolates. *Mol. Biol. Evol.* 16: 1484–1495.

Lafay, B., A. T. Lloyd, M. J. McLean, K. M. Devine, P. M. Sharp and K. H. Wolfe. 1999. Proteome composition and codon usage in spirochaetes: Species-specific and DNA strand-specific mutational biases. *Nucleic Acids Res.* 27: 1642–1649.

Laforest, M. J., C. E. Bullerwell, L. Forget and B. F. Lang. 2004. Origin, evolution, and mechanism of 5′ tRNA editing in chytridiomycete fungi. *RNA* 10: 1191–1199.

Lagercrantz, U. 1998. Comparative mapping between *Arabidopsis thaliana* and *Brassica nigra* indicates that *Brassica* genomes have evolved through extensive genome replication accompanied by chromosome fusions and frequent rearrangements. *Genetics* 150: 1217–1228.

Lagercrantz, U., M. Kruskopf Osterberg and M. Lascoux. 2002. Sequence variation and haplotype structure at the putative flowering-time locus *COL1* of *Brassica nigra*. *Mol. Biol. Evol.* 19: 1474–1482.

Lahn, B. T. and D. C. Page. 1999. Four evolutionary strata on the human X chromosome. *Science* 286: 964–967.

Lake, J. A. and M. C. Rivera. 1994. Was the nucleus the first endosymbiont? *Proc. Natl. Acad. Sci. USA* 91: 2880–2881.

Lake, J. A., R. Jain and M. C. Rivera. 1999. Mix and match in the tree of life. *Science* 283: 2027–2028.

Lal, S. K. and L. C. Hannah. 1999. Maize transposable element Ds is differentially spliced from primary transcripts in endosperm and suspension cells. *Biochem. Biophys. Res. Commun.* 261: 798–801.

Lamb, B. C. 1985. The effects of mispair and nonpair correction in hybrid DNA on base ratios (G + C content) and total amounts of DNA. *Mol. Biol. Evol.* 2: 175–188.
Lamb, B. C. 1998. Gene conversion disparity in yeast: Its extent, multiple origins, and effects on allele frequencies. *Heredity* 80: 538–552.
Lambert, J. D. and N. A. Moran. 1998. Deleterious mutations destabilize ribosomal RNA in endosymbiotic bacteria. *Proc. Natl. Acad. Sci. USA* 95: 4458–4462.
Lambowitz, A. M. and S. Zimmerly. 2004. Mobile group II introns. *Annu. Rev. Genet.* 38: 1–35.
Lambowitz, A. M., M. G. Caprara, S. Zimmerly and P. S. Perlman. 1999. Group I and group II ribozymes as RNPs: Clues to the past and guides to the future. In R. F. Gesteland, T. R. Cech and J. F. Atkins, eds., *The RNA World*, 2nd ed., 451–486. Cold Spring Harbor Laboratory Press, Cold Spring Harbor, NY.
Lampe, D. J., D. J. Witherspoon, F. N. Soto-Adames and H. M. Robertson. 2003. Recent horizontal transfer of *mellifera* subfamily *mariner* transposons into insect lineages representing four different orders shows that selection acts only during horizontal transfer. *Mol. Biol. Evol.* 20: 554–562.
Lampson, B. C., M. Inouye and S. Inouye. 2005. Retrons, msDNA, and the bacterial genome. *Cytogenet. Genome Res.* 110: 491–499.
Lande, R. 1975. The maintenance of genetic variability by mutation in a polygenic character with linked loci. *Genet. Res.* 26: 221–235.
Lande, R. 1976. Natural selection and random genetic drift in phenotypic evolution. *Evolution* 30: 314–334.
Lande, R. 1979. Effective deme sizes during long-term evolution estimated from rates of chromosomal rearrangement. *Evolution* 33: 234–251.
Lande, R. 1994. Risk of population extinction from fixation of new deleterious mutations. *Evolution* 48: 1460–1469.
Lander, E. S., L. M. Linton, B. Birren, C. Nusbaum, M. C. Zody, J. Baldwin, K. Devon, K. Dewar, M. Doyle, W. FitzHugh et al. 2001. Initial sequencing and analysis of the human genome. *Nature* 409: 860–921.
Landweber, L. F. and W. Gilbert. 1993. RNA editing as a source of genetic variation. *Nature* 363: 179–182.
Lang, B. F., M. W. Gray and G. Burger. 1999. Mitochondrial genome evolution and the origin of eukaryotes. *Annu. Rev. Genet.* 33: 351–397.
Langham, R. J., J. Walsh, M. Dunn, C. Ko, S. A. Goff and M. Freeling. 2004. Genomic duplication, fractionation and the origin of regulatory novelty. *Genetics* 166: 935–945.
Langkjaer, R. B., P. F. Cliften, M. Johnston and J. Piskur. 2003. Yeast genome duplication was followed by asynchronous differentiation of duplicated genes. *Nature* 421: 848–852.
Langley, C. H., J. F. Y. Brookfield and N. Kaplan. 1983. Transposable elements in Mendelian populations. I. Theory. *Genetics* 104: 457–471.
Langley, C. H., E. Montgomery, R. Hudson, N. Kaplan and B. Charlesworth. 1988. On the role of unequal exchange in the containment of transposable element copy number. *Genet. Res.* 52: 223–235.
Laporte, V. and B. Charlesworth. 2002. Effective population size and population subdivision in demographically structured populations. *Genetics* 162: 501–519.
Laporte, V., D. A. Filatov, E. Kamau and D. Charlesworth. 2005. Indirect evidence from DNA sequence diversity for genetic degeneration of the Y-chromosome in dioecious species of the plant *Silene*: The *SlY4/SlX4* and *DD44-X/DD44*-Y gene pairs. *J. Evol. Biol.* 18: 337–347.
Larson, A., E. M. Prager and A. C. Wilson. 1984. Chromosomal evolution, speciation and morphological change in vertebrates: The role of social behaviour. *Chromosomes Today* 8: 215–228.
Laski, F. A., D. C. Rio and G. M. Rubin. 1986. Tissue specificity of *Drosophila P* element transposition is regulated at the level of mRNA splicing. *Cell* 44: 7–19.
Lathe, W. C. III, B. Snel and P. Bork. 2000. Gene context conservation of a higher order than operons. *Trends Biochem Sci.* 25: 474–479.
Latorre, A., R. Gil, F. J. Silva and A. Moya. 2005. Chromosomal stasis versus plasmid plasticity in aphid endosymbiont *Buchnera aphidicola*. *Heredity* 95: 339–347.
Lavner, Y. and D. Kotlar. 2005. Codon bias as a factor in regulating expression via translation rate in the human genome. *Gene* 345: 127–138.
Lavrov, D. V., W. M. Brown and J. L. Boore. 2000. A novel type of RNA editing occurs in the mitochondrial tRNAs of the centipede *Lithobius forficatus*. *Proc. Natl. Acad. Sci. USA* 97: 13738–13742.
Lawrence, J. G. and H. Ochman. 1998. Molecular archaeology of the *Escherichia coli* genome. *Proc. Natl. Acad. Sci. USA* 95: 9413–9417.
Lawrence, J. G. and J. R. Roth. 1996. Selfish operons: Horizontal transfer may drive the evolution of gene clusters. *Genetics* 143: 1843–1860.
Lawrence, J. G., R. W. Hendrix and S. Casjens. 2001. Where are the pseudogenes in bacterial genomes? *Trends Microbiol.* 9: 535–540.
Lawson, F. S., R. L. Charlebois and J. A. Dillon. 1996. Phylogenetic analysis of carbamoylphosphate synthetase genes: Complex evolutionary history includes an internal duplication within a gene which can root the tree of life. *Mol. Biol. Evol.* 13: 970–977.
Lawson, L. J. and G. M. Hewitt. 2002. Comparison of substitution rates in *ZFX* and *ZFY* introns of sheep and goat related species supports the hypothesis of male-biased mutation rates. *J. Mol. Evol.* 54: 54–61.
LeBlanc, P., S. Desset, F. Giorgi, A. R. Taddei, A. M. Fausto, M. Mazzini, B. Dastugue and C. Vaury. 2000. Life cycle of an endogenous retrovirus, *ZAM*, in *Drosophila melanogaster*. *J. Virol.* 74: 10658–10669.
Lee, C. C., E. L. Beall and D. C. Rio. 1998. DNA binding by the KP repressor protein inhibits P-element transposase activity in vitro. *EMBO J.* 17: 4166–4174.
Lee, H. R., W. Zhang, T. Langdon, W. Jin, H. Yan, Z. Cheng and J. Jiang. 2005. Chromatin immunopre-

cipitation cloning reveals rapid evolutionary patterns of centromeric DNA in *Oryza* species *Proc. Natl. Acad. Sci. USA* 102: 11793–11798.

Lee, J. M. and E. L. Sonnhammer. 2003. Genomic gene clustering analysis of pathways in eukaryotes. *Genome Res.* 13: 875–882.

Lee, K. Z. and R. J. Sommer. 2003. Operon structure and *trans*-splicing in the nematode *Pristionchus pacificus*. *Mol. Biol. Evol.* 20: 2097–2103.

Lee, S. J. 1991. Expression of growth/differentiation factor 1 in the nervous system: Conservation of a bicistronic structure. *Proc. Natl. Acad. Sci. USA* 88: 4250–4254.

Lee, S. R. and K. Collins. 2006. Two classes of endogenous small RNAs in *Tetrahymena thermophila*. *Genes Dev.* 20: 28–33.

Leeds, P., J. M. Wood, B. S. Lee and M. R. Culbertson. 1992. Gene products that promote mRNA turnover in *Saccharomyces cerevisiae*. *Mol. Cell. Biol.* 12: 2165–2177.

Leipe, D. D., L. Aravind and E. V. Koonin. 1999. Did DNA replication evolve twice independently? *Nucleic Acids Res.* 27: 3389–3401.

Leister, D. 2005. Origin, evolution and genetic effects of nuclear insertions of organelle DNA. *Trends Genet.* 21: 655–663.

Lemos, B., B. R. Bettencourt, C. D. Meiklejohn and D. L. Hartl. 2005. Evolution of proteins and gene expression levels are coupled in *Drosophila* and are independently associated with mRNA abundance, protein length, and number of protein-protein interactions. *Mol. Biol. Evol.* 22: 1345–1354.

Lenormand, T., T. Guillemaud, D. Bourguet and M. Raymond. 1998. Appearance and sweep of a gene duplication: Adaptive response and potential for new functions in the mosquito *Culex pipiens*. *Evolution* 52: 1705–1712.

Lerat, E., V. Daubin, H. Ochman and N. A. Moran. 2005. Evolutionary origins of genomic repertoires in bacteria. *PLoS Biol.* 3: 807–814.

Lercher, M. J. and L. D. Hurst. 2002. Human SNP variability and mutation rate are higher in regions of high recombination. *Trends Genet.* 18: 337–340.

Lercher, M. J., A. O. Urrutia and L. D. Hurst. 2002. Clustering of housekeeping genes provides a unified model of gene order in the human genome. *Nature Genet.* 31: 180–183.

Lercher, M. J., T. Blumenthal and L. D. Hurst. 2003. Coexpression of neighboring genes in *Caenorhabditis elegans* is mostly due to operons and duplicate genes. *Genome Res.* 13: 238–243.

Lercher, M. J., A. O. Urrutia and L. D. Hurst. 2003. Evidence that the human X chromosome is enriched for male-specific but not female-specific genes. *Mol. Biol. Evol.* 20: 1113–1116.

Lesage, P. and A. L. Todeschini. 2005. Happy together: The life and times of *Ty* retrotransposons and their hosts. *Cytogenet. Genome Res.* 110: 70–90.

Lespinet, O., Y. I. Wolf, E. V. Koonin and L. Aravind. 2002. The role of lineage-specific gene family expansion in the evolution of eukaryotes. *Genome Res.* 12: 1048–1059.

Lester, L., A. Meade and M. Pagel. 2006. The slow road to the eukaryotic genome. *BioEssays* 28: 57–64.

Letunic, I., R. R. Copley and P. Bork. 2002. Common exon duplication in animals and its role in alternative splicing. *Hum. Mol. Genet.* 11: 1561–1567.

Levin, D. A., B. G. Palestis, R. N. Jones and R. Trivers. 2005. Phyletic hot spots for B chromosomes in angiosperms. *Evolution* 59: 962–969.

Levin, H. L. 1995. A novel mechanism of self-primed reverse transcription defines a new family of retroelements. *Mol. Cell. Biol.* 15: 3310–3317.

Levine, A. and R. Durbin. 2001. A computational scan for U12-dependent introns in the human genome sequence. *Nucleic Acids Res.* 29: 4006–4013.

Levis, R. W., R. Ganesan, K. Houtchens, L. A. Tolar and F. M. Sheen. 1993. Transposons in place of telomeric repeats at a *Drosophila* telomere. *Cell* 75: 1083–1093.

Lev-Maor, G., R. Sorek, N. Shomron and G. Ast. 2003. The birth of an alternatively spliced exon: 3′ splice-site selection in *Alu* exons. *Science* 300: 1288–1291.

Lewis, B. P., R. E. Green and S. E. Brenner. 2003. Evidence for the widespread coupling of alternative splicing and nonsense-mediated mRNA decay in humans. *Proc. Natl. Acad. Sci. USA* 100: 189–192.

Lewontin, R. C. 1974. *The Genetic Basis of Evolutionary Change*. Columbia University Press, New York.

Li, B., C. Wachtel, E. Miriami, G. Yahalom, G. Friedlander, G. Sharon, R. Sperling and J. Sperling. 2002. Stop codons affect 5′ splice site selection by surveillance of splicing. *Proc. Natl. Acad. Sci. USA* 99: 5277–5282.

Li, Q., K. R. Peterson, X. Fang and G. Stamatoyannopoulos. 2002. Locus control regions. *Blood* 100: 3077–3086.

Li, S., T. Nosenko, J. D. Hackett and D. Bhattacharya. 2006. Phylogenomic analysis identifies red algal genes of endosymbiotic origin in the chromalveolates. *Mol. Biol. Evol.* 23: 663–674.

Li, W. and J. E. Shaw. 1993. A variant Tc4 transposable element in the nematode *C. elegans* could encode a novel protein. *Nucleic Acids Res.* 21: 59–67.

Li, W.-H. 1985. Accelerated evolution following gene duplication and its implication for the neutralist-selectionist controversy. In T. Ohta and K. Aoki, eds., *Population Genetics and Molecular Evolution*, 333–352. Springer-Verlag, Berlin.

Li, W.-H. 1987. Models of nearly neutral mutations with particular implications for nonrandom usage of synonymous codons. *J. Mol. Evol.* 24: 337–345.

Li, W.-H. 1997. *Molecular Evolution*. Sinauer Associates, Sunderland, MA.

Li, W.-H. and M. Nei. 1972. Total number of individuals affected by a single deleterious mutation in a finite population. *Amer. J. Hum. Genet.* 24: 667–679.

Li, W.-H., Z. Gu, H. Wang and A. Nekrutenko. 2001. Evolutionary analyses of the human genome. *Nature* 409: 847–849.

Lieber, M. R., Y. Ma, U. Pannicke and K. Schwarz. 2003. Mechanism and regulation of human non-homologous DNA end-joining. *Nature Rev. Mol. Cell. Biol.* 4: 712–720.

Lim, L. P., M. E. Glasner, S. Yekta, C. B. Burge and D. P. Bartel. 2003a. Vertebrate microRNA genes. *Science* 299: 1540.
Lim, L. P., N. C. Lau, E. G. Weinstein, A. Abdelhakim, S. Yekta, M. W. Rhoades, C. B. Burge and D. P. Bartel. 2003b. The microRNAs of *Caenorhabditis elegans*. *Genes Dev.* 17: 991–1008.
Lim, L. P., N. C. Lau, P. Garrett-Engele, A. Grimson, J. M. Schelter, J. Castle, D. P. Bartel, P. S. Linsley and J. M. Johnson. 2005. Microarray analysis shows that some microRNAs downregulate large numbers of target mRNAs. *Nature* 433: 769–773.
Lin, C. T., W. H. Lin, Y. L. Lyu and J. Whang-Peng. 2001. Inverted repeats as genetic elements for promoting DNA inverted duplication: Implications in gene amplification. *Nucleic Acids Res.* 29: 3529–3538.
Lin, H., W. Zhu, J. C. Silva, X. Gu and C. R. Buell. 2006. Intron gain and loss in segmentally duplicated genes in rice. *Genome Biol.* 7(5): R41.
Lin, S., H. Zhang, D. F. Spencer, J. E. Norman and M. W. Gray. 2002. Widespread and extensive editing of mitochondrial mRNAS in dinoflagellates. *J. Mol. Biol.* 320: 727–739.
Lin, Y. and A. S. Waldman. 2001a. Capture of DNA sequences at double-strand breaks in mammalian chromosomes. *Genetics* 158: 1665–1674.
Lin, Y. and A. S. Waldman. 2001b. Promiscuous patching of broken chromosomes in mammalian cells with extrachromosomal DNA. *Nucleic Acids Res.* 29: 3975–3981.
Lin, Z., H. Kong, M. Nei and H. Ma. 2006. Origins and evolution of the *recA/RAD51* gene family: Evidence for ancient gene duplication and endosymbiotic gene transfer. *Proc. Natl. Acad. Sci. USA* 103: 10328–10333.
Lindahl, T. 1993. Instability and decay of the primary structure of DNA. *Nature* 362: 709–715.
Lindsay, M. R., R. I. Webb, M. Strous, M. S. Jetten, M. K. Butler, R. J. Forde and J. A. Fuerst. 2001. Cell compartmentalisation in planctomycetes: Novel types of structural organisation for the bacterial cell. *Arch. Microbiol.* 175: 413–429.
Ling, F., H. Morioka, E. Ohtsuka and T. Shibata. 2000. A role for *MHR1*, a gene required for mitochondrial genetic recombination, in the repair of damage spontaneously introduced in yeast mtDNA. *Nucleic Acids Res.* 28: 4956–4963.
Lipatov, M., P. F. Arndt, T. Hwa and D. A. Petrov. 2006. A novel method distinguishes between mutation rates and fixation biases in patterns of single-nucleotide substitution. *J. Mol. Evol.* 62: 168–175.
Lippman, Z., B. May, C. Yordan, T. Singer and R. Martienssen. 2003. Distinct mechanisms determine transposon inheritance and methylation via small interfering RNA and histone modification. *PLoS Biol.* 1: 420–428.
Lipson, H., J. B. Pollack and N. P. Such. 2002. On the origin of modular variation. *Evolution* 56: 1549–1556.
Lister, J. A., J. Close and D. W. Raible. 2001. Duplicate *mitf* genes in zebrafish: Complementary expression and conservation of melanogenic potential. *Dev. Biol.* 237: 333–344.
Liston, D. R. and P. J. Johnson. 1999. Analysis of a ubiquitous promoter element in a primitive eukaryote: Early evolution of the initiator element. *Mol. Cell. Biol.* 19: 2380–2388.
Liu, H. X., M. Zhang and A. R. Krainer. 1998. Identification of functional exonic splicing enhancer motifs recognized by individual SR proteins. *Genes Dev.* 12: 1998–2012.
Liu, R. Z., M. K. Sharma, Q. Sun, C. Thisse, B. Thisse, E. M. Denovan-Wright and J. M. Wright. 2005. Retention of the duplicated cellular retinoic acid-binding protein 1 genes (*crabp1a* and *crabp1b*) in the zebrafish genome by subfunctionalization of tissue-specific expression. *FEBS J.* 272: 3561–3571.
Liu, Y., S. Kuersten, T. Huang, A. Larsen, M. MacMorris and T. Blumenthal. 2003. An uncapped RNA suggests a model for *Caenorhabditis elegans* polycistronic pre-mRNA processing. *RNA* 9: 677–687.
Liu, Z., P. H. Moore, H. Ma, C. M. Ackerman, M. Ragiba, Q. Yu, H. M. Pearl, M. S. Kim, J. W. Charlton, J. I. Stiles et al. 2004. A primitive Y chromosome in papaya marks incipient sex chromosome evolution. *Nature* 427: 348–352.
Lizama, L., L. Holuigue and X. Jordana. 1994. Transcription initiation sites for the potato mitochondrial gene coding for subunit 9 of ATP synthase (atp9). *FEBS Lett.* 349: 243–248.
Llopart, A., J. M. Comeron, F. G. Brunet, D. Lachaise and M. Long. 2002. Intron presence-absence polymorphism in *Drosophila* driven by positive Darwinian selection. *Proc. Natl. Acad. Sci. USA* 99: 8121–8126.
Lo, K. and S. T. Smale. 1996. Generality of a functional initiator consensus sequence. *Gene* 182: 13–22.
Loayza, D. and T. de Lange. 2003. POT1 as a terminal transducer of TRF1 telomere length control. *Nature* 423: 1013–1018.
Lobry, J. R. 1995. Properties of a general model of DNA evolution under no-strand-bias conditions. *J. Mol. Evol.* 40: 326–330.
Lobry, J. R. 1996. Asymmetric substitution patterns in the two DNA strands of bacteria. *Mol. Biol. Evol.* 13: 660–665.
Lobry, J. R. 1997. Influence of genomic G+C content on average amino-acid composition of proteins from 59 bacterial species. *Gene* 205: 309–316.
Lobry, J. R. and N. Sueoka. 2002. Asymmetric directional mutation pressures in bacteria. *Genome Biol.* 3: RESEARCH0058.
Locascio, A., M. Manzanares, M. J. Blanco and M. A. Nieto. 2002. Modularity and reshuffling of Snail and Slug expression during vertebrate evolution. *Proc. Natl. Acad. Sci. USA* 99: 16841–16846.
Locke, D. P., R. Segraves, L. Carbone, N. Archidiacono, D. G. Albertson, D. Pinkel and E. E. Eichler. 2003. Large-scale variation among human and great ape genomes determined by array comparative genomic hybridization. *Genome Res.* 13: 347–357.
Loewe, L. 2006. Quantifying the genomic decay paradox due to Muller's ratchet in human mitochondrial DNA. *Genet. Res.* 87: 133–159.
Logsdon, J. M., Jr., M. G. Tyshenko, C. Dixon, J. D-Jafari, V. K. Walker and J. D. Palmer. 1995. Seven newly

discovered intron positions in the triose-phosphate isomerase gene: Evidence for the introns-late theory. *Proc. Natl. Acad. Sci. USA* 92: 8507–8511.

Logsdon, J. M., Jr., A. Stoltzfus and W. F. Doolittle. 1998. Molecular evolution: Recent cases of spliceosomal intron gain? *Curr. Biol.* 8: R560–R563.

Lonergan, K. M. and M. W. Gray. 1993. Editing of transfer RNAs in *Acanthamoeba castellanii* mitochondria. *Science* 259: 812–816.

Long, M. 2001. Evolution of novel genes. *Curr. Opin. Genet. Dev.* 11: 673–680.

Long, M. and M. Deutsch. 1999. Association of intron phases with conservation at splice site sequences and evolution of spliceosomal introns. *Mol. Biol. Evol.* 16: 1528–1534.

Long, M., C. Rosenberg and W. Gilbert. 1995. Intron phase correlations and the evolution of the intron/exon structure of genes. *Proc. Natl. Acad. Sci. USA* 92: 12495–12499.

Long, M., S. J. de Souza, C. Rosenberg and W. Gilbert. 1998. Relationship between "proto-splice sites" and intron phases: Evidence from dicodon analysis. *Proc. Natl. Acad. Sci. USA* 95: 219–223.

Longman, D., I. L. Johnstone and J. F. Caceres. 2000. Functional characterization of SR and SR-related genes in *Caenorhabditis elegans*. *EMBO J.* 19: 1625–1637.

Lopato, S., A. Mayeda, A. R. Krainer and A. Barta. 1996. Pre-mRNA splicing in plants: Characterization of Ser/Arg splicing factors. *Proc. Natl. Acad. Sci. USA* 93: 3074–3079.

Lopez, C. C., L. Nielsen and J. E. Edstrom. 1996. Terminal long tandem repeats in chromosomes form *Chironomus pallidivittatus*. *Mol. Cell. Biol.* 16: 3285–3290.

Lopez, J. V., S. Cevario and S. J. O'Brien. 1996. Complete nucleotide sequences of the domestic cat (*Felis catus*) mitochondrial genome and a transposed mtDNA tandem repeat (Numt) in the nuclear genome. *Genomics* 33: 229–246.

Lopez-Bigas, N., B. Audit, C. Ouzounis, G. Parra and R. Guigo. 2005. Are splicing mutations the most frequent cause of hereditary disease? *FEBS Lett.* 579: 1900–1903.

Lopez-Garcia, P. and D. Moreira. 2006. Selective forces for the origin of the eukaryotic nucleus. *BioEssays* 28: 525–533.

Lorković, Z. J., D. A. Wieczorek Kirk, M. H. Lambermon and W. Filipowicz. 2000. Pre-mRNA splicing in higher plants. *Trends Plant Sci.* 5: 160–167.

Lorković, Z. J., R. Lehner, C. Forstner and A. Barta. 2005. Evolutionary conservation of minor U12-type spliceosome between plants and humans. *RNA* 11: 1095–1107.

Louis, E. J. and A. V. Vershinin. 2005. Chromosome ends: Different sequences may provide conserved functions. *BioEssays* 27: 685–697.

Lozano, E., A. G. Saez, A. J. Flemming, A. Cunha and A. M. Leroi. 2006. Regulation of growth by ploidy in *Caenorhabditis elegans*. *Curr. Biol.* 16: 493–498.

Lu, J. and C. I. Wu. 2005. Weak selection revealed by the whole-genome comparison of the X chromosome and autosomes of human and chimpanzee. *Proc. Natl. Acad. Sci. USA* 102: 4063–4067.

Lu, R., M. Maduro, F. Li, H.-W. Li, G. Broitman-Maduro, W. X. Li and S. W. Ding. 2005. Animal virus replication and RNAi-mediated antiviral silencing in *Caenorhabditis elegans*. *Nature* 436: 1040–1043.

Luan, D. D., M. H. Korman, J. L. Jakubczak and T. H. Eickbush. 1993. Reverse transcription of *R2Bm* RNA is primed by a nick at the chromosomal target site: A mechanism for non-LTR retrotransposition. *Cell* 72: 595–605.

Lucattini, R., V. A. Likic and T. Lithgow. 2004. Bacterial proteins predisposed for targeting to mitochondria. *Mol. Biol. Evol.* 21: 652–658.

Lucchesi, J. C. 1998. Dosage compensation in flies and worms: The ups and downs of X-chromosome regulation. *Curr. Opin. Genet. Dev.* 8: 179–184.

Ludwig, A., N. M. Belfiore, C. Pitra, V. Svirsky and I. Jenneckens. 2001. Genome duplication events and functional reduction of ploidy levels in sturgeon (*Acipenser, Huso* and *Scaphirhynchus*). *Genetics* 158: 1203–1215.

Ludwig, M. Z., C. Bergman, N. H. Patel and M. Kreitman. 2000. Evidence for stabilizing selection in a eukaryotic enhancer element. *Nature* 403: 564–567.

Ludwig, M. Z., A. Palsson, E. Alekseeva, C. M. Bergman, J. Nathan and M. Kreitman. 2005. Functional evolution of a *cis*-regulatory module. *PLoS Biol.* 3: 588–598.

Luehrsen, K. R. and V. Walbot. 1994. Intron creation and polyadenylation in maize are directed by AU-rich RNA. *Genes Dev.* 8: 1117–1130.

Luehrsen, K. R., S. Taha and V. Walbot. 1994. Nuclear pre-mRNA processing in higher plants. *Prog. Nucleic Acid Res. Mol. Biol.* 47: 149–193.

Lund, M., T. O. Tange, H. Dyhr-Mikkelsen, J. Hansen and J. Kjems. 2000. Characterization of human RNA splice signals by iterative functional selection of splice sites. *RNA* 6: 528–544.

Lundblad, V. 2000. DNA ends: Maintenance of chromosome termini versus repair of double strand breaks. *Mutat. Res.* 451: 227–240.

Lundgren, M., A. Andersson, L. Chen, P. Nilsson and R. Bernander. 2004. Three replication origins in *Sulfolobus* species: Synchronous initiation of chromosome replication and asynchronous termination. *Proc. Natl. Acad. Sci. USA* 101: 7046–7051.

Luo, H. R., G. A. Moreau, N. Levin and M. J. Moore. 1999. The human Prp8 protein is a component of both U2- and U12-dependent spliceosomes. *RNA* 5: 893–908.

Luo, M. J. and R. Reed. 1999. Splicing is required for rapid and efficient mRNA export in metazoans. *Proc. Natl. Acad. Sci. USA* 96: 14937–14942.

Luscombe, N. M., J. Qian, Z. Zhang, T. Johnson and M. Gerstein. 2002. The dominance of the population by a selected few: Power-law behaviour applies to a wide variety of genomic properties. *Genome Biol.* 3: RESEARCH0040.

Lutsenko, E. and A. S. Bhagwat. 1999. Principal causes of hot spots for cytosine to thymine mutations at sites of cytosine methylation in growing cells: A model, its experimental support and implications. *Mutat. Res.* 437: 11–20.

Lykke-Andersen, J. 2001. mRNA quality control: Marking the message for life or death. *Curr. Biol.* 11: R88–R91.
Lyman, R. F., F. Lawrence, S. V. Nuzhdin and T. F. Mackay. 1996. Effects of single *P*-element insertions on bristle number and viability in *Drosophila melanogaster. Genetics* 143: 277–292.
Lynch, M. 1990. The rate of morphological evolution in mammals from the standpoint of the neutral expectation. *Amer. Natur.* 136: 727–741.
Lynch, M. 1994. The neutral theory of phenotypic evolution. In L. Real, ed., *Ecological Genetics*, 86–108. Princeton University Press, Princeton, NJ.
Lynch, M. 1997. Mutation accumulation in nuclear, organelle, and prokaryotic transfer RNA genes. *Mol. Biol. Evol.* 14: 914–925.
Lynch, M. 1998. Mutation accumulation in transfer RNAs: Molecular evidence for Muller's ratchet in mitochondrial genomes. *Mol. Biol. Evol.* 13: 209–220.
Lynch, M. 1999. The age and relationships of the major animal phyla. *Evolution* 53: 319–325.
Lynch, M. 2002a. Chromosomal repatterning by gene duplication. *Science* 297: 945–947.
Lynch, M. 2002b. Intron evolution as a population-genetic process. *Proc. Natl. Acad. Sci. USA* 99: 6118–6123.
Lynch, M. 2006a. The origins of eukaryotic gene structure. *Mol. Biol. Evol.* 23: 450–468.
Lynch, M. 2006b. Streamlining and simplification of microbial genome architecture. *Annu. Rev. Microbiol.* 60: 327–349.
Lynch, M. and J. L. Blanchard. 1998. Deleterious mutation accumulation in organelle genomes. *Genetica* 102–103: 29–39.
Lynch, M. and J. S. Conery. 2000. The evolutionary fate and consequences of duplicate genes. *Science* 290: 1151–1154.
Lynch, M. and J. S. Conery. 2003a. The evolutionary demography of duplicate genes. *J. Struc. Func. Genomics* 3: 35–44.
Lynch, M. and J. S. Conery. 2003b. The origins of genome complexity. *Science* 302: 1401–1404.
Lynch, M. and A. Force. 2000a. Gene duplication and the origin of interspecific genomic incompatibility. *Amer. Natur.* 156: 590–605.
Lynch, M. and A. Force. 2000b. The probability of duplicate-gene preservation by subfunctionalization. *Genetics* 154: 459–473.
Lynch, M. and W. Gabriel. 1990. Mutation load and the survival of small populations. *Evolution* 44: 1725–1737.
Lynch, M. and W. G. Hill. 1986. Phenotypic evolution by neutral mutation. *Evolution* 40: 915–935.
Lynch, M. and V. Katju. 2004. The altered evolutionary trajectories of gene duplicates. *Trends Genet.* 20: 544–549.
Lynch, M. and A. Kewalramani. 2003. Messenger RNA surveillance and the evolutionary proliferation of introns. *Mol. Biol. Evol.* 20: 563–571.
Lynch, M. and A. O. Richardson. 2002. The evolution of spliceosomal introns. *Curr. Opin. Genet. Dev.* 12: 701–710.
Lynch, M. and J. B. Walsh. 1998. *Genetics and Analysis of Quantitative Traits*. Sinauer Associates, Sunderland, MA.
Lynch, M., R. Bürger, D. Butcher and W. Gabriel. 1993. Mutational meltdowns in asexual populations. *J. Hered.* 84: 339–344.
Lynch, M., J. Conery and R. Bürger. 1995a. Mutation accumulation and the extinction of small populations. *Amer. Natur.* 146: 489–518.
Lynch, M., J. Conery and R. Bürger. 1995b. Mutational meltdowns in sexual populations. *Evolution* 49: 1067–1080.
Lynch, M., J. Blanchard, D. Houle, T. Kibota, S. Schultz, L. Vassilieva and J. Willis. 1999. Spontaneous deleterious mutation. *Evolution* 53: 645–663.
Lynch, M., M. O'Hely, B. Walsh and A. Force. 2001. The probability of fixation of a newly arisen gene duplicate. *Genetics* 159: 1789–1804.
Lynch, M., D. G. Scofield and X. Hong. 2005. The evolution of transcription-initiation sites. *Mol. Biol. Evol.* 22: 1137–1146.
Lynch, M., B. Koskella and S. Schaack. 2006a. Mutation pressure and the evolution of organelle genomic architecture. *Science* 311: 1727–1730.
Lynch, M., X. Hong and D. G. Scofield. 2006b. Nonsense-mediated decay and the evolution of eukaryotic gene structure. In L. E. Maquat, ed., *Nonsense-Mediated mRNA Decay*, 197–211. Landes Bioscience, Georgetown, TX.
Lyubomirskaya, N. V., J. B. Smirnova, O. V. Razorenova, N. N. Karpova, S. A. Surkov, S. N. Avedisov, A. I. Kim and Y. V. Ilyin. 2001. Two variants of the *Drosophila melanogaster* retrotransposon *gypsy* (*mdg4*): Structural and functional differences, and distribution in fly stocks. *Mol. Genet. Genomics* 265: 367–374.

Ma, J. and J. L. Bennetzen. 2006. Recombination, rearrangement, reshuffling, and divergence in a centromeric region of rice. *Proc. Natl. Acad. Sci. USA* 103: 383–388.
Ma, J., K. M. Devos and J. L. Bennetzen. 2004. Analyses of LTR-retrotransposon structures reveal recent and rapid genomic DNA loss in rice. *Genome Res.* 14: 860–869.
MacAlpine, D. M., P. S. Perlman and R. A. Butow. 1998. The high mobility group protein Abf2p influences the level of yeast mitochondrial DNA recombination intermediates in vivo. *Proc. Natl. Acad. Sci. USA* 95: 6739–6743.
MacAlpine, D. M., H. K. Rodriguez and S. P. Bell. 2004. Coordination of replication and transcription along a *Drosophila* chromosome. *Genes Dev.* 18: 3094–3105.
MacArthur, S. and J. F. Brookfield. 2004. Expected rates and modes of evolution of enhancer sequences. *Mol. Biol. Evol.* 21: 1064–1073.
Macdonald, P. 2001. Diversity in translational regulation. *Curr. Opin. Cell Biol.* 13: 326–331.
Machado, C. A., R. M. Kliman, J. A. Markert and J. Hey. 2002. Inferring the history of speciation from multilocus DNA sequence data: The case of *Drosophila pseudoobscura* and close relatives. *Mol. Biol. Evol.* 19: 472–488.

Maciejowski, J., J. H. Ahn, P. G. Cipriani, D. J. Killian, A. L. Chaudhary, J. I. Lee, R. Voutev, R. C. Johnsen, D. L. Baillie, K. C. Gunsalus et al. 2005. Autosomal genes of autosomal/X-linked duplicated gene pairs and germ-line proliferation in *Caenorhabditis elegans*. *Genetics* 169: 1997–2011.

Mackay, T. F., R. F. Lyman and M. S. Jackson. 1992. Effects of *P* element insertions on quantitative traits in *Drosophila melanogaster. Genetics* 130: 315–332.

Maere, S., S. De Bodt, J. Raes, T. Casneuf, M. Van Montagu, M. Kuiper and Y. Van de Peer. 2005. Modeling gene and genome duplications in eukaryotes. *Proc. Natl. Acad. Sci. USA* 102: 5454–5459.

Magni, G. E. and R. C. von Borstel. 1962. Different rates of spontaneous mutation during mitosis and meiosis in yeast. *Genetics* 47: 1097–1108.

Mai, Z., S. Ghosh, M. Frisardi, B. Rosenthal, R. Rogers and J. Samuelson. 1999. Hsp60 is targeted to a cryptic mitochondrion-derived organelle ("crypton") in the microaerophilic protozoan parasite *Entamoeba histolytica*. *Mol. Cell. Biol.* 19: 2198–2205.

Maicas, E., M. Shago and J. D. Friesen. 1990. Translation of the *Saccharomyces cerevisiae tcm1* gene in the absence of a 5′-untranslated leader. *Nucleic Acids Res.* 18: 5823–5828.

Maier, R. M., P. Zeltz, H. Kossel, G. Bonnard, J. M. Gualberto and J. M. Grienenberger. 1996. RNA editing in plant mitochondria and chloroplasts. *Plant Mol. Biol.* 32: 343–365.

Maisnier-Patin, S., J. R. Roth, A. Fredriksson, T. Nystrom, O. G. Berg and D. I. Andersson. 2005. Genomic buffering mitigates the effects of deleterious mutations in bacteria. *Nature Genet.* 37: 1376–1379.

Maizels, N. 2006. Regulated and unregulated recombination of G-rich genomic regions. In L. H. Caporale, ed., *The Implicit Genome*, 191–207. Oxford University Press, Oxford.

Majewski, J. and F. M. Cohan. 1999. DNA sequence similarity requirements for interspecific recombination in *Bacillus*. *Genetics* 153: 1525–1533.

Majewski, J. and J. Ott. 2002. Distribution and characterization of regulatory elements in the human genome. *Genome Res.* 12: 1827–1836.

Majewski, J., P. Zawadzki, P. Pickerill, F. M. Cohan and C. G. Dowson. 2000. Barriers to genetic exchange between bacterial species: *Streptococcus pneumoniae* transformation. *J. Bacteriol.* 182: 1016–1023.

Makova, K. D. and W.-H. Li. 2002. Strong male-driven evolution of DNA sequences in humans and apes. *Nature* 416: 624–626.

Makova, K. D., S. Yang and F. Chiaromonte. 2004. Insertions and deletions are male biased too: A whole-genome analysis in rodents. *Genome Res.* 14: 567–573.

Maldonado, R., J. Jimenez and J. Casadesus. 1994. Changes of ploidy during the *Azotobacter vinelandii* growth cycle. *J. Bacteriol.* 176: 3911–3919.

Malek, O. and V. Knoop. 1998. Trans-splicing group II introns in plant mitochondria: The complete set of cis-arranged homologs in ferns, fern allies, and a hornwort. *RNA* 4: 1599–1609.

Malek, O., K. Lattig, R. Hiesel, A. Brennicke and V. Knoop. 1996. RNA editing in bryophytes and a molecular phylogeny of land plants. *EMBO J.* 15: 1403–1411.

Malik, H. S. 2005. Ribonuclease H evolution in retrotransposable elements. *Cytogenet. Genome Res.* 110: 392–401.

Malik, H. S. and T. H. Eickbush. 2000. *NeSL-1*, an ancient lineage of site-specific non-LTR retrotransposons from *Caenorhabditis elegans*. *Genetics* 154: 193–203.

Malik, H. S. and T. H. Eickbush. 2001. Phylogenetic analysis of ribonuclease H domains suggests a late, chimeric origin of LTR retrotransposable elements and retroviruses. *Genome Res.* 11: 1187–1197.

Malik, H. S. and S. Henikoff. 2001. Adaptive evolution of Cid, a centromere-specific histone in *Drosophila. Genetics* 157: 1293–1298.

Malik, H. S. and S. Henikoff. 2005. Positive selection of Iris, a retroviral envelope-derived host gene in *Drosophila melanogaster*. *PLoS Genet.* 1: 429–443.

Malik, H. S., W. D. Burke and T. H. Eickbush. 1999. The age and evolution of non-LTR retrotransposable elements. *Mol. Biol. Evol.* 16: 793–805.

Malik, H. S., S. Henikoff and T. H. Eickbush. 2000. Poised for contagion: Evolutionary origins of the infectious abilities of invertebrate retroviruses. *Genome Res.* 10: 1307–1318.

Malinsky, S., A. Bucheton and I. Busseau. 2000. New insights on homology-dependent silencing of *I* factor activity by transgenes containing ORF1 in *Drosophila melanogaster*. *Genetics* 156: 1147–1155.

Malkova, A., J. Swanson, M. German, J. H. McCusker, E. A. Housworth, F. W. Stahl and J. E. Haber. 2004. Gene conversion and crossing over along the 405-kb left arm of *Saccharomyces cerevisiae* chromosome VII. *Genetics* 168: 49–63.

Mamoun, C. B., I. Y. Gluzman, S. M. Beverley and D. E. Goldberg. 2000. Transposition of the *Drosophila* element *mariner* within the human malaria parasite *Plasmodium falciparum*. *Mol. Biochem. Parasitol.* 110: 405–407.

Mango, S. E. 2001. Stop making nonSense: The *C. elegans smg* genes. *Trends Genet.* 17: 646–653.

Maniatis, T. and R. Reed. 2002. An extensive network of coupling among gene expression machines. *Nature* 416: 499–506.

Maniatis, T. and B. Tasic. 2002. Alternative pre-mRNA splicing and proteome expansion in metazoans. *Nature* 418: 236–243.

Mank, J. E. and J. C. Avise. 2006. Cladogenetic correlates of genomic expansions in the recent evolution of actinopterygiian fishes. *Proc. Biol. Sci.* 273: 33–38.

Mansky, L. M. and H. M. Temin. 1995. Lower in vivo mutation rate of human immunodeficiency virus type 1 than that predicted from the fidelity of purified reverse transcriptase. *J. Virol.* 69: 5087–5094.

Maquat, L. E. 2004. Nonsense-mediated mRNA decay: A comparative analysis of different species. *Curr. Genomics* 5: 175–190.

Maquat, L. E. (ed.). 2006. *Nonsense-Mediated mRNA Decay*. Landes Bioscience, Georgetown, TX.

Maquat, L. E. and G. G. Carmichael. 2001. Quality control of mRNA function. *Cell* 104: 173–176.

Maquat, L. E. and X. Li. 2001. Mammalian heat shock p70 and histone H4 transcripts, which derive from naturally intronless genes, are immune to nonsense-mediated decay. *RNA* 7: 445–456.

Marais, G. 2003. Biased gene conversion: Implications for genome and sex evolution. *Trends Genet.* 19: 330–338.

Marais, G. and G. Piganeau. 2002. Hill-Robertson interference is a minor determinant of variations in codon bias across *Drosophila melanogaster* and *Caenorhabditis elegans* genomes. *Mol. Biol. Evol.* 19: 1399–1406.

Marais, G., D. Mouchiroud and L. Duret. 2001. Does recombination improve selection on codon usage? Lessons from nematode and fly complete genomes. *Proc. Natl. Acad. Sci. USA* 98: 5688–92.

Marais, G., T. Domazet-Loso, D. Tautz and B. Charlesworth. 2004. Correlated evolution of synonymous and nonsynonymous sites in *Drosophila*. *J. Mol. Evol.* 59: 771–779.

Marais, G., P. Nouvellet, P. D. Keightley and B. Charlesworth. 2005. Intron size and exon evolution in *Drosophila*. *Genetics* 170: 481–485.

Marashi, S. A. and Z. Ghalanbor. 2004. Correlations between genomic GC levels and optimal growth temperatures are not "robust." *Biochem. Biophys. Res. Commun.* 325: 381–383.

Marc, P., A. Margeot, F. Devaux, C. Blugeon, M. Corral-Debrinski and C. Jacq. 2002. Genome-wide analysis of mRNAs targeted to yeast mitochondria. *EMBO Rep.* 3: 159–164.

Marchington, D. R., G. M. Hartshorne, D. Barlow and J. Poulton. 1997. Homopolymeric tract heteroplasmy in mtDNA from tissues and single oocytes: Support for a genetic bottleneck. *Amer. J. Hum. Genet.* 60: 408–416.

Marechal-Drouard, L., P. Guillemaut, A. Cosset, M. Arbogast, F. Weber, J. H. Weil and A. Dietrich. 1990. Transfer RNAs of potato (*Solanum tuberosum*) mitochondria have different genetic origins. *Nucleic Acids Res.* 18: 3689–3696.

Margulis, L. 1981. *Symbiosis in Cell Evolution*. Freeman, San Francisco, CA.

Margulis, L. 1996. Archaeal-eubacterial mergers in the origin of Eukarya: Phylogenetic classification of life. *Proc. Natl. Acad. Sci. USA* 93: 1071–1076.

Margulis, L., M. F. Dolan and R. Guerrero. 2000. The chimeric eukaryote: Origin of the nucleus from the karyomastigont in amitochondriate protists. *Proc. Natl. Acad. Sci. USA* 97: 6954–6959.

Marino-Ramírez, L., K. C. Lewis, D. Landsman and I. K. Jordan. 2005. Transposable elements donate lineage-specific regulatory sequences to host genomes. *Cytogenet. Genome Res.* 110: 333–341.

Marques, A. C., I. Dupanloup, N. Vinckenbosch, A. Reymond and H. Kaessmann. 2005. Emergence of young human genes after a burst of retroposition in primates. *PLoS Biol.* 3: 1970–1979.

Marshall, C. R., E. C. Raff and R. A. Raff. 1994. Dollo's law and the death and resurrection of genes. *Proc. Natl. Acad. Sci. USA* 91: 12283–12287.

Martienssen, R. A. 2003. Maintenance of heterochromatin by RNA interference of tandem repeats. *Nature Genet.* 35: 213–214.

Martienssen, R., Z. Lippman, B. May, M. Ronemus and M. Vaughn. 2004. Transposons, tandem repeats, and the silencing of imprinted genes. *Cold Spring Harbor Symp. Quant. Biol.* 69: 371–379.

Martin, A. P. 1999. Increasing genomic complexity by gene duplication and the origin of vertebrates. *Amer. Natur.* 154: 111–128.

Martin, A. P. and S. R. Palumbi. 1993. Body size, metabolic rate, generation time, and the molecular clock. *Proc. Natl. Acad. Sci. USA* 90: 4087–4091.

Martin, G., S. P. Otto and T. Lenormand. 2006. Selection for recombination in structured populations. *Genetics* 172: 593–609.

Martin, W. and E. V. Koonin. 2006. Introns and the origin of nucleus-cytosol compartmentalization. *Nature* 440: 41–45.

Martin, W. and M. Müller. 1998. The hydrogen hypothesis for the first eukaryote. *Nature* 392: 37–41.

Martin, W. and M. J. Russell. 2003. On the origins of cells: A hypothesis for the evolutionary transitions from abiotic geochemistry to chemoautotrophic prokaryotes, and from prokaryotes to nucleated cells. *Phil. Trans. Roy. Soc. Lond.* B *Biol. Sci.* 358: 59–85.

Martin, W., B. Stoebe, V. Goremykin, S. Hansmann, M. Hasegawa and K. V. Kowallik. 1998. Gene transfer to the nucleus and the evolution of chloroplasts. *Nature* 393: 162–165.

Martin, W., T. Rujan, E. Richly, A. Hansen, S. Cornelsen, T. Lins, D. Leister, B. Stoebe, M. Hasegawa and D. Penny. 2002. Evolutionary analysis of *Arabidopsis*, cyanobacterial, and chloroplast genomes reveals plastid phylogeny and thousands of cyanobacterial genes in the nucleus. *Proc. Natl. Acad. Sci. USA* 99: 12246–12251.

Maruyama, T. and C. W. Birky, Jr. 1991. Effects of periodic selection on gene diversity in organelle genomes and other systems without recombination. *Genetics* 127: 449–451.

Maruyama, T. and M. Kimura. 1974. A note on the speed of gene frequency changes in reverse direction in a finite population. *Evolution* 28: 161–163.

Maside, X., S. Assimacopoulos and B. Charlesworth. 2000. Rates of movement of transposable elements on the second chromosome of *Drosophila melanogaster*. *Genet. Res.* 75: 275–284.

Maside, X., C. Bartolome, S. Assimacopoulos and B. Charlesworth. 2001. Rates of movement and distribution of transposable elements in *Drosophila melanogaster: In situ* hybridization vs Southern blotting data. *Genet. Res.* 78: 121–136.

Maside, X., A. W. Lee and B. Charlesworth. 2004. Selection on codon usage in *Drosophila americana*. *Curr. Biol.* 14: 150–154.

Masly, J. P., C. D. Jones, M. A. Noor, J. Locke and H. A. Orr. 2006. Gene transposition as a cause of hybrid sterility in *Drosophila*. *Science* 313: 1448–1450.

Mason, P. A. and R. N. Lightowlers. 2003. Why do mammalian mitochondria possess a mismatch repair activity? *FEBS Lett.* 554: 6–9.

Mason, P. A., E. C. Matheson, A. G. Hall and R. N. Lightowlers. 2003. Mismatch repair activity in mammalian mitochondria. *Nucleic Acids Res.* 31: 1052–1058.

Matsunaga, F., P. Forterre, Y. Ishino and H. Myllykallio. 2001. In vivo interactions of archaeal Cdc6/Orc1 and minichromosome maintenance proteins with the replication origin. *Proc. Natl. Acad. Sci. USA* 98: 11152–11157.

Matsunaga, F., C. Norais, P. Forterre and H. Myllykallio. 2003. Identification of short "eukaryotic" Okazaki fragments synthesized from a prokaryotic replication origin. *EMBO Rep.* 4: 154–158.

Matsuzaki, M., O. Misumi, T. Shin-I, S. Maruyama , M. Takahara , S. Y. Miyagishima, T. Mori, K. Nishida, F. Yagisawa, K. Nishida et al. 2004. Genome sequence of the ultrasmall unicellular red alga *Cyanidioschyzon merolae* 10D. *Nature* 428: 653–657.

Mattick, J. S. 2003. Challenging the dogma: The hidden layer of non-protein-coding RNAs in complex organisms. *BioEssays* 25: 930–939.

Mattick, J. S. and M. J. Gagen. 2001. The evolution of controlled multitasked gene networks: The role of introns and other noncoding RNAs in the development of complex organisms. *Mol. Biol. Evol.* 18: 1611–1630.

Mauro, V. P. and G. M. Edelman. 2002. The ribosome filter hypothesis. *Proc. Natl. Acad. Sci. USA* 99: 12031–12036.

May, B. P., Z. B. Lippman, Y. Fang, D. L. Spector and R. A. Martienssen. 2005. Differential regulation of strand-specific transcripts from *Arabidopsis* centromeric satellite repeats. *PLoS Genet.* 1: 705–714.

Mayeda, A., G. R. Screaton, S. D. Chandler, X. D. Fu and A. R. Krainer. 1999. Substrate specificities of SR proteins in constitutive splicing are determined by their RNA recognition motifs and composite premRNA exonic elements. *Mol. Cell. Biol.* 19: 1853–1863.

Maynard Smith, J. 1978. *The Evolution of Sex*. Cambridge University Press, Cambridge.

Maynard Smith, J. and J. Haigh. 1974. The hitch-hiking effect of a favourable gene. *Genet. Res.* 23: 23–35.

Maynard Smith, J. and E. Szathmáry. 1995. *The Major Transitions in Evolution*. W. H. Freeman, San Francisco, CA.

Mayr, E. 1998. Two empires or three? *Proc. Natl. Acad. Sci. USA* 95: 9720–9723.

Mazumder, B., V. Seshadri and P. L. Fox. 2003. Translational control by the 3′-UTR: The ends specify the means. *Trends Biochem. Sci.* 28: 91–98.

McArthur, A.G., L. A. Knodler, J. D. Silberman, B. J. Davids, F. D. Gillin and M. L. Sogin. 2001. The evolutionary origins of eukaryotic protein disulfide isomerase domains: New evidence from the amitochondriate protist *Giardia lamblia*. *Mol. Biol. Evol.* 18: 1455–1463.

McCartney, M. A., G. Keller and H. A. Lessios. 2000. Dispersal barriers in tropical oceans and speciation in Atlantic and eastern Pacific sea urchins of the genus *Echinometra*. *Mol. Ecol.* 9: 1391–1400.

McClintock, B. 1956. Controlling elements and the gene. *Cold Spring Harbor Symp. Quant. Biol.* 21: 197–216.

McClintock, B. 1984. The significance of responses of the genome to challenge. *Science* 226: 792–802.

McClintock, J. M., M. A. Kheirbek and V. E. Prince. 2002. Knockdown of duplicated zebrafish *hoxb1* genes reveals distinct roles in hindbrain patterning and a novel mechanism of duplicate gene retention. *Development* 129: 2339–2354.

McCracken, S., M. Lambermon and B. J. Blencowe. 2002. SRm160 splicing coactivator promotes transcript 3′-end cleavage. *Mol. Cell. Biol.* 22: 148–160.

McLysaght, A., A. J. Enright, L. Skrabanek and K. H. Wolfe. 2000. Estimation of synteny conservation and genome compaction between pufferfish (*Fugu*) and human. *Yeast* 17: 22–36.

McLysaght, A., K. Hokamp and K. H. Wolfe. 2002. Extensive genomic duplication during early chordate evolution. *Nature Genet.* 31: 200–204.

McVean, G. A. 2001. What do patterns of genetic variability reveal about mitochondrial recombination? *Heredity* 87: 613–620.

McVean, G. A and B. Charlesworth. 2000. The effects of Hill-Robertson interference between weakly selected mutations on patterns of molecular evolution and variation. *Genetics* 155: 929–944.

McVean, G. A. and J. Vieira. 2001. Inferring parameters of mutation, selection and demography from patterns of synonymous site evolution in *Drosophila*. *Genetics* 157: 245–257.

McVean, G. A., S. R. Myers, S. Hunt, P. Deloukas, D. R. Bentley and P. Donnelly. 2004. The fine-scale structure of recombination rate variation in the human genome. *Science* 304: 581–584.

Meadows, J. R., R. J. Hawken and J. W. Kijas. 2004. Nucleotide diversity on the ovine Y chromosome. *Anim. Genet.* 35: 379–385.

Medghalchi, S. M., P. A. Frischmeyer, J. T. Mendell, A. G. Kelly, A. M. Lawler and H. C. Dietz. 2001. *Rent1*, a trans-effector of nonsense-mediated mRNA decay, is essential for mammalian embryonic viability. *Hum. Mol. Genet.* 10: 99–105.

Meeks, J. C., E. L. Campbell, M. L. Summers and F. C. Wong. 2002. Cellular differentiation in the cyanobacterium *Nostoc punctiforme*. *Arch. Microbiol.* 178: 395–403.

Meijer, H. A. and A. A. Thomas. 2002. Control of eukaryotic protein synthesis by upstream open reading frames in the 5′-untranslated region of an mRNA. *Biochem J.* 367: 1–11.

Mejlumian, L., A. Pelisson, A. Bucheton and C. Terzian. 2002. Comparative and functional studies of *Drosophila* species invasion by the *gypsy* endogenous retrovirus. *Genetics* 160: 201–209.

Meland, S., S. Johansen, T. Johansen, K. Haugli and F. Haugli. 1991. Rapid disappearance of one parental mitochondrial genotype after isogamous mating in the myxomycete *Physarum polycephalum*. *Curr. Genet.* 19: 55–59.

Mellone, B. G. and R. C. Allshire. 2003. Stretching it: Putting the CEN(P-A) in centromere. *Curr. Opin. Genet. Dev.* 13: 191–198.

Mendell, J. T., S. M. Medghalchi, R. G. Lake, E. N. Noensie and H. C. Dietz. 2000. Novel Upf2p orthologues suggest a functional link between translation initiation and nonsense surveillance complexes. *Mol. Cell. Biol.* 20: 8944–8957.

Mendell, J. T., N. A. Sharifi, J. L. Meyers, F. Martinez-Murillo and H. C. Dietz. 2004. Nonsense surveillance regulates expression of diverse classes of mammalian transcripts and mutes genomic noise. *Nature Genet.* 36: 1073–1078.

Mendez, R. and J. D. Richter. 2001. Translational control by CPEB: A means to the end. *Nature Rev. Mol. Cell. Biol.* 2: 521–529.

Mereschkowsky, C. 1905. Uber Natur und Ursprung der Chromatophoren im Pflanzenreiche. *Biol. Zentralbl.* 25: 593–604.

Meunier, J. and L. Duret. 2004. Recombination drives the evolution of GC-content in the human genome. *Mol. Biol. Evol.* 21: 984–990.

Meyerowitz, E. M. 2002. Plants compared to animals: The broadest comparative study of development. *Science* 295: 1482–1485.

Mézard, C., D. Pompon and A. Nicolas. 1992. Recombination between similar but not identical DNA sequences during yeast transformation occurs within short stretches of identity. *Cell* 70: 659–670.

Mi, S., X. Lee, X. Li, G. M. Veldman, H. Finnerty, L. Racie, E. LaVallie, X. Y. Tang, P. Edouard, S. Howes et al. 2000. Syncytin is a captive retroviral envelope protein involved in human placental morphogenesis. *Nature* 403: 785–789.

Michalakis, Y. and M. Slatkin. 1996. Interaction of selection and recombination in the fixation of negative-epistatic genes. *Genet. Res.* 67: 257–269.

Michel, F. and J.-L. Ferat. 1995. Structure and activities of group II introns. *Annu. Rev. Biochem.* 64: 435–461.

Michod, R. E. 1999. *Darwinian Dynamics: Evolutionary Transitions in Fitness and Individuality.* Princeton University Press, Princeton, NJ.

Michod, R. E. and D. Roze. 2001. Cooperation and conflict in the evolution of multicellularity. *Heredity* 86: 1–7.

Mikkelsen, T. S., L. W. Hillier, E. E. Eichler, M. C. Zody, D. B. Jaffe, S.-P. Yang, W. Enard, I. Hellmann, K. Lindblad-Toh, T. K. Altheide et al. 2005. Initial sequence of the chimpanzee genome and comparison with the human genome. *Nature* 437: 69–87.

Milhausen, M., R. G. Nelson, S. Sather, M. Selkirk and N. Agabian. 1984. Identification of a small RNA containing the trypanosome spliced leader: A donor of shared 5′ sequences of trypanosomatid mRNAs? *Cell* 38: 721–729.

Millen, R. S., R. G. Olmstead, K. L. Adams, J. D. Palmer, N. T. Lao, L. Heggie, T. A. Kavanagh, J. M. Hibberd, J. C. Gray, C. W. Morden et al. 2001. Many parallel losses of *infA* from chloroplast DNA during angiosperm evolution with multiple independent transfers to the nucleus. *Plant Cell* 13: 645–658.

Miller, J. H. 1996. Spontaneous mutators in bacteria: Insights into pathways of mutagenesis and repair. *Annu. Rev. Microbiol.* 50: 625–643.

Mills, R. E., E. A. Bennett, R. C. Iskow, C. T. Luttig, C. Tsui, W. S. Pittard and S. E. Devine. 2006. Recently mobilized transposons in the human and chimpanzee genomes. *Amer. J. Hum. Genet.* 78: 671–679.

Mira, A., H. Ochman and N. A. Moran. 2001. Deletional bias and the evolution of bacterial genomes. *Trends Genet.* 17: 589–596.

Miriami, E., U. Motro, J. Sperling and R. Sperling. 2002. Conservation of an open-reading frame as an element affecting 5′ splice site selection. *J. Struct. Biol.* 140: 116–122.

Mirkin, E. V. and S. M. Mirkin. 2005. Mechanisms of transcription-replication collisions in bacteria. *Mol. Cell. Biol.* 25: 888–895.

Misra, S., M. A. Crosby, C. J. Mungall, B. B. Matthews, K. S. Campbell, P. Hradecky, Y. Huang, J. S. Kaminker, G. H. Millburn, S. E. Prochnik et al. 2002. Annotation of the *Drosophila melanogaster* euchromatic genome: A systematic review. *Genome Biol.* 3(12): RESEARCH0083.

Mitchell, A. and D. Graur. 2005. Inferring the pattern of spontaneous mutation from the pattern of substitution in unitary pseudogenes of *Mycobacterium leprae* and a comparison of mutation patterns among distantly related organisms. *J. Mol. Evol.* 61: 795–803.

Mitchell, P. J. and R. Tjian. 1989. Transcriptional regulation in mammalian cells by sequence-specific DNA binding proteins. *Science* 245: 371–378.

Mitchison, G. 2005. The regional rule for bacterial base composition. *Trends Genet.* 21: 440–443.

Mitrovich, Q. M. and P. Anderson. 2000. Unproductively spliced ribosomal protein mRNAs are natural targets of mRNA surveillance in *C. elegans. Genes Dev.* 14: 2173–2184.

Mitrovich, Q. M. and P. Anderson. 2005. mRNA surveillance of expressed pseudogenes in *C. elegans. Curr. Biol.* 15: 963–967.

Miura, A., S. Yonebayashi, K. Watanabe, T. Toyama, H. Shimada and T. Kakutani. 2001. Mobilization of transposons by a mutation abolishing full DNA methylation in *Arabidopsis. Nature* 411: 212–214.

Miyamoto, T., J. Obokata and M. Sugiura. 2004. A site-specific factor interacts directly with its cognate RNA editing site in chloroplast transcripts. *Proc. Natl. Acad. Sci. USA* 101: 48–52.

Miyata, T., H. Hayashida, K. Kuma, K. Mitsuyasu and T. Yasunaga. 1987. Male-driven molecular evolution: A model and nucleotide sequence analysis. *Cold Spring Harbor Symp. Quant. Biol.* 52: 863–867.

Modrek, B. and C. J. Lee. 2003. Alternative splicing in the human, mouse and rat genomes is associated with an increased frequency of exon creation and/or loss. *Nature Genet.* 34: 177–180.

Mohr, G. and A. M. Lambowitz. 2003. Putative proteins related to group II intron reverse transcriptase/maturases are encoded by nuclear genes in higher plants. *Nucleic Acids Res.* 31: 647–652.

Moll, I., S. Grill, C. O. Gualerzi and U. Bläsi. 2002. Leaderless mRNAs in bacteria: Surprises in ribosomal recruitment and translational control. *Mol. Microbiol.* 43: 239–246.

Montchamp-Moreau, C., S. Ronsseray, M. Jacques, M. Lehmann and D. Anxolabehere. 1993. Distribution and conservation of sequences homologous to the

1731 retrotransposon in *Drosophila*. *Mol. Biol. Evol.* 10: 791–803.

Montgomery, E. A., S. M. Huang, C. H. Langley and B. H. Judd. 1991. Chromosome rearrangement by ectopic recombination in *Drosophila melanogaster*: Genome structure and evolution. *Genetics* 129: 1085–1098.

Moore, J. K. and J. E. Haber. 1996a. Capture of retrotransposon DNA at the sites of chromosomal double-strand breaks. *Nature* 383: 644–646.

Moore, J. K. and J. E. Haber. 1996b. Cell cycle and genetic requirements of two pathways of nonhomologous end-joining repair of double-strand breaks in *Saccharomyces cerevisiae*. *Mol. Cell. Biol.* 16: 2164–2173.

Moore, M. J. 2000. Intron recognition comes of AGe. *Nature Struct. Biol.* 7: 14–16.

Moore, M. J. 2005. From birth to death: The complex lives of eukaryotic mRNAs. *Science* 309: 1514–1518.

Moran, J. V. and N. Gilbert. 2002. Mammalian *LINE-1* retrotransposons and related elements. In N. L. Craig, R. Craigie, M. Gellert and A. M. Lambowitz, eds., *Mobile DNA II*, 836–869. ASM Press, Washington, DC.

Moran, J. V., R. J. DeBerardinis and H. H. Kazazian, Jr. 1999. Exon shuffling by *L1* retrotransposition. *Science* 283: 1530–1534.

Moran, N. A. 1996. Accelerated evolution and Muller's ratchet in endosymbiotic bacteria. *Proc. Natl. Acad. Sci. USA* 93: 2873–2878.

Moran, N. A. 2002. Microbial minimalism: Genome reduction in bacterial pathogens. *Cell* 108: 583–586.

Moran, N. A. and G. R. Plague. 2004. Genomic changes following host restriction in bacteria. *Curr. Opin. Genet. Dev.* 14: 627–633.

Morden, C. W. and A. R. Sherwood. 2002. Continued evolutionary surprises among dinoflagellates. *Proc. Natl. Acad. Sci. USA* 99: 11558–11560.

Moreira, D., H. Le Guyader and H. Philippe. 2000. The origin of red algae and the evolution of chloroplasts. *Nature* 405: 69–72.

Morgan, T. H. 1925. *Evolution and Genetics*. 2nd ed. Princeton University Press, Princeton, NJ.

Morin, G. B. and T. R. Cech. 1988. Mitochondrial telomeres: Surprising diversity of repeated telomeric DNA sequences among six species of *Tetrahymena*. *Cell* 52: 367–374.

Moriyama, E. N. and J. R. Powell. 1997. Codon usage bias and tRNA abundance in *Drosophila*. *J. Mol. Evol.* 45: 514–523.

Moriyama, E. N., D. A. Petrov and D. L. Hartl. 1998. Genome size and intron size in *Drosophila*. *Mol. Biol. Evol.* 15: 770–773.

Morrell, P. L., D. M. Toleno, K. E. Lundy and M. T. Clegg. 2006. Estimating the contribution of mutation, recombination, and gene conversion in the generation of haplotypic diversity. *Genetics* 173: 1705–1723.

Morris, D. R. and A. P. Geballe. 2000. Upstream open reading frames as regulators of mRNA translation. *Mol. Cell. Biol.* 20: 8635–8642.

Morris, R. M., M. S. Rappe, S. A. Connon, K. L. Vergin, W. A. Siebold, C. A. Carlson and S. J. Giovannoni. 2002. SAR11 clade dominates ocean surface bacterioplankton communities. *Nature* 420: 806–810.

Morrish, T. A., N. Gilbert, J. S. Myers, B. J. Vincent, T. D. Stamato, G. E. Taccioli, M. A. Batzer and J. V. Moran. 2002. DNA repair mediated by endonuclease-independent LINE-1 retrotransposition. *Nature Genet.* 31: 159–165.

Morton, B. R., I. V. Bi, M. D. McMullen and B. S. Gaut. 2006. Variation in mutation dynamics across the maize genome as a function of regional and flanking base composition. *Genetics* 172: 569–577.

Mounsey, A., P. Bauer and I. A. Hope. 2002. Evidence suggesting that a fifth of annotated *Caenorhabditis elegans* genes may be pseudogenes. *Genome Res.* 12: 770–775.

Mount, S. M., C. Burks, G. Hertz, G. D. Stormo, O. White and C. Fields. 1992. Splicing signals in *Drosophila*: Intron size, information content, and consensus sequences. *Nucleic Acids Res.* 20: 4255–4262.

Mourier, T. and D. C. Jeffares. 2003. Eukaryotic intron loss. *Science* 300: 1393.

Muckenfuss, H., M. Hamdorf, U. Held, M. Perkovic, J. Lower, K. Cichutek, E. Flory, G. G. Schumann and C. Munk. 2006. APOBEC3 proteins inhibit human LINE-1 retrotransposition. *J. Biol. Chem.* 281: 22161–22172.

Muller, H. J. 1932. Some genetic aspects of sex. *Amer. Natur.* 66: 118–138.

Muller, H. J. 1938. The remaking of chromosomes. *The Collecting Net*, Woods Hole 13: 181–198.

Muller, H. J. 1940. Bearing of the *Drosophila* work on systematics. In J. S. Huxley, ed., *The New Systematics*, 185–268. Clarendon Press, Oxford.

Muller, H. J. 1964. The relation of recombination to mutational advance. *Mutat. Res.* 106: 2–9.

Müller, W. E., M. Bohm, V. A. Grebenjuk, A. Skorokhod, I. M. Müller and V. Gamulin. 2002. Conservation of the positions of metazoan introns from sponges to humans. *Gene* 295: 299–309.

Murphy, T. D. and G. H. Karpen. 1995. Localization of centromere function in a *Drosophila* minichromosome. *Cell* 82: 599–609.

Murray, A. 1990. Telomeres: All's well that ends well. *Nature* 346: 797–798.

Murti, K. G. and D. M. Prescott. 1999. Telomeres of polytene chromosomes in a ciliated protozoan terminate in duplex DNA loops. *Proc. Natl. Acad. Sci. USA* 96: 14436–14439.

Mushegian, A. R., J. R. Garey, J. Martin and L. X. Liu. 1998. Large-scale taxonomic profiling of eukaryotic model organisms: A comparison of orthologous proteins encoded by the human, fly, nematode, and yeast genomes. *Genome Res.* 8: 590–598.

Myers, J. S., B. J. Vincent, H. Udall, W. S. Watkins, T. A. Morrish, G. E. Kilroy, G. D. Swergold, J. Henke, L. Henke, J. V. Moran et al. 2002. A comprehensive analysis of recently integrated human Ta *L1* elements. *Amer. J. Hum. Genet.* 71: 312–326.

Myers, S., L. Bottolo, C. Freeman, G. McVean and P. Donnelly. 2005. A fine-scale map of recombination rates and hotspots across the human genome. *Science* 310: 321–324.

Myllykallio, H., P. Lopez, P. Lopez-Garcia, R. Heilig, W. Saurin, Y. Zivanovic, H. Philippe and P. Forterre. 2000. Bacterial mode of replication with eukaryotic-like machinery in a hyperthermophilic archaeon. *Science* 288: 2212–2215.

Myllykallio, H., G. Lipowski, D. Leduc, J. Filée, P. Forterre and U. Liebl. 2002. An alternative flavin-dependent mechanism for thymidylate synthesis. *Science* 297: 105–107.

Nachman, M. W. 1997. Patterns of DNA variability at X-linked loci in *Mus domesticus*. *Genetics* 147: 1303–1316.

Nachman, M. W. 2001. Single nucleotide polymorphisms and recombination rate in humans. *Trends Genet.* 17: 481–485.

Nachman, M. W. 2002. Variation in recombination rate across the genome: Evidence and implications. *Curr. Opin. Genet. Dev.* 12: 657–663.

Nachman, M. W. and S. L. Crowell. 2000. Estimate of the mutation rate per nucleotide in humans. *Genetics* 156: 297–304.

Nachman, M. W., S. N. Boye and C. F. Aquadro. 1994. Nonneutral evolution at the mitochondrial NADH dehydrogenase subunit 3 gene in mice. *Proc. Natl. Acad. Sci. USA* 91: 6364–6368.

Nachman, M. W., W. M. Brown, M. Stoneking and C. F. Aquadro. 1996. Nonneutral mitochondrial DNA variation in humans and chimpanzees. *Genetics* 142: 953–963.

Nagaki, K., Z. Cheng, S. Ouyang, P. B. Talbert, M. Kim, K. M. Jones, S. Henikoff, C. R. Buell and J. Jiang. 2004. Sequencing of a rice centromere uncovers active genes. *Nature Genet.* 36: 138–145.

Nagawa, F. and G. R. Fink. 1985. The relationship between the "TATA" sequence and transcription initiation sites at the *HIS4* gene of *Saccharomyces cerevisiae*. *Proc. Natl. Acad. Sci. USA* 82: 8557–8561.

Nagy, E. and L. E. Maquat. 1998. A rule for termination-codon position within intron-containing genes: When nonsense affects RNA abundance. *Trends Biochem. Sci.* 6: 198–199.

Nagylaki, T. 1983. Evolution of a finite population under gene conversion. *Proc. Natl. Acad. Sci. USA* 80: 6278–6281.

Nagylaki, T. and T. D. Petes. 1982. Intrachromosomal gene conversion and the maintenance of sequence homogeneity among repeated genes. *Genetics* 100: 315–337.

Nakaar, V., D. Bermudes, K. R. Peck and K. A. Joiner. 1998. Upstream elements required for expression of nucleoside triphosphate hydrolase genes of *Toxoplasma gondii*. *Mol. Biochem. Parasitol.* 92: 229–239.

Nakabachi, A., A. Yamashita, H. Toh, H. Ishikawa, H. E. Dunbar, N. A. Moran and M. Hattori. 2006. The 160-kilobase genome of the bacterial endosymbiont *Carsonella*. *Science* 314: 267.

Nakazono, M. and A. Hirai. 1993. Identification of the entire set of transferred chloroplast DNA sequences in the mitochondrial genome of rice. *Mol. Gen. Genet.* 236: 341–346.

Nalbantoglu, J., G. Phear and M. Meuth. 1987. DNA sequence analysis of spontaneous mutations at the *aprt* locus of hamster cells. *Mol. Cell. Biol.* 7: 1445–1449.

Nanda, I., M. Kondo, U. Hornung, S. Asakawa, C. Winkler, A. Shimizu, Z. Shan, T. Haaf, N. Shimizu, A. Shima et al. 2002. A duplicated copy of DMRT1 in the sex-determining region of the Y chromosome of the medaka, *Oryzias latipes*. *Proc. Natl. Acad. Sci. USA* 99: 11778–11783.

Nanda, I., T. Haaf, M. Schartl, M. Schmid and D. W. Burt. 2002. Comparative mapping of Z-orthologous genes in vertebrates: Implications for the evolution of avian sex chromosomes. *Cytogenet. Genome Res.* 99: 178–184.

Navarro, A. and N. H. Barton. 2003. Accumulating postzygotic isolation genes in parapatry: A new twist on chromosomal speciation. *Evolution* 57: 447–459.

Nei, M. 1969. Linkage modifications and sex difference in recombination. *Genetics* 63: 681–699.

Nei, M. 1983. Genetic polymorphism and the role of mutation in evolution. In M. Nei and R. K. Koehn, eds., *Evolution of Genes and Proteins*, 165–190. Sinauer Associates,, Sunderland, MA.

Nei, M. 1987. *Molecular Evolutionary Genetics*. Columbia University Press, New York.

Nei, M. 2005. Selectionism and neutralism in molecular evolution. *Mol. Biol. Evol.* 22: 2318–2342.

Nei, M. and S. Kumar. 2000. *Molecular Evolution and Phylogenetics*. Oxford University Press, Oxford.

Nei, M. and A. P. Rooney. 2005. Concerted and birth-and-death evolution of multigene families. *Annu. Rev. Genet.* 39: 121–152.

Nei, M., P. Xu and G. Glazko. 2001. Estimation of divergence times from multiprotein sequences for a few mammalian species and several distantly related organisms. *Proc. Natl. Acad. Sci. USA* 98: 2497–2502.

Nembaware, V., K. Crum, J. Kelso and C. Seoighe. 2002. Impact of the presence of paralogs on sequence divergence in a set of mouse-human orthologs. *Genome Res.* 12: 1370–1376.

Nesbø, C. L. and W. F. Doolittle. 2003. Active self-splicing group I introns in 23S rRNA genes of hyperthermophilic bacteria, derived from introns in eukaryotic organelles. *Proc. Natl. Acad. Sci. USA* 100: 10806–10811.

Nesbø, C. L., M. Dlutek and W. F. Doolittle. 2006. Recombination in *Thermotoga*: Implications for species concepts and biogeography. *Genetics* 172: 759–769.

Neu-Yilik, G., N. H. Gehring, R. Thermann, U. Frede, M. W. Hentze and A. E. Kulozik. 2001. Splicing and 3′ end formation in the definition of nonsense-mediated decay-competent human β-globin mRNPs. *EMBO J.* 20: 532–540.

Newlon, C. S. and J. F. Theis. 2002. DNA replication joins the revolution: Whole-genome views of DNA replication in budding yeast. *BioEssays* 24: 300–304.

Newton, K. J. 1988. Plant mitochondrial genomes: Organization, expression, and variation. *Annu. Rev. Plant Physiol. Plant Mol. Biol.* 39: 503–532.

Ng, R. and J. Carbon. 1987. Mutational and in vitro protein-binding studies on centromere DNA from *Saccharomyces cerevisiae*. *Mol. Cell. Biol.* 7: 4522–4534.

Nguyen, H. D., M. Yoshihama and N. Kenmochi. 2005. New maximum likelihood estimators for eukaryotic intron evolution. *PLoS Comput Biol.* 1: 631–638.

Nguyen, H. D., M. Yoshihama and N. Kenmochi. 2006. Phase distribution of spliceosomal introns: Implications for intron origin. *BMC Evol. Biol.* 6: 69.

Nicolaides, N. C., K. W. Kinzler and B. Vogelstein. 1995. Analysis of the 5′ region of PMS2 reveals heterogeneous transcripts and a novel overlapping gene. *Genomics* 29: 329–334.

Nicolas, M., G. Marais, V. Hykelova, B. Janousek, V. Laporte, B. Vyskot, D. Mouchiroud, I. Negrutiu, D. Charlesworth and F. Moneger. 2005. A gradual process of recombination restriction in the evolutionary history of the sex chromosomes in dioecious plants. *PLoS Biol.* 3: 47–56.

Nieduszynski, C. A., Y. Knox and A. D. Donaldson. 2006. Genome-wide identification of replication origins in yeast by comparative genomics. *Genes Dev.* 20: 1874–1879.

Niehrs, C. and N. Pollet. 1999. Synexpression groups in eukaryotes. *Nature* 402: 483–487.

Nielsen, C. B., B. Friedman, B. Birren, C. B. Burge and J. E. Galagan. 2004. Patterns of intron gain and loss in fungi. *PLoS Biol.* 2: 2234–2242.

Nielsen, R., C. Bustamante, A. G. Clark, S. Glanowski, T. B. Sackton, M. J. Hubisz, A. Fledel-Alon, D. M. Tanenbaum, D. Civello, T. J. White et al. 2005. A scan for positively selected genes in the genomes of humans and chimpanzees. *PLoS Biol.* 3: 976–985.

Niemi, J. B., J. D. Raymond, R. Patrek and M. J. Simmons. 2004. Establishment and maintenance of the *P* cytotype associated with telomeric *P* elements in *Drosophila melanogaster*. *Genetics* 166: 255–264.

Niimura, Y., M. Terabe, T. Gojobori and K. Miura. 2003. Comparative analysis of the base biases at the gene terminal portions in seven eukaryote genomes. *Nucleic Acids Res.* 31: 5195–5201.

Nikolaev, S. I., C. Berney, J. F. Fahrni, I. Bolivar, S. Polet, A. P. Mylnikov, V. V. Aleshin, N. B. Petrov and J. Pawlowski. 2004. The twilight of Heliozoa and rise of Rhizaria, an emerging supergroup of amoeboid eukaryotes. *Proc. Natl. Acad. Sci. USA* 101: 8066–8071.

Nilsen, T. W. 2001. Evolutionary origin of SL-addition *trans*-splicing: Still an enigma. *Trends Genet.* 17: 678–680.

Nilsson, A. I., S. Koskiniemi, S. Eriksson, E. Kugelberg, J. C. Hinton and D. I. Andersson. 2005. Bacterial genome size reduction by experimental evolution. *Proc. Natl. Acad. Sci. USA* 102: 12112–12116.

Ninio, J. 1991. Transient mutators: A semiquantitative analysis of the influence of translation and transcription errors on mutation rates. *Genetics* 129: 957–962.

Nisbet, E. 2000. The realms of Archaean life. *Nature* 405: 625–626.

Nisbet, E. G. and N. H. Sleep. 2001. The habitat and nature of early life. *Nature* 409: 1083–1091.

Nissim-Rafinia, M. and B. Kerem. 2002. Splicing regulation as a potential genetic modifier. *Trends Genet.* 18: 123–127.

Niu, D. K., W. R. Hou and S. W. Li. 2005. mRNA-mediated intron losses: Evidence from extraordinarily large exons. *Mol. Biol. Evol.* 22: 1475–1481.

Niwa, M., S. D. Rose and S. M. Berget. 1990. In vitro polyadenylation is stimulated by the presence of an upstream intron. *Genes Dev.* 4: 1552–1559.

Niwa, M., C. C. MacDonald and S. M. Berget. 1992. Are vertebrate exons scanned during splice-site selection? *Nature* 360: 277–280.

Nixon, J. E., A. Wang, H. G. Morrison, A. G. McArthur, M. L. Sogin, B. J. Loftus and J. Samuelson. 2002. A spliceosomal intron in *Giardia lamblia*. *Proc. Natl. Acad. Sci. USA* 99: 3701–3705.

Noor, M. A., K. L. Grams, L. A. Bertucci and J. Reiland. 2001. Chromosomal inversions and the reproductive isolation of species. *Proc. Natl. Acad. Sci. USA* 98: 12084–12088.

Noordam, M. J. and S. Repping. 2006. The human Y chromosome: A masculine chromosome. *Curr. Opin. Genet. Dev.* 16: 225–232.

Nornes, S., M. Clarkson, I. Mikkola, M. Pedersen, A. Bardsley, J. P. Martinez, S. Krauss and T. Johansen. 1998. Zebrafish contains two *Pax6* genes involved in eye development. *Mechs. Dev.* 77: 185–196.

Nosek, J and L. Tomáška. 2003. Mitochondrial genome diversity: Evolution of the molecular architecture and replication strategy. *Curr. Genet.* 44: 73–84.

Nouaud, D., B. Boeda, L. Levy and D. Anxolabehere. 1999. A P element has induced intron formation in *Drosophila*. *Mol. Biol. Evol.* 16: 1503–1510.

Novina, C. D. and P. A. Sharp. 2004. The RNAi revolution. *Nature* 430: 161–164.

Nowak, M. A., M. C. Boerlijst, J. Cooke and J. Maynard Smith. 1997. Evolution of genetic redundancy. *Nature* 388: 167–170.

Nusinow, D. A. and B. Panning. 2005. Recognition and modification of sex chromosomes. *Curr. Opin. Genet. Dev.* 15: 206–213.

Nuzhdin, S. V. 1999. Sure facts, speculations, and open questions about the evolution of transposable element copy number. *Genetica* 107: 129–137.

Nuzhdin, S. V. and T. F. Mackay. 1995. The genomic rate of transposable element movement in *Drosophila melanogaster*. *Mol. Biol. Evol.* 12: 180–181.

O'Hely, M. 2006. A diffusion approach to approximating preservation probabilities for gene duplicates. *J. Math. Biol.* 53: 215–230.

Obado, S. O., M. C. Taylor, S. R. Wilkinson, E. V. Bromley and J. M. Kelly. 2005. Functional mapping of a trypanosome centromere by chromosome fragmentation identifies a 16-kb GC-rich transcriptional "strand-switch" domain as a major feature. *Genome Res.* 15: 36–43.

Ochman, H. 2003. Neutral mutations and neutral substitutions in bacterial genomes. *Mol. Biol. Evol.* 20: 2091–2096.

Ochman, H. and L. M. Davalos. 2006. The nature and dynamics of bacterial genomes. *Science* 311: 1730–1733.

Ochman, H., J. G. Lawrence and E. A. Groisman. 2000. Lateral gene transfer and the nature of bacterial innovation. *Nature* 405: 299–304.

Ohler, U., G. C. Liao, H. Niemann and G. M. Rubin. 2002. Computational analysis of core promoters in the *Drosophila* genome. *Genome Biol.* 3: RESEARCH0087.

Ohnishi, G., K. Endo, A. Doi, A. Fujita, Y. Daigaku, P. Nunoshiba and K. Yamamoto. 2004. Spontaneous mutagenesis in haploid and diploid *Saccharomyces cerevisiae*. *Biochem. Biophys. Res. Commun.* 325: 928–933.

Ohno, S. 1967. *Sex Chromosomes and Sex-Linked Genes.* Springer-Verlag, Berlin.

Ohno, S. 1970. *Evolution by Gene Duplication*. Springer-Verlag, Berlin.

Ohta, T. 1973. Slightly deleterious mutant substitutions in evolution. *Nature* 246: 96–98.

Ohta, T. 1974. Mutational pressure as the main cause of molecular evolution and polymorphism. *Nature* 252: 351–354.

Ohta, T. 1986. Population genetics of an expanding family of mobile genetic elements. *Genetics* 113: 145–159.

Ohta, T. 1997. *Selected Papers on Theoretical Population Genetics and Molecular Evolution*. Department of Population Genetics, National Institute of Genetics, Mishima, Japan.

Ohta, T. and M. Kimura. 1971. Linkage disequilibrium between two segregating nucleotide sites under the steady flux of mutations in a finite population. *Genetics* 68: 571–580.

Okazaki, Y., M. Furuno, T. Kasukawa, J. Adachi, H. Bono, S. Kondo, I. Nikaido, N. Osato, R. Saito, H. Suzuki et al. 2002. Analysis of the mouse transcriptome based on functional annotation of 60,770 full-length cDNAs. *Nature* 420: 563–573.

Oldenburg, D. J. and A. J. Bendich. 2004. Changes in the structure of DNA molecules and the amount of DNA per plastid during chloroplast development in maize. *J. Mol. Biol.* 344: 1311–1330.

Olsen, G. J. and C. R. Woese. 1997. Archaeal genomics: An overview. *Cell* 89: 991–994.

Ometto, L., W. Stephan and D. De Lorenzo. 2005. Insertion/deletion and nucleotide polymorphism data reveal constraints in *Drosophila melanogaster* introns and intergenic regions. *Genetics* 169: 1521–1527.

Omilian, A. R., M. E. Cristescu, J. L. Dudycha and M. Lynch. 2006. Ameiotic recombination in asexual lineages of *Daphnia*. *Proc. Natl. Acad. Sci. USA* 103: 18638–18643.

O'Neil, M. T. and J. M. Belote. 1992. Interspecific comparison of the transformer gene of *Drosophila* reveals an unusually high degree of evolutionary divergence. *Genetics* 131: 113–128.

O'Neill, C. M. and I. Bancroft. 2000. Comparative physical mapping of segments of the genome of *Brassica oleracea* var. *alboglabra* that are homoologous to sequenced regions of chromosomes 4 and 5 of *Arabidopsis thaliana*. *Plant J.* 23: 233–243.

O'Neill, R. J. W., F. E. Brennan, M. L. Delbridge, R. H. Crozier and J. A. M. Graves. 1998. *De novo* insertion of an intron into the mammalian sex determining gene, *SRY*. *Proc. Natl. Acad. Sci. USA* 95: 1653–1657.

O'Neill, R. J., M. J. O'Neill and J. A. Graves. 1998. Undermethylation associated with retroelement activation and chromosome remodelling in an interspecific mammalian hybrid. *Nature* 393: 68–72.

Ophir, R. and D. Graur. 1997. Patterns and rates of indel evolution in processed pseudogenes from humans and murids. *Gene* 205: 191–202.

Opperman, R., E. Emmanuel and A. A. Levy. 2004. The effect of sequence divergence on recombination between direct repeats in *Arabidopsis*. *Genetics* 168: 2207–2215.

Oren, A. 2004. Prokaryote diversity and taxonomy: Current status and future challenges. *Phil. Trans. Roy. Soc. Lond.* B *Biol. Sci.* 359: 623–638.

Orgel, L. E. and F. H. Crick. 1980. Selfish DNA: The ultimate parasite. *Nature* 284: 604–607.

Orphanides, G. and D. Reinberg. 2002. A unified theory of gene expression. *Cell* 108: 439–451.

Orr, H. A. 1996. Dobzhansky, Bateson, and the genetics of speciation. *Genetics* 144: 1331–1335.

Orr, H. A. 1997. Haldane's rule. *Annu. Rev. Ecol. Syst.* 28: 195–218.

Orr, H. A. 2000a. Adaptation and the cost of complexity. *Evolution* 54: 13–20.

Orr, H. A. 2000b. The rate of adaptation in asexuals. *Genetics* 155: 961–968.

Orr, H. A. and Y. Kim. 1998. An adaptive hypothesis for the evolution of the Y chromosome. *Genetics* 150: 1693–1698.

Osawa, S., T. H. Jukes, K. Watanabe and A. Muto. 1992. Recent evidence for evolution of the genetic code. *Microbiol. Rev.* 56: 229–264.

Osborn, T. C., J. C. Pires, J. A. Birchler, D. L. Auger, Z. J. Chen, H. S. Lee, L. Comai, A. Madlung, R. W. Doerge, V. Colot and R. A. Martienssen. 2003. Understanding mechanisms of novel gene expression in polyploids. *Trends Genet.* 19: 141–147.

Osheim, Y. N., M. L. Sikes and A. L. Beyer. 2002. EM visualization of Pol II genes in *Drosophila*: Most genes terminate without prior 3′ end cleavage of nascent transcripts. *Chromosoma* 111: 1–12.

Ostertag, E. M. and H. H. Kazazian, Jr. 2001. Biology of mammalian *L1* retrotransposons. *Annu. Rev. Genet.* 35: 501–538.

Ostheimer, G. J., R. Williams-Carrier, S. Belcher, E. Osborne, J. Gierke and A. Barkan. 2003. Group II intron splicing factors derived by diversification of an ancient RNA-binding domain. *EMBO J.* 22: 3919–3929.

Otake, L. R., P. Scamborova, C. Hashimoto and J. A. Steitz. 2002. The divergent U12-type spliceosome is required for pre-mRNA splicing and is essential for development in *Drosophila*. *Mol. Cell* 9: 439–446.

Otto, S. P. 2003. The advantages of segregation and the evolution of sex. *Genetics* 164: 1099–1118.

Otto, S. P. and N. H. Barton. 2001. Selection for recombination in small populations. *Evolution* 55: 1921–1931.

Otto, S. P. and M. C. Whitlock. 1997. The probability of fixation in populations of changing size. *Genetics* 146: 723–733.

Otto, S. P. and J. Whitton. 2000. Polyploid incidence and evolution. *Annu. Rev. Genet.* 34: 401–437.

Otto, S. P. and P. Yong. 2002. The evolution of gene duplicates. *Adv. Genet.* 46: 451–483.

Ovchinnikov, I., A. Rubin and G. D. Swergold. 2002. Tracing the LINEs of human evolution. *Proc. Natl. Acad. Sci. USA* 99: 10522–10527.

Overbeek, R., M. Fonstein, M. D'Souza, G. D. Pusch and N. Maltsev. 1999. The use of gene clusters to infer functional coupling. *Proc. Natl. Acad. Sci. USA* 96: 2896–2901.

Pace, N. R., G. J. Olsen and C. R. Woese. 1986. Ribosomal RNA phylogeny and the primary lines of evolutionary descent. *Cell* 45: 325–326.

Pagani, F., M. Raponi and F. E. Baralle. 2005. Synonymous mutations in CFTR exon 12 affect splicing and are not neutral in evolution. *Proc. Natl. Acad. Sci. USA* 102: 6368–6372.

Page, B. T., M. K. Wanous and J. A. Birchler. 2001. Characterization of a maize chromosome 4 centromeric sequence: Evidence for an evolutionary relationship with the B chromosome centromere. *Genetics* 159: 291–302.

Pagel, M. and R. A. Johnstone. 1992. Variation across species in the size of the nuclear genome supports the junk-DNA explanation for the C-value paradox. *Proc. Roy. Soc. Lond.* B *Biol. Sci.* 249: 119–124.

Pain, A., H. Renauld, M. Berriman, L. Murphy, C. A. Yeats, W. Weir, A. Kerhornou, M. Aslett, R. Bishop, C. Bouchier et al. 2005. Genome of the host-cell transforming parasite *Theileria annulata* compared with *T. parva*. *Science* 309: 131–133.

Paland, S. and M. Lynch. 2006. Transitions to asexuality result in excess amino acid substitutions. *Science* 311: 990–992.

Palestis, B. G., R. Trivers, A. Burt and R. N. Jones. 2004. The distribution of B chromosomes across species. *Cytogenet. Genome Res.* 106: 151–158.

Palmer, J. D. 1997. The mitochondrion that time forgot. *Nature* 387: 454–455.

Palmer, J. D. and J. M. Logsdon. 1991. The recent origins of introns. *Curr. Opin. Genet. Dev.* 1: 470–477.

Palmer, J. D. and C. R. Shields. 1984. Tripartite structure of the *Brassica campestris* mitochondrial genome. *Nature* 307: 437–440.

Palmer, J. D., K. L. Adams, Y. Cho, C. L. Parkinson, Y. L. Qiu and K. Song. 2000. Dynamic evolution of plant mitochondrial genomes: Mobile genes and introns and highly variable mutation rates. *Proc. Natl. Acad. Sci. USA* 97: 6960–6966.

Pamilo, P., M. Nei and W. H. Li. 1987. Accumulation of mutations in sexual and asexual populations. *Genet. Res.* 49: 135–146.

Pan, Q., A. L. Saltzman, Y. K. Kim, C. Misquitta, O. Shai, L. E. Maquat, B. J. Frey and B. J. Blencowe. 2006. Quantitative microarray profiling provides evidence against widespread coupling of alternative splicing with nonsense-mediated mRNA decay to control gene expression. *Genes Dev.* 20: 153–158.

Panopoulou, G., S. Hennig, D. Groth, A. Krause, A. J. Poustka, R. Herwig, M. Vingron and H. Lehrach. 2003. New evidence for genome-wide duplications at the origin of vertebrates using an amphioxus gene set and completed animal genomes. *Genome Res.* 13: 1056–1066.

Papadopoulou, B. and C. Dumas. 1997. Parameters controlling the rate of gene targeting frequency in the protozoan parasite *Leishmania*. *Nucleic Acids Res.* 25: 4278–4286.

Papke, R. T., J. E. Koenig, F. Rodriguez-Valera and W. F. Doolittle. 2004. Frequent recombination in a saltern population of *Halorubrum*. *Science* 306: 1928–1929.

Papp, B., C. Pal and L. D. Hurst. 2003a. Dosage sensitivity and the evolution of gene families in yeast. *Nature* 424: 194–197.

Papp, B., C. Pal and L. D. Hurst. 2003b. Evolution of cis-regulatory elements in duplicated genes of yeast. *Trends Genet.* 19: 417–422.

Paquin, B., M. J. Laforest, L. Forget, I. Roewer, Z. Wang, J. Longcore and B. F. Lang. 1997. The fungal mitochondrial genome project: Evolution of fungal mitochondrial genomes and their gene expression. *Curr. Genet.* 31: 380–395.

Pardo, B. and S. Marcand. 2005. Rap1 prevents telomere fusions by nonhomologous end joining. *EMBO J.* 24: 3117–3127.

Pardo-Manuel de Villena, F. and C. Sapienza. 2001. Female meiosis drives karyotypic evolution in mammals *Genetics* 159: 1179–1189.

Pardue, M. L. and P. G. DeBaryshe. 2003. Retrotransposons provide an evolutionarily robust non-telomerase mechanism to maintain telomeres. *Annu. Rev. Genet.* 37: 485–511.

Pardue, M. L., S. Rashkova, E. Casacuberta, P. G. DeBaryshe, J. A. George and K. L. Traverse. 2005. Two retrotransposons maintain telomeres in *Drosophila*. *Chromosome Res.* 13: 443–453.

Parisi, M., R. Nuttall, D. Naiman, G. Bouffard, J. Malley, J. Andrews, S. Eastman and B. Oliver. 2003. Paucity of genes on the *Drosophila* X chromosome showing male-biased expression. *Science* 299: 697–700.

Parker, H. R., D. P. Philipp and G. S. Whitt. 1985. Relative developmental success of interspecific *Lepomis* hybrids as an estimate of gene regulatory divergence between species. *J. Exp. Zool.* 233: 451–466.

Parkinson, C. L., J. P. Mower, Y. L. Qiu, A. J. Shirk, K. Song, N. D. Young, C. W. DePamphilis and J. D. Palmer. 2005. Multiple major increases and decreases in mitochondrial substitution rates in the plant family Geraniaceae. *BMC Evol. Biol.* 5: 73.

Parmley, J. L., J. V. Chamary and L. D. Hurst. 2006. Evidence for purifying selection against synonymous mutations in mammalian exonic splicing enhancers. *Mol. Biol. Evol.* 23: 301–309.

Parsch, J. 2003. Selective constraints on intron evolution in *Drosophila*. *Genetics* 165: 1843–1851.

Parsons, T. J. and J. A. Irwin. 2000. Questioning evidence for recombination in human mitochondrial DNA. *Science* 288: 1931.

Partridge, L. and N. H. Barton. 2000. Evolving evolvability. *Nature* 407: 457–458.

Pasyukova, E. G., S. V. Nuzhdin and D. A. Filatov. 1998. The relationship between the rate of transposition and transposable element copy number for *copia* and *Doc* retrotransposons of *Drosophila melanogaster*. *Genet. Res.* 72: 1–11.

Pasyukova, E. G., S. V. Nuzhdin, T. V. Morozova and T. F. Mackay. 2004. Accumulation of transposable elements in the genome of *Drosophila melanogaster* is associated with a decrease in fitness. *J. Hered.* 95: 284–290.

Patel, A. A. and J. A. Steitz. 2003. Splicing double: Insights from the second spliceosome. *Nature Rev. Mol. Cell. Biol.* 4: 960–970.

Patel, A. A., M. McCarthy and J. A. Steitz. 2002. The splicing of U12-type introns can be a rate-limiting step in gene expression. *EMBO J.* 21: 3804–3815.

Paterson, A. H., J. E. Bowers and B. A. Chapman. 2004. Ancient polyploidization predating divergence of the cereals, and its consequences for comparative genomics. *Proc. Natl. Acad. Sci. USA* 101: 9903–9908.

Patthy, L. 1999a. Genome evolution and the evolution of exon-shuffling—a review. *Gene* 238: 103–114.

Patthy, L. 1999b. *Protein Evolution*. Blackwell Science Ltd., Oxford.

Pavlov, Y. I., I. M. Mian and T. A. Kunkel. 2003. Evidence for preferential mismatch repair of lagging strand DNA replication errors in yeast. *Curr. Biol.* 13: 744–748.

Pearson, M. N. and G. F. Rohrmann. 2006. Envelope gene capture and insect retrovirus evolution: The relationship between errantivirus and baculovirus envelope proteins. *Virus Res.* 118: 7–15.

Peck, J. R. 1994. A ruby in the rubbish: Beneficial mutations, deleterious mutations and the evolution of sex. *Genetics* 137: 597–606.

Peichel, C. L., J. A. Ross, C. K. Matson, M. Dickson, J. Grimwood, J. Schmutz, R. M. Myers, S. Mori, D. Schluter and D. M. Kingsley. 2004. The master sex-determination locus in threespine sticklebacks is on a nascent Y chromosome. *Curr. Biol.* 14: 1416–1424.

Pelisson, A., L. Mejlumian, V. Robert, C. Terzian and A. Bucheton. 2002. *Drosophila* germline invasion by the endogenous retrovirus *gypsy*: Involvement of the viral env gene. *Insect Biochem. Mol. Biol.* 32: 1249–1256.

Penton, E. H., B. W. Sullender and T. J. Crease. 2002. *Pokey*, a new DNA transposon in *Daphnia* (Cladocera: Crustacea). *J. Mol. Evol.* 55: 664–673.

Pereira, V. 2004. Insertion bias and purifying selection of retrotransposons in the *Arabidopsis thaliana* genome. *Genome Biol.* 5: R79.

Perez, G. I., A. M. Trbovich, R. G. Gosden and J. L. Tilly. 2000. Mitochondria and the death of oocytes. *Nature* 403: 500–501.

Pérez-Brocal, V., R. Gil, S. Ramos, A. Lamelas, M. Postigo, J. M. Michelena, F. J. Silva, A. Moya and A. Latorre. 2006. A small microbial genome: The end of a long symbiotic relationship? *Science* 314: 312–313.

Pérez-González, C. E. and T. H. Eickbush. 2001. Dynamics of *R1* and *R2* elements in the rDNA locus of *Drosophila simulans*. *Genetics* 158: 1557–1567.

Pérez-González, C. E. and T. H. Eickbush. 2002. Rates of *R1* and *R2* retrotransposition and elimination from the rDNA locus of *Drosophila melanogaster*. *Genetics* 162: 799–811.

Pérez-González, C. E., W. D. Burke and T. H. Eickbush. 2003. *R1* and *R2* retrotransposition and deletion in the rDNA loci on the X and Y chromosomes of *Drosophila melanogaster*. *Genetics* 165: 675–685.

Peri, S. and A. Pandey. 2001. A reassessment of the translation initiation codon in vertebrates. *Trends Genet.* 17: 685–687.

Perron, K., M. Goldschmidt-Clermont and J. D. Rochaix. 1999. A factor related to pseudouridine synthases is required for chloroplast group II intron trans-splicing in *Chlamydomonas reinhardtii*. *EMBO J.* 18: 6481–6490.

Perron, K., M. Goldschmidt-Clermont and J. D. Rochaix. 2004. A multiprotein complex involved in chloroplast group II intron splicing. *RNA* 10: 704–711.

Perry, J. and A. Ashworth. 1999. Evolutionary rate of a gene affected by chromosomal position. *Curr. Biol.* 9: 987–989.

Pesole, G., C. Gissi, G. Grillo, F. Licciulli, S. Liuni and C. Saccone. 2000. Analysis of oligonucleotide AUG start codon context in eukaryotic mRNAs. *Gene* 261: 85–91.

Pesole, G., G. Grillo, A. Larizza and S. Liuni. 2000. The untranslated regions of eukaryotic mRNAs: Structure, function, evolution and bioinformatic tools for their analysis. *Brief Bioinform.* 1: 236–249.

Pesole, G., F. Mignone, C. Gissi, G. Grillo, F. Licciulli and S. Liuni. 2001. Structural and functional features of eukaryotic mRNA untranslated regions. *Gene* 276: 73–81.

Pesole, G., S. Liuni, G. Grillo, F. Licciulli, F. Mignone, C. Gissi and C. Saccone. 2002. UTRdb and UTRsite: Specialized databases of sequences and functional elements of 5′ and 3′ untranslated regions of eukaryotic mRNAs. Update 2002. *Nucleic Acids Res.* 30: 335–340.

Peters, A. D. and P. D. Keightley. 2000. A test for epistasis among induced mutations in *Caenorhabditis elegans*. *Genetics* 156: 1635–1647.

Peterson, K. J., J. B. Lyons, K. S. Nowak, C. M. Takacs, M. J. Wargo and M. A. McPeek. 2004. Estimating metazoan divergence times with a molecular clock. *Proc. Natl. Acad. Sci. USA* 101: 6536–6541.

Petes, T. D. 2001. Meiotic recombination hot spots and cold spots. *Nature Rev. Genet.* 2: 360–369.

Petrov, D. A. 2001. Evolution of genome size: New approaches to an old problem. *Trends Genet.* 17: 23–28.

Petrov, D. A. 2002a. DNA loss and evolution of genome size in *Drosophila*. *Genetica* 115: 81–91.

Petrov, D. A. 2002b. Mutational equilibrium model of genome size evolution. *Theor. Pop. Biol.* 61: 531–544.

Petrov, D. A. and D. L. Hartl. 1999. Patterns of nucleotide substitution in *Drosophila* and mammalian genomes. *Proc. Natl. Acad. Sci. USA* 96: 1475–1479.

Petrov, D. A., J. L. Schutzman, D. L. Hartl and E. R. Lozovskaya. 1995. Diverse transposable elements are mobilized in hybrid dysgenesis in *Drosophila virilis*. *Proc. Natl. Acad. Sci. USA* 92: 8050–8054.

Petrov, D. A., E. R. Lozovskaya and D. L. Hartl. 1996. High intrinsic rate of DNA loss in *Drosophila*. *Nature* 384: 346–349.

Petrov, D. A., T. A. Sangster, J. S. Johnston, D. L. Hartl and K. L. Shaw. 2000. Evidence for DNA loss as a determinant of genome size. *Science* 287: 1060–1062.

Petrov, D. A., Y. T. Aminetzach, J. C. Davis, D. Bensasson and A. E. Hirsh. 2003. Size matters: Non-LTR retrotransposable elements and ectopic recombination in *Drosophila*. *Mol. Biol. Evol.* 20: 880–892.

Pevzner, P. and G. Tesler. 2003. Human and mouse genomic sequences reveal extensive breakpoint reuse in mammalian evolution. *Proc. Natl. Acad. Sci. USA* 100: 7672–7677.

Pfeiffer, T. and S. Bonhoeffer. 2003. An evolutionary scenario for the transition to undifferentiated multicellularity. *Proc. Natl. Acad. Sci. USA* 100: 1095–1098.

Phadnis, N., R. Mehta, N. Meednu and E. A. Sia. 2006. Ntg1p, the base excision repair protein, generates mutagenic intermediates in yeast mitochondrial DNA. *DNA Repair* 5: 829–839.

Philippe, H. and P. Forterre. 1999. The rooting of the universal tree of life is not reliable. *J. Mol. Evol.* 49: 509–523.

Philippe, H., P. Lopez, H. Brinkmann, K. Budin, A. Germot, J. Laurent, D. Moreira, M. Muller and H. Le Guyader. 2000. Early-branching or fast-evolving eukaryotes? An answer based on slowly evolving positions. *Proc. Roy. Soc. Lond.* B *Biol. Sci.* 267: 1213–1221.

Philips, A. V. and T. A. Cooper. 2000. RNA processing and human disease. *Cell. Mol. Life Sci.* 57: 235–249.

Phillips, P. C. 1996. Waiting for a compensatory mutation: Phase zero of the shifting-balance process. *Genet. Res.* 67: 271–283.

Phreaner, C. G., M. A. Williams and R. M. Mulligan. 1996. Incomplete editing of rps12 transcripts results in the synthesis of polymorphic polypeptides in plant mitochondria. *Plant Cell.* 8: 107–117.

Piatigorsky, J. and G. Wistow. 1991. The recruitment of crystallins: New functions precede gene duplication. *Science* 252: 1078–1079.

Picard, G. and P. L'Heritier. 1971. A maternally inherited factor inducing sterility in *Drosophila melanogaster. Drosophila Information Service* 46: 54.

Pidoux, A. L. and R. C. Allshire. 2005. The role of heterochromatin in centromere function. *Phil. Trans. Roy. Soc. Lond.* B *Biol. Sci.* 360: 569–579.

Pierron, G. and M. Bénard. 1996. DNA replication in *Physarum*. In M. L. DePamphilis, ed., *DNA Replication in Eukaryotic Cells*, 933–946. Cold Spring Harbor Laboratory Press, Plainview, NY.

Piganeau, G., M. Gardner and A. Eyre-Walker. 2004. A broad survey of recombination in animal mitochondria. *Mol. Biol. Evol.* 21: 2319–2325.

Pinsker, W., E. Haring, S. Hagemann and W. J. Miller. 2001. The evolutionary life history of *P* transposons: From horizontal invaders to domesticated neogenes. *Chromosoma* 110: 148–158.

Plasterk, R. H. A. and H. G. A. M. van Luenen. 2002. The *Tc1/mariner* family of transposable elements. In N. L. Craig, R. Craigie, M. Gellert and A. M. Lambowitz, eds., *Mobile DNA II*, 519–532. ASM Press, Washington, DC.

Plasterk, R. H., Z. Izsvak and Z. Ivics. 1999. Resident aliens: The *Tc1/mariner* superfamily of transposable elements. *Trends Genet.* 15: 326–332.

Podlutsky, A., A. M. Osterholm, S. M. Hou, A. Hofmaier and B. Lambert. 1998. Spectrum of point mutations in the coding region of the hypoxanthine-guanine phosphoribosyltransferase (*hprt*) gene in human T-lymphocytes in vivo. *Carcinogenesis* 19: 557–566.

Pont-Kingdon, G., N. A. Okada, J. L. Macfarlane, C. T. Beagley, C. D. Watkins-Sims, T. Cavalier-Smith, G. D. Clark-Walker and D. R. Wolstenholme. 1998. Mitochondrial DNA of the coral *Sarcophyton glaucum* contains a gene for a homologue of bacterial *MutS*: A possible case of gene transfer from the nucleus to the mitochondrion. *J. Mol. Evol.* 46: 419–431.

Pont-Kingdon, G., C. G. Vassort, R. Warrior, R. Okimoto, C. T. Beagley and D. R. Wolstenholme. 2000. Mitochondrial DNA of *Hydra attenuata* (Cnidaria): A sequence that includes an end of one linear molecule and the genes for l-rRNA, tRNA(f-Met), tRNA(Trp), COII, and ATPase8. *J. Mol. Evol.* 51: 404–415.

Poole, A. M., D. C. Jeffares and D. Penny. 1998. The path from the RNA world. *J. Mol. Evol.* 46: 1–17.

Poole, A. M., D. C. Jeffares and D. Penny. 1999. Early evolution: Prokaryotes, the new kids on the block. *BioEssays* 21: 880–889.

Popescu, C. E., T. Borza, J. P. Bielawski and R. W. Lee. 2006. Evolutionary rates and expression level in *Chlamydomonas*. *Genetics* 172: 1567–1576.

Posada, D., K. A. Crandall, M. Nguyen, J. C. Demma and R. P. Viscidi. 2000. Population genetics of the *porB* gene of *Neisseria gonorrhoeae*: Different dynamics in different homology groups. *Mol. Biol. Evol.* 17: 423–436.

Postlethwait, J. H., I. G. Woods, P. Ngo-Hazelett, Y. L. Yan, P. D. Kelly, F. Chu, H. Huang, A. Hill-Force and W. S. Talbot. 2000. Zebrafish comparative genomics and the origins of vertebrate chromosomes. *Genome Res.* 10: 1890–1902.

Pouchkina-Stantcheva, N. N. and A. Tunnacliffe. 2005. Spliced leader RNA-mediated *trans*-splicing in phylum Rotifera. *Mol. Biol. Evol.* 22: 1482–1489.

Powell, J. R. and E. N. Moriyama. 1997. Evolution of codon usage bias in *Drosophila*. *Proc. Natl. Acad. Sci. USA* 94: 7784–7790.

Prachumwat, A. and W.-H. Li. 2006. Protein function, connectivity, and duplicability in yeast. *Mol. Biol. Evol.* 23: 30–39.

Prachumwat, A., L. DeVincentis and M. F. Palopoli. 2004. Intron size correlates positively with recombination rate in *Caenorhabditis elegans*. *Genetics* 166: 1585–1590.

Pramateftaki, P. V., V. N. Kouvelis, P. Lanaridis and M. A. Typas. 2006. The mitochondrial genome of the wine yeast *Hanseniaspora uvarum*: A unique genome organization among yeast/fungal counterparts. *FEMS Yeast Res.* 6: 77–90.

Preer, J. R., Jr., L. B. Preer, B. Rudman and A. Jurand. 1971. Isolation and composition of bacteriophage-like particles from kappa of killer *Paramecia*. *Mol. Gen. Genet.* 111: 202–208.

Preiss, T. and M. W. Hentze. 1999. From factors to mechanisms: Translation and translational control in eukaryotes. *Curr. Opin. Genet. Dev.* 9: 515–521.

Presgraves, D. C. 2002. Patterns of postzygotic isolation in Lepidoptera. *Evolution* 56: 1168–1183.

Presgraves, D. C. 2006. Intron length evolution in *Drosophila*. *Mol. Biol. Evol.* 23: 2203–2213.

Prêt, A. M. and L. L. Searles. 1991. Splicing of retrotransposon insertions from transcripts of the *Drosophila melanogaster vermilion* gene in a revertant. *Genetics* 129: 1137–1145.

Price, M. N., E. J. Alm and A. P. Arkin. 2005. Interruptions in gene expression drive highly expressed operons to the leading strand of DNA replication. *Nucleic Acids Res.* 33: 3224–3234.

Price, M. N., K. H. Huang, A. P. Arkin and E. J. Alm. 2005. Operon formation is driven by co-regulation and not by horizontal gene transfer. *Genome Res.* 15: 809–819.

Price, T. D. and M. M. Bouvier. 2002. The evolution of F1 postzygotic incompatibilities in birds. *Evolution* 56: 2083–2089.

Promislow, D. E., I. K. Jordan and J. F. McDonald. 1999. Genomic demography: A life-history analysis of transposable element evolution. *Proc. Roy. Soc. Lond.* B *Biol. Sci.* 266: 1555–1560.

Proudfoot, N. J. 1989. How RNA polymerase II terminates transcription in higher eukaryotes. *Trends Biochem. Sci.* 14: 105–10.

Proudfoot, N. J., A. Furger and M. J. Dye. 2002. Integrating mRNA processing with transcription. *Cell* 108: 501–512.

Proulx, S. R. and P. C. Phillips. 2005. The opportunity for canalization and the evolution of genetic networks. *Amer. Natur.* 165: 147–162.

Provine, W. B. 1971. *The Origins of Theoretical Population Genetics*. University of Chicago Press, Chicago.

Provine, W. B. 1986. *Sewall Wright and Evolutionary Biology*. University of Chicago Press, Chicago.

Prud'homme, N., M. Gans, M. Masson, C. Terzian and A. Bucheton. 1995. *Flamenco*, a gene controlling the *gypsy* retrovirus of *Drosophila melanogaster*. *Genetics* 139: 697–711.

Pryde, F. E., H. C. Gorham and E. J. Louis. 1997. Chromosome ends: All the same under their caps. *Curr. Opin. Genet. Dev.* 7: 822–828.

Ptak, S. E., K. Voelpel and M. Przeworski. 2004. Insights into recombination from patterns of linkage disequilibrium in humans. *Genetics* 167: 387–397.

Ptashne, M. and A. Gann. 2002. *Genes and Signals*. Cold Spring Harbor Laboratory Press, Cold Spring Harbor, NY.

Puchta, H. 2005. The repair of double-strand breaks in plants: Mechanisms and consequences for genome evolution. *J. Exp. Bot.* 56: 1–14.

Puig, M., M. Cáceres and A. Ruiz. 2004. Silencing of a gene adjacent to the breakpoint of a widespread *Drosophila* inversion by a transposon-induced antisense RNA. *Proc. Natl. Acad. Sci. USA* 101: 9013–9018.

Pulak, R. and P. Anderson. 1993. mRNA surveillance by the *Caenorhabditis elegans smg* genes. *Genes Dev.* 7: 1885–1897.

Purugganan, M. and S. Wessler. 1992. The splicing of transposable elements and its role in intron evolution. *Genetica* 86: 295–303.

Pyne, S., S. Skiena and B. Futcher. 2005. Copy correction and concerted evolution in the conservation of yeast genes. *Genetics* 170: 1501–1513.

Qian, J., N. M. Luscombe and M. Gerstein. 2001. Protein family and fold occurrence in genomes: Power-law behaviour and evolutionary model. *J. Mol. Biol.* 313: 673–681.

Qin, H., W. B. Wu, J. M. Comeron, M. Kreitman and W.-H. Li. 2004. Intragenic spatial patterns of codon usage bias in prokaryotic and eukaryotic genomes. *Genetics* 168: 2245–2260.

Qiu, W. G., N. Schisler and A. Stoltzfus. 2004. The evolutionary gain of spliceosomal introns: Sequence and phase preferences. *Mol. Biol. Evol.* 21: 1252–1263.

Que, X., S. G. Svard, T. C. Meng, M. L. Hetsko, S. B. Aley and F. D. Gillin. 1996. Developmentally regulated transcripts and evidence of differential mRNA processing in *Giardia lamblia*. *Mol. Biochem. Parasitol.* 81: 101–110.

Quint, E., T. Zerucha and M. Ekker. 2000. Differential expression of orthologous *Dlx* genes in zebrafish and mice: Implications for the evolution of the *Dlx* homeobox gene family. *J. Exp. Zool.* 288: 235–241.

Quiros, C. F., F. Grellet, J. Sadowski, T. Suzuki, G. Li and T. Wroblewski. 2001. *Arabidopsis* and *Brassica* comparative genomics: Sequence, structure and gene content in the *ABI-Rps2-Ck1* chromosomal segment and related regions. *Genetics* 157: 1321–1330.

Radford, A. and J. H. Parish. 1997. The genome and genes of *Neurospora crassa*. *Fungal Genet. Biol.* 21: 258–266.

Radman, M. 1998. DNA replication: One strand may be more equal. *Proc. Natl. Acad. Sci. USA* 95: 9718–9719.

Raff, R. A. 1996. *The Shape of Life*. University of Chicago Press, Chicago.

Raghuraman, M. K., E. A. Winzeler, D. Collingwood, S. Hunt, L. Wodicka, A. Conway, D. J. Lockhart, R. W. Davis, B. J. Brewer and W. L. Fangman. 2001. Replication dynamics of the yeast genome. *Science* 294: 115–121.

Raible, F., K. Tessmar-Raible, K. Osoegawa, P. Wincker, C. Jubin, G. Balavoine, D. Ferrier, V. Benes, P. de Jong, J. Weissenbach et al. 2005. Vertebrate-type intron-rich genes in the marine annelid *Platynereis dumerilii*. *Science* 310: 1325–1326.

Rajavel, K. S. and E. F. Neufeld. 2001. Nonsense-mediated decay of human *HEXA* mRNA. *Mol. Cell. Biol.* 21: 5512–5519.

Rajkovic, A., R. E. Davis, J. N. Simonsen and F. M. Rottman. 1990. A spliced leader is present on a subset of mRNAs from the human parasite *Schistosoma mansoni*. *Proc. Natl. Acad. Sci. USA* 87: 8879–8883.

Rajkowitsch, L., C. Vilela, K. Berthelot, C. V. Ramirez and J. E. McCarthy. 2004. Reinitiation and recycling are distinct processes occurring downstream of translation termination in yeast. *J. Mol. Biol.* 335: 71–85.

Ralph, S. A., B. J. Foth, N. Hall and G. I. McFadden. 2004. Evolutionary pressures on apicoplast transit peptides. *Mol. Biol. Evol.* 21: 2183–2194.

Ramesh, M. A., S. B. Malik and J. M. Logsdon, Jr. 2005. A phylogenomic inventory of meiotic genes; evidence for sex in *Giardia* and an early eukaryotic origin of meiosis. *Curr. Biol.* 15: 185–191.

Ramsey, J. and D. W. Schemske. 1998. Pathways, mechanisms, and rates of polyploid formation in flowering plants. *Annu. Rev. Ecol. Syst.* 29: 467–501.

Rand, D. M. 2001. The units of selection on mitochondrial DNA. *Annu. Rev. Ecol. Syst.* 32: 415–448.

Rand, D. M. and R. G. Harrison. 1986. Mitochondrial DNA transmission genetics in crickets. *Genetics* 114: 955–970.

Rand, D. M., M. Dorfsman and L. M. Kann. 1994. Neutral and non-neutral evolution of *Drosophila* mitochondrial DNA. *Genetics* 138: 741–756.

Ranea, J. A., A. Grant, J. M. Thornton and C. A. Orengo. 2005. Microeconomic principles explain an optimal genome size in bacteria. *Trends Genet.* 21: 21–25.

Ranz, J. M., F. Casals and A. Ruiz. 2001. How malleable is the eukaryotic genome? Extreme rate of chromosomal rearrangement in the genus *Drosophila*. *Genome Res.* 11: 230–239.

Ranz, J. M., J. González, F. Casals and A. Ruiz. 2003. Low occurrence of gene transposition events during the evolution of the genus *Drosophila*. *Evolution* 57: 1325–1335.

Rappsilber, J., U. Ryder, A. I. Lamond and M. Mann. 2002. Large-scale proteomic analysis of the human spliceosome. *Genome Res.* 12: 1231–1245.

Rasmussen, B. 2000. Filamentous microfossils in a 3,235-million-year-old volcanogenic massive sulphide deposit. *Nature* 405: 676–679.

Ratner, V. A., S. A. Zabanov, O. V. Kolesnikova and L. A. Vasilyeva. 1992. Induction of the mobile genetic element *Dm-412* transpositions in the *Drosophila* genome by heat shock treatment. *Proc. Natl. Acad. Sci. USA* 89: 5650–5654.

Rattray, A. J., B. K. Shafer, C. B. McGill and J. N. Strathern. 2002. The roles of REV3 and RAD57 in double-strand-break-repair-induced mutagenesis of *Saccharomyces cerevisiae*. *Genetics* 162: 1063–1077.

Raup, D. M. 1978. Cohort analysis of generic survivorship. *Paleobiology* 4: 1–15.

Rayssiguier, C., D. S. Thaler and M. Radman. 1989. The barrier to recombination between *Escherichia coli* and *Salmonella typhimurium* is disrupted in mismatch-repair mutants. *Nature* 342: 396–401.

Raz, E., H. G. van Luenen, B. Schaerringer, R. H. Plasterk and W. Driever. 1998. Transposition of the nematode *Caenorhabditis elegans Tc3* element in the zebrafish *Danio rerio*. *Curr. Biol.* 8: 82–88.

Reboud, X. and C. Zeyl. 1994. Organelle inheritance in plants. *Heredity* 72: 132–140.

Redmond, D. L. and D. P. Knox. 2001. *Haemonchus contortus* SL2 *trans*-spliced RNA leader sequence. *Mol. Biochem. Parasitol.* 117: 107–110.

Redon, R., S. Ishikawa, K. R. Fitch, L. Feuk, G. H. Perry, T. D. Andrews, H. Fiegler, M. H. Shapero, A. R. Carson, W. Chen et al. 2006. Global variation in copy number in the human genome. *Nature* 444: 444–454.

Reed, R. and E. Hurt. 2002. A conserved mRNA export machinery coupled to pre-mRNA splicing. *Cell* 108: 523–531.

Reenan, R. A. 2001. The RNA world meets behavior: A→I pre-mRNA editing in animals. *Trends Genet.* 17: 53–56.

Reich, D. E., M. Cargill, S. Bolk, J. Ireland, P. C. Sabeti, D. J. Richter, T. Lavery, R. Kouyoumjian, S. F. Farhadian, R. Ward and E. S. Lander. 2001. Linkage disequilibrium in the human genome. *Nature* 411: 199–204.

Reines, D., J. W. Conaway and R. C. Conaway. 1996. The RNA polymerase II general elongation factors. *Trends Biochem Sci.* 21: 351–355.

Reinhart, B. J. and D. P. Bartel. 2002. Small RNAs correspond to centromere heterochromatic repeats. *Science* 297: 1831.

Reinke, V., H. E. Smith, J. Nance, J. Wang, C. Van Doren, R. Begley, S. J. Jones, E. B. Davis, S. Scherer, S. Ward and S. K. Kim. 2000. A global profile of germline gene expression in *C. elegans*. *Mol. Cell* 6: 605–616.

Ren, X.-Y., O. Vorst, M. W. E. J. Fiers, W. J. Stiekema and J. P. Nap. 2006. In plants, highly expressed genes are the least compact. *Trends Genet.* 22: 528–532.

Rest, J. S. and D. P. Mindell. 2003. Retroids in archaea: Phylogeny and lateral origins. *Mol. Biol. Evol.* 20: 1134–1142.

Reyes, A., C. Gissi, G. Pesole and C. Saccone. 1998. Asymmetrical directional mutation pressure in the mitochondrial genome of mammals. *Mol. Biol. Evol.* 15: 957–966.

Reznikoff, W. S. 2002. Tn5 transposition. In N. L. Craig, R. Craigie, M. Gellert and A. M. Lambowitz, eds., *Mobile DNA II*, 403–422. ASM Press, Washington, DC.

Rhounim, L., J. L. Rossignol and G. Faugeron. 1992. Epimutation of repeated genes in *Ascobolus immersus*. *EMBO J.* 11: 4451–4457.

Ribeiro-dos-Santos, G., A. C. Schenberg, D. C. Gardner and S. G. Oliver. 1997. Enhancement of *Ty* transposition at the ADH4 and ADH2 loci in meiotic yeast cells. *Mol. Gen. Genet.* 254: 555–561.

Ricchetti, M., C. Fairhead and B. Dujon. 1999. Mitochondrial DNA repairs double-strand breaks in yeast chromosomes. *Nature* 402: 96–100.

Rice, W. R. 1984. Sex chromosomes and the evolution of sexual dimorphism. *Evolution* 38: 735–742.

Rice, W. R. 1994. Degeneration of a nonrecombining chromosome. *Science* 263: 230–232.

Rice, W. R. 1998. Male fitness increases when females are eliminated from gene pool: Implications for the Y chromosome. *Proc. Natl. Acad. Sci. USA* 95: 6217–6221.

Richards, T. A. and T. Cavalier-Smith. 2005. Myosin domain evolution and the primary divergence of eukaryotes. *Nature* 436: 1113–1118.

Richly, E. and D. Leister. 2004. NUPTs in sequenced eukaryotes and their genomic organization in relation to NUMTs. *Mol. Biol. Evol.* 21: 1972–1980.

Richly, E., P. F. Chinnery and D. Leister. 2003. Evolutionary diversification of mitochondrial proteomes: Implications for human disease. *Trends Genet.* 19: 3563–3562.

Rieseberg, L. H. 2001. Chromosomal rearrangements and speciation. *Trends Ecol. Evol.* 16: 351–358.

Rifkin, S. A., D. Houle, J. Kim and K. P. White. 2005. A mutation accumulation assay reveals a broad capacity for rapid evolution of gene expression. *Nature* 438: 220–223.

Rio, D. C. 2002. *P* transposable elements in *Drosophila melanogaster*. In N. L. Craig, R. Craigie, M. Gellert and A. M. Lambowitz, eds., *Mobile DNA II*, 484–518. ASM Press, Washington, DC.

Riska, B. and W. R. Atchley. 1985. Genetics of growth predict patterns of brain-size evolution. *Science* 253: 669–673.

Rivera, M. C. and J. A. Lake. 2004. The ring of life provides evidence for a genome fusion origin of eukaryotes. *Nature* 431: 152–155.

Rivera, M. C., R. Jain, J. E. Moore and J. A. Lake. 1998. Genomic evidence for two functionally distinct gene classes. *Proc. Natl. Acad. Sci. USA* 95: 6239–6244.

Rivier, C., M. Goldschmidt-Clermont and J. D. Rochaix. 2001. Identification of an RNA-protein complex involved in chloroplast group II intron trans-splicing in *Chlamydomonas reinhardtii*. *EMBO J.* 20: 1765–1773.

Rizzon, C., G. Marais, M. Gouy and C. Biémont. 2002. Recombination rate and the distribution of transposable elements in the *Drosophila melanogaster* genome. *Genome Res.* 12: 400–407.

Robart, A. R. and S. Zimmerly. 2005. Group II intron retroelements: Function and diversity. *Cytogenet. Genome Res.* 110: 589–597.

Roberts, G. C. and C. W. Smith. 2002. Alternative splicing: Combinatorial output from the genome. *Curr. Opin. Chem. Biol.* 6: 375–383.

Robertson, H. M. 1993. The *mariner* transposable element is widespread in insects. *Nature* 362: 241–245.

Robertson, H. M. 1997. Multiple *Mariner* transposons in flatworms and hydras are related to those of insects. *J. Hered.* 88: 195–201.

Robertson, H. M. 2000. The large *srh* family of chemoreceptor genes in *Caenorhabditis* nematodes reveals processes of genome evolution involving large duplications and deletions and intron gains and losses. *Genome Res* 10: 192–203.

Robertson, H. M. and K. L. Zumpano. 1997. Molecular evolution of an ancient mariner transposon, *Hsmar1*, in the human genome. *Gene* 205: 203–217.

Robin, S., S. Chambeyron, C. Brun, A. Bucheton and I. Busseau. 2002. Trans-complementation of an endonuclease-defective tagged *I* element as a tool for the study of retrotransposition in *Drosophila melanogaster. Mol. Genet. Genomics* 267: 829–834.

Robinson-Rechavi, M., B. Boussau and V. Laudet. 2004. Phylogenetic dating and characterization of gene duplications in vertebrates: The cartilaginous fish reference. *Mol. Biol. Evol.* 21: 580–586.

Rocha, E. P. 2004. Codon usage bias from tRNA's point of view: Redundancy, specialization, and efficient decoding for translation optimization. *Genome Res.* 14: 2279–2286.

Rocha, E. P. and A. Danchin. 2002. Base composition bias might result from competition for metabolic resources. *Trends Genet.* 18: 291–294.

Rocha, E. P. and A. Danchin. 2003. Essentiality, not expressiveness, drives gene-strand bias in bacteria. *Nature Genet.* 34: 377–378.

Rockman, M. V. and G. A. Wray. 2002. Abundant raw material for *cis*-regulatory evolution in humans. *Mol. Biol. Evol.* 19: 1991–2004.

Rockman, M. V., M. W. Hahn, N. Soranzo, F. Zimprich, D. B. Goldstein and G. A. Wray. 2005. Ancient and recent positive selection transformed opioid *cis*-regulation in humans. *PLoS Biol.* 3: 2208–2219.

Rodin, S. N. and A. D. Riggs. 2003. Epigenetic silencing may aid evolution by gene duplication. *J. Mol. Evol.* 56: 718–729.

Roger, A. J., S. G. Svard, J. Tovar, C. G. Clark, M. W. Smith, F. D. Gillin and M. L. Sogin. 1998. A mitochondrial-like chaperonin 60 gene in *Giardia lamblia*: Evidence that diplomonads once harbored an endosymbiont related to the progenitor of mitochondria. *Proc. Natl. Acad. Sci. USA* 95: 229–234.

Rogers, J. H. 1990. The role of introns in evolution. *FEBS Lett.* 268: 339–343.

Rogozin, I. B., A. V. Kochetov, F. A. Kondrashov, E. V. Koonin and L. Milanesi. 2001. Presence of ATG triplets in 5′ untranslated regions of eukaryotic cDNAs correlates with a "weak" context of the start codon. *Bioinformatics* 17: 890–900.

Rogozin, I. B., K. S. Makarova, D. A. Natale, A. N. Spiridonov, R. L. Tatusov, Y. I. Wolf, J. Yin and E. V. Koonin. 2002. Congruent evolution of different classes of non-coding DNA in prokaryotic genomes. *Nucleic Acids Res.* 30: 4264–4271.

Rogozin, I. B., Y. I. Wolf, A. V. Sorokin, B. G. Mirkin and E. V. Koonin. 2003. Remarkable interkingdom conservation of intron positions and massive, lineage-specific intron loss and gain in eukaryotic evolution. *Curr. Biol.* 13: 1512–1517.

Rokas, A., D. Kruger and S. B. Carroll. 2005. Animal evolution and the molecular signature of radiations compressed in time. *Science* 310: 1933–1938.

Rolfe, D. F. and G. C. Brown. 1997. Cellular energy utilization and molecular origin of standard metabolic rate in mammals. *Physiol. Rev.* 77: 731–758.

Romero, P. R., S. Zaidi, Y. Y. Fang, V. N. Uversky, P. Radivojac, C. J. Oldfield, M. S. Cortese, M. Sickmeier, T. LeGall, Z. Obradovic and A. K. Dunker. 2006. Alternative splicing in concert with protein intrinsic disorder enables increased functional diversity in multicellular organisms. *Proc. Natl. Acad. Sci. USA* 103: 8390–8395.

Rose, A. B. 2004. The effect of intron location on intron-mediated enhancement of gene expression in *Arabidopsis*. *Plant J.* 40: 744–751.

Roselius, K., W. Stephan and T. Stadler. 2005. The relationship of nucleotide polymorphism, recombination rate and selection in wild tomato species. *Genetics* 171: 753–763.
Rosing, M. T. 1999. 13C-depleted carbon microparticles in >3700-Ma sea-floor sedimentary rocks from west Greenland. *Science* 283: 674–676.
Ross, M. T., D. V. Grafham, A. J. Coffey, S. Scherer, K. McLay, D. Muzny, M. Platzer, G. R. Howell, C. Burrows, C. P. Bird et al. 2005. The DNA sequence of the human X chromosome. *Nature* 434: 325–337.
Rousset, F. 2003. Effective size in simple metapopulation models. *Heredity* 91: 107–111.
Roy, P. J., J. M. Stuart, J. Lund and S. K. Kim. 2002. Chromosomal clustering of muscle-expressed genes in *Caenorhabditis elegans*. *Nature* 418: 975–979.
Roy, S. W. and W. Gilbert. 2005a. Complex early genes. *Proc. Natl. Acad. Sci. USA* 102: 1986–1991.
Roy, S. W. and W. Gilbert. 2005b. The pattern of intron loss. *Proc. Natl. Acad. Sci. USA* 102: 713–718.
Roy, S. W. and W. Gilbert. 2005c. Rates of intron loss and gain: Implications for early eukaryotic evolution. *Proc. Natl. Acad. Sci. USA* 102: 5773–5778.
Roy, S. W. and W. Gilbert. 2006. The evolution of spliceosomal introns: Patterns, puzzles and progress. *Nature Rev. Genet.* 7: 211–221.
Roy, S. W. and D. L. Hartl. 2006. Very little intron loss/gain in *Plasmodium*: Intron loss/gain mutation rates and intron number. *Genome Res.* 16: 750–756.
Roy, S. W. and D. Penny. 2006. Large-scale intron conservation and order-of-magnitude variation in intron loss/gain rates in apicomplexan evolution. *Genome Res.* 16: 1270–1275.
Roy, S. W., M. Nosaka, S. J. de Souza and W. Gilbert. 1999. Centripetal modules and ancient introns. *Gene* 238: 85–91.
Roy-Engel, A. M., M. L. Carroll, E. Vogel, R. K. Garber, S. V. Nguyen, A. H. Salem, M. A. Batzer and P. L. Deininger. 2001. Alu insertion polymorphisms for the study of human genomic diversity. *Genetics* 159: 279–290.
Roy-Engel, A. M., M. El-Sawy, L. Farooq, G. L. Odom, V. Perepelitsa-Belancio, H. Bruch, O. O. Oyeniran and P. L. Deininger. 2005. Human retroelements may introduce intragenic polyadenylation signals. *Cytogenet. Genome Res.* 110: 365–371.
Rozen, S., H. Skaletsky, J. D. Marszalek, P. J. Minx, H. S. Cordum, R. H. Waterston, R. K. Wilson and D. C. Page. 2003. Abundant gene conversion between arms of palindromes in human and ape Y chromosomes. *Nature* 423: 873–876.
Rubin, E. J., B. J. Akerley, V. N. Novik, D. J. Lampe, R. N. Husson and J. J. Mekalanos. 1999. In vivo transposition of *mariner*-based elements in enteric bacteria and mycobacteria. *Proc. Natl. Acad. Sci. USA* 96: 1645–1650.
Rubin, G. M., M. D. Yandell, J. R. Wortman, G. L. Gabor Miklos, C. R. Nelson, I. K. Hariharan, M. E. Fortini, P. W. Li, R. Apweiler, W. Fleischmann et al. 2000. Comparative genomics of the eukaryotes. *Science* 287: 2204–2215.
Rubnitz, J. and S. Subramani. 1984. The minimum amount of homology required for homologous recombination in mammalian cells. *Mol. Cell. Biol.* 4: 2253–2258.
Rudd, M. K., G. A. Wray and H. F. Willard. 2006. The evolutionary dynamics of α-satellite. *Genome Res.* 16: 88–96.
Ruiz-Echevarria, M. J. and S. W. Peltz. 2000. The RNA binding protein Pub1 modulates the stability of transcripts containing upstream open reading frames. *Cell* 101: 741–751.
Ruiz-Echevarria, M. J., C. I. González and S. W. Peltz. 1998. Identifying the right stop: Determining how the surveillance complex recognizes and degrades an aberrant mRNA. *EMBO J.* 17: 575–589.
Ruiz-Pesini, E., A. C. Lapena, C. Diez-Sanchez, A. Perez-Martos, J. Montoya, E. Alvarez, M. Diaz, A. Urries, L. Montoro, M. J. Lopez-Perez and J. A. Enriquez. 2000. Human mtDNA haplogroups associated with high or reduced spermatozoa motility. *Amer. J. Hum. Genet.* 67: 682–696.
Rujan, T. and W. Martin. 2001. How many genes in *Arabidopsis* come from cyanobacteria? An estimate from 386 protein phylogenies. *Trends Genet.* 17: 113–120.
Runnegar, B. 2000. Loophole for snowball Earth. *Nature* 405: 403–404.
Rushforth, A. M. and P. Anderson. 1996. Splicing removes the *Caenorhabditis elegans* transposon Tc1 from most mutant pre-mRNAs. *Mol. Cell. Biol.* 16: 422–429.
Russell, A. G., T. E. Shutt, R. F. Watkins and M. W. Gray. 2005. An ancient spliceosomal intron in the ribosomal protein L7a gene (Rpl7a) of *Giardia lamblia*. *BMC Evol. Biol.* 5: 45.
Russell, A. G., J. M. Charette, D. F. Spencer and M. W. Gray. 2006. An early evolutionary origin for the minor spliceosome. *Nature* 443: 863–866.
Russo, C. A., N. Takezaki and M. Nei. 1995. Molecular phylogeny and divergence times of drosophilid species. *Mol. Biol. Evol.* 12: 391–404.
Rutherford, S. L. and S. Lindquist. 1998. Hsp90 as a capacitor for morphological evolution. *Nature* 396: 336–342.
Ruvinsky, A., S. T. Eskesen, F. N. Eskesen and L. D. Hurst. 2005. Can codon usage bias explain intron phase distributions and exon symmetry? *J. Mol. Evol.* 60: 99–104.
Ruvinsky, I. and G. Ruvkun. 2003. Functional tests of enhancer conservation between distantly related species. *Development* 130: 5133–5142.
Ruvolo, M. 1997. Molecular phylogeny of the hominoids: Inferences from multiple independent DNA sequence data sets. *Mol. Biol. Evol.* 14: 248–265.
Rycovska, A., M. Valach, L. Tomaska, M. Bolotin-Fukuhara and J. Nosek. 2004. Linear versus circular mitochondrial genomes: Intraspecies variability of mitochondrial genome architecture in *Candida parapsilosis*. *Microbiol.* 150: 1571–1580.

Sachidanandam, R., D. Weissman, S. C. Schmidt, J. M. Kakol, L. D. Stein, G. Marth, S. Sherry, J. C. Mullikin, B. J. Mortimore, D. L. Willey et al. 2001. A map of human genome sequence variation containing 1.42 million single nucleotide polymorphisms. *Nature* 409: 928–933.

Sachs, A. B. and G. Varani. 2000. Eukaryotic translation initiation: There are (at least) two sides to every story. *Nature Struct. Biol.* 7: 356–361.

Sadowski, M., B. Dichtl, W. Hubner and W. Keller. 2003. Independent functions of yeast Pcf11p in pre-mRNA 3′ end processing and in transcription termination. *EMBO J.* 22: 2167–2177.

Sadusky, T., A. J. Newman and N. J. Dibb. 2004. Exon junction sequences as cryptic splice sites: Implications for intron origin. *Curr. Biol.* 14: 505–509.

Sahara, K., F. Marec and W. Traut. 1999. TTAGG telomeric repeats in chromosomes of some insects and other arthropods. *Chromosome Res.* 7: 449–460.

Saifi, G. M. and H. S. Chandra. 1999. An apparent excess of sex- and reproduction-related genes on the human X chromosome. *Proc. Biol. Sci.* 266: 203–209.

Sainz, A., J. A. Wilder, M. Wolf and H. Hollocher. 2003. *Drosophila melanogaster* and *D. simulans* rescue strains produce fit offspring, despite divergent centromere-specific histone alleles. *Heredity* 91: 28–35.

Saito, R. and M. Tomita. 1999. On negative selection against ATG triplets near start codons in eukaryotic and prokaryotic genomes. *J. Mol. Evol.* 48: 213–217.

Sakurai, A., S. Fujimori, H. Kochiwa, S. Kitamura-Abe, T. Washio, R. Saito, P. Carninci, Y. Hayashizaki and M. Tomita. 2002. On biased distribution of introns in various eukaryotes. *Gene* 300: 89–95.

Salem, A. H., G. E. Kilroy, W. S. Watkins, L. B. Jorde and M. A. Batzer. 2003. Recently integrated *Alu* elements and human genomic diversity. *Mol. Biol. Evol.* 20: 1349–1361.

Salzberg, S. L., O. White, J. Peterson and J. A. Eisen. 2001. Microbial genes in the human genome: Lateral transfer or gene loss? *Science* 292: 1903–1906.

Samanta, M. P., W. Tongprasit, H. Sethi, C. S. Chin and V. Stolc. 2006. Global identification of noncoding RNAs in *Saccharomyces cerevisiae* by modulating an essential RNA processing pathway. *Proc. Natl. Acad. Sci. USA* 103: 4192–4197.

Samonte, R. V. and E. E. Eichler. 2002. Segmental duplications and the evolution of the primate genome. *Nature Rev. Genet.* 3: 65–72.

Sánchez-Gracia, A., X. Maside and B. Charlesworth. 2005. High rate of horizontal transfer of transposable elements in *Drosophila*. *Trends Genet.* 21: 200–203.

Sandmeyer, S. B., M. Aye and T. Menees. 2002. *Ty3*, a position specific, *gypsy*-like element in *Saccharomyces cerevisiae*. In N. L. Craig, R. Craigie, M. Gellert and A. M. Lambowitz, eds., *Mobile DNA II*, 663–683. ASM Press, Washington, DC.

Sandoval, P., G. Leon, I. Gomez, R. Carmona, P. Figueroa, L. Holuigue, A. Araya and X. Jordana. 2004. Transfer of *RPS14* and *RPL5* from the mitochondrion to the nucleus in grasses. *Gene* 324: 139–147.

San Miguel, P., B. S. Gaut, A. Tikhonov, Y. Nakajima and J. L. Bennetzen. 1998. The paleontology of intergene retrotransposons of maize. *Nature Genet.* 20: 43–45.

San Miguel, P. J., W. Ramakrishna, J. L. Bennetzen, C. S. Busso and J. Dubcovsky. 2002. Transposable elements, genes and recombination in a 215-kb contig from wheat chromosome 5A(m). *Func. Integr. Genomics* 2: 70–80.

Santos, C., R. Montiel, B. Sierra, C. Bettencourt, E. Fernandez, L. Alvarez, M. Lima, A. Abade and M. P. Aluja. 2005. Understanding differences between phylogenetic and pedigree-derived mtDNA mutation rate: A model using families from the Azores Islands (Portugal). *Mol. Biol. Evol.* 22: 1490–1505.

Santoyo, G., J. M. Martinez-Salazar, C. Rodriguez and D. Romero. 2005. Gene conversion tracts associated with crossovers in *Rhizobium etli*. *J. Bacteriol.* 187: 4116–4126.

Sanyal, K., M. Baum and J. Carbon. 2004. Centromeric DNA sequences in the pathogenic yeast *Candida albicans* are all different and unique. *Proc. Natl. Acad. Sci. USA* 101: 11374–11379.

Sargentini, N. J. and K. C. Smith. 1994. DNA sequence analysis of spontaneous and gamma-radiation (anoxic)-induced lacId mutations in *Escherichia coli* umuC122::Tn5: Differential requirement for umuC at G.C vs. A.T sites and for the production of transversions vs. transitions. *Mutat. Res.* 311: 175–189.

Sarkar, N. 1997. Polyadenylation of mRNA in prokaryotes. *Annu. Rev. Biochem.* 66: 173–197.

Sarkar, S. F. and D. S. Guttman. 2004. Evolution of the core genome of *Pseudomonas syringae*, a highly clonal, endemic plant pathogen. *Appl. Environ. Microbiol.* 70: 1999–2012.

Sarot, E., G. Payen-Groschene, A. Bucheton and A. Pelisson. 2004. Evidence for a *piwi*-dependent RNA silencing of the *gypsy* endogenous retrovirus by the *Drosophila melanogaster flamenco* gene. *Genetics* 166: 1313–1321.

Sasa, M. M., P. T. Chippindale and N. A. Johnson. 1998. Patterns of postzygotic isolation in frogs. *Evolution* 52: 1811–1820.

Satoh, M. and T. Kuroiwa. 1991. Organization of multiple nucleoids and DNA molecules in mitochondria of a human cell. *Exp. Cell Res.* 196: 137–40.

Satou, Y., M. Hamaguchi, K. Takeuchi, K. E. Hastings and N. Satoh. 2006. Genomic overview of mRNA 5′-leader *trans*-splicing in the ascidian *Ciona intestinalis*. *Nucleic Acids Res.* 34: 3378–3388.

Saville, B. J., Y. Kohli and J. B. Anderson. 1998. mtDNA recombination in a natural population. *Proc. Natl. Acad. Sci. USA* 95: 1331–1335.

Sawamura, K. and M. T. Yamamoto. 1993. Cytogenetical localization of Zygotic hybrid rescue (Zhr), a *Drosophila melanogaster* gene that rescues interspecific hybrids from embryonic lethality. *Mol. Gen. Genet.* 239: 441–449.

Sawyer, S. A., D. E. Dykhuizen, R. F. DuBose, L. Green, T. Mutangadura-Mhlanga, D. F. Wolczyk and D. L. Hartl. 1987. Distribution and abundance of insertion sequences among natural isolates of *Escherichia coli*. *Genetics* 115: 51–63.

Scannell, D. R., K. P. Byrne, J. L. Gordon, S. Wong and K. H. Wolfe. 2006. Rapid speciation associated with reciprocal gene loss in polyploidy yeasts. *Nature* 440: 341–345.

Schaal, T. D. and T. Maniatis. 1999. Selection and characterization of pre-mRNA splicing enhancers: Identification of novel SR protein-specific enhancer sequences. *Mol. Cell. Biol.* 19: 1705–1719.

Schaaper, R. M. and R. L. Dunn. 1991. Spontaneous mutation in the *Escherichia coli lacI* gene. *Genetics* 129: 317–326.

Schaeffer, S. W. 2002. Molecular population genetics of sequence length diversity in the Adh region of *Drosophila pseudoobscura*. *Genet. Res.* 80: 163–175.

Schaetzlein, S., A. Lucas-Hahn, E. Lemme, W. A. Kues, M. Dorsch, M. P. Manns, H. Niemann and K. L. Rudolph. 2004. Telomere length is reset during early mammalian embryogenesis. *Proc. Natl. Acad. Sci. USA* 101: 8034–8038.

Schlenke, T. A. and D. J. Begun. 2004. Strong selective sweep associated with a transposon insertion in *Drosophila simulans*. *Proc. Natl. Acad. Sci. USA* 101: 1626–1631.

Schmid, P. E., M. Tokeshi and J. M. Schmid-Araya. 2000. Relation between population density and body size in stream communities. *Science* 289: 1557–1560.

Schmidt, T. R., J. W. Doan, M. Goodman and L. I. Grossman. 2003. Retention of a duplicate gene through changes in subcellular targeting: An electron transport protein homologue localizes to the golgi. *J. Mol. Evol.* 57: 222–228.

Schneider, C., C. L. Will, O. V. Makarova, E. M. Makarov and R. Lührmann. 2002. Human U4/U6.U5 and U4atac/U6atac.U5 tri-snRNPs exhibit similar protein compositions. *Mol. Cell. Biol.* 22: 3219–3229.

Schneider, C., C. L. Will, J. Brosius, M. J. Frilander and R. Lührmann. 2004. Identification of an evolutionarily divergent U11 small nuclear ribonucleoprotein particle in *Drosophila*. *Proc. Natl. Acad. Sci. USA* 101: 9584–9589.

Schopf, J. W. 1993. Microfossils of the early Archean Apex chert: New evidence of the antiquity of life. *Science* 260: 640–646.

Schopf, J. W., A. B. Kudryavtsev, D. G. Agresti, T. J. Wdowiak and A. D. Czaja. 2002. Laser-Raman imagery of Earth's earliest fossils. *Nature* 416: 73–76.

Schouten, G. J., H. G. van Luenen, N. C. Verra, D. Valerio and R. H. Plasterk. 1998. Transposon *Tc1* of the nematode *Caenorhabditis elegans* jumps in human cells. *Nucleic Acids Res.* 26: 3013–3017.

Schueler, M. G., A. W. Higgins, M. K. Rudd, K. Gustashaw and H. F. Willard. 2001. Genomic and genetic definition of a functional human centromere. *Science* 294: 109–115.

Schueler, M. G., J. M. Dunn, C. P. Bird, M. T. Ross, L. Viggiano, NISC Comparative Sequencing Program, M. Rocchi, H. F. Willard and E. D. Green. 2005. Progressive proximal expansion of the primate X chromosome centromere. *Proc. Natl. Acad. Sci. USA* 102: 10563–10568.

Schultz, D. W. and M. Yarus. 1994. tRNA structure and ribosomal function. II. Interaction between anticodon helix and other tRNA mutations. *J. Mol. Biol.* 235: 1395–1405.

Schultz, S. T. and M. Lynch. 1997. Deleterious mutation and extinction: Effects of variable mutational effects, synergistic epistasis, beneficial mutations, and degree of outcrossing. *Evolution* 51: 1363–1371.

Schuster, W., B. Wissinger, M. Unseld and A. Brennicke. 1990. Transcripts of the NADH-dehydrogenase subunit 3 gene are differentially edited in *Oenothera* mitochondria. *EMBO J.* 9: 263–269.

Sebat, J., B. Lakshmi, J. Troge, J. Alexander, J. Young, P. Lundin, S. Maner, H. Massa, M. Walker, M. Chi et al. 2004. Large-scale copy number polymorphism in the human genome. *Science* 305: 525–528.

Seffens, W. and D. Digby. 1999. mRNAs have greater negative folding free energies than shuffled or codon choice randomized sequences. *Nucleic Acids Res.* 27: 1578–1584.

Segurado, M., A. de Luis and F. Antequera. 2003. Genome-wide distribution of DNA replication origins at A+T-rich islands in *Schizosaccharomyces pombe*. *EMBO Rep.* 4: 1048–1053.

Selinger, D. W., K. J. Cheung, R. Mei, E. M. Johansson, C. S. Richmond, F. R. Blattner, D. J. Lockhart and G. M. Church. 2000. RNA expression analysis using a 30 base pair resolution *Escherichia coli* genome array. *Nature Biotechnol.* 18: 1262–1268.

Selker, E. U. 1990. Premeiotic instability of repeated sequences in *Neurospora crassa*. *Annu. Rev. Genet.* 24: 579–613.

Selosse, M.-A., A. Béatrice and B. Godelle. 2001. Reducing the genome size of organelles favours gene transfer to the nucleus. *Trends Ecol. Evol.* 16: 135–141.

Sémon, M. and L. Duret. 2006. Evolutionary origin and maintenance of coexpressed gene clusters in mammals. *Mol. Biol. Evol.* 23: 1715–1723.

Seoighe, C., N. Federspiel, T. Jones, N. Hansen, V. Bivolarovic, R. Surzycki, R. Tamse, C. Komp, L. Huizar, R. W. Davis et al. 2000. Prevalence of small inversions in yeast gene order evolution. *Proc. Natl. Acad. Sci. USA* 97: 14433–14437.

Seoighe, C., C. R. Johnston and D. C. Shields. 2003. Significantly different patterns of amino acid replacement after gene duplication as compared to after speciation. *Mol. Biol. Evol.* 20: 484–490.

Seoighe, C., C. Gehring and L. D. Hurst. 2005. Gametophytic selection in *Arabidopsis thaliana* supports the selective model of intron length reduction. *PLoS Genet.* 1: 154–158.

Sezutsu, H., E. Nitasaka and T. Yamazaki. 1995. Evolution of the *LINE*-like *I* element in the *Drosophila melanogaster* species subgroup. *Mol. Gen. Genet.* 249: 168–178.

Shabalina, S. A., A. Y. Ogurtsov, V. A. Kondrashov and A. S. Kondrashov. 2001. Selective constraint in intergenic regions of human and mouse genomes. *Trends Genet.* 17: 373–376.

Shackelton, L. A. and E. C. Holmes. 2004. The evolution of large DNA viruses: Combining genomic information of viruses and their hosts. *Trends Microbiol.* 12: 458–465.

Shakirov, E. V., Y. V. Surovtseva, N. Osbun and D. E. Shippen. 2005. The *Arabidopsis* Pot1 and Pot2 proteins function in telomere length homeostasis and chromosome end protection. *Mol. Cell. Biol.* 25: 7725–7733.

Shankar, R., D. Grover, S. K. Brahmachari and M. Mukerji. 2004. Evolution and distribution of RNA polymerase II regulatory sites from RNA polymerase III dependant mobile *Alu* elements. *BMC Evol. Biol.* 4: 37.

Shao, C., P. J. Stambrook and J. A. Tischfield. 2001. Mitotic recombination is suppressed by chromosomal divergence in hybrids of distantly related mouse strains. *Nature Genet.* 28: 169–172.

Shao, Z., S. Graf, O. Y. Chaga and D. V. Lavrov. 2006. Mitochondrial genome of the moon jelly *Aurelia aurita* (Cnidaria, Scyphozoa): A linear DNA molecule encoding a putative DNA-dependent DNA polymerase. *Gene* 381: 92–101.

Shapira, S. K. and V. G. Finnerty. 1986. The use of genetic complementation in the study of eukaryotic macromolecular evolution: Rate of spontaneous gene duplication at two loci of *Drosophila melanogaster*. *J. Mol. Evol.* 23: 159–167.

Sharakhov, I. V., A. C. Serazin, O. G. Grushko, A. Dana, N. Lobo, M. E. Hillenmeyer, R. Westerman, J. Romero-Severson, C. Costantini, N. Sagnon et al. 2002. Inversions and gene order shuffling in *Anopheles gambiae* and *A. funestus*. *Science* 298: 182–185.

Sharp, P. A. 1985. On the origin of RNA splicing and introns. *Cell* 42: 397–400.

Sharp, P. M. and W.-H. Li. 1987. The rate of synonymous substitution in enterobacterial genes is inversely related to codon usage bias. *Mol. Biol. Evol.* 4: 222–230.

Sharp, P. M., E. Bailes, R. J. Grocock, J. F. Peden and R. E. Sockett. 2005. Variation in the strength of selected codon usage bias among bacteria. *Nucleic Acids Res.* 33: 1141–1153.

Shaw, P. J., N. S. Wratten, A. P. McGregor and G. A. Dover. 2002. Coevolution in bicoid-dependent promoters and the inception of regulatory incompatibilities among species of higher Diptera. *Evol. Dev.* 4: 265–277.

Shaw, R. J., N. D. Bonawitz and D. Reines. 2002. Use of an in vivo reporter assay to test for transcriptional and translational fidelity in yeast. *J. Biol. Chem.* 277: 24420–24426.

She, X., J. E. Horvath, Z. Jiang, G. Liu, T. S.Furey, L. Christ, R. Clark, T. Graves, C. L. Gulden, C. Alkan et al. 2004. The structure and evolution of centromeric transition regions within the human genome. *Nature* 430: 857–864.

Shearer, T. L., M. J. Van Oppen, S. L. Romano and G. Worheide. 2002. Slow mitochondrial DNA sequence evolution in the Anthozoa (Cnidaria). *Mol. Ecol.* 11: 2475–2487.

Sheen, F. M. and R. W. Levis. 1994. Transposition of the LINE-like retrotransposon *TART* to *Drosophila* chromosome termini. *Proc. Natl. Acad. Sci. USA* 91: 12510–12514.

Shen, P., F. Wang, P. A. Underhill, C. Franco, W. H. Yang, A. Roxas, R. Sung, A. A. Lin, R. W. Hyman, D. Vollrath et al. 2000. Population genetic implications from sequence variation in four Y chromosome genes. *Proc. Natl. Acad. Sci. USA* 97: 7354–7359.

Sherratt, D. J. 2003. Bacterial chromosome dynamics. *Science* 301: 780–785.

Sheth, U. and R. Parker. 2006. Targeting of aberrant mRNAs to cytoplasmic processing bodies. *Cell* 125: 1095–1109.

Sheveleva, E. V. and R. B. Hallick. 2004. Recent horizontal intron transfer to a chloroplast genome. *Nucleic Acids Res.* 32: 803–810.

Shields, D. C. 1990. Switches in species-specific codon preferences: The influence of mutation biases. *J. Mol. Evol.* 31: 71–80.

Shields, D. C. and K. H. Wolfe. 1997. Accelerated evolution of sites undergoing mRNA editing in plant mitochondria and chloroplasts. *Mol. Biol. Evol.* 14: 344–349.

Shigenobu, S., H. Watanabe, M. Hattori, Y. Sakaki and H. Ishikawa. 2000. Genome sequence of the endocellular bacterial symbiont of aphids *Buchnera* sp. APS. *Nature* 407: 81–86.

Shikanai, T. 2006. RNA editing in plant organelles: Machinery, physiological function and evolution. *Cell. Mol. Life Sci.* 63: 698–708.

Shine, J. and L. Dalgarno. 1974. The 3′-terminal sequence of *Escherichia coli* 16S ribosomal RNA: Complementarity to nonsense triplets and ribosome binding sites. *Proc. Natl. Acad. Sci. USA* 17: 1342–1346.

Shiu, S. H., J. K. Byrnes, R. Pan, P. Zhang and W.-H. Li. 2006. Role of positive selection in the retention of duplicate genes in mammalian genomes. *Proc. Natl. Acad. Sci. USA* 103: 2232–2236.

Shixing, Z. and C. Huineng. 1995. Megascopic multicellular organisms form the 1700-million-year-old Tuanshanzi formation in the Jixian area, north China. *Science* 270: 620–622.

Shpak, M. 2005. The role of deleterious mutations in allopatric speciation. *Evolution* 59: 1389–1399.

Shukla, G. C. and R. A. Padgett. 2002. A catalytically active group II intron domain 5 can function in the U12-dependent spliceosome. *Mol. Cell* 9: 1145–1150.

Shuter, B. J., J. E. Thomas, W. D. Taylor, A. M. Zimmerman. 1983. Phenotypic correlates of genomic DNA content in unicellular eukaryotes and other cells. *Amer. Natur.* 122: 26–44.

Sickmann, A., J. Reinders, Y. Wagner, C. Joppich, R. Zahedi, H. E. Meyer, B. Schonfisch, I. Perschil, A. Chacinska, B. Guiard et al. 2003. The proteome of *Saccharomyces cerevisiae* mitochondria. *Proc. Natl. Acad. Sci. USA* 100: 13207–13212.

Siepel, A. and D. Haussler. 2004. Phylogenetic estimation of context-dependent substitution rates by maximum likelihood. *Mol. Biol. Evol.* 21: 468–488.

Siepel, A., G. Bejerano, J. S. Pedersen, A. S. Hinrichs, M. Hou, K. Rosenbloom, H. Clawson, J. Spieth, L. W. Hillier, S. Richards et al. 2005. Evolutionarily conserved elements in vertebrate, insect, worm, and yeast genomes. *Genome Res.* 15: 1034–1050.

Sijen, T. and R. H. Plasterk. 2003. Transposon silencing in the *Caenorhabditis elegans* germ line by natural RNAi. *Nature* 426: 310–314.

Silberman, J. D., A. G. Simpson, J. Kulda, I. Cepicka, V. Hampl, P. J. Johnson and A. J. Roger. 2002. Retortamonad flagellates are closely related to diplomonads—implications for the history of mitochondrial function in eukaryote evolution. *Mol. Biol. Evol.* 19: 777–786.

Silliker, M. E., M. R. Liotta and D. J. Cummings. 1996. Elimination of mitochondrial mutations by sexual reproduction: Two *Podospora anserina* mitochondrial mutants yield only wild-type progeny when mated. *Curr. Genet.* 30: 318–324.

Silva, C., P. Vinuesa, L. E. Eguiarte, V. Souza and E. Martinez-Romero. 2005. Evolutionary genetics and biogeographic structure of *Rhizobium gallicum sensu lato*, a widely distributed bacterial symbiont of diverse legumes. *Mol. Ecol.* 14: 4033–4050.

Silva, J. C. and M. G. Kidwell. 2004. Evolution of *P* elements in natural populations of *Drosophila willistoni* and *D. sturtevanti*. *Genetics* 168: 1323–1335.

Silva-Filho, M. C. 2003. One ticket for multiple destinations: Dual targeting of proteins to distinct subcellular locations. *Curr. Opin. Plant Biol.* 6: 589–595.

Simillion, C., K. Vandepoele, M. C. Van Montagu, M. Zabeau and Y. Van de Peer. 2002. The hidden duplication past of *Arabidopsis thaliana*. *Proc. Natl. Acad. Sci. USA* 99: 13627–13632.

Simmons, M. J., K. J. Haley and S. J. Thompson. 2002. Maternal transmission of *P* element transposase activity in *Drosophila melanogaster* depends on the last *P* intron. *Proc. Natl. Acad. Sci. USA* 99: 9306–9309.

Simon, D. K. and D. R. Johns. 1999. Mitochondrial disorders: Clinical and genetic features. *Annu. Rev. Med.* 50: 111–127.

Simpson, A. G., E. K. MacQuarrie and A. J. Roger. 2002. Eukaryotic evolution: Early origin of canonical introns. *Nature* 419: 270.

Simpson, C. G., G. Thow, G. P. Clark, S. N. Jennings, J. A. Watters and J. W. Brown. 2002. Mutational analysis of a plant branchpoint and polypyrimidine tract required for constitutive splicing of a mini-exon. *RNA* 8: 47–56.

Simpson, L., O. H. Thiemann, N. J. Savill, J. D. Alfonzo and D. A. Maslov. 2000. Evolution of RNA editing in trypanosome mitochondria. *Proc. Natl. Acad. Sci. USA* 97: 6986–6993.

Singh, N. D., P. F. Arndt and D. A. Petrov. 2005. Genomic heterogeneity of background substitutional patterns in *Drosophila melanogaster*. *Genetics* 169: 709–722.

Singh, U., J. B. Rogers, B. J. Mann and W. A. Petri, Jr. 1997. Transcription initiation is controlled by three core promoter elements in the hgl5 gene of the protozoan parasite *Entamoeba histolytica*. *Proc. Natl. Acad. Sci. USA* 94: 8812–8817.

Sironi, M., G. Menozzi, G. P. Comi, R. Cagliani, N. Bresolin and U. Pozzoli. 2005. Analysis of intronic conserved elements indicates that functional complexity might represent a major source of negative selection on non-coding sequences. *Hum. Mol. Genet.* 14: 2533–2546.

Skaletsky, H., T. Kuroda-Kawaguchi, P. J. Minx, H. S. Cordum, L. Hillier, L. G. Brown, S. Repping, T. Pyntikova, J. Ali, T. Bieri et al. 2003. The male-specific region of the human Y chromosome is a mosaic of discrete sequence classes. *Nature* 423: 825–837.

Slamovits, C. H. and P. J. Keeling. 2006. A high density of ancient spliceosomal introns in oxymonad excavates. *BMC Evol. Biol.* 6: 34.

Slatkin, M. 1985. Genetic differentiation of transposable elements under mutation and unbiased gene conversion. *Genetics* 110: 145–158.

Slattery, J. P. and S. J. O'Brien. 1998. Patterns of Y and X chromosome DNA sequence divergence during the Felidae radiation. *Genetics* 148: 1245–1255.

Sleep, N. H., K. J. Zahnle, J. F. Kasting and H. J. Morowitz. 1989. Annihilation of ecosystems by large asteroid impacts on the early Earth. *Nature* 342: 139–142.

Slupska, M. M., A. G. King, S. Fitz-Gibbon, J. Besemer, M. Borodovsky and J. H. Miller. 2001. Leaderless transcripts of the crenarchaeal hyperthermophile *Pyrobaculum aerophilum*. *J. Mol. Biol.* 309: 347–360.

Smale, S. T. and J. T. Kadonaga. 2003. The RNA polymerase II core promoter. *Annu. Rev. Biochem.* 72: 449–479.

Smalheiser, N. R. and V. I. Torvik. 2006. Alu elements within human mRNAs are probable microRNA targets. *Trends Genet.* 22: 532–536.

Smit, A. F. 1999. Interspersed repeats and other mementos of transposable elements in mammalian genomes. *Curr. Opin. Genet. Dev.* 9: 657–663.

Smit, A. F. and A. D. Riggs. 1996. Tiggers and DNA transposon fossils in the human genome. *Proc. Natl. Acad. Sci. USA* 93: 1443–1448.

Smith, C. A. (ed.). 2001. *The Ribosome*. Cold Spring Harbor Laboratory Press, Woodbury, NY.

Smith, C. W. and J. Valcárcel. 2000. Alternative pre-mRNA splicing: The logic of combinatorial control. *Trends Biochem. Sci.* 25: 381–388.

Smith, H. C. 2006. Editing informational content of expressed DNA sequences and their transcripts. In L. H. Caporale, ed., *The Implicit Genome*, 248–265. Oxford University Press, Oxford.

Smith, N. G. and A. Eyre-Walker. 2002. Adaptive protein evolution in *Drosophila*. *Nature* 415: 1022–1024.

Sniegowski, P. D. and H. A. Murphy. 2006. Evolvability. *Curr. Biol.* 16: R831–R834.

Sogin, M. L. 1991. Early evolution and the origin of eukaryotes. *Curr. Opin. Genet. Dev.* 1: 457–463.

Solignac, M., M. Monnerot and J. C. Mounolou. 1983. Mitochondrial DNA heteroplasmy in *Drosophila mauritiana*. *Proc. Natl. Acad. Sci. USA* 80: 6942–6946.

Söll, D. and U. L. RajBhandary. 1995. *tRNA: Structure, Biosynthesis, and Function*. ASM Press, Washington, DC.

Sommer, R. J. 1997. Evolution and development—the nematode vulva as a case study. *BioEssays* 19: 225–231.

Sontheimer, E. J., P. M. Gordon and J. A. Piccirilli. 1999. Metal ion catalysis during group II intron self-splicing: Parallels with the spliceosome. *Genes Dev.* 13: 1729–1741.

Soppa, J. 1999a. Normalized nucleotide frequencies allow the definition of archaeal promoter elements for different archaeal groups and reveal base-specific TFB contacts upstream of the TATA box. *Mol. Microbiol.* 31: 1589–1592.

Soppa, J. 1999b. Transcription initiation in Archaea: Facts, factors and future aspects. *Mol. Microbiol.* 31: 1295–1305.

Sorek, R. and G. Ast. 2003. Intronic sequences flanking alternatively spliced exons are conserved between human and mouse. *Genome Res.* 13: 1631–1637.

Sorek, R., G. Ast and D. Graur. 2002. *Alu*-containing exons are alternatively spliced. *Genome Res.* 12: 1060–1067.

Sorek, R., R. Shamir and G. Ast. 2004. How prevalent is functional alternative splicing in the human genome? *Trends Genet.* 20: 68–71.

Spellman, P. T. and G. M. Rubin. 2002. Evidence for large domains of similarly expressed genes in the *Drosophila* genome. *J. Biol.* 1: 5.

Spitz, F., F. Gonzalez and D. Duboule. 2003. A global control region defines a chromosomal regulatory landscape containing the HoxD cluster. *Cell* 113: 405–417.

Spofford, J. B. 1969. Heterosis and the evolution of duplications. *Amer. Natur.* 103: 407–432.

Springer, B., P. Sander, L. Sedlacek, W. D. Hardt, V. Mizrahi, P. Schar and E. C. Bottger. 2004. Lack of mismatch correction facilitates genome evolution in mycobacteria. *Mol. Microbiol.* 53: 1601–1609.

Städler, T. and L. F. Delph. 2002. Ancient mitochondrial haplotypes and evidence for intragenic recombination in a gynodioecious plant. *Proc. Natl. Acad. Sci. USA* 99: 11730–11735.

Stahley, M. R. and S. A. Strobel. 2005. Structural evidence for a two-metal-ion mechanism of group I intron splicing. *Science* 309: 1587–1590.

Stajich, J. E. and F. S. Dietrich. 2006. Evidence of mRNA-mediated intron loss in the human-pathogenic fungus *Cryptococcus neoformans*. *Eukaryot. Cell* 5: 789–793.

Stamm, S., S. Ben-Ari, I. Rafalska, Y. Tang, Z. Zhang, D. Toiber, T. A. Thanaraj and H. Soreq. 2005. Function of alternative splicing. *Gene* 344: 1–20.

Staněk, D. and K. M. Neugebauer. 2006. The Cajal body: A meeting place for spliceosomal snRNPs in the nuclear maze. *Chromosoma* 115: 343–354.

Stanier, R. Y. 1970. Some aspects of the biology of cells and their possible evolutionary significance. *Symp. Soc. Gen. Microbiol.* 20: 1–38.

Stanley, S. M. 1979. *Macroevolution*. W. H. Freeman, New York.

Stanley, S. M. 1985. Rates of evolution. *Paleobiology* 11: 13–26.

Stechmann, A. and T. Cavalier-Smith. 2002. Rooting the eukaryote tree by using a derived gene fusion. *Science* 297: 89–91.

Steenkamp, E. T., J. Wright and S. L. Baldauf. 2006. The protistan origins of animals and fungi. *Mol. Biol. Evol.* 23: 93–106.

Stein, L. D., Z. Bao, D. Blasiar, T. Blumenthal, M. R. Brent, N. Chen, A. Chinwalla, L. Clarke, C. Clee, A. Coghlan et al. 2003. The genome sequence of *Caenorhabditis briggsae*: A platform for comparative genomics. *PLoS Biol.* 1: E45.

Steinberg, S. and R. Cedergren. 1994. Structural compensation in atypical mitochondrial tRNAs. *Nature Struct. Biol.* 1: 507–510.

Steinberg, S., D. Gautheret and R. Cedergren. 1994. Fitting the structurally diverse animal mitochondrial tRNAs(Ser) to common three-dimensional constraints. *J. Mol. Biol.* 236: 982–989.

Steinemann, M. and S. Steinemann. 1997. The enigma of Y chromosome degeneration: TRAM, a novel retrotransposon is preferentially located on the neo-Y chromosome of *Drosophila miranda*. *Genetics* 145: 261–266.

Steinemann, S. and M. Steinemann. 2005a. Retroelements: Tools for sex chromosome evolution. *Cytogenet. Genome Res.* 110: 134–143.

Steinemann, S. and M. Steinemann. 2005b. Y chromosomes: Born to be destroyed. *BioEssays* 27: 1076–1083.

Stephan, W. 1996. The rate of compensatory evolution. *Genetics* 144: 419–426.

Stephan, W. and C. H. Langley. 1992. Evolutionary consequences of DNA mismatch inhibited repair opportunity. *Genetics* 132: 567–574.

Stephan, W. and C. H. Langley. 1998. DNA polymorphism in *Lycopersicon* and crossing-over per physical length. *Genetics* 150: 1585–1593.

Stern, D. L. 2000. Evolutionary developmental biology and the problem of variation. *Evolution* 54: 1079–1091.

Stevens, S. W., D. E. Ryan, H. Y. Ge, R. E. Moore, M. K. Young, T. D. Lee and J. Abelson. 2002. Composition and functional characterization of the yeast spliceosomal penta-snRNP. *Mol. Cell* 9: 31–44.

Stevenson, R. D., M. F. Hill and P. J. Bryant. 1995. Organ and cell allometry in Hawaiian *Drosophila*: How to make a big fly. *Proc. Roy. Soc. Lond.* B *Biol. Sci.* 259: 105–110.

Stiller, J. W., E. C. Duffield and B. D. Hall. 1998. Amitochondriate amoebae and the evolution of DNA-dependent RNA polymerase II. *Proc. Natl. Acad. Sci. USA* 95: 11769–11774.

Stillman, B. 1996. Cell cycle control of DNA replication. *Science* 274: 1659–1664.

Stolc, V., Z. Gauhar, C. Mason, G. Halasz, M. F. van Batenburg, S. A. Rifkin, S. Hua, T. Herreman, W. Tongprasit, P. E. Barbano et al. 2004. A gene expression map for the euchromatic genome of *Drosophila melanogaster*. *Science* 306: 655–660.

Stoltzfus, A. 1999. On the possibility of constructive neutral evolution. *J. Mol. Evol.* 49: 169–181.

Stoltzfus, A. 2006a. Mutation-biased adaptation in a protein NK model. *Mol. Biol. Evol.* 23: 1852–1862.

Stoltzfus, A. 2006b. Mutationism and the dual causation of evolutionary change. *Evol. Dev.* 8: 304–317.

Stoltzfus, A., J. M. Logsdon, Jr., J. D. Palmer and W. F. Doolittle. 1997. Intron "sliding" and the diversity of intron positions. *Proc. Natl. Acad. Sci. USA* 94: 10739–10744.

Stone, A. C., R. C. Griffiths, S. L. Zegura and M. F. Hammer. 2002. High levels of Y-chromosome nucleotide diversity in the genus *Pan*. *Proc. Natl. Acad. Sci. USA* 99: 43–48.

Stone, J. R. and G. A. Wray. 2001. Rapid evolution of *cis*-regulatory sequences via local point mutations. *Mol. Biol. Evol.* 18: 1764–1770.

Stover, N. A. and R. E. Steele. 2001. *Trans*-spliced leader addition to mRNAs in a cnidarian. *Proc. Natl. Acad. Sci. USA* 98: 5693–5698.

Strathern, J. N., B. K. Shafer and C. B. McGill. 1995. DNA synthesis errors associated with double-strand-break repair. *Genetics* 140: 965–972.

Struhl, K. 1989. Molecular mechanisms of transcriptional regulation in yeast. *Annu. Rev. Biochem.* 58: 1051–1077.

Stupar, R. M., J. W. Lilly, C. D. Town, Z. Cheng, S. Kaul, C. R. Buell and J. Jiang. 2001. Complex mtDNA constitutes an approximate 620-kb insertion on *Arabidopsis thaliana* chromosome 2: Implication of potential sequencing errors caused by large-unit repeats. *Proc. Natl. Acad. Sci. USA* 98: 5099–5103.

Sturm, N. R., M. C. Yu and D. A. Campbell. 1999. Transcription termination and 3′-end processing of the spliced leader RNA in kinetoplastids. *Mol. Cell. Biol.* 19: 1595–1604.

Su, Z., J. Wang, J. Yu, X. Huang and X. Gu. 2006. Evolution of alternative splicing after gene duplication. *Genome Res.* 16: 182–189.

Subramanian, S. and S. Kumar. 2004. Gene expression intensity shapes evolutionary rates of the proteins encoded by the vertebrate genome. *Genetics* 168: 373–381.

Sueoka, N. 1962. On the genetic basis of variation and heterogeneity of DNA base composition. *Proc. Natl. Acad. Sci. USA* 48: 582–592.

Sueoka, N. 1988. Directional mutation pressure and neutral molecular evolution. *Proc. Natl. Acad. Sci. USA* 85: 2653–2657.

Sueoka, N. 1995. Intrastrand parity rules of DNA base composition and usage biases of synonymous codons. *J. Mol. Evol.* 40: 318–325.

Sugawara, N. and J. E. Haber. 1992. Characterization of double-strand break-induced recombination: Homology requirements and single-stranded DNA formation. *Mol. Cell. Biol.* 12: 563–575.

Sullivan, B. A. and G. H. Karpen. 2004. Centromeric chromatin exhibits a histone modification pattern that is distinct from both euchromatin and heterochromatin. *Nature Struct. Mol. Biol.* 11: 1076–1083.

Sullivan, K. F. 2001. A solid foundation: Functional specialization of centromeric chromatin. *Curr. Opin. Genet. Dev.* 11: 182–188.

Sun, X., H. D. Le, J. M. Wahlstrom and G. H. Karpen. 2003. Sequence analysis of a functional *Drosophila* centromere. *Genome Res.* 13: 182–194.

Sutak, R., P. Dolezal, H. L. Fiumera, I. Hrdy, A. Dancis, M. Delgadillo-Correa, P. J. Johnson, M. Muller and J. Tachezy. 2004. Mitochondrial-type assembly of FeS centers in the hydrogenosomes of the amitochondriate eukaryote *Trichomonas vaginalis*. *Proc. Natl. Acad. Sci. USA* 101: 10368–10373.

Suttle, C. A. 2005. Viruses in the sea. *Nature* 437: 356–361.

Suzuki, Y., D. Ishihara, M. Sasaki, H. Nakagawa, H. Hata, T. Tsunoda, M. Watanabe, T. Komatsu, T. Ota, T. Isogai et al. 2000. Statistical analysis of the 5′ untranslated region of human mRNA using "oligo-capped" cDNA libraries. *Genomics* 64: 286–297.

Suzuki, Y., H. Taira, T. Tsunoda, J. Mizushima-Sugano, J. Sese, H. Hata, T. Ota, T. Isogai, T. Tanaka, S. Morishita et al. 2001. Diverse transcriptional initiation revealed by fine, large-scale mapping of mRNA start sites. *EMBO Rep.* 2: 388–393.

Suzuki, Y., T. Tsunoda, J. Sese, H. Taira, J. Mizushima-Sugano, H. Hata, T. Ota, T. Isogai, T. Tanaka, Y. Nakamura et al. 2001. Identification and characterization of the potential promoter regions of 1031 kinds of human genes. *Genome Res.* 11: 677–684.

Sverdlov, A. V., I. B. Rogozin, V. N. Babenko and E. V. Koonin. 2003. Evidence of splice signal migration from exon to intron during intron evolution. *Curr. Biol.* 13: 2170–2174.

Sverdlov, E. D. 2000. Retroviruses and primate evolution. *BioEssays* 22: 161–171.

Swire, J., O. P. Judson and A. Burt. 2005. Mitochondrial genetic codes evolve to match amino acid requirements of proteins. *J. Mol. Evol.* 60: 128–139.

Symer, D. E., C. Connelly, S. T. Szak, E. M. Caputo, G. J. Cost, G. Parmigiani and J. D. Boeke. 2002. Human *L1* retrotransposition is associated with genetic instability in vivo. *Cell* 110: 327–338.

Szczepanik, D., P. Mackiewicz, M. Kowalczuk, A. Gierlik, A. Nowicka, M. R. Dudek and S. Cebrat. 2001. Evolution rates of genes on leading and lagging DNA strands. *J. Mol. Evol.* 52: 426–33.

Szostak, J. W., T. L. Orr-Weaver, R. J. Rothstein and F. W. Stahl. 1983. The double-strand-break repair model for recombination. *Cell* 33: 25–35.

Taanman, J. W. 1999. The mitochondrial genome: Structure, transcription, translation and replication. *Biochim. Biophys. Acta* 1410: 103–123.

Tabara, H., M. Sarkissian, W. G. Kelly, J. Fleenor, A. Grishok, L. Timmons, A. Fire and C. C. Mello. 1999. The *rde-1* gene, RNA interference, and transposon silencing in *C. elegans*. *Cell* 99: 123–132.

Tachezy, J., L. B. Sanchez and M. Muller. 2001. Mitochondrial type iron-sulfur cluster assembly in the amitochondriate eukaryotes *Trichomonas vaginalis* and *Giardia intestinalis*, as indicated by the phylogeny of IscS. *Mol. Biol. Evol.* 18: 1919–1928.

Takahara, T., D. Kasahara, D. Mori, S. Yanagisawa and H. Akanuma. 2002. The *trans*-spliced variants of *Sp1* mRNA in rat. *Biochem. Biophys. Res. Commun.* 298: 156–162.

Takahashi, I. and J. Marmur. 1963. Replacement of thymidylic acid by deoxyuridylic acid in the deoxyribonucleic acid of a transducing phage for *Bacillus subtilis*. *Nature* 197: 794–795.

Takahashi, Y., S. Urushiyama, T. Tani and Y. Ohshima. 1993. An mRNA-type intron is present in the *Rhodotorula hasegawae* U2 small nuclear RNA gene. *Mol. Cell. Biol.* 13: 5613–5619.

Takahashi, Y., T. Tani and Y. Ohshima. 1996. Spliceosomal introns in conserved sequences of U1 and U5 small nuclear RNA genes in yeast *Rhodotorula hasegawae*. *J. Biochem.* (Tokyo) 120: 677–683.

Takahata, N. and M. Slatkin. 1983. Evolutionary dynamics of extranuclear genes. *Genet. Res.* 42: 257–265.

Takebayashi, S. I., E. M. Manders, H. Kimura, H. Taguchi and K. Okumura. 2001. Mapping sites where replication initiates in mammalian cells using DNA fibers. *Exp. Cell Res.* 271: 263–268.

Talbert, P. B., T. D. Bryson and S. Henikoff. 2004. Adaptive evolution of centromere proteins in plants and animals. *J. Biol.* 3: 18.

Tamas, I., L. Klasson, B. Canbäck, A. K. Naslund, A. S. Eriksson, J. J. Wernegreen, J. P. Sandstrom, N. A. Moran and S. G. Andersson. 2002. 50 million years of genomic stasis in endosymbiotic bacteria. *Science* 296: 2376–2379.

Tang, T. H., J. P. Bachellerie, T. Rozhdestvensky, M. L. Bortolin, H. Huber, M. Drungowski, T. Elge, J. Brosius and A. Huttenhofer. 2002. Identification of 86 candidates for small non-messenger RNAs from the archaeon *Archaeoglobus fulgidus*. *Proc. Natl. Acad. Sci. USA* 99: 7536–7541.

Tanguay, R. L. and D. R. Gallie. 1996a. The effect of the length of the 3′-untranslated region on expression in plants. *FEBS Lett.* 394: 285–288.

Tanguay, R. L. and D. R. Gallie. 1996b. Translational efficiency is regulated by the length of the 3′ untranslated region. *Mol. Cell. Biol.* 16: 146–156.

Tani, T. and Y. Ohshima. 1991. mRNA-type introns in U6 small nuclear RNA genes: Implications for the catalysis in pre-mRNA splicing. *Genes Dev.* 5: 1022–1031.

Tarrío, R., F. Rodriguez-Trelles and F. J. Ayala. 1998. New *Drosophila* introns originate by duplication. *Proc. Natl. Acad. Sci. USA* 95: 1658–1662.

Tautz, D. 1992. Redundancies, development and the flow of information. *BioEssays* 14: 263–266.

Taylor, D. R., C. Zeyl and E. Cooke. 2002. Conflicting levels of selection in the accumulation of mitochondrial defects in *Saccharomyces cerevisiae*. *Proc. Natl. Acad. Sci. USA* 99: 3690–3694.

Taylor, J. S. and J. Raes. 2004. Duplication and divergence: The evolution of new genes and old ideas. *Annu. Rev. Genet.* 38: 615–643.

Taylor, J. S., Y. Van de Peer, I. Braasch and A. Meyer. 2001a. Comparative genomics provides evidence for an ancient genome duplication event in fish. *Phil. Trans. Roy. Soc. Lond.* B *Biol. Sci.* 356: 1661–1679.

Taylor, J. S., Y. van de Peer and A. Meyer. 2001b. Genome duplication, divergent resolution and speciation. *Trends Genet.* 17: 299–301.

Taylor, J. S., I. Braasch, T. Frickey, A. Meyer and Y. Van de Peer. 2003. Genome duplication, a trait shared by 22000 species of ray-finned fish. *Genome Res.* 13: 382–390.

Taylor, J., S. Tyekucheva, M. Zody, F. Chiaromonte and K. D. Makova. 2006. Strong and weak male mutation bias at different sites in the primate genomes: Insights from the human-chimpanzee comparison. *Mol. Biol. Evol.* 23: 565–573.

Technau, U., S. Rudd, P. Maxwell, P. M. Gordon, M. Saina, L. C. Grasso, D. C. Hayward, C. W. Sensen, R. Saint, T. W. Holstein et al. 2005. Maintenance of ancestral complexity and non-metazoan genes in two basal cnidarians. *Trends Genet.* 21: 633–639.

Teixeira, M. T. and E. Gilson. 2005. Telomere maintenance, function and evolution: The yeast paradigm. *Chromosome Res.* 13: 535–548.

Temin, H. M. 1980. Origin of retroviruses from cellular moveable genetic elements.*Cell* 21: 599–600.

Temin, H. M. 1991. Sex and recombination in retroviruses. *Trends Genet.* 7: 71–74.

Templeton, A. R. 1996. Gene lineages and human evolution. *Science* 272: 1363–1364.

Templeton, T. J., L. M. Iyer, V. Anantharaman, S. Enomoto, J. E. Abrahante, G. M. Subramanian, S. L. Hoffman, M. S. Abrahamsen and L. Aravind. 2004. Comparative analysis of apicomplexa and genomic diversity in eukaryotes. *Genome Res.* 14: 1686–1695.

Tenaillon, M. I., J. U'Ren, O. Tenaillon and B. S. Gaut. 2004. Selection versus demography: A multilocus investigation of the domestication process in maize. *Mol. Biol. Evol.* 21: 1214–1225.

Teng, S. C. and V. A. Zakian. 1999. Telomere-telomere recombination is an efficient bypass pathway for telomere maintenance in *Saccharomyces cerevisiae*. *Mol. Cell. Biol.* 19: 8083–8093.

Teng, S. C., B. Kim and A. Gabriel. 1996. Retrotransposon reverse-transcriptase-mediated repair of chromosomal breaks. *Nature* 383: 641–644.

Terzian, C., C. Ferraz, J. Demaille and A. Bucheton. 2000. Evolution of the *Gypsy* endogenous retrovirus in the *Drosophila melanogaster* subgroup. *Mol. Biol. Evol.* 17: 908–914.

Terzian, C., A. Pelisson and A. Bucheton. 2001. Evolution and phylogeny of insect endogenous retroviruses. *BMC Evol. Biol.* 1: 3.

Tessier, L. H., M. Keller, R. L. Chan, R. Fournier, J. H. Weil and P. Imbault. 1991. Short leader sequences may be transferred from small RNAs to pre-mature mRNAs by *trans*-splicing in *Euglena*. *EMBO J.* 10: 2621–2625.

Tessier, L. H., F. Paulus, M. Keller, C. Vial and P. Imbault. 1995. Structure and expression of *Euglena gracilis* nuclear rbcS genes encoding the small subunits of the ribulose 1,5-bisphosphate carboxylase/oxygenase: A novel splicing process for unusual intervening sequences? *J. Mol. Biol.* 245: 22–33.

Thanaraj, T. A., F. Clark and J. Muilu. 2003. Conservation of human alternative splice events in mouse. *Nucleic Acids Res.* 31: 2544–2552.

Thomas, C. A. 1971. The genetic organization of chromosomes. *Annu. Rev. Genet.* 5: 237–256.

Thomas, E. E., N. Srebro, J. Sebat, N. Navin, J. Healy, B. Mishra and M. Wigler. 2004. Distribution of short paired duplications in mammalian genomes. *Proc. Natl. Acad. Sci. USA* 101: 10349–10354.

Thomas, J. H. 2006. Concerted evolution of two novel protein families in *Caenorhabditis* species. *Genetics* 172: 2269–2281.

Thomm, M. 1996. Archaeal transcription factors and their role in transcription initiation. *FEMS Microbiol. Rev.* 18: 159–171.

Thornton, K., D. Bachtrog and P. Andolfatto. 2006. X chromosomes and autosomes evolve at similar rates in *Drosophila*: No evidence for faster-X protein evolution. *Genome Res.* 16: 498–504.

Thorsness, P. E. and T. D. Fox. 1990. Escape of DNA from mitochondria to the nucleus in *Saccharomyces cerevisiae*. *Nature* 346: 376–379.

Thorsness, P. E. and T. D. Fox. 1993. Nuclear mutations in *Saccharomyces cerevisiae* that affect the escape of

DNA from mitochondria to the nucleus. *Genetics* 134: 21–28.
Thyagarajan, B., R. A. Padua and C. Campbell. 1996. Mammalian mitochondria possess homologous DNA recombination activity. *J. Biol. Chem.* 271: 27536–275343.
Tian, M. and T. Maniatis. 1993. A splicing enhancer complex controls alternative splicing of doublesex pre-mRNA. *Cell* 74: 105–114.
Tice, M. M. and D. R. Lowe. 2004. Photosynthetic microbial mats in the 3,416-Myr-old ocean. *Nature* 431: 549–552.
Tillich, M., P. Lewahrk, B. R. Morton and U. G. Maier. 2006. The evolution of chloroplast RNA editing. *Mol. Biol. Evol.* 23: 1912–1921.
Tillier, E. R. and P. A. Collins. 2000. Replication orientation affects the rate and direction of bacterial gene evolution. *J. Mol. Evol.* 51: 459–463.
Timmis, J. N., M. A. Ayliffe, C. Y. Huang and W. Martin. 2004. Endosymbiotic gene transfer: Organelle genomes forge eukaryotic chromosomes. *Nature Rev. Genet.* 5: 123–135.
Timmons, L., H. Tabara, C. C. Mello and A. Z. Fire. 2003. Inducible systemic RNA silencing in *Caenorhabditis elegans*. *Mol. Biol. Cell* 14: 2972–2983.
Ting, C. T., S. C. Tsaur, S. Sun, W. E. Browne, Y. C. Chen, N. H. Patel and C.-I. Wu. 2004. Gene duplication and speciation in *Drosophila*: Evidence from the *Odysseus* locus. *Proc. Natl. Acad. Sci. USA* 101: 12232–12235.
Tjaden, J., I. Haferkamp, B. Boxma, A. G. Tielens, M. Huynen and J. H. Hackstein. 2004. A divergent ADP/ATP carrier in the hydrogenosomes of *Trichomonas gallinae* argues for an independent origin of these organelles. *Mol. Microbiol.* 51: 1439–1446.
Toh, H., B. L. Weiss, S. A. Perkin, A. Yamashita, K. Oshima, M. Hattori and S. Aksoy. 2005. Massive genome erosion and functional adaptations provide insights into the symbiotic lifestyle of *Sodalis glossinidius* in the tsetse host. *Genome Res.* 16: 149–156.
Tollervey, D. 2004. Termination by torpedo. *Nature* 432: 456–457.
Tolstrup, N., C. W. Sensen, R. A. Garrett and I. G. Clausen. 2000. Two different and highly organized mechanisms of translation initiation in the archaeon *Sulfolobus solfataricus*. *Extremophiles* 4: 175–179.
Tomita, K., T. Ueda and K. Watanabe. 1996. RNA editing in the acceptor stem of squid mitochondrial tRNA(Tyr). *Nucleic Acids Res.* 24: 4987–4991.
Tomitani, A., K. Okada, H. Miyashita, H. C. Matthijs, T. Ohno and A. Tanaka. 1999. Chlorophyll b and phycobilins in the common ancestor of cyanobacteria and chloroplasts. *Nature* 400: 159–162.
Toor, N., G. Hausner and S. Zimmerly. 2001. Coevolution of group II intron RNA structures with their intron-encoded reverse transcriptases. *RNA* 7: 1142–1152.
Tordai, H. and L. Patthy. 2004. Insertion of spliceosomal introns in proto-splice sites: The case of secretory signal peptides. *FEBS Lett.* 575: 109–111.
Torgerson, D. G. and R. S. Singh. 2003. Sex-linked mammalian sperm proteins evolve faster than autosomal ones. *Mol. Biol. Evol.* 20: 1705–1709.
Toro, N. 2003. Bacteria and Archaea Group II introns: Additional mobile genetic elements in the environment. *Environ. Microbiol.* 5: 143–151.
Touchon, M., S. Nicolay, B. Audit, E. B. Brodie of Brodie, Y. d'Aubenton-Carafa, A. Arneodo and C. Thermes. 2005. Replication-associated strand asymmetries in mammalian genomes: Toward detection of replication origins. *Proc. Natl. Acad. Sci. USA* 102: 9836–9841.
Tovar, J., G. Leon-Avila, L. B. Sanchez, R. Sutak, J. Tachezy, M. van der Giezen, M. Hernandez, M. Muller and J. M. Lucocq. 2003. Mitochondrial remnant organelles of *Giardia* function in iron-sulphur protein maturation. *Nature* 426: 172–176.
Trachtulec, Z. and J. Forejt. 1999. Transcription and RNA processing of mammalian genes in *Saccharomyces cerevisiae*. *Nucleic Acids Res.* 27: 526–531.
Tranque, P., M. C. Hu, G. M. Edelman and V. P. Mauro. 1998. rRNA complementarity within mRNAs: A possible basis for mRNA-ribosome interactions and translational control. *Proc. Natl. Acad. Sci. USA* 95: 12238–12243.
Traut, W. and B. Wollert. 1998. An X/Y DNA segment from an early stage of sex chromosome differentiation in the fly *Megaselia scalaris*. *Genome* 41: 289–294.
Tristem, M. 2000. Identification and characterization of novel human endogenous retrovirus families by phylogenetic screening of the human genome mapping project database. *J. Virol.* 74: 3715–3730.
True, H. L. and S. L. Lindquist. 2000. A yeast prion provides a mechanism for genetic variation and phenotypic diversity. *Nature* 407: 477–483.
True, J. R. and S. B. Carroll. 2002. Gene co-option in physiological and morphological evolution. *Annu. Rev. Cell. Dev. Biol.* 18: 53–80.
True, J. R. and E. S. Haag. 2001. Developmental system drift and flexibility in evolutionary trajectories. *Evol. Dev.* 3: 109–119.
Tsong, A. E., B. B. Tuch, H. Li and A. D. Johnson. 2006. Evolution of alternative transcriptional circuits with identical logic. *Nature* 443: 415–420.
Tsudzuki, T., T. Wakasugi and M. Sugiura. 2001. Comparative analysis of RNA editing sites in higher plant chloroplasts. *J. Mol. Evol.* 53: 327–332.
Tucker, P. K., R. M. Adkins and J. S. Rest. 2003. Differential rates of evolution for the *ZFY*-related zinc finger genes, *Zfy*, *Zfx*, and *Zfa* in the mouse genus *Mus*. *Mol. Biol. Evol.* 20: 999–1005.
Tuzun, E., A. J. Sharp, J. A. Bailey, R. Kaul, V. A. Morrison, L. M. Pertz, E. Haugen, H. Hayden, D. Albertson, D. Pinkel et al. 2005. Fine-scale structural variation of the human genome. *Nature Genet.* 37: 727–732.
Tvrdik, P. and M. R. Capecchi. 2006. Reversal of *Hox1* gene subfunctionalization in the mouse. *Dev. Cell* 11: 239–250.
Tyler-Smith, C. and G. Floridia. 2000. Many paths to the top of the mountain: Diverse evolutionary solutions to centromere structure. *Cell* 102: 5–8.

Ueda, T., T. Ohta and K. Watanabe. 1985. Large scale isolation and some properties of AGY-specific serine tRNA from bovine heart mitochondria. *J. Biochem.* 98: 1275–1284.
Unniraman, S., R. Prakash and V. Nagaraja. 2002. Conserved economics of transcription termination in eubacteria. *Nucleic Acids Res.* 30: 675–684.
Urrutia, A. O. and L. D. Hurst. 2001. Codon usage bias covaries with expression breadth and the rate of synonymous evolution in humans, but this is not evidence for selection. *Genetics* 159: 1191–1199.
Urrutia, A. O. and L. D. Hurst. 2003. The signature of selection mediated by expression on human genes. *Genome Res.* 13: 2260–2264.

Valadkhan, B. and J. L. Manley. 2002. Intrinsic metal binding by a spliceosomal RNA. *Nature Struct. Biol.* 9: 498–499.
Valentine, J. W., D. Jablonski and D. H. Erwin. 1999. Fossils, molecules and embryos: New perspectives on the Cambrian explosion. *Development* 126: 851–859.
Vallender, E. J. and B. T. Lahn. 2004. How mammalian sex chromosomes acquired their peculiar gene content. *BioEssays* 26: 159–169.
Vaňáčova, S., W. Yan, J. M. Carlton and P. J. Johnson. 2005. Spliceosomal introns in the deep-branching eukaryote *Trichomonas vaginalis. Proc. Natl. Acad. Sci. USA* 102: 4430–4435.
Vandenberghe, A. E., T. H. Meedel and K. E. Hastings. 2001. mRNA 5′-leader *trans*-splicing in the chordates. *Genes Dev.* 15: 294–303.
van den Heuvel, J. J., R. J. Bergkamp, R. J. Planta and H. A. Raue. 1989. Effect of deletions in the 5′-noncoding region on the translational efficiency of phosphoglycerate kinase mRNA in yeast. *Gene* 79: 83–95.
Van de Peer, Y., S. A. Rensing, U. G. Maier and R. De Wachter. 1996. Substitution rate calibration of small subunit ribosomal RNA identifies chlorarachniophyte endosymbionts as remnants of green algae. *Proc. Natl. Acad. Sci. USA* 93: 7732–7736.
Van de Peer, Y., S. L. Baldauf, W. F. Doolittle and A. Meyer. 2000. An updated and comprehensive rRNA phylogeny of (crown) eukaryotes based on rate-calibrated evolutionary distances. *J. Mol. Evol.* 51: 565–576.
Van de Peer, Y., J. S. Taylor, I. Braasch I and A. Meyer. 2001. The ghost of selection past: Rates of evolution and functional divergence of anciently duplicated genes. *J. Mol. Evol.* 53: 436–446.
Vandepoele, K., C. Simillion and Y. Van de Peer. 2003. Evidence that rice and other cereals are ancient aneuploids. *Plant Cell* 15: 2192–2202.
Vandepoele, K., W. De Vos, J. S. Taylor, A. Meyer and Y. Van de Peer. 2004. Major events in the genome evolution of vertebrates: Paranome age and size differ considerably between ray-finned fishes and land vertebrates. *Proc. Natl. Acad. Sci. USA* 101: 1638–1643.
van der Giezen, M., D. J. Slotboom, D. S. Horner, P. L. Dyal, M. Harding, G. P. Xue, T. M. Embley and E. R. Kunji. 2002. Conserved properties of hydrogenosomal and mitochondrial ADP/ATP carriers: A common origin for both organelles. *EMBO J.* 21: 572–579.
van Ham, R. C., J. Kamerbeek, C. Palacios, C. Rausell, F. Abascal, U. Bastolla, J. M. Fernandez, L. Jimenez, M. Postigo, F. J. Silva et al. 2003. Reductive genome evolution in *Buchnera aphidicola. Proc. Natl. Acad. Sci. USA* 100: 581–586.
Van Holde, K. E. 1989. *Chromatin*. Springer-Verlag, New York.
van Hoof, A. 2005. Conserved functions of yeast genes support the duplication, degeneration and complementation model for gene duplication. *Genetics* 171: 1455–1461.
van Hoof, A. and P. J. Green. 1996. Premature nonsense codons decrease the stability of phytohemagglutinin mRNA in a position-dependent manner. *Plant J.* 10: 415–424.
van Hoof, A. and P. J. Green. 2006. NMD in plants. In L. E. Maquat, ed., *Nonsense-Mediated mRNA Decay*, 167–172. Landes Bioscience, Georgetown, TX.
van Noort, V. and M. A. Huynen. 2006. Combinatorial gene regulation in *Plasmodium falciparum. Trends Genet.* 22: 73–78.
van Ommen, G. J. 2005. Frequency of new copy number variation in humans. *Nature Genet.* 37: 333–334.
van Oppen, M. J., J. Catmull, B. J. McDonald, N. R. Hislop, P. J. Hagerman and D. J. Miller. 2002. The mitochondrial genome of *Acropora tenuis* (Cnidaria; Scleractinia) contains a large group I intron and a candidate control region. *J. Mol. Evol.* 55: 1–13.
Van't Hof, J. and C. A. Bjerknes. 1981. Similar replicon properties of higher plant cells with different S periods and genome sizes. *Exp. Cell Res.* 136: 461–465.
Van Valen, L. 1985. Why and how do mammals evolve unusually rapidly? *Evol. Theory* 7: 127–132.
Venkatesh, B., Y. Ning and S. Brenner. 1999. Late changes in spliceosomal introns define clades in vertebrate evolution. *Proc. Natl. Acad. Sci. USA* 96: 10267–10271.
Venter, J. C., M. D. Adams, E. W. Myers, P. W. Li, R. J. Mural, G. G. Sutton, H. O. Smith, M. Yandell, C. A. Evans, R. A. Holt et al. 2001. The sequence of the human genome. *Science* 291: 1304–1351.
Ventura, M., N. Archidiacono and M. Rocchi. 2001. Centromere emergence in evolution. *Genome Res.* 11: 595–599.
Ventura, M., S. Weigl, L. Carbone, M. F. Cardone, D. Misceo, M. Teti, P. D'Addabbo, A. Wandall, E. Bjorck, P. J. de Jong et al. 2004. Recurrent sites for new centromere seeding. *Genome Res.* 14: 1696–1703.
Vieira, C., C. Nardon, C. Arpin, D. Lepetit and C. Biémont. 2002. Evolution of genome size in *Drosophila*. Is the invader's genome being invaded by transposable elements? *Mol. Biol. Evol.* 19: 1154–1161.
Vilela, C. and J. E. McCarthy. 2003. Regulation of fungal gene expression via short open reading frames in the mRNA 5′ untranslated region. *Mol. Microbiol.* 49: 859–867.
Villanueva, M. S., S. P. Williams, C. B. Beard, F. F. Richards and S. Aksoy. 1991. A new member of a family of site-specific retrotransposons is present in

the spliced leader RNA genes of *Trypanosoma cruzi*. *Mol. Cell. Biol.* 11: 6139–6148.
Villarreal, L. P. 2005. *Viruses and the Evolution of Life*. ASM Press, Washington, DC.
Vinogradov, A. E. 2003. Selfish DNA is maladaptive: Evidence from the plant Red List. *Trends Genet.* 19: 609–614.
Vinogradov, A. E. 2004. Genome size and extinction risk in vertebrates. *Proc. Roy. Soc. Lond.* B *Biol. Sci.* 271: 1701–1705.
Vinogradov, A. E. 2005. Noncoding DNA, isochores and gene expression: Nucleosome formation potential. *Nucleic Acids Res.* 33: 559–563.
Vinogradov, A. E. and O. V. Anatskaya. 2006. Genome size and metabolic intensity in tetrapods: A tale of two lines. *Proc. Roy. Soc. Biol. Sci.* B *Biol. Sci.* 273: 27–32.
Vinuesa, P., C. Silva, D. Werner and E. Martinez-Romero. 2005. Population genetics and phylogenetic inference in bacterial molecular systematics: The roles of migration and recombination in *Bradyrhizobium* species cohesion and delineation. *Mol. Phylogenet. Evol.* 34: 29–54.
Vision, T. J., D. G. Brown and S. D. Tanksley. 2000. The origins of genomic duplications in *Arabidopsis*. *Science* 290: 2114–2117.
Vitte, C. and O. Panaud. 2003. Formation of solo-LTRs through unequal homologous recombination counterbalances amplifications of LTR retrotransposons in rice *Oryza sativa* L. *Mol. Biol. Evol.* 20: 528–540.
Vogel, J., T. Borner and W. R. Hess. 1999. Comparative analysis of splicing of the complete set of chloroplast group II introns in three higher plant mutants. *Nucleic Acids Res.* 27: 3866–3874.
Voight, B. F., S. Kudaravalli, X. Wen and J. K. Pritchard. 2006. A map of recent positive selection in the human genome. *PLoS Biol.* 4: 446–458.
Volff, J. N. 2005. Genome evolution and biodiversity in teleost fish. *Heredity* 94: 280–294.
Volpe, T. A., C. Kidner, I. M. Hall, G. Teng, S. I. Grewal and R. A. Martienssen. 2002. Regulation of heterochromatic silencing and histone H3 lysine-9 methylation by RNAi. *Science* 297: 1833–1837.
von Dassow, G. and E. Monro. 1999. Modularity in animal development and evolution: Elements of a conceptual framework for evo-devo. *J. Exp. Zool.* 285: 307–325.
von Dohlen, C. D., S. Kohler, S. T. Alsop and W. R. McManus. 2001. Mealybug beta-proteobacterial endosymbionts contain gamma-proteobacterial symbionts. *Nature* 412: 433–436.
von Heijne, G. 1986. Why mitochondria need a genome. *FEBS Lett.* 198: 1–4.
Voytas, D. F. 1996. Retroelements in genome organization. *Science* 274: 737–738.
Voytas, D. F. and J. D. Boeke. 2002. *Ty1* and *Ty5* of *Saccharomyces cerevisiae*. In N. L. Craig, R. Craigie, M. Gellert and A. M. Lambowitz, eds., *Mobile DNA II*, 631–662. ASM Press, Washington, DC.
Vulić, M., F. Dionisio, F. Taddei and M. Radman. 1997. Molecular keys to speciation: DNA polymorphism and the control of genetic exchange in enterobacteria. *Proc. Natl. Acad. Sci. USA* 94: 9763–9767.
Vulić, M., R. E. Lenski and M. Radman. 1999. Mutation, recombination, and incipient speciation of bacteria in the laboratory. *Proc. Natl. Acad. Sci. USA* 96: 7348–7351.

Wada, H., M. Kobayashi, R. Sato, N. Satoh, H. Miyasaka and Y. Shirayama. 2002. Dynamic insertion-deletion of introns in deuterostome EF-1alpha genes. *J. Mol. Evol.* 54: 118–128.
Wagner, A. 1999. Redundant gene functions and natural selection. *J. Evol. Biol.* 12: 1–16.
Wagner, A. 2000. The role of population size, pleiotropy and fitness effects of mutations in the evolution of overlapping gene functions. *Genetics* 154: 1389–1401.
Wagner, A. 2001. Birth and death of duplicated genes in completely sequenced eukaryotes. *Trends Genet.* 17: 237–239.
Wagner, A. 2006. Periodic extinctions of transposable elements in bacterial lineages: Evidence from intragenomic variation in multiple genomes. *Mol. Biol. Evol.* 23: 723–733.
Wagner, E. G. and K. Flärdh. 2002. Antisense RNAs everywhere? *Trends Genet.* 18: 223–226.
Wagner, G. P. 1996. Homologues, natural kinds and the evolution of modularity. *American Zoologist* 36: 36–43.
Wagner, G. P. and L. Altenberg. 1996. Complex adaptations and the evolution of evolvability. *Evolution* 50: 967–976.
Wahle, E. and U. Rüegsegger. 1999. 3′-end processing of pre-mRNA in eukaryotes. *FEMS Microbiol. Rev.* 23: 277–295.
Waldman, A. S. and R. M. Liskay. 1988. Dependence of intrachromosomal recombination in mammalian cells on uninterrupted homology. *Mol. Cell. Biol.* 8: 5350–5357.
Wall, J. D. and J. K. Pritchard. 2003. Haplotype blocks and linkage disequilibrium in the human genome. *Nature Rev. Genet.* 4: 587–597.
Wallace, D. C. 1994. Mitochondrial DNA sequence variation in human evolution and disease. *Proc. Natl. Acad. Sci. USA* 91: 8739–8746.
Walsh, B. 2003. Population-genetic models of the fates of duplicate genes. *Genetica* 118: 279–294.
Walsh, J. B. 1982. Rate of accumulation of reproductive isolation by chromosome rearrangements. *Amer. Natur.* 120: 510–532.
Walsh, J. B. 1983. Role of biased gene conversion in one-locus neutral theory and genome evolution. *Genetics* 105: 461–468.
Walsh, J. B. 1992. Intracellular selection, conversion bias, and the expected substitution rate of organelle genes. *Genetics* 130: 939–946.
Walsh, J. B. 1995. How often do duplicated genes evolve new functions? *Genetics* 110: 345–364.
Walters, M. C., S. Fiering, J. Eidemiller, W. Magis, M. Groudine and D. I. Martin. 1995. Enhancers increase the probability but not the level of gene expression. *Proc. Natl. Acad. Sci. USA* 92: 7125–7129.
Wang, D., D. A. Kreutzer and J. M. Essigmann. 1998. Mutagenicity and repair of oxidative DNA dam-

age: Insights from studies using defined lesions. *Mutat. Res.* 400: 99–115.
Wang, D. Y., S. Kumar and S. B. Hedges. 1999. Divergence time estimates for the early history of animal phyla and the origin of plants, animals and fungi. *Proc. Roy. Soc. Lond.* B *Biol. Sci.* 266: 163–171.
Wang, D., M. Hsieh and W. H. Li. 2005. A general tendency for conservation of protein length across eukaryotic kingdoms. *Mol. Biol. Evol.* 22: 142–147.
Wang, J., R. Tokarz and C. Savage-Dunn. 2002. The expression of TGFbeta signal transducers in the hypodermis regulates body size in *C. elegans. Development* 129: 4989–4998.
Wang, P. J., J. R. McCarrey, F. Yang and D. C. Page. 2001. An abundance of X-linked genes expressed in spermatogonia. *Nature Genet.* 27: 422–426.
Wang, W., H. Yu and M. Long. 2004. Duplication-degeneration as a mechanism of gene fission and the origin of new genes in *Drosophila* species. *Nature Genet.* 36: 523–527.
Wang, X. and J. Zhang. 2004. Rapid evolution of mammalian X-linked testis-expressed homeobox genes. *Genetics* 167: 879–888.
Wang, X., W. E. Grus and J. Zhang. 2006. Gene losses during human origins. *PLoS Biol.* 4: 366–377.
Wang, Z., M. E. Rolish, G. Yeo, V. Tung, M. Mawson and C. B. Burge. 2004. Systematic identification and analysis of exonic splicing silencers. *Cell* 119: 831–845.
Ward, B. L., R. S. Anderson and A. J. Bendich. 1981. The mitochondrial genome is large and variable in a family of plants (Cucurbitaceae). *Cell* 25: 793–803.
Washietl, S., I. L. Hofacker, M. Lukasser, A. Huttenhofer and P. F. Stadler. 2005. Mapping of conserved RNA secondary structures predicts thousands of functional noncoding RNAs in the human genome. *Nature Biotechnol.* 23: 1383–1390.
Watanabe, J., M. Sasaki, Y. Suzuki and S. Sugano. 2002. Analysis of transcriptomes of human malaria parasite *Plasmodium falciparum* using full-length enriched library: Identification of novel genes and diverse transcription start sites of messenger RNAs. *Gene* 291: 105–113.
Watanabe, K. I. and T. Ohama. 2001. Regular spliceosomal introns are invasive in *Chlamydomonas reinhardtii*: 15 introns in the recently relocated mitochondrial *cox2* and *cox3* genes. *J. Mol. Evol.* 53: 333–339.
Watanabe, K. I., Y. Bessho, M. Kawasaki and H. Hori. 1999. Mitochondrial genes are found on minicircle DNA molecules in the mesozoan animal *Dicyema*. *J. Mol. Biol.* 286: 645–650.
Watanabe, N., Y. Nagamatsu, K. Gengyo-Ando and S. Mitani and Y. Ohshima. 2005. Control of body size by SMA-5, a homolog of MAP kinase BMK1/ERK5, in *C. elegans*. *Development* 132: 3175–3184.
Watanabe, Y., H. Tsurui, T. Ueda, R. Furushima, S. Takamiya, K. Kita, K. Nishikawa and K. Watanabe. 1994. Primary and higher order structures of nematode (*Ascaris suum*) mitochondrial tRNAs lacking either the T or D stem. *J. Biol. Chem.* 269: 22902–22906.
Waterston, R. H., K. Lindblad-Toh, E. Birney, J. Rogers, J. F. Abril, P. Agarwal, R. Agarwala, R. Ainscough, M. Alexandersson, P. An et al. 2002. Initial sequencing and comparative analysis of the mouse genome. *Nature* 420: 520–562.
Watson, J. D. and F. H. Crick. 1953. Molecular structure of nucleic acids; a structure for deoxyribose nucleic acid. *Nature* 171: 737–738.
Watt, V. M., C. J. Ingles, M. S. Urdea and W. J. Rutter. 1985. Homology requirements for recombination in *Escherichia coli*. *Proc. Natl. Acad. Sci. USA* 82: 4768–4772.
Watterson, G. A. 1983. On the time for gene silencing at duplicate loci. *Genetics* 105: 745–766.
Webb, B. A., M. R. Strand, S. E. Dickey, M. H. Beck, R. S. Hilgarth, W. E. Barney, K. Kadash, J. A. Kroemer, K. G. Lindstrom, W. Rattanadechakul et al. 2005. Polydnavirus genomes reflect their dual roles as mutualists and pathogens. *Virology* 347: 160–174.
Webb, C. J., C. M. Romfo, W. J. van Heeckeren and J. A. Wise. 2005. Exonic splicing enhancers in fission yeast: Functional conservation demonstrates an early evolutionary origin. *Genes Dev.* 19: 242–254.
Webb, C. T., S. A. Shabalina, A. Y. Ogurtsov and A. S. Kondrashov. 2002. Analysis of similarity within 142 pairs of orthologous intergenic regions of *Caenorhabditis elegans* and *Caenorhabditis briggsae*. *Nucleic Acids Res.* 30: 1233–1239.
Webster, M. T. and N. G. Smith. 2004. Fixation biases affecting human SNPs. *Trends Genet.* 20: 122–126.
Webster, M. T., N. G. Smith and H. Ellegren. 2003. Compositional evolution of noncoding DNA in the human and chimpanzee genomes. *Mol. Biol. Evol.* 20: 278–286.
Webster, M. T., E. Axelsson and H. Ellegren. 2006. Strong regional biases in nucleotide substitution in the chicken genome. *Mol. Biol. Evol.* 23: 1203–1216.
Wei, W., N. Gilbert, S. L. Ooi, L. F. Lawler, E. M. Ostertag, H. H. Kazazian, J. D. Boeke and J. V. Moran. 2001. Human *L1* retrotransposition: *cis* preference versus *trans* complementation. *Mol. Cell. Biol.* 21: 1429–1439.
Weil, C. F. and R. Kunze. 2000. Transposition of maize Ac/Ds transposable elements in the yeast *Saccharomyces cerevisiae*. *Nature Genet.* 26: 187–190.
Weiner, A. M. 2002. SINEs and LINEs: The art of biting the hand that feeds you. *Curr. Opin. Cell. Biol.* 14: 343–350.
Weiner, J. III, R. Herrmann and G. F. Browning. 2000. Transcription in *Mycoplasma pneumoniae*. *Nucleic Acids Res.* 28: 4488–4496.
Welch, J. J. and D. Waxman. 2003. Modularity and the cost of complexity. *Evolution* 57: 1723–1734.
Wendel, J. F. 2000. Genome evolution in polyploids. *Plant Mol. Biol.* 42: 225–249.
Wernegreen, J. J. and D. J. Funk. 2004. Mutation exposed: A neutral explanation for extreme base composition of an endosymbiont genome. *J. Mol. Evol.* 59: 849–858.
Werth, C. R. and M. D. Windham. 1991. A model for divergent, allopatric speciation of polyploid pteridophytes resulting from silencing of duplicate-gene expression. *Amer. Natur.* 137: 515–526.

Wessler, S. R. 1989. The splicing of maize transposable elements from pre-mRNA—a minireview. *Gene* 82: 127–133.
Wessler, S. R. 2006. Eukaryotic transposable elements: Teaching old genomes new tricks. In L. H. Caporale, ed., *The Implicit Genome*, 138–162. Oxford University Press, Oxford.
West-Eberhard, M. J. 2003. *Developmental Plasticity and Evolution*. Oxford University Press, Oxford.
Westin, J. and M. Lardelli. 1997. Three novel *notch* genes in zebrafish: Implications for vertebrate *Notch* gene evolution and function. *Dev. Genes Evol.* 207: 51–63.
Whamond, G. S. and J. M. Thornton. 2006. An analysis of intron positions in relation to nucleotides, amino acids, and protein secondary structure. *J. Mol. Biol.* 359: 238–247.
Whitfield, L. S., R. Lovell-Badge and P. N. Goodfellow. 1993. Rapid sequence evolution of the mammalian sex-determining gene *SRY*. *Nature* 364: 713–715.
Whitlock, M. C. 2003. Fixation probability and time in subdivided populations. *Genetics* 164: 767–779.
Whitlock, M. C. and N. H. Barton. 1997. The effective size of a subdivided population. *Genetics* 146: 427–441.
Whitman, W. B., D. C. Coleman and W. J. Wiebe. 1998. Prokaryotes: The unseen majority. *Proc. Natl. Acad. Sci. USA* 95: 6578–6583.
Whittle, C. A. and M. O. Johnston. 2002. Male-driven evolution of mitochondrial and chloroplastidial DNA sequences in plants. *Mol. Biol. Evol.* 19: 938–949.
Wicky, C., A. M. Villeneuve, N. Lauper, L. Codourey, H. Tobler and F. Muller. 1996. Telomeric repeats (TTAGGC)n are sufficient for chromosome capping function in *Caenorhabditis elegans*. *Proc. Natl. Acad. Sci. USA* 93: 8983–8988.
Wilkie, D. and D. Y. Thomas. 1973. Mitochondrial genetic analysis by zygote cell lineages in *Saccharomyces cerevisiae*. *Genetics* 73: 367–377.
Wilkins, A. S. 2002. *The Evolution of Developmental Pathways*. Sinauer Associates, Sunderland, MA.
Wilkins, A. S. 2005. Recasting developmental evolution in terms of genetic pathway and network evolution... and the implications for comparative biology. *Brain Res. Bull.* 66: 495–509.
Wilkins, C., R. Dishongh, S. C. Moore, M. A. Whitt, M. Chow and K. Machaca. 2005. RNA interference is an antiviral defence mechanism in *Caenorhabditis elegans*. *Nature* 436: 1044–1047.
Will, C. L., C. Schneider, R. Reed and R. Lührmann. 1999. Identification of both shared and distinct proteins in the major and minor spliceosomes. *Science* 284: 2003–2005.
Will, C. L., C. Schneider, A. M. MacMillan, N. F. Katopodis, G. Neubauer, M. Wilm, R. Lührmann and C. C. Query. 2001. A novel U2 and U11/U12 snRNP protein that associates with the pre-mRNA branch site. *EMBO J.* 20: 4536–4546.
Will, C. L., C. Schneider, M. Hossbach, H. Urlaub, R. Rauhut, S. Elbashir, T. Tuschl and R. Lührmann. 2004. The human 18S U11/U12 snRNP contains a set of novel proteins not found in the U2-dependent spliceosome. *RNA* 10: 929–941.
Williams, B. A., R. P. Hirt, J. M. Lucocq and T. M. Embley. 2002. A mitochondrial remnant in the microsporidian *Trachipleistophora hominis*. *Nature* 418: 865–869.
Williams, B. A., C. H. Slamovits, N. J. Patron, N. M. Fast and P. J. Keeling. 2005. A high frequency of overlapping gene expression in compacted eukaryotic genomes. *Proc. Natl. Acad. Sci. USA* 102: 10936–10941.
Williams, E. J. and D. J. Bowles. 2004. Coexpression of neighboring genes in the genome of *Arabidopsis thaliana*. *Genome Res.* 14: 1060–1067.
Williams, G. C. 1966. *Adaptation and Natural Selection*. Princeton University Press, Princeton, NJ.
Williamson, D. 2002. The curious history of yeast mitochondrial DNA. *Nature Rev. Genet.* 3: 475–481.
Willie, E. and J. Majewski. 2004. Evidence for codon bias selection at the pre-mRNA level in eukaryotes. *Trends Genet.* 20: 534–538.
Wilquet, V. and M. Van de Casteele. 1999. The role of the codon first letter in the relationship between genomic GC content and protein amino acid composition. *Res. Microbiol.* 150: 21–32.
Wilson, A. C., G. L. Bush, S. M. Case and M. C. King. 1975. Social structuring of mammalian populations and rate of chromosomal evolution. *Proc. Natl. Acad. Sci. USA* 72: 5061–5065.
Wilson, D. S. and J. W. Szostak. 1999. In vitro selection of functional nucleic acids. *Annu. Rev. Biochem.* 68: 611–47.
Wilson, E. O. 1975. *Sociobiology: The New Synthesis*. Belknap Press, Cambridge, MA.
Wilusz, C. J., W. Wang and S. W. Peltz. 2001. Curbing the nonsense: The activation and regulation of mRNA surveillance. *Genes Dev.* 15: 2781–2785.
Wise, C. A., M. Sraml and S. Easteal. 1998. Departure from neutrality at the mitochondrial NADH dehydrogenase subunit 2 gene in humans, but not in chimpanzees. *Genetics* 148: 409–421.
Witherspoon, D. J. and H. M. Robertson. 2003. Neutral evolution of ten types of mariner transposons in the genomes of *Caenorhabditis elegans* and *C. briggsae*. *J. Mol. Evol.* 56: 751–769.
Wiuf, C. 2001. Recombination in human mitochondrial DNA? *Genetics* 159: 749–756.
Woese, C. R. 1983. The primary lines of descent and the universal ancestor. In D. S. Bendall, ed., *Evolution: From Molecules to Men*, 209–233. Cambridge University Press, Cambridge.
Woese, C. R. 1998. Default taxonomy: Ernst Mayr's view of the microbial world. *Proc. Natl. Acad. Sci. USA* 95: 11043–11046.
Woese, C. R. and G. E. Fox. 1977. Phylogenetic structure of the prokaryotic domain: The primary kingdoms. *Proc. Natl. Acad. Sci. USA* 74: 5088–5090.
Woese, C. R., O. Kandler and M. L. Wheelis. 1990. Towards a natural system of organisms: Proposal for the domains Archaea, Bacteria, and Eucarya. *Proc. Natl. Acad. Sci. USA* 87: 4576–4579.
Wolf, J. B., E. D. Brodie III and M. J. Wade. 2000. *Epistasis and the Evolutionary Process*. Oxford University Press, New York.

Wolf, P. G., C. A. Rowe and M. Hasebe. 2004. High levels of RNA editing in a vascular plant chloroplast genome: Analysis of transcripts from the fern *Adiantum capillus-veneris*. *Gene* 339: 89–97.

Wolf, Y. I., I. B. Rogozin and E. V. Koonin. 2004. Coelomata and not Ecdysozoa: Evidence from genome-wide phylogenetic analysis. *Genome Res.* 14: 29–36.

Wolfe, K. H. 2001. Yesterday's polyploids and the mystery of diploidization. *Nature Rev. Genet.* 2: 333–341.

Wolfe, K. H. and D. C. Shields. 1997. Molecular evidence for an ancient duplication of the entire yeast genome. *Nature* 387: 708–713.

Wolfe, K. H., W. H. Li and P. M. Sharp. 1987. Rates of nucleotide substitution vary greatly among plant mitochondrial, chloroplast, and nuclear DNAs. *Proc. Natl. Acad. Sci. USA* 84: 9054–9058.

Wolfe, K. H., P. M. Sharp and W.-H. Li. 1989. Mutation rates differ among regions of the mammalian genome. *Nature* 337: 283–285.

Wolff, E., M. Kim, K. Hu, H. Yang and J. H. Miller. 2004. Polymerases leave fingerprints: Analysis of the mutational spectrum in *Escherichia coli* rpoB to assess the role of polymerase IV in spontaneous mutation. *J. Bacteriol.* 186: 2900–2905.

Wolstenholme, D. R. 1992. Animal mitochondrial DNA: Structure and evolution. In D. R. Wolstenholme and K. W. Jeon, eds., *Mitochondrial Genomes*, 173–216. Academic Press, New York.

Wolstenholme, D. R., R. Okimoto and J. L. Macfarlane. 1994. Nucleotide correlations that suggest tertiary interactions in the TV-replacement loop-containing mitochondrial tRNAs of the nematodes, *Caenorhabditis elegans* and *Ascaris suum*. *Nucleic Acids Res.* 22: 4300–4306.

Wong, L. H. and K. H. Choo. 2004. Evolutionary dynamics of transposable elements at the centromere. *Trends Genet.* 20: 611–616.

Wood, D. W., J. C. Setubal, R. Kaul, D. E. Monks, J. P. Kitajima, V. K. Okura,Y. Zhou, L. Chen,G. E. Wood, N. F. Almeida, Jr. et al. 2001. The genome of the natural genetic engineer *Agrobacterium tumefaciens* C58. *Science* 294: 2317–2323.

Wood, V., R. Gwilliam, M. A. Rajandream, M. Lyne, R. Lyne, A. Stewart, J. Sgouros, N. Peat, J. Hayles, S. Baker et al. 2002. The genome sequence of *Schizosaccharomyces pombe*. *Nature* 415: 871–880.

Woolfe, A., M. Goodson, D. K. Goode, P. Snell, G. K. McEwen, T. Vavouri, S. F. Smith, P. North, H. Callaway, K. Kelly et al. 2005. Highly conserved non-coding sequences are associated with vertebrate development. *PLoS Biol.* 3: 116–130.

Worning, P., L. J. Jensen, P. F. Hallin, H. H. Staerfeldt and D. W. Ussery. 2006. Origin of replication in circular prokaryotic chromosomes. *Environ. Microbiol.* 8: 353–361.

Woyke, T., H. Teeling, N. N. Ivanova, M. Huntemann, M. Richter, F. O. Gloeckner, D. Boffelli, I. J. Anderson, K. W. Barry, H. J. Shapiro et al. 2006. Symbiosis insights through metagenomic analysis of a microbial consortium. *Nature* 443: 950–955.

Wray, G. A., J. S. Levinton and L. H. Shapiro. 1996. Molecular evidence for deep Precambrian divergences among metazoan phyla. *Science* 274: 568–573.

Wray, G. A., M. W. Hahn, E. Abouheif, J. P. Balhoff, M. Pizer, M. V. Rockman and L. A. Romano. 2003. The evolution of transcriptional regulation in eukaryotes. *Mol. Biol. Evol.* 20: 1377–1419.

Wright, S. 1934a. An analysis of variability in number of digits in an inbred strain of guinea pigs. *Genetics* 19: 506–536.

Wright, S. 1934b. The results of crosses between inbred strains of guinea pigs, differing in numbers of digits. *Genetics* 19: 537–551.

Wright, S. I., Q. H. Le, D. J. Schoen and T. E. Bureau. 2001. Population dynamics of an *Ac*-like transposable element in self- and cross-pollinating *Arabidopsis*. *Genetics* 158: 1279–1288.

Wright, S. I., N. Agrawal and T. E. Bureau. 2003a. Effects of recombination rate and gene density on transposable element distributions in *Arabidopsis thaliana*. *Genome Res.* 13: 1897–1903.

Wright, S. I., B. Lauga and D. Charlesworth. 2003b. Subdivision and haplotype structure in natural populations of *Arabidopsis lyrata*. *Mol. Ecol.* 12: 1247–1263.

Wright, S. I., C. B. Yau, M. Looseley and B. C. Meyers. 2004. Effects of gene expression on molecular evolution in *Arabidopsis thaliana* and *Arabidopsis lyrata*. *Mol. Biol. Evol.* 21: 1719–1726.

Wu, C. I. and C. T. Ting. 2004. Genes and speciation. *Nature Rev. Genet.* 5: 114–122.

Wu, C. I. and E. Y. Xu. 2003. Sexual antagonism and X inactivation—the SAXI hypothesis. *Trends Genet.* 19: 243–247.

Wyles, J. S., J. G. Kunkel and A. C. Wilson. 1983. Birds, behavior, and anatomical evolution. *Proc. Natl. Acad. Sci. USA* 80: 4394–4397.

Xia, X., M. S. Hafner and P. D. Sudman. 1996. On transition bias in mitochondrial genes of pocket gophers. *J. Mol. Evol.* 43: 32–40.

Xiao, S., Y. Zhang, A. H. Knoll. 1998. Three-dimensional preservation of algae and animal embryos in a Neoproterozoic phosphorite. *Nature* 391: 553–558.

Xie, X., J. Lu, E. J. Kulbokas, T. R. Golub, V. Mootha, K. Lindblad-Toh, E. S. Lander and M. Kellis. 2005. Systematic discovery of regulatory motifs in human promoters and 3′ UTRs by comparison of several mammals. *Nature* 434: 338–345.

Xing, Y. and C. J. Lee. 2004. Negative selection pressure against premature protein truncation is reduced by alternative splicing and diploidy. *Trends Genet.* 20: 472–475.

Xing, Y. and C. J. Lee. 2005. Protein modularity of alternatively spliced exons is associated with tissue-specific regulation of alternative splicing. *PLoS Genet.* 1: 323–328.

Xing, Y. and C. Lee. 2006. Alternative splicing and RNA selection pressure—evolutionary consequences for eukaryotic genomes. *Nature Rev. Genet.* 7: 499–509.

Xiong, Y. and T. H. Eickbush. 1990. Origin and evolution of retroelements based upon their reverse transcriptase sequences. *EMBO J.* 9: 3353–3362.

Xu, Q., B. Modrek and C. Lee. 2002. Genome-wide detection of tissue-specific alternative splicing in the human transcriptome. *Nucleic Acids Res.* 30: 3754–3766.

Xu, Z., Y. Yu, J. L. Schwartz, M. L. Meltz and A. W. Hsie. 1995. Molecular nature of spontaneous mutations at the hypoxanthine-guanine phosphoribosyltransferase (*hprt*) locus in Chinese hamster ovary cells. *Environ. Mol. Mutagen.* 26: 127–138.

Yamamura, E., T. Nunoshiba, M. Kawata and K. Yamamoto. 2000. Characterization of spontaneous mutation in the oxyR strain of *Escherichia coli*. *Biochem. Biophys. Res. Commun.* 2000 279: 427–432.

Yamamura, E., E. H. Lee, A. Kuzumaki, N. Uematsu, T. Nunoshiba, M. Kawata and K. Yamamoto. 2002. Characterization of spontaneous mutation in the delta soxR and SoxS overproducing strains of *Escherichia coli*. *J. Radiat. Res.* (Tokyo) 43: 195–203.

Yamauchi, K., M. Mukai, T. Ochiai and I. Usuki. 1992. Molecular cloning of the cDNA for the major hemoglobin component from *Paramecium caudatum*. *Biochem. Biophys. Res. Commun.* 182: 195–200.

Yampolsky, L. Y. and A. Stoltzfus. 2001. Bias in the introduction of variation as an orienting factor in evolution. *Evol. Dev.* 3: 73–83.

Yanai, I., C. J. Comacho and C. DeLisi. 2000. Predictions of gene family distributions in microbial genomes: Evolution by gene duplication and modification. *Phys. Rev. Lett.* 85: 2641–2644.

Yang, F., B. Fu, P. C. O'Brien, W. Nie, O. A. Ryder and M. A. Ferguson-Smith. 2004. Refined genome-wide comparative map of the domestic horse, donkey and human based on cross-species chromosome painting: Insight into the occasional fertility of mules. *Chromosome Res.* 12: 65–76.

Yang, H. P. and S. V. Nuzhdin. 2003. Fitness costs of *Doc* expression are insufficient to stabilize its copy number in *Drosophila melanogaster*. *Mol. Biol. Evol.* 20: 800–804.

Yang, H. P., A. Y. Tanikawa, W. A. Van Voorhies, J. C. Silva and A. S. Kondrashov. 2001. Whole-genome effects of ethyl methanesulfonate-induced mutation on nine quantitative traits in outbred *Drosophila melanogaster*. *Genetics* 157: 1257–1265.

Yang, J., S. Zimmerly, P. S. Perlman and A. M. Lambowitz. 1996. Efficient integration of an intron RNA into double-stranded DNA by reverse splicing. *Nature* 381: 332–335.

Yang, J., Z. Gu and W. H. Li. 2003. Rate of protein evolution versus fitness effect of gene deletion. *Mol. Biol. Evol.* 20: 772–774.

Yang, J. W., C. Pendon, J. Yang, N. Haywood, A. Chand and W. R. Brown. 2000. Human mini-chromosomes with minimal centromeres. *Hum. Mol. Genet.* 9: 1891–1902.

Yang, Y. W., K. N. Lai, P. Y. Tai and W. H. Li. 1999. Rates of nucleotide substitution in angiosperm mitochondrial DNA sequences and dates of divergence between *Brassica* and other angiosperm lineages. *J. Mol. Evol.* 48: 597–604.

Ye, J., C. E. Pérez-González,, D. G. Eickbush and T. H. Eickbush. 2005. Competition between *R1* and *R2* transposable elements in the 28S rRNA genes of insects. *Cytogenet. Genome Res.* 110: 299–306.

Yeadon, P. J. and D. E. Catcheside. 1999. Polymorphism around *cog* extends into adjacent structural genes. *Curr. Genet.* 35: 631–637.

Yean, S. L., G. Wuenschell, J. Termini and R. J. Lin. 2000. Metal-ion coordination by U6 small nuclear RNA contributes to catalysis in the spliceosome. *Nature* 408: 881–884.

Yee, J., M. R. Mowatt, P. P. Dennis and T. E. Nash. 2000. Transcriptional analysis of the glutamate dehydrogenase gene in the primitive eukaryote, *Giardia lamblia*. Identification of a primordial gene promoter. *J. Biol. Chem.* 275: 11432–11439.

Yeo, G., S. Hoon, B. Venkatesh and C. B. Burge. 2004. Variation in sequence and organization of splicing regulatory elements in vertebrate genes. *Proc. Natl. Acad. Sci. USA* 101: 15700–15705.

Yi, S. and J. T. Streelman. 2005. Genome size is negatively correlated with effective population size in ray-finned fishes. *Trends Genet.* 21: 643–646.

Yokobori, S and S. Pääbo. 1995. Transfer RNA editing in land snail mitochondria. *Proc. Natl. Acad. Sci. USA* 92: 10432–10435.

Yoon, H. S., J. D. Hackett and D. Bhattacharya. 2002. A single origin of the peridinin- and fucoxanthin-containing plastids in dinoflagellates through tertiary endosymbiosis. *Proc. Natl. Acad. Sci. USA* 99: 11724–11729.

Yoon, H. S., J. D. Hackett, C. Ciniglia, G. Pinto and D. Bhattacharya. 2004. A molecular timeline for the origin of photosynthetic eukaryotes. *Mol. Biol. Evol.* 21: 809–818.

Yoshihama, M., A. Nakao, H. D. Nguyen and N. Kenmochi. 2006. Analysis of ribosomal protein gene structures: Implications for intron evolution. *PLoS Genet.* 2: 237–242.

Yoshiyama, K. and H. Maki. 2003. Spontaneous hotspot mutations resistant to mismatch correction in *Escherichia coli*: Transcription-dependent mutagenesis involving template-switching mechanisms. *J. Mol. Biol.* 327: 7–18.

Yu, N., M. I. Jensen-Seaman, L. Chemnick, O. Ryder and W.-H. Li. 2004. Nucleotide diversity in gorillas. *Genetics* 166: 1375–1383.

Yu, X. and A. Gabriel. 1999. Patching broken chromosomes with extranuclear cellular DNA. *Mol. Cell* 4: 873–881.

Yu, Y. S., Y. Suzuki, K. Yoshitomo, M. Muramatsu, N. Yamaguchi and S. Sugano. 1996. The promoter structure of TGF-β type II receptor revealed by "oligo-capping" method and deletion analysis. *Biochem. Biophys. Res. Commun.* 225: 302–306.

Yuh, C. H., H. Bolouri and E. H. Davidson. 1998. Genomic *cis*-regulatory logic: Experimental and computational analysis of a sea urchin gene. *Science* 279: 1896–1902.

Zahler, A. M., K. M. Neugebauer, W. S. Lane and M. B. Roth. 1993. Distinct functions of SR proteins in alternative pre-mRNA splicing. *Science* 260: 219–222.

Zamore, P. D. and B. Haley. 2005. Ribo-gnome: The big world of small RNAs. *Science* 309: 1519–1524.

Zayas, R. M., T. D. Bold and P. A. Newmark. 2005. Spliced-leader *trans*-splicing in freshwater planarians. *Mol. Biol. Evol.* 22: 2048–2054.
Zechner, E. L., C. A. Wu and K. J. Marians. 1992. Coordinated leading- and lagging-strand synthesis at the *Escherichia coli* DNA replication fork. II. Frequency of primer synthesis and efficiency of primer utilization control Okazaki fragment size. *J. Biol. Chem.* 267: 4045–4053.
Zeiner, G. M., N. R. Sturm and D. A. Campbell. 2003. The *Leishmania tarentolae* spliced leader contains determinants for association with polysomes. *J. Biol. Chem.* 278: 38269–38275.
Zerbib, D., P. Polard, J. M. Escoubas, D. Galas and M. Chandler. 1990. The regulatory role of the IS1-encoded InsA protein in transposition. *Mol. Microbiol.* 4: 471–477.
Zeyl, C., G. Bell and D. M. Green. 1996. Sex and the spread of retrotransposon Ty3 in experimental populations of *Saccharomyces cerevisiae*. *Genetics* 143: 1567–1577.
Zhang, H. and S. Lin. 2005. Mitochondrial cytochrome b mRNA editing in dinoflagellates: Possible ecological and evolutionary associations? *J. Eukaryot. Microbiol.* 52: 538–545.
Zhang, J. 2003. Evolution by gene duplication: An update. *Trends Ecol. Evol.* 18: 292–298.
Zhang, J. 2004. Evolution of *DMY*, a newly emergent male sex-determination gene of medaka fish. *Genetics* 166: 1887–1895.
Zhang, J., X. Sun, Y. Qian, J. P. LaDuca and L. E. Maquat. 1998. At least one intron is required for the nonsense-mediated decay of triosephosphate isomerase mRNA: A possible link between nuclear splicing and cytoplasmic translation. *Mol. Cell. Biol.* 18: 5272–5283.
Zhang, J., D. M. Webb and O. Podlaha. 2002. Accelerated protein evolution and origins of human-specific features: Foxp2 as an example. *Genetics* 162: 1825–1835.
Zhang, K., J. M. Akey, N. Wang, M. Xiong, R. Chakraborty and L. Jin. 2003. Randomly distributed crossovers may generate block-like patterns of linkage disequilibrium: An act of genetic drift. *Hum. Genet.* 113: 51–59.
Zhang, L., T. J. Vision and B. S. Gaut. 2002. Patterns of nucleotide substitution among simultaneously duplicated gene pairs in *Arabidopsis thaliana*. *Mol. Biol. Evol.* 19: 1464–1473.
Zhang, L., H. H. Lu, W. Y. Chung, J. Yang and W.-H. Li. 2005. Patterns of segmental duplication in the human genome. *Mol. Biol. Evol.* 22: 135–141.
Zhang, L. H., H. Vrieling, A. A. van Zeeland and D. Jenssen. 1992. Spectrum of spontaneously occurring mutations in the *hprt* gene of V79 Chinese hamster cells. *J. Mol. Biol.* 223: 627–635.
Zhang, P., S. Chopra and T. Peterson. 2000. A segmental gene duplication generated differentially expressed *myb*-homologous genes in maize. *Plant Cell* 12: 2311–2322.
Zhang, P., Z. Gu, W.-H. Li. 2003. Different evolutionary patterns between young duplicate genes in the human genome. *Genome Biol.* 4(9): R56.
Zhang, Q., J. Arbuckle and S. R. Wessler. 2000. Recent, extensive, and preferential insertion of members of the miniature inverted-repeat transposable element family *Heartbreaker* into genic regions of maize. *Proc. Natl. Acad. Sci. USA* 97: 1160–1165.
Zhang, X., N. Jiang, C. Feschotte and S. R. Wessler. 2004. *PIF*- and *Pong*-like transposable elements: Distribution, evolution and relationship with *Tourist*-like miniature inverted-repeat transposable elements. *Genetics* 166: 971–986.
Zhang, Z. and M. Gerstein. 2003. Patterns of nucleotide substitution, insertion and deletion in the human genome inferred from pseudogenes. *Nucleic Acids Res.* 31: 5338–5348.
Zhang, Z. and M. Gerstein. 2004. Large-scale analysis of pseudogenes in the human genome. *Curr. Opin. Genet. Dev.* 14: 328–335.
Zhang, Z., B. R. Green and T. Cavalier-Smith. 1999. Single gene circles in dinoflagellate chloroplast genomes. *Nature* 400: 155–159.
Zhang, Z., B. R. Green and T. Cavalier-Smith. 2000. Phylogeny of ultra-rapidly evolving dinoflagellate chloroplast genes: A possible common origin for sporozoan and dinoflagellate plastids. *J. Mol. Evol.* 51: 26–40.
Zhang, Z., T. Cavalier-Smith and B. R. Green. 2002. Evolution of dinoflagellate unigenic minicircles and the partially concerted divergence of their putative replicon origins. *Mol. Biol. Evol.* 19: 489–500.
Zhang, Z., P. Harrison and M. Gerstein. 2002. Identification and analysis of over 2000 ribosomal protein pseudogenes in the human genome. *Genome Res.* 12: 1466–1482.
Zhang, Z., N. Carriero and M. Gerstein. 2004. Comparative analysis of processed pseudogenes in the mouse and human genomes. *Trends Genet.* 20: 62–67.
Zhao, D. and M. Bownes. 1998. The RNA product of the *Doc* retrotransposon is localized on the *Drosophila* oocyte cytoskeleton. *Mol. Gen. Genet.* 257: 497–504.
Zhao, J., L. Hyman and C. Moore. 1999. Formation of mRNA 3′ ends in eukaryotes: Mechanism, regulation, and interrelationships with other steps in mRNA synthesis. *Microbiol. Mol. Biol. Rev.* 63: 405–445.
Zheng, W., K. Khrapko, H. A. Coller, W. G. Thilly and W. C. Copeland. 2006. Origins of human mitochondrial point mutations as DNA polymerase gamma-mediated errors. *Mutat. Res.* 599: 11–20.
Zhong, J. and A. M. Lambowitz. 2003. Group II intron mobility using nascent strands at DNA replication forks to prime reverse transcription. *EMBO J.* 22: 4555–4565.
Zhou, Z., L. J. Licklider, S. P. Gygi and R. Reed. 2002. Comprehensive proteomic analysis of the human spliceosome. *Nature* 419: 182–185.
Zhu, J., A. Mayeda and A. R. Krainer. 2001. Exon identity established through differential antagonism between exonic splicing silencer-bound hnRNP A1 and enhancer-bound SR proteins. *Mol. Cell* 8: 1351–1361.
Zhu, W. and V. Brendel. 2003. Identification, characterization and molecular phylogeny of U12-dependent

introns in the *Arabidopsis thaliana* genome. *Nucleic Acids Res.* 31: 4561–4572.
Zhuang, Y. and A. M. Weiner. 1989. A compensatory base change in human U2 snRNA can suppress a branch site mutation. *Genes Dev.* 3: 1545–1552.
Zigler, K. S., M. A. McCartney, D. R. Levitan and H. A. Lessios. 2005. Sea urchin bindin divergence predicts gamete compatibility. *Evolution* 59: 2399–2404.
Zilberman, D., X. Cao and S. E. Jacobsen. 2003. Argonaute4 control of locus-specific siRNA accumulation and DNA and histone methylation. *Science* 299: 716–719.
Zillig, W. 1991. Comparative biochemistry of Archaea and Bacteria. *Curr. Opin. Genet. Dev.* 1: 544–551.
Zimmerly, S., H. Guo, P. S. Perlman and A. M. Lambowitz. 1995. Group II intron mobility occurs by target DNA-primed reverse transcription. *Cell* 82: 545–554.
Zimmerly, S., G. Hausner and X. Wu. 2001. Phylogenetic relationships among group II intron ORFs. *Nucleic Acids Res.* 29: 1238–1250.
Zipf, G. K. 1949. *Human Behavior and the Principle of Least Effort*. Addison-Wesley, Boston, MA.
Zou, S., J. M. Kim and D. F. Voytas. 1996. The *Saccharomyces* retrotransposon *Ty5* influences the organization of chromosome ends. *Nucleic Acids Res.* 24: 4825–4831.
Zsurka, G., Y. Kraytsberg, T. Kudina, C. Kornblum, C. E. Elger, K. Khrapko and W. S. Kunz. 2005. Recombination of mitochondrial DNA in skeletal muscle of individuals with multiple mitochondrial DNA heteroplasmy. *Nature Genet.* 37: 873–877.
Zuckerkandl, E. 2001. Intrinsically driven changes in gene interaction complexity. I. Growth of regulatory complexes and increase in number of genes. *J. Mol. Evol.* 53: 539–554.
Zuegge, J., S. Ralph, M. Schmuker, G. I. McFadden and G. Schneider. 2001. Deciphering apicoplast targeting signals—feature extraction from nuclear-encoded precursors of *Plasmodium falciparum* apicoplast proteins. *Gene* 280: 19–26.
Župunski, V., F. Gubensek and D. Kordis. 2001. Evolutionary dynamics and evolutionary history in the *RTE* clade of non-LTR retrotransposons. *Mol. Biol. Evol.* 18: 1849–1863.

Author Index

Subject Index